Islamic Philosophy, Theology and Science

Texts and Studies

Edited by

H. Daiber

VOLUME LXXV

Prof. Dr. Hans Daiber

Islamic Thought in the Middle Ages

Studies in Text, Transmission and Translation, in Honour of Hans Daiber

Edited by

Anna Akasoy and Wim Raven

BRILL

LEIDEN • BOSTON
2008

This book is printed on acid-free paper.

Library of Congress Cataloging-in-Publication Data

Islamic thought in the Middle Ages : studies in text, transmission and translation, in honour of Hans Daiber / edited by Anna Akasoy and Wim Raven.
p. cm. — (Islamic philosophy, theology and science. text and studies)
Includes bibliographical references and index.
ISBN 978-90-04-16565-6 (hardback : alk. paper) 1. Philosophy, Islamic—Greek influences. 2. Civilization, Islamic—Greek influences. 3. Science and civilization. 4. Translations. I. Raven, Wim. II. Akasoy, Anna. III. Daiber, Hans, 1942–
B744.3.I85 2008
181'.07—dc22

2008006512

ISSN 0169-8729
ISBN 978 90 04 16565 6

PRINTED IN THE NETHERLANDS

CONTENTS

History of Science

PREFACE

An often quoted dictum of the prophet has it that one should seek knowledge even as far as China. It is often referred to as a principle which allowed medieval Islamic intellectual culture to develop its stunning intercultural traits, first and foremost in the famous translation movement under the ʿAbbāsids. This spirit prevailed among medieval Muslim philosophers, starting with al-Kindī, the 'philosopher of the Arabs', who presented the first fusion of Greek philosophy and Islamic religion in ninth-century Baghdad. The giants of medieval thought, Ibn Sīnā, al-Ghazālī and Ibn Rushd, were to share the same open-minded attitude. The afterlife of their works led to equally prodigious developments in both East and West. The works of Mollā Ṣadrā, for example, in the Safawid Empire, show the impact of Avicennian metaphysics. In the extreme West of the Muslim world, the Catalan missionary Raimundus Lullus was inspired by the works of al-Ghazālī and Ibn Sīnā and fought intellectual battles with the Averroists in Paris.

Hans Daiber is one of the pioneers in the study of this field—a field which stretches from the eighth until the seventeenth century and even further, and from al-Andalus to Central Asia. What distinguishes Daiber in his approach to this massive area is his broad vision which combines the philological study of texts and their translation and transmission in different languages with a philosophical analysis and a multi-layered contextualisation involving, among others, theology, science, history and literature. Indeed, if there is a common methodological link between the classes Hans Daiber has taught at the Free University in Amsterdam and the Johann Wolfgang Goethe-University in Frankfurt, it is that, even though they dealt with a great variety of subjects, there was always an attempt to trace the development of an idea as it travelled through periods and regions, fields of thought and cultures.

In his teaching as well as in his research, Hans Daiber is living up to the principle to travel as far as you can in pursuit of knowledge. He has been and still is travelling to regions not many Western scholars set foot in. His journey to India in order to explore Arabic manuscripts has opened new perspectives in the study of the Aristotelian tradition. He has shared his knowledge with students in Bosnia, Japan and

Malaysia and did not hesitate to participate in conferences in Iran and Uzbekistan.

In his teaching and research he has also displayed the openness and flexibility of mind which distinguished the *falāsifa*. His colleagues and his students in particular have always enjoyed the tolerance with which Hans Daiber responded to opinions and methods that he did not share. Furthermore, when asked about their experience of having Hans Daiber as their supervisor, most students—former as well as present—will respond referring to the personal qualities which distinguish their teacher: his constant support, his patience and gentleness. His colleagues will add his great generosity, demonstrated, for example, in the willingness to share valuable manuscript material. They will equally highlight that despite his achievements Hans Daiber has always remained a modest and unpretentious man with a great sense of humour.

The articles collected in this volume reflect Hans Daiber's interests as well as his methods. And, in accordance with the complexity of his own approach, the borders between the sections are blurred. The publication of these articles marks the occasion of Hans Daiber's retirement as a professor at Frankfurt University, a position he had taken up in 1995 after having spent the previous eighteen years at the Free University in Amsterdam. During his time in Frankfurt, he has been part of several interdisciplinary networks concerning the history of science, where his contributions have been much appreciated.

Anna Akasoy (Oxford)
Wim Raven (Marburg)

HANS DAIBER, A BIBLIOGRAPHY

I. Monographs

1) *Die arabische Übersetzung der Placita Philosophorum*, doctoral thesis (University of Saarbrücken, 1968). For a revised version cf. no. 5.
2) *Ein Kompendium der aristotelischen Meteorologie in der Fassung des Ḥunain Ibn Isḥāq* (Aristoteles Semitico-Latinus. Prolegomena et parerga, 1; Amsterdam, 1975).
3) *Das theologisch-philosophische System des Muʿammar Ibn ʿAbbād as-Sulamī* (gest. 830 n. Chr.) (Beiruter Texte und Studien, 19; Beirut, 1975).
4) *Gott, Natur und menschlicher Wille im frühen islamischen Denken.* Rede, uitgesproken in het Nederlands bij de aanvaarding van het ambt van gewoon hoogleraar in het Arabisch en de Islam aan de Faculteit der Letteren van de Vrije Universiteit te Amsterdam op 15 juni 1978 (Amsterdam, 1978).
5) *Aetius Arabus. Die Vorsokratiker in arabischer Überlieferung* (Akademie der Wissenschaften und der Literatur Mainz, Veröffentlichungen der Orientalischen Kommission, 33; Wiesbaden, 1980).
6) *The Ruler as Philosopher. A New Interpretation of Fārābī's View* (Mededelingen der Koninklijke Nederlandse Akademie van Wetenschappen, afd. Letterkunde. Nieuwe reeks, d. 49, no. 4; Amsterdam, 1986). Persian translation in *Dowwomīn yādnāma-i ʿAllāma Ṭabāṭabāʾī. The Second Commemoration Volume.* The Second Cultural Congress on the Occasion of the Death Anniversary of the Late Allamah Seyyed Muhammad Husayn Ṭabāṭabīʾī, Tehran, 15–18 November 1983 (Tehran, 1363/1985), pp. 353–69.
7) *Wāṣil Ibn ʿAṭāʾ als Prediger und Theologe. Ein neuer Text aus dem 8. Jahrhundert n. Chr.* (Islamic Philosophy and Theology, Texts and Studies, 2; Leiden, 1988).
8) *Catalogue of the Arabic Manuscripts in the Daiber Collection, Institute of Oriental Culture, University of Tokyo* (Tokyo, 1988).
9) *Naturwissenschaft bei den Arabern im 10. Jahrhundert n. Chr. Briefe des Abū l-Faḍl Ibn al-ʿAmīd (gest. 360/970) an ʿAḍudaddaula*, Mit Einleitung, kommentierter Übersetzung und Glossar herausgegeben (Islamic Philosophy, Theology and Science, Texts and Studies, 13; Leiden, 1993).

10) *Neuplatonische Pythagorica in arabischem Gewande. Der Kommentar des Iamblichus zu den* Carmina aurea (Koninklijke Nederlandse Akademie van Wetenschappen, afd. Letterkunde, nieuwe reeks, deel 161; Amsterdam, 1995).
11) *The Islamic Concept of Belief in the 4th/10th Century. Abū l-Lait as-Samarqandī's Commentary on Abū Ḥanīfa (died 150/767)*, al-Fiqh al-absaṭ, Introduction, Text and Commentary (Studia culturae islamicae, 52; Tokyo, 1995).
12) *Catalogue of the Arabic Manuscripts in the Daiber Collection II, Institute of Oriental Culture, University of Tokyo* (Tokyo, 1996).
13) *Bibliography of Islamic Philosophy*, 3 vols. (Leiden, 1999–2007).
14) *The Struggle for Knowledge in Islam. Some Historical Aspects/Borba za znanje u Islamu: Neki historijski aspekti* (Sarajevo, 2004).
15) *Catalogue of the Arabic Manuscripts in the Daiber Collection III* (in press).
16) *Islamic Thought in the Dialogue of Cultures. Innovation and Mediation between Antiquity and Middle Ages* (Sarajevo, in press).

II. Articles

1) 'Die arabische Handschrift Chester Beatty 4183 und das Kitāb Sirr al-asrār,' *Der Islam* 42 (1965), pp. 71–8 (together with H. Gätje).
2) 'Ein bisher unbekannter pseudoplatonischer Text über die Tugenden der Seele in arabischer Überlieferung,' *Der Islam* 47 (1971), pp. 25–42 and 49 (1972), pp. 122–3.
3) 'Zur Erstausgabe von al-Māturīdī, Kitāb at-Tauḥīd,' *Der Islam* 52 (1975), pp. 299–313.
4) 'Neue Handschriftenfunde aus dem Jemen und aus der Südosttürkei. Die Erstversion des Tashīl al-manāfiʿ fi ṭ-ṭibb wa-l-ḥikma,' in *19. Deutscher Orientalistentag vom 28. September bis 4. Oktober 1975 in Freiburg im Breisgau, Vorträge* (= *ZDMG*, Suppl. III/1), pp. 661–8.
5) 'Anfänge und Entstehung der Wissenschaft im Islam,' *Saeculum* 29 (1978), pp. 356–66.
6) 'Briefe des Abū l-Faḍl Ibn al-ʿAmīd an ʿAḍudaddaula,' *Der Islam* 56 (1979), pp. 106–16.
7) 'The Creed (ʿAqīda) of the Ḥanbalite Ibn Qudāma al-Maqdisī. A Newly Discovered Text,' in W. al-Qāḍī (ed.), *Studia Arabica et Islamica. Festschrift for Iḥsān ʿAbbās* (Beirut, 1981), pp. 105–25.

8) 'Vom Nutzen der Metaphysik, auch in der Philologie,' *Der Islam* 58 (1981), pp. 310–20.
9) 'De zendeling Raimundus Lullus en zijn kritiek op de Islam,' *Historische betrekkingen tussen moslims en christenen* (Nijmegen, 1982) = *Midden Oosten en Islampublicatie* 9, pp. 21–32.
10) 'Fārābīs Abhandlung über das Vakuum: Quellen und Stellung in der islamischen Wissenschaftsgeschichte,' *Der Islam* 60 (1983), pp. 37–47. (Cf. the Turkish newspaper *Milliyet*, 20 May 1986, p. 6).
11) 'The Qurʾān as Stimulus of Science in Early Islam,' in *International Conference on Science in Islamic Polity, its Past, Present and Future*, Islamabad, 19–24 November 1983, i (Islamabad, 1983), pp. 122–30.
12) 'Abū l-Faradj Ibn aṭ-Ṭayyib on Plants: an Inquiry into his Sources,' in *The Third International Symposium on the History of Arab Science*, December 10–14 1983, Kuwait, Xerox print. A revised version appeared as n. 68.
13) 'Der Missionar Raimundus Lullus und seine Kritik am Islam,' *Estudios Lulianos* 25 (1981–83), pp. 47–57. A revised version appeared as n. 81.
14) 'The Qurʾān as a Matter of Dispute. New Material on the History of the Ḥanbalite-Ashʿarite Controversy in the 12/13th Century A.D.,' in *Proceedings of the 31. International Congress of Human Sciences in Asia and North Africa*, Tokyo and Kyoto, 31.8.–7.9.1983, i (Tokyo, 1984), pp. 282–3.
15) 'Der Ṣiwān al-ḥikma und Abū Sulaimān al-Manṭiqī as-Siǧistānī in der Forschung,' *Arabica* 31 (1984), pp. 36–68.
16) 'Democritus in Arabic and Syriac Tradition,' in *Proceedings of the 1st International Congress on Democritus*, Xanthi 6–9 October 1983, B (Xanthi, 1984), pp. 251–65.
17) Article 'Dialog (Prosa),' in *Lexikon des Mittelalters*, iii (Munich, 1985), cols. 964–5.
18) 'Die Kreuzzüge im Lichte islamischer Theologie. Theologische Interpretamente bei Abū Šāma,' in *Orientalische Kultur und europäisches Mittelalter* (Berlin, 1985), pp. 77–85.
19) 'Ein vergessener syrischer Text: Bar Zoʿbī über die Teile der Philosophie,' *Oriens Christianus* 69 (1985), pp. 73–80.
20) 'A Survey of Theophrastean Texts and Ideas in Arabic. Some New Material,' in *Rutgers University Studies in Classical Humanities* (New Brunswick, 1985), ii, pp. 103–14.
21) See monographs, n. 6.

22) 'Orosius' *Historiae adversus paganos* in arabischer Überlieferung,' in J.W. van Henten et al. (eds.), *Tradition and Re-Interpretation in Jewish and Early Christian Literature. Essays in honour of Jürgen C.H. Lebram* (Leiden, 1986), pp. 202–49.
23) 'Semitische Sprachen als Kulturvermittler zwischen Antike und Mittelalter. Stand und Aufgaben der Forschung,' *ZDMG* 136 (1986), pp. 292–313.
24) 'Prophetie und Ethik bei Fārābī (gest. 339/950),' in C. Wenin (ed.), *L'homme et son univers au moyen âge*. Actes du septième congrès international de philosophie médiévale, 30 août–4 septembre 1982 (Louvain, 1986), ii, pp. 729–53.
25) 'Anfänge muslimischer Theologie im Lichte eines neuen Textes von Wāṣil Ibn ʿAṭāʾ,' in *Islão e Arabismo na peninsula iberica*. Actas do XI congresso da União europeia de arabistas e islamologos (Evora, 1986), pp. 383–91.
26) 'Die Kritik des Ibn Ḥazm an Kindīs Metaphysik,' *Der Islam* 63 (1986), pp. 284–302. English summary in:
27) 'Al-Kindī in Andalus. Ibn Ḥazm's Critique of his Metaphysics,' in *Actas del XII congreso de la U.E.A.I.*, Malaga 1984 (Madrid, 1986), pp. 229–35.
28) 'New Manuscript Findings from Indian Libraries,' *Manuscripts of the Middle East* 1 (1986), pp. 26–48.
29) 'Helmut Gätje (1927–1986). Ein Nachruf,' *ZDMG* 138 (1988), pp. 1–16.
30) 'An Arabic Manuscript Library: Some Important Discoveries,' *Manuscripts of the Middle East* 2 (1987, publ. Leiden 1988), pp. 18–36.
31) Article 'al-Masāʾil wa-l-adjwiba' in *EI*[2].
32) Article 'Bahmanyār,' in *Encyclopaedia Iranica*, iii (London, 1989), col. 501b–503a.
33) 'Abū Ḥātim ar-Rāzī (10th Century A.D.) on the Unity and Diversity of Religions,' in G. Gort (ed.), *Dialogue and Syncretism. An Interdisciplinary Approach* (Grand Rapids, 1989), pp. 87–104.
34) 'Theophrasts Meteorologica in syrisch-arabischer Überlieferung,' in *Symposium Graeco-Arabicum II.* Akten des Zweiten Symposium Graeco-Arabicum Ruhr-Universität Bochum, 3.–5. März 1987 (Amsterdam, 1989), pp. 10–18.
35) 'Die Technik im Islam,' in A. Stöcklein and M. Rassem (eds.), *Technik und Religion* (Düsseldorf, 1990), pp. 102–16.

36) 'Al-Islām wa-ʿalāqatuhu bi'l-ʿilm wa'l-tiqniyya. Naẓratun fī 'l-jadal al-muʿāṣir ḥawl taʾrīkh al-ʿilm wa'l-thaqāfa al-taʾrīkhiyya,' *Fikrun wa Fann* 50 (1990), pp. 78–85.
37) 'Die Autonomie der Philosophie im Islam,' in *Knowledge and the Sciences in Medieval Philosophy*, i (Helsinki, 1990), pp. 228–49.
38) 'The Ismaili Background of Fārābī's Political Philosophy.—Abū Ḥātim ar-Rāzī as a Forerunner of Fārābī,' in U. Tworuschka (ed.), *Gottes ist der Orient, Gottes ist der Okzident. Festschrift für Abdoljavad Falaturi zum 65. Geburtstag* (Cologne, 1991), pp. 143–50.
39) 'Lateinische Übersetzungen arabischer Texte zur Philosophie und ihre Bedeutung für die Scholastik des Mittelalters. Stand und Aufgaben der Forschung,' *Rencontres de cultures dans la philosophie médiévale. Traductions et traducteurs de l'antiquité tardive au XIVe siècle* (Louvain, 1990 [1991]), pp. 203–50.
40) 'The Qurʾan as Stimulus of Science in Early Islam,' in *Islamic Thought and Scientific Creativity* (Islamabad, 1991), ii/2, pp. 29–42.
41) Article 'Muʿammar Ibn ʿAbbād as-Sulamī,' in *EI*^{2}.
42) 'Qosṭā Ibn Lūqā (9. Jh.) über die Einteilung der Wissenschaften,' *ZGAIW* 6 (1990 [1991]), pp. 94–129.
43) Article 'Muḥammad B. Aṣbagh,' in *EI*^{2}.
44) 'Griechische Ethik in islamischem Gewande. Das Beispiel von Rāġib al-Iṣfahānī (11. Jh.),' in B. Mojsisch und O. Pluta (eds.), *Historia philosophiae medii aevi. Studien zur Geschichte der Philosophie des Mittelalters*. Festschrift für Kurt Flasch zu seinem 60. Geburtstag (Amsterdam, 1991 [1992]), pp. 181–92.
45) Article 'Al-Murdār, Abū Mūsā ʿĪsā B. Ṣubayḥ,' in *EI*^{2}.
46) 'Doxographie und Geschichtsschreibung über griechische Philosophen in islamischer Zeit,' *Medioevo* 16 (1990, published Padua, 1992), pp. 1–21.
47) 'The *Meteorology* of Theophrastus in Syriac and Arabic Translation,' in W.W. Fortenbaugh and D. Gutas (eds.), *Theophrastus. His Psychological, Doxographical, and Scientific Writings* (New Brunswick, 1992), pp. 166–293.
48) 'Sharing Religious Experience as a Problem in Early Islamic Mysticism,' in J.D. Gort et al. (eds.), *On Sharing Religious Experience. Possibilities of Interfaith Mutuality* (Amsterdam, 1992), pp. 245–53.
49) 'In memoriam prof. dr. D.S. Attema,' in Vrije Universiteit Amsterdam. Jaarboek 1989/1990 (Amsterdam, 1992), pp. 85–6.
50) Revised Article 'Naẓar' (first version by T. de Boer), in *EI*^{2}.

51) 'Science and Technology versus Islam: a Controversy from Renan and Afghānī to Nasr and Needham and its Historical Background,' *Annals of Japan Association for Middle East Studies* 8 (1993), pp. 169–87.
52) 'Nestorians of 9th Century Iraq as a Source of Greek, Syriac and Arabic. A Survey of Some Unexploited Sources,' *ARAM* 3/1–2 (1991, published Oxford, 1993), pp. 45–52.
53) 'Al-Wazīr al-buwaihī Abū l-Faḍl Ibn al-ʿAmīd—Leonardo da Vinci min al-qarn al-ʿāshir al-mīlādī,' *Fikrun wa Fann* 57 (1993), pp. 58–62.
54) 'Das Studium der Islamwissenschaft in Japan,' *ZDMG* 143 (1993), pp. *1*–*6*.
55) 'The Reception of Islamic Philosophy at Oxford in the 17th Century: The Pococks' (Father and Son) Contribution to the Understanding of Islamic Philosophy in Europe,' in C.E. Butterworth and B.A. Kessel (eds.), *The Introduction of Arabic Philosophy into Europe* (Leiden, 1994), pp. 65–82.
56) Articles 'Alfārābī', 'Alkindī', 'Arabisch-islamische Theologie u. Philosophie' and 'Avicebron,' in *Lexikon für Theologie und Kirche*, i/3 (Freiburg/Br., 1993), col. 383, 395, 907–10 and 1313–14.
57) 'Hellenistisch-kaiserzeitliche Doxographie und philosophischer Synkretismus in islamischer Zeit,' in *Aufstieg und Niedergang der römischen Welt* (Berlin, 1994), ii/36.7, pp. 4974–92.
58) 'Recent Islamic and Arabic Studies in the Netherlands,' *Asian Research Trends* 4 (1994), pp. 1–24. (together with W. Raven)
59) 'The Qurʾān as a "Shibboleth" of Varying Conceptions of the Godhead. A 12th Century Ḥanbalite-Ashʿarite Discussion and its Theological Sequel in the Protocol of Ibn Qudāma al-Maqdisī,' *IOS* 14 (1994), pp. 249–95.
60) Article 'Ruʾyā', 2. In its philosophical-mystical meaning, in *EI*^{2}.
61) Article 'Saʿāda', in *EI*^{2}.
62) 'L'incontro con la filosofia islamica,' in P. Rossi and C.A. Viano (eds.), *Storia della filosofia*, 2: Il medioevo (Rome, 1994), pp. 180–95.
63) 'Science and Technology versus Islam: a Controversy from Renan and Afghānī to Nasr and Needham and its Historical Background,' *Journal for the History of Arabic Science* 10/1–2 (1992–93–94), pp. 119–33.
64) 'New Manuscript Material Relevant to the History of Arabic Literature,' *Manuscripts of the Middle East* 6 (1994), pp. 101–19.

65) 'Literarischer Prozess zwischen Fiktion und Wirklichkeit—Ein Beispiel aus der klassisch-arabischen Erzählliteratur,' *Annals of Japan Association for Middle East Studies* 10 (1995), pp. 27–67.
66) 'Political Philosophy', in S.H. Nasr and O. Leaman (eds.), *History of Islamic Philosophy* (London, 1996), ii, pp. 841–85.
67) Article 'Shaḳāwa,' in *EI*[2].
68) 'Abū 'l-Farağ Ibn Aṭ-Ṭayyib on Plants. An Inquiry into his Sources,' *Erdem*, cilt 9, sayı 26 (Ankara, 1996), pp. 629–42.
69) 'Salient Trends of the Arabic Aristotle,' in G. Endress and R. Kruk (eds.), *The Ancient Tradition in Christian and Islamic Hellenism.* Studies on the Transmission of Greek Philosophy and Sciences dedicated to H.J. Drossaart Lulofs on his ninetieth birthday (Leiden, 1997), pp. 29–41.
70) 'What is the Meaning of and to What End do we Study the History of Islamic Philosophy?,' *Proceedings of the Israel Academy of Sciences and Humanities* 7/5 (Jerusalem, 1998), pp. 177–99.
71) 'Rebellion gegen Gott. Formen atheistischen Denkens im frühen Islam,' in F. Niewöhner and O. Pluta (eds.), *Atheismus im Mittelalter und in der Renaissance* (Wiesbaden, 1999), pp. 23–44.
72) 'Hendrik Joan Drossaart Lulofs (22 april 1906–21 oktober 1998). Herdenking,' in Koninklijke Nederlandse Akademie van Wetenschappen, *Levensberichten en herdenkingen 2000* (Amsterdam, 2000), pp. 11–16.
73) Article 'Al-Ṭūsī, Naṣīr Al-Dīn', 1. Life. 2. As Theologian and Philosopher, in *EI*[2].
74) 'Mullā Ṣadrā on the Problem of Creation and the Role of Greek Philosophers. New Light on Mullā Ṣadrā as Historian of Greek Philosophers,' *Spektrum Iran* 13/1 (Bonn, 2000), pp. 7–22.
75) 'Gottesliebe und menschliche Größe im frühen Islam. Das Beispiel von Rābiʿa al-ʿAdawīya (717–801),' *Spektrum Iran* 13/2 (Bonn, 2000), pp. 5–25.
76) 'Ibn Khaldūn—Leben und Werk,' in B. Schefold (ed.), *Ibn Khaldūn. Ökonomie aus der "Muqaddima"*, Textauswahl von Hans Daiber (Düsseldorf, 2000), pp. 33–54.
77) 'Die Aristotelesrezeption in der syrischen Literatur,' in D. Kuhn and H. Stahl (eds.), *Die Gegenwart des Altertums. Formen und Funktionen des Altertumsbezugs in den Hochkulturen der Alten Welt* (Heidelberg, 2001), pp. 327–45.
78) 'Der Aristoteleskommentator Alexander von Aphrodisias (2./3. Jh. n. Chr.) und der samaritanische Gelehrte Levi über die Ewigkeit

der Welt. Der Bericht des Abū l-Fatḥ as-Sāmirī (14. Jh.) über ein Streitgespräch vor dem römischen Kaiser Commodus,' *JSAI* 27 (2002), pp. 305–47.

79) 'Von der Weisheit Gottes zur Wissenschaft. Wege des Wissens im Islam,' *Evangelium und Wissenschaft. Beiträge zum interdisziplinären Gespräch* 42 (2003), pp. 3–13. Slightly extended version of an article published in *Neue Zürcher Zeitung*, 2 March 2002.

80) 'The Limitations of Knowledge According to Ibn Sīnā: Epistemological and Theological Aspects and the Consequences,' in M. Lutz-Bachmann, A. Fidora and P. Antolic (eds.), *Erkenntnis und Wissenschaft. Probleme der Epistemologie in der Philosophie des Mittelalters* (Berlin, 2004), pp. 25–34.

81) 'Raimundus Lullus in der Auseinandersetzung mit dem Islam. Eine philosophiegeschichtliche Analyse des "Liber disputationis Raimundi Christiani et Homeri Saraceni". Juden, Christen und Muslime,' in M. Lutz-Bachmann and A. Fidora (eds.), *Religionsdialoge im Mittelalter* (Darmstadt, 2004), pp. 136–72. A revised and expanded version of n. 13.

82) 'Die Faḍā'il-Literatur als Quelle der Wissenschaftsgeschichte. Das Beispiel des ägyptischen Historikers Ibn Zūlāq (10. Jh.),' in R. Arnzen and J. Thielmann (eds.), *Words, Texts and Concepts Cruising the Mediterranean Sea. Studies on the Sources, Contents and Influences of Islamic Civilization and Arabic Philosophy and Science.* Dedicted to Gerhard Endress on his sixty-fifth birthday (Leuven, 2004), pp. 355–85.

83) 'Annemarie Schimmel. 7 April 1922–26 Januari 2003. Levensbericht,' in Koninklijke Nederlandse Akademie van Wetenschappen, *Levensberichten en herdenkingen 2005* (Amsterdam, 2005), pp. 104–9.

84) 'Raimundus Lullus' Dispute with Homer Saracenus in the Year 1307. An Inquiry into their Theological Positions,' in M.I. Ripoll Perelló (ed.), *Actes de les Jornades Internacionals Lullianes "Ramon Llull al s. XXI".* Palma, 1, 2 i 3 d'abril de 2004 (Palma, 2005), pp. 259–64.

85) 'Das Fārābī-Bild des Maimonides. Ideentransfer als hermeneutischer Weg zu Maimonides' Philosophie,' in G. Tamer (ed.), *The Trias of Maimonides/Die Trias des Maimonides. Jewish, Arabic, and Ancient Culture of Knowledge/Jüdische, arabische und antike Wissenskultur* (Berlin, 2005), pp. 199–209.

86) 'Ibn Khaldūn ('Abd al-Rahman Ibn Mohammed),' in D. Herz and V. Weinberger (eds.), *Lexikon ökonomischer Werke. 650 wegweisende*

Schriften von der Antike bis ins 20. Jahrhundert (Stuttgart, 2006), pp. 215–16.

87) 'Maqāla fī l-radd ʿalā l-Mujbira (MS Najaf) Attributed to ʿAbd al-Jabbār al-Hamadhānī,' in C. Adang, S. Schmidtke and D. Sklare (eds.), *A Common Rationality. Muʿtazilism in Islam and Judaism* (Würzburg, 2007), pp. 101–25.

88) 'Frühe islamische Diskussionen über die Willensfreiheit des Menschen,' in *Hat der Mensch einen freien Willen? Die Antworten der großen Philosophen* (Stuttgart, 2007), pp. 324–39.

89) 'A New Manuscript of Jāḥiẓ's Works and its Importance for their Transmission,' in *Al-Jahiz: a Muslim Humanist for our Time* (Proceedings of an International Conference, Beirut, 20–22 January 2005) (in press).

90) 'Die Stellung des Theologen Abū l-Lait as-Samarqandī in der Schule des Māturīdī,' in *S. Teoman DURALI Armağan*, Istanbul (in press).

91) 'Weltgeschichte als Unheilsgeschichte. Die arabische Übersetzung von Orosius' Historiae adversus paganos als Warnung an die Muslime Spaniens,' in A. Fidora and M. Tischler (eds.), *Christlicher Norden—Muslimischer Süden. Die iberische Halbinsel im Kontext kultureller, religiöser und politischer Veränderungen zwischen dem 11. und 15. Jahrhundert* (in press).

92) 'The Contribution of Islam to Science in the 4th/10th Century. The Case of Abū l-Faḍl Ibn al-ʿAmīd,' *Malaysian Journal of Science and Technology Studies* (Kuala Lumpur) (in press).

93) 'Die griechisch-arabische Wissenschaftsüberlieferung in der arabisch-islamischen Kultur in Übersetzungen des 8.–10. Jahrhunderts,' in Kittel et al. (eds.), *Übersetzung, Translation, Traduction* (Berlin, 2007), pp. 1206–17.

94) 'Pythagoras and Pythagoreanism [addendum 2: Pythagoras in Arabic-Jewish transmission],' in *Encyclopaedia of Philosophy* (second edition; in press).

95) 'Science Connecting Scholars and Cultures in Khwārazm. The Case of Khwārazmshāh Maʾmūn Ibn Maʾmūn,' in Festschrift Naquib Al-Attas (Kuala Lumpur; in press).

96) 'Die philosophische Tradition bei den syrischen Christen. Abschnitt A: Die Fortsetzung der hellenischen Tradition in frühislamischer Zeit,' in U. Rudolph (ed.), *Grundriss der Geschichte der Philosophie. Die Philosophie in islamischer Zeit* (in press).

97) 'Die philosophische Tradition bei den syrischen Christen. Abschnitt B: Die Öffnung gegenüber islamischen Einflüssen,' in U. Rudolph (ed.), *Grundriss der Geschichte der Philosophie. Die Philosophie in islamischer Zeit* (in press).
98) 'Abū Bakr ar-Rāzī (gest. 925 o. 935 n. Chr.),' in U. Rudolph (ed.), *Grundriss der Geschichte der Philosophie. Die Philosophie in islamischer Zeit* (in press).
99) 'Al-Fārābīs Aristoteles. Grundlagen seiner Erkenntnislehre,' in A. Vrolijk and J.P. Hogendijk (eds.), *O ye Gentlemen. Arabic Studies on Science and Literary Culture in Honor of Remke Kruk* (Leiden, 2007), pp. 99–112.
100) 'Ibn Sīnā in Christian Transmission? A Meteorological Treatise by 'Abdallāh Ibn al-Faḍl (11th c. A.D.)' (in preparation).
101) 'Islamic Roots of Knowledge in Europe,' in *Islam and Europe* (Kuala Lumpur, in press).
102) 'Islamwissenschaft zwischen Philologie und Kulturwissenschaft. Einheit und Vielheit einer Disziplin,' *Spektrum Iran* (in press).

III. Reviews

1) F. Rosenthal, *Das Fortleben der Antike im Islam* (Zurich, 1965).—In *ZDMG* 117 (1967), pp. 203–4.
2) N. Rescher, *Temporal Modalities in Arabic Logic* (Dordrecht, 1967).—In *Indogermanische Forschungen* 72 (1967), pp. 310–11.
3) G.E. von Grunebaum, *Der Islam im Mittelalter* (Zurich, 1963).—In *ZDMG* 118 (1968), pp. 176–7.
4) N. Rescher, *Galen and the Syllogism* (Pittsburgh, 1966).—In *Der Islam* 44 (1968), pp. 284–5.
5) J. van Ess, *Die Erkenntnislehre des 'Aḍudaddīn al-Īcī* (Wiesbaden, 1966).—In *Der Islam* 45 (1969), pp. 137–9.
6) F.E. Peters, *Aristoteles Arabus* (Leiden, 1968).—In *Gnomon* 42 (1970), pp. 538–47.
7) F. Kholeif, *A Study on Fakhr Al-Dīn Al-Rāzī and his Controversies in Transoxiana* (Beirut, 1966).—In *OLZ* 66 (1971), col. 378–83.
8) C. Zurayk, *The Refinement of Character. A Translation from the Arabic of Aḥmad Ibn Muḥammad Miskawayh's* Tahdhīb al-Akhlāq (Beirut, 1968).—In *OLZ* 67 (1972), col. 370–73.
9) Lucien, *Dialogues des morts et Menippe. Traduction arabe avec introduction et notes par Elias Saad Ghali* (Beirut, 1967).—In *OLZ* 67 (1972), col. 465–7.

10) T. Izutsu, *Concept and Reality of Existence* (Tokyo, 1971).—In *Der Islam* 50 (1973), pp. 176–8.
11) *Imām Rāzī's 'Ilm Al-Akhlāq. English translation of his Kitāb al-Nafs Wa'l-Rūḥ wa Sharḥ Quwāhumā*, trans. M.Ṣ.Ḥ. Maʿṣūmī (Islamabad, 1969).—In *Der Islam* 50 (1973), pp. 174–5.
12) F. Rosenthal, *Knowledge Triumphant* (Leiden, 1970).—In *ZDMG* 123 (1973), pp. 410–21.
13) J. van Ess, *Das* Kitāb an-Nakṯ *des Naẓẓām und seine Rezeption im* Kitāb al-Futyā *des Ǧāḥiẓ* (Göttingen, 1972).—In *Der Islam* 51 (1974), pp. 155–60.
14) T. Nagel, *Frühe Ismailiya und Fatimiden im Lichte der Risālat Iftitāḥ ad-daʿwā* (Bonn, 1972).—In *OLZ* 73 (1978), col. 57–61.
15) M. Ghaleb (ed.), *Die Ismailitische Theologie des Ibrāhīm Ibn al-Ḥusain al-Ḥāmidī* (Wiesbaden, 1971).—In *OLZ* 73 (1978), col. 61–3.
16) Ibn Abī d-Dunyā, *The Noble Qualities of Characters*, ed. J.A. Bellamy (Wiesbaden, 1973).—In *OLZ* 75 (1980), col. 561–2.
17) E.A. Gruber, *Verdienst und Rang* (Freiburg/Br., 1975).—In *OLZ* 75 (1980), col. 562–4.
18) S. Diwald, *Arabische Philosophie und Wissenschaft in der Enzyklopädie Kitāb Ihwān aṣ-Ṣafā' (III)* (Wiesbaden, 1975).—In *OLZ* 76 (1981), col. 46–52.
19) B. Radtke, *Al-Ḥakīm at-Tirmidī* (Freiburg/Br., 1980).—In *BiOr* 39 (1982), col. 455–6.
20) *Das Vermächtnis des Islams*, ed. C.E. Bosworth (Zurich, 1980).—In *Der Islam* 59 (1982), pp. 319–20.
21) U. Weisser, *Das 'Buch über das Geheimnis der Schöpfung' von Pseudo-Apollonius von Tyana* (Berlin, 1980).—In *Der Islam* 59 (1982), pp. 326–32.
22) A.Y. Al-Hassan, Gh. Karmi and N. Manum (eds.), *Proceedings of the First International Symposium for the History of Arabic Science (April 5–12, 1976)*, ii *Papers in European Languages* (Aleppo, 1978).—In *Archives Internationales d'Histoire des Sciences* 109 (1982), pp. 287–9.
23) M. Asín Palacios, *The Mystical Philosophy of Ibn Masarra and his Followers* (Leiden, 1978).—In *OLZ* 78 (1983), col. 573–5.
24) D.M. Dunlop (ed.), *The* Muntakhab Ṣiwān Al-Ḥikmah *of Abū Sulaimān As-Sijistānī* (The Hague, 1979).—In *Der Islam* 60 (1983), pp. 321–2.
25) R.B. Serjeant, *Studies in Arabian History and Civilisation* (London, 1981).—In *Der Islam* 60 (1983), pp. 332–3.
26) I.R. Netton, *Muslim Neoplatonists* (London, 1982).—In *Der Islam* 62 (1985), pp. 139–41.

27) M.M. Qāsim, C.E. Butterworth and A.ʿA. Harīdī (eds.), *Averrois Cordubensis in Librum Aristotelis De interpretatione* (Cairo, 1981); C.E. Butterworth (trans.), *Averroes' Middle Commentaries on Aristotle's Categories and De Interpretatione* (Princeton, 1983).—In *Der Islam* 62 (1985), pp. 146–7.
28) E. Booth, *Aristotelian Aporetic Ontology in Islamic and Christian Thinkers* (Cambridge, 1983).—In *Der Islam* 62 (1985), pp. 143–5.
29) Kh. Samir and S. Nwiya (eds.), *Une correspondence islamo-chretienne entre Ibn al-Munağğim, Ḥunayn ibn Isḥāq et Qusṭā ibn Lūqā* (Turnhout, 1981).—In *ZDMG* 135 (1985), pp. 193–4.
30) J. Needham, *Science and Civilisation in China*, v (Chemistry and Chemical Technology, part. IV; Cambridge, 1980).—In *BiOr* 41 (1984, Leiden, 1985), col. 785–6.
31) M. Cook, *Early Islamic Dogma* (Cambridge, 1981).—In *BiOr* 41 (1984, Leiden, 1985), col. 764–6.
32) Ibn Sīnā, *Remarks and Admonitions, Part One: Logic*, trans. S.C. Inati (Toronto, 1984).—In *Der Islam* 62 (1985), pp. 338–9.
33) R.M. Frank, *Beings and their Attributes* (Albany, 1978).—In *OLZ* 82 (1987), col. 164–5.
34) Hazrat Inayat Khan, *Das Lied in allen Dingen. Sufi-Erzählungen und Gleichnisse vom Glück der Harmonie* (Freiburg, 1985).—In *Der Islam* 64 (1987), p. 172.
35) G. Vajda (ed. and trans.), *Al-Kitāb al-Muḥtawī de Yūsuf al-Baṣīr* (Leiden, 1985).—In *ZDMG* 137 (1987), p. 188.
36) E.L. Ormsby, *Theodicy in Islamic Thought* (Princeton, 1984).—In *Der Islam* 64 (1987), pp. 298–9.
37) G.F. Hourani, *Reason and Tradition in Islamic Ethics* (Cambridge, 1985).—In *Der Islam* 64 (1987), pp. 299–302.
38) *Ismail Hakki Bursevi's Translation of and Commentary on* Fusus Al-Hikam *by Muhyiddin Ibn ʿArabi*, trans. B. Rauf with R. Brass and H. Tollemache (Oxford, 1986).—In *Der Islam* 64 (1987), pp. 305–6.
39) A. Zimmermann (ed.), *Aristotelisches Erbe im arabisch-lateinischen Mittelalter. Übersetzungen, Kommentare, Interpretationen* (Berlin, 1986).—In *Der Islam* 64 (1987), pp. 308–10.
40) B.S. Kogan, *Averroes and the Metaphysics of Causation* (Albany, 1985).—In *Der Islam* 64 (1987), pp. 310–11.
41) W. Madelung (ed.), *Die Streitschrift* (Kitāb an-Nağāt) *des Zaiditenimams Aḥmad an-Nāṣir wider die ibaditische Prädestinationslehre* (Wiesbaden, 1985).—In *ZDMG* 138 (1988), p. 190.

42) J. Kraye, W.F. Ryan and C.B. Schmitt (eds.), *Pseudo-Aristotle in the Middle Ages. The Theology and Other Texts* (London, 1986).—In *Der Islam* 65 (1988), pp. 130–34.
43) Ibn aṭ-Ṭayyib, *Proclus' Commentary on the Pythagorean Verses*, ed. and trans. N. Linley (Buffalo, 1984).—In *Der Islam* 65 (1988), pp. 134–7.
44) *La Hidāya de Al-Rayrāyī. Un espejo de príncipes medieval*, ed. and trans. B.J. Calabozo (Madrid, 1983).—In *BiOr* 44 (1987, Leiden, 1988), col. 555.
45) *Al-Farabi on the Perfect State*, ed. and trans. R. Walzer (Oxford, 1985).—In *WI* 29 (1989), pp. 209–11.
46) J.L. Kraemer, *Philosophy in the Renaissance of Islam* (Leiden, 1986).—In *Der Islam* 66 (1989), pp. 374–5.
47) Ibn Rushd, *Averroes' Middle Commentary on Aristotle's Poetics*, trans. C.E. Butterworth (Princeton, 1986).—In *Der Islam* 66 (1989), pp. 376–8.
48) D. Gimaret, *Les noms divins en Islam* (Paris, 1988).—In *ZDMG* 140 (1990), p. 198.
49) Abū Ḥāmid Muḥammad al-Ghazālī, *Der Erretter aus dem Irrtum*, trans. A.A. Elschazli (Hamburg, 1988); id., *Die Nische der Lichter*, trans. A.A. Elschazli (Hamburg, 1987).—In *ZDMG* 140 (1990), pp. 159–60.
50) P. Kraus, *Jābir Ibn Ḥayyān* (Paris, 1986).—In *BiOr* 47/1–2 (1990), col. 236–7.
51) O. Leaman, *An Introduction to Medieval Islamic Philosophy* (Cambridge, 1985).—In *BiOr* 47/1–2 (1990), col. 237–9.
52) B. Holmberg (ed.), *A Treatise on the Unity and Trinity of God by Israel of Kashkar (d. 872)* (Lund, 1989).—In *ZDMG* 141 (1991 [1992]), pp. 401–2.
53) D. Gimaret, *La doctrine d'al-Ashʿarī* (Paris, 1990).—In *ZDMG* 142 (1992), p. 216.
54) G. Levi della Vida, *Pitagora, Bardesane e altri studi siriaci*, ed. R. Contini (Rome, 1989).—In *ZDMG* 142/2 (1992 [1993]), p. 380.
55) S.B. Abed, *Aristotelian Logic and the Arabic Language in Alfārābī* (Albany, 1991).—In *ZDMG* 142/2 (1992 [1993]), pp. 382–4.
56) F. Daftary (ed.), *Mediaeval Ismaʿili History and Thought* (Cambridge, 1996).—In *OLZ* 93 (1998), pp. 55–7.
57) T. Nagel, *Geschichte der islamischen Theologie. Von Mohammed bis zur Gegenwart* (Munich, 1994).—In *Der Islam* 76 (1999), pp. 158–60.

58) D. Gutas, *Greek Thought, Arabic Culture. The Graceo-Arabic Translation Movement in Baghdad and Early ʿAbbāsid Sopciety (2nd–4th/8th–10th Centuries)* (New York, 1998).—In *Classical World* 94/2 (New York, 2001), 215–6.
59) M. Ullmann, *Wörterbuch zu den griechisch-arabischen Übersetzungen des 9. Jahrhunderts* (Wiesbaden, 2002).—In *Oriens Christianus* 88 (2004), pp. 271–2.

Hans Daiber is co-editor of *Aristoteles Semitico-Latinus*, a project of the Union Académique Internationale in cooperation with the Royal Dutch Academy of Arts and Sciences, and of *Islamic Philosophy, Theology and Science. Texts and Studies.*

ABBREVIATIONS

AHDLMA	*Archives d'histoire doctrinale et littéraire du Moyen Age*
AI	*Annales Islamologiques*
ArScPh	*Arabic Sciences and Philosophy*
ASL	*Aristoteles Semitico-Latinus*
BEO	*Bulletin des Études Orientales*
BiOr	*Bibliotheca Orientalis*
BSOAS	*Bulletin of the School of Oriental and African Studies*
CCAA	*Corpus commentariorum Averrois in Aristotelem*
CC CM	*Corpus christianorum continuatio medievalis*
EI²	*Encyclopaedia of Islam*, CD Rom Edition (Leiden, 2007).
EIr	E. Yarshater (ed.), *Encylcopaedia Iranica*, available online under http://iranica.com/index.html
EQ	J. Dammen McAuliffe (ed.), *Encyclopaedia of the Qurʾān*, 6 vols. (Leiden, 2001–2006).
GAL	C. Brockelmann, *Geschichte der arabischen Litteratur*, 2 + 3 vols. (Leiden, ²1937–1946).
GAS	F. Sezgin, *Geschichte des arabischen Schrifttums*, 12 vols. (i–ix Leiden, 1967–2000; x–xii Frankfurt).
HdO	*Handbuch der Orientalistik*
IC	*Islamic Culture*
IJMES	*International Journal of Middle Eastern Studies*
IOS	*Israel Oriental Studies*
IPTS	*Islamic Philosophy, Theology and Science*
IQ	*Islamic Quarterly*
JA	*Journal Asiatique*
JAOS	*Journal of the American Oriental Society*
JIS	*Journal of Islamic Studies*
JJS	*Journal of Jewish Studies*
JNES	*Journal of Near Eastern Studies*
JQR	*Jewish Quarterly Review*
JRAS	*Journal of the Royal Asiatic Society*
JSAI	*Jerusalem Studies in Arabic and Islam*
JSS	*Journal of Semitic Studies*
MIDEO	*Mélanges de l'Institut Dominicain d'Études Orientales*
MUSJ	*Mélanges de l'Université Saint-Joseph*

OLZ	*Orientalistische Literaturzeitung*
REI	*Revue des Études Islamiques*
REJ	*Revue des Études Juives*
RSO	*Rivista degli studi orientali*
SI	*Studia Islamica*
WI	*Die Welt des Islam*
ZDMG	*Zeitschrift der Deutschen Morgenländischen Gesellschaft*
ZGAIW	*Zeitschrift für die Geschichte der Arabisch-Islamischen Wissenschaften*

ISLAMIC PHILOSOPHY AND THEOLOGY

SOME NOTES ON THE NOTION OF *NASKH* IN THE *KALĀM*

Binyamin Abrahamov

I) There are two doctrines of abrogation (*naskh*) in Islam: a. abrogation of certain qurʾānic commandments by other qurʾānic commandments, or of certain commandments of the *sunna* by qurʾānic commandments and vice versa;[1] and b. abrogation of previous divine laws (especially those of the Jews and the Christians) by the revelation of the Qurʾān.[2] The doctrine of *naskh* seems to imply the affirmation of the Shīʿite theory of *badāʾ* (literally: appearance), namely, God's revision of His knowledge, or will or command, for if God abrogates a previous law He gave, He changes His mind.[3] Perceived in its extreme form, *badāʾ* is incompatible with God's immutable and eternal knowledge.[4] Consequently, Muslim theologians have tried to elucidate *naskh* as a doctrine which does not entail God's changing His mind. Their efforts, however, have not been made only for the sake of maintaining consistency in

---

[1] J. Burton, 'Naskh,' in *EI*². Idem, 'The Exegesis of Q. 2:106 and the Islamic Theories of *Naskh*: *mā nansakh min āya aw nansahā naʾti bi khairin minhā aw mithlihā*,' *BSOAS* 48 (1985), pp. 452–69.

[2] R. Bell and W.M. Watt, *Introduction to the Qurʾān* (Edinburgh, 1977), pp. 86–9; Al-Suyūṭī, *al-Itqān fī ʿulūm al-Qurʾān* (Cairo, 1951), ii, pp. 20–27; Al-Baghdādī, *Uṣūl al-dīn* (Istanbul, 1928); Al-Juwaynī, *Kitāb al-irshād ilā qawāṭiʿ al-adilla fī uṣūl al-iʿtiqād*, ed. M.Y. Mūsā and ʿA.ʿA. ʿAbd al-Ḥamīd (Cairo, 1950), pp. 338–44; Al-Shahrastānī, *Nihāyat al-aqdām fī ʿilm al-kalām*, ed. A. Guillaume (Oxford, 1931), pp. 499–503.

[3] Shīʿite views on this doctrine differed. Some early theologians affirmed absolute *badāʾ*, using as their point of departure the doctrine of the Shīʿite theologian Hishām ibn al-Ḥakam (d. 179/795–796), that is, God's knowledge follows the existence of its objects. Some others of the Imamiyya moderate school of theology like al-Shaykh al-Mufīd (d. 413/1022) tried to exclude or at least minimise the possibility of change in God's knowledge; I. Goldziher and A.S. Tritton, 'Badāʾ,' in *EI*². Al-Ashʿarī, *Maqālāt al-islāmiyyīn wa-ikhtilāf al-muṣallīn*, ed. H. Ritter (Wiesbaden, 1963), p. 39, l. 4–15; Al-Shaykh al-Mufīd, *Awāʾil al-maqālāt fī ʾl-madhāhib al-mukhtārāt* (Tabriz, 1371 H), p. 43f; idem, *Taṣḥīḥ al-iʿtiqād* (Tabriz, 1371 H), pp. 24–6; W. Madelung, 'The Shīʿite and Khārijite Contribution to Pre-Ashʿarite *Kalām*,' in P. Morewedge (ed.), *Islamic Philosophical Theology* (New York, 1979), p. 123f; idem, 'Imamism and Muʿtazilite Theology,' in *Le Shiʿism Imamite, Colloque de Strasbourg* (Mai 1968) (Paris, 1970), p. 23; M.J. McDermott, *The Theology of al-Shaikh al-Mufīd* (Beirut, 1978), pp. 329–39.

[4] R. Brunschvig, 'L'Argumentation d'un théologien Musulman du X siècle contre le Judaisme,' in *Homenaje a Millás-Vallicrosa* (Barcelona, 1954), i, p. 234f.

Islamic dogma, but also in order to refute the Jews who rejected *naskh* because it necessitates *badāʾ*.[5] In fact, the question of whether God can change His previous law is expressed, as far as I know, for the first time in Islamic theology in a discussion between a Jew named Manasseh ibn Ṣāliḥ and the Muʿtazilite al-Naẓẓām (d. between 220–230/835–845).[6] In the following literature of *kalām* and *uṣūl al-fiqh* (the foundations of Islamic jurisprudence), the explanation of *naskh* has been repeatedly taken up mainly in the light of Jewish arguments. Religious polemics with the Jews undoubtedly contributed to the refinement of the doctrine of *naskh* in Islam.[7]

The well-known Muʿtazilite ʿAbd al-Jabbār[8] (d. 414–416/1023–1025) is a good example of a theologian who thoroughly summed up the teaching of the Baṣrian school of the Muʿtazila on the question of *naskh* and who approached it in terms of polemics with the Jews. In his vast theological encyclopedia[9] he devotes a considerable number of sections to a detailed discussion of the problem.[10] What follows is an examination and analysis of ʿAbd al-Jabbār's basic principles of the theory of *naskh*.[11] The views of certain Ashʿarites will also be put forward in order to attain a better understanding of Islamic theological views on *naskh*.

II) In a chapter entitled 'On the explanation of the meaning of abrogation and its reality' (*fī bayān fāʾidat al-naskh wa-ḥaqīqatihi*),[12] ʿAbd al-Jabbār sets forth the conditions by which *naskh* operates. He states that an act

---

[5] J. van Ess, *Frühe Muʿtazilitische Häresiographie* (Beirut, 1971), p. 64 and pp. 74–6 of the Arabic text; Ibn al-Murtaḍā, *al-Baḥr al-zakhkhār* (Cairo, 1975), i, p. 75; Brunschvig, 'L'Argumentation,' p. 225–41.

[6] A.S. Tritton, '"Debate" between a Muslim and a Jew,' *Islamic Studies* (Karachi) 1 (1962), pp. 60–64.

[7] J. Waardenburg, 'World Religions as Seen in the Light of Islam,' in A.T. Welch and P. Cachia (eds.), *Islam: Past Influence and Present Challenge* (Edinburgh, 1979), p. 255.

[8] See J.R.T.M. Peters, *God's Created Speech* (Leiden, 1976), pp. 8–25.

[9] ʿAbd al-Jabbār, *Al-Mughnī fī abwāb al-tawḥīd wa'l-ʿadl*, 16 vols. (Cairo, 1960–69) (henceforth *Mughnī*).

[10] M. Perlmann's note ('The Medieval Polemics between Islam and Judaism,' in S.D. Goitein [ed.], *Religion in a Religious Age* [Cambridge, Mass., 1974], p. 130, n. 4), stating that ʿAbd al-Jabbār discusses (vol. v) Christianity and dualism and ignores Judaism is not correct. On the contrary, ʿAbd al-Jabbār refutes Judaism in volumes xv, xvi of *Mughnī* which treat prophecy.

[11] ʿAbd al-Jabbār's theory occurs in volume xvi of *Mughnī*, pp. 49–142. Since his writing is imbued with repetitions, I shall try to introduce his main principles without necessarily following the order of the chapters.

[12] For the meanings of *fāʾida* and *ḥaqīqa*, see Peters, *God's Created Speech*, index.

of devotion (*ʿibāda*) becomes binding on man on account of an indication (*dalīl*). This indication is of two kinds: a. an indication which deals with one act of devotion. *Naskh* is not applicable to this case; and b. an indication which deals with the repetition and continuity of an act of devotion whose aspects are necessitated by the indication. By the term aspect the author means a certain time or condition.

This second class of indication is further sub-divided into three parts: The first is an indication obliging a law which is generally known by intuition (*ʿaql*) before the coming of revelation, such as showing gratitude in return for a favor (*shukr al-niʿma*). In this case the continuity of the imposition of obligations (*taklīf*) may cease by the perception of intuition.[13] Here *naskh* is not applicable, because *naskh* operates only through an indication of revealed law (*dalīl sharʿī*).[14] The second indication affirms a revealed law and is accompanied by several conditions, such as a certain time of performing the law or the physical ability of the person under obligation (*mukallaf*). In this case cessation of continuity is expected in a certain time or circumstance, but not total cessation. Thus also *naskh* does not apply to this case. The third indication applies neither to a certain time nor to a certain circumstance, but rather to the performance of an act of devotion when man is able and no hindrances prevent him from doing the act. So long as no indication necessitating the abrogation of the continuity of an act of devotion ensues, man is obliged to perform this act. But such an indication of abrogation is expected as the expected coming of the wind which effaces man's traces. The wind may come or not. So long as it does not come the traces exist, but the moment the wind comes, the traces are no longer existent.[15]

ʿAbd al-Jabbār explains this principle as follows:

---

[13] *Mughnī*, xvi, p. 92. For the explanation of this case, see below p. 7. For the approach of the Basrian Muʿtazilites to precepts obliged by intuition, see R.M. Frank, 'Several Fundamental Assumptions of the Baṣra School of the Muʿtazila,' *SI* 33 (1971), pp. 5–18. On the question of whether useful acts are prohibited, permitted or obligatory, see A.K. Reinhart, *Before Revelation. The Boundaries of Muslim Moral Thought* (New York, 1995).

[14] Mānkdīm, *Sharḥ al-uṣūl al-khamsa*, ed. ʿA. ʿUthmān (as a treatise of ʿAbd al-Jabbār) (Cairo, 1965) (henceforth Mānkdīm), p. 584. Here the author distinguishes between four kinds of indications: one which derives from the human intuition and the other three which derive from the divine revelation, namely, the Qurʾān, the *sunna* and the *ijmāʿ*; Peters, *God's Created Speech*, p. 67f.

[15] *Mughnī*, xvi, pp. 92–4.

> [...] Therefore we say concerning a messenger: 'If he summoned [the people to perform] a revealed law (*law da'ā ilā sharī'a*) for one year, a second messenger would not abrogate this law, for the *mukallaf* does not expect the second messenger because the continuity [of the revealed law] of the first ceases, but he expects a messenger only as intelligent people expect messengers and knows that [the obligation of] the first messenger's revealed law will be interrupted after the expiring of one year, whether another messenger will come or not. It is said concerning the second messenger that he will abrogate the first messenger's revealed law by his own law only when the first messenger summons [the people] to make [the performance of] this act continue (*idāmat dhālika al-fi'l*) without connecting it (namely, the continuity of performance) with a certain time. The logical possibility (*jawāz... min jihat al-'aql*) of the coming of the second messenger necessitates that whenever he comes and indicates (*dalla*) the cessation of the continuity of the first revealed law (*zawāl takrār al-shar' al-awwal*), he will be [considered] an abrogator (*nāsikh*).'[16]

'Abd al-Jabbār emphasizes that the second revealed law abrogates the first because it necessitates the cessation of its continuity. A substitute law, when contradicting a former law, has a twofold function: that of affirmation of itself (*ithbāt al-ḥukm*) as well as the cessation of the continuity of the law contradicted.[17] Thus abrogation, according to 'Abd al-Jabbār, denotes the cessation of the continuity of a previous law caused by a new law. This definition of *naskh* is connected, as we shall see, with 'Abd al-Jabbār's theory of *maṣlaḥa* (literally: benefit), that is, each law is good for a certain time.[18]

As we have seen, *naskh*, in 'Abd al-Jabbār's thought, applies to repetitive acts obliged by the revealed law. It does not apply to acts known generally by intuition before the coming of revelation. This point needs further explanation. 'Abd al-Jabbār distinguishes between the acts of the limbs (*af'āl al-jawāriḥ*) and the acts of the intelligence (literally: hearts) (*af'āl al-qulūb*).[19] By the acts of the intelligence he means man's knowledge of God (*ma'rifat allāh*), his knowledge of the obligatory force of obligations perceived by intuition (*al-ma'rifa bi-wujūb al-wājibāt al-'aqliyya*),

---

[16] Ibid., p. 95, l. 1–8.

[17] Ibid., l. 9–12.

[18] See below p. 7.

[19] Cf. 'Abd al-Jabbār, *Kitāb al-majmū' fī 'l-muḥīṭ bi'l-taklīf*, ed. J.J. Houben (Beirut, 1962), i, p. 366; Peters, *God's Created Speech*, p. 127. The distinction between the acts of the limbs and the acts of the intelligence already appeared in the writings of the theologian and mystic al-Ḥārith ibn Asad al-Muḥāsibī (d. 243/857). A. Goldreich, 'Possible Arabic Sources for the Distinction between "Duties of the Heart" and "Duties of the Limbs",' (Hebrew) *Te'udah* 6 (1988), pp. 179–208.

man's decision to fulfil what is necessary and the like. Whoever is *compos mentis* (*ʿāqil*) and capable of carrying out the acts of the intelligence, and no obstacles prevent him from fulfilling these acts, is obliged to fulfil them.[20] Concerning the acts of intelligence, there is no possibility of different ways of fulfilment, for the way these acts become a divine grace (*luṭf*)[21] is not connected with a certain time or with a certain *mukallaf*. Therefore these obligations must continue to exist so long as God's imposition (*taklīf*) exists, that is, these acts are obligatory on man in any time or state.[22] With regard to matters of human reason (*ʿaqliyyāt*), then, there is no difference in the states of whoever is *compos mentis*, for neither the way they are obligatory is specified, nor the way they are to be stopped. But with regard to matters of divine legislation (*sharʿiyyāt*), the states of the *mukallafs* may be different in different times, for *sharʿiyyāt* are based on man's benefits (*maṣāliḥ*)[23] which reason cannot attain (*lā ṭarīq lahā bi'l-ʿaql*).[24]

According to ʿAbd al-Jabbār, God's informing man of his obligation or making his obligation known (*ījāb*)[25] necessitates—concerning things lacking the quality of obligatoriness (*ṣifat al-wujūb*) known by reason—its being benefit (*maṣlaḥa*). If we do not describe God's *ījāb* as such, it means that He obliges man a thing concerning which *ījāb* is not good. But in ʿAbd al-Jabbār's view, all God's acts must be good.[26] The same holds true with regard to God's prohibition (*taḥrīm*) and making man

[20] *Mughnī*, xvi, p. 84, l. 4–8.

[21] For the theory of *luṭf* see my "Abd al-Jabbār's Theory of Divine Assistance (*luṭf*),' *JSAI* 16 (1993), pp. 41–57. ʿAbd al-Jabbār devotes a whole volume (xiii) of his *Mughnī* to a discussion of this theory. He defines *luṭf* as 'what motivates (*yadʿū*) man to carry out acts of obedience in a way according to which he has free choice (to carry out these acts or not). *Mughnī*, xiii, p. 9, l. 3–4. Man is led to know God by a motive of fear which He implants in him urging him to speculate on the world and consequently to attain knowledge about God. Ibid., xii, pp. 487–9; xvi, p. 85, l. 13–20. When this motive is absent, man is not obliged to know God; Ibid.; Cf. J. van Ess, 'Early Islamic Theologians on the Existence of God,' in K.I. Semaan (ed.), *Islam and the Medieval West. Aspects of Intercultural Relations* (Albany, 1980), pp. 64–81, especially p. 74ff; Frank, 'Several Fundamental Assumptions,' p. 16, n. 3.

[22] *Mughnī*, xvi, p. 84, l. 9–11.

[23] Most of the Muʿtazila held that God does what is best for man (*aṣlaḥ*). *Maṣlaḥa* (pl. *maṣāliḥ*) is derived from the same root (*ṣlḥ*) and can be rendered as advantage or benefit; cf. Peters, *God's Created Speech*, p. 90f; for a discussion of the term *aṣlaḥ* and other forms of the same root, see R. Brunschvig, 'Muʿtazilisme et Optimum (*al-aṣlaḥ*),' *SI* 39 (1974), pp. 5–23.

[24] *Mughnī*, xvi, p. 88, l. 9–11.

[25] G.F. Hourani, *Islamic Rationalism. The Ethics of ʿAbd al-Djabbār* (Oxford, 1971), p. 119f.

[26] Ibid., p. 109f.

desirous of acting (*targhīb fī 'l-fiʿl*). Since man's benefits may change from time to time, God's obligations must follow them accordingly. Therefore prayers and fasts are obligatory in certain states and on certain *mukallafs*. Consequently, it is possible that they should be abrogated or substituted by other laws. All other religious laws (*sharāʾiʿ*) follow the same principle.[27] The author realizes that such a principle would imply an uninterrupted chain of revealed laws abrogating each other.[28] His solution to the problem points to an indication to be found in revelation. If there is an indication which shows that God's law will be neither abrogated nor changed, abrogation and change will not take place. If not, the case will be as before the coming of revelation, namely, it will be possible for a new law to to be brought forth.[29]

The last point that is important to our discussion is the distinction ʿAbd al-Jabbār makes between two aspects of abrogation: a. emission and elimination (*isqāṭ* and *izāla*); and b. an opposite substitute (*badal muḍādd*). Abrogation in its two aspects does not apply to acts of devotion whose continuity is unceasing, such as one's knowledge of God.[30] Abrogation in the first aspect applies to acts whose continuity is stopped, but the opposite does not take place.[31] As for acts whose continuity

[27] *Mughnī*, xvi, p. 75f, 79; Mānkdīm, p. 577. ʿAbd al-Jabbār's theory of *naskh* is based on his ethics. The answer to the question of whether a certain act is evil (*qabīḥ*) or good (*ḥasan*) depends on circumstances, or the total character of the act. It is possible, for instance, that one's entrance into his friend's house should be good, because his friend permits him to enter, but if the entrance occurs without permission it is evil; *Mughnī*, xvi, p. 87; Mānkdīm, p. 577f; Hourani, *Islamic Rationalism*, pp. 64, 69–109, 105 n. 60.

[28] See Saadia's argument below p. 17f, n. 68.

[29] *Mughnī*, xvi, p. 90f.

[30] In his work on *uṣūl al-fiqh* entitled *al-Mustaṣfā min ʿilm al-uṣūl*, a summa of the science of jurisprudence ([Cairo, 1322 H], p. 122f), the Ashʿarite theologian and mystic al-Ghazālī (d. 505/1111) states that 'there is no unabrogatable religious law (*ḥukm sharʿī*) in contradiction to the Muʿtazilites, for they have said that it is inadmissible to abrogate acts having intrinsic qualities (*ṣifāt nafsiyya*) which necessitate their being good or evil. These are acts such as one's knowledge of God, may He be exalted, doing just acts (*ʿadl*) and showing gratitude towards a benefactor (*shukr al-munʿim*). It is inadmissible to abrogate the obligatoriness of these acts, and the prohibition of acts such as unbelief, wrong-doing (*ẓulm*) and lying (*kadhb*) as well. They (the Muʿtazilites) have based this (theory) on intuition which declares acts to be either good or evil, and on the duty of doing the best (*al-aṣlaḥ*) incumbent upon God...' Cf. idem, *Kitāb al-iqtiṣād fī 'l iʿtiqād* (Cairo, n.d.) (henceforth *Iqtiṣād* 1), p. 55f; idem, *idem*, ed. I. Agâh Çubukçu and H. Atay (Ankara, 1962) (henceforth *Iqtiṣād* 2), p. 184.

[31] The example ʿAbd al-Jabbār brings is showing gratitude for a favor (*shukr al-niʿma*). The benefactor may spoil his good graces and deal badly with the person whom he benefits. As a result, the benefited person is not obliged now to thank the benefactor. Thus showing gratitude may be stopped; it is no longer obligatory, but ingratitude is not permitted. Thus showing gratitude is different from man's obligation to know

may be stopped and their opposites take place, it is possible that they should be abrogated in line with these two aspects.[32]

Now that we have laid out the basic elements of ʿAbd al-Jabbār's theory of *naskh*, it is appropriate to show his answer to the question: does *naskh* entail *badā*ʾ? He states that abrogation of a law or a commandment necessitates *badā*ʾ only when the abrogating law deals with the same *mukallaf*, act, time and aspect of the abrogated law. To illustrate this statement he gives the following example: A master says to his servant: 'When sun sets and you enter the market, buy meat.' Soon afterwards he says to him: 'When sun sets and you enter the market, don't buy meat.' This change of decision is called *badā*ʾ, because something appears to the master, concerning the buying of the meat, of which he knew nothing before. The second order directed to the same *mukallaf* is not to do the same act (buying) at the same time (sunset) and according to the same aspect (in the market).[33] Correctly applied, the term *badā*ʾ refers only to him who does not know a thing, then knows it, or to him who no longer knows what he has known, or to one who may undergo a change of assumption (*ẓann*) and convictions (*iʿtiqādāt*).[34]

ʿAbd al-Jabbār's thesis of *naskh* leads him to the refutation of the Jews. Here his point of departure is the principle of *al-aṣlaḥ* (or *ṣalāḥ*) referred to above. God always does what is best for man. God's imposition of duties on mankind is one of His best acts. Since what is best for man may change according to circumstances and times, it is possible that a revealed law will change. Moreover, continuity of a certain law means for ʿAbd al-Jabbār evilness of God's imposition of precepts (*qubḥ al-taklīf*).[35] Thus *naskh* of a revealed law is admissible, and as we

---

God, which does not change so long as imposition is valid. Man's knowledge of God resembles showing gratitude only generally (*ʿalā ṭarīq al-jumla*), that is, the knowledge of the obligatoriness of both, but speaking in terms of circumstances, each case is different; *Mughnī*, xvi, p. 86, l. 1–7.

[32] Ibid., p. 88, l. 16–p. 89, l. 3.

[33] *Mughnī*, xvi, pp. 62–4, 109f, Mānkdīm, p. 584f.

[34] *Mughnī*, ibid., p. 65f; Cf. Abū 'l-Ḥusayn al-Baṣrī, *Kitāb al-muʿtamad fī uṣūl al-fiqh*, ed. M. Ḥamidullah (Damascus, 1964), i, p. 398f.

[35] *Mughnī*, xvi, pp. 79, 81, 101. Here ʿAbd al-Jabbār seems to undermine his own thesis, for it is also possible that Islamic law would be liable to change because of the change of circumstances and times. His solution mentioned above (p. 5) to the effect that if there comes an indication telling people that a certain law will not cease, it will not cease, does not solve the difficulty, because it implies that change no longer occurs, and this contradicts of course what man observes in the world. However, ʿAbd al-Jabbār does not ignore this difficulty stating that a commandment to carry out an act forever is accompanied by an announcement of its benefit. See below p. 10f.

have seen does not entail *badāʾ*. This contradicts the arguments of the overwhelming majority of the Jews that *naskh* is impossible because it necessitates *badāʾ*, and also Moses said that his Law obliges forever.[36] ʿAbd al-Jabbār answers the last argument by asserting that a commandment or a prohibition accompanied by words denoting 'always' or 'forever' means fixing a time or an aim. 'When one says to his fellow: "attach to someone always (*abadan*) and learn always and come to me always", the meaning is only fixing the time (*tawqīt*)...it is as if whoever expresses this phrase indicates that he means by this attachment "until one attains his wish", in order to distinguish between this case and an attachment for one measure [of time] or for special measures [of time].'[37] Now, if such words are said by God, it means that God says to man: 'do it until you attain your wish.' If one does not know when his wish can be attained, then God must indicate to him either by an affirmative report (*khabar*) or by a prohibition (*nahy*).[38]

ʿAbd al-Jabbār asserts that man will perform his act forever only if God says: 'This act will be from this *mukallaf* in his states of being obliged only benefit'[39] meaning that the commandment is accompanied with an announcement which indicates its benefit. *Naskh* does not apply

---

[36] Ibid., pp. 97, 99. The wording 'do it (the law) forever' (*ifʿalū abadan*) expressed in the text by the Jews (p. 99, l. 14) does not occur in the Bible; cf. al-Bāqillānī, *Kitāb al-tamhīd*, ed. R.J. McCarthy (Beirut, 1957), p. 176, n. 9. It is rather implied in verses stating that the covenant between God and the Children of Israel exists forever; see, for example, *Deuteronomy* 29:13–14; *Judges* 2: 2. Cf. Saadia, *Kitāb al-amānāt waʾl-iʿtiqādāt*, ed. J. Kafiḥ (Jerusalem, 1970), p. 132; trans. S. Rosenblatt (New Haven, 1967), p. 157f.

[37] *Mughnī*, xvi, p. 99, l. 21–p. 100, l. 4.

[38] Ibid., p. 100, l. 4–7; Here an adversary might have raised an objection: each language differs from another in the usage of words and phrases. Thus one cannot bring an example from one language to serve as a general true statement (ʿAbd al-Jabbār does not even bring pieces of evidence in Arabic [*shawāhid*] to prove that 'do it always' means 'do it up to the attainment of your aim'). Furthermore, as we know, he rejects the thesis that language originally derives from divine guidance (*tawqīf*; *Mughnī*, xv, p. 106; R.M. Frank, *Beings and their Attributes. The Teaching of the Baṣrian School of the Muʿtazila in the Classical Period* [Albany, 1978], p. 29f, n. 10; Peters, *God's Created Speech*, p. 387, n. 8; B.G. Weiss, 'Medieval Muslim Discussions of the Origin of Language,' *ZDMG* 124 [1974], p. 39). ʿAbd al-Jabbār generally holds that language is meaningful only through convention (*muwāḍaʿa*; *Mughnī*, vii, p. 101; Weiss, ibid., n. 15). Here, however, he states that with regard to this question the contextual use or meaning of a phrase (*taʿāruf*; *Mughnī*, xvi, p. 100, l. 8–11; Frank, *Beings*, index) is preferable over convention because a word according to *taʿāruf* is immediately understood (literally: immediately present: *waʾl-taʿāruf aḥḍaru min al-muwāḍaʿa*. Read *aḥḍaru* instead of the *akhṣaru* in the text, which does not make sense). Evidently *taʿāruf* is particular to each language, therefore what ʿAbd al-Jabbār states may be true only in Arabic, but not in other languages. Consequently, refutation by such an argument seems to be inefficient.

[39] *Mughnī*, xvi, p. 106, l. 7–8.

to this possibility. He understands the phrase 'making something act forever' (*lafẓat al-taʾbīd*) as a repetitive commandment.[40] This meaning of the phrase is demonstrated by the following analogy: Just as ability is a condition of performing an act, so benefit may also serve as a condition of carrying out an act.[41] Just as when a man is unable, he is not obliged to carry out the act imposed upon him, also when the act is harmful to him. Our author, then, distinguishes between a commandment to perform a law successively (*taʾbīd*) on the one hand, and a commandment to perform a law accompanied by an announcement which states that the law will not be abrogated.

> And His saying, may He be exalted: "I shall not abrogate My law" is like the announcement which we have mentioned, for it is an announcement that this commandment will not disappear so long as imposition exists. This necessitates that the situation of the act will not change. Likewise if the messenger says: "my law is binding always, and it is nothing but benefit, and the prophethood was completed by me,[42] and I have been sent to all the *mukallafs* and similar other phrases".[43]

The comparison just mentioned between a commandment accompanied with a condition of benefit and an act whose performance is conditioned by the ability of the *mukallaf* leads ʿAbd al-Jabbār to another argument against the Jews. According to Moses' Law, ʿAbd al-Jabbār says, acts of devotion may not be performed by reason of excuses such as disease which allows man not to fast even if he is capable. So imposition disappears if a difficulty is involved. Now, he argues,

---

[40] Saadia brings forward five kinds of possible laws to prove that a law promulgated by God cannot be abrogated: a. A law promulgated by God for performance forever (*sharʿ muʾabbad*) cannot be abrogated; b. A law promulgated by God for a certain time cannot be abrogated in this time, and after this time abrogation does not apply to it, for its time has expired; c. A law promulgated by God whose performance is restricted to a certain place. It cannot be abrogated in this place. Another law which contradicts the former but given for performance in another place cannot be considered as abrogation of the former; d. A law promulgated by God whose performance is connected with a certain cause. So long as the cause exists this law cannot be abrogated; and e. A law which is promulgated without fixing a time limit of its performance. So man goes on performing it until he is commanded to do otherwise. The time, however, of its performance must be known by God before, and by man when he is informed of a second law. Thus, abrogation does not apply here either, for the time of the law is known. Saadia, *Kitāb al-amānāt*, p. 132f; trans. Rosenblatt, p. 158f; this is exactly what ʿAbd al-Jabbār regards as cessation of continuity which he interprets as abrogation! See al-Ghazālī's view below.

[41] *Mughnī*, xvi, pp. 103, 104, l. 1–3, 106–107.

[42] Qurʾān 33:40; *Tafsīr Ibn Kathīr* (Beirut, 1970), v, pp. 469–73.

[43] *Mughnī*, xvi, p. 107, l. 5–9.

if a commandment necessitated continuity so long as the *mukallaf* is capable, it would be impossible that one should not fast because of a disease, for he is still capable, even if the difficulty is great. They (the Jews), says the author, could not say in this case that an indication of intuition (*dalīl al-ʿaql*) is connected with the announcement (*khiṭāb*) as they say concerning inability (*ʿajz*).[44] The conclusion implied in ʿAbd al-Jabbār's argument is that the Jewish Law itself does not preserve its continuity when allowing a capable person not to carry out his duties; it acknowledges *naskh* within itself.[45]

In sum, *naskh*, according to ʿAbd al-Jabbār, applies to acts of devotion imposed by the religious law alone. Every act of this kind was promulgated by God on account of its benefit to human beings. *Naskh* does not imply *badāʾ*, for *naskh* does not apply to the same aspects of an act obliged by God. ʿAbd al-Jabbār's refutation of the Jews is mainly based on his theory of benefit.

In the following passages I shall introduce the views of some Ashʿarites on *naskh*.

III) The Ashʿarite theologian al-Bāqillānī (d. 404/1013)[46] was influenced by the Muʿtazilites on this issue. He bases his theory of *naskh*—which is put forward within the range of his refutation of the Jews—on the Muʿtazilite notion of *maṣlaḥa*: A commandment given in a certain time

[44] Ibid., p. 108, l. 4–10. As we have seen, *naskh*, according to ʿAbd al-Jabbār, applies only to an indication of a religious law.

[45] A Jewish adversary might have answered ʿAbd al-Jabbār's argument by pointing out that Jewish Law establishes that man should not endanger himself when performing acts of devotion. So the duty of saving human life overrides the Sabbath laws, the fast laws and other laws. It is not a question of how difficult it is to carry out the duty, but of danger to life (even in a doubtful case of saving a life). When such a circumstance arises the law should be disregarded for the benefit of the person whose life is imperiled; *Encyclopaedia Judaica* (Jerusalem, 1971), xiii, pp. 509–10. This apparent abrogation forms a part of the Sabbath laws themselves. It operates under special conditions and at a certain time and does not apply to every person. Moreover, a person whose life is imperiled cannot be always spoken of as a capable person.

It should be noted that the ancient sages of Israel (the second and third centuries C.E.), having been disputed by the Christians, held that the Revelation at Sinai—which is eternal and cannot be changed even by God Himself—consists of the written Law (Torah), the Oral Law (*Mishna* and *Talmud*), all the legends and prophecies, and even what scholars would teach. All these elements have existed from the time of the Revelation, but have been revealed in different times; E.E. Urbach, *The Sages, their Concepts and Beliefs* (Jerusalem, 1975) (Hebrew), p. 270ff; B.J. Bamberger, 'Revelations of Torah after Sinai,' *Hebrew Union College Annual* 16 (1941), p. 97.

[46] See R.J. McCarthy, 'Al-Bāqillānī,' in *EI*^{2}.

is beneficial and good but in another time, may be harmful and evil. Eating and drinking are good for the hungry and the thirsty, respectively, but evil for the satiated and well watered. Likewise, revealed acts of devotion such as fast, prayer, turning to Jerusalem in prayer and rest on Saturday are good in one time but evil in another.[47]

It is interesting to note that al-Bāqillānī does not refer to the Ashʿarite theory of attributes—as one might have expected—in answering the following argument against *naskh*. His opponents argue that giving a commandment (God's promulgating of a law) implies that whoever commands something wills it, while prohibition (God's abrogation of this law) implies that whoever prohibits something does not will it. Thus it is impossible that the object willed by God (*murād*, that is the law) is willed and not willed by Him. The opponents' point of departure is based on the identification of God's will with His commandment, a thesis propounded by the Muʿtazilites.[48] Al-Bāqillānī answers by referring to the principle mentioned above, namely, "the object willed in one time is not the same object willed in another time".[49] Thus a Muʿtazilite answer is given to a Muʿtazilite question by an Ashʿarite theologian. The same answer is given by the author when he is confronted with the question of *naskh* which entails *badāʾ*. He even goes farther stating that *naskh* may occur before the performance of the act commanded by God. If whoever commands knows that the commandment is no longer a benefit to man, he may abrogate the commandment. He also emphasizes that the prohibition does not come so long as the commandment exists but with its passage. Thus, in order to abstain from the joining of contraries, namely, commandment and prohibition, he fixes a time in which a commanded act must be carried out and another time in which the same act is prohibited by God.[50]

---

[47] Al-Bāqillānī, *Kitāb al-tamhīd*, ed. R.J. McCarthy (Beirut, 1957) (henceforth *Tamhīd*), p. 184f; Brunschvig, 'L'Argumentation,' p. 234f. Nowhere do the *mutakallimūn* explain why these acts are evil, or why they were good in times past.

[48] *Mughnī*, xvi, p. 58. L. Gardet, *Dieu et la destiné de l'homme* (Paris, 1967), p. 217; M. Schwarz, 'Some Notes on the Notion of *iljāʾ* (Constraint) in Muʿtazilite Kalām,' *IOS* 2 (1972), p. 414.

[49] *Tamhīd*, p. 185, art. 317.

[50] Ibid., p. 186, art. 319. Al-Baṣrī, i, p. 406.

IV) Also al-Ghazālī's (d. 505/1111) explanation of *naskh* and *badā'* occurs in connection with the Jews.[51] Al-Ghazālī defines the term *naskh* as an announcement (*khiṭāb*) indicating the cancellation of a fixed rule (*irtifāʿ al-ḥukm al-thābit*). It is not impossible that a master should command his servant unrestrictedly: Stand up! without stating the time of his standing up. The master knows how long it is beneficial for the servant to stand up, but does not reveal this to him. The servant understands that he is obliged to stand up until such time as his master commands him to sit down. One must not suppose that a benefit unknown before appears (*badā*) to the master, but it is admissible that the master should know three things of benefit concerning the servant: a. the time of standing up; b. the time of sitting down; and c. concealment of the knowledge of the two times from the servant. In this manner, al-Ghazālī concludes, one must understand the difference in the rules of revelations (*aḥkām al-sharāʾiʿ*), which is due to the difference of times and situations. *Naskh* indicates no change or knowledge after ignorance or self-contradiction (*tanāquḍ*) on the part of God. It just indicates that the performance of a law is limited by God to a certain time known to Him alone. The Jews, according to al-Ghazālī, could not disregard the Laws of Noah and Abraham which came before Moses' Law, that is, they could not deny that Moses' Law abrogated what had been promulgated before.[52]

In sum, al-Ghazālī adopts the principle of the Muʿtazilite theory of *naskh*; that is, divine law brings benefits to human beings which are known to God from eternity. Each benefit is restricted to a certain time. But unlike the Muʿtazilites, he holds that there is no distinction between revelational laws (*samʿiyyāt*) and intellectual laws (*ʿaqliyyāt*). When admitting that *naskh* does not apply to the majority of the laws but only to certain laws such as the changing of the point toward which Muslims turn in praying (*qibla*) and making the forbidden lawful (*taḥlīl muḥarram*) and the like,[53] al-Ghazālī, however, may hint, contrary to what he said elsewhere (see note 30 above), that other principal duties (for instance, man's knowledge of God) cannot be abrogated.

---

[51] *Iqtiṣād* 1, p. 91f; *Iqtiṣād* 2, p. 202f. On al-Ghazālī see W. Montgomery Watt s.v. in *EI*².

[52] *Iqtiṣād* 1, p. 92. *Iqtiṣād* 2, p. 203f. Saadia answers this argument saying that Moses' Law did not abrogate Abraham's Law; the former is the same as the latter with the addition of some precepts promulgated by God as a consequence of events which befell the children of Israel. Saadia, p. 135; trans. Rosenblatt, p. 162f.

[53] *Iqtiṣād* 1, p. 92, l. 13–14. *Iqtiṣād* 2, p. 204, l. 4–6.

V) Al-Ghazālī's teacher al-Juwaynī (d. 478/1085)[54] states that the Muʿtazilites hold that *naskh* means the termination of the period of a law (*intihāʾ muddat sharīʿa*). As we have seen above it is the view al-Ghazālī also adopted. Al-Juwaynī mentions that some orthodox scholars inclined to this view.[55] However, in his opinion *naskh* means 'an announcement (*khiṭāb*) indicating the elimination of a fixed law (*irtifāʿ al-ḥukm al-thābit*) by another announcement in such a way without which the abrogated law may continue [to be valid]'.[56] He asserts that the *naskh* of an act of devotion is possible before the termination of its period, but it is impossible in spite of this notion that *naskh* is an announcement of the termination of the period of devotion, for it is impossible that God should fix a time for an act of devotion in which this act cannot not be performed; there can be no commandment without time, thus the abrogation applies to the commandment itself, not to its time. Although al-Juwaynī uses the same phrase (*irtifāʿ al-ḥukm al-thābit*) used by al-Ghazālī, he means something different by it. Whereas al-Ghazālī considers the cause of abrogation to be *maṣlaḥa* (thus *naskh* means *irtifāʿ muddat al-sharīʿa*), al-Juwaynī holds that it is God's eternal will and knowledge unperceived by man's intuition.

Al-Juwaynī rejects the Muʿtazilite notion of *naskh* as the termination of the period of a law because, I suppose, he rejects the theory of *maṣlaḥa*, which is connected with times and circumstances. He holds that God's omnipotence and omniscience cannot be understood by man's intuition. Accordingly, a commandment may be changed before it is performed by man. Abraham was commanded first to slaughter his son, but before he performed the act this commandment was abrogated. The essence of what he was commanded to do was slaughter (*ʿayn al-maʾmūr bihi huwa al-dhabḥ*).[57] It was not a series of continual different acts so that the commandment turned to one act[58] and the abrogation to another. So, al-Juwaynī sums up, if *naskh* turns to the essence of what was commanded, it is real elimination (*rafʿ al-ḥukm ʿalā al-taḥqīq*).[59]

Subsequently, al-Juwaynī explains the basis of his method. He states that there is no rational stage between logical possibility (*jawāz*) and

---

54 See on him C. Brockelmann and L. Gardet, 'Al-D̲juwaynī,' in *EI*².
55 Al-Juwaynī, *Kitāb al-irshād*, p. 339, l. 6–10. Cf. al-Baghdādī, *Uṣūl al-dīn*, p. 226.
56 Al-Juwaynī, *Kitāb al-irshād*, ibid., l. 3–4.
57 For the term *ʿayn* see S. van den Bergh, 'ʿAyn,' in *EI*².
58 Literally: 'to one thing' (*ilā shayʾin*) according to the reading of ms. m.
59 Al-Juwaynī, *Kitāb al-irshād*, p. 340. Cf. Tritton, '"Debate" between a Muslim and a Jew,' p. 62.

logical impossibility (*istiḥāla*). Transformation of genera (*inqilāb al-ajnās*) and joining of contraries (*ijtimāʿ al-ḍiddayn*) are considered by him impossible by virtue of themselves. A commandment to do what was forbidden is not impossible by virtue of itself. If such a commandment is not impossible by virtue of itself, it does not make others' acts impossible. Thus, it is not impossible for God to command to carry out an act after He forbade its carrying out. That is, because this change does not entail the change of an essential attribute of God (*ṣifa nafsiyya*).[60] Making a law is not an essential attribute of God.[61]

Now al-Juwaynī lays himself open to the charge mentioned before, that is, *naskh* is impossible for it implies the description of God by *badāʾ*. His doctrine of attributes serves, however, to meet this charge.[62] *Badāʾ*, says al-Juwaynī, is knowledge coming after ignorance or regretting a command after it is willed. God's knowledge and will are both eternal. Moreover, God's will is not identical with His commandments; He commands what He does not will and wills what He does not command.[63] Consequently, God has known from eternity whatever would be, so one cannot ascribe to Him knowledge after ignorance, neither can one ascribe to Him regret at His commandments, for there is no connection between His will and His commandments. In sum, there is no way to maintain *badāʾ*.

VI) Generally al-Shahrastānī (d. 548/1153)[64] seems to follow al-Juwaynī. He begins his discussion by mentioning the definitions of *naskh* as abrogation of a valid law (*rafʿ al-ḥukm baʿda thubūtihi*) and as the termination of the period of a law (*intihāʾ muddat al-ḥukm*), and succeeds in stating that neither *badāʾ* nor *nadam* (regret) can be ascribed to Him who knows even the smallest particle in the world. Like al-Juwaynī, he states that *naskh* of a commandment is not impossible by virtue of itself nor does it bring about something impossible.[65]

---

[60] For *ṣifa nafsiyya* see al-Juwaynī, *Kitāb al-irshād*, p. 30. According to al-Juwaynī, this term denotes a positive attribute of the essence inseparable from the latter so long as the essence exists. Contrary to *ṣifa maʿnawiyya*, it is not explained by causes inherent in the qualified thing. Cf. M. Allard, *Le problème des attributes divins dans la doctrine d'Al-Ašʿarī et de ses premiers grands disciples* (Beirut, 1965), p. 387; ʿA. ʿUthmān, *Naẓariyyat al-taklīf. Ārāʾ ʿAbd al-Jabbār al-kalāmiyya* (Cairo, 1971), pp. 169–96.

[61] Al-Juwaynī, *Kitāb al-irshād*, p. 340f.

[62] Ibid., pp. 94–9.

[63] Ibid., p. 347, l. 7–14 and above note 48.

[64] See G. Monnot, 'Al-Shahrastānī,' in *EI*[2].

[65] Al-Shahrastānī, *Nihāyat al-aqdām*, p. 499f.

Al-Shahrastānī expressly rejects the Muʿtazilite notion of *maṣlaḥa*. Since the causes of God's sayings and acts cannot be known (*aqwāluhu wa-afʿāluhu lā tuʿallalu*), one cannot ascribe to Him action according to an aim or a benefit. If this is so, *badāʾ* does not apply to Him, because *badāʾ* means to regret what was done according to an aim or a benefit.[66]

Contrary to the Muʿtazilites, he explains that prohibition (*ḥaẓr*) and obligatoriness (*wujūb*) are not attributes of acts nor do acts themselves have attributes of good or evil affirmed by the revealed law. The fact that an act is prohibited or obligatory, namely its judgment (*ḥukm*), has its basis in the Lawgiver's statements (*aqwāl al-shāriʿ*), neither in an intrinsic attribute nor in an attribute acquired through the Lawgiver's statements. Acts are qualified only through these statements which can be abrogated.[67]

The fact that al-Shahrastānī regards the judgments of acts as merely God's statements devoid of any intrinsic value seems to aim at two targets: a. to disconnect God's acts from the perception of human intuition, contrary to the Muʿtazilite view; and b. to emphasize that *naskh* is not impossible by virtue of itself, because it deals with statements, not with attributes (see below al-Āmidī's view).

Al-Shahrastānī brings examples to prove that there are contradictions and changes in God's acts. First, in the sphere of law we see that marriage can be abrogated by a definite repudiation (*ṭalāq mubīn*), and that laws applying to man who stays in one place are different from laws applying to a traveler. Examples regarding man's condition are also given. God gives life and puts man to death, He makes man ill and cures him and so on. Likewise, plants and animals are liable to change. The stages of man's creation from the sperm to a perfect creature are compared to the different laws which have existed before Islam. Just as each stage in man's creation abrogates the stage before, the same is true of God's laws which abrogated one another until Islam, the perfect law, arrives.[68]

---

[66] Ibid., p. 500, l. 8–14.

[67] Ibid., p. 501, l. 18, p. 502, l. 3.

[68] Ibid., p. 502. This comparison seems to be faulty, since each law in its time has been a complete law, while there is no completeness in each stage of the stages of man's creation. Furthermore, some of the stages (flesh, bones) are never 'abrogated'; they remain part of the human being throughout life.

The arguments al-Shahrastānī employs were already well known in Saadia's era. Saadia refutes them in ch. 3, art. 7 of his *Book of Beliefs and Opinions*. He begins by referring to the argument drawn from analogy, that is, just as it is possible for God to

VII) In accordance with al-Bāqillānī and al-Ghazālī and in opposition to al-Juwaynī and al-Shahrastānī, the Ashʿarite *mutakallim* Sayf al-Dīn al-Āmidī (d. 631/1233),[69] who was well versed in philosophy, held the Muʿtazilite notion of *maṣlaḥa* concerning abrogation. God commands man unrestrictedly to do something knowing that at a certain time He will abrogate this commandment for man's benefit.[70] Like ʿAbd al-Jabbār, al-Āmidī rejects the notion advanced by al-Bāqillānī that God's commandment may be abrogated before its being carried out by man.[71]

The term *rafʿ* (literally: elimination)—used by some Ashʿarites to denote abrogation—is, according to al-Āmidī, nothing but the cessation of continuity; it is neither the abrogation of whatever exists nor of whatever does not exist. Al-Āmidī adds a principle, which other theologians pay no attention to, similar to the one advanced by al-Shahrastānī: God's cessation of the continuity of a law must not be ascribed to His speech, His attribute, concerning which non-existence is impossible. *Naskh*, then, means the interruption of the connection of God's speech to human beings.[72] But this last notion may raise another question: Is it possible that within God's attribute of speech there should be two self-contradictory announcements, one affirming a commandment and

give life and then put to death, so it is possible for him to legislate laws and then to abrogate them. Saadia refutes this argument by stating that God gives life in order to put to death only because death is the means to life in the world to come, which is the ultimate goal of man. But laws were not given by God for the purpose of their future abrogation. If a law had been given for the purpose of its abrogation, there would have been an endless series of laws, therefore this is a false notion; Saadia, p. 133, trans. Rosenblatt, p. 160. Furthermore, Saadia says, a law subject to abrogation has always an inner antithesis and contradiction, because it serves as a means to an end which is the following abrogating law; Ibid.

Saadia refutes also the argument drawn from analogy to the effect that just as God changes man's situation, that is, makes him rich or poor, seeing or blind, so He changes the laws. There is a difference, according to Saadia, between these two categories of action. Whereas the former can be explained as a consequence of God's reward or punishment, the latter cannot be explained so; Saadia, p. 134; trans. Rosenblatt, p. 161.

Likewise, the change of a law cannot be compared to changes in the nature of things such as the change of the color of a date to red when it ripens, for changes in nature take place either because of the inner constitution of a thing or habit, both of which cannot be ascribed to God's laws. If God's laws had followed such a pattern, every law would have been abrogated; Saadia, p. 134f; trans. Rosenblatt, p. 161.

[69] See D. Sourdel, 'Al-Āmidī,' in *EI*[2].

[70] Sayf al-Dīn al-Āmidī, *Ghāyat al-marām fī ʿilm al-kalām*, ed. Ḥ. Maḥmūd ʿAbd al-Laṭīf (Cairo, 1971), p. 358. Idem, *al-Iḥkām fī uṣūl al-aḥkām* (Cairo n. d.), iii, p. 101.

[71] Al-Āmidī, *Ghāyat al-marām*, p. 359, l. 1–3.

[72] Ibid., l. 3–10.

the other denying it? That is, also, the abrogating law, as a qurʾānic commandment, derives from God's attribute of speech, even if the act of abrogation does not belong to this attribute.

VIII) Concluding remarks: a. We have seen that out of five eminent Ashʿarite theologians, three (al-Bāqillānī, al-Ghazālī and al-Āmidī) were partly influenced by Muʿtazilite notions. The fact that Muʿtazilism not only influenced Ashʿarism regarding methods of argumentation but also respecting certain notions and terms is well known, but still needs further investigation.[73] This influence was not limited to issues of *uṣūl al-dīn*.[74]

b. Both al-Juwaynī and al-Shahrastānī are remote from Muʿtazilite influence. Their point of departure is the thesis that God can do every act except an act which is logically impossible. Thus a law can be abrogated even before its being performed. They also use the Ashʿarite doctrine of attributes to show that *naskh* does not imply *badāʾ*.

c. We have seen the important position of intuition in Muʿtazilite thought; only what is necessitated by intuition is not abrogated so long as imposition exists.

d. The arguments of the *mutakallimūn* are not irrefutable. This fact, which has been shown here and elsewhere, has led some Muslim scholars to hold *takāfuʾ al-adilla* (equality of contradictory proofs),[75] while some others, like al-Ghazālī and Ibn Taymiyya, have shown the danger involved in using *kalām*.[76]

---

[73] Cf. B. Abrahamov, *Islamic Theology, Traditionalism and Rationalism* (Edinburgh, 1998), especially ch. 2.

[74] W.M. Watt, 'Ashʿariyya,' in *EI*².

[75] Cf. J. van Ess, 'Skepticism in Islamic Religious Thought,' *Al-Abḥāth* 21 (1968), i, p. 7.

[76] H. Lazarus-Yafeh, *Studies in al-Ghazālī* (Jerusalem, 1975), pp. 373–90; B. Abrahamov, 'Ibn Taymiyya on the Agreement of Reason with Tradition,' *MW* 82 (1992), pp. 256–72.

AVICENNA'S DOCTRINE OF THE PRIMARY NOTIONS AND ITS IMPACT ON MEDIEVAL PHILOSOPHY

Jan A. Aertsen

The reception of Arabic philosophy in the Latin West is an important but still fragmentarily known chapter in the history of Western thought. In an erudite essay, Hans Daiber has stated that the influence of the Latin translations of Arabic philosophical texts on Scholastic thought 'has as yet by no means been exhaustively discussed.'[1] This conclusion motivated me when choosing the theme of my contribution to this volume. It is concerned with the medieval reception of a well-known doctrine of Avicenna's philosophy, whose significance for the history of Western metaphysics is generally underestimated. His original teaching on the primary notions of the intellect had an immense impact on Latin philosophy; it captivated medieval thinkers like Thomas Aquinas, Henry of Ghent and John Duns Scotus. They not only adopt Avicenna's doctrine, but develop it in a critical and productive way. The reasons for this fascination as well as the nature of their adoption and critical transformation of the doctrine deserve closer scrutiny.[2] To that end we shall first (1) examine Avicenna's motive for the introduction of primary concepts. Since he lists a plurality of such notions, we next consider them separately: (2) the concepts 'thing' and 'being', and (3) the status of the concept 'one'. By way of conclusion (4) we attempt to characterize Avicenna's doctrine as a whole in the light of the medieval reception.

---

[1] H. Daiber, 'Lateinische Übersetzungen arabischer Texte zur Philosophie und ihre Bedeutung für die Scholastik des Mittelalters. Stand und Aufgaben der Forschung,' in J. Hamesse and J. Fattori (eds.), *Rencontres de cultures dans la philosophie médiévale* (Leuven, 1990), pp. 203–50, p. 204.

[2] Some aspects of this reception were discussed by E. Gilson, 'Avicenne en Occident au Moyen Age,' *AHDLMA* 44 (1969), pp. 89–121. R. Schönberger, *Die Transformation des klassischen Seinsverständnisses. Studien zur Vorgeschichte des neuzeitlichen Seinsbegriffs im Mittelalter* (Berlin, 1986), pp. 95–121.

(1) The Beginning of Thought: 'The First Impressions in the Soul'

Avicenna's doctrine of the primary notions has a place in his *Metaphysics* that suits its primacy. In the first treatise of this work, the Arab philosopher first deals with what one could call 'the Prolegomena' of his metaphysics: its subject-matter (ch. 1–2), utility (ch. 3) and contents (ch. 4).[3] The constructive part of his account in fact starts in the fifth chapter, entitled 'On Indicating the Existent (*ens*), the Thing (*res*) and their First Divisions', in which he develops the doctrine of the primary notions of the intellect.[4] The chapter begins with the following statement: '"Thing" (*res*), "being" (*ens*) and "the necessary" (*necesse*) are such notions that they are impressed immediately in the soul by a first impression (*prima impressio*) and are not acquired from other and better known notions.'[5] This programmatic statement from Avicenna is probably the text from his *Metaphysics* most frequently cited by medieval authors. Two aspects are noteworthy in his doctrine.

(i) Why is it necessary to accept primary notions? The structure of Avicenna's argument does not become very transparent in his exposition; its force rests on an analogy between two orders of knowledge, the order of 'assent' (*taṣdīq*; in the Latin translation *credulitas*) and that of 'conception' (*taṣawwur*; in the Latin translation *imaginatio*). He does not explain these terms which have been called 'the cornerstones of medieval Arabic epistemology'.[6] But Algazel, who in the Middle Ages

---

[3] Cf. A. Bertolacci, 'The Structure of Metaphysical Science in the *Ilāhiyyāt* (Divine Science) of Avicenna's *Kitāb al-Shifā'* (*Book of the Cure*),' *Documenti e Studi sulla tradizione filosofica medievale* 13 (2002), pp. 1–69, in particular p. 22.

[4] For the Latin translation of the text: Avicenna latinus, *Liber de philosophia prima sive scientia divina* I, c. 5 (ed. S. Van Riet [Leuven, 1977], pp. 31–42). On this chapter, see the 'Introduction doctrinale' by G. Verbeke in the same volume of the Avicenna latinus, pp. 31*–39*. M.E. Marmura, 'Avicenna on Primary Concepts in the Metaphysics of his al-Shifā',' in R.M. Savory and D.A. Agius (eds.), *Logos Islamikos. Studia Islamica in Honorem Georgii Michaelis Wickens* (Toronto, 1984), pp. 219–39 (with an Engl. trans.). J. Jolivet, 'Aux origines de l'ontologie d'Ibn Sina,' in J. Jolivet and R. Rashed (eds.), *Études sur Avicenne* (Paris, 1984), pp. 11–28 (repr. in J. Jolivet, *Philosophie médiévale arabe et latine* [Paris, 1995], pp. 221–36). A. de Libera, *L'art des généralités. Théories de l'abstraction* (Paris, 1999), pp. 579–90 (with a French trans. of Avicenna's text, pp. 645–53).

[5] Avicenna latinus, *Liber de philosophia prima* I, c. 5 (ed. Van Riet, pp. 31–2): 'Dicemus igitur quod res et ens et necesse talia sunt quod statim imprimuntur in anima prima impressione, quae non acquiritur ex aliis notioribus se'.

[6] D.L. Black, *Logic and Aristotle's 'Rhetoric' and 'Poetics' in Medieval Arabic Philosophy* (Leiden, 1990), p. 71. Cf. H.A. Wolfson, 'The Terms *Tasawwur* and *Tasdiq* in Arabic Philosophy and their Greek, Latin and Hebrew Equivalents,' in idem, *Studies in the*

was regarded as Avicenna's faithful student, gives in his *Logic* a description of 'the first two parts of science', according to which the order of *credulitas* concerns the domain of propositions, the order of *imaginatio* that of concepts.[7]

The analogy Avicenna has in mind consists in the claim that in both orders a reduction is necessary to first principles that are known *per se*. Just as there are first principles, known through themselves, in the realm of assent, so also in the realm of conception there are principles that are conceived *per se* and do not require any prior conception. If one desires to indicate them to somebody, his doing so would thus not make an unknown thing known, but would merely draw attention to them or bring them to mind through the use of a sign.[8]

In Avicenna's argument, the first member of the analogy, the order of *credulitas*, is the better known, because he takes the Aristotelian analysis of the structure of demonstrative knowledge (*scientia*) for granted. Since 'science' is grounded knowledge, i.e. a *habitus* that is produced by demonstration, what is scientifically knowable in the proper sense are the conclusions of a demonstration, for these propositions meet the demand for foundation. From this it follows that science is always derived from something prior, insofar as the conclusion is deduced from propositions previously known. This structure, however, raises the problem of the ultimate foundation of science, for the reduction (Gr. 'analysis', Lat. 'resolutio') to something prior seems to lead to an infinite regress. Aristotle solves this problem by concluding that the first principles of science cannot be demonstrated, since the search for a foundation would imply either an infinite regress, which is impossible,

---

History of Philosophy and Religion, i (Cambridge, Mass., 1973), pp. 478–92. The medievals were aware of the Arabic background of the terminology. See, for instance, Thomas Aquinas, *De spiritualibus creaturis* a. 9 ad 6.

[7] Algazel, *Tractatus de Logica* 1 (ed. C.H. Lohr, *Traditio* 21 [1965], p. 239): 'Scientiarum, quamvis multi sint rami, duae tamen sunt primae partes, imaginatio et credulitas. *Imaginatio* est apprehensio rerum, quas significant singulae dictiones ad intelligendum et certificandum eas. Sicut est apprehensio significationis huius nominis, "lapis", "arbor", "angelus" "spiritus" et similium. *Credulitas* vero est sicut hoc quod dicitur, quia "Mundus coepit", et "Obedientia remunerabitur".'

[8] Avicenna latinus, *Liber de philosophia prima* I, c. 5 (ed. Van Riet, p. 32): 'Sicut credulitas quae habet prima principia, ex quibus ipsa provenit per se, et est alia ab eis, sed propter ea (...). Similiter in imaginationibus sunt multa quae sunt principia imaginandi, quae imaginantur per se, sed, cum voluerimus ea significare, non faciemus per ea certissime cognoscendi ignotum, sed fiet assignatio aliqua transitus ille per animam nomine vel signo quod aliquando in se erit minus notum quam illud, sed per aliquam rem vel per aliquam dispositionem fiet notius in significatione'.

or a circular argument. Therefore the first principles of science are not derived from something else, but are immediately known.[9] In his *Metaphysics*, Aristotle shows that the first principle of demonstration, which he calls the *anhypotheton* of thought, is the principle that 'the same thing cannot at the same time belong and not belong to the same thing in the same respect'.[10]

Avicenna's originality consists in his application of the finite structure of *scientia* to the order of concepts as well. For Aristotle, the ascent to the most general predicate of which a definition is constituted terminates in the ten *genera generalissima*, the categorial diversity of being. For Avicenna, it is a logical complement of the Aristotelian analysis that the impossibility of an infinite regress and the reduction to a first likewise holds for the order of concepts. Just as propositions must be reduced to first indemonstrable principles, so too in the order of *imaginatio* there must be primary notions. 'If every conception requires a prior conception, then this state of affairs would lead either to an infinite regress or to circularity.'[11]

In his argument, Avicenna aims at a systematic beginning of human thought. His discovery of first notions realizes an ambition of metaphysics in its search for a first: our knowledge starts from first principles through which all subsequent knowledge has to be gathered. It is this aspect that may explain the strong medieval interest in his doctrine. *Secundum Avicennam* and *secundum rei veritatem* there are 'firsts' in what is conceived by the intellect, which are referred to as the *primae intentiones*, *primae conceptiones* or *prima intelligibilia* in the thirteenth century.[12] Most authors avoid Avicenna's expression *primae impressiones* because of its

---

9 Aristotle, *Post. Anal.* I, c. 3.

10 Aristotle, *Metaphysics* iv, c. 3, 1005b 14–20. Avicenna deals with this principle in the final chapter of the first treatise: *Liber de philosophia prima* I, c. 8 (ed. Van Riet, p. 56): 'Prima vero omnium dictionum certorum ad quam perducitur quicquid est per resolutionem (...) est cum inter affirmationem et negationem non est medium'.

11 Avicenna latinus, *Liber de philosophia prima* I, c. 5 (ed. Van Riet, p. 33): 'Si autem omnis imaginatio egeret alia praecedente imaginatione, procederet hoc in infinitum vel circulariter'.

12 Henry of Ghent uses the expression '*secundum Avicennam et secundum rei veritatem*' in the *Summa quaestionum ordinarium* a. 22, q. 5 (ed. Paris, 1520, fol. 134) and a. 25, q. 3 (156rS). Some examples of the terminology: *primae intentiones*: Philip the Chancellor, *Summa de bono* q. 9 (ed. N. Wicki [Bern, 1985], p. 30); Henry of Ghent, *Summa* a. 24, q. 7 (fol. 144rH); *primae conceptiones*: Thomas Aquinas, *Super Boethium De trinitate* q. 6, a. 4 (ed. Leonina, vol. 50 [Rome, 1992] p. 170); id., *Quodlibet* viii, q. 2, a. 2 (ed. Leonina, xxv, 1 [Rome, 1996], p. 59); *primum intelligibile*: Alexander of Hales, *Summa theologica* I, n. 72 (ed. Quaracchi, p. 113); Thomas Aquinas, *Summa theologiae* I, q. 5, a. 2.

epistemological implications; his terminology reflects an extrinsic view of the origin of the primary notions, insofar as they are seen as the direct impressions by the cosmic active Intellect. The term *conceptio*, by contrast, expresses the inner activity of the human intellect in the forming of these notions.[13]

The impact of Avicenna's doctrine can be seen in two accounts of the transcendentals, those of Aquinas in *De veritate* q. 1, a. 1 and Henry of Ghent in his *Summa* a. 34, q. 3. Both thinkers prepare their accounts by adopting the Avicennian analogy between the two orders of intellectual knowledge. Aquinas describes them as the order of demonstrable propositions and the order of the investigation into what something is. Henry indicates them with the terms *intellectus complexus*, which connects concepts in a proposition, and *intellectus incomplexus*. In both orders a reduction (*reductio*) is necessary to a first that is known *per se* and therefore *notissimum*.[14]

(ii) Avicenna's analogy argument is strictly formal; it does not say which concepts are the primary notions of the intellect. He twice presents a list of them. In the opening statement of chapter 5 he mentions 'thing', 'being' and 'the necessary'. Later in the same chapter, immediately after the analogy argument, he establishes that 'what is most suited to be conceived through itself is that which is common to all things (*ea quae communia sunt omnibus rebus*), as are "thing", "being" and "one".'[15] Between the two lists there exist some differences, which, however, Avicenna does not discuss.

In the first list, the conceptual primacy is accounted for by the impossibility of acquiring these notions from other and better known notions. In this idea it is implied that the primary concepts cannot be

---

[13] Cf. Aquinas's critique of Avicenna in *De veritate* q. 10, a. 6.

[14] Thomas Aquinas, *De veritate* q. 1, a. 1 (ed. Leonina, xxii [Rome, 1970], pp. 4–5): 'Dicendum quod sicut in demonstrabilibus oportet fieri reductionem in aliqua principia per se intellectui nota ita investigando quid est unumquodque, alias utrobique in infinitum iretur, et sic periret omnino scientia et cognitio rerum'. Henry of Ghent, *Summa quaestionum ordinarium* a. 34, q. 3 (ed. R. Macken, *Opera Omnia*, xxvii [Leuven, 1991], p. 190): 'Sciendum quod, sicut in intellectu complexo oportet fieri reductionem in aliquod primum principium complexum omnino intellectui per se notum—aliter enim procederetur in infinitum—, sic in conceptu intellectus incomplexi de eo quod quid est (...), oportet omnes huiusmodi conceptus reducere ad aliquem conceptum incomplexum primum et notissimum, in quem omnes alii habent reduci, et qui inclditur in omnibus aliis'.

[15] Avicenna latinus, *Liber de philosophia prima* I, c. 5 (ed. Van Riet, p. 33): 'Quae autem promptiora sunt ad imaginandum per seipsa, sunt ea quae communia sunt omnibus rebus, sicut res et ens et unum, et cetera'.

defined. Any attempt in that direction, Avicenna observes, rather conceals these notions. If someone were to say, for example, 'the reality of 'being' (*ens*) consists in being either active (*agens*) or passive (*patiens*)', that person uses a division of being that is less known than 'being'. All men conceive 'being' without knowing at all that it must be either active or passive.[16] Another implication is that the primary notions are the condition for all further conceptual knowledge. Henry of Ghent clearly expresses this priority: Nothing can be known and understood as such, for instance, as 'man' or 'white', when it is not first known and understood under the notion of 'being' and 'one', that is, as 'being' or 'one'. These notions are necessarily conceived to belong to a thing by a first impression, at least according to a logical priority—, before that thing is conceived as 'man' or 'white'.[17]

In Avicenna's second list, the conceptual firstness is related to the commonness of these notions. That may be the reason that 'the necessary' is absent here, because it rather belongs to the first division of 'being'. In accordance with the title of the chapter that announces such a division, Avicenna, at the end of the chapter, deals with the modal concepts 'necessary', 'possible' and 'impossible'. It is striking that, when the medievals quote the opening statement of the fifth chapter, they always leave out 'the necessary'.

The primary notions are the *communissima*;[18] because of their universal predicability they transcend the categories of being that Aristotle had distinguished. So medieval texts refer to Avicenna's primary notions as *transcendentia*.[19] This reference illustrates an important development in

[16] Avicenna latinus, *Liber de philosophia prima* I, c. 5 (ed. Van Riet, p. 33): 'Unde quisquis voluerit discurrere de illis incidet in volucrum, sicut ille qui dixit quod certitudo entis est quod vel est agens vel patiens, quamvis haec divisio sit entis, sed tamen ens notius est quam agens vel patiens. Omnes enim homines imaginant certitudinem entis, sed ignorant an debeat esse agens vel patiens'.

[17] Henry of Ghent, *Summa quaestionum ordinarium* a. 24, q. 7 (ed. Paris, 1520, fol. 144rH): 'Nihil enim talium cognoscitur in creatura aut intelligitur ut tale nisi prius cognoscendo ei intelligendo ipsum sub intentione entis et unius et caeterarum primarum intentionum, ut quod sit ens aut unum, quae necessario prima impressione saltem prioritate naturae concipiuntur de quolibet antequam concipitur aliquid eorum quia album aut quia homo'.

[18] Cf. Duns Scotus, *Quaestiones super libros Metaphysicam* I, q. 10, n. 6 (ed. The Franciscan Institute, p. 182): '...secundum Avicennam, "communissima prima impressione imprimuntur"'.

[19] See, for instance, Anonymus (quidam Scotista), *Duae quaestiones ordinariae de conceptibus transcendentibus*, q. 1 (ed. S.F. Brown and S.D. Dumont, *Mediaeval Studies* 51 [1989], p. 39): '...per Avicennam, I *Metaphysicae* suae, ubi dicit quod ens et res et huiusmodi transcendentia prima impressione imprimuntur in anima'.

medieval philosophy: Avicenna's identification of the primary notions with the most common concepts was incorporated into the doctrine of the transcendentals that was formed in the thirteenth century. *Transcendentia* are the 'firsts' (*prima*) in a cognitive respect, the first conceptions of the intellect.[20] Does this mean that Avicenna's chapter on the primary notions essentially was a doctrine of the transcendentals? An answer to this question should be postponed until the conclusion of our essay.

(2) 'Thing' and 'Being'

Avicenna's exposition in the fifth chapter of the first treatise is focused on the two basic notions that are mentioned in both lists, *ens* and *res*. The introduction of the latter term is surprising, because in the preceding chapters he had shown that being-as-being is the proper subject of metaphysics without making any reference to *res*.[21] The introduction of *res* is also remarkable, since the term does not have an antecedent in Aristotle's *Metaphysics*. The Latin Avicenna was the origin of the career of *res* in medieval philosophy. But why did this concept enter philosophy and what was its significance for our understanding of 'reality'?[22] Some clues as to why 'thing' was introduced are provided by Avicenna's analysis of the relation between *ens* and *res*: how they differ from each other and how they are identical.

He first shows that they have different meanings. In all languages, he states, *res* signifies something different from *ens*. Every 'thing' has a 'stable

---

[20] Cf. J.A. Aertsen, 'What is First and Most Fundamental? The Beginnings of Transcendental Philosophy,' in J.A. Aertsen and A. Speer (eds.), *Was ist Philosophie im Mittelalter?* (Berlin, 1998), pp. 305–21.

[21] On the concept of *res* in Avicenna, R. Wisnovsky, 'Notes on Avicenna's Concept of Thingness (*šay'iyya*),' *ArScPh* 10 (2000), pp. 181–222. T.-A. Druart, '"Shay" or "Res" as Concomitant of "Being" in Avicenna,' *Documenti e studi sulla tradizione filosofica medievale* 12 (2001), pp. 125–42.

[22] There exists no comprehensive study on *res* as a philosophical concept. A good overview is offered by J.F. Courtine, 'Res,' *Historisches Wörterbuch der Philosophie*, viii (Basel, 1992), pp. 892–901. The volume *Res. Atti del iii° Colloquio internazionale del Lessico intellettuale europeo*, ed. by M. Fattori and M. Bianchi (Rome, 1982), contains two relevant contributions: J. Hamesse, 'Res chez les auteurs philosophiques du 12e et 13e siècles ou le passage de la neutralité à la spécificité,' pp. 91–104; L. Oeing-Hanhoff, 'Res comme concept transcendental et sur-transcendental,' pp. 285–96. See also R. Darge, 'Suarez' Analyse der Transzendentalien "Ding" und "Etwas" im Kontext der scholastischen Metaphysiktradition,' *Theologie und Philosophie* 75 (2000), pp. 339–58; J.A. Aertsen, '"Res" as Transcendental: its Introduction and Significance,' in G. Federici Vescovini (ed.), *Le problème des transcendantaux du xive au xviie siècle* (Paris, 2002), pp. 139–57.

nature' (*certitudo*) through which it is what it is. Thus the *certitudo* of a triangle is that whereby it is a triangle, that of whiteness that whereby it is white. The 'certitude' of a thing may be called its 'proper being' (*esse proprium*). This is different from the 'affirmed being' (*esse affirmativum*), signified by the term *ens*, which is synonymous with *aliquid*—in the Latin translation *aliquid* stands for two Arabic terms that mean 'what is established' and 'what is realized'. Avicenna is hesitant with respect to the phrase *esse proprium*—an expression that attracted the attention of medieval readers; in the concluding summary he describes the *certitudo* of a thing as its 'whatness' (*quidditas*).[23]

What was Avicenna's motive for introducing *res*? The surprising answer is that this notion in itself does not contain anything new. The Avicennian 'thing' is related to the *certitudo* of a thing, it signifies its 'whatness'. *Res* expresses the Greek tradition of intelligibility, which centers on the quiddity of a thing by posing the question as to *what* it is. What is new in Avicenna's account is not the introduction of *res*, but rather the conceptual differentiation between *res* and *ens*, which signifies *that* something is. The basis for this differentiation is an ontological distinction unknown to Aristotle, but fundamental to Arabic metaphysics, namely the distinction between 'essence' and 'existence'.[24] As Avicenna observes, each thing has a *certitudo* proper to it, which is 'something other than the *esse* that is synonymous with what is affirmed (*aliquid*)'.[25]

After having shown the conceptual difference between *res* and *ens*, Avicenna emphasizes their extensional identity. *Ens* is a necessary concomitant (*concomitantia*) of *res*: 'The concept of *ens* is always concomitant with *res*, because the thing has being either in the singulars or in the estimation and intellect. If it were not so, it would not be a

---

[23] Avicenna latinus, *Liber de philosophia prima* i, c. 5 (ed. Van Riet, pp. 34–5): 'Dico ergo quod intentio entis et intentio rei imaginantur in animabus duae intentiones; ens vero et aliquid sunt nomina multivoca unius intentionis nec dubitabis quin intentio istorum non sit iam impressa in anima legentis hunc librum. Sed res et quicquid aequipollet ei, significat etiam aliquid aliud in omnibus linguis; unaquaeque enim res habet certitudinem qua est id quod est, sicut triangulus habet certitudinem qua est triangulus, et albedo habet certitudinem qua est albedo. Et hoc est quod fortasse appellamus esse proprium, nec intendimus per illud [nisi (must be deleted in accordance with manuscript A)] intentionem esse affirmativi. (...) Redeamus igitur et dicamus quod (...) est hoc quod unaquaeque res habet certitudinem propriam quae est eius quidditas'.

[24] Cf. A.M. Goichon, *La distinction de l'essence et de l'existence d'après Ibn Sina (Avicenne)* (Paris, 1937).

[25] Avicenna latinus, *Liber de philosophia prima* I, c. 5 (ed. Van Riet, p. 35): 'Et notum est quod certitudo cuiuscumque rei quae propria est ei, est praeter esse quod multivocum est cum aliquid'.

thing.'[26] There follows an extensive refutation of the view that denies the extensional identity of *res* and *ens* and holds that *res* is the more comprehensive notion, insofar as there are non-existing things.

Avicenna's account of the primary notions left a question open, which turns out to be of importance for the history of its reception. He distinguishes a plurality of first notions, states their difference and identity, but does not discuss their order. Does the notion of *res* precede that of *ens*? Avicenna does not give an explicit answer; it is, however, significant that he uses the term *concomitans* to express the relation of 'being' (and also, as we shall see, of 'one') to 'thing'. The term suggests a conceptual priority of *res*, insofar as notions that 'accompany' 'thing' are later than than that which is 'accompanied'.[27]

Avicenna's account was read in this sense by several medieval authors. But when *res* has a certain priority, another question presents itself: Should First Philosophy then not be the science of 'thing as thing'? As we shall see, a fourteenth-century thinker, Francis of Marchia, did not hesitate to draw this conclusion. The medieval reception of Avicenna's exposition on 'thing' and 'being' was various. We give three examples that represent different manners of interpretation: the priority of *ens* to *res*, their identification and the priority of *res* to *ens*.

(i) A first witness of the penetration of Avicenna's doctrine into the Latin West is *Aquinas*'s account of the transcendentals in *De veritate* q. 1, a. 1. After having adopted the Avicennian analogy between the order of demonstration and the order of concepts, he states: 'That which the intellect first conceives, as best known, and into which it resolves all its conceptions, is "being" (*ens*)'.[28] Thomas supports this thesis by a reference to Avicenna's *Metaphysics*, but in fact he modifies Avicenna's exposition. Whereas the Arabic philosopher names a plurality of primary notions, Aquinas reduces them to one single concept. He does not deny the plurality of first notions, but sees a conceptual order among them, in which *ens* has a clear priority. 'Being' is the first

---

[26] Avicenna latinus, *Liber de philosophia prima* i, c. 5 (ed. Van Riet, p. 36): 'Nec separabitur a comitantia intelligendi ens cum illa ullo modo, quoniam intellectus de ente semper comitabitur illam, quia illa habet esse vel in singularibus vel in aestimatione vel intellectu. Si autem non esset ita, tunc non esset res'.

[27] Contra T.-A. Druart (n. 21), in particular the claim on p. 130 that Avicenna 'always gives precedence to "being" over "thing"'.

[28] Thomas Aquinas, *De veritate* q. 1, a. 1 (ed. Leonina, p. 5): 'Illud autem quod primo intellectus concipit quasi notissimum et in quod conceptiones omnes resolvit est ens, ut Avicenna dicit in principio suae Metaphysicae'.

transcendental and *primum intelligibile*, because something is knowable only insofar as it *is* in act.

Aquinas gives a place to the Avicennian notion of *res* in his derivation of the other transcendentals. These add something to 'being' in the sense that they express a general mode of being that is not yet expressed by the term 'being' itself. What is expressed by the name *res* is a mode of being that pertains to every being in itself and can be said positively of every being, namely that it has an essence. Thomas explains the distinction between *ens* and *res* by another reference to Avicenna's *Metaphysics*: the name *ens* is taken from 'the act of being' (*actus essendi*), while the name *res* expresses the 'quiddity' or 'essence' of something.[29] He interprets Avicenna's concept of *ens* in terms of the 'act of being', a phrase that is typical of his own understanding of being as 'actuality'.

Has Aquinas really managed to incorporate *res* into the doctrine? It seems that the Avicennian 'thing' does not fit well into his systematization. The tension is due to the fact that he explains the convertibility of the transcendentals, not in a purely extensional manner but in a more intrinsic way, namely from the inner modes of the first transcendental, 'being'. *Res*, however, is the only transcendental that is not based on the act of being, but on the other component in the ontological structure of things, the essence or quiddity, that is really different from their *esse*.[30] In this sense, *res* possesses a peculiar position within Aquinas's doctrine.

(ii) Exemplary for another model of interpretation is *Henry of Ghent*, the most influential thinker in the generation after Aquinas. In his account of the transcendentals, he establishes, like Thomas, 'being' as the first concept of the intellect. But he determines that which is first known not only as *ens*, but as *ens inquantum ens*, the traditional expression for the subject-matter of metaphysics. He refers to Avicenna for this phrase, but, although the latter adopts the Aristotelian formula, the Arab philosopher does not employ it in his exposition of the primary

---

[29] Thomas Aquinas, *De veritate* q. 1, a. 1 (ed. Leonina, p. 5): '(...) non autem invenitur aliquid affirmative dictum absolute quod possit accipi in omni ente nisi essentia eius secundum quam esse dicitur, et sic imponitur hoc nomen res, quod in hoc differt ab ente, secundum Avicennam in principio Metaphysicae, quod ens sumitur ab actu essendi sed nomen rei exprimit quiditatem vel essentiam entis'.

[30] Thomas Aquinas, *In I Sent.* d. 8, q. 1, a. 1; d. 25, q. 1, a. 4: 'Secundum Avicennam (...) hoc nomen "ens" et "res" differunt secundum quod est duo considerare in re, scilicet quidditatem et rationem ejus, et esse ipsius; et a quidditate sumitur hoc nomen "res"'.

notions. The identification of the first concept of the intellect with the proper 'subject' (*subiectum*) of metaphysics is Henry's innovation. This science is called 'First Philosophy', because it deals with that which is first known.[31]

A significant difference from Aquinas consists in Henry's explanation of the *ratio entis*. '"Being" signifies the *certitudo*, whereby everything is what it is in its nature and essence *absolute*, without any qualification (*condicio*) or addition.'[32] It is striking that Henry determines the meaning of *ens* through the concept of *certitudo*, because it is the term used by Avicenna in his description of the notion *res*. Henry identifies the concept of 'being' with Avicenna's notion of 'thing'. How was this step possible?

Henry knows very well that in Avicenna's *Metaphysics* the notions of *res* and *ens* are distinct. In his third *Quodlibet*, he paraphrases the fifth chapter to this effect, that the concept of *res*, signifying the proper *certitudo* of a thing, is different from the concept whereby *esse* is ascribed to it. Henry's reading even assigns priority to the former notion: 'Thing' is the absolutely first concept (*prima simpliciter*); its concomitant is 'being', by reason of which the *certitudo* of a thing has existence in the mind or in the extra-mental world.[33]

In this *Quodlibet*, the key to Henry's identification of 'thing' and 'being' becomes perceptible. He understands the *esse proprium*, a concept that Avicenna tentatively suggested in his description of the *certitudo* of a thing, as a distinctive kind of being, the *esse essentiae*.[34] The 'essential being' of a thing is its proper reality that has to be distinguished from its 'actual existence' (*esse existentiae*), since the essence is indeterminate

---

[31] Henry of Ghent, *Summa* a. 34, q. 3 (ed. R. Macken, p. 190): 'Et est iste ‚conceptus entis in quantum ens est', secundum Avicennam in I° *Metaphysicae*'. On Henry's identification, M. Pickavé, 'Heinrich von Gent über das Subjekt der Metaphysik als Ersterkanntes,' *Documenti e Studi sulla tradizione filosofica medievale* 12 (2001), pp. 493–522 (in particular p. 512).

[32] Henry of Ghent, *Summa* a. 34, q. 3 (ed. R. Macken, p. 190): 'Et hoc est quod significat certitudinem, qua est unumquodque id quod est in natura et essentia sua absolute, absque omni conditione et additione'.

[33] Henry of Ghent, *Quodlibet* III, q. 9 (ed. Badius, fol. 61rO): 'Quoniam intentio de re est intentio prima simpliciter, ad quam concomitatur intentio de esse, ex hoc scilicet quod certitudo rei qua est id quod est, secundum se habet esse in anima (...) aut in singularibus extra, et sequitur secundum rationem intelligendi intentio de esse intentionem de re'.

[34] Cf. P. Porro, 'Possibilità ed *esse essentiae* in Enrico di Gand,' in W. Vanhamel (ed.), *Henry of Ghent. Proceedings of the International Colloquium on the Occasion of the 700th Anniversary of his Death (1293)* (Leuven, 1996), pp. 211–53.

and prior to any existence. A thing can thus be called 'being' (*ens*), not because it actually exists, but insofar as it has an essence or a quiddity capable to extramental existence. Since 'essential being' is considered prior to 'actual existence', the concept of 'being' referring to the *esse essentiae* is what is first conceived by the intellect. According to Henry it is this primary notion of 'being', of which Avicenna is speaking.[35] In contrast to Aquinas, Henry clearly understands 'being' in a quidditative sense: 'reality' versus 'actuality' of being.

(iii) A third manner of interpretation was presented by the Franciscan *Francis of Marchia*, a highly original thinker who was active at the University of Paris around 1320. In his *Questions on Metaphysics*, he raises as the first question: 'Whether *res secundum quod res* is the subject of metaphysics or something else?'[36] Its phrasing is noteworthy: in the traditional formulation of the subject-matter of metaphysics, *ens* is replaced by *res*.

Francis's reply begins with a comparison of the concept of 'thing' to other concepts that are above, equal to (*aequale*) and under 'thing'. What matters is the second group, to which the notions convertible with 'thing' pertain, like the transcendentals 'being', 'one', 'true' and 'good'. None of these transcendental notions is the first subject of metaphysics. Francis advances the argument that a property cannot be the subject of this science, because the subject is that to which the properties are attributed. 'Entity' and 'unity', that is, 'being' and 'one' considered abstractly, are properties of 'thing', taken abstractly. Therefore, neither 'being' nor 'one' can be the subject of metaphysics.[37] Francis does not

---

[35] Henry of Ghent, *Summa* a. 21, q. 3 (ed. Badius, fol. 126rE): 'Loquendo autem de esse quod convenit rei ratione suae essentiae (...), de quo Avicenna dicit quod est illud quod naturaliter primo de re concipitur'. *Summa* a. 24, q. 3 (ed. Badius, fol. 138vP): 'Dubitatio de re quacumque an sit in esse essentiae natura aliqua an non, debet determinari in principio cuiuslibet cognitionis scientialis (...). Et est istud scire de primo et simplicissimo conceptu incomplexo entis, qui (...) ut dicit Avicenna (...) prima impressione in anima imprimitur'.

[36] Franciscus de Marchia, *Quaestiones in Metaphysicam* I, q. 1 (ed. A. Zimmermann, *Ontologie oder Metaphysik? Die Diskussion über den Gegenstand der Metaphysik im 13. und 14. Jahrhundert, Texte und Untersuchungen* [Leuven, ²1998], p. 84–98); analysis of the question on p. 348ff.

[37] Franciscus de Marchia, *Quaestiones in Metaphysicam* i, q. 1 (ed. A. Zimmermann, p. 86): 'Secundo dico, quod nihil convertibile cum ipsa re simpliciter vel distinctum est primum subiectum metaphysicae. Quod patet: Quia nulla proprietas (...) est primum subiectum primae scientiae, quia subiectum primae scientiae est illud, cui attribuuntur primae passiones, non autem aliqua istarum passionum. Entitas autem et unitas sunt proprietates ipsius rei abstractae. Ergo nec ens nec unum nec quodcumque simile potest esse subiectum metaphysicae'.

explain the concept of 'being' that he regards as a property of 'thing', but in the continuation of the question it becomes clear that *ens* means the actual existence of a thing.

Francis elaborates upon his conclusion only with respect to the relation between 'thing' and 'being'. He appeals to the authority of Avicenna: *secundum intentionem Avicennae*, the concept of *ens* is concomitant with the concept of *res*. Now the expression 'concomitant' implies a relation of 'posterior' and 'prior'. Thus 'being' is not the first concept and consequently cannot be the subject of metaphysics.[38] Only 'thing as thing' meets the condition of its subject. Yet, later in the same *quaestio*, Francis also designates the subject of this science by the term *ens*; 'being in general' then means, as in Henry of Ghent, the 'essential' being of a thing. The Avicennian 'reality' found acceptance in the medieval understanding of being.

(3) The Accidentality of the 'One'

Beside 'thing' and 'being' Avicenna lists the 'one' among the primary notions. He does not deal, however, with this concept in the fifth chapter of the first treatise, but in the third treatise of his *Metaphysics*, devoted to the nine accidental categories. Because this exposition sheds light on the relation between the primary notions, it is necessary to take it into consideration for a complete picture of his doctrine. Particularly because Avicenna's view of unity was highly controversial among medieval authors. In their reading, his account seems to contain an equivocation, or at least a lack of clarity. At the beginning of the third treatise, Avicenna mentions two reasons for dealing with unity in this place: unity has much in agreement with the being that is the subject of this science, and it is, in a certain way, the beginning of quantity, that is, of number.[39] But why is it appropriate to treat of the metaphysical 'one' in the context of the accidents?

[38] Franciscus de Marchia, *Quaestiones in Metaphysicam* i, q. 1 (ed. A. Zimmermann, p. 86): 'Ex quo patet secundum intentionem Avicennae, quod intentio entis concomitatur intentionem rei. Sed intentio posterior concomitatur intentionem prioris (...) Ergo intentio entis, cum non sit prima intentio, non erit primum subiectum metaphysicae'.

[39] Avicenna latinus, *Liber de philosophia prima* iii, c. 1 (ed. Van Riet, p. 108). For Avicenna's concept of unity, see R.E. Houser, *Thomas Aquinas on Transcendenal Unity: Scholastic and Aristotelian Predecessors*, PhD thesis (University of Toronto, 1980), pp. 120–98.

With respect to the relation between 'one' and 'being' the Arabic philosopher makes three statements. First, the extension of the 'one' is equal (*parificatur*) to that of 'being' because *unum*, like *ens*, is said of each of the categories. Avicenna recognizes the transcendental character of the 'one' and its convertibility with 'being'; they are identical according to their supposition or subject (*unum subiecto*). Second, this does not mean, however, that their concepts are identical. The 'one' signifies 'what is not divided in act'. Avicenna illustrates the conceptual difference by pointing out that the many as many is not one, but is nevertheless a being. A third feature spells out the agreement between 'being' and 'one': neither signifies the substance of any thing.[40] The first two features are adopted from Aristotle's account in book iv of his *Metaphysics* (ch. 2, 1003b 22–25): 'Being and the one are the same and one single nature (*physis*) in the sense that they follow upon each other (...) but not in the sense that they are determined by one concept (*logos*).' The third characteristic is non-Aristotelian; Avicenna's distinction between essentiality and existentiality also appears to have consequences for his determination of the place of unity.

Avicenna clarifies the third feature by means of a proof for the accidentality of unity. Unity is either said of accidents or of substance. When it belongs to the domain of accidents—the term must here refer to accidents in the 'categorial' or 'predicamental' sense—, then substantial unity would be excluded, and that is 'doubtful' according to Avicenna. When unity belongs to the domain of substance, it is not said of substance as a genus or difference, for unity does not enter into the definition that determines the essence of substance. It is rather a

[40] Avicenna latinus, *Liber de philosophia prima* iii, c. 2 (ed. Van Riet, p. 114): 'Unum autem parificatur ad esse, quia unum dicitur de unoquoque praedicamentorum, sicut ens, sed intellectus eorum, sicut nosti, diversus est. Conveniunt autem in hoc quod nullum eorum significat substantiam alicuius rei'. See also the instructive summary in vii,1 (ed. Van Riet, p. 349): 'Scias autem quod unum et ens iam parificantur in praedicatione sui de rebus, ita quod, de quocumque dixeris quod est ens uno respectu, illud potest esse unum alio respectu. Nam quicquid est, unum est, et ideo fortasse putatur quia id quod intelligitur de utroque sit unum et idem, sed non est ita; sunt autem unum subiecto, scilicet quia, in quocumque est hoc, est et illud. Si enim id quod intelligitur de uno omnino esset id quod intelligitur per ens, tunc multum, secundum quod est multum, non esset ens sicut non est unum, quamvis accidat ei etiam esse unum; dicitur enim quod multitudo est una, sed non inquantum est multitudo'.

'concomitant' (*comitans*) of the substance; consequently unity is said of it as an accident.[41]

Avicenna's thesis of the accidentality of the 'one' is equivocal, since the term 'accident' has different meanings. His thesis concerns the place of the predicate 'one' within the Porphyrian order of the predicables. He establishes that unity is neither a genus nor a difference but an accident. 'Accident' in the Avicennian sense refers to all that does not belong to the essence of a thing. He underlines that unity is not an accident separable from the subject of which it is said, for it 'belongs to the universality of that which is concomitant with thing'.[42] Avicenna does not maintain that the 'one' is an accident in the predicamental sense, which is the proper subject-matter of the third treatise. Yet some passages in this treatise could suggest that Avicenna's thesis also has an ontological import. So, from the accidentality of unity, he infers the accidentality of number that belongs to the category of quantity.[43]

The medieval reading of the *Avicenna latinus* was strongly influenced by the severe criticism of 'the Commentator', that is, Averroes, who accuses Avicenna of having confused two distinct types of unity, the metaphysical 'one' and the mathematical 'one'. In his 'Long Commentary' on *Metaphysics* iv, ch. 2, he portrays Avicenna as a thinker who 'sinned much in holding that 'the one' and 'being' signify dispositions added to the essence of a thing'.[44] Next Averroes indicates the reasons which led Avicenna astray. (i) If 'being' and 'one' signify the same, it would be a useless repetition (*nugatio*) to say that 'being is one'. In order to avoid such a 'nugatory' predication, Averroes observes, it is not necessary to suppose that 'being' and 'one' signify diverse dispositions added to the

[41] Avicenna latinus, *Liber de philosophia prima* iii, c. 3 (ed. Van Riet, p. 117): 'Dico igitur quod unitas vel dicitur de accidentibus vel dicitur de substantia; cum autem dicitur de accidentibus, non est substantia, et hoc est dubium; cum vero dicitur de substantiis, non dicitur de eis sicut genus nec siut differentiaullo modo: non enim recipitur in certificatione quidditatis alicuius substantiarum, sed est quiddam comitans substantiam, sicut iam nosti. Non ergo dicitur de eis sicut genus vel sicut differentiam, sed sicut accidens'.

[42] Avicenna latinus, *Liber de philosophia prima* iii, c. 3 (ed. Van Riet, p. 121): 'Certitudo unitatis est intentio accidentis et est de universitate eorum quae concomitantur res'.

[43] Avicenna latinus, *Liber de philosophia prima* iii, 3 (ed. Van Riet, p. 122): 'Postquam igitur unitas est accidens, tunc numerus qui necessario provenit ex unitate accidens est'.

[44] Averroes, *In iv Metaph.*, comm. 3, *Aristotelis Opera cum Averrois Commentariis*, viii (Venice, 1562–74; repr. Frankfurt, 1962), fol. 67rB: 'Avicenna autem peccavit multum in hoc, quod existimavit, quod unum et ens significant dispositiones additas essentiae rei'.

essence. 'We, however, said that they signify the same essence, but in diverse ways (*modis diversis*)'. (ii) The name 'one' belongs to the *nomina denominativa*, and these names signify an accident and substance—what is meant by this brief argument will be explained by the medieval commentators. (iii) Avicenna identified the one that is predicated of all categories with the one that is the principle of number. Number is an accident; 'therefore, this man held that the one signifies an accident.'[45] The last mistake was unmistakably Avicenna's basic error: he fails to distinguish the transcendental 'one' from the quantitative 'one'.

Avicenna's theory of the *one* was discussed in the Middle Ages in two different genres of texts. The first setting is logical: Avicenna's thesis was examined in sophismatic exercises, conducted in the Faculty of Arts. A 'sophisma' is a proposition that presents logical problems because it appears to be possible to prove both its falsity and its truth. A standard *sophisma* in the thirteenth century was the proposition 'Only one (thing) is' (*Tantum unum est*).[46] In the analysis of this proposition the different meanings of the 'one' were inquired into and the convertibility of being and one discussed. One of these sophisms, composed before *c.*1270, revives the controversy between Avicenna and Averroes. Arguments against the convertibility of being and the one are presented as the *rationes Avicennae*—in fact, they are the arguments that Averroes had adduced for Avicenna's position; their refutations are based on the 'Commentator's' criticisms.[47]

Avicenna's thesis of the accidentality of the 'one' was also discussed in commentaries on Aristotle's *Metaphysics*. In his commentary on book iv, ch. 2, Albert the Great inserts a 'digression' (*digressio*) in which he

[45] Averroes, *In iv Metaph.*, comm. 3 (fol. 67rC–E): 'Et iste homo ratiocinatur ad suam opinionem, dicendo quod, si unum et ens significant idem, tunc dicere ens est unum esset nugatio, quasi dicere unum est unum, aut ens est ens (...). Nos autem diximus, quod significant eandem essentiam, sed modis diversis, non dispositiones diversas essentiae additas (...). Et fecerunt errare illum hominem res, quarum quaedam est, quia innuit hoc nomen unumde genere nominum denominativorum, et ista nomina significant accidens, et substantiam (...). Et etiam, quia existimavit, quod unum dictum de omnibus praedicamentis, est illud unum quod est principium numerorum. Numerus autem est accidens. Unde opinatus fuit iste, quod hoc nomen unum significat accidens in entibus'.

[46] S. Ebbesen, '*Tantum unum est.* 13th-Century Sophismatic Discussions around the Parmenidean Thesis,' *The Modern Schoolman* 72 (1995), pp. 175–99.

[47] Cf. A. de Libera, 'D'Avicenne à Averroès, et retour. Sur les sources arabes de la théorie scolastique de l'un transcendental,' *ArScPh* 4 (1994), pp. 141–79, in particular pp. 156–7. De Libera discusses the anonymous sophism '*Tantum unum est*,' contained in the *Collectio secunda* of the Bibliothèque Nationale, ms. lat. 16135.

disputes arguments of *sophistae* against the view that *being* and the *one* are one and the same nature.[48] It seems plausible that Albert is referring here to the sophismatic texts. Like the *sophistae*, he ascribes to Avicenna (*obicit enim contra hoc Avicenna*) a series of objections against the convertibility of being and one, among which the *nugatio* argument, the argument that *one* added to *being* is a denominative noun (that signifies a form added to what is denominated) and that the *one* is the principle of number and thus an accident. These objections are, however, in Albert's view, easy to refute. At the conclusion of his excursus he tempers his criticism of Avicenna. The Arabic philosopher is to be excused (*excusare*). Closer scrutiny of his statements makes clear he intended to say the same as Aristotle, namely, that *being* and the *one* signify one and the same nature.[49]

Thomas Aquinas, in his Commentary on the *Metaphysics*, follows Averroes's reading and is less reserved in his critique of Avicenna.[50] Avicenna held that the 'one' adds something real to the essence of a thing because he identified the one that is convertible with being with the one that is the principle of number. But this view is false, since in general it holds that what is transcendental cannot belong to a determinate category. 'Nothing that is in a determinate category is consequent upon all beings'. The one that belongs to the category of quantity cannot therefore be convertible with being in general. Moreover, Avicenna confuses two distinct kinds of unity. Aquinas distinguishes more clearly than his contemporaries between the metaphysical *one* and the mathematical *one*. The transcendental 'one' that is convertible with 'being' adds something to it which is merely conceptual, namely the negation of division. The 'one' signifies that which is undivided. These two features, real identity and conceptual difference, constitute

[48] Albert the Great, *Metaphysica* iv, tract. 1, c. 5 (*Opera omnia* xvi/1, ed. B. Geyer [Münster, 1960], p. 166–7): 'Et est digressio declarans solutionem rationum sophistarum inductarum ad hoc quod ens et unum non sint natura una et eadem'. Cf. A. Bertolacci, 'Albert the Great, Metaphysica iv, 1, 5: From the *refutatio* to the *excusatio* of Avicenna's Theory of Unity,' in J.A. Aertsen and A. Speer (eds.), *Was ist Philosophie im Mittelalter?* (Berlin, 1998), pp. 881–7.

[49] Albert the Great, *Metaphysica* iv, tract. 1, c. 5 (ed. Geyer, p. 167): 'Et facile est per haec quae hic dicta sunt, excusare dicta Avicennae, quia pro certo, si quis subtiliter dicta sua respiciat, dicere intendit quod hic dictum est'.

[50] Thomas Aquinas, *In iv Metaph.*, lect. 2, pp. 556–60. Cf. T. O' Shaughnessy, 'St. Thomas and Avicenna on the Nature of the One,' *Gregorianum* 41 (1960), pp. 665–79. J.A. Aertsen, *Medieval Philosophy and the Transcendentals. The Case of Thomas Aquinas* (Leiden, 1996), pp. 212–18.

the thirteenth-century model for interpreting the relation between the transcendentals. The 'one' that is the principle of number, by contrast, adds something real to that of which it is predicated; because of its negation of quantitative division, this *one* has the property of being a measure (*mensura*).[51] Avicenna, Aquinas concludes, was deceived by 'the equivocation of the one'.[52]

Against the background of the critical trend in the commentaries, the novelty of Duns Scotus's approach stands out in sharp relief. His approach can be described as a return to Avicenna. Sympathy for Avicenna's position runs like a thread through his entire treatment of the second question in his Commentary on book iv of the *Metaphysics*: 'Do being and one signify the same nature?'[53] This *quaestio* is a long and complicated text—it covers more than forty pages in the new critical edition—, in which it is not actually Aristotle's account of the relation between being and the one that is central, but Avicenna's view.

Scotus presents the *opinio Avicennae* by means of four arguments, in which we easily recognize Averroes's exposition of Avicenna's motives. (i) 'Being' and 'one', although predicated of all things, are not identical according to their nature, but as to their subject. The reason for this is that if the 'one' were essentially the same as 'being', then a multitude would be not-being. (ii) It would be a useless repetition of the same (*nugatio*) to say 'one being' if there were an essential identity between the two. (iii) If the 'one' signifies the same as 'being', then it would be predicated *in quid* of being. That, however, is not the case: the one is

---

[51] Thomas Aquinas, *In iv Metaph.*, lect. 2, p. 559: 'De uno autem non videtur esse verum, quod sit idem quod convertitur cum ente, et quod est principium numeri. Nihil enim quod est in determinato genere videtur consequi omnia entia. Unde unum quod determinatur ad special genus entis, scilicet ad genus quantitatis discretae, non videtur posse cum ente universali converti (...). 560: Unum igitur quod est principium numeri, aliud est ab eo quod cum ente convertitur. Unum enim quod cum ente convertitur, ipsum ens designat, superaddens indivisionis rationem, quae, cum sit negatio vel privatio, non ponit aliquam naturam enti additam. Et sic in nullo differt ab ente secundum rem, sed solum ratione (...). Unum vero quod est principium numeri addit supra substantiam, rationem mensurae, quae est propria passio quantitatis, et primo invenitur in unitate'.

[52] Thomas Aquinas, *In x Metaph.*, lect. 3, 1981: 'Hoc autem non considerans Avicenna posuit quod unum et ens sunt praedicata accidentalia, et quod significant naturam additam supra ea de quibus dicuntur. Deceptus enim fuit ex aequivocatione unius'.

[53] Duns Scotus, *Quaestiones super libros Metaphysicorum* iv, q. 2 (ed. The Franciscan Institute, St. Bonaventure [New York, 1997], pp. 321–63). Cf. J.A. Aertsen, 'Being and One: The Doctrine of the Convertible Transcendentals in Duns Scotus,' *Franciscan Studies* 56 (1998) [= G.A. Wilson and T.B. Noone (eds.), *Essays in Honor of Girard Etzkorn*], pp. 47–64.

said denominatively of being. What is predicated *per modum denominantis* is always predicated *in quale*, that is, as a further modification or qualification of the essence of the subject. (iv) Every unity belongs to the category of quantity. Therefore the one is not convertible with being.[54]

Scotus's further procedure in the *quaestio* is to discuss Aristotle's arguments for the convertibility from Avicenna's perspective (*sustinendo opinionem Avicennae*), adduce objections to Avicenna's arguments, formulate answers to these objections, and finally present his own solution.

Scotus concludes that 'being' and 'one' are not simply convertible. He explains this conclusion by referring to Avicenna's first argument.[55] But does the denial of the convertibility not conflict with Avicenna's statement that 'being and the one are on a par (*parificantur*) in predication'?[56] In his reply Scotus stresses that the Arabic philosopher does recognize a convertibility, but not an 'essential' convertibility. Avicenna's point is that the 'one' and 'being' are not 'essentially' (*essentialiter*) identical, since otherwise it would be 'nugatory' to say 'one being' (his second argument), and 'one' would be predicated of being *in quid* and not denominatively (his third argument). In his interpretation of Avicenna, Scotus distinguishes two types of convertibility. The type that Avicenna recognizes with respect to 'being' and the 'one' presupposes not an essential identity but an identity according to subject.[57] This type of convertibility means that 'being' and the 'one' are related to each other as *subiectum* and *passio*.

But how is the transcendentality of the 'one' to be reconciled with Avicenna's (fourth) argument that the one belongs 'to a determinate category'? Scotus subscribes to this argument. 'Being' and the 'one' are related to each other as *subiectum* and *passio*. No substance is ever the *passio* of another substance or of itself. The property is therefore an accident. 'I concede that every "one" belongs to a determinate

[54] Duns Scotus, *Quaestiones super libros Metaphysicorum* iv, q. 2, nn. 12–21 (ed. The Franciscan Institute, p. 323–5).

[55] Ibid. iv, q. 2, n. 66 (ed. The Franciscan Institute, p. 335): 'Dicendum est ad quaestionem, quod non convertuntur propter rationem tactam, quae (...) fuit prima pro opinione Avicennae'.

[56] Ibid. iv, q. 2, n. 41 (ed. The Franciscan Institute, p. 330): 'Contra primam rationem: sequitur ex illa quod ens et unum non convertuntur. Hoc haberi potest ex prima ratione pro opinione sua: quia non praedicantur de eisdem. Contradicit sibi ipsi, quia dicit in vii Metaphysicae a: "Ens et unum parificantur in praedicatione"'.

[57] Ibid. iv, q. 2, n. 56 (ed. The Franciscan Institute, p. 333): 'Ad rationes contra opinionem Avicennae, dicendum quod concedit convertibilitatem, non essentialem, sed quod idem sunt subiecto, non secundum essentiam'.

category, namely, to that of quantity.' For all that, Scotus maintains that the transcendental character of the one is not thereby eliminated. He endeavors to clarify this by making a comparison with the notion of *creatio*. 'Creation' in the passive sense signifies the createdness of things, which belongs to everything that is not God. *Creatio* is therefore a property of being, but it still belongs to a determinate category, that of relation. In the same way, the one can belong to a determinate category and still to the whole of being.[58]

What is the reason for Scotus's positive evaluation of Avicenna's theory of the 'one'? The explanation has to be sought in the new model he develops for the relation between being and the convertible transcendentals. According to Scotus there is more than a merely conceptual difference between 'being' and 'one'. There exists a non-identity not caused by the intellect; the difference between them must, however, not be understood as a 'real diversity', with which Averroes mistakenly charged Avicenna.[59] Scotus calls it a 'formal' distinction, because it exists between different *formalitates*. which are not distinct things (*res*) but quiddities independent of the intellect.

(4) Conclusion

Avicenna's doctrine contains several non-Aristotelian aspects. Its major innovations are the introduction of primary notions, irreducible to other, better known notions, and the introduction of the concept *res* that is different from 'being'. Their difference is based on the distinction between essence and existence which also determines the accidental status of the 'one'.

Avicenna relates the firstness of these notions to their commonness; they are predicated of all categories. For this reason they 'transcend'

[58] Ibid. iv, q. 2, n. 70 (ed. The Franciscan Institute, p. 336): 'Concedo tunc quod omne unum est determinati generis, scilicet quantitatis. Sicut enim omne aliud a Deo dicitur creatum, ita quod creatio-passio est proprietas entis, et tamen creatio est unius generis, ut relationis, et determinata species in illo genere, sic unum potest esse determinati generis, et tamen convenire toti enti simpliciter vel secundum quid. (...) Unde hoc nihil concludit contra Avicennam.'

[59] Ibid. iv, q. 2, n. 142 (ed. The Franciscan Institute, p. 354): 'Nota etiam quod opinionem istam de diversitate reali non oportet imponere Avicennae, licet Averroes videatur ei imponere. Quaecumque enim dicit Avicenna iii Metaphysicae suae vel vii de hac materia, exponi possunt: quod accidens est quidquid est extra per se intellectum quiditatis'.

the individual categories Aristotle had distinguished and could thus be called 'transcendentals'. Understandably, some scholars hold that Avicenna's doctrine really is a doctrine of the transcendentals, and even see his teaching as the basis of the Scholastic doctrine.[60] What is usually credited to medieval philosophy should rather be regarded as an achievement of Arabic thought.

It cannot be denied that Avicenna's metaphysics played a decisive role in the formation of a transcendental systematic in the Middle Ages.[61] But it seems to me that his doctrine of the primary notions is not a theory of the transcendentals in the medieval sense. The essential difference concerns the analysis of the relations between the primary notions. According to Avicenna 'being' and 'one' are on a par: they agree in that neither signifies the essence of a thing; they add something to 'thing' in the manner of an accident. The distinction between, on the one hand, the mutual relation of *being* and *one*, and, on the other hand, the relation of either of them to 'thing' comes to expression in Avicenna's terminology. To express the relation between 'being' and 'one', he uses the term *parificatur*; to express the relation of 'being' and 'one' to 'thing', he uses the term *concomitans*. *Ens* accompanies *res*, and the same holds for *unum*. Both terms refer to a purely *extensional* identity.

Characteristic of the medieval accounts in, for instance, Thomas Aquinas and Henry of Ghent, is the attempt to derive the other transcendentals from the first transcendental, 'being'; they are an inner explication of this first, they signify general modes of being, not yet expressed by the term 'being' itself. The *intensional* understanding of transcendentality belongs to the core of the medieval doctrine. A telling illustration of that is a brief text by Albert the Great. After having mentioned Avicenna's triad, *unum*, *res* and *aliquid* (which he regards as

[60] A. Bertolacci, *Avicenna ed Averroè come fonti del Commento di Alberto Magno alla 'Metafisica' di Aristotele: la dottrina dei trascendentali nei commentatori arabi di Aristotele e nel xiii secolo latino*, thesis (University of Florence, 1998), cap. 2 : 'La dottrina dei trascendentali in Avicenna,' pp. 65–107. I am grateful to the author for sending me a copy of this chapter. Cf. J. Janssens, 'Elements of Avicennian Metaphysics in the *Summa*,' in G. Guldentops and C. Steel (eds.), *Henry of Ghent and the Transformation of Scholastic Thought. Studies in Memory of Jos Decorte* (Leuven, 2003), pp. 41–59. The author claims (p. 52) that 'the *Avicenna Latinus* formed the direct basis of the Scholastic doctrine of the four transcendentals, usually designated by the terms *ens, unum, bonum* and *verum*' by showing that the Avicennian notion of *necesse* can be linked with *bonum*, the notion of *res* with *verum*.

[61] I agree with J.F. Courtine, *Suarez et le système de la métaphysique* [Paris, 1990], p. 349: '...la tradition issue d'Avicenne nous parait décisive por comprendre la constitution d'une systématique transcendantale...'.

a distinct transcendental), he derives them from the first notion *ens* in a non-Avicennian way: *unum* adds 'indivision' to 'being', *aliquid* 'particularized being' (*ens particularizatum*) and *res* 'stable being' (*ens ratum*).[62]

The difference between Avicenna's doctrine and the medieval account can be expressed in yet another way. Avicenna aims at the beginning of human thought, but focuses on one side of it, the reduction to primary notions. What is lacking is the corresponding synthetic part, the derivation or 'deduction' from the first. The medievals, on the other hand, elaborated this aspect of the idea of a systematic beginning. We are not belittling Avicenna's genius when we maintain that the elaboration of a systematic doctrine of the transcendentals is an original achievement of Latin philosophy.

[62] Albert the Great, *In I Perih.*, tract. ii, c. 5 (*Opera omnia* i, ed. Borgnet, p. 395): 'Similiter autem est de aliis nominibus, unum, res, et aliquid, ut dicit Avicenna (...) "unum" addit super ens indivisionem, et "aliquid" addit super ens particularizatum, et "res" addit super ens ratum.'

ON THE NATURE AND FATE OF CHAPTER V OF IBN RUSHD'S *EPITOME* OF ARISTOTLE'S *METAPHYSICS**

Rüdiger Arnzen

> One beginning and one ending for a book was a thing I did not agree with. A good book may have three openings (...) inter-related only in the prescience of the author, or for that matter one hundred times as many endings.
>
> Flann O'Brien, *At Swim-Two-Birds*

In 554/1159 Ibn Rushd completed his compendium of the Aristotelian natural sciences (other than psychology and biology), which consisted of epitomes of four works, *Physica, De caelo, De generatione et corruptione*, and *Meteorologica.* Presumably not much later, he decided to append to this compendium a fifth epitome, the *Jawāmiʿ Kitāb mā baʿd al-ṭabīʿa* (*Epitome of the Metaphysics*, henceforth *EM*). The text is preserved in 17 Arabic manuscripts dating from the thirteenth to the nineteenth centuries. During the first half of the thirteenth century it was translated into Hebrew twice: the translation prepared by Mosheh ben Shemuel ibn Tibbon is preserved in at least 14 manuscripts, while another, anonymous, translation is quoted nearly in its entirety in books IX and X of Shem Tov ibn Falaquera's *Deʿot ha-Filosofim*, of which we have two manuscripts. The Hebrew version of Ibn Tibbon, in turn, was translated into Latin in 1523 by Iacob Mantinus.

In all these versions the text is divided into four chapters (*maqālāt*, in the Arabic tradition). However, Ibn Rushd declares shortly after the beginning of the text that he had divided his work into five chapters, and this declaration is likewise unanimously transmitted in all versions. This raises the question whether *EM*, as we know it today from the manuscript tradition and the editions based thereon, contains the

* It is a great pleasure and honour to dedicate this article to Prof. Hans Daiber who has constantly followed, encouraged and supported my work for many years. My gratitude goes to Prof. Dimitri Gutas, Prof. Amos Bertolacci and Prof. Hinrich Biesterfeldt for their corrections and erudite comments on a first draft of this paper. I would also like to thank Prof. Mauro Zonta who kindly shared his knowledge of the Hebrew tradition of Ibn Rushd's works.

final version intended by Ibn Rushd for circulation—i.e., whether the absence of Chapter V is due to the fact that Ibn Rushd changed his mind about it (he either never composed it or did so, but decided later to withhold it from publication), or to an omission in the manuscript tradition. In either case, the question of what this fifth chapter was about (or was supposed to be about) deserves study, either in order to gain a complete picture of Ibn Rushd's basic division of metaphysics or to learn what part of the *Metaphysics* it is that he changed his mind about and possibly discarded, either because of its minor importance or for other reasons.[1]

Having explained that the contents of the Aristotelian *Metaphysics* basically fall into three sections, Ibn Rushd provides some introductory notes on the topics of the first two sections, and then proceeds to the third section, as follows:[2]

والقسم الثالث ينظر فيه في موضوعات العلوم الجزئية و يزيل الأغاليط الواقعة فيها لمن سلف من القدماء وذلك في صناعة المنطق وفي الصناعتين الجزئيتين أعني العلم الطبيعي والتعليمي وإنما كان ذلك كذلك لأنه ليس من شأن العلوم الجزئية أن تصحح[1] مبادئها ولا أن تزيل[2] الغلط الواقع فيها على ما تبين في كتاب البرهان وإنما ذلك إلى[3] صناعة عامة وذلك إما هذه الصناعة وإما صناعة الجدل إلا أن صناعة الجدل إنما تبطل تلك الآراء بأقاويل مشهورة ليس يؤمن أن ينطوي فيها

[1] The scholarly literature on *EM* provides no detailed investigation into this question, but only three conflicting statements: I. Husik ('Averroes on the Metaphysics of Aristotle,' *The Philosophical Review* 18 [1909], pp. 416–28; repr. in F. Sezgin in collabor. with M. Amawi, C. Ehrig-Eggert, E. Neubauer [eds.], *Abu l-Walīd Muḥammad Ibn Rushd: Texts and Studies*, vi: *Ibn Rushd and his Commentaries on Plato and Aristotle* [Frankfurt, 1999], pp. 136–48) maintains that Chapter V was about 'the subjects of the special sciences' (p. 422/142) and that Ibn Rushd's reference to former thinkers and their errors on this topic is related to 'the philosophical and theological sects of his days' (ibid.). S. van den Bergh (*Die Epitome der Metaphysik des Averroes*, trans. S. van den Bergh [Leiden, 1924], p. II), on the other hand, holds that Chapter V dealt with the axioms treated in *Metaph.* IV (Γ) and Aristotle's polemic against his predecessors, while J. Puig Montada remarks that book IV (Γ) was not taken into consideration in *EM* and believes that Chapter V provided a correction of the principles of the particular sciences or the foundation of physics and mathematics through metaphysics (cf. Ibn Rushd, *Compendio de Metafísica*, ed. and trans. C. Quirós Rodríguez, introd. J. Puig Montada [Madrid, 1919; repr. Seville, 1998], pp. xvii–xviii).

[2] The Arabic text is based on my examination of various manuscripts and roughly corresponds to the text in the three most current editions: Ibn Rushd, *Compendio de Metafísica*, ed. and trans. C. Quirós Rodríguez, p. 9sq. Ibn Rushd, *Talkhīṣ Mā baʿd al-ṭabīʿa*, ed. ʿU. Amīn (Cairo, 1958; repr. Tehran, 1377/1999), p. 5sq. Ibn Rushd, *Risālat Mā baʿd al-ṭabīʿa*, ed. J. Jihāmī, introd. R. al-ʿAjam (Rasāʾil Ibn Rushd al-falsafiyya, 6; Beirut, 1994), p. 33sq. N.B.: This edition is a reprint of *Rasāʾil Ibn Rushd* (Hyderabad: Maṭbaʿat Dāʾirat al-Maʿārif al-ʿUthmāniyya, 1366/1947).

كذب وهذه بأقاويل صادقة وإن كان يلحقها أن تكون[٤] مشهورة فلهذا ما[٥] كان من ضرورة[٦] هذا العلم تصحيح مبادئ الصنائع الجزئية وتبين من هذا أن الأجزاء الضرورية من هذا العلم إنما هي الجزءان الأولان فقط وأما الجزء الثالث فعلى جهة الأفضل إذ[٧] كان وجود أكثر موضوعات العلوم الجزئية وجهة وجودها من الأمور البينة بنفسها وإنما وقع فيها غلط لمن سلف من القدماء فكان من تمام المعرفة بها حل تلك المغالطات بمنزلة ما تكون حل الشكوك الواقعة في الشيء من[٨] تمام المعرفة به مع حصول المعرفة بجوهره لكن رأينا نحن أن نجعل هذا الكتاب خمس مقالات المقالة الأولى نذكر فيها الصدر الذي نحن بسبيله ونشرح فيها الأسماء المستعملة في هذه الصناعة والمقالة الثانية نذكر فيها الأمور التي تتنزل[٩] من الجزء الأول من هذه الصناعة منزلة الأنواع والمقالة الثالثة نذكر فيها اللواحق العامة لها والمقالة[١٠] الرابعة تتضمن القول فيما يشتمل عليه الجزء الثاني من هذا العلم والمقالة الخامسة تحتوي على ما تضمنه الجزء الثالث من هذه الصناعة

١ تصحح] يصحح *DGQ* ٢ تزيل] يزيل *DQα* ٣ إلى] على *om. Q: DPT* ٤ تكون] يكون *Dα* ٥ ما] *om. Mantinus* ٦ ضرورة] صورة *Q* ٧ إذ] إذا *MG* ٨ من] في *ACDα* ٩ تتنزل] تنزل *APT*: ينزل *Q:D s.p.*: نتنزل *G* ١٠ والمقالة] و *MQα Mantinus*

In the third part [of metaphysics] he (*i.e. Aristotle*) examines the fundamentals[3] of the particular sciences and eliminates the mistakes committed by the ancients on this subject, namely in the discipline of logic and in the two particular disciplines, i.e. physics and mathematics. The reason why he acted like this is that it is inappropriate for the particular sciences to establish the truth of their principles and to eliminate errors occurring about them (as is shown in the *Book of the Apodeictic Proof*).[4] This is rather the task of a general discipline, which is either this discipline [of metaphysics] or the discipline of dialectics. However, the discipline of dialectics reduces such opinions to absurdity merely through generally accepted statements for which there is no guarantee that they do not contain any falsehood, while this [discipline of metaphysics does so] through true statements, even if they might also serve as generally accepted [statements]. Accordingly, the verification of the principles of the particular sciences is not a necessary part of this science.[5] From this it is clear that only the

[3] I choose this term to translate *al-mawḍūʿāt* in order to avoid prejudging its significance. The specific interpretation of the Arabic term is crucial for the entire evaluation of this section of Ibn Rushd's introduction. Quirós Rodríguez, p. 12, translates it by 'materias proprias', Van den Bergh, p. 4, by 'Grundsätze', and Horten by 'Postulate' (Ibn Rushd, *Die Metaphysik des Averroes (1198†)*, trans. M. Horten [Halle an der Saale, 1912; repr. Frankfurt, 1960], p. 6).

[4] I.e. *Analytica posteriora.*

[5] That *mā* serves in this clause as a particle of negation and not as a pronoun, as translated by Van den Bergh, p. 5, and Horten, p. 6, is clear from the immediately following sentence. For Ibn Rushd's use of *mā* as negation in similar syntactic constructions see Ibn Rushd, *Tahāfut al-Tahāfut (L'incohérence de l'incohérence)*, ed. M. Bouyges (Beirut, 1930), Index E [Lexique grammatical], p. 672 s.n. *mā*; also Ibn Rushd, *Averrois Cordubensis <Commentarium medium> in librum Aristotelis De Interpretatione. Recensum textis arabicis initiavit M.M. Kassem, complevit […] C.E. Butterworth adjuv. A.A. Haridi* (Cairo, 1981),

> first two parts are necessary parts of this science, whereas [this] third part [is included] because that is the best [method],[6] since the existence of most of the fundamentals of the particular sciences and their mode of existence belong to what is self-evident, and only the ancients lapsed into mistakes in this matter. Then, however, the complete knowledge of these [fundamentals] solved these aporiai as any uncertainty occurring in a certain thing is resolved through completing one's knowledge of this [thing] together with acquiring knowledge of its substance.[7] Nevertheless, we decided to compose this treatise in five chapters. In Chapter I we present the introduction,[8] which we are now in the midst of, and explain the terms employed in this science. In Chapter II we set forth the things which hold the position of species of what belongs to the first part of this discipline. In Chapter III we set forth the general concomitants of these [species]. Chapter IV contains a lecture on what the second part of this science includes. Chapter V comprises what the third part of this discipline contains.

This section of the introduction leaves no doubt that Ibn Rushd intended to include a fifth chapter dealing with what he describes as the "third part" of Aristotelian metaphysics. As he says at the end of his concise table of contents, he planned to treat in this chapter exclusively the topics of this part of metaphysics. These consist, as explained at the

p. 78, l. 12, p. 122, l. 6; as well as Ibn Rushd, *Talkhīṣ Kitāb al-maqūlāt (Paraphrase du Livre des catégories)*, ed. M. Bouyges (Beirut, 1932), p. 44 ult.

[6] The Arabic is very concise and not quite clear. One might also think about translating this phrase through "because [metaphysics] is the highest/most excellent [discipline]", thus referring to *Metaph.* IV (Γ) 3, 1005a 33–b 1. Quirós Rodríguez's translation ('en razón de mejoría', p. 13) is rather vague. The translation 'zur Vervollständigung' (i.e. of metaphysics) provided by Horten and Van den Bergh is in my view not supported by the Arabic wording. Ibn Rushd often uses *al-afḍal* in a methodological context (*al-afḍal fī tartīb al-ʿilm, al-afḍal fī 'l-taʿlīm*, etc.), and does so also with reference to metaphysics, cf. Ibn Rushd, *Tafsīr Mā baʿd al-ṭabīʿa*, ed. M. Bouyges, 3 vols. (Beirut, 1938–42), i, p. 167, l. 5–10, p. 168, l. 7, ii, p. 476, l. 3.

[7] The first part of this sentence cannot be interpreted as a general epistemological statement as in the translations by Van den Bergh and Horten, but only as a sort of historical report referring to Aristotle's decisive solution of any aporia and error concerning the principles and fundamentals of the sciences. Otherwise it would make no sense to open the following sentence with the adversative conjunction *lākin*, which specifies the relation between the fact that the solution of these problems is not a necessary part of metaphysics and has already been completed by Aristotle and the explicit intention to include a separate chapter on this topic anyhow. (Such an interpretation is additionally supported by the historical approach Ibn Rushd displays in the general introduction to the four epitomes of natural sciences with regard to the doubts and errors of other thinkers; cf. J. al-ʿAlawī, *al-Matn al-rushdī. Madkhal li-qirāʾa jadīda* [Casablanca, 1986], p. 161sq., right column [the Cairo version].)

[8] Pace Van den Bergh, p. 275, the Cairo manuscript reads, as all other manuscripts, *al-ṣadr*.

beginning of the passage quoted, of the examination of the fundamentals of the particular disciplines of theoretical science and the removal of any unsettled questions concerning these fundamentals. The inclusion of such an examination is vindicated through the argument that it is inappropriate for the particular sciences to verify their principles. It is, therefore, not entirely out of place to relate Ibn Rushd's "third part" of metaphysics partly or entirely to *Metaph.* IV (Γ) 3–8, as suggested by Van den Bergh,[9] and, implicitly, Horten.

In book IV (Γ) 3 Aristotle solves the second aporia of book III (B) (cf. 995b 7–10 and 996b 26–997a 15) and subsequently performs the metaphysician's task entailed by this solution (IV [Γ] 4–8).[10] Aristotle presents the second aporia in book III (B) in a form slightly different from that of its counterpart, book XI (K). While he asks in book III (B) whether one and the same science studies both substance and the first principles of demonstration, the question in book XI (K) is whether it is one or more than one science that studies these first principles. However, this difference affects primarily the structure of the subsequent unfolding of the aporia and its implied objections. The solutions presented in books III (B) 3 and XI (K) 4 are more or less congruent. It is the metaphysician who (also) considers the first principles of demonstration, for these apply to all existing things, while the particular sciences make use of them by applying them to their special subject-matter, e.g. physics by relating them to being *qua* being in motion, mathematics to being *qua* continuous being, etc. Aristotle performs this metaphysical task in the immediately following chapters—i.e., chapters 4–8 of book IV (Γ) and chapters 5–6 of book XI (K)—by defending the laws of contradiction and of the excluded middle, refuting arguments for their denial, and indicating the fundamental errors of previous philosophers which caused their critique or opposition.

---

[9] Cf. Van den Bergh, p. II.

[10] The 14 (or 15) aporiai are first enumerated in *Metaphysics* III (B) 1, then discussed in greater detail in book III (B) 2–6. They reappear in book XI (K) 1–2 and are (partly) discussed and solved in the remaining chapters of book XI (K). According to Jaeger and Ross the first part of book XI (K) is an earlier and shorter version of the contents of books III (B)–IV (Γ) and book VI (E); cf. W. Jaeger, *Aristoteles. Grundlegung einer Geschichte seiner Entwicklung* (Berlin, ²1955), pp. 215–21; W.D. Ross, *Aristotle's Metaphysics*, 2 vols. (Oxford, 1924; repr. Oxford, 1997), ii, p. 305sqq. For a conspectus of corresponding passages of the list of *aporiai* cf. Ross, i, p. 224; for a rough survey on the question of which aporia is discussed and (more or less) solved in what part of Aristotle's *Metaphysics* cf. ibid., p. xxiii sq. and p. 223.

In his *Tafsīr* (*Long Commentary*) on Aristotle's *Metaphysics*, as we have it today, Ibn Rushd neither quotes nor comments on books XI (K), XIII (M) and XIV (N). We know, however, from some brief remarks in this work and especially from his *Talkhīṣ* (*Middle Commentary* or *Paraphrasis*) on the *Metaphysics* that he was well acquainted with these books.[11] In case it is really the solution of the second aporia what is announced for Chapter V of *EM*, Ibn Rushd probably did not restrict himself to book IV (Γ) 3, but also took into consideration book XI (K) 4. Having stated that no particular science inquires into the truth of its own principles, he refers to *Posterior Analytics* just as Aristotle does in the same context in book IV (Γ) 3, 1005b 4 (without correspondence in book XI [K]). On the other hand, he considers in the immediately following sentences the option that this might be the task of another general discipline, i.e. dialectics, thereby possibly referring to the second aporia as phrased in book XI (K).

Be that as it may, the section quoted above raises another problem of greater relevance: Ibn Rushd uses twice the term *mawḍūʿāt* and twice the term *mabādiʾ* ("principles") without making clear what he is referring to and whether the terms are semantically distinct. Since Aristotle likewise employs in both relevant passages two different terms in order to refer to the first principles of a science, i.e. ἀρχαί ("principles") and ἀξιώματα ("axioms"), one might suppose that each Arabic term is related to one of these Greek terms.[12] But while *mabādiʾ* corresponds perfectly to ἀρχαί and is confirmed in this meaning by various Græco-Arabic texts,[13] this certainly does not apply to *mawḍūʿāt* and ἀξιώματα: in none of the nine instances where ἀξίωμα occurs in *Analytica posteriora*, *Topica* and *Metaphysics* in the epistemological context

[11] A critical edition of Qalonymos ben Qalonymos' translation, one of two extant Hebrew versions of the *Middle Commentary*, is being prepared by Mauro Zonta. For Ibn Rushd's knowledge of books XI, XIII and XIV displayed in this work cf. M. Zonta, 'Il Commento medio di Averroè alla Metafisica nella tradizione ebraica: alcuni problemi testuali,' in C. Baffioni (ed.), *Averroes and the Aristotelian Heritage* (Naples, 2004), pp. 189–99, esp. pp. 190–92; also M. Steinschneider, 'Die Metaphysik des Aristoteles in jüdischen Bearbeitungen. Ein Versuch,' in Curatorium der Zunz-Stiftung (ed.), *Jubelschrift zum neunzigsten Geburtstag des Dr. Leopold Zunz* (Berlin, 1884; repr. Hildesheim, 1974), pp. 1–35, there pp. 13–16. For evidence of Ibn Rushd's knowledge concerning these books in his *Long Commentary*, cf. A. Bertolacci, 'On the Arabic Translations of Aristotle's *Metaphysics*,' *ArScPhil* 15 (2005), pp. 241–75, cf. pp. 247–51, esp. p. 250, note 22.

[12] As Horten and Van den Bergh obviously did when translating *mawḍūʿāt* through "Postulate" or "Grundsätze" (cf. above, note 3).

[13] Cf. G. Endress and D. Gutas (eds.), *A Greek and Arabic Lexicon (GALex). Fasc. 8:* ب *to* بدل (Leiden, 2007), pp. 114–16; also M. Ullmann, *Wörterbuch zu den griechisch-arabischen Übersetzungen des 9. Jahrhunderts. Suppl. I: A–O* (Wiesbaden, 2006), pp. 175sq.

in question do the medieval Arabic translations render it by *mawḍūʿ* or any related term.[14] In his comments on the relevant passages of the *Metaphysics*, Ibn Rushd himself sometimes employs the expression *ārāʾ ʿāmmiyya*,[15] which he found in the translation, and sometimes coins his own periphrastic expressions, e.g. *awāʾil al-taṣdīq al-ʿāmmiyya li-jamīʿ al-ʿulūm* ("general principles of assent for all sciences"),[16] yet he nowhere uses the term *mawḍūʿāt* for that purpose.

Instead, the term *mawḍūʿ* is usually employed in medieval philosophical texts (including Ibn Rushd's works) in order to denote what Aristotle called ὑποκείμενον both (*a*) in its logical and ontological meanings, none of which was ever employed by Aristotle in order to refer to the principles or fundamentals of a science,[17] and (*b*) in its epistemological meaning, referring to the subject-matter of a science,[18] which is likewise semantically clearly distinct from ἀξίωμα. Since *EM*, unlike

---

[14] Instead, we find the following correspondences: *Analytica posteriora* 72a 17: "aksiyūmā aʿnī 'l-mutaʿārif", *Manṭiq Arisṭū*, ed. ʿA. Badawī, 3 vols. (Kuwait, [2]1980), ii, p. 335, l. 5–6; *Analytica posteriora* 75a 41 and 76b 14: "ʿulūm mutaʿārifa", ed. Badawī, ii, p. 353, l. 2–3, and p. 359, l. 14; *Topica* 156a 23: "qaḍāyā wājibun qabūluhā", ed. Badawī, iii, p. 729, l. 12 (N.B.: the latter two translations are also found in the translation of *An. post.* quoted by Ibn Rushd in his *Sharḥ al-Burhān li-Arisṭū*); *Topica* 159a 4: "qaḍiyya", ed. Badawī, iii, p. 742, l. 10; *Metaphysics* 997a 7: "ārāʾ ʿāmmiyya bayyina", ed. Bouyges, i, p. 193, l. 3; *Metaphysics* 997a 11: "ārāʾ ʿāmmiyya wāḍiḥa", ed. Bouyges, i, p. 193, l. 7; *Metaphysics* 1005a 20: "umūr ʿāmmiyya", ed. Bouyges, i, p. 335, l. 16; *Metaphysics* 1005b 33: "ārāʾ ʿāmmiyya mushtaraka", ed. Bouyges, i, p. 347, l. 1.

[15] Cf. ed. Bouyges, i, p. 197, ll. 5, 9 and 11.

[16] Ibid., i, p. 337, l. 11sq. Other expressions used by Ibn Rushd in his *Long Commentary on the Metaphysics* are "awāʾil al-maʿrifa al-ʿāmma", "al-awāʾil/umūr al-maʿrūfa bi-nafsihā", "al-muqaddimāt al-ʿāmma al-uwal allatī hiya mabdaʾ kull burhān" and the like; cf. ibid., i, p. 196, ll. 6, 11 and 17, p. 337, ll. 4sq., 13sq. etc. In his *Sharḥ al-Burhān li-Arisṭū* Ibn Rushd employs the terms he found in his translation (see note 14), while we find again other periphrastic expressions in his *Middle Commentary*, e.g. "al-umūr al-maʿlūma bi'l-ṭabʿ", Ibn Rushd, *Talkhīṣ Manṭiq Arisṭū*, ed. J. Jihāmī, 3 vols. (Beirut, 1982), ii, p. 394, or "muqaddimāt wājibun qabūluhā", ibid., p. 398.

[17] The only instance where Aristotle comes close to such a terminology is *Metaphysics* I (A) 2, 982a 23: οὗτος (scil. ὁ μάλιστα ἔχων τὴν καθόλου ἐπιστήμην) γὰρ οἶδέ πως πάντα τὰ ὑποκείμενα. But even there τὰ ὑποκείμενα does not mean general principles, but what is subordinate to the universals.

[18] For examples and further literature cf. A. Zimmermann, *Ontologie oder Metaphysik. Die Diskussion über den Gegenstand der Metaphysik im 13. und 14. Jahrhundert* (Leuven, 1998), p. 132, n. 45. An extensive list of passages, in which Ibn Rushd uses *mawḍūʿ* in this meaning, is quoted in Ibn Rushd, *Talkhīṣ Manṭiq Arisṭū*, ed. Jihāmī, iii, p. 1000sq. Another specific epistemological use of *mawḍūʿ*, hardly to be mistaken for ἀξίωμα, is attributive *mawḍūʿ* in *aṣl mawḍūʿ* = ὑπόθεσις (cf. Aristotle, *Analytica posteriora* I 2, 72a 20–23; I 10, 76b 23–77a 4, and the translation and corresponding commentaries in Ibn Rushd, *Sharḥ al-Burhān li-Arisṭū* wa-*Talkhīṣ al-Burhān*, ed. ʿA. Badawī [Kuwait, 1405/1984], p. 192, l. 9–p. 194, l. 22, and p. 313, l. 17–p. 319, l. 17).

the *Talkhīṣ* and the *Tafsīr*, offers an independent restructuring of the contents of the *Metaphysics*, it is possible to relate the Arabic term to the Aristotelian concept of ὑποκείμενον in this latter meaning, even if it occurs in a context which deals (among other things) with the first principles of all sciences. But if this is indeed the case in the present passage of *EM*, the scope of Ibn Rushd's "third part" of metaphysics would be significantly widened, if not completely shifted. Since Ibn Rushd mentions immediately afterwards Aristotle's efforts to eliminate the mistakes committed by previous logicians, physicists and mathematicians concerning these *mawḍūʿāt*, it would no longer be possible to restrict the scope of this "third part" to *Metaph.* IV (Γ) 3–8 and the corresponding chapters in book XI (K). For what Aristotle treats there are not errors concerning the specific subjects of particular sciences but those concerning the status and validity of the first principles of demonstration and his solution of the second aporia. In other words, the "third part" of metaphysics, according to Ibn Rushd, would also include those sections of the *Metaphysics* which deal with problems and errors concerning the subject-matters of the particular sciences.

Above all, this would concern large parts of books XIII (M) and XIV (N), in which Aristotle deals with the subject-matter of mathematics and errors committed by Plato and the Pythagoreans on this subject. That Ibn Rushd did indeed intend (also) this meaning of *mawḍūʿāt* is corroborated by a parallel passage in Chapter II of *EM* which likewise refers to the verification of the *mawḍūʿāt* of the particular sciences and especially to mathematics:

> Whether there is a separate quantity which differs in its being from that of this sensible quantity and forms the subject-matter (*al-mawḍūʿ*) of mathematics, as the Pythagoreans thought, we will discuss in connection with the verification of the fundamentals (*mawḍūʿāt*) of the particular sciences.[19]

Obviously, Ibn Rushd refers here to *Metaph.* XIII (M) 1–3, Aristotle's discussion of the question of an immaterial existence or separability of mathematical entities and their relation to Platonic Ideas, as one of the topics of the "third part" of metaphysics to be treated in Chapter V.

---

[19] Cf. Quirós Rodríguez, p. 42, l. 6–9 ≈ Amīn, p. 39, l. 4–6 ≈ Jihāmī, p. 64, l. 4–7:
فأما هل هاهنا كم مفارق وجوده غير وجود هذا الكم المحسوس هو الموضوع لصناعة التعاليم على ما كان يرى ذلك آل فوثاغورس فسنفحص عنه عند تصحيح موضوعات الصنائع الجزئية

In addition, this chapter would have to include those sections of the *Metaphysics* which deal with the threefold division of theoretical sciences according to their different subject-matters, i.e. books VI (E) 1 and XI (K) 7, and perhaps also XI (K) 8–12, in which Aristotle treats basic concepts of physics such as change and movement.

The interpretation proposed here has the following three advantages over that provided by Horten and Van den Bergh: (*i*) It is in harmony with Ibn Rushd's terminology in this and in other works as well as with the common terminology of medieval Græco-Arabic texts. (*ii*) It avoids a certain discrepancy between Ibn Rushd's concept of a "third part" of metaphysics and the Aristotelian text itself, which does not discuss specific postulates or axioms of any particular science—as would be suggested by taking '*mawḍūʿāt*' to refer to such postulates—but only first principles of demonstration common to all sciences. (*iii*) It is supported by the second occurrence in the passage quoted above (p. 45, l. 3sq.). Ibn Rushd explains there that for most of these *mawḍūʿāt*, their 'That' (*wujūduhā*) and their 'What' (*jihat wujūdihā*) are self-evident. In doing so he seems to have in mind the following passage of Chapter 10 of the first book of Aristotle's *Posterior Analytics*:[20]

> For every demonstrative science has to do with three things: what it posits to be (these form the genus of what it considers the attributes that belong to it in itself); and what are called the common axioms, the primitives from which it demonstrates; and thirdly the attributes, of which it assumes what each signifies. Nothing, however, prevents some sciences from overlooking some of these—e.g. from not supposing that its genus is, if it is evident that it is (for it is not equally clear that number is and that hot and cold are), and from not assuming what the attributes signify, if they are clear—just as in the case of the common items it does not assume what to take equals from equals signifies, because it is familiar. But none the less there are by nature these three things, that about which the science proves, what it proves, and the things from which it proves.[21]

Just like Ibn Rushd with reference to the *mawḍūʿāt* of the particular sciences, Aristotle states here that the 'That' and the 'What' of the subject-matter of some sciences need not be presupposed because they are well known. But this is not stated (and would make no sense to

[20] That Ibn Rushd explicitly refers to this work in the present context has already been mentioned above. In addition, his reference to dialectics in the previous sentence clearly echoes *Analytica posteriora* I 11, 77a 26–31.

[21] Aristotle, *Analytica posteriora* I 10, 76b 11–22, trans. J. Barnes in *The Complete Works of Aristotle. The revised Oxford translation*, 2 vols. (Princeton, 1984), i, p. 124.

be stated) of the third class of principles, the self-evident principles of demonstration (cf. also *An. post.* I 10, 76a 37sq.). In all probability, Ibn Rushd refers here with '*mawḍūʿāt*' to the two other classes of principles, i.e. the subject-matter of particular sciences and its essential attributes, the 'That' and the 'What' of which are either self-evident or taken for granted by these sciences.[22]

Summing up the preceding considerations we have the following result with regard to Ibn Rushd's introductory note on the so-called "third part" of metaphysics: this part, which forms the sole topic of Chapter V of *EM*, dealt or was supposed to deal in all likelihood with two closely related, though distinct problems of metaphysics. First, the verification of the first principles of demonstration and refutation of those who deny their truth or their fundamental status, and second, the discussion of problems and errors concerning the subject-matter of the particular sciences, especially that of mathematics and physics. Accordingly, Chapter V will have taken into consideration the following parts of the *Metaphysics* not discussed in any other part of *EM*: IV (Γ) 3–8, VI (E) 1, XI (K) 4–7 (or 4–12), and (parts of) books XIII (M) and XIV (N).

But was this chapter ever written?—Three Arabic manuscripts transmit a statement on this point after the end of Chapter IV:

[1] Ms. Cairo, Dār al-kutub al-qawmiyya, Taymūr Ḥikma 117, fol. 217r	[2] Ms. Hyderabad, al-Maktaba al-Āṣafiyya, printed in *Rasāʾil Ibn Rushd*, Hyderabad 1366/1947, p. 172sq. (= Jihāmī 1994, p. 173sq.), and [3] Ms. Rampur, Raẓā Library, ʿA 3609, fol. 257v
قال بعض من شافه المؤلف انه لم يلتفت الى اتمام الكتاب بالمقالة الخامسة الذي وعد بها لانها	كان في الاصل المنتسخ منه هذه المقالة الرابعة هى آخر هذا الكتاب وما وعد ان يتكلم به في الخامسة

[22] Cf. *Analytica posteriora* I 10, 76b 3–9 (trans. Barnes, p. 124): 'Proper too are the things which are assumed to be, about which the science considers what belongs to them in themselves—as e.g. arithmetic is about units, and geometry is about points and lines. For they assume these to be and to be *this*. As to what are attributes of these in themselves, they assume what each signifies—e.g. arithmetic assumes what odd or even or quadrangle or cube signifies, and geometry what irrational or inflection or verging signifies…'.

تشتمل على اكثر امور غير مهمة كتصحيح مبادى العلوم والمقدمات اليقينية ورأى انه يكفى بذلك ما اشير اليه

Somebody who personally talked to the author reported that he (*i.e.* the author) did not attend to completing the book by [adding] *the promised fifth chapter*—because it would contain *predominantly irrelevant things such as the verification of the principles of the sciences and the indisputable premises*—but considered what he had pointed out [in the previous chapters] to be sufficient.

فانه بعد ذلك لم يعول عليه لانه رأى ان الذي بقى القول فيه من هذا العلم غير مهم اذ كان اكثر ذلك انما هو في اعطاء مبادى العلوم وفي تصحيح المقدمات اليقينية بالاقاويل المشهورة اى الجدلية ولما كان هذا غير ضروري جزم القول في آخر هذه المقالة الرابعة وهذا قول من شافه المؤلف رضى (Hyderabad رضى: Rampur عفو) الله عنه

In the manuscript from which this manuscript was copied [the following] was [written:] This fourth chapter is the last one of this book. [The author] decided not to add *what he had promised* to discuss *in the fifth [chapter]*, because he considered that [part] of this science which remains to be treated *to be irrelevant*, for *most of it consists* only in providing *the principles of the sciences and verifying the indisputable premises* through generally accepted—i.e. dialectical—arguments. Since this is not necessary, he broke off his teaching at the end of Chapter IV. This *was reported by somebody who personally talked to the author* (may God be pleased with him).

As can be gathered from the *corresponding* phrases, both versions of the colophon ultimately must go back to one and the same source. There is good reason to doubt the authenticity of this source. The author of the original version of this colophon claims authenticity by referring to somebody personally acquainted with Ibn Rushd as his source of information. He must have been acquainted, therefore, directly or indirectly, with this informant. Taking into consideration the crucial relevance attached in medieval Muslim society to the name and identity of an informant in connection with the reliability of a report, the fact that the author of the colophon failed to identify his source is highly suspicious. Furthermore, this informant must have been a contemporary of Ibn Rushd, but the colophon does not appear in the textual tradition before the sixteenth century.[23] If the colophon was originally

[23] The Cairo manuscript is dated 936 A.H. = 1529 C.E., ms. Āṣafiyya dates in all likelihood from Safavid times, the Rampur manuscript is dated 1266 A.H. = 1850 C.E.

appended to the text in the early thirteenth century one would expect (*a*) to find its traces also in earlier stages of the transmission (i.e., in the two early Arabic manuscripts Madrid, Biblioteca Nacional 5000, and Cairo, Ḥikma wa-falsafa 5, or the two thirteenth century Hebrew translations mentioned above), and (*b*) to come across a much broader transmission in the later tradition.

Additional doubt is raised by the content of the colophon itself. Ibn Rushd certainly did not consider the finding and verification of the first principles of demonstration and of the principles of the particular sciences as 'irrelevant' (*ghayr muhimm[a]*), as stated in both versions of the colophon. His *Long Commentary* (*Tafsīr*) on the subject, i.e. on *Metaph.* IV (Γ) 3–8, covers no less than 137 pages (ed. Bouyges, i, pp. 335–472), while his *Long Commentary* on the corresponding passages of book I 10 of Aristotle's *Analytica posteriora* is likewise very detailed.[24] What Ibn Rushd says in the relevant section of the introduction of *EM* is not that the study of the first principles is irrelevant but that it is not a necessary part of metaphysics, i.e. that metaphysics as such consists necessarily of the two parts mentioned in the preceding passage of the introduction, namely the study of being *qua* material, sensible being, its species and its essential attributes (part 1), and the study of being *qua* separate and immaterial being, i.e. the principles and causes of all being (part 2). These two parts together constitute metaphysics as science of being *qua* being. In this respect neither the finding and confirmation of the first principles of demonstration nor the establishment of the subject-matters of the particular sciences is a necessary part of metaphysics. However, since all particular sciences make use of these principles in a specific way appropriate to each specific science, only metaphysics as the most general and the only universal science can perform the task of studying these principles *qua* apodeictic principles of all sciences. Hence, the "third part", although not a necessary, i.e. constitutive, part of metaphysics, is nevertheless a part of metaphysics which cannot be entrusted to any other science, and certainly anything but 'irrelevant'.

Of course, the probable inauthenticity of the colophon does not allow us simply to conclude that the promised Chapter V was ever actually

[24] Cf. Ibn Rushd, *Sharḥ al-Burhān li-Arisṭū* wa-*Talkhīṣ al-Burhān*, ed. Badawī, pp. 302–19.

written. There are three possible explanations for the absence of this chapter: (*i*) Ibn Rushd never composed it; (*ii*) Ibn Rushd did compose it, but it got lost in an early stage of the manuscript tradition; or (*iii*) Ibn Rushd composed Chapter V, but later (possibly in a thorough revision of the work) discarded what he had written earlier and decided to withhold it from publication.

Explanation (*i*) is rather improbable.[25] *EM* is an early work of Ibn Rushd, and it was succeeded by many others. Biographical obstacles to its completion, such as death or severe and protracted illness of its author, can be excluded. Even if Chapter V was not yet finished by the time it was announced in the introduction, Ibn Rushd had plenty of time after the completion of Chapters I–IV either to finish it or to revise accordingly the introductory notes concerning the structure and contents of his work—especially since he indeed found the time to revise *EM* after having completed the *Long Commentary on the Metaphysics.*[26] The introduction of *EM* is not the only place where reference is made to Chapter V. In the subsequent chapters of *EM* there are at least three further references to this chapter, which demonstrate not only its importance but also the fact that Chapter V, if not entirely completed, at least must have been in an advanced state during the composition of Chapters I–IV. The first of these references, which relates the discussion of the subject-matter of mathematics to the verification of the fundamentals of the particular sciences, has been mentioned earlier.[27] A further reference follows some pages later. In connection with considerations concerning the nature of universals (*al-kulliyyāt*) and their mode of being (dealing with *Metaph.* VII [Z] 14–16), Ibn Rushd mentions that many contemporary *mutakallimūn* totally deny their existence and then refers to a discussion of this point as follows:

> A controversy with these and others on this [point] will be held in [the chapter on] the verification of the principles of logic and other particular sciences.[28]

---

[25] Nevertheless favoured, without further arguments, by Quirós Rodríguez, p. xxxiv.

[26] Cf. al-ʿAlawī, p. 164.

[27] Cf. above, p. 50 and note 19. Such a discussion does not appear in any section of Chapters II–IV but fits perfectly in what has been outlined as topic of Chapter V.

[28] Cf. Quirós Rodríguez, p. 59, l. 22–4 ≈ Amīn, p. 56, l. 7–9 ≈ Jihāmī, p. 79, l. 6–8:

وسيأتي القول في هذا معهم ومع غيرهم في تصحيح مبادئ صناعة المنطق وغيرها من الصنائع الجزئية

This reference clearly corresponds to the outline of Chapter V presented in the introduction. The fact that Ibn Rushd announces that he will discuss there the logical and epistemological consequences of a general denial of universals bears out the assumption, mentioned above, that Chapter V also included parts of *Metaph.* XIII (M) and XIV (N), for such consequences are indeed examined, among others, in book XIII (M) 9–10.[29] Finally, Ibn Rushd announces at the end of Chapter IV that he will clarify in the following chapter questions raised by the theory that the Good as such is different from the First Principle of being (probably referring to chapter 4 of book XIV [N]):

> God willing, we will devote ourselves to showing the reprehensibility of what follows from these [opinions] in the chapter which follows this [fourth] one.[30]

Apart from the detailed references themselves, the mere fact that Ibn Rushd revised *EM* about three decades after its composition without deleting these references to Chapter V—a minor undertaking that could have been carried out in a few minutes—is a further strong argument against explanation (*i*).

Explanation (*ii*) cannot be excluded with certainty. However, we have to take into consideration that the manuscript tradition of *EM* split up into at least three branches already about one century after Ibn Rushd's revision of the text—these three branches being represented by the two Hebrew versions and the Arabic manuscript Cairo, Ḥikma wa-falsafa 5, all of which occasionally contain independent readings.[31] We would thus have to assume, if we are to entertain explanation (*ii*), that these three branches trace back all together to one and the same defective Arabic manuscript, which is possible but not very likely.

The most plausible reason for the absence of Chapter V is therefore explanation (*iii*). And yet a simple deletion of this chapter during the process of revision must be excluded for the very reason already given

---

[29] E.g., *Metaphysics* XIII 9, 1086b 5sqq.: δηλοῖ δ'ἐκ τῶν ἔργων· ἄνευ μὲν γὰρ τοῦ καθόλου οὐκ ἔστιν ἐπιστήμην λαβεῖν… (etc.).

[30] Cf. Quirós Rodríguez, p. 171, l. 3sq. ≈ Amīn, p. 164, l. 15sq. ≈ Jihāmī, p. 173, l. 11sq.:

وسنفرغ لبيان ما يلحقها من الشناعة في المقالة التي تلي هذه إن شاء الله تعالى

[31] Cf. B. Chiesa, 'Note su al-Fārābī, Averroè e Ibn Bāğğa (Avempace) in traduzione ebraica,' *Henoch* 8 (1986), pp. 79–86, esp. p. 83sq., also I. Husik, 'Averroes on the Metaphysics of Aristotle,' in S.W. Baron and A. Marx (eds.), *Jewish Studies in Memory of George A. Kohut 1874–1933* (New York, 1935), pp. 370–78; repr. in Sezgin (et al.) (eds.), *Abu l-Walīd Muḥammad Ibn Rushd: Texts and Studies*, vi, pp. 150–58.

for explanation (*i*): if that were so, it would have been easy to revise the introduction and eliminate the references. That Ibn Rushd refrained from implementing such slight revisions suggests that he never intended to delete Chapter V without substitution, but rather separated it from the first four chapters, already provisionally revised, in order to perform a more thorough revision or to replace it with an entirely new version. Such a revision or new composition of Chapter V obviously was never finished, and this, unlike explanation (*i*), may very well have its cause in the events of Ibn Rushd's life. Since the revision of Chapters I–IV of *EM* must have taken place after the completion of the *Tafsīr* on the *Metaphysics*, this work (intended or undertaken) on Chapter V of *EM* is to be dated to the last four years of Ibn Rushd's life,[32] years which were marked by banishment, the burning of his works, and his emigration to Marrakech.[33]

As for the motives which might have caused Ibn Rushd's decision to revise or re-write Chapter V, one might take into consideration his lifelong grappling with the problem of how to acquire the first principles of demonstration and of identifying the science which is to verify and confirm such principles. Only recently, C.E. Butterworth has shown that Ibn Rushd failed to offer any attempt to settle this problem in his early *Compendium of Logic* (*al-Ḍarūrī fī 'l-manṭiq* or *Mukhtaṣar al-manṭiq*).[34] Regarding the question of how this problem can be integrated into metaphysics, the view Ibn Rushd takes in his *Tafsīr* is different from that found in *EM*. In the latter he defends, as we have seen, a clear-cut threefold division of metaphysical topics: (*i*) being *qua* sensible being, its species and its essential attributes, (*ii*) being *qua* separate, immaterial being or principle and cause of sensible being, and (*iii*) problems and errors concerning the first principles of all sciences and the subject-matter of the particular sciences. In the *Long Commentary* the aporetic character of the "third part" of metaphysics stressed in *EM* is discarded in favour of an ontological approach to these principles. Starting with a paraphrase of *Metaph.* IV (Γ) 3, 1005a 21sq., Ibn Rushd explains there:

[32] According to al-ʿAlawī, p. 109sq., the *Long Commentary* on the *Metaphysics* was completed around 590 A.H. = 1194 C.E.

[33] Cf. M. Geoffroy, 'Averroè,' in C. D'Ancona (ed.), *Storia della filosofia nell'Islam medievale*, 2 vols. (Torino, 2005), ii, pp. 723–82, there p. 768sq.

[34] C.E. Butterworth, 'Finding First Principles, Possibility or Impasse?,' in R. Arnzen and J. Thielmann (eds.), *Words, Texts and Concepts Cruising the Mediterranean Sea* (Leuven, 2004), pp. 211–22, esp. 219–22.

> The examination of these principles belongs to one and the same science, i.e. this science which is the science of the philosopher, because
> *[i]* these principles are common to all genera of the entities studied by the theoretical sciences, and
> *[ii]* that which is common to all existing genera is one of the concomitants of being *qua* being, and
> *[iii]* everything which is essentially a concomitant of being is examined by him who examines being as such, and this is the philosopher.[35]

Hence, the examination of the first principles of demonstration is no longer exiled from the study of being *qua* being and its species and essential attributes, as outlined in the introduction of *EM*, but repatriated into this area of metaphysics ([iii]) through inclusion of the first principles under the essential concomitants of being ([i]–[ii]).

It is a matter of speculation whether that which caused the actual absence of Chapter V was in fact the unrealized plan for a revision—motivated either by this shift concerning the epistemological location of the study of the first principles or by some other consideration—or a mere omission or codicological mishap in an early stage of the transmission. We are, however, quite safe in assuming that its absence was not caused by Ibn Rushd's deliberate decision either never to compose it or to delete an early version without substitute. The version of the *Epitome of the Metaphysics* we know today is, therefore, not the one intended by Ibn Rushd for circulation, but rather either an unfinished work or a work preserved incompletely.

---

[35] Ed. Bouyges, i, p. 337, l. 18–p. 338, l. 6.

ON THE MANUSCRIPTS OF THE *ILĀHIYYĀT* OF AVICENNA'S *KITĀB AL-SHIFĀʾ**

Amos Bertolacci

In a pioneering article published in 1986, Hans Daiber has masterly shown that Indian libraries represent extremely rich, and largely unexplored, depositories of manuscripts of Arabic philosophical works.[1] The Arabic manuscripts discovered by Daiber in Indian libraries regard also some of Avicenna's most important works on philosophy.[2]

Several manuscripts of Avicenna's masterpiece on metaphysics, the *Ilāhiyyāt* ([*Science of*] *Divine Things*) of the *Kitāb al-shifāʾ* (*Book of the Cure*), are preserved in India: further manuscripts of the same work will predictably be discovered in other libraries worldwide by scholars that will follow the example of Daiber's ground-breaking research. In the footstep of Daiber's aforementioned study, the present contribution tries to collect, order and update the information on the manuscript tradition of Avicenna's *Ilāhiyyāt* available in secondary literature.

The first section of the present essay will provide an inventory of the known codices of the *Ilāhiyyāt*, adding a few items to the list that can be drawn from the bibliographies of C. Brockelmann, G.C. Anawati and Y. Mahdavī,[3] and showing the extremely wide circulation of Avicenna's

---

* I wish to thank sincerely Dr. Anna Akasoy (Oriental Institute, Oxford) for the information kindly provided on the Arabic manuscripts of the Bodleian Library of Oxford and the British Library of London, and Father René-Vincent du Grandlaunay (IDEO, Cairo) for having provided me with a digital reproduction of mss. 16–18 below. My sincere gratitude goes also to Dr. Heidrun Eichner (Universität Halle) who brought to my attention mss. 38, 41 below, and provided abundant information on the Istanbul codices of Avicenna, as well as to Dr. Hamed Naji Isfahani (Islamic Azad University, Dehagan [Iran]-Freie Universität, Berlin) who informed me of the existence, date and content of mss. 61–2, 77–8 below. I am also deeply indebted to Dr. Rüdiger Arnzen (Thomas-Institut, Universität zu Köln) for his remarks on a first draft of this article.

[1] H. Daiber, 'New Manuscript Findings from Indian Libraries,' *Manuscripts of the Middle East* 1 (1986), pp. 26–48.

[2] J.L. Janssens, *An Annotated Bibliography on Ibn Sînâ (1970–1989)* (Leuven, 1991), p. 84, remarks that Daiber's article "offers a valuable complement to the bibliographies [of Avicenna's works] of Anawati, Mahdavī and Ergin".

[3] *GAL* i, 592 and *S* i, 815; G.C. Anawati, *Essai de bibliographie avicennienne* (Cairo, 1950), pp. 69–78, 430–34; Y. Mahdavī, *Fihrist-i nuskhahā-yi muṣannafāt-i Ibn-i Sīnā* (Tehran, 1954), pp. 170–72. The data concerning the libraries of Istanbul in Anawati are mainly derived from O. Ergin, *İbni Sina Bibliografyası* (Istanbul, [2]1956) (see G.C. Anawati, 'La

work in the Islamic world. In the second section, the chronology of these manuscripts will be taken into account, pointing to another remarkable feature displayed by the *Ilāhiyyāt*, namely the extremely long period of time during which this work was copied. The final section will evaluate the editions of Avicenna's work presently available in light of the data discussed in the previous two sections, pointing to the necessity of a new, better grounded and more precise, critical edition.

In some cases, the manuscripts containing the *Ilāhiyyāt* preserve also all or some of the other parts of the *Shifā'*. Incidentally, thus, the information provided here will shed light also on the manuscript transmission of the sections on logic, natural philosophy and mathematics of Avicenna's philosophical *magnum opus*.

I. Inventory

Excluding the manuscripts of private collections,[4] and without taking the fragments into account,[5] more than one hundred manuscripts of Avicenna's *Ilāhiyyāt* are recorded by Brockelmann, Anawati and Mahdavī; to these, some others (one manuscript preserved in Cairo—below, no. 17—and the ones in the libraries of Ankara, Damascus, Khoy and Princeton) can now be added. The available data can be provisionally arranged in the following list, with some *caveat*. First, the indications provided by the aforementioned bibliographies are not always clear and correct: some manuscripts reported as containing the *Ilāhiyyāt*, for example, do not in fact contain it.[6] Second, with the progress of research new manuscripts of the *Ilāhiyyāt* from other geographical areas of the Middle East and Europe will probably be brought to light. Thus, the next step in this path of research, i.e. the systematic scrutiny of the catalogues of libraries preserving Arabic manuscripts, and of the funds

---

tradition manuscrite orientale de l'œuvre d'Avicenne,' *Revue Thomiste* 51 [1951], pp. 407–40; also in idem, *Études de philosophie musulmane* [Paris, 1974], pp. 229–62).

[4] See the manuscripts of this kind mentioned below, n. 72, and the ones used by Ḥasanzādah al-Āmulī for his printed version of the *Ilāhiyyāt* (below, section III). Also Hans Daiber's personal library contains an Arabic manuscript of the *Ilāhiyyāt*, copied in 865/1461 (provisionally catalogued as Daiber Collection III, 131; I wish to thank Prof. Daiber for having kindly put at my disposal a copy of this codex).

[5] Ms. Istanbul, Köprülü 1604, fol. 62r, l. 11–62v, l. 6, for example, is an almost literal quotation of *Ilāhiyyāt* viii, 6, p. 357, 4–9: this fragment occurs in the *Faṣl min kalām al-shaykh*, fol. 62r–v, comprising passages from Avicenna's *Ta'līqāt*, ed. 'A. Badawī (Cairo, 1973), p. 78, 24–8 (I owe this information to D.C. Reisman).

[6] See, for example, below, n. 23, 56, 62.

of the libraries themselves, will probably in a way slightly reduce, and in another substantially enlarge, the proposed list.[7]

Aligarh (Uttar Pradesh, India) (5 mss.)
1) Maulana Azad Library, Aligarh Muslim University 80/25–31 (+ m, ṭ, r) (B, A)
2) Maulana Azad Library, Aligarh Muslim University 81/41, 50–57 (+ m, ṭ, r) (B, A)
3) Maulana Azad Library, Aligarh Muslim University 84/26 (+ m, ṭ, r) (B)
4) Maulana Azad Library, Aligarh Muslim University 110/30 (+ m, ṭ, r) (M)
5) Maulana Azad Library, Aligarh Muslim University 110/40; 53 (M)

Ankara (1 ms.)
6) Millī Kütüphane (National Library) B 153 (694/1294)[8]

Berlin (1 ms.)
7) Deutsche Staatsbibliothek 5045, Minutoli 229 (Iṣfahān, 1083/1672) (M)[9]

Bihar, Patna (India) (5 mss.)[10]
8) Khuda Bakhsh Oriental Public Library (Bankipur) I, 213 (+ m, ṭ, r) (B)
9) Khuda Bakhsh Oriental Public Library (Bankipur) I, 904/6 (+ m, ṭ, r) (B)

[7] Sigla: (A) = ms. recorded in Anawati; (B) = ms. recorded in Brockelmann; (M) = ms. recorded in Mahdavī; (+ m, ṭ, r) = ms. containing all the *Shifā'* (i.e. besides the *Ilāhiyyāt*, also *Manṭiq* [= logic], *Ṭabī'iyyāt* [= natural philosophy] and *Riyāḍiyyāt* [= mathematics]); (+ m, ṭ) = ms. containing *Ilāhiyyāt*, *Manṭiq* and *Ṭabī'iyyāt*; (+ m) = ms. containing *Ilāhiyyāt* and *Manṭiq*; (+ ṭ) = ms. containing *Ilāhiyyāt* and *Ṭabī'iyyāt*.

[8] See *World Survey of Islamic Manuscripts*, ed. G. Roper, 4 vols. (London, 1992–94), iii, p. 302.

[9] Datation in Ibn Rushd, *Die Metaphysik Avicennas enthaltend die Metaphysik, Theologie, Kosmologie und Ethik*, trans. M. Horten (Leipzig, 1907; repr. Frankfurt, 1960), p. ix and n. 1. The ms. Deutsche Staatsbibliothek 5044, reported as containing the entire *Shifā'* in *GAL*, 592, is reported as containing part of one section of logic (*Qiyās*) by Anawati, p. 70.

[10] Brockelmann refers to this library as both "Patna" and "Bank"; Anawati as both "Bankibūr" and "Būhār".

10) Khuda Bakhsh Oriental Public Library (Bankipur) II, 525,$_{2822}$ (B)
11) Khuda Bakhsh Oriental Public Library VII, 284 (Bankipur) (+ m, ṭ, r) (A)
12) Khuda Bakhsh Oriental Public Library (Bankipur) XXI, 2226 (B, A, M)[11]

Cairo (9 mss.)[12]
13) Azhar 331 *khuṣūṣiyya*, 2415 Bakhīt (684/1285) (+ m, ṭ, r) (A, M)[13]
14) Dār al-Kutub 51 *falsafa* (+ m, ṭ, r) (A)
15) Dār al-Kutub 144 *ḥikma* (684/1285) (A, M)[14]
16) Dār al-Kutub 262 *ḥikma wa-falsafa* (1337/1914, copy of a ms. of 992/1584) (+ m, ṭ) (A, M)[15]
17) Dār al-Kutub, 349 *ḥikma wa-falsafa* (13th c./19th c.) [incompl.: I, 1–3 pp. 3, 4–23, 16 (fol. 68–86, 13); descendant of ms. 42].[16]
18) Dār al-Kutub, 826 *ḥikma wa-falsafa* (1084/1673) (A, M)[17]

[11] According to Brockelmann, *GAL*, 815, the *Ilāhiyyāt* is contained in ms. xxi, 2226. Mahdavī, p. 170, reports mss. xxi, 2223–6 as containing the entire *Shifāʾ*. Anawati, p. 70, reports mss. xxi, 2223 and xxi, 2226 as containing the entire *Shifāʾ*.

[12] Photostatic reproductions of two mss. of uncertain provenance (possibly from Cairo libraries) are recorded by D.C. Reisman, 'Avicenna at the ARCE,' in R. Wisnovsky (ed.), *Aspects of Avicenna* (Princeton, 2001), pp. 131–82, ##133 and 232 (the former ms. is reported as containing mathematics and *Ilāhiyyāt*, the latter an incomplete version of the *Shifāʾ*).

[13] Description in Avicenna, *La Métaphysique du Shifāʾ*. Livres i à v, trans. G.C. Anawati (Paris, 1978) (henceforth: Anawati [1978]), pp. 18–19. Anawati, p. 70, reports this ms. as Azhar 331, Bakhīt 44988. Detailed table of contents in Anawati, pp. 30–66 (see Anawati, 'La tradition,' p. 418). The date of 684 A.H. is provided by F. Sayyid, *Ibn Sīnā: muʾallafātuhū* (catalogue of Avicenna's works held in Dār al-Kutub Library of Cairo; I owe this information to D.C. Reisman), whereas Anawati [1978] takes it to be written, generically, in the seventh century A.H. The *Ilāhiyyāt* covers fol. 390–427. See Maʿhad Iḥyāʾ al-Makhṭūṭāt al-ʿArabiyya, *Fihris al-makhṭūṭāt al-muṣawwara* (henceforth: *Fihris*), p. 226, n. 261.

[14] Description and datation in Anawati [1978], p. 20. Anawati, p. 72, names this ms. 144 *falsafa* and dates it to 1083 A.H. Mahdavī, p. 170, reports this ms. as containing the entire *Shifāʾ*.

[15] Description in Anawati [1978], pp. 19–20, who names this ms. 262 *ḥikma*. Anawati, p. 72, reports only the date of 1919 [sic] C.E.

[16] Description and datation in Reisman, *The Making of the Avicennan Tradition. The Transmission, Contents, and Structure of Ibn Sīnā's* al-Mubāḥaṯāt (The Discussions) (Leiden, 2002), p. 76, who names this ms. 349 *falsafa*. Mcrf. F 511 [399] at the Maʿhad al-Makhṭūṭāt bi-Jāmiʿa al-Duwal al-ʿArabiyya (henceforth: Maʿhad, see Anawati, p. 425).

[17] Description and datation in Anawati [1978], p. 20, who names this ms. 826 *falsafa*.

19) Dār al-Kutub, 894 *falsafa* (+ m, ṭ, r) [incompl.: the first 20 folia of the *Ilāhiyyāt* are missing] (A, M)[18]
20) Dār al-Kutub, Ṭalʿat 363 (1105/1693) (A, M)[19]
21) Dār al-Kutub, Taymūr 140 (535/1140) (+ ṭ) (A, M)[20]

Damascus (1 ms.)
22) Ẓāhiriyya (now Maktabat al-Asad al-Waṭaniyya, Asad National Library) 2905 (+ m, ṭ, r)[21]

Istanbul (37 mss.)[22]
23) Atıf Efendi Kütüphanesi 1596 (1098/1686) (A, M)[23]
24) Beyazit Kütüphanesi (form.: ʿUmūmī) 3966 (+ m, ṭ, r) (A)
25) Beyazit Kütüphanesi (form.: ʿUmūmī) 3967 (+ m, ṭ, r) (A)
26) Beyazit Kütüphanesi (form.: ʿUmūmī) 3969 (+ m, ṭ, r) (M)
27) Çoban 572 (A)[24]
28) Istanbul Üniversitesi Kütüphanesi, Üniversite 760 (1060/1650) (A, M)[25]
29) Istanbul Üniversitesi Kütüphanesi, Üniversite 766 (1060/1650) (A)[26]
30) Köprülü Kütüphanesi, Fazil Ahmet Paşa 894 (+ m, ṭ) (A, M)[27]
31) Millet Kütüphanesi, Feyzullah Efendi 1445 (1132/1719) (A, M)[28]
32) Nuruosmaniye Kütüphanesi 2708 (+ m, ṭ, a fragment of r) (A, M)[29]

---

[18] Description in Anawati [1978], p. 19. Mcf. F 1030 and 1031 at the Maʿhad (see Anawati, p. 423); cf. *Fihris*, p. 226, n. 260.
[19] Datation in Anawati, p. 74.
[20] Datation in Anawati, p. 71, and Mahdavī, p. 171.
[21] See S. al-Khīmī, 'Manuscripts of Avicennian Works in the National Library al-Ẓāhiriyya,' *Al-Turāth al-ʿarabī* 2 (1981), pp. 91–112, p. 105 (the incipit and explicit of this manuscript, reported in Al-Khīmī's catalogue under the heading '*al-Shifāʾ, al-jumla al-ūlā*', make clear that it contains the entire *Shifāʾ*).
[22] Anawati, p. 70, maintains that he has not been able to identify a ms. Istanbul, Beşīr reported by Ergin (see below, ms. 41).
[23] Datation in Anawati, p. 74.
[24] I have not been able to identify the library or collection of Istanbul to which Anawati is referring.
[25] Datation in Anawati, p. 78. The ms. Atif Efendi Kütüphanesi 1597, reported by Anawati as containing the entire *Shifāʾ*, does in fact contain only natural philosophy.
[26] Datation in Anawati, p. 78.
[27] Anawati reports this ms. as containing also mathematics.
[28] Datation in Anawati, p. 76; Anawati remarks that this ms. is not recorded by Ergin.
[29] Mcrf. F 822 [1] at the Maʿhad (see Anawati p. 423); cf. *Fihris*, p. 226, n. 263.

33) Nuruosmaniye Kütüphanesi 2709 (886/1481–897/1481) (+ m, ṭ, r) (A, M)[30]
34) Nuruosmaniye Kütüphanesi 2710 (666/1267) (+ m, ṭ, r) (A, M)[31]
35) Nuruosmaniye Kütüphanesi 2711 (+ ṭ) (A, M)
36) Nuruosmaniye Kütüphanesi 4894 [incompl.: I, 1–3] (A, M)[32]
37) Ragıp Paşa Kütüphanesi 865 (M)
38) Ragıp Paşa Kütüphanesi 910 (+ m, ṭ, r)[33]
39) Süleymaniye Kütüphanesi, Ayasofya 2389 (9th c./15th c.) [incompl.: I, 1–3, p. 3, 4–23, 16] (A)[34]
40) Süleymaniye Kütüphanesi, Ayasofya 2442 (671/1272) (+ m, ṭ, r) (A, M)[35]
41) Süleymaniye Kütüphanesi, Beşir Ağa (Eyüp) 102.
42) Süleymaniye Kütüphanesi, Carullah 1332 (882/1477) (+ ṭ) (A, M)[36]
43) Süleymaniye Kütüphanesi, Carullah 1333 (+ ṭ) [incompl.: it starts from III,7] (A, M)
44) Süleymaniye Kütüphanesi, Carullah 1424 (693/1293) (+ m, ṭ, r) (A, M)[37]
45) Süleymaniye Kütüphanesi, Carullah 1425 (1115/1703) (+ m) (A, M)[38]
46) Süleymaniye Kütüphanesi, Damat Ibrahim Paşa 822 (8th c./14th c.) (+ m, ṭ, r) (M) [incompl.: it contains X, 4, p. 447, 14–end; X, 5 (fol. 355r)] (A, M)[39]

---

[30] The right datation (Mahdavī, p. 170) regards logic; the second datation (Mahdavī, p. 77) mathematics. The *Ilāhiyyāt* was copied in 894/1488.

[31] Datation in Anawati, p. 77, and Mahdavī, p. 170.

[32] Anawati, pp. 246–7, 452, records this ms. under Avicenna's work no. 189 (*Al-Falsafa al-ūlā*). Mahdavī, p. 172, states that this ms. contains the entire *Ilāhiyyāt*. According to G.C. Anawati, 'Avicenniana. Le manuscrit Nour Osmaniyye 4894,' *MIDEO* 3 (1956), pp. 381–6, p. 383, as the part of the *Ilāhiyyāt* preserved in this ms. is i, 1,3. On this ms. see also D. Gutas, *Avicenna and the Aristotelian Tradition. Introduction to Reading Avicenna's Philosophical Works* (Leiden, 1988), p. 87, n. 2, p. 120, n. 18, and Reisman, *The Making*, p. 44 ("Anawati produced a seriously flawed contents list").

[33] The ms. Ragip Paşa Kütüphanesi 1461 reported by Anawati as containing the entire *Shifā'*, contains in fact only a short excerpt of one section of logic (*Burhān*).

[34] Description and datation in Reisman, *The Making*, pp. 71–4. Anawati, p. 69, reports this ms. as complete.

[35] Datation in Mahdavī, p. 170.

[36] Datation in Anawati, p. 71.

[37] Datation in Anawati, p. 71, and Mahdavī, p. 170.

[38] Datation in Anawati, p. 71.

[39] Contrary to Mahdavī, p. 170, who contends that this ms. contains the entire *Shifā'*, Anawati, p. 72, maintains that it does not contain the *Ilāhiyyāt*. Anawati refers to the

47) Süleymaniye Kütüphanesi, Damat Ibrahim Paşa 823 (697/1297) (+ m, ṭ, r)[40]
48) Süleymaniye Kütüphanesi, Eyüp Câmii Hz. Hâlid 883 (A)[41]
49) Süleymaniye Kütüphanesi, Hâlet Efendi 513 (A, M)
50) Süleymaniye Kütüphanesi, Hamidiye 795 (1066/1655) (+ m, ṭ) (A, M)[42]
51) Süleymaniye Kütüphanesi, Hekimoğlu Ali Paşa 857 (1102/1690) (+ m, ṭ, r) (A, M)[43]
52) Süleymaniye Kütüphanesi, Kılıç Ali Paşa 673 or 689 (1185/1771) (A, M)[44]
53) Süleymaniye Kütüphanesi, Lâleli 2546 (A, M)
54) Süleymaniye Kütüphanesi, Lâleli 2550 (1023/1614) (+ m) (A)[45]
55) Süleymaniye Kütüphanesi, Şehid Ali Paşa 1748 (879/1474) (+ m, ṭ) (A, M)[46]
56) Süleymaniye Kütüphanesi, Yeni Câmi 770 (888/1483) (+ m, ṭ, r) (B, A, M)[47]
57) Topkapı Sarayı Müzesi Kütüphanesi, Ahmed III 3262 (+ m, ṭ, r) (A, M)[48]
58) Topkapı Sarayı Müzesi Kütüphanesi, Ahmed III 3263 (+ ṭ, r) (A, M)[49]
59) Topkapı Sarayı Müzesi Kütüphanesi, Ahmed III 3445 (+ m, ṭ, r) (A)

---

study of this ms. by H. Ritter and others, who do not take as authentic the *qirā'a* and *mubālagha* note, allegedly from Avicenna's hand, dated 3 Rajab 422, occurring at the end of the section on mathematics (music, fol. 354v), and date this ms. to the beginning of the eighth/fourteenth century. Cp. Mahdavī, p. 170, bottom of page.

[40] Datation in Anawati, p. 72, and Mahdavī, p. 170.

[41] The siglum *Ayub* employed by Anawati with regard to this ms. does not occur in the list of libraries at the beginning of his bibliography. It might refer also to the collection Beşir Ağa (Eyüp) of the Süleymaniye Library.

[42] Datation in Anawati, p. 71.

[43] Datation in Anawati, p. 71.

[44] Datation in Anawati, p. 71. Anawati, who records it as 689, remarks that this ms. is recorded as Kılıç Ali Paşa 673 by Ergin.

[45] According to Anawati, this ms. does not contain logic.

[46] Datation in Anawati, p. 74.

[47] Datation in Anawati, p. 77, and Mahdavī, p. 170.

[48] Mahdavī, p. 171, reports this ms. as containing, besides the *Ilāhiyyāt*, only logic and natural philosophy.

[49] Mahdavī, p. 170, reports this ms. as containing the entire *Shifā'*.

60) Topkapı Sarayı Müzesi Kütüphanesi, Ahmed III 3447 (866/1462) [incompl.: I, 1–3, p. 3, 4–23, 16 (fol. 396v–404v); descendant of ms. 42] (A, M)[50]

Khvoy (Khoy, Khuy) (Iran) (2 mss.)

61) Madrasa Nimāzī 248 (+ ṭ) [incompl.: the end of the *Ilāhiyyāt* is missing] (7th c./13th c.)[51]
62) Madrasa Nimāzī: a further ms. of the *Ilāhiyyāt* (9th c./15th c.) is reported as preserved in this library[52]

Leiden (2 mss.)

63) Universiteitsbibliotheek 1444, Golius 4 (+ m, ṭ, r) (B, A, M)[53]
64) Universiteitsbibliotheek 1445, Golius 84 (881–882/1476–1477) (+ m, ṭ, r) (B, A, M)[54]

---

[50] Description in Reisman, 'Avicenna,' pp. 136–9, and Reisman, *The Making*, pp. 74–5. Anawati, p. 247 (cf. p. 69), records this ms. under work no. 189 of Avicenna (*Al-Falsafa al-ūlā*). Mahdavī, p. 172, states that this ms. contains the entire *Ilāhiyyāt*. Mcrf. F 653 [732] at the Maʿhad (see Anawati p. 426); cf. *Fihris*, p. 206, n. 73; photographic reproduction at the American Research Center in Cairo.

[51] Dr. Hamed Naji Isfahani has kindly reported to me that, according to the catalogue of the Madrasa Nimāzī, p. 13, this manuscript (in *naskh* writing, with some folios not in order) was originally from Istanbul, and passed later into the possession of Naṣīr al-Dīn al-Ṭūsī, who stamped it with the date of 670 A.H.

[52] I owe this information to Dr. Hamed Naji Isfahani.

[53] Mahdavī, p. 171, reports this ms. as containing logic, natural philosophy and metaphysics. See C. Landberg, R.P.A. Dozy, P. de Jong, and M.J. de Goeje, *Catalogus Codicum Orientalium Bibliothecae Academiae Lugduno-Batavae* (Leiden, 1865), iii, pp. 315–19. Date uncertain. An owner's note from X c. A.H. The *Ilāhiyyāt* is contained from fol. 275 until the end. Used in Horten, *Die Metaphysik* (see p. ix and n. 1).

[54] Logic: 881 A.H.; other parts: 882 A.H.; the *Ilāhiyyāt* precedes the other parts (see Landberg, Dozy, de Jong, and de Goeje, *Catalogus*, iii, p. 319). Anawati, p. 76, dates the entire ms. to 881 A.H.; Mahdavī, p. 170, to 882 A.H. Anawati, p. 76, states that this ms. contains the entire *Shifāʾ* "except the three [first?] sections of the first part", and that "parts of this manuscript can be found in the Bodleian [Library of Oxford]"; at the end of the list of the Bodleian mss., possibly referring to the last codices mentioned (II, 281ff), he writes: "All these manuscripts are simple reproductions ([?] *mujarrad ṣuwar*) of the manuscripts Leiden 1444–1445" (p. 71). Used by Horten, *Die Metaphysik* (see p. ix and n. 1). Brockelmann, p. 815, apparently reports mss. 63–64 as 144/5.

London (4 mss.)[55]

65) India Office (now: British Library, Oriental and India Office Collections) 477 (+ r) (M)[56]
66) British Museum (now: British Library, Oriental and India Office Collections) Suppl. 711 (+ m, ṭ, r) (B, A)[57]
67) British Museum Or. 5592_1 (+ m, ṭ, r) (B, A)[58]
68) British Museum Or. 7500 (+ m, ṭ) [incompl.: I–IX] (B, A, M)[59]

Manchester (1 ms.)

69) John Rylands University Library 378 (B, A, M)

Mashhad (Iran) (9 mss.)

70) Āstān-i Quds-i Razavī Library I,1/172 (B, A, M)[60]
71) Āstān-i Quds-i Razavī Library I,1/182 (M)
72) Āstān-i Quds-i Razavī Library I,1/183 (M)
73) Āstān-i Quds-i Razavī Library I,1/184 (M)
74) Āstān-i Quds-i Razavī Library IV,1/876 (M)
75) Āstān-i Quds-i Razavī Library IV,1/877 (M)
76) Āstān-i Quds-i Razavī Library IV,1/878 (M)
77) Āstān-i Quds-i Razavī Library 5662 (1082/1671)
78) Āstān-i Quds-i Razavī Library XI, 7347 [incompl.: I–IX] (5th c./11th c.)

---

[55] A fifth ms. (India Office 114), reported by Brockelmann as containing the *Ilāhiyyāt*, is mentioned by Mahdavī, p. 171, as containing logic and mathematics.

[56] The *Ilāhiyyāt* is contained at fol. 179–264. Mahdavī, p. 170, mentions together with mss. 475–7 (containing, respectively, logic, natural philosophy, and mathematics plus *Ilāhiyyāt*), also ms. 474, which, however, does not contain Avicenna's *Shifā'*, but the *Rasā'il* of the Ikhwān al-ṣafā'. Brockelmann, p. 592, reports mss. 475–6 as containing the entire *Shifā'*. Anawati, p. 76, refers exclusively to ms. 475, which he reports as identical to British Museum Or. 7500 (this latter, however, contains not only logic, but also natural philosophy and *Ilāhiyyāt*; see below, n. 59). See *Fihris*, p. 226, n. 262.

[57] Anawati, p. 76, apparently reports this ms. as British Museum Suppl. 7.

[58] Brockelmann, p. 815, reports this ms. as containing only the *Ilāhiyyāt*.

[59] Brockelmann, p. 815, reports this manuscript as containing only the *Ilāhiyyāt*. Anawati, p. 76, reports it as containing only logic and as identical to ms. London, India Office 475 (see above, n. 56). Description in Anawati [1978], pp. 20–21; see *Fihris*, p. 226, n. 264.

[60] Brockelmann, p. 815, and Anawati, p. 76, report this manuscript as I, $52_{,172}$.

Mosul (Iraq) (2 mss.)
79) University Library 43.226 (M)
80) University Library 189.16 (+ m, ṭ, r) (B, A)[61]

Oxford (3 mss.)[62]
81) Bodleian, Pococke 125 (Uri's catalogue I, 435) (561/1166 or 571/1175) (+ ṭ) (B, A, M)[63]
82) Bodleian, Pococke 110 (Uri's catalogue I, 477) (601/1204 or 604/1207) [incompl.: V–VIII, 6] (B, A, M)[64]
83) Bodleian, Pococke 117 (Uri's catalogue I, 482) (601/1204 or 604/1207) [incompl.: I–IV] (B, A, M)[65]

Paris (1 ms.)
84) Bibliothèque Nationale 6829 (+ m) (A, M)[66]

Peshawar (1 ms.)
85) Dār al-ʿUlūm al-Islāmiyya 1672 (+ m, ṭ, r) (B, A)

Princeton (2 mss.)
86) Princeton University Library 2(769) (1082/1671)[67]
87) Princeton University Library 5(2092) (12th c./18th c.)[68]

---

[61] The precise content of this ms. is uncertain: at p. 592, Brockelmann reports it as containing all the *Shifāʾ*, whereas at p. 815 as containing only the *Ilāhiyyāt*.

[62] The other Bodleian mss. reported by Brockelmann, p. 592, and Anawati, p. 71 (Uri's catalogue i, 452; i, 467/8; i, 471/3; i, 485/7; i, 490; i, 493; i, 495; i, 893; ii, 581ff in Brockelmann, ii, 281ff in Anawati) do not contain the *Ilāhiyyāt*, but other parts of the *Shifāʾ*; ms. i, 483 in Uri's catalogue, reported by Brockelmann, ibid., does not contain any part of the *Shifāʾ*.

[63] The date of 571 A.H. is provided by Mahdavī, p. 171. The card catalogue of the Bodleian Library reports the dates of 1166 or 1175 C.E. (corrected from 1369).

[64] The date 601 A.H. of mss. i, 109–24 is reported by Mahdavī, p. 170. The card catalogue of the Bodleian Library reports the date of 1207 C.E. for mss. i, 477, 482.

[65] Mahdavī, p. 170, reports mss. i, 477 and i, 482 as containing, together, the entire *Ilāhiyyāt*.

[66] This is possibly the Paris ms. reported by Anawati, p. 70—with no indication of library and number, and the notation *nuskha fāʾidan* (possibly meaning "manuscript with marginal annotations", see A. Gacek, *The Arabic Manuscript Tradition. A Glossary of Technical Terms and Bibliography* [Leiden, 2001], p. 111)—as containing only the *Ilāhiyyāt*.

[67] See R. Mach and J. Ormsby, *Handlist of Arabic Manuscripts (New Series) in the Princeton University Library* (Princeton, 1987), p. 307, n. 1363. This ms. has 205 folios. No manuscript of the *Shifāʾ* is reported in R. Mach, *Catalogue of Arabic Manuscripts (Yahuda Section) in the Garrett Collection, Princeton University Library* (Princeton, 1977).

[68] See Mach and Ormsby, *Handlist*, p. 307, n. 1363. This ms. has 168 folios; the date of the copy (12th c. A.H.) is estimated.

Rampur (India) (1 ms.)
88) Raza Library 397/1 (+ m, ṭ, r) (A)

Tehran (20 mss.)
89) Kitābkhānah-i Majlis-i Shūrā-yi Millī (now: Majlis-i Shūrā-yi Islāmī Library) 135 (871/1466) (+ m, ṭ) (M)[69]
90) Kitābkhānah-i Majlis-i Shūrā-yi Millī 136 (M)
91) Kitābkhānah-i Majlis-i Shūrā-yi Millī 144/6 (+ m, ṭ, r) (B)
92) Kitābkhānah-i Majlis-i Shūrā-yi Millī, Ṭabāṭabā'ī 862 (M)
93) Kitābkhānah-i Majlis-i Shūrā-yi Millī, Ṭabāṭabā'ī 865 (+ ṭ) (M)
94) Kitābkhānah-i Majlis-i Shūrā-yi Millī, Ṭabāṭabā'ī 929 (M)
95) Kitābkhānah-i Majlis-i Shūrā-yi Millī, Ṭabāṭabā'ī 1300 (M)
96) Kitābkhānah-i Majlis-i Shūrā-yi Millī, Ṭabāṭabā'ī 1344 (M)
97) Millī Malik Library 714 (M)
98) Millī Malik Library 715 (M)
99) Millī Malik Library 745 (M)
100) Millī Malik Library 1085 (509/1115) (M)[70]
101) Kitābkhānah-i Markazī va-Markaz-i Asnād-i Dānishgāh-i Tihrān (Central Library of Tehran University), Mishkāt 242 (949/1542) (A, M)[71]
102) Kitābkhānah-i Markazī va-Markaz-i Asnād-i Dānishgāh-i Tihrān (Central Library of Tehran University), Mishkāt 243 (1075/1664) (+ m, ṭ r) (A, M)[72]
103) Millī Library 580 (+ m, ṭ, r) (M)

[69] Datation in Mahdavī, p. 171. Microfilms of several Majlis manuscripts (among others, Majlis 593 and Majlis 4547) are preserved in the Central Library of Tehran University (I take this information from Samaw'al al-Maghribī [d. 570/1175], *Ifḥām al-yahūd. The Early Recension*, ed. I. Marazka, R. Pourjavady and S. Schmidtke [Wiesbaden, 2006], pp. 9–10).

[70] Datation in Mahdavī, p. 172.

[71] Anawati, p. 432, Mahdavī, p. 172, n. ††, and Y. al-Khachab, 'Avicenne et les manuscripts d'Iran', *Revue du Caire* 27 (1951), pp. 172–83, p. 182, report that the "Third Master" Mīr Muḥammad Bāqir Dāmād Astarabādī (Mīr Dāmād, d. 1041/1631), teacher of Ṣadrā al-Dīn al-Shīrāzī (Mullā Ṣadrā), corrected and annotated this manuscript. Al-Khachab and Anawati provide the date and place (Iṣfahān) where the copy was accomplished.

[72] According to Anawati, p. 432, and Al-Khachab, 'Avicenne', p. 182, this ms. comprehends logic, natural philosophy and *Ilāhiyyāt*. Al-Khachab and Anawati report that this manuscript was written in 1075/1664 in Shīrāz by Muḥammad Ṣāliḥ al-Urdistānī (al-Ardistānī [?]), and corrected and annotated by Mīrzā Abū 'l-Ḥasan Jīlwa (1238/1823–1314/1897) and his disciple Mīrzā Ṭāhir Tankabni [?]. The fate of a further complete ms. of the *Shifā'* owned by Prof. Muḥammad Mishkāt of Tehran University (see Anawati, p. 432, and Al-Khachab, 'Avicenne', p. 182) is uncertain.

104) Kitābkhānah-i Madrasah-i ʿĀlī-i Sipahsālār (now: Madrasah-i ʿĀlī-i Shahīd Muṭahharī Library) 1435 (M)
105) Kitābkhānah-i Madrasah-i ʿĀlī-i Sipahsālār 1437 (M)
106) Kitābkhānah-i Madrasah-i ʿĀlī-i Sipahsālār 1438 (+ m, ṭ, r) (M)
107) Kitābkhānah-i Madrasah-i ʿĀlī-i Sipahsālār 1439 (+ m, ṭ, r) (M)
108) Kitābkhānah-i Madrasah-i ʿĀlī-i Sipahsālār 8331 (+ m, ṭ, r) (M)

Tunis (2 mss.)
109) Ahmadiyya 5217 (+ m, ṭ, r) (A)
110) Ahmadiyya 5218 (+ m, ṭ, r) (A)

Uppsala (1 ms.)
111) Universitetsbibliotek 344 (+ m, ṭ, r) (B, A)

Only a small part (15) of the entire set of manuscripts is preserved in Europe (Berlin, 1 ms.; Leiden, 2 mss.; London, 4 mss.; Manchester, 1 ms.; Oxford, 3 mss.; Paris, 1 ms.; Uppsala, 1 ms.) and U.S.A. (Princeton, 2 mss.). Surprising—and, if confirmed, instructive—is the absence of manuscripts from Spain, where the Latin medieval translation of the *Ilāhiyyāt* was accomplished, but where Avicenna's metaphysics was also fiercely opposed by Averroes. The majority of manuscripts is spread in libraries of the Near East and Central Asia. They come from all the main areas of the Islamic world: Tunisia (Tunis), Egypt (Cairo), Turkey (Istanbul; Ankara), Syria (Damascus), Iraq (Mosul), Iran (Tehran; Mashhad; Khoy), Pakistan (Peshawar) and India (Aligarh; Bihar, Patna; Rampur). The highest concentration of manuscripts is in Istanbul (37 mss.), followed by Tehran (20 mss.), Cairo and Mashhad (9 mss. each), Aligarh and Patnā (5 mss. each), Mosul and Tunis (2 mss. each), Ankara, Damascus, Khoy, Peshawar and Rampur (1 ms. each).

The high number of manuscripts, and their wide spread in Islamic countries (the place of origin of the manuscripts preserved nowadays in Europe and U.S.A.), attest the immense impact of the *Ilāhiyyāt* on subsequent Arabic philosophy.

II. Datation

Some of the above manuscripts are dated. In chronological order, they are:

Ms. Mashhad, Āstān-i Quds-i Razavī Library XI, 7347 (5th c./11th c.)
Ms. Tehran, Millī Malik Library 1085 (509/1115)
Ms. Cairo, Dār al-Kutub, Taymūr 140 (535/1140)
Ms. Oxford, Bodleian, Pococke 125 (561/1166 or 571/1175)
Ms. Oxford, Bodleian, Pococke 110 (601/1204 or 604/1207)
Ms. Oxford, Bodleian, Pococke 117 (601/1204 or 604/1207)
Ms. Istanbul, Nuruosmaniye Kütüphanesi 2710 (666/1267)
Ms. Istanbul, Süleymaniye Kütüphanesi, Ayasofya 2442 (671/1272)
Ms. Cairo, Azhar 331 *khuṣūṣiyya*, 2415 Bakhīt (684/1285)
Ms. Cairo, Dār al-Kutub 144 *ḥikma* (684/1285)
Ms. Istanbul, Süleymaniye Kütüphanesi, Carullah 1424 (693/1293)
Ms. Ankara, Millī Kütüphane B 153 (694/1294)
Ms. Istanbul, Süleymaniye Kütüphanesi, Damat Ibrahim Paşa 823 (697/1297)
Ms. Khvoy Madrasa Nimāzī 248 (7th c./13th c.)
Ms. Istanbul, Süleymaniye Kütüphanesi, Damat Ibrahim Paşa 822 (8th c./14th c.)
Ms. Tehran, Kitābkhānah-i Majlis-i Shūrā-yi Millī 135 (871/1466)
Ms. Istanbul, Süleymaniye Kütüphanesi, Şehid Ali Paşa 1748 (879/1474)
Ms. Leiden, Golius 84 (881–882/1476–1477)
Ms. Istanbul, Süleymaniye Kütüphanesi, Carullah 1332 (882/1477)
Ms. Istanbul, Nuruosmaniye Kütüphanesi 2709 (886/1481 or 894/1488)
Ms. Istanbul, Süleymaniye Kütüphanesi, Yeni Cami 770 (888/1483)
Ms. Istanbul, Süleymaniye Kütüphanesi, Ayasofya 2389 (9th c./15th c.)
Ms. Tehran, Kitābkhānah-i Markazī va-Markaz-i Asnād-i Dānishgāh-i Tihrān, Mishkāt 242 (949/1542)
Ms. Istanbul, Süleymaniye Kütüphanesi, Lâleli 2550 (1023/1614)
Ms. Istanbul, Istanbul Üniversitesi Kütüphanesi, Üniversite 760 (1060/1650)
Ms. Istanbul, Istanbul Üniversitesi Kütüphanesi, Üniversite 766 (1060/1650)
Ms. Tehran, Kitābkhānah-i Markazī va-Markaz-i Asnād-i Dānishgāh-i Tihrān, Mishkāt 243 (1075/1664)
Ms. Princeton University Library 2(769) (1082/1671)

Ms. Mashhad, Āstān-i Quds-i Razavī Library 5662 (1082/1671)
Ms. Berlin 5045 (1083/1672)
Ms. Istanbul, Atif Efendi Kütüphanesi 1596 (1098/1686)
Ms. Istanbul, Süleymaniye Kütüphanesi, Hekimoğlu Ali Paşa 857 (1102/1690)
Ms. Cairo, Dār al-Kutub, Ṭalʿat 363 (1105/1693)
Ms. Istanbul, Süleymaniye Kütüphanesi, Carullah 1425 (1115/1703)
Ms. Istanbul, Millet Kütüphanesi, Feyzullah Efendi 1445 (1132/1719)
Ms. Istanbul, Süleymaniye Kütüphanesi, Kılıç Ali Paşa 689 (1185/1771)
Ms. Princeton University Library 5(2092) (12th c./18th c.)
Ms. Cairo, Dār al-Kutub, 349 *falsafa* (13th c./19th c.)
Ms. Cairo, Dār al-Kutub 262 *ḥikma* (1337/1914, copy of a ms. of 992/1584)

The temporal distance separating the first from the last of the manuscripts in the above list is impressive: the *Ilāhiyyāt* has kept being copied for nine centuries, from the fifth until the fourteenth century of the Islamic era, that is to say: from a few decades after Avicenna's death until a few decades ago. This aspect, which mirrors an equally prolonged exegetical activity regarding this work,[73] shows clearly that the subsequent impact of the *Ilāhiyyāt* was not only wide (as the previous section has documented), but also lasting.

III. Towards a Critical Edition

When compared with the manuscript tradition described in the previous two sections, the available printed editions of the *Ilāhiyyāt* appear

[73] See G. Endress, 'Philosophische Ein-Band-Bibliotheken aus Isfahan,' *Oriens* 36 (2001), pp. 10–58; D. Gutas, 'The Heritage of Avicenna: the Golden Age of Arabic Philosophy, 1000–ca. 1350,' in J. Janssens and D. De Smet (eds.), *Avicenna and his Heritage* (Leuven, 2002), pp. 81–97; R. Arnzen, 'Mapping Philosophy and Science in Ṣafawid Iran and Mughal India: the Case of Niẓāmaddīn Aḥmad Gīlānī and ms. Khudā Bakhsh 2641,' *MUSJ* 56 (1999–2003), pp. 107–60; R. Wisnovsky, 'The Nature and Scope of Arabic Philosophical Commentary in Post Classical (ca. 1100–1900 A.D.) Islamic Intellectual History: Some Preliminary Observations,' in P. Adamson, H. Baltussen and M.W.F. Stone (eds.), *Philosophy, Science and Exegesis in Greek, Arabic and Latin Commentaries* (London, 2004), ii, pp. 149–91; H. Ziai, 'Recent Trends in Arabic and Persian Philosophy,' in P. Adamson and R. Taylor (eds.), *The Cambridge Companion to Arabic Philosophy* (Cambridge, 2005), pp. 405–25.

unreliable.[74] The first of them (the lithograph published in Tehran in 1885, encompassing, besides the *Ilāhiyyāt*, also the second part of the *Shifā'* on natural philosophy) is very likely the reproduction of one manuscript, whose identity is still to be ascertained.[75] The still standard "edition" of the *Ilāhiyyāt*, published in Cairo in 1960, is based on five manuscripts (13, 15, 18–19, 68 in the above list), whose selection is limited not only numerically, but also geographically, since four of them are preserved in Cairo libraries.[76] The most ancient dated manuscripts of this selection go back to the seventh/thirteenth century. Finally, the printed version of the *Ilāhiyyāt* published in Qum in 1997/8 by Ḥasanzādah al-Āmulī is based on an unspecified number of manuscripts belonging to the private collection of the curator (in all likelihood, five codices, as it results from some annotations and the photographic reproductions at the end of the book), one of which appears to have been chosen as the basis of the edition.[77]

The collation of the Cairo edition of the *Ilāhiyyāt* with manuscripts 62, 81 (571/1175), and 82–3 (601/1204) of the above list, the Tehran lithograph (carefully inspected),[78] the printed version by al-Āmulī, the Latin Medieval translation (accomplished in the third quarter of the twelfth century, thus one of most ancient witnesses of the text), and the parallel passages in the metaphysics of Avicenna's *Kitāb al-najāt* (*Book of the Salvation*), has allowed more than 750 (!) emendations of the edited text, as I have documented elsewhere.[79]

---

[74] See the detailed account in A. Bertolacci, *The Reception of Aristotle's* Metaphysics *in Avicenna's* Kitāb al-Shifā': *A Milestone of Western Metaphysical Thought* (Leiden, 2006), pp. 483–5 ("Appendix A: Towards a Critical Edition of the *Ilāhiyyāt*: List of Corrections of the Cairo Printed Text").

[75] Ibn Sīnā, *Al-Ilāhiyyāt min al-Shifā' li-Shaykh al-Ra'īs Abū 'Alī Ḥusayn Ibn 'Abd Allāh Ibn Sīnā ma'a ta'līqāt* (Tehran, 1885), i, pp. 266–567. Like a manuscript, the Tehran lithograph contains interlinear and marginal corrections and explanatory glosses.

[76] A sixth manuscript (16 of the list) is mentioned in the introduction of Anawati's French translation of the *Ilāhiyyāt* (Anawati [1978], pp. 17, 19–21); it might correspond to the siglum H that appears occasionally in the critical apparatus, but is not included in the list of manuscripts of the edition.

[77] Ibn Sīnā, *Al-Ilāhiyyāt min Kitāb al-Shifā'*, ed. Ḥasanzādah al-Āmulī (Qum, 1418 A.H. q., 1376 A.H.sh. [= 1997–98]). In lack of a critical apparatus, variants are recorded by al-Āmulī occasionally in the footnotes.

[78] Some readings of the Tehran lithograph are omitted, others are wrongly recorded in the critical apparatus of the Cairo edition.

[79] See Bertolacci, *The Reception*, pp. 483–558.

Besides showing the need of a new critical edition,[80] the scrutiny of the manuscript tradition of the *Ilāhiyyāt* is useful in three further respects. First, it allows some emendations of the critical text of the *Najāt* established by M.T. Dānishpazūh in 1985.[81] In *Ilāhiyyāt* IX, 4, p. 407, 1, for instance, the reading *al-kathratu* ("the multiplicity")—reported in the Cairo edition with no variants, and transmitted by an interlinear correction of the Tehran lithograph and by the printed version of al-Āmulī—has to be corrected in *al-kuratu* ("the sphere"), on account of mss. 81 and 62, the original reading of the Tehran lithograph,[82] and the Latin translation (*sphaera*). The same correction has to be made also in the critical edition of the *Najāt*, which reports the reading *al-kathratu* (p. 606, 14), with no variants, in the parallel passage.[83]

Second, on account of the manuscript tradition of the *Ilāhiyyāt* some further emendations of the Latin Medieval translation can be added to the ones listed by the editor herself (S. van Riet) in the *corrigenda et addenda*.[84] In VII, 3, p. 317, 7, for example, the second occurrence of *takhayyulihā* ("their imagination") in the edited text—a repetition of the term occurring shortly before in the same line, transmitted by mss. 13 (B in the edition), 15 (J), 18 (Ṣ) and 68 (M)—has to be emended in *takhayyulinā* ("our imagination"), on the basis of mss. 81, 82, 62, 19 (D in the edition) and the Tehran lithograph[85] (see also the printed version of al-Āmulī). Thus, the reading of the Latin translation (*naturae*

[80] Hamed Naji Isfahani's in progress edition of some of the most significant commentaries on the *Ilāhiyyāt* contains a new version of Avicenna's text, established on the basis of mss. 61, 77–8, 82–3, 100–101, of the above list. See Ibn Sīnā, *al-Shifā' (al-Ilāhiyyāt), with Marginal Glosses by Mullā Ṣadrā and Others*, ed. H. Naji Isfahani (Tehran, 1383 A.H.sh/2004). Dr. Naji Isfahani is planning to publish this version of the *Ilāhiyyāt* also as an independent book.

[81] Ibn Sīnā, *Al-Najāt min al-gharq fī baḥr al-ḍalālāt*, ed. M.T. Dānishpazūh (Tehran, 1985).

[82] This reading is not reported in the Cairo edition.

[83] Other emendations of the *Najāt* (i.e. other cases in which the edited text of this work agrees with a reading of the Cairo edition of the *Ilāhiyyāt* that the collation shows to be wrong) in Bertolacci, *The Reception*, pp. 537–56 *passim*.

[84] Avicenna, *Liber de Philosophia prima sive Scientia divina, i–iv*, ed. S. Van Riet (Louvain, 1977); Avicenna, *Liber de Philosophia prima sive Scientia divina, v–x*, ed. S. Van Riet (Louvain, 1980); Avicenna, *Liber de Philosophia prima sive Scientia divina, i–x*, Lexiques par S. Van Riet (Louvain, 1983) (the *corrigenda and addenda* are at pp. 6*–13* of the third volume).

[85] In the apparatus of the Cairo edition, the Tehran lithograph is mistakenly reported as having the edited reading.

imaginationis, p. 367, 74–75), which presents no variants in the Latin apparatus, has probably to be emended in *nostrae imaginationis*.[86]

Third, some readings of the Cairo edition that the collation has shown to be questionable present signs of "Ghazalian" influence. In *Ilāhiyyāt* I, 1, pp. 4, 16, for example, the edited expression *musabbib al-asbāb* ("Causer of causes", i.e. God)—transmitted by mss. 13, 18, 19, 68, and the Tehran lithograph (see also the printed version of al-Āmulī)[87]—is replaced by *sabab al-asbāb* ("Cause of causes") in mss. 15, 16, 62, 81, 82,[88] and is possibly witnessed also by the Latin translation (*causa causarum*).[89] The expression *musabbib al-asbāb* designates frequently God in al-Ghazālī's works.[90] Thus, if the reading *musabbib* has to be retained, this passage of the *Ilāhiyyāt* deserves to be included among the Avicennian sources of al-Ghazālī's thought; if, on the other hand, it has to be rejected and replaced by *sabab* (as I am inclined to think, on the basis of the principle of the *lectio brevior*), it can be regarded as a reading inserted in Avicenna's original text by a scribe familiar with al-Ghazālī's writings.

Thus, a new critical edition of the *Ilāhiyyāt*, besides finally providing a reliable Arabic text of this fundamental work of Avicenna, will also shed new light on Avicenna's philosophical works related to the *Ilāhiyyāt* (like the *Najāt*), the Latin transmission of the *Ilāhiyyāt* itself, and the doctrinal tendencies of the reception of Avicenna's thought in subsequent Arabic philosophy.

---

[86] See also the proposed emendation of the punctuation of iii, 10, p. 152, 7–8, in Bertolacci, *The Reception*, p. 506.

[87] This reading is present in ms. Daiber Collection iii, 131, fol. 1v15 (see above, n. 4). It is retained by Naji Isfahani in the work mentioned above, n. 80. The presence of the expression *musabbib al-asbāb* in Avicenna's *Risāla ʿarshiyya* (ed. Hilāl, p. 26, 8–9) and *Taʿlīqāt* (ed. Badawī, p. 152, 9) deserves a more thorough codicological investigation (see J. Janssens, 'Ibn Sīnā's Ideas of Ultimate Realities. Neoplatonism and the Qurʾān as Problem-Solving Paradigms in the Avicennian System,' *Ultimate Reality and Meaning* 9 [1986], pp. 252–71, especially pp. 265–6).

[88] Ms. 17 has *sabab musabbib al-asbāb*.

[89] The expression *causa causarum* renders *musabbib al-asbāb*, rather than *sabab al-asbāb*, in the Latin translation of al-Ghazālī's *Maqāṣid al-falāsifa* (*Algazel's Metaphysics. A Mediaeval Translation*, ed. J.T. Muckle [Toronto, 1933], p. 119, 5, corresponding to *Maqāṣid al-falāsifa*, ed. S. Dunyā [Cairo, 1961], p. 288, 5).

[90] To the list of occurrences in R.M. Frank, *Creation and the Cosmic System: Al-Ghazālī and Avicenna* (Heidelberg, 1992), p. 18 and n. 18, also the place of the *Maqāṣid al-falāsifa* mentioned in the previous footnote can be added.

UNE CLASSIFICATION ISMAÉLIENNE DES SCIENCES. L'APPORT D'ABŪ YAʿQŪB AL-SIJISTĀNĪ À LA 'TRADITION D'AL-KINDĪ' ET SES LIENS AVEC ABŪ 'L-ḤASAN AL-ʿĀMIRĪ

Daniel De Smet

Al-Kindī, 'le premier philosophe arabe', a introduit dans la pensée arabo-musulmane une réflexion sur la philosophie, sa définition, ses objectifs et ses différentes disciplines, qui se situe dans le prolongement direct de la tradition alexandrine tardo-antique et de son héritier, l'aristotélisme syriaque.[1] La philosophie y est divisée en deux branches, l'une théorique et l'autre pratique: la philosophie théorique comprend la métaphysique, les mathématiques et la physique, alors que la philosophie pratique englobe l'éthique, l'économie et la politique. Cette division, admise par la plupart des commentateurs alexandrins d'Aristote (dont Ammonius, David et Élias), ainsi que par leurs successeurs syriaques, apparaît pour la première fois en arabe dans le *Compendium de logique* d'Ibn al-Muqaffaʿ (milieu du 8e siècle),[2] quelques décennies avant son adoption par al-Kindī.[3]

Si les philosophes arabo-musulmans, fidèles à leurs modèles grecs et syriaques, présentent 'la division des sciences' (*taqsīm al-ʿulūm*) dans le cadre d'une introduction générale à la philosophie qui, avant d'introduire les *Catégories* d'Aristote par le biais de l'*Isagoge* de Porphyre, se propose de classer les écrits du Stagirite et d'en déterminer l'ordre d'étude (allant des *Catégories* à la *Métaphysique*),[4] elle leur ouvre à la fois

---

[1] Bien que son traité sur la division des sciences profanes, le *Kitāb aqsām al-ʿilm al-unsī*, soit perdu, al-Kindī a abordé ce thème à plusieurs reprises dans son œuvre conservée; on en trouvera une étude d'ensemble dans A. Cortabarria Beita, 'La classification des sciences chez al-Kindī,' *MIDEO* 11 (1972), pp. 49–76.

[2] Pour un aperçu des antécédents grecs et syriaques d'al-Kindī, voir C. Hein, *Definition und Einteilung der Philosophie. Von der spätantiken Einleitungsliteratur zur arabischen Enzyklopädie* (Frankfurt, 1985), pp. 146–51, 163–70, 226–9.

[3] Voir en particulier al-Kindī, *Rasāʾil al-Kindī al-falsafiyya*, éd. M. Abū Rīda (Le Caire, 1950), i, p. 97; ii, pp. 8–10.

[4] En témoigne l'épître d'al-Kindī intitulée *Risāla fī kammiyyat kutub Arisṭūṭālīs wa-mā yuḥtāj ilayhi fī taḥṣīl al-falsafa* ('Sur le nombre des livres d'Aristote et ce dont on a besoin pour acquérir la philosophie'), éd. Abū Rīda in *Rasāʾil*, i, pp. 363–84; M. Guidi et R. Walzer, 'Studi su al-Kindī. I. Uno scritto introduttivo allo studio di Aristotele,'

une perspective nouvelle: harmoniser l'enseignement coranique avec celui des philosophes 'païens' ou, ce qui revient au même, les sciences religieuses avec les sciences profanes héritées des Grecs. Ce souci d'harmonisation est déjà bien présent dans la pensée d'al-Kindī: il constitue même la démarche principale de sa philosophie.[5]

Le thème de la 'division des sciences' sera repris et développé dans cette même optique par une série d'auteurs qui présentent tous des liens plus ou moins étroits avec al-Kindī: son contemporain, le chrétien Qusṭā ibn Lūqā (m. vers 912);[6] Abū Zayd al-Balkhī (m. 934), disciple d'al-Kindī qui a introduit sa philosophie en Iran;[7] Ibn Farīghūn[8] et Abū 'l-Ḥasan al-ʿĀmirī (m. 992),[9] tous deux disciples d'al-Balkhī; al-Khwārizmī, un contemporain d'Ibn Farīghūn et d'al-ʿĀmirī;[10] Abū 'l-Faraj ibn Hindū (m. 1029), qui dépendrait d'al-ʿĀmirī;[11] et enfin le 'philosophe littérateur' Abū Ḥayyān al-Tawḥīdī (m. 1023), qui avait fréquenté avec al-ʿĀmirī le 'cercle' d'Abū Sulaymān al-Sijistānī.[12]

Tous ces auteurs véhiculent une conception de la philosophie, de ses objectifs, divisions et rapports avec les sciences religieuses, qui présente de nombreux traits communs, de sorte qu'on a pu parler d'une 'école'

Memorie della Accademia Nazionale dei Lincei. Classe di scienze morali, storiche e filologiche, série 6, vol. 6 (Rome, 1940), pp. 375–419.

[5] Comme l'a bien démontré G. Endress, 'The Defense of Reason: the Plea for Philosophy in the Religious Community,' *ZGAIW* 6 (1990), pp. 3–13.

[6] H. Daiber, 'Qosṭā Ibn Lūqā (9. Jh.). Über die Einteilung der Wissenschaften,' *ZGAIW* 6 (1990), pp. 93–129.

[7] Son *Kitāb aqsām al-ʿulūm* est malheureusement perdu; cf. H.H. Biesterfeldt, 'Medieval Arabic Encyclopedias of Science and Philosophy,' dans S. Harvey (éd.), *The Medieval Hebrew Encyclopedia of Science and Philosophy* (Dordrecht, 2000), p. 83.

[8] Biesterfeldt, 'Medieval Arabic Encyclopedias,' p. 84; W. Heinrichs, 'The Classification of the Sciences and the Consolidation of Philology in Classical Islam,' dans J.W. Drijvers et A.A. Mc Donald (éds.), *Centres of Learning. Learning and Location in Pre-Modern Europe and the Near East* (Leiden, 1995), pp. 129–30. Les passages du *Jawāmiʿ al-ʿulūm* d'Ibn Farīghūn concernant la division des sciences ont été traduits par D. Gutas, *Avicenna and the Aristotelian Tradition* (Leiden, 1988), pp. 245–9.

[9] H.H. Biesterfeldt, 'Abū l-Ḥasan al-ʿĀmirī und die Wissenschaften,' dans W. Voigt (éd.), *XIX. Deutscher Orientalistentag* (Wiesbaden, 1977), pp. 335–41; Heinrichs, 'Classification,' pp. 130–36.

[10] Biesterfeldt, 'Medieval Arabic Encyclopedias,' pp. 86–7; Heinrichs, 'Classification,' pp. 128–9.

[11] M. Türker-Küyel, 'La classification des sciences selon *Cumal al-falsafa* d'Ibn Hindī,' *Araştırma* 5 (1967), pp. 47–53.

[12] M. Bergé, 'Épître sur les sciences (*Risāla fī l-ʿulūm*) d'Abū Ḥayyān at-Tawhīdī. Introduction, traduction, glossaire technique, manuscrit et édition critique,' *BEO* 18 (1963–64), pp. 241–98; cf. Heinrichs, 'Classification,' pp. 136–7.

ou d'une 'tradition kindienne'.[13] Celle-ci se démarque nettement de la division des sciences élaborée par al-Fārābī dans son *Iḥṣā' al-ʿulūm*[14] qui remonterait en partie à d'autres sources (notamment le philosophe syriaque Paul le Perse).[15]

Par la publication d'un écrit sans titre de Qusṭā ibn Lūqā sur la division des sciences, contenu dans le ms. Aya Sofya 4855 (où il est introduit seulement par la mention *min kalām Qusṭā ibn Lūqā*), sa traduction allemande et sa savante annotation, Hans Daiber a livré une contribution majeure à la connaissance de cette 'tradition d'al-Kindī'.[16] En ce volume d'hommages qui lui est dédié, nous nous proposons d'y contribuer à notre tour, d'une façon bien plus modeste certes, en présentant un texte d'un auteur ismaélien du 10e siècle, Abū Yaʿqūb al-Sijistānī, qui jusqu'ici a échappé à l'attention des chercheurs. Pourtant, il s'agit d'une pièce importante à verser au dossier kindien, d'autant plus qu'il s'avère très proche d'Abū 'l-Ḥasan al-ʿĀmirī, sans toutefois dépendre directement de lui. Il soulève en même temps la question du rapport entre les philosophes ismaéliens de la dite 'École persane' et les disciples iraniens d'al-Kindī.

Dans son *Kitāb ithbāt al-nubuwwāt*, Abū Yaʿqūb al-Sijistānī—auteur de tendance carmathe, mort vers 971, qui joua un rôle majeur dans l'élaboration du néoplatonisme ismaélien au sein de l''École persane'[17]—se propose de démontrer par des arguments rationnels la nécessité de la

---

[13] Endress, 'The Defense of Reason,' p. 25; Gutas, *Avicenna*, pp. 243–8; P. Adamson, 'The Kindian Tradition. The Structure of Philosophy in Arabic Neoplatonism,' dans C. d'Ancona (éd.), *The Libraries of the Neoplatonists. Proceedings of the Meeting of the European Science Foundation Network 'Late Antiquity and Arabic Thought. Patterns in the Constitution of European Culture" held in Strasbourg, March 12–14, 2004* (Leiden, 2007), pp. 351–70.

[14] Biesterfeldt, 'Abū l-Ḥasan al-ʿĀmirī,' p. 337; Hein, *Definition*, pp. 230–31; Gutas, *Avicenna*, pp. 248–9; Endress, 'The Defense of Reason,' pp. 28–9; Heinrichs, 'Classification,' pp. 122–6.

[15] D. Gutas, 'Paul the Persian on the Classification of the Parts of Aristotle's Philosophy: a Milestone between Alexandria and Baġdād,' *Der Islam* 60 (1983), pp. 231–67; cf. P. Vallat, *Farabi et l'École d'Alexandrie. Des prémisses de la connaissance à la philosophie politique* (Paris, 2004), pp. 39–42.

[16] Voir l'article cité *supra*, n. 6.

[17] On trouvera un état de la question sur al-Sijistānī, l''École persane' et le néoplatonisme ismaélien en général dans notre article 'Die ismaʿilitischen Denker des 10. und frühen 11. Jhs. (von Nasafī bis Kirmānī),' à paraître dans U. Rudolph (éd.), *Philosophie in der islamischen Welt*, i (Überweg, Grundriss der Geschichte der Philosophie; Bâle, 2008).

prophétie et de l'imamat.[18] Il consacre un chapitre de son ouvrage à l'harmonie entre *al-ḥikma* et *al-risāla*, la 'sagesse' et la 'mission', termes par lesquels il entend respectivement les sciences profanes héritées des Grecs, en particulier la philosophie, et les sciences religieuses de l'islam. C'est à cette occasion qu'il aborde le thème de la 'division des sciences' (*taqsīm al-ʿulūm*).[19]

Par le choix du terme *ḥikma* pour désigner la philosophie (au lieu de *falsafa*, transcrit du grec), al-Sijistānī se rattache d'emblée à la tradition d'al-Kindī. Ce terme coranique traduit en effet l'essence même du projet philosophique kindien: légitimer la philosophie dans un contexte musulman en montrant que l'enseignement des Anciens relève d'une 'sagesse' universelle qui s'harmonise parfaitement avec les données du Coran.[20] Dès lors, fidèle à l'optique kindienne, al-Sijistānī soutient qu'il n'existe aucune contradiction entre *ḥikma* et *risāla*. La similitude entre ces deux domaines du savoir s'articule autour de trois points de convergence: (1) ils poursuivent le même but; (2) ils présentent la même structure; (3) ils ont la même source et atteignent leur perfection dans la même personne. Examinons ces trois points de plus près.

1. *Ḥikma* et *Risāla* poursuivent un but identique

Selon al-Sijistānī, la *ḥikma* permet à ceux qui la détiennent de choisir ce qui est le meilleur, le plus approprié, le plus utile (*ikhtiyār al-aṣlaḥ li'l-kull wa'l-ajdar li'l-nafʿ ʿalayhim*) et, en même temps, de connaître tous les degrés de la réalité, du plus humble au plus élevé. Elle présente ainsi un double aspect, théorique et pratique: la *ḥikma* est à la fois une science

[18] Pour un aperçu du contenu de l'ouvrage, voir P. Walker, *Abu Yaʿqub al-Sijistani. Intellectual Missionary* (Londres, 1996), pp. 110–12; voir en outre P. Walker, *Early Philosophical Shiism. The Ismaili Neoplatonism of Abū Yaʿqūb al-Sijistānī* (Cambridge, 1993), p. 22; Y. Marquet, 'La pensée d'Abū Yaʿqūb as-Sijistānī à travers l''Ithbāt an-Nubuwwāt' et la 'Tuḥfat al-Mustajībīn',' *SI* 54 (1981), pp. 95–128.

[19] Abū Yaʿqūb al-Sijistānī, *Kitāb ithbāt al-nubuwwāt*, éd. ʿĀ. Tāmir (Beyrouth, 1966), pp. 119–23. Comme le remarque avec raison Walker, *Abu Yaʿqub al-Sijistani*, p. 107, cette édition est médiocre. Le passage qui nous intéresse ici offre par endroits un texte corrompu, qu'il faudrait revoir sur les manuscrits.

[20] Endress, 'The Defense of Reason,' p. 4; J. Jolivet, 'L'idée de la Sagesse et sa fonction dans la philosophie des 4e et 5e siècles,' *ArScPhil* 1 (1991), pp. 38–9, 42–4. Le prototype du *ḥakīm* dans le Coran est Luqmān, que les disciples d'al-Kindī, en particulier Abū 'l-Ḥasan al-ʿĀmirī, suivis par les Ismaéliens, présentent comme le maître des premiers philosophes grecs: Thalès, Pythagore et surtout Empédocle; voir notre livre *Empedocles Arabus. Une lecture néoplatonicienne tardive* (Bruxelles, 1998), pp. 38–45.

(*ʿilm*) et une pratique (*ʿamal*).[21] La philosophie pratique interdit à l'homme de nombreuses actions qu'il serait par nature tenté d'accomplir et lui commande de faire des choses auxquelles il répugnerait par nature: 'il appartient à l'essence de la *ḥikma* d'inciter l'homme à l'ascétisme (*zuhd*) en ce monde, de lui montrer la nature éphémère et relative des jouissances et des biens terrrestres, et de susciter en lui le désir de l'Au-delà, de lui révéler sa noblesse et sa permanence, ainsi que la permanence des formes (*ṣuwar*, à savoir les formes intelligibles) qui s'y trouvent'.[22]

Cette conception de la philosophie comme *ḥikma*, qui allie la connaissance (*ʿilm*) de la vérité à la pratique (*ʿamal*) de la vertu, est un élément majeur de la tradition kindienne. Héritier d'une longue tradition antique qui remonte en fin de compte à Aristote, al-Kindī écrit dans son 'Traité sur la philosophie première' (*Risāla fī 'l-falsafa al-ūlā*): 'le but poursuivi par le philosophe (*gharaḍ al-faylasūf*) consiste à connaître la vérité (*al-ḥaqq*) et à agir en conformité avec elle'.[23] À plusieurs reprises, il souligne que la *ḥikma* montre la voie menant à la science universelle, à une vie vertueuse et à la félicité (*saʿāda*),[24] thèse qui sera adoptée par la plupart de ses disciples.[25]

Or, selon al-Sijistānī, la *risāla* poursuit ce même double objectif: 'Les philosophes divisent la *ḥikma* en deux parties: la connaissance et la pratique (*al-ʿilm wa'l-ʿamal*); de même, la *risāla* est divisée en deux parties: la connaissance et la pratique'.[26] Si cette division de la *ḥikma* repose sur l'autorité des philosophes, celle de la *risāla* prend racine dans le Coran. En effet, l'auteur ismaélien retrouve ces deux domaines du savoir clairement distingués dans le texte coranique: 'Il m'est révélé que votre Dieu est un Dieu unique et que celui qui espère la rencontre de son Seigneur doit accomplir de bonnes actions (*fal-yaʿmal ʿamalan*

[21] Sijistānī, *Ithbāt*, p. 119.

[22] Ibid., p. 121.

[23] Kindī, *Rasāʾil*, i, p. 97. C. d'Ancona, 'Al-Kindī on the Subject-Matter of the First Philosophy. Direct and Indirect Sources of *Falsafa al-ūlā*, Chapter one,' dans J.A. Aertsen et A. Speer (éds.), *Was ist Philosophie im Mittelalter?* (Berlin, 1998), pp. 844–5, a relevé un passage parallèle dans la préface de la dite 'Théologie d'Aristote', éd. ʿA. Badawī, *Iflūṭīn ʿinda l-ʿArab. Plotinus apud Arabes. Theologia Aristotelis et fragmenta quae supersunt* (Kuwayt, ^{3}1977), p. 4.

[24] Voir, p. ex., Kindī, *Rasāʾil*, i, p. 177, ii, p. 8; cf. Endress, 'The Defense of Reason,' p. 5; Cortabarria, 'La classification des sciences,' p. 70. Les Arabes ont en partie repris les riches développements que connut ce thème chez les commentateurs alexandrins d'Aristote; voir Hein, *Definition*, pp. 149–51.

[25] Dont Qusṭā b. Lūqā (Daiber, 'Qosṭā Ibn Lūqā,' pp. 106–7, 117–19) et al-Khwārizmī, *Kitāb mafātīḥ al-ʿulūm*, éd. G. van Vloten (Leiden, 1895), pp. 131–2.

[26] Sijistānī, *Ithbāt*, p. 119.

ṣāliḥan) (18:110); 'Ceux qui savent (*alladhīna yaʿlamūna*) et les ignorants sont-ils égaux? Les hommes doués d'intelligence sont les seuls qui réfléchissent' (39:9).[27] Le Coran exhorte ainsi les croyants à acquérir la science théorique et pratique, afin qu'ils entrevoient par la raison la perfection de la création et qu'ils mènent une vie vertueuse conforme aux injonctions divines. Tout comme la *ḥikma* des philosophes *interdit* les mauvaises actions et *commande* le bien, la Loi divine (la *sharīʿa*) a pour objectif de 'commander le bien et d'interdire ce qui est répréhensible': *al-amr bi'l-maʿrūf wa'l-nahy ʿan al-munkar.*

Pour al-Sijistānī, la conclusion s'impose: loin d'être hostile à la philosophie, le Coran en fait un devoir imposé par Dieu aux hommes. Cette obligation de pratiquer la philosophie est énoncée de façon explicite: 'Appelle les hommes dans le chemin de ton Seigneur, par la Sagesse (*bi'l-ḥikma*) et une belle exhortation' (16:125).[28]

En présentant la philosophie comme une obligation religieuse, notre auteur se rallie à une des thèses majeures de la tradition kindienne. Al-Kindī soutient en effet dans sa 'Philosophie première' que quiconque refuse d'acquérir 'la science de la nature réelle des choses' (*ʿilm al-ashyāʾ bi-ḥaqāʾiqihā*),[29] s'oppose à la religion, car 'la science de la nature réelle des choses englobe la science de Dieu et de son unicité, la science de la vertu, la connaissance de tout ce qui est utile et de la voie qui permet de l'acquérir, et de s'abstenir de tout ce qui est nuisible et répréhensible. C'est précisément cette science que les messagers véridiques ont apportée de la part de Dieu. En effet, les messagers véridiques ont exhorté (les hommes) à attester l'existence de Dieu et à professer son unicité; ils leur ont montré la nécessité de poursuivre les vertus qui lui sont agréables et à se détourner des vices qui sont contraires à ces vertus'.[30] Par conséquent, il n'existe aucune contradiction entre l'enseignement des Anciens et celui du Coran ou, dans les termes employés par al-Sijistānī, entre *ḥikma* et *risāla*.

---

[27] Ibid. Nous citons les versets coraniques dans la traduction de D. Masson, *Le Coran* (Paris, 2001).
[28] Sijistānī, *Ithbāt*, p. 122.
[29] Il s'agit là d'une définition traditionnelle de la philosophie.
[30] Kindī, *Rasāʾil*, i, p. 104.

2. *Ḥikma* et *Risāla* présentent une structure identique

Cette convergence se reflète également dans la division des différentes branches du savoir.

Après avoir remarqué que les philosophes divisent la *ḥikma* en une philosophie théorique et une philosophie pratique, al-Sijistānī poursuit: 'Les philosophes disent que la philosophie théorique se divise en trois parties: (1) la théologie (*ʿilm al-lāhūt*), ce qu'ils appellent la science supérieure (*al-ʿilm al-aʿlā*), à savoir la connaissance de Dieu et de ses anges; (2) la science médiane (*al-ʿilm al-awsaṭ*), à savoir l'astronomie (*ʿilm al-tanjīm*), la connaissance des mouvements des corps célestes; (3) la science inférieure (*al-ʿilm al-adnā*), à savoir la médecine et les arts. Ils divisent également la philosophie pratique en trois parties: le gouvernement de la communauté (*siyāsat al-ʿāmma*), le gouvernement de l'individu (*siyāsat al-khāṣṣa*) et le gouvernement de l'équité (*siyāsat al-ḥāqqa*)'.[31]

La classification générale des sciences ou disciplines philosophiques attribuée par al-Sijistānī aux 'philosophes' (*ḥukamāʾ*), relève de la tradition aristotélicienne. Elle est devenue un lieu commun dans les introductions à la philosophie produites par l'École d'Alexandrie (Ammonius, David, Élias). La philosophie théorique se ramifie en trois branches: la métaphysique, les mathématiques et la physique; il en va de même de la philosophie pratique qui comprend la politique, l'économie et l'éthique.[32] Cette division fut reprise par les philosophes syriaques et leurs continuateurs arabes, tout comme la hiérarchisation des disciplines théoriques en une science supérieure (la métaphysique), médiane (les mathématiques) et inférieure (la physique).[33] L'identification de la métaphysique avec la théologie, introduite dans la pensée arabe par al-Kindī,[34] est caractéristique de toute la tradition kindienne ultérieure, dont al-Fārābī se démarquera clairement.[35] Ce qui frappe toutefois dans la présentation d'al-Sijistānī est l'absence des mathématiques: la science médiane s'y réduit à l'astronomie, qui selon la conception tardo-antique adoptée par la plupart des Arabes n'est qu'une des quatre disciplines mathématiques

---

[31] Sijistānī, *Ithbāt*, p. 120; cf. ibid., p. 122.

[32] Hein, *Definition*, pp. 28, 146–7, 163, 226; Gutas, 'Paul the Persian,' p. 261.

[33] Voir les références dans Hein, *Definition*, pp. 166–9; pour al-Kindī, voir Cortabarria, 'La classification des sciences,' pp. 63–4.

[34] Kindī, *Rasāʾil*, i, pp. 104, 384; cf. Cortabarria, 'La classification des sciences,' p. 72; Gutas, *Avicenna*, pp. 243–4.

[35] Gutas, *Avicenna*, pp. 245–9.

(avec l'arithmétique, la musique et la géométrie). En outre, al-Sijistānī réduit la physique à la médecine et aux 'arts' (*ṣināʿāt*).

Quant à la philosophie pratique, l'auteur ismaélien s'inscrit également dans la tradition kindienne qui, à la suite de ses modèles syriaques (Sévère Bar Shakkō, Paul le Perse) et arabes (Ibn al-Muqaffaʿ), désigne les trois disciplines par un même terme générique—*siyāsa*, 'gouvernement'—la politique devenant alors 'le gouvernement général' (*siyāsa ʿāmma*), l'économie 'le gouvernement particulier' (*siyāsa khāṣṣa*) et l'éthique 'le gouvernement le plus particulier' (*siyāsat khāṣṣat al-khāṣṣa*).[36] Tout en se montrant tributaire de cette terminologie traditionnelle, al-Sijistānī la transforme: il désigne la politique comme 'le gouvernement de la généralité ou de la communauté' (*siyāsat al-ʿāmma*), l'éthique comme 'le gouvernement du particulier ou de l'individu' (*siyāsat al-khāṣṣa*) et l'économie comme 'le gouvernement de l'équité ou de la distribution équitable' (*siyāsat al-ḥāqqa*). Nous ignorons où al-Sijistānī a puisé cette dénomination de l'économie, car nous n'en avons trouvé aucun équivalent ailleurs. L'auteur a-t-il mal compris le sens des termes figurant dans sa source? Le texte dont nous disposons a-t-il été corrompu au cours de sa transmission ou par son éditeur moderne?

Malgré ces réserves, la division de la philosophie présentée par al-Sijistānī se rattache à la tradition d'al-Kindī. Ainsi, elle suit de près le canevas tracé par ce dernier[37] et élaboré par son contemporain Qusṭā ibn Lūqā.[38]

Bien que la philosophie, avec toutes ses ramifications, soit une science 'profane' héritée des Grecs, al-Sijistānī en retrouve les traces dans le Coran. La pratique de chaque discipline y est, selon lui, prescrite aux croyants: (1) la révélation coranique impose aux hommes de connaître Dieu et les anges, et de croire en eux [= la théologie]; (2) elle prescrit

[36] Voir les références dans Hein, *Definition*, pp. 227–9.

[37] En témoigne la citation d'al-Kindī dans le *Kitāb al-mūsīqā* d'un certain Ḥasan ibn Aḥmad ibn ʿAlī, contenu dans un manuscrit d'Istanbul découvert par Rosenthal; voir F. Rosenthal, 'From Arabic Books and Manuscripts. VI: Istanbul Materials for al-Kindī and as-Sarakhsī,' *JAOS* 76 (1956), p. 27, ainsi que Biesterfeldt, 'Medieval Arabic Encyclopedias,' p. 81.

[38] Les rapprochements les plus marquants sont: (1) la distinction d'une philosophie théorique (*ʿilm*) et d'une philosophie pratique (*ʿamal*); (2) la tripartition de la philosophie théorique en une science supérieure, médiane et inférieure; (3) l'identification de la science supérieure à la théologie; (4) la tripartition de la philosophie pratique, la dénomination de chaque discipline étant introduite par le terme générique *siyāsa* et spécifiée par les termes *khāṣṣa* ou *ʿāmma*; voir Daiber, 'Qosṭā Ibn Lūqā,' pp. 106–11.

la connaissance des cieux, des astres et de la terre, selon 3:191: 'pour ceux qui pensent à Dieu debout, assis ou couchés et qui méditent sur la création des cieux et de la terre' [= l'astronomie]; (3) elle recommande la connaissance des corps, de ce qui leur est utile et nuisible, selon 30:8: 'N'ont-ils pas réfléchi en eux-mêmes?' (*aw lam yatafakkarū fī anfusihim*) [= la médecine]; (4) elle établit les règles qui déterminent le gouvernement de la communauté, notamment la célébration des fêtes religieuses, la prière du vendredi, le rassemblement à La Mecque [= la politique]; (5) elle trace les principes selon lesquels l'homme doit organiser sa famille avec équité, à savoir l'éducation qu'il doit donner aux membres de sa famille et la punition qu'il doit leur infliger lorsqu'ils se montrent désobéissants, selon 66:6: 'Ô vous les croyants! Préservez vos personnes et vos familles d'un Feu dont les hommes et les pierres sont l'aliment' [= l'économie]; (6) la Loi divine prescrit le gouvernement de l'individu, c'est-à-dire les mœurs et les principes qui doivent régler sa conduite: la façon de se couper les cheveux, de se tailler les ongles, de se laver, de se purifier, de se parfumer, de s'abstenir de certains aliments et boissons [= l'éthique]. Puisque la division aristotélicienne de la philosophie se retrouve dans le Coran, la conclusion s'impose: il n'y a aucune contradiction entre la *ḥikma* et l'enseignement coranique.[39]

Par cette légitimation coranique des différentes disciplines philosophiques, al-Sijistānī s'inscrit une fois de plus dans la tradition kindienne. Tout en présentant une division de la philosophie similaire mais distincte de celle adoptée par al-Sijistānī, Abū 'l-Ḥasan al-ʿĀmirī retrouve lui aussi l'obligation d'étudier les sciences aristotéliciennes énoncée dans le Coran, en particulier l'arithmétique (19:94 et 72:28) et l'astronomie (5:188 et 30:7).[40]

À l'image de la classification aristotélico-kindienne des sciences 'profanes', al-Sijistānī divise les sciences religieuses (*risāla*) en disciplines théoriques et pratiques. 'La branche théorique, précise-t-il, se divise en trois parties: (1) l'exégèse (*taʾwīl*)[41] qui correspond (*shabīh*) à la science supérieure; (2) la science du *kalām* (*ʿilm al-kalām*), qui correspond à la

---

[39] Sijistānī, *Ithbāt*, p. 120.

[40] Abū 'l-Ḥasan al-ʿĀmirī, *Kitāb al-iʿlām bi-manāqib al-islām*, éd. A.ʿA. Ghurāb (Le Caire, 1967), pp. 88–90; traduction allemande de ce passage dans F. Rosenthal, *Das Fortleben der Antike im Islam* (Zürich, 1965), pp. 94–5.

[41] Il va de soi qu'il faut prendre ce terme dans son sens chiite, ismaélien: c'est l''exégèse ésotérique', enseignée par les Imāms, qui permet de dégager le sens caché (*bāṭin*) du sens obvie (*ẓāhir*) du texte révélé.

science médiane; (3) la science du *fiqh* (*ʿilm al-fiqh*), qui correspond à la science inférieure. De même, la branche pratique se divise en deux (!): (1) la pratique concernant les biens (*ʿamal fī ʾl-amwāl*, sans doute la gestion des finances et des biens publics), qui correspond au gouvernement de la communauté; (2) la pratique concernant les corps, qui correspond au gouvernement de l'individu'.[42] Face à chaque discipline philosophique se dresse ainsi une science religieuse, à cette seule exception près que l'économie ne semble pas avoir d'équivalent 'islamique'. Nous ignorons pourquoi al-Sijistānī rompt sur ce point le parallélisme.

Il résulte de ce texte une homologie de structure entre *ḥikma* et *risāla*, que l'on peut représenter de la façon suivante:

Ḥikma	***Risāla***
(a) *al-ʿilm*	(a) *al-ʿilm*
(1) *al-ʿilm al-aʿlā* = *ʿilm al-lāhūt*	(1) *al-taʾwīl*
(2) *al-ʿilm al-awsaṭ* = *ʿilm al-tanjīm*	(2) *ʿilm al-kalām*
(3) *al-ʿilm al-adnā* = *ʿilm al-ṭibb wa'l-ṣināʿāt*	(3) *ʿilm al-fiqh*
(b) *al-ʿamal*	(b) *al-ʿamal*
(1) *siyāsat al-ʿāmma* [= la politique]	(1) *al-ʿamal fī ʾl-amwāl*
(2) *siyāsat al-ḥāqqa* [= l'économie]	(2) ???
(3) *siyāsat al-khāṣṣa* [= l'éthique]	(3) *al-ʿamal ʿalā ʾl-abdān*

Al-Kindī et ses disciples iraniens, dont Ibn Farīghūn et al-Khwārizmī, ont élaboré un concept de la science qui englobe à la fois les sciences profanes et les sciences islamiques.[43] Toutefois, ils laissent subsister une certaine dichotomie entre les deux branches du savoir, sans vraiment

---

[42] Sijistānī, *Ithbāt*, p. 122.

[43] Al-Khwārizmī aurait introduit la distinction, promise à un long avenir dans la pensée arabo-musulmane, entre les 'sciences islamiques et arabes' (*ʿulūm al-sharīʿa wa-mā yaqtarin bihā min al-ʿulūm al-ʿarabiyya*) et les 'sciences étrangères héritées des Grecs et des autres nations' (*ʿulūm al-ʿajam min al-yūnāniyyīn wa-ghayrihim min al-umam*) (al-Khwārizmī, *Mafātīḥ al-ʿulūm*, pp. 5–6); cf. Heinrichs, 'Classification,' pp. 128–9.

établir de rapport entre elles. En d'autres termes, il y a chez eux une juxtaposition théorique, mais pas d'harmonie véritable.[44] Un tel souci d'harmonisation est en revanche bien présent dans l'exposé d'al-Sijistānī.

Nous retrouvons une démarche tout à fait analogue chez un autre disciple iranien d'al-Kindī, contemporain d'al-Sijistānī, bien qu'il ne fut pas ismaélien: Abū 'l-Ḥasan al-ʿĀmirī. Le projet qu'il présente dans son *Kitāb al-iʿlām bi-manāqib al-Islām* ressemble à bien des égards à celui d'al-Sijistānī.[45] Lui aussi divise la science (*ʿilm*) en deux branches: *al-ʿulūm al-ḥikmiyya*, les sciences profanes des philosophes (*ḥukamāʾ*), et *al-ʿulūm al-milliyya*, les sciences religieuses dans lesquelles excellent les prophètes. Chaque branche se subdivise en trois disciplines, selon le mode de connaisance qui leur est propre. Les sciences philosophiques englobent: (1) l'art des métaphysiciens (*ṣināʿat al-ilāhiyyīn*),[46] basé sur la raison (elle est *ʿaqliyya*); l'art des mathématiciens (*ṣināʿat al-riyāḍiyyīn*), basé sur la raison et les sens (*al-ʿaql wa'l-ḥiss*); (3) l'art des physiciens (*ṣināʿat al-ṭabīʿiyyīn*), basé sur les sens (elle est *ḥissiyya*). Parallèlement, les sciences religieuses comprennent: (1) l'art des théologiens (*ṣināʿat al-mutakallimīn*), basé sur la raison; (2) l'art des juristes (*ṣināʿat al-fuqahāʾ*), basé sur la raison et les sens; (3) l'art des traditionnistes (*ṣināʿat al-muḥaddithīn*), basé sur les sens.[47]

Si les détails de la division et la terminologie diffèrent,[48] la manière dont les deux auteurs établissent un rapport de correspondance—une

---

[44] Ce que remarque avec raison Gutas, *Avicenna*, pp. 247–8, au sujet d'Ibn Farīghūn et al-Khwārizmī.

[45] ʿĀmirī, *Manāqib*, pp. 83–97; trad. Rosenthal, *Fortleben*, pp. 91–101; cf. Biesterfeldt, 'Abū l-Ḥasan al-ʿĀmirī,' pp. 335–41.

[46] Littéralement: 'l'art des théologiens", mais nous avons vu (*supra*, p. 83) que la tradition kindienne assimile la métaphysique à la théologie. Les *ilāhiyyūn* se distinguent alors des *mutakallimūn*, le *kalām* étant considéré comme une science religieuse.

[47] ʿĀmirī, *Manāqib*, pp. 84–5; trad. Rosenthal, *Fortleben*, pp. 91–2. Sur la classification des sciences chez al-ʿAmirī, voir en outre les schémas dans Heinrichs, 'Classification,' pp. 131–2; M. Arkoun, 'Logocentrisme et vérité religieuse dans la pensée islamique d'après *al-Iʿlām bi-manāqib al-Islām* d'al-ʿĀmirī,' dans idem, *Essais sur la pensée islamique* (Paris, 1973), p. 211.

[48] Ces divergences s'expliquent en partie par l'orientation religieuse différente des deux auteurs. Pour l'ismaélien al-Sijistānī, le *taʾwīl*, dont l'enseignement est le privilège des Imāms chiites, représente la science religieuse la plus éminente, de loin supérieure au *kalām*, tandis que l'étude du *ḥadīth*, propre au sunnisme, n'apparaît pas sous cette forme dans l'islam chiite. En revanche, il va de soi que le sunnite al-ʿĀmirī ne peut inclure le *taʾwīl* dans sa classification.

'relation miroir-image'[49]—entre les disciplines profanes et les disciplines religieuses est très similaire. En outre, al-ʿĀmirī légitime la pratique des sciences profanes (métaphysique, physique, astronomie, médecine) à l'aide de versets coraniques, tout comme le fait al-Sijistānī. Aucune contradiction ne peut opposer les deux branches du savoir, puisque leur contenu et leurs objectifs sont identiques. Par la connaissance qui mène à la vérité et la pratique qui mène à la vertu, l'enseignement des philosophes et des prophètes conduit au même but ultime: la béatitude éternelle (*al-saʿāda al-abadiyya*). Pour al-Sijistānī et al-ʿĀmirī, un lien très étroit unit dès lors le philosophe et le prophète.

3. *Ḥikma* et *Risāla* ont une source commune et atteignent leur perfection dans la même personne

Selon al-Sijistānī, le philosophe accompli, celui qui maîtrise la *ḥikma* en son intégralité et d'une façon parfaite, est 'un homme pur qui est inspiré par l'esprit de sainteté' (*al-insān al-ṣāfī al-muʾayyad bi-rūḥ al-quds*).[50] Outre qu'elle reflète l'idée kindienne selon laquelle l'acquisition de la connaissance suppose la vertu ou, en d'autres termes, la purification de l'âme par la pratique du bien, cette expression nous révèle la source de la *ḥikma*: 'l'esprit de sainteté'. Dans la philosophie d'al-Sijistānī, la notion coranique du *rūḥ al-quds* se réfère à l'Intellect,[51] le *Noûs* plotinien qui procède de l'Un transcendant et accomplit le rôle de l'Intellect agent des *falāsifa*: par son 'inspiration' (*taʾyīd*), l'intellect humain passe de la puissance à l'acte.[52]

La philosophie, au sens le plus noble du terme, advient donc à l'homme dont l'intellect s'est pleinement actualisé sous l'influence de l'Intellect agent. Al-Sijistānī inclut parmi ces *muʾayyadūn bi-rūḥ al-quds* les 'envoyés' (*rusul*), à savoir les Prophètes et les Imāms, en tant que

---

[49] Heinrichs, 'Classification,' p. 132: 'a mirror-image relationship between religious and philosophical sciences'.

[50] Sijistānī, *Ithbāt*, p. 119.

[51] Walker, *Early Philosophical Shiism*, p. 117.

[52] Sous l'influence d'al-Fārābī, le philosophe ismaélien Ḥamīd al-Dīn al-Kirmānī (m. vers 1021) identifie explicitement le *rūḥ al-quds* à l'Intellect agent (*al-ʿaql al-faʿʿāl*), le dixième Intellect; voir notre article 'Ḥamīd al-Dīn al-Kirmānī's Theory of Intellect and Soul. A Critical Re-examination of its Supposed Fārābian Background,' à paraître dans P. Adamson (éd.), *In the Age of al-Fārābī. Arabic Thought in the 4th/10th Century* (Londres, 2008).

'vicaires' (*khulafā'*) de l'Intellect dans le monde corporel.[53] Un lien très étroit s'instaure ainsi entre les 'philosophes' (*ḥukamā'*) et les 'envoyés' (*rusul*), et de là entre *ḥikma* et *risāla*, les sciences 'profanes' et les sciences 'religieuses', pour autant que ces deux branches du savoir ont une source commune—les 'grâces' (*faḍā'il*) qui procèdent de la lumière de la Parole divine (*kalimat Allāh* = l'Intelllect)[54]—et atteignent leur complétude dans une même catégorie d'hommes.

Déjà al-Kindī avait établi une distinction entre la science humaine (*'ilm insānī*), qui englobe l'apport de la philosophie grecque, et la science divine (*'ilm ilāhī*), qui est l'apanage des 'envoyés' (*rusul*): si la première s'acquiert par l'étude préalable des sciences propédeutiques (dont la logique) et nécessite une longue application morale et intellectuelle, la seconde est obtenue sans effort par les 'intellects lumineux et purs' (*al-'uqūl al-nayyira al-ṣāfiyya*) des prophètes, grâce à une inspiration (*ta'yīd*!) divine. Par ailleurs, malgré leur mode d'acquisition différente, les deux types de sciences se complètent et tendent vers un but identique.[55]

Cette doctrine, reprise par la plupart des disciples d'al-Kindī, inspirera à al-'Āmirī la phrase devenue célèbre: 'chaque prophète est philosophe, mais chaque philosophe n'est pas prophète' (*kull nabī ḥakīm wa-laysa kull ḥakīm nabiyyan*).[56] Le prophète est un philosophe parfait et le philosophe accompli s'assimile au prophète à condition que, par l'étude et la pratique de la vertu, il ait disposé son intellect à recevoir l'action de l'Intellect agent, qui le fera passer de la puissance à l'acte. Al-Sijistānī souscrit lui aussi à cette thèse, dans laquelle on pourrait voir une transposition islamique de la définition antique de la philosophie comme *théôsis*: 'la philosophie est assimilation à la divinité, pour autant que la nature de l'homme le permette'.[57] Le philosophe 'se divinise' en s'assimilant aux envoyés de Dieu: les Prophètes et les Imāms.

Notre analyse de ce chapitre du *Kitāb ithbāt al-nubuwwāt* consacré à la division des sciences montre qu'al-Sijistānī s'inscrit dans la tradition d'al-Kindī, qui fut introduite en Iran par son disciple Abū Zayd al-Balkhī

---

[53] Sijistānī, *Ithbāt*, p. 119; cf. Walker, *Early Philosophical Shiism*, p. 117.

[54] Sijistānī, *Ithbāt*, p. 119. *Faḍā'il* est un terme assez flou, qui se réfère souvent dans les textes néoplatoniciens et ismaéliens aux formes intelligibles (*ṣuwar rūḥāniyya*).

[55] Kindī, *Rasā'il*, i, pp. 372–6; cf. Cortabarria, 'La classification des sciences,' pp. 59–60; Jolivet, 'L'idée de la Sagesse,' pp. 34–7.

[56] 'Āmirī, *Manāqib*, p. 84.

[57] Sur cette définition de la philosophie, la quatrième selon la classification traditionnelle, voir Hein, *Definition*, pp. 99–101.

et élaborée par les élèves de ce dernier: Ibn Farīghūn et Abū 'l-Ḥasan al-ʿĀmirī. Ces mêmes auteurs semblent avoir joué un rôle prépondérant dans la diffusion en Iran des paraphrases arabes de Plotin et de Proclus, qui furent élaborées à Bagdad par un 'cercle' de traducteurs réunis autour d'al-Kindī.[58] Si la postérité immédiate du philosophe al-Kindī en Iran reste encore très imparfaitement connue,[59] il en est encore davantage de ses rapports avec le néoplatonisme ismaélien de l''École persane'—dont al-Sijistānī faisait partie—qui se développe en ce même dixième siècle, dans un même milieu géographique et intellectuel. Ainsi, Abū Zayd al-Balkhī, disciple direct d'al-Kindī, était probablement un chiite duodécimain, mais il passa une grande partie de sa vie au Khurassan sous la protection de l'émir ismaélien al-Marwāzī, dans l'entourage duquel il semble avoir côtoyé certains représentants de l''École persane', en particulier Muḥammad al-Nasafī.[60]

Nous espérons avoir démontré, par l'exemple de la division des sciences et de l'harmonie entre les sciences philosophiques et religieuses, l'existence d'un rapport étroit entre Abū Yaʿqūb al-Sijistānī et son contemporain et compatriote Abū 'l-Ḥasan al-ʿĀmirī. D'autres chercheurs ont occasionnellement relevé des liens entre la tradition kindienne et le néoplatonisme ismaélien,[61] mais il faudra attendre un dépouillement systématique des ouvrages d'al-Sijistānī et des autres auteurs de 'l'École persane' (Muḥammad al-Nasafī, Abū Tammām, Abū Ḥātim al-Rāzī) avant de pouvoir déterminer la nature et l'ampleur de cette relation. Or un tel dépouillement, qui permettrait selon nous de recueillir de nombreux éléments à verser au dossier de la 'tradition kindienne' en Iran, n'en est encore qu'à ses débuts.

---

[58] G. Endress, 'The Circle of al-Kindī. Early Arabic Translations from the Greek and the Rise of Islamic Philosophy,' dans G. Endress et R. Kruk (éds.), *The Ancient Tradition in Christian and Islamic Hellenism. Studies on the Transmission of Greek Philosophy and Sciences dedicated to H.J. Drossaart Lulofs on his ninetieth birthday* (Leiden, 1997), pp. 43–76.

[59] Le livre récent d'Elvira Wakelnig, *Feder, Tafel, Mensch. Al-ʿĀmirī's Kitāb al-fuṣūl fī l-maʿālim al-ilāhīya und die arabische Proklos-Rezeption im 10. Jh.* (Leiden, 2006), apporte une contribution fondamentale en ce domaine.

[60] Walker, *Early Philosophical Shiism*, pp. 14–15.

[61] Ainsi, p. ex., Walker, *Early Philosophical Shiism*, index, s.v. 'al-ʿĀmirī' et 'al-Kindī,' Jolivet, 'L'idée de la Sagesse,' pp. 41–2.

'I WAS A HIDDEN TREASURE'. SOME NOTES ON A COMMENTARY ASCRIBED TO MULLĀ ṢADRĀ SHĪRĀZĪ: *SHARḤ ḤADĪTH: 'KUNTU KANZAN MAKHFIYYAN…'*

Armin Eschraghi

Ṣadr al-Dīn Muḥammad ibn Ibrāhīm Shīrāzī (d. 1640), better known as 'Ṣadr al-Muta'allihīn' or 'Mullā Ṣadrā', is generally considered one of the most popular and influential thinkers of Shiite Islam.[1] It is often claimed that after him the philosophical tradition of Islam ceased to produce anything original. The larger part of Ṣadrā's voluminous oeuvre is published. His major works are available in many editions and have been commented upon. However, there are also a certain number of lesser known pieces, one of them a short commentary on a celebrated *ḥadīth qudsī*. The text has been published in a compilation of philosophical treatises as part of a longer collection of *fawā'id*.[2]

Ṣadrā's actual authorship of it is not certain. In the introduction to his edition Iṣfahānī explains that it has at times also been ascribed to Muḥyī al-Dīn ibn al-ʿArabī and ʿAlā' al-Dawla Simnānī, although, he states, it is mostly found in manuscript collections of works belonging or ascribed to Mullā Ṣadrā. He is reluctant to judge the matter conclusively but believes that the fact that this commentary is part of a longer collection of *fawā'id*, which contains internal evidence of Ṣadrā's authorship, 'strengthens probability of its belonging to Ṣadrā.'[3] Whether or not its style and contents conform to Ṣadrā's other writings needs to be decided by experts who have intimate knowledge of his works. As shall presently be seen, the fact that a very similar passage can be found in Ṣadrā's *magnum opus* might strengthen probability of his authorship.

---

[1] There is a vast literature on Mullā Ṣadrā. For a general introduction to his life and works see D. MacEoin, 'Mullā Ṣadrā Shīrāzī,' in *EI*[2]; H. Ziai, 'Mullā Ṣadrā: his Life and Works,' in S.H. Nasr and O. Leaman (eds.), *History of Islamic Philosophy*, i (London, 1996), pp. 635–42; S.H. Nasr, 'Mullā Ṣadrā: his Teachings,' in ibid., pp. 643–62.

[2] *Majmūʿiy-i rasā'il-i falsafiy-i Ṣadru'l-Muta'allihīn*, ed. Ḥ.N. Iṣfahānī (Teheran, 1999), p. 356–8 (based on two manuscripts and a lithograph [Teheran, 1315]).

[3] Ibid., pp. 37–8.

Background of the 'Hidden Treasure'-Tradition

Whatever the case, the contents of this text are interesting enough to grant it a second look. That is mainly because the *ḥadīth* commented upon has a long tradition in Sufi literature and can in fact be considered as one of the best known and most widely celebrated apocryphal traditions of mysticism. Its earliest known occurrence seems to be in ʿAbdallāh Anṣārī's (d. 1089) *Ṭabaqāt al-ṣūfiyya*.[4] However, from the eleventh century onwards references to it multiply and there seem to be very few mystical works that do not quote it. Jalāl al-Dīn Rūmī (d. 1273), for example, alludes to it in his *Mathnawī*,[5] even to an extent that Nicholson considered 'certain motifs, such as that of the "hidden treasure"' to be 'overworked.'[6] In his commentary on the *Mathnawī* Furūzānfar quotes the following from the *Luʾluʾ al-marṣūʿ*:[7]

> Ibn Taymiyya said, this [*ḥadīth*] was not uttered by the prophet (ṣ), and no *isnād* for it, either solid or weak, is known. He was followed [in this] by al-Zarkashī and Ibn Ḥajar. But its content is correct (*wa-lākin maʿnāhu ṣaḥīḥ*) and clear and it circulates among the Ṣūfīs.[8]

Its authenticity has thus been debated, not so much because of its content, but rather due to a lack of formal criteria of transmission, i.e. absence of an *isnād*, as well as, in other cases, due to alleged violations of Arabic grammar.[9] Sometimes the words are said to have been revealed to the prophet Muhammad, sometimes David is named as the addressee.

But this criticism by traditional scholars hardly concerns mystics who share a large corpus of non-canonical prophetic traditions from which they quote extensively and to which they apply criteria of authenticity, different from those of traditional *ḥadīth*-scholarship. Rūmī's contem-

---

[4] Cf. M. Afnānī, 'Mulāḥaẓātī dar-bāriy-i Lawḥ wa Ḥadīth-i Kuntu Kanz,' in *Safīniy-i ʿIrfān* (Darmstadt, 2001), iv, p. 162.

[5] E.g. Book 1, verse 2863.

[6] *The Mathnawí of Jalálu'ddín Rúmí*, trans. R.A. Nicholson (London, 1934), vi, p. xi.

[7] This is a reference to Muḥammad ibn Khalīl al-Qāwijī's *al-Luʾluʾ al-marṣūʿ fī-mā qīla lā aṣla lahu aw bi-aṣlihi mawḍūʿ fī 'l-ḥadīth.* A manuscript of this work can be accessed via http://www.sahab.org/books/book.php?id=1392 [16 June 2006]. The passage here referred to is on p. 22 of that manuscript.

[8] *Aḥādīth wa-qiṣaṣ-i Mathnawī* (Teheran, 1376), pp. 120–21.

[9] Iṣfahānī, *Majmūʿiy-i rasāʾil*, p. 38 (n. 84), cf. M. Afnani, 'Mulāḥaẓātī,' pp. 166–7.

porary, the celebrated and most influential mystic Muḥyī al-Dīn ibn al-ʿArabī (d. 1240), thus writes of it in his *magnum opus*:

> It has been transmitted as correct *ḥadīth*, through unveiling—not confirmed through transmission (*al-ghayr al-thābit naqlan*)—that the Messenger of God, on behalf of His Lord, said something to this effect: I was a hidden Treasure...[10]

He also alludes to the *ḥadīth* on numerous other occasions throughout his major works.[11]

It is obvious that the *ḥadīth* was known to Mullā Ṣadrā and he apparently did not have any objections against it. He quotes it at least twice in his major work, the *Asfār*.[12] In one instance he uses it to refute a specific concept of 'unity of being' taught by some 'misguided Sufis'. He quotes it to illustrate God's self-disclosure which takes place on several stages. Thus, the level of the 'Hidden Treasure' is the one of divine essence in the state of 'pure Oneness' (*aḥadiyya*), whereas the second station, i.e. the 'being known' (*maʿrūfiyya*), follows from self-disclosure of the essence in different degrees (*ṭawran baʿda ṭawrin*). The divine essence now sees itself in creation as in a mirror.[13] The main argument of this passage seems to be partly reflected in replies two and four of the commentary presently under consideration.

Another passage in the *Asfār* is equally relevant—and similar in purport—to parts two and four of the commentary. Here Ṣadrā deals with another celebrated *ḥadīth*, according to which 'God is behind seventy thousand veils of light and darkness, were He to lift them, the rays of His face would burn everything His sight reaches.'[14] The veils of light are

[10] *Al-Futūḥāt al-Makkiyya*, 4 vols. (Beirut, n.d.) ii, p. 399. Cf. W. Chittick, *The Sufi Path of Knowledge. Ibn ʿArabī's Metaphysics of Imagination* (Albany, NY, 1989), pp. 131 and 391. Some of Ibn ʿArabī's later followers and commentators also make ample reference to this *ḥadīth*. See for example ʿAbd al-Razzāq al-Kāshānī, *Laṭaʾif al-aʿlām fī ishārāt ahl al-ilhām* (Teheran, 2000), pp. 486–7; Haydar Āmolī, *Jāmiʿ al-asrār* (Teheran, ^{2}1368), pp. 102, 144, and many more; Shams al-Dīn Lāhījī, *Sharḥ-i golshan-i rāz* (Teheran, 1374), p. 94.

[11] *Fuṣūṣ al-ḥikam*, ed. A. ʿAfīfī (Beirut, n.d.), p. 203; *al-Futūḥāt al-Makkiyya*, ii, pp. 112, 232, 310, 322, 331, and many more.

[12] The full title of this voluminous work is: *Al-Ḥikma al-mutaʿāliyya fī 'l-asfār al-ʿaqliyya al-arbaʿa*. All references in this article are to the 1981 Beirut edition, printed in nine volumes by Dār Iḥyāʾ al-Turāth al-ʿArabī.

[13] *Al-Asfār*, ii, pp. 345–7.

[14] Ibid., vi, 299–301.

the pure intellects (*al-ʿuqūl al-mujarrada*), which, unlike essence and nature (*nufūs wa-ṭabāʾiʿ*), are not temporary. The latter are the veils of darkness. However, all of them, as Ṣadrā points out, are archetypes and part of God's knowledge. They possess two types of existence: collective and individual. Out of grace God wished to be known. Thus, He granted all things individual existence, i.e. transferred them from the state of existence in knowledge (*wujūd ʿilmī*) to actual/individual existence (*ʿaynī*). This process is one of completion (*takmīlī*)—not of God's immutable essence which necessarily remains unaltered, but rather of knowledge of God (*ʿilm bi'llāh*). True knowledge derives from existence (*al-maʿrifa farʿ al-wujūd*), which is granted through God's self-disclosure. Since all existence eventually belongs to God, it is in truth He who revealed Himself to Himself for Himself (*tajallā bi-nafsihi li-nafsihi*). This is the gist of Ṣadrā's more complicated and detailed analysis in the *Asfār*, the main points of which are closely reflected especially in the second part of the commentary.

Summary of the Commentary

The first paragraph of the *Sharḥ ḥadīth al-kanz al-makhfī* repeats or rather summarizes the problem raised by the addressee. The question essentially deals with the well-known philosophical problem of the pre-existence of creation. The *ḥadīth* states that God was hidden before He brought creation into being. Strictly speaking this is not possible, because if God was alone, how and by whom could He be considered 'hidden'? Also, if God were alone, He would be manifest, because obviously He is aware of Himself and cannot thus be considered hidden from Himself. There would be nobody else in existence to know or not know Him. He thus could not be considered hidden. It is only through the creation of another that God can actually be in 'hiding'. But this again, runs contrary to the text of the *ḥadīth* which seems to say the opposite, namely that God brought forth creation in order *not* to be hidden anymore. This apparent contradiction the author is asked to resolve.

The reply begins with the author's statement that he received what is to follow through direct inspiration from God. This should not be understood as a claim to some sort of spiritual authority. It might merely be a common expression of modesty, in the sense that nothing can be accomplished without God's assistance. But most probably it is a reference to the Ishrāqī concept of intuitive knowledge, which was also adopted by Mullā Ṣadrā.[15] Similar statements can be found in his other works, e.g. in his *al-Asfār*.[16]

The author then begins the actual commentary by stating that there are different possible solutions to the problem. Out of the four replies that he continues to present afterwards the first and the third deal with the outward form of the text and explain the semantics of certain keywords. No prior specific knowledge is needed to come to the understanding of the text suggested in those two replies. However, the second and fourth replies draw on concepts which lie outside the text of the *ḥadīth* itself and are of a mystical and philosophical nature.

In the first reply a less literal reading of the *ḥadīth* is encouraged, which harmonizes its contents with the generally acknowledged interpretation. The author suggests that not too much emphasis should be placed on the word 'hidden' and its implications, because the meaning of the text is obvious. Hidden here does not necessarily—in contrast to what the addressee suggests—involve a 'hiding' and a 'hidden-from.' In this context it merely refers to the fact, that there was no created being to know God and that He thus wished to be known.

The second reply revolves around questions of ontology and epistemology. The relation of God to His creation needs to be seen from different perspectives. Each being has two kinds of existence, temporal and archetypal. The archetypes (*al-aʿyān al-thābita*) feature prominently in the works of authors like Ibn ʿArabī and Mullā Ṣadrā and are thought to exist beyond any conception of time in God's immutable

[15] For a general overview of this concept see: H. Ziai, 'Shihāb al-Dīn Suhrawardī: Founder of the Illuminationist School,' in Nasr and Leaman, *History of Islamic Philosophy*, pp. 434–64, particularly pp. 451–7. For a more detailed discussion see M.A. Razavi, *Suhrawardi and the School of Illumination* (Richmond, 1997). Claims to intuitive knowledge or divine inspiration can also be found outside the Suhrawardī tradition, i.e. in the works of many mystics. Ibn ʿArabī, who has been quoted earlier, is another example. In the introduction to his *Fuṣūṣ al-ḥikam* (i, p. 47) he states that the book was directly inspired by the prophet Muḥammad himself. Even a thinker like Abū Ḥāmid al-Ghazālī (d. 1111), who is generally perceived of as rather orthodox, described the source of his knowledge as a light, which God cast into his breast. *Al-Munqidh min al-ḍalāl*, ed. M.M. Jābir (Cairo, n.d.), pp. 10–11; here quoted from *Der Erretter aus dem Irrtum*, ed. and trans. A.A. Elschazlī (Hamburg, 1988), pp. 9–10.

[16] See for example: i, pp. 7–9.

knowledge,[17] although they themselves do not have knowledge of their Creator. This is, according to the author of the commentary, due to the fact that knowledge of something requires the knower's outward or temporal existence. When God wished to be known by the archetypes, He thus 'had' to bestow outward existence on them. On the level of pre-existence and archetypes God remains hidden from the archetypes, who here are the 'hidden from', yet at the same time are also not 'entirely' created and existent. Only after God calls them into outward existence do they become capable of knowing Him. This is obviously a simplification and a summary of the more detailed explanations in the *Asfār*, referred to earlier.

The third reply is the simplest and does not require much of an elaboration. The author points to the dictionary of al-Jawharī[18] where the word *makhfī* is named amongst the *aḍdād*, namely words that have two contradictory meanings, in this case 'hidden' and 'manifest'. If the latter meaning be taken in this context, there will no longer be a problem. Interestingly, the point raised by some scholars, that *makhfī* here is a violation of the rules of grammar, is not dealt with in this commentary. The author seems not to object to the word itself, but rather tries to re-define its meaning.

The fourth and final reply involves a concept that is well-known to mystics although it might sound paradoxical at first, i.e. God's hiding in His own manifestation. The sun, which is referred to here, is a popular example which serves to illustrate this point. According to our author, God was so openly manifest to Himself, that He was almost blinded by His own light. He therefore brought forth creation as a cover, or, in terms of the example referred to earlier, as a hand to cover his eyes. Now He can perceive Himself clearly in creation, which at the same time serves as both a veil, as well as a tool to observe God's light. In other words, creation is a channel for God's overwhelming light. Also, as follows from our author's commentary, through this process God grants His creation the ability to know Him by means of His self-revelation

[17] They have to be perceived of as existent, because otherwise a change in God's knowledge would be implied, which is not permissible. On the other hand, eternal existence of anything beside God poses different theological problems. The archetypes are an attempt at a solution of this dilemma.

[18] The title of this work is *Al-Ṣiḥāḥ fī 'l-lugha wa'l-ʿulūm*.

in them. Here the reader wishes the commentary would go on and explain in more detail and more explicitly, whether the fact that 'God knows Himself through creation' also implies that God Himself takes part in the process of knowledge, in other words, whether God and His creation, knower and known are here thought to be essentially one. Finding an answer to this question requires sufficient knowledge of Mullā Ṣadrā's other writings, particularly the concept of 'the union of intellect and intelligible' (*ittiḥād al-ʿāqil wa'l-maʿqūl*) revived by Ṣadrā, as well as general expertise in the mystical and philosophical tradition of Islam. The aim of this short piece is merely to draw attention to Ṣadrā's lesser known but obviously significant works.

What follows is an attempt to translate this short text into English:

Translation of the Text According to Iṣfahānī's Edition

Know—may God grant you success in whatever He loves and pleases—that one of the learned presented me with a very complicated problem (*ishkāl ʿaẓīm*) in the *ḥadīth qudsī*, where He says: 'I was a hidden Treasure. Then I loved to be known. Therefore I created Creation that I may be known.' He also mentioned that he had asked several of the learned of our times, but had not received a satisfying and sufficient reply. When I thus contemplated the question, God Most High inspired me with four answers. I will now quote what he [the addressee] presented me with and enclose the answers which God Most High bestowed on me.

The problem: Hiding is a relative matter and of necessity involves a 'hidden', and a 'hidden from'. It is neither permissible to deem God the 'hidden from' since He is manifest in Himself [to/for Himself] (*ẓāhirun bi-nafsihi [li-nafsihi]*) and knows His own Essence in pre- and post-eternity (*al-azal wa'l-abad*). Nor is it permissible [to think] that it is the 'Creation', since they did not exist in pre-eternity in order for God Most-High to be hidden from them. And in the *ḥadīth* [it is said]: 'God was, and no thing was beside Him'. Being hidden requires [the existence of] creation. Thus, creation would become the cause of hiding, not of manifestation. Now this would contradict the purport of the *ḥadīth*. For this *ḥadīth* seems to say that God Most High was hidden in pre-eternity during the non-existence of creation. This is the question.

I say: A reply to this question [can be given] from [different] perspectives:

One: The meaning of 'hiding' could be that there was no knower but/beside Him. And when He wished to multiply those who know Him He created creation. And He expressed non-existence of [another] knower through [the word] 'hiding'. It is as if He had said: 'I was a mighty (*ʿazīzan*) treasure, and a noble jewel (*jawhar*). But nobody knew about Me (*ʿālim bī*) but Me, and no one was aware of My existence (*ʿārif bi-wujūdī*) besides Me.' So He called it 'hidden' and intended by that its necessary corollaries (?) (*lāzimuhu*), namely that there were no others aware of Him. So the meaning is this: 'I was a good Lord and a beneficent and bestowing (*munʿiman mufīḍan*) God. There was no one to know Me nor to be aware of My perfection and beauty. Then I loved to be known. Therefore I created Creation that I may be known.' And this meaning is correct and there is no problem with it.

Two: All things have two kinds of existence: existence in knowledge and outward existence (*wujūd khārijī*). The existence in knowledge is called archetypes (*aʿyān thābita*) and they are pre-eternal (*azaliyya qadīma*), whereas the outward existence is created in time (*muḥdath*). Thus God's hiding is, in relation to the archetypes, in pre-eternity, since they exist with God but have no knowledge of Him. Therefore God is hidden in relation to them. When He wished for the Archetypes to know Him He transferred them from 'existence in knowledge' to the 'outward existence', so that God Most-High may be known. For God Most-High is not known except through outward existence.

Three: In the *Ṣiḥāḥ* it is reported from al-Aṣmaʿī: '*Khafaytu al-shayʾa* can mean "I hid it" and it can also mean "I displayed/manifested it". It is one of the *aḍdād*.' [In] His word: 'I was a hidden treasure', *khafāʾ* can thus also mean 'manifestation'. Then the meaning of the *ḥadīth* would be: 'I was a treasure manifest before Myself and there was no one else to know Me. I wished to be known by others and therefore created [creation].'

Four: It is possible that the meaning is: I was manifest, hidden in the utmost degree of manifestation (*kuntu ẓāhiran fī ghāyat al-ẓuhūr khafiyyan*). As if He had said: 'Through the utmost degree of manifestation My own self almost became hidden from Myself, let alone others. Thus I created creation as a veil for my manifestation and to cover my light. So that part of my manifestation may be hidden and creation would be able to apprehend Me.' Have you not seen how someone who wants to look directly at the sun puts his hand on his forehead and covers some of its light that he may be able to apprehend [other] parts of its light. So He created creation as a veil for His light and yet made it a cause/means for His apprehension. 'Therefore I created Creation that I may be known.' Praised be He, who made manifestation prevent apprehension, and turned cover and veil into a means for manifestation and apprehension. He is the knower of truths (*ḥaqāʾiq*).

AL-GHAZĀLĪ OR AL-GHAZZĀLĪ? ON A LIVELY DEBATE AMONG AYYŪBID AND MAMLŪK HISTORIANS IN DAMASCUS

Frank Griffel

How to spell and pronounce the *nisba*, the adjective of relation or the family name of the great Muslim theologian and jurist Abū Ḥāmid Muḥammad ibn Muḥammad al-Ghazālī, who died in 505/1111 in his birthplace Ṭābarān in the district of Ṭūs in Khorasan, is a contested subject among scholars in contemporary Islamic studies. While in Iran, in past and present, he is widely known as *Ghazzālī*, the Arab world is probably equally split between those who pronounce his name with a *tashdīd*, a gemination, of the *zāy* and those who omit the *shadda*. In the West there is now a clear tendency to write his name with just one z. While some Western scholars write *al-Ghazzālī*, both two major works of reference in this field of study published, during the second half of the twentieth century, the second edition of the *Encyclopaedia of Islam* and the *Encyclopaedia Iranica*,[1] opted for the spelling with one z. In the past, the matter was much more confusing. Carl Brockelmann in his *Geschichte der arabischen Litteratur* of 1898 initially wrote *al-Ghazālī* and defended his choice by quoting evidence from classical Arabic literature.[2] Yet, in the Supplement-volumes to this work, published in 1937, and in the second edition of 1943, Brockelmann writes the name with a double-z.[3] This change of opinion was prompted by conflicting evidence to the one quoted earlier, taken both from classical Arabic literature and Persian

[1] Cf. G. Böwering et al., 'Ġazālī,' in *EIr*, x, pp. 358–77.

[2] C. Brockelmann, *Geschichte der arabischen Litteratur* (= *GAL*), 2 vols. (Weimar, 1898–1902), i, p. 419. Brockelmann quotes Ibn Khallikān's (d. 681/1282) claim that al-Samʿānī (d. 562/1166) in his *Kitāb al-ansāb* opted for *al-Ghazālī*. Al-Samʿānī, writes Brockelmann, was the greatest authority on Arabic family names (*ansāb*). Yet, Brockelmann disregards Ibn Khallikān's balancing conclusion, which is uncommitted to either spelling; cf. *Wafayāt al-aʿyān wa-anbāʾ abnāʾ al-zamān*, ed. I. ʿAbbās, 8 vols. (Beirut, 1968–77), i, p. 98. Al-Samʿānī's *Kitāb al-ansāb* was not available to Brockelmann. When it was first published in 1912, the fact that the quoted comment is lacking became one of several reasons for Brockelmann to change his mind.

[3] C. Brockelmann, *Geschichte der arabischen Litteratur*, 2 vols. (Leiden, ²1943–49), i, p. 535 and *Supplement*, 3 vols. (Leiden, 1937–42), i, p. 744.

poetry, where the meter of some poems dictates the reading *Ghazzālī*. The amount of confusion about this matter might be illustrated by the fact that while the first edition of the *Encyclopaedia of Islām*, which was published between 1908 and 1936, writes the scholar's name with one z, the *Shorter Encyclopaedia of Islam*, which was published in 1953 and which is just a selection of articles from the former, makes an editorial change in the very same text published earlier and writes the name with two z.[4]

The subsequent tendency towards a spelling with one z seems to be a result of William M. Watt's argument from the early 1960s: since the classical Arab sources attest both forms of spelling with equal right, the one with the more complex explanation—meaning: the one that seems more farfetched—should be accepted 'on an analogy with the principle of *difficilior lectio potius*.'[5] Watt supposed that while the reasons for the two spellings of al-Ghazālī's name are known, the information about which one is correct is shrouded in history and will probably never become available to us. Yet, Watt assumed that one of the two explanations, namely the one that leads to a spelling with two z, is so obvious that it begs the question why a second one had been put forward. The fact that there were always Muslim scholars who defended the reading with one *zāy*, which according to Watt is much more obscure than its alternative, should lead us to accept this one as initially correct. A brief version of the argument was included in Watt's article on al-Ghazālī in the second edition of the *Encyclopaedia of Islam* and spread widely.[6]

During the seventh/thirteenth and eighth/fourteenth centuries, a group of Muslim historians in Ayyūbid and Mamlūk Damascus debated the spelling of al-Ghazālī's name and—given that they did not share Watt's method adopted from Latin textual criticism—came to a different

---

[4] Cf. D.B. Macdonald, 'al-Ghazālī,' in *Encyclopaedia of Islām*, 4 vols. (Leiden, 1908–36), ii, pp. 146–9. Cf. with idem, article 'al-Ghazzālī,' in H.A.R. Gibb and J.H. Kramers (eds.), *Shorter Encyclopaedia of Islam* (Leiden, 1953), pp. 111–14. The *Shorter Encyclopaedia of Islam* takes its cues from the earlier German *Handwörterbuch des Islam*, ed. A.J. Wensinck and J.H. Kramers (Leiden, 1941), pp. 140–44, where the name is also written with two z. The same article in the full edition of the *Enzyklopaedie des Islām*, 4 vols. (Leiden, 1908–36), ii, pp. 154–7 has the name with one z. In his brief article 'The Name al-Ghazzālī,' *JRAS* (1902), pp. 18–22, Duncan B. Macdonald discusses the evidence for each alternative and remains uncommitted.

[5] 'The more difficult reading should be preferred,' W.M. Watt, *Muslim Intellectual. A Study of al-Ghazālī* (Edinburgh, 1963), pp. 181–3.

[6] W.M. Watt, 'al-Ghazālī,' in *Encyclopaedia of Islam*, 11 vols. (Leiden, ^{2}1954–2002), ii, p. 1038b.

conclusion. They gathered and weighted the evidence pro and against each of the two versions of the name. In their discussions they also offer a number of interesting insights into al-Ghazālī's family background.

The Wool-Spinner versus the Man from Ghazāla

In 1889, Wilhelm Ahlwardt published a brief excerpt from a biography of al-Ghazālī by the Egyptian historian of the Shāfiʿite school of law, al-Sharqāwī (d. 1227/1812), where the author discusses al-Ghazālī's family name (*nisba*):

> His father used to spin wool and to sell it in his shop. This would require that 'al-Ghazzālī' is with the gemination [of the *zāy*] as it is a reference to the spinning (*al-ghazl*). It is the habit of the people of Khwārazm and Jurjān to say '*al-qaṣṣārī*' and '*al-khabbāzī*' and similar words like these with a *yāʾ* [at the end] in the meaning of 'the fuller' (*al-qaṣṣār*) or 'the baker' (*al-khabbāz*) and similar. Thus, the people would make a reference to the spinning and they would say '*al-ghazzālī*' and mean 'the spinner' (*al-ghazzāl*). Yaḥyā al-Nawawī mentions in his book *Daqāʾiq al-Rawḍa*[7] that the gemination [of the *zāy*] is well known and has already been mentioned by Ibn al-Athīr.
>
> From al-Ghazālī himself it has reached us that he said: 'I am connected to Ghazāla without geminization, one of the villages of Ṭūs.'[8]

The question of whether 'al-Ghazālī' should be written with a double-*zāy* or simply with a single depends on whether his father or an earlier relative was a spinner of wool. The Iraqi historian Ibn al-Athīr

[7] Al-Nawawī's *Fine Points of the 'Garden'* is apparently a supplementary work to his own legal handbook *Rawḍat al-ṭālibīn wa-ʿumdat al-muftīn*. The *Daqāʾiq al-Rawḍa* is not available to us. A similar work of al-Nawawī, the *Daqāʾiq al-Minhāj* (Beirut, 1996) is extant and offers explanations of 'fine points' in his *Minhāj al-ṭālibīn*. Yet this is not the book al-Sharqawī refers to. Cf. the same quote from al-Nawawī's *Daqāʾiq al-Rawḍa* in al-Murtaḍā al-Zabīdī, *Itḥāf al-ṣāda al-muttaqīn bi-sharḥ asrār Iḥyāʾ ʿulūm al-dīn*, 10 vols. (Cairo, 1311 [1894]), i, p. 18, l. 12–14.

[8] W. Ahlwardt, *Die Handschriften-Verzeichnisse der Königlichen Bibliothek zu Berlin. Verzeichnis der arabischen Handschriften*, 10 vols. (Berlin, 1887–99), ii, p. 306. The text is taken from al-Sharqāwī, *al-Tuḥfa al-bahiyya fī ṭabaqāt al-Shāfiʿiyya*, ms. Berlin, Orientabteilung der Staatsbibliothek Preussischer Kulturbesitz, Landberg 115, fol. 122b. This book is yet unedited. On the book and its author see *GAL* ii, p. 479f, *S* ii, p. 729; Ahlwardt, *Verzeichnis*, ix, p. 449 (no. 10041). A second manuscript at Yale University, Beinecke Memorial Library, Landberg 459, fol. 54b, has only the first sentence of this passage on the profession of al-Ghazālī's father and not the rest.

(d. 630/1233), who is quoted by al-Sharqāwī as the earliest authority for this explanation, was hardly decisive in his judgment. In his book on common family names (*ansāb*), he writes on 'al-Ghazzālī':

> I believe this is a reference to 'the spinner' (*al-ghazzāl*) according to the usage of the people of Jurjān and Khwārazm (...). The best-known bearer of this name was Abū Ḥāmid al-Ghazzālī. I also heard people saying that it is with just one *zāy* and that it refers to Ghazāla, which is one of the villages of Ṭūs. This is a well-known dispute.[9]

A few years after Ibn al-Athīr wrote these lines, the dispute was picked up by a number of Damascene legal scholars of the seventh/thirteenth century. Their debate was triggered by the interest of Ayyūbid legal commentators such as Ibn al-Ṣalāḥ al-Shahrazūrī (d. 643/1245) and his student Yaḥyā al-Nawawī, (d. 676/1277) for al-Ghazālī's works. Both scholars produced voluminous commentaries either directly on al-Ghazālī's books, like Ibn al-Ṣalāḥ on his *al-Wasīṭ fī 'l-madhhab al-Shāfiʿī*,[10] or indirectly as super-commentaries like al-Nawawī's *Rawḍat al-ṭālibīn wa-ʿumdat al-muftīn*.[11] Al-Nawawī's work became a very popular handbook of Shāfiʿite *fiqh* and is based on *Fatḥ al-ʿazīz fī sharḥ al-Wajīz* by ʿAbd al-Karīm al-Rāfiʿī (d. 623/1226) of Qazwīn, a book that is itself a commentary on al-Ghazālī's *al-Wajīz fī 'l-fiqh*.[12] Although al-Ghazālī's three works on the individual judgments (*furūʿ*) of the Shāfiʿite tradition, the very voluminous *al-Basīṭ fī 'l-furūʿ*, the mid-size excerpt *al-Wasīṭ* and the short excerpt *al-Wajīz*, may by themselves not have been much studied by later Shāfiʿites, these commentaries, particular the one by al-Nawawī, made sure that al-Ghazālī's opinions had a lasting impact on the rulings of the Shāfiʿite school of law.

The eminent historian al-Dhahabī (d. 748/1348), who wrote two generations after al-Nawawī in Mamlūk Damascus, presents the results of the discussions on al-Ghazālī's name in his book on the lives of the

[9] Ibn al-Athīr, *al-Lubāb fī tahdhīb al-ansāb*, ed. Ḥ. al-Qudsī, 3 vols. (Cairo, 1357–69 [1938–50]), ii, p. 170.

[10] *Sharḥ mushkil al-Wasīṭ*, now available in al-Ghazālī, *al-Wasīṭ fī 'l-madhhab wa-bi-hāmishihi: al-Tanqīh fī sharḥ al-Wasīṭ li-Muḥyī al-Dīn ibn Sharaf al-Nawawī, Sharḥ mushkil al-Wasīṭ li-Abī ʿAmr ʿUthmān ibn al-Ṣalāḥ, Sharḥ mushkilāt al-Wasīṭ li-Muwaffaq al-Dīn Hamza ibn Yūsuf al-Hamawī. Taʿlīqa mūjaza ʿalā 'l-Wasīṭ li-Ibrāhīm ibn ʿAbdallāh ibn Abī 'l-Dam*, ed. A.M. Ibrāhīm, 7 vols. (Cairo, 1997).

[11] Al-Nawawī, *Rawḍat al-ṭālibīn wa-ʿumdat al-muftīn*, ed. ʿA.ʿU. al-Bārādī (Beirut, 1995).

[12] Al-Rāfiʿī, *Fatḥ al-ʿazīz fī sharḥ al-wajīz*, ms. Yale University, Beinecke Memorial Library, Landberg 639. Cf. *GAL* i, pp. 393 and 424, no. 50, *S* i, p. 753.

eminent scholars in Islam. He informs his reader that both al-Nawawī and Ibn al-Ṣalāḥ showed a keen interest in this question.[13] Both spellings of al-Ghazālī's name have arguments in their favor. That al-Ghazālī's father was a spinner of wool is reported on the authority of the historian of Baghdad Ibn al-Najjār (d. 643/1245), whose work is not fully available to us.[14] Al-Dhahabī cites a passage from ʿAbd al-Ghāfir al-Fārisī's (d. 529/1134) book on the biographies of the scholars of Nishapur, a book of which we also do not have a full copy. ʿAbd al-Ghāfir is quoted as saying that in the usage of the people of Ṭūs words like *ghazzālī*, *ʿaṭṭārī*, or *khabbāzī* refer to the member of the profession of spinners, perfumer, or bakers.[15] ʿAbd al-Ghāfir al-Fārisī was a well-known authority on al-Ghazālī's life. He was his earliest biographer who interviewed the great scholar in person. Yet, he did not comment specifically on al-Ghazālī's name and al-Dhahabī's quote seems to be a general observation and not made in the context of explaining al-Ghazālī's name.[16]

Al-Dhahabī also brings strong evidence for the second spelling, the one with one *zāy*. He claims that he saw a text by al-Nawawī where he reports on the authority of a chain of five transmitters that a student of al-Ghazālī, namely Tāj al-Islām ibn Khamīs (d. 552/1157) of Mosul,[17] heard him saying:

[13] Al-Dhahabī, *Siyar aʿlām al-nubalāʾ*, ed. Sh. al-Arnaʾūt, 25 vols. (Beirut, 1981–88), xix, p. 343, l. 13. The *tarjama* on al-Ghazālī (ibid., xix, pp. 322–46) is slightly different from to the one in al-Dhahabī's *Taʾrīkh al-Islām wa-wafayāt al-mashāhīr wa'l-aʿlām*, ed. ʿU.ʿA. Tadmurī (Beirut, 1407–/1987–), vol. 501–520 A.H., pp. 115–29.

[14] Al-Dhahabī, *Siyar aʿlām al-nubalāʾ*, xix, p. 335, l. 9–10. Ibn al-Najjār's *Dhayl ʿalā Taʾrīkh Baghdād* is lost. The excerpts of this book by al-Dimyāṭī, *al-Mustafād min Dhayl Taʾrīkh Baghdād*, ed. Q. Abū Farah (Haydarabad, 1399/1979), p. 37f, do not mention this point.

[15] Al-Dhahabī, *Siyar aʿlām al-nubalāʾ*, xix, p. 343, l. 8–9 (*Taʾrīkh al-Islām*, p. 126, l. 6–7). Only one of two parts of ʿAbd al-Ghāfir al-Fārisī's *al-Siyāq li-Taʾrīkh Nīsābūr* is preserved. It is edited as a facsimile in R. Frye, *The Histories of Nishapur* (London, 1965), text 2. An excerpt of the whole work exists in al-Ṣarīfīnī (d. 541/1243), *al-Muntakhab min al-Siyāq li-Taʾrīkh Nīsābūr*, ed. M.K. al-Maḥmūdī (Qum, 1362 [1983]).

[16] ʿAbd al-Ghafīr al-Fārisī's *tarjama* on al-Ghazālī is extensively quoted in al-Subkī, *Ṭabaqāt al-Shāfiʿiyya al-kubrā*, ed. M.M. al-Ṭanāḥī and ʿA.M. al-Ḥilw, 10 vols. (Cairo, 1964–76), vi, p. 204, l. 6–p. 214, l. 3 and Ibn ʿAsākir, *Tabyīn kadhib al-muftarī fī-mā ansaba al-Imām Abī 'l-Ḥasan al-Ashʿarī* (Damascus, 1347/1928), p. 291, l. 15–p. 296, l. 17. Obviously, these quotations might not be the complete *tarjama*.

[17] Al-Ḥusayn ibn Naṣr ibn Khamīs al-Juhānī; al-Subkī, *Ṭabaqāt*, vii, p. 81; *GAL S* i, p. 776.

> The people call me 'al-Ghazzālī', but I am not 'al-Ghazzālī'. I am rather al-Ghazālī, related to a village that is called Ghazāla.[18]

Another important authority on the life of al-Ghazālī was the Khorasanian historian al-Samʿānī (d. 562/1166), who studied with many of the great scholar's students and asked them about his life.[19] Al-Samʿānī is often quoted as a source for the information that people in 'Khwārazm and Jurjān', i.e. in the northern province bordering on Khorasan, simply add a *yāʾ* to Arab words for professions such as *ghazzāl*, 'spinner'. Yet al-Samʿānī's comments on al-Ghazālī's life are lost and can only be reconstructed from quotations. One such quotation appears in al-Murtaḍā al-Zabīdī's (d. 1205/1791) commentary on al-Ghazālī's *Iḥyāʾ ʿulūm al-dīn*, which was written in the second half of the eighteenth century in Cairo. Al-Murtaḍā al-Zabīdī says that regarding the village of Ghazāla, al-Samʿānī wrote in one of his works:

> I asked the people of Ṭūs about this village and they did not know anything about it.[20]

But the association with this place found new fodder during the Mamlūk period from one of al-Ghazālī's descendents. The lexicographer al-Fayyūmī (d. after 770/1368), who was born in Egypt but who was active in Syria, reports that in the year 710/1310–11 he met someone in Baghdad who could credibly trace his lineage to one of al-Ghazālī's daughters and who informed him:

> The people are wrong when they pronounce the name of our forefather with a geminization of the *zāy*. The name is just with one *zāy*. It is in relation to Ghazāla, one of the villages of Ṭūs.[21]

What all this tells us is that by the time historians became interested in al-Ghazālī's family name, it was simply too late to get conclusive information about it. None of the two conflicting explanations of the name of al-Ghazālī was handed down with enough certainty to settle

---

18 Al-Dhahabī, *Siyar aʿlām al-nubalāʾ*, xix, p. 343, l. 17.

19 On al-Samʿānī's works see Rudolf Sellheim's article in *EI*[2] and *GAL* i, p. 329f, *S* i, p. 564f. On his position among the scholars of Khorasan see H. Halm, *Die Ausbreitung der šāfiʿitischen Rechtsschule von den Anfängen bis zum 8./14. Jahrhundert* (Wiesbaden, 1974), pp. 84–6.

20 Al-Murtaḍā al-Zabīdī, *Itḥāf*, i, p. 18, l. 16.

21 Al-Fayyūmī, *Miṣbāh al-munīr fī gharīb al-Sharḥ al-kabīr*, ed. ʿA. al-Shannawī (Cairo, 1977), p. 447 (sub gh-z-l). The work is a dictionary of difficult words that appear in al-Rāfiʿī's commentary to al-Ghazālī's *al-Wajīz*.

the matter. The second explanation was reported on a single authority, a *khabar al-wāḥid*, of a student. A clear identification of the village Ghazāla was impossible. That could not trump the equally plausible explanation via the profession of al-Ghazālī's father, which seemed to have had the backing of ʿAbd al-Ghāfir al-Fārisī, al-Ghazālī's most authoritative biographer. The experienced historian al-Ṣafadī (d. 766/1363), a teacher and colleague of those who took sides in this dispute, admitted that the matter cannot be decided and acknowledged that God knows best which of the two spellings is right.[22] Almost a century earlier, Ibn Khallikān (d. 681/1282) had already taken the same position.[23]

AL-GHAZĀLĪ 'THE ELDER'

Yet there is more historical evidence than what al-Dhahabī and al-Ṣafadī discuss. This appears in the work of their student Tāj al-Dīn al-Subkī (d. 771/1370). Another battleground of the dispute about al-Ghazālī's name—and thus his family background—was the issue of whether there existed a prominent scholar by that name in a generation before the great theologian. Abū Isḥāq al-Shīrāzī (d. 476/1083) of Baghdad, who was the most influential Shāfiʿī scholar during the middle of the fifth/eleventh century, includes the name 'al-Ghazālī' in a list of great members of his school who had taught in Khorasan and Transoxania.[24] Al-Shīrāzī wrote this at a time when our al-Ghazālī was either yet unborn or still in the early years of his education. The list contains only scholars who at the time of writing were already deceased and it does not mention, for instance, Abū 'l-Maʿālī al-Juwaynī (d. 478/1085), who was the most famous Shāfiʿī teacher in Nishapur during this time.[25] Rather, the name 'al-Ghazālī' is followed by that of Abū Muḥammad al-Juwaynī (d. 438/1047),[26] the father of the great theologian, which suggests that this al-Ghazālī lived somewhat around his time or even earlier.

---

[22] Al-Ṣafadī, *al-Wāfī bi'l-wafayāt*, ed. H. Ritter et al. (Istanbul, 1931–), i, p. 277, l. 15.
[23] Ibn Khallikān, *Wafayāt al-aʿyān*, i, p. 98, l. 16.
[24] Al-Shīrāzī, *Ṭabaqāt al-fuqahāʾ*, ed. I. ʿAbbās (Beirut, 1970), p. 133; Halm, *Ausbreitung*, p. 94.
[25] Al-Shīrāzī met Abū 'l-Maʿālī al-Juwaynī when he traveled to Nishapur in a diplomatic mission of the caliph shortly before his death.
[26] *GAL* i, p. 385, *S* i, p. 667; al-Subkī, *Ṭabaqāt*, v, pp. 73–93.

Al-Ghazālī 'the elder' is mentioned a second time by the historian of Nishapur ʿAbd al-Ghāfir al-Fārisī. He reports that an elder 'Abū Ḥāmid al-Ghazālī' was a teacher of Abū ʿAlī al-Fāramadhī (d. 477/1085), who taught the younger al-Ghazālī in Sufism.[27] A generation later this information is repeated by al-Samʿānī, who gives the elder al-Ghazālī's full name: Abū Ḥāmid Muḥammad ibn Aḥmad al-Ghazālī.[28] We hear nothing about this scholar until the seventh/thirteenth century when Ibn al-Ṣalāḥ writes about the elder al-Ghazālī, whom he names Abū Ḥāmid Aḥmad ibn Muḥammad. He says, he died in Ṭābarān, al-Ghazālī's hometown, in 435/1043–44.[29] Ibn al-Ṣalāḥ received his information from the relatively little known Khorasanian historian Abū Ḥafṣ al-Muṭṭawiʿī, who wrote in the middle of the fifth/eleventh century and whose works on this subject are lost.[30]

Apparently, Ibn al-Ṣalāḥ's report triggered a debate among the Damascene historians in the generations after him. Did the elder al-Ghazālī really exist or has the great theologian's name simply cast a shadow on the historiography of the Shāfiʿite school before him? Ibn al-Ṣalāḥ suggested that the elder al-Ghazālī was the theologian's paternal uncle. One of the greatest authorities of Muslim historiography, al-Dhahabī, decided that there was not enough evidence in favor of the elder al-Ghazālī's existence. Al-Dhahabī was unaware of al-Samʿānī's report and believed that the only testimony he had, namely the appearance of the name 'al-Ghazālī' in al-Shīrāzī's list, was an anachronism triggered by a scribe's mistake in a manuscript.[31] For al-Dhahabī, 'the uncle of al-Ghazālī' was no famous scholar. Al-Dhahabī's student al-Subkī reports a dispute with his teacher about the existence of al-Ghazālī 'the elder'. Their discussion gives an interesting glimpse into the workshop tools

[27] Al-Faḍl ibn Muḥammad al-Fāramadhī; on him see al-Ṣarīfīnī, *al-Muntakhab min al-Siyāq*, p. 629 = Frye, *The Histories of Nishapur*, text 3, fol. 121b; al-Subkī, *Ṭabaqāt*, v, pp. 304–6; Halm, *Ausbreitung*, p. 94. Fāramadh is one of the villages of Ṭūs.

[28] Al-Samʿānī, *Kitāb al-ansāb*, 13 vols. (Haydarabad, 1382–1402/1962–83), x, p. 125.

[29] Al-Isnawī (d. 772/1370), *Ṭabaqāt al-Shāfiʿiyya*, ed. ʿA. al-Jabūrī, 2 vols. (Baghdad, 1390–91 [1970–71]), ii, p. 246f, reports this information on the authority of Ibn al-Ṣalāḥ's lost book *Fawāʾid riḥlatihi* (which might be identical to his *al-Muntakhab min al-Mudhhab*, mentioned below in note 32).

[30] Regarding al-Muṭṭawiʿī see below note 32. A text from the margins of one ms. of al-Subkī, *Ṭabaqāt*, iv, p. 90, note 1 says that Ẓāhir al-Dīn ibn Funduq al-Bayhaqī (d. 565/1169–70) repeated al-Muṭṭawiʿī's information on al-Ghazālī the elder and thus provided an additional channel of transmission.

[31] Al-Subkī, *Ṭabaqāt*, iv, p. 88, l. 4f.

of Arab historians. This is how al-Subkī's exchange with his teacher is quoted by al-Murtaḍā al-Zabīdī:

> [During one of al-Dhahabī's teaching sessions] the subject of this man was brought up and most people were ignorant about him. I asked our master al-Dhahabī about him since I had read about him in the *Ṭabaqāt* of the master Abū Isḥāq al-Shīrāzī where he mentions the preceding masters. Al-Dhahabī said: 'This is an addition of a copyist and we do not know people by the name of Ghazālī other than Ḥujjat al-Islām and his brother. It is most farfetched that there was another [with that name].' I told him: 'There is conclusive evidence (*dalīl qāṭiʿ*) that al-Shīrāzī did not mean Ḥujjat al-Islām.' 'What is it?' he asked. 'Al-Shīrāzī's statement that their date of death was unknown to him. That is evidence (*dalīl*) for the fact that he did not mean Ḥujjat al-Islām because [these scholars were already dead when al-Shīrāzī wrote] and he [Ḥujjat al-Islām] lived after the death of al-Shīrāzī.' Al-Dhahabī said: 'That is correct.'
>
> I mentioned this exchange to my father [the historian Ṭaqī 'l-Dīn al-Subkī] but he tended to agree with al-Dhahabī on this matter until I found [a passage] in al-Samʿānī's *Kitāb al-ansāb* where he writes about the ascetic Abū ʿAlī al-Fāramadhī and mentions that he studied with Abū Ḥāmid al-Ghazālī the elder. Then I read in the book of al-Muṭṭawiʿī[32] about the teachers of Abū ʿAlī al-Fāramadhī and found he mentions this Abū Ḥāmid who is described as someone who preceded Abū ʿAlī. There it said that [this al-Ghazālī] had a son by the name of Aḥmad and his *kunya* was [also] Abū Ḥāmid. He surpassed his father in knowledge. I understood that he was a relative of Ḥujjat al-Islām, he was the paternal uncle of his father or the brother of his grandfather. [Finally], Jamāl al-Dīn Muḥammad ibn Muḥammad al-Jamālī told me that such a grave is well-known in the graveyard of Ṭūs and that they call him al-Ghazālī 'the elder' and believe God will respond to prayers issued at this grave.[33]

Al-Subkī found the evidence in al-Samʿānī and Ibn al-Ṣalāḥ's source al-Muṭṭawiʿī only after he had his discussion with al-Dhahabī. Subsequently

---

[32] In his *Ṭabaqāt*, iv, p. 89, l. 14–15, al-Subkī clarifies that this is the history of Shāfiʿī scholars *al-Mudhhab fī dhikr shuyūkh al-madhhab* by Abū Ḥafṣ ʿUmar ibn ʿAlī ibn Muḥammad al-Muṭṭawiʿī (d. *c.* 440/1048) of Nishapur which he read in the abbreviation of Ibn al-Ṣalāḥ, *al-Muntakhab min al-Mudhhab*. Both works are lost and neither al-Muṭṭawiʿī nor Ibn al-Ṣalaḥ's abbreviation are mentioned in *GAL*. On this author and Ibn al-Ṣalāḥ's abbreviation cf. the editor's introduction to al-Muṭṭawiʿī, *Darj al-ghurar wa-darj al-durar*, ed. Kh. al-ʿAṭiyya (Beirut, 1406/1986), pp. 5–8.

[33] Al-Murtaḍā al-Zabīdī, *Itḥāf*, i, p. 18. ult.–p. 19, l. 9. In his *Ṭabaqāt al-Shāfiʿiyya al-kubrā*, iv, pp. 87–90, al-Subkī offers a more detailed discussions of all the motives and arguments in this dispute. Al-Murtaḍā al-Zabīdī's text seems to come from one of al-Subkī's own abbreviations of that book, the *Ṭabaqāt al-Shāfiʿiyya al-wusṭā* or the *Ṭabaqāt al-Shāfiʿiyya al-ṣughrā*, which are both not edited.

al-Subkī came to a different conclusion than his teacher. He decided that the elder al-Ghazālī did, in fact, exist.

Al-Subkī leaves open how this judgment is to be reconciled with his verdict, put down at a different place in his dictionary, that al-Ghazālī's father was a spinner of wool.[34] If the name 'al-Ghazālī' or 'al-Ghazzālī' was inherited through the generations, it does not convey any information about the great scholar's immediate family and their occupation. The information given by al-Shīrāzī, al-Muṭṭawiʿī, ʿAbd al-Ghāfir al-Fārisī and al-Samʿānī paints a relatively consistent picture of an 'elder' al-Ghazālī.[35] He was a Shāfiʿite jurist of Ṭūs, who taught two or three generations before al-Ghazālī during the second quarter of the fifth/eleventh century and who died in 435/1043–44 about fifteen years before the great theologian was born. Although there is a dispute about his full name, both reported versions make him a member of al-Ghazālī's family. He might have been, for instance, a cousin of al-Ghazālī's paternal grandfather or his brother. There was an educational connection between the two al-Ghazālīs as the elder was a teacher of two of the younger's instructors. The *faqīh* Aḥmad al-Rādhakānī and the Sufi Abū ʿAlī al-Fāramadhī, who were both among the students of al-Ghazālī 'the elder', were also teachers of the great theologian.[36] Fame in scholarship and the high social stature that comes with it was probably not unknown in al-Ghazālī's family.

Like many other Muslim historians before him, al-Sharqawī, who wrote in the nineteenth century, assumed that al-Ghazālī had a humble family background. Al-Sharqawī implies that because al-Ghazālī may have felt ashamed of the profession of his father, a poor spinner of wool, he chose to associate himself to a village by the name of Ghazāla.[37] Such dissociation from his father would also explain the lack of infor-

---

[34] Al-Subkī, *Ṭabaqāt*, vi, p. 193, l. 10.

[35] In addition, al-Isnawī, *Ṭabaqāt*, ii, p. 246 claims that the elder al-Ghazālī is mentioned by Muḥammad ibn Aḥmad al-ʿAbbādī (d. 458/1066) of Herat in his *Ṭabaqāt al-fuqahāʾ al-Shāfiʿiyya. Das Klassenbuch der Gelehrten Šāfiʿiten*. ed. G. Vitestam (Leiden, 1964). Al-ʿAbbādī would be a contemporary of the elder al-Ghazālī. In Vitestam's criticial edition of this book, however, 'al-Ghazālī' does not appear.

[36] Al-Subkī, *Ṭabaqāt*, iv, p. 91, v, p. 305, l. 6. On al-Rādhakānī see al-Subkī, *Ṭabaqāt*, v, p. 204, l. 9; al-Ṣarīfīnī, *al-Muntakhab min al-Siyāq*, p. 83 = Frye, *Histories*, text 3, fol. 20a. I am currently preparing a more detailed study of al-Ghazālī's teachers and the evidence about his family. This will be part of a forthcoming monograph on the great theologian.

[37] *GAL S* i, p. 744 translates this motive into Western literature: 'Die Aussprache mit [einem] z sollte die Erinnerung an die wenig vornehme Herkunft verwischen (...).'

mation on him, as al-Ghazālī apparently avoided to write about him or his profession in his autobiography *al-Munqidh min al-ḍalāl*.[38] This explanation, which appears in numerous biographies of al-Ghazālī, has many problems and the lack of conclusive evidence for the profession of al-Ghazālī's father is not the least of it. It is, in fact, unclear why al-Ghazālī would be ashamed of his father's profession as a craftsman. Diligent and honest work is one of the backbones of al-Ghazālī's ethics for the un-educated people (*al-ʿawāmm*). Equally, he often stresses the transparency of the Seljuq society that allows those who are able, to become members of the intellectual elite. Curiosity and doubt are the main vehicles to reach the level of the educated elite (*khawāṣṣ*). A poor and uneducated family background would tally well with al-Ghazālī's grand narrative in the *Munqidh min al-ḍalāl*, where such a fact is, however, not mentioned.[39]

Conclusion

Al-Ghazālī or al-Ghazzālī was most probably a *nisba* that was used by members of this family from Ṭābarān in Ṭūs for at least three or four generations before our scholar. If that is the case, there is no reason to assume that our scholar's father had anything to do with the spinning of wool. Maybe an earlier member of the family had. It is impossible to determine what the *nisba* initially referred to. Maybe our al-Ghazālī himself did not know its origin. There seems to be some indication that the family itself preferred 'al-Ghazālī' while the double-z seems an attempt of philologists to make sense out of a name whose origin is shrouded in mystery. The most talented Arab historians of this period—Ibn Khallikān, al-Dhahabī, and al-Ṣafadī—understood that the matter could not be settled and remained uncommitted: *Allāhu aʿlam*.

Watt's application of the *difficilior lectio potius*-principle includes a commitment towards one explanation, namely the one that he did not adopt. Watt assumed that the explanation via the word *ghazzāl* is more plausible than the one via the village Ghazāla, which should make us adopt the latter. Yet, in light of the discussions of the Muslim historians, I remain unconvinced that any of the two alternatives is more

[38] Al-Ghazālī, *al-Munqidh min al-ḍalāl/Erreur et deliverance*, ed. and trans. F. Jabre (Beirut, [3]1969).

[39] Cf. Watt, *Muslim Intellectual*, p. 182.

likely than its opposite. Still, like Watt, I would argue that we should use al-Ghazālī with one z.

Adapting the spelling 'al-Ghazālī' should not be viewed as an indication that one accepts the explanation via a place called 'Ghazāla' any more than the one that associates it to the occupation of a wool-spinner. As long as there is no more conclusive evidence about the name's origin this is simply a case for Ockham's razor: given that both spellings are equally possible, we should remain uncommitted and use the spelling that saves effort, ink, and paper: *al-Ghazālī.*

AVERROES' UNDERSTANDING OF THE PHILOSOPHER'S ROLE IN SOCIETY[1]

Alfred L. Ivry

One may say of Averroes, as one might of any committed philosopher, that he viewed his responsibilities as a philosopher to be those that brought him to the truth. 'The truth,' *al-ḥaqq*, is also the name—one of the names—of the True One, God, in Arabic, and the connection between these two dimensions of the truth was evident to Muslim philosophers from the ninth-century Yaʿqūb ibn Isḥāq al-Kindī, the 'philosopher of the Arabs,' on.[2]

The pursuit of the truth along philosophical lines thus had a religious dimension for all Muslim philosophers, however divorced it was from more conventional religious practices, as it was for Averroes. Philosophy had for him the earmarks of a spiritual calling, and he gave himself to it as fully as circumstances permitted. He was, after all, actively engaged and an authority as well in both the judicial and medical professions, disciplines that brought him into contact with other persons and with society at large.

In these capacities, Averroes may be said to have fulfilled his civic obligations. His philosophical activity, on the other hand, was more of a private affair, limited to those relatively few in society who were prepared for it. It is true that Averroes' commentaries on Aristotle's corpus were commissioned by the Almohad caliph Abū Yaʿqūb Yūsuf (reigned 1163–84) in order allegedly to help him understand Aristotle's work,[3] and in that sense Averroes may be seen as having rendered a public service qua philosopher. Yet the public for which the commentaries were written was a very select one, and one to which Averroes barely

[1] This is a revised version of an article that appeared originally in Arabic, in the proceedings of an international symposium titled 'Actualité d'Averroes,' ed. M.A. Mansiyya (Tunis, 1999), i, pp. 319–32.

[2] See al-Kindī, *Rasāʾil al-Kindī al-falsafiyya*, ed. M. Abū Rīda (Cairo, 1950), i, pp. 160–62, and cf. the discussion of al-Kindī's usage of *al-wāḥid al-ḥaqq*, 'the true One,' in the introduction to A. Ivry's translation, *Al-Kindi's Metaphysics* (Albany, 1974), pp. 15–17.

[3] Cf. ʿAbd al-Wāḥid al-Marrākushī, *Al-Muʿjib fī talkhīṣ akhbār al-Maghrib*, ed. R. Dozy (Leiden, 1847), pp. 174–5, and see L. Gauthier, *Ibn Thofail, sa vie, ses œuvres* (Paris, 1909), pp. 15–17.

condescended. Nor would these commentaries, not even the politically oriented commentary on Plato's *Republic*, which served in lieu of Aristotle's *Politics*, have assisted the caliph discharge his responsibilities towards his subjects.

Each of the three types of commentaries—the Epitome or Short (*jawāmiʿ*), Middle (*talkhīṣ*) and Long (*sharḥ*)—which Averroes composed (and he wrote thirty eight commentaries in all) assumes some philosophical sophistication in the reader. The five Long Commentaries—to the *Posterior Analytics, Physics, De caelo, De anima* and *Metaphysics*,—are very detailed and demanding, while the more numerous Short Commentaries, or Epitomes, often range beyond Aristotle's text to survey the subject at hand in an abbreviated but eminently non-elementary manner.[4]

It is only Averroes' Middle Commentaries that make an attempt to clarify Aristotle's text for the non-cognoscenti, through paraphrase and relatively brief comment and explanation. The Middle Commentaries also render Aristotle's text more accessible to the Muslim reader in various subtle ways, as I have attempted to show elsewhere.[5] These Middle Commentaries are called *talākhīṣ* in Arabic, the same term we are told the Almohad caliph had used in requesting Averroes' assistance. These *talākhīṣ* may be seen, therefore, as evidence of Averroes' having fulfilled the responsibilities of a philosopher, as required by his society.

That society, though, was the privileged one of the court and of other philosophers. In his independent works, such as the *Tahāfut al-Tahāfut* ('The Incoherence of the Incoherence') and the *Faṣl al-maqāl* (paraphrased as 'On the Harmony of Religion and Philosophy'), Averroes engages al-Ghazālī and the theologians or *mutakallimūn* of Islam whom al-Ghazālī represents, but only to distinguish himself from them. Among their many errors, they make the mistake, he believes, of reaching out to the masses in order to refine their beliefs, i.e., to disabuse them of

[4] Cf. the inventory of Averroes' commentaries compiled by Harry A. Wolfson in his 'Plan for the Publication of a Corpus Commentariorum Averroes in Aristotelem,' reprinted in Wolfson's *Studies in the History of Philosophy and Religion*, ed. I. Twersky and G.H. Williams (Cambridge, Mass., 1973), i, pp. 433–40. See now the description of these commentaries given by Jamāl al-Dīn al-ʿAlawī, *Al-Matn al-Rushdī. Madkhal li-qirāʾa jadīda* (Casablanca, 1986), pp. 14–51. For the current state of publications of this corpus, cf. the introduction to J. Puig's Spanish translation of Averroes' *Epitome De Fisica* (Madrid, 1987), pp. 20–24.

[5] Cf. A.L. Ivry, 'Averroes' Middle and Long Commentaries on De Anima,' *ArScPhil* 5 (1995), pp. 84–6.

the anthropomorphic and polytheistic entailments of their convictions, based on a literal understanding of the Qurʾān.[6]

Averroes would agree, of course, with the purpose of the theologians' work, but believes their effort is fundamentally unsound. The masses, for him, are not prepared for the hermeneutical sophistication of the *mutalkallimūn*, as expressed in their allegorical interpretations of God's book.[7] *Kalām* treatises should not undermine the literal understanding of the Qurʾān and popular formulations of the faith. Such works are relatively comprehensible to simple readers or semi-educated auditors, their dialectical form of argument relatively comprehensible (if of limited value, in Averroes' opinion).

It is only the philosophers who are entitled to use allegorical techniques to explain the faith, Averroes asserts. The demonstrative nature of their argument establishes the credibility of their allegorical readings and offers strong grounds for accepting their interpretation as correct.[8] Moreover, we are to understand, the rigorous syllogistic style of demonstrative argument serves to discourage the average, untrained reader or auditor from paying attention to the philosophers' remarks.[9]

For Averroes, then, the philosopher and only the philosopher may use, and in fact must use, allegory whenever reason/philosophy conflicts with belief/Scripture (since the public is not listening to him anyway?). Only thus may truth be spoken out loud. The philosopher cannot enlighten the masses and is thus not responsible for bringing them the truth.

Consequently, Averroes does not demand that Muslims accept philosophical formulations of the faith, as many *mutakallimūn* had required,

---

[6] Cf. Averroës, *Kitāb faṣl al-maqāl wa-taqrīr mā bayna 'l-sharīʿa wa'l-ḥikma min al-ittiṣāl*, ed. G. Hourani (Leiden, 1959), pp. 24–5, and Hourani's translation, *Averroes on the Harmony of Religion and Philosophy* (London, 1967), p. 61. See too now the Arabic edition and English translation of Charles Butterworth, *Averroës: Decisive Treatise & Epistle Dedicatory* (Provo, Utah, 2001), p. 21.

[7] Averroës, *Faṣl al-maqāl*, pp. 32–4; Hourani, *Harmony*, p. 66; Butterworth, *Decisive Treatise*, p. 26.

[8] Jorge Gracia has argued that even the philosophers cannot claim their interpretations of allegorical symbols are definitively (i.e., necessarily) true, though they can conclusively refute false interpretations. Cf. J. Gracia, 'Interpretation and the Law: Averroes's Contribution to the Hermeneutics of Sacred Texts,' *History of Philosophy Quarterly* 14 (1997), pp. 147–9; see too idem, 'The Philosopher and the Understanding of the Law,' in M. Wahba and M. Abousenna (eds.), *Averroës and the Enlightenment* (Amherst, NY, 1996), pp. 243–51.

[9] As Averroes says in *Faṣl al-maqāl*, p. 27 (Hourani, *Harmony*, p. 61; Butterworth, *Decisive Treatise*, p. 21), allegorical interpretations (*al-taʾwīlāt*) should occur only in demonstrative books, because only people familiar with demonstrative logic (*ahl al-burhān*) will read such works.

and as his contemporary Moses Maimonides wished to impose on his fellow Jews.[10] On the contrary, Averroes is adamantly opposed to acquainting the masses with the truth as philosophers understand it, and with exposing them to the kinds of logical argument and allegorical interpretation which philosophers adopt. He believes such misguided attempts only confuse the masses and weaken their faith; a faith which may anyway, with proper interpretation, correspond in its essentials with the teachings of philosophy.[11]

Averroes is not, however, opposed in principle to inculcating in the public true beliefs and moral practices. As he says in his commentary on Plato's *Republic*, such beliefs ought to be taught on the level that the masses can comprehend, by means of rhetorical and poetic techniques.[12] The public, that is, is to be brought towards the truth through popular forms of imaginative discourse, rather than by rigorous, if only dialectical, argument. It is understood, of course, that the philosopher as philosopher (i.e., when not engaged in political philosophy) is far removed from these techniques, which are often captured by the very ideas that they presume to co-opt.

In endorsing this approach, Averroes in effect is passing favorable judgment on the Qur'ān and the classical oral traditions (*ḥadīth*) of Islam.[13] Averroes sees this canonical literature as employing these means to establish the beliefs and virtues fundamental to Islamic society, and these beliefs must remain intact in their original formulations, for the public good.

It is the theologians, not the philosophers, Averroes contends, who have been irresponsible towards society in their attempt to use allegory in support of the faith. They are also to be faulted for accepting non-apodictic forms of reasoning which Averroes (and all other philosophers) regarded as inferior. Moreover, the *mutakallimūn* worked from premises that he would have called (today) counter-intuitive. For *kalām*, Occa-

---

[10] Cf. Maimonides' thirteen principles of faith in his *Mishnah Commentary* to *Sanhedrin*, ch. 10 (*ḥeleq*); found in English translation in I. Twersky (ed.), *A Maimonides Reader* (New York, 1972), pp. 417–23.

[11] This is the inverse of Averroes' oft-repeated teaching that the essential teachings of philosophy agree with those of Islam. As he says in *Faṣl al-maqāl*, p. 13 (*Harmony*, pp. 50/9; all as in n. 6), 'truth does not oppose truth' (*al-ḥaqq lā yuḍādd al-ḥaqq*).

[12] Cf. *Averroes on Plato's Republic*, trans. R. Lerner (Ithaca, 1974), p. 10.

[13] Ibid., pp. 12 and 48; see too Averroes, *Faṣl al-maqāl*, pp. 23–4; Hourani, *Harmony*, p. 59; Butterworth, *Decisive Treatise*, p. 19.

sionalism defied in principle any appeal to natural events and to the principle of causal efficacy that Averroes held essential.[14]

The scientific approach which Aristotelian philosophy proclaimed and to which Averroes was pledged was, he knew, foreign to the vast majority of his fellow men, and therefore Averroes felt no compulsion to acquaint them with it. Though he responded to the challenges to philosophy from its critics, and to what he regarded as distortions of the philosophical heritage from within, he would, I am sure, have preferred to avoid public debate with non-philosophers. Such debate could only let the allegorical cat out of the bag and expose the philosophers' exclusive, elitist natures. As a philosopher, therefore, Averroes would have preferred not to be a public figure at all. His polemical defense of the religious obligation to do philosophy, as given in his *Faṣl al-maqāl* and towards the end of his *Tahāfut al-Tahāfut*, is not particularly impressive philosophically.[15]

Philosophy attracted Averroes as a form of personal gratification, in which a person undergoes a sort of apotheosis. *Iʿraf dhātaka, taʿrif khāliqaka*, 'know thyself, know thy Creator,' he says in the *Epitome to De anima*, averting to a popular maxim current as well in philosophical circles.[16]

'Personal gratification' is better understood, in medieval terms, as the beatitude or felicity (*saʿāda*) consequent upon conjunction (*ittiṣāl*) of the individual intellect with the Agent (or Active) Intellect, *al-ʿaql al-faʿʿāl*.[17] The Agent Intellect is that eternal and hence divine, universal immaterial substance that, it was believed, served as the facilitating

[14] This is a main point of contention between Averroes and al-Ghazālī, as described in Averroes' *Tahāfut al-Tahāfut* and elsewhere. Cf. *Tahafot at-Tahafot*, ed. M. Bouyges (Beirut, 1930), pp. 512–42; English translation by S. van den Bergh, *Averroes' Tahafut al-Tahafut (The Incoherence of the Incoherence)* (London, 1969), i, pp. 312–33. See too the discussion of this controversy in A.L. Ivry, 'Averroes on Causation,' in *Studies in Jewish Religious and Intellectual History* (Alabama, 1979), pp. 143–56.

[15] Cf. now the analysis of Averroes' argument as given by Butterworth, *Decisive Treatise*, pp. xxii–xxx, and see his translation of the closing argument in Averroes' *Tahāfut al-Tahāfut*, given as an appendix, pp. 43–6.

[16] Cf. A.F. al-Ahwānī, *Talkhīṣ (!) Kitāb al-nafs* (Cairo, 1950), p. 93; Spanish translation by S. Gómez Nogales, *La Psicología de Averroes. Commentario al libro sobre el alma de Aristóteles* (Madrid, 1987), p. 218. Cf. the variations of this remark and widespread usage in A. Altmann, 'The Delphic Oracle in Medieval Islam and Judaism,' in *Von der mittelalterlichen zur modernen Aufklärung* (Tübingen, 1987), pp. 1–33.

[17] The doctrine of conjunction is at the heart of medieval theories of cognition, and has been discussed by many authors. Cf. H. Davidson, *Alfarabi, Avicenna, and Averroes, On Intellect. Their Cosmologies, Theories of the Active Intellect, and Theories of the Human Intellect* (New York, 1992), pp. 220–340 in particular.

agent in human intellection.[18] Speculative thought, in this scheme, is not only an epistemic act, it has ontological significance as well, uniting an individual intelligible with its universal paradigm.

The quest for conjunction is thus the quest for truth, and the quest as well for intellectual and hence psychic and existential perfection. Averroes understood that this is the challenge facing the philosopher, and it is the responsibility of the philosopher to meet this challenge.

Ultimately, therefore, the task of the philosopher is a lonely one, occurring between himself and the object of his knowledge, the goal being to unite with it mentally. Each such act brings him (or her) into experiencing, however briefly, the world of eternal truth. This is, for Averroes, our only chance at participating in eternal being, in experiencing immortality of some kind.[19]

This experience is contingent upon possession of universal truth, as articulated in propositions that fall within the province of that domain to which the theoretical intellect alone has access. While in principle every human being may exercise his or her theoretical intellect, in practice very few persons actually do so. It is the rare individual who penetrates beyond the truths of individual sciences to the 'mother of sciences,' metaphysics, there to achieve direct knowledge of and therefore conjunction with the Agent Intellect itself. This is the ultimate perfection to which human beings may aspire, and for which only the philosopher is eligible. It is the philosopher's responsibility to pursue

[18] How the Agent Intellect facilitated conjunction is a difficult question, discussed by Deborah L. Black, 'Conjunction and the Identity of Knower and Known in Averroes,' *American Catholic Philosophical Quarterly* 73 (1999), pp. 171–83. As Davidson has shown, Averroes held diverse opinions on the nature of the Agent Intellect, just as he changed his mind repeatedly on the character of the material intellect. Viewing the Agent Intellect initially as an emanating cause of forms on earth, Averroes moved to a view that distinguished between the formal content of a material object and its intelligibility; the former inherent in its matter, the latter a function of the Agent Intellect. However, as all material objects are potentially intelligible, and something cannot actualize that which it does not itself have in actuality, Averroes must have thought that the Agent Intellect possesses the intelligible essence of all material forms on earth. It is thus a repository of the universal species that order our world. Conjunction is a linking up with this eternal order and a partial identification with it. Cf., however, for a different understanding of the Agent Intellect, one that denies its possession of forms, R. Taylor, 'The Agent Intellect as 'Form for us' and Averroes's Critique of al-Fārābī,' *Topicos* 29 (2005), pp. 29–51; repr. in *Proceedings of the Society for Medieval Logic and Metaphysics* 5 (2005). See http://www.fordham.edu/gsas/phil/klima/SMLM.

[19] Cf. A.L. Ivry, 'Averroes on Intellection and Conjunction,' *JAOS* 86 (1966), pp. 76–85.

this goal, to realize the perfection that defines his species, that of being a 'rational animal,' *zoon logistikon*.

Though it is the ultimate goal of knowledge and conjunction, the Agent Intellect is not absent from the normative cognitive process as well, in Averroes' understanding of its function.[20] It is present at every stage of intellection, involved in practical as well as theoretical issues, helping to formulate our thoughts and to articulate them. Not only the force which activates our mind to think, the Agent Intellect, through its relation to the potential or material intellect, *al-'aql al-hayūlānī*, establishes in human beings the potentiality to think, laying the groundwork for acquiring individual intelligence. The presence of the Agent Intellect is thus ubiquitous, but manifested indirectly, through the activity of individual intellects in their various stages of development.

For Averroes, therefore, every stage of intellectual endeavor has its merit, or as we may call it, its relative perfection, and the philosopher is not successful, and does not experience happiness or felicity only upon reaching ultimate truth and total union with the Agent Intellect itself. Indeed, Averroes believes in the value of partial truth as well, it too enhances and ennobles the character of the person who acquires it.

This is brought out strikingly, if obliquely, in Averroes' discussion of the faculty of taste and the possible role as a medium to that faculty which saliva could offer, in his *Epitome to De anima*. Aristotle thought that the faculty of taste did not have a medium (*De anima* 422a 16), and Averroes was familiar with the various views of his Greek and Muslim predecessors on this issue. In the *Epitome* he vacillates in his views of the function of saliva, concluding partially with Aristotle that taste does not necessarily require a medium.

Averroes is aware, however, that his examination of this issue is incomplete, and he mentions his intention, God willing, to clarify the issue at greater length and thoroughness at another opportunity. An opportunity, incidentally, which apparently never came, as his two later and larger *De anima* commentaries silently testify. Nevertheless, despite Averroes' awareness of the very partial state of his knowledge of this issue, he says that what he has written is sufficient, being the amount of information necessary for human perfection, *al-kamāl al-insānī*. What

[20] Cf. Black, 'Conjunction,' pp. 181–2.

he has written, he avers, will enable the person knowing this to reach the 'top rank of mankind,' *awwalu marātib al-insān*.[21]

As brought out in this rather hyperbolic way, Averroes' notion of *istikmāl*, the *entelecheia* or perfected realization towards which we aspire, is not a remote, practically inaccessible treasure, even if few ever possess it fully. It is rather something that we can all experience, in varying degrees, once we commit ourselves to the philosophical discipline. Complete knowledge of a subject is not therefore necessary for reaping the psychic rewards of philosophical investigation.

Naturally, it is critical for Averroes that one has true knowledge of a subject, and follows the correct path to that knowledge. Such a path for him is well defined, excluding in his opinion Neoplatonic and allegorical formulations of the sort Avicenna propounded. Yet within the philosophical enterprise as conceived, essentially, by Aristotle, Averroes was relatively tolerant of diverse ideas.[22] The very examination of an issue, when correctly formulated, was one that he clearly cherished, the philosophical act itself obviously regarded as virtuous.

This understanding may help account for Averroes' preoccupation with the Aristotelian corpus, as well as for his paraphrase of Plato's *Republic*. These works were written not for the caliph only, but I believe also, and especially, for himself, his students, and those few philosophical colleagues he may have had. For this audience, Averroes was not obliged to resolve all the issues that he addressed, or attempt to find practical applications for them. Just as he had inherited a rich tradition of theoretical philosophical inquiry, so he was passing it on to those who came after him.[23]

The practical application of this philosophy was indeed something he could do little about, at most he could write a theoretical treatise about such practical topics as ethics and politics.[24] His *Commentary on Plato's Republic* is such a work. In it he informs the caliph of Plato's views on how to build and defend a virtuous, even an ideal, state. The

[21] Al-Ahwānī, *Talkhīṣ*, p. 44; Gómez Nogales, *Psicología*, p. 152.

[22] Averroes, *Faṣl al-maqāl*, p. 11; Hourani, *Harmony*, p. 48; Butterworth, *Decisive Treatise*, p. 6.

[23] Cf. Averroes' remarks in his Long Commentary to *De anima*, *Averrois Cordubensis Commentarium Magnum in Aristoteles De Anima Libros*, ed. F. Stuart Crawford (Cambridge, Mass., 1953), p. 399, l. 362–9.

[24] Cf. C. Butterworth, 'Ethics and Classical Islamic Philosophy: A Study of Averroes' Commentary on Plato's *Republic*,' in R.G. Hovannisian (ed.), *Ethics in Islam* (Malibu, Cal., 1985), pp. 17–45.

Platonic practices which Averroes does not disavow in this work, such as egalitarianism between the sexes, communal marriages and property, could never have been taken seriously by the caliph or anyone else in a position of political responsibility, as Averroes had to know.[25] He would have viewed these positions as utopian prescriptions, representing an ideal state that Averroes knew was unrealistic.

Nor should Averroes' apparent advocacy in this commentary of coercive, aggressive warfare, in contrast to Plato's approval of defensive warfare only, be seen as necessarily conveying a realistic or full view of Averroes' views on this subject. The *jihād* that Averroes espouses here is in the name of general philosophical virtues, and not of any particular religious or Islamic creed; it is a war on behalf of civilization, as it were, and not for Islam *per se*. It is only by asserting that Islamic law, the *sharīʿa*, agrees with human laws, that Averroes can be seen as giving the caliph a philosophical rationale, should he need it, for an expansionist foreign policy.[26] As we know from his legal compendium on this subject, Averroes' full views on *jihād* are more nuanced, and prudent, than appear in this commentary, which appears politically—or is it intentionally?—naive.[27]

The *Commentary on Plato's Republic* thus does little to change our impression of Averroes' sense of the responsibility of the philosopher. He offers his views on political philosophy, as on everything else, from a theoretical perspective that only tangentially addresses the practical needs and concerns of his society. The philosopher's life is the life of the mind, and the pursuit of the ideal, and the philosophers' obligation to society is to be true to this vocation.[28]

Still, Averroes was aware that as a member of his society, the philosopher must conform to its mores and rituals, performing the deeds that society deems virtuous. In the closing chapters of his *Epistle on the Possibility of Conjunction with the Active Intellect*, Averroes addresses this

---

[25] *Averroes on Plato's Republic*, trans. Lerner, pp. 57–66.

[26] Ibid., pp. 11–21, and see Plato, *Republic* v: 469–71.

[27] Cf. Averroes' *Bidāyat al-mujtahid* (Cairo, 1966), i, pp. 390–400, particularly p. 398; *Jihad in Medieval and Modern Islam*, trans. R. Peters (Leiden, 1977), p. 22. See too R. Brunschvig, 'Averroès Juriste,' in *Études d'Orientalisme dédiées à la Mèmoire de Lévi-Provençal* (Paris, 1962), i, p. 67; A.L. Ivry, 'The Toleration of Ethics and the Ethics of Tolerance in Judaism and Islam,' in W. Brinner and S. Ricks (eds.), *Studies in Islamic and Judaic Traditions* (Atlanta, 1986), p. 174.

[28] Cf. *The Epistle on the Possibility of Conjunction with the Active Intellect by Ibn Rushd with the Commentary of Moses Narboni*, Hebrew ed. and English trans. K. Bland (New York, 1982), p. 108.

issue and endorses various ascetic practices and ritual performances peculiar to Islam. Yet he recognizes them to be essentially personal, a-social acts, the attainment of felicity remaining thus a 'regimen of the solitary' (*hanhagat hamityaḥed*).[29]

Political science demands, Averroes knows, that a person needs to be part of his society to achieve perfection, or felicity. Theoretically, therefore, he accepts Plato's view of the philosopher's responsibility towards involving himself in society; but on condition that society itself should not set up impediments towards the attainment of this felicity. This may have been realized in ancient Greece, Averroes surmises, but it certainly is not the case in his society. The social contract, as it were, does not exist in his time, and thus the philosopher best keep at a distance from society, not spurning it necessarily, but not investing in it particularly either. His responsibility transcends a particular time and place, as well as any particular people or faith. Given this attitude, Averroes' historical eclipse is not surprising, even as the renascence of interest in him in portions of the Arab world today may prove significant.

[29] Ibid., Hebrew, pp. 137–9 and 146; English, pp. 103–4 and 108.

AL-GHAZĀLĪ'S *MĪZĀN AL-ʿAMAL*: AN ETHICAL SUMMA BASED ON IBN SĪNĀ AND AL-RĀGHIB AL-IṢFAHĀNĪ

Jules Janssens

Thanks to the studies of Madelung[1] and Sherif,[2] since years now the scholarly community knows almost with certainty that al-Ghazālī, in his *Mīzān al-ʿamal*, used Ibn Sīnā's *Najāt* and al-Rāghib al-Iṣfahānī's *Kitāb al-dharīʿa ilā makārim al-sharīʿa*. Both authors offer a general significant outline that at the same time provides a valuable basis for a more detailed analysis. However, one looks in vain for a precise account of the respective influence of these two works. Neither do they state that al-Ghazālī, as usual, not always copies his sources *verbatim*. Al-Ghazālī is never a slavish follower of whatever source he is using, hence important terminological modifications do sometimes appear. However, a difference in wording does not necessarily imply a difference in thought, and even if the latter is the case, the 'new idea' may still have been inspired by a formulation in the source. In what follows, I will try to present an overview of all passages that deserve to be characterized as a 'source' for the *Mīzān*. Of course, a systematic in-depth comparison between all the fragments and their respective sources largely surpasses the limits of the present study since it would clearly require a monograph. I will therefore confine myself to indicating similarities while roughly qualifying their nature.

Before presenting a list of comparison, I would like to stress that the work of Ibn Sīnā that influenced al-Ghazālī in the *Mīzān* is not really *al-Najāt*, but rather *Aḥwāl al-nafs*. Despite the close resemblance of the fragments concerned in both of Ibn Sīnā's works, it is clear that al-Ghazālī's wording is in most respects much closer to the formulation of the *Aḥwāl* than to the one of the *Najāt*. Furthermore, special mention has to be made of the recent discovery of a new ethical tract

---

[1] W. Madelung, 'al-Rāġib al-Iṣfahānī und die Ethik al-Ġazālīs,' in R. Gramlich (ed.), *Islamwissenschaftliche Abhandlungen* (Wiesbaden, 1974), pp. 152–63.

[2] M.A. Sherif, *Ghazali's Theory of Virtue* (Albany, 1975).

of Ibn Sīnā,[3] which turned out to be the direct source of al-Ghazālī's discussion of the virtues and vices. Before this tract became known, it seemed plausible enough that al-Ghazāli had somewhat modified the ethics of al-Rāghib al-Iṣfahānī, or Miskawayh, while remaining faithful to a Neo-Platonic basis.[4] Finally, it is obvious that al-Rāghib al-Iṣfahānī and Ibn Sīnā have many ideas in common and that one of them clearly influenced the other in several respects. Since both thinkers were more or less contemporaneous, it is difficult to decide who influenced whom, at least based on the data that we have actually at our disposal.[5] However, for the present research this issue is devoid of any importance, and hence will not be dealt with.

We have not been able to trace any direct source for the first three chapters of the *Mīzān*.[6] After having underlined that no effort is too great in order to reach a happy life in the hereafter (ch. 1), al-Ghazālī mentions four major concepts regarding life in the hereafter: the common religious belief in bodily resurrection; an intellectual interpretation by some of the *ilāhiyyūn* (divine philosophers), which however leaves room for sensual experiences in an 'imaginary'—or should one say: imaginally?[7]—way; a purely intellectual interpretation (sufis and the majority of the *ilāhiyyūn*); the simple denial of life in the hereafter (ch. 2). Then he stresses that for all serious scholars, whatever their ideological background, both genuine knowledge (*ʿilm*) and action (*ʿamal*) have to be developed in view of the achievement of true happiness (ch. 3).

---

[3] See B. Karliga, 'Un nouveau traité d'éthique d'Ibn Sīnā inconnu jusqu'à nos jours,' in J. Janssens and D. De Smet (eds.), *Avicenna and His Heritage* (Leuven, 2002), pp. 21–35.

[4] See H. Daiber, 'Griechische Ethik in islamischem Gewande. Das Beispiel von Rāġib al-Iṣfahānī,' in B. Mojsisch and O. Pluta (eds.), *Historia philosophiae medii aevi. Studien zur Geschichte der Philosophie des Mittelalters* (Amsterdam, 1991), pp. 181–92. In absence of any knowledge of Ibn Sīnā's tract, Daiber's hypothesis of a reworking of the ethics of Rāghib al-Iṣfahānī or Miskawayh was undoubtedly the best possible.

[5] R. Wisnovsky, *Avicenna's Metaphysics in Context* (London, 2003), pp. 242–3 rightly stresses that in view of the uncertainty of the data of Rāghib al-Iṣfahānī's life, one has to remain 'agnostic' about the fact whether he influenced Ibn Sīnā, or whether the reverse is the case.

[6] There are several editions of this work, see H. Daiber, *Bibliography of Islamic Philosophy*, i (Leiden, 1999), p. 351, n. 3519. To the ones mentioned there, one may add two other: 1. one by M. al-Ṣabrī, ʿA. Maʿrūf and M.Ḥ. Nuʿaymī (Cairo, 1328 A.H.); 2. another by ʿA. Shams al-Dīn (Beirut, 1989). When offering a precise pagination, I refer to this latter edition, but I always indicate the number of the chapters so that the reader may easily identify the passages in other editions.

[7] For the exact meaning of this notion in Ibn Sīnā's system, see J. Michot, *La destinée de l'homme selon Avicenne* (Leuven, 1986), *passim*.

In the very beginnings of chapter 4 (22,1–17), al-Ghazālī, in a manner apparently characteristic of him, emphasises the necessity to avoid blind imitation (*taqlīd*). Thereafter (22,18–23,15), he insists on the importance of self-knowledge by invoking variants of the Delphic maxim. In this passage, he is much indebted to al-Rāghib al-Iṣfahānī's *Kitāb al-dharī'a*, i, 2 (pp. 16,20–17,19).[8] Then (23,18–28,10), a systematic survey follows of the animal and human souls, and their respective faculties. This section, despite changes in the order of presentation, has been inspired by Ibn Sīnā's *Aḥwāl al-nafs*, ch. 2, more specifically the part covering pp. 58,5–65,12.[9] A short presentation of the different degrees one encounters regarding the human intellects, with a particular attention for the highest of them, i.e., the prophetic (28,11–18) shows many affinities with the opening lines of the fourteenth chapter of Ibn Sīnā's work (122,6–123,9). In the final part of the chapter—which, as far as I can see, is proper to al-Ghazālī (28,10–30,10), he formulates a warning against all those who in one way or another divinize the human intellect. Furthermore, he invites his reader to humbly open himself to the divine Light and Revelation.

Chapter 5 firstly (31–32,2) indicates that man constitutes an intermediary level between animal and angel. This section reminds one of al-Rāghib al-Iṣfahānī's *Kitāb al-dharī'a*, i, 6 (23,3–14) although it rather deserves to be characterized as a rewording than as a literal copy. Then (32,3–33,7) the hierarchy and mutual dependence between the diverse faculties of the soul is presented, based on Ibn Sīnā's *Aḥwāl al-nafs*, ch. 2, final part (67,12–68,15). Afterwards (33,8–34,2) a comparison between a few major faculties of the soul and different degrees of responsibility in a kingdom is made. This time, the *Kitāb al-dharī'a* is used again, i.e., i, 4 (20,11–21,8—with only minor modifications). In terms reminiscent of the Qur'ān, al-Ghazālī, in the final part of the chapter (34,3–35,8) insists that the complex nature of the human soul offers strong evidence for the existence of God, the Creator. I looked in vain for a source. Given its outspoken religious tonality, this last section might well constitute a personal addition of al-Ghazālī.

---

[8] I refer to the edition by Ṭ. 'Abd al-Ra'ūf Sa'd (Cairo, 1973). For other editions, see Daiber, *Bibliography of Islamic Philosophy*, i, p. 753, n. 7337. Note that when a specific pagination is given after the mentioning of the chapter, this means that only that part of the chapter is involved.

[9] Ibn Sīnā, *Aḥwāl al-nafs. Risāla fī 'l-nafs wa-baqā'ihā wa-ma'ādihā*, in idem, *Aḥwāl al-nafs li-shaykh al-ra'īs Ibn Sīnā*, ed. A.F. Ahwani (Cairo, 1952), pp. 45–152.

The chapters 6–8 deal with action in its relation to science and the difference between the mystical and the 'scientific' ways of life. They show, that for most people the way of action has to be preferred over that of science although the latter is truly superior. The main stress lies on the purification of the soul, which is essentially described as a mirror that, once polished, reflects the highest realities. Also striking is the strongly emphasized distinction between masses and elite. All such ideas are consistent with al-Ghazālī's general thought and hence, insofar as a source lacks, as I believe is the case, they seem to have been introduced and developed by him.[10]

Chapter 9 (49–50,17) starts with a division of the theoretical and practical sciences inspired by the Aristotelian-Avicennian tradition. Since there are many languages, theoretical science has not so much to concentrate on their particularities, but has to look for the essential knowledge, i.e., the one about God, the angels and the prophets (a *kalāmic* rewording of the object of metaphysics as defined by Ibn Sīnā). Then a tri-division of practical science is proposed, once more in Aristotelian-Avicennian terms: ethics, domestics and politics. It is stressed that the present book first and foremost concentrates on the former of these three practical sciences. The rest of the chapter (50,18–52,16) affirms the necessity to educate the faculties of thought (*fikr*), desire (*shahwa*) and anger (*ghaḍab*), each according to its own rank, and such in order to acquire the all-encompassing virtue of justice. This part almost *verbatim* reproduces *Kitāb al-dharīʿa*, i, 16, (37,15–38,14). It may be worthwhile to note that the idea of justice as the highest of all virtues has of course already been expressed by Plato, in his *Republic*, iv, 443 Cff, and by Aristotle, in his *Nicomachean Ethics*, v, 3.

In chapter 10 (53–6) the relation between reason and the faculties of envy and anger is clarified by three examples: 1. the relation existing between a good counsellor and two servants, i.e., one who supplies livelihood and another who functions as a policeman—in the present case, the soul is compared to a good counsellor (*wālī*) and the body to a kingdom;[11] 2. the relation one finds between a king and his enemies—

[10] The ideas that action and thought are complimentary and that the human soul is in need of purification are certainly present in al-Rāghib al-Iṣfahānī's *Kitāb al-dharīʿa ilā makārim al-sharīʿa*, but the actual wording used by al-Ghazālī in the present chapters seems to have no direct source in that work, unless I have overlooked some similarities.

[11] The comparison between body and kingdom is already present in al-Fārābī, but the formulation of al-Ghazālī has in fact more in common with that of the Ikhwān al-ṣafāʾ, see Daiber, 'Griechische Ethik in islamischem Gewande,' pp. 187–8.

the human being is here characterized as intentionally a macrocosm and extensionally a microcosm;[12] 3. the relation between a horseman and his horse and dog. Regarding this latter idea, one easily detects a Platonic(-Galenic) inspiration.[13] But al-Ghazālī's immediate source is once more the *Kitāb al-dharī'a*, i.e., i, 13 (31,14–33,18), which he quotes with a few minor additions.

Chapter 11 first (57–58,14) distinguishes three possible states in the human being: one where passion predominates; another where there is a permanent struggle between passion and reason and a third where man is master of his passions. *Kitāb al-dharī'a*, i, 13 (33,19–34,7) clearly determines the basic articulation, notwithstanding some modifications and, above all, a larger addition (58,1–14) that warns against a too great confidence in one's own capacities to control one's passions. Then the problem (58,15–60,13) is discussed how one, as a simple human being, can make the right choices in terms very similar to those of the *Kitāb al-dharī'a*, i, 14 (34,9–35,19), although a major difference in wording comes to the fore in the very beginnings of the discussion. Finally (60,14–22), the difference between passion (*hawā*) and envy (*shahwa*) is elucidated—to this end *Kitāb al-dharī'a*, i, 14 (36,7–14) is quoted in a slightly modified way.

Every human being is able to improve his moral character: this forms the basic affirmation of the first part of chapter 12 (61–62,11). Despite changes and additions, one can easily see that this part has been directly inspired by *Kitāb al-dharī'a*, i, 18 (40–41,11). The second part of the chapter (62,11–63,11) presents four categories of men who are in need of moral education, although in case of the lowest of them this is almost an impossible task. This constitutes an elaborated version of *Kitāb al-dharī'a*, i, 32 (66,9–20).

Chapter 13 (64–6) points to the reciprocal influence between body and soul, while emphasizing the necessity of a continuously sustained effort in view of the purification of the soul. Since 'habit' is awarded a central role in the acquisition of virtue, the main inspiration is certainly Greek, most probably Aristotle's *Nicomachean Ethics*, ii, 1, where

[12] Al-Ghazālī presents a kind of anthropocentric microcosmism, which might ultimately derive from Plato, especially his *Timaeus*, see R. Allers, 'Microcosmos. From Anaximandros to Paracelsus,' *Traditio* 2 (1944), pp. 319–407, 323 and 351.

[13] See Plato, *Phaedrus*, 246 a–d; 253 d–254, but, above all, Galen (Jālīnūs), *Mukhtaṣar min Kitāb al-akhlāq*, in 'A. Badawī (ed.), *Dirāsāt wa-nuṣūṣ fī 'l-falsafa wa'l-'ulūm 'inda 'l-'Arab* (Beirut, 1981), p. 192 (compare al-Sijistānī, *Muntakhab ṣiwān al-ḥikma*, ed. D.M. Dunlop [The Hague, 1979], pp. 109–10, §233).

Aristotle, inspired by Plato's *Laws*, 792E, affirms that ἦθος has been derived from ἔθος.[14] It was a common idea in the ethical writings of the *falāsifa*, but the present chapter shows a particular affinity with a part of Ibn Sīnā's *Risāla fī 'l-ʿahd, Treatise of the Pact*, i.e., p. 146.[15] Two major ideas are common to both texts: on the one hand, 'habit' is a multiple repetition over a long period of time; and the acquisition of virtue resembles that of other arts, on the other. But the discussions in the *Mīzān* on the arts of writing and of *fiqh* are not extant in Ibn Sīnā. The idea that 'habit' facilitates the right action may have been derived from *Kitāb al-dharīʿa*, i, 17 (39,16–20)—note that the *ḥadīth* that closes this chapter is also mentioned in this latter work, i, 20 (43,5–7).

Chapter 14 first (67,2–12) insists that a virtuous life requires a genuine discernment and a good behaviour. Then (67,13–68,16), particular emphasis is put on the necessity of an attitude of obedience and of spontaneity in the practice of cultural obligations as well as virtues. Finally (68,17–69,12), it is said that the virtues are obtained either by learning and effort, or by divine grace. The first part of the chapter offers a rewording of *Kitāb al-dharīʿa*, i, 20 (43,8–13). The beginning of the second (67,15–68,2) is similar to the same work, i, 23 (47,8–11), while its end (68,13–16) is almost identical with i, 20 (44,7–11); between these two fragments, there is a consideration on the longitude of life. The final part constitutes an almost *verbatim* copy of chapter 21 of book i of the very same *Kitāb al-dharīʿa*.

The formation, maintenance and restoration of the moral character are compared with the physical care and development of the body in chapter 15 (70–73). Special attention is paid to the need of finding, or restoring the just equilibrium illustrated by striking examples of such cases. Ibn Sīnā's *New Ethical Tract* (38,15–39,16)[16] constitutes undoubtedly a direct source of inspiration, a rather close resemblance coming to the fore regarding the examples given pp. 71,17–72,5, which have much in common with *New Ethical Tract* (39,5–16). It may be worthwhile

[14] Galen (Jālīnūs), *Mukhtaṣar min Kitāb al-akhlāq*, p. 208, mentions this explanation, ascribing it vaguely to *baʿḍ al-falāsifa* (a philosopher, or some philosophers).

[15] Ibn Sīnā, *Risāla fī 'l-ʿahd*, in idem, *Tisʿ rasāʾil* (Cairo, 1326 A.H.), pp. 142–51.

[16] All references are to the (provisory) edition of B. Karliga, *Un nouveau traité de morale d'Ibn Sīnā inconnu jusqu'à maintenant* (Istanbul, 1995) (I thank the author to have put a copy at my disposal) (besides the edition, an introduction and a French translation are given).

to mention that the latter ultimately derive from Pseudo-Nicolaus' *al-Madkhal ilā 'ilm al-akhlāq*.[17]

The chapters 16–19 deal with the different virtues, their divisions and the opposite vices. The basic description of each of them is sometimes almost *verbatim* the same as that given by Ibn Sīnā in his *New Ethical Tract*, and is in the other cases at least inspired by the latter. One may establish the following similarities:[18] 76,1–12 = 27,8–28,1; 78,10–15 ~ 28,2–7; 80,14–81,10 ~ 28,8–16; 82–83 = 29,2–30,15; 84–86 = 31–33,10; 87,2–8 = 33,12–17; 88,5–91,2 = 33,18–37,9. Note that the first three parts figure in chapter 16 and the last two in chapter 19. The fourth and the fifth part cover the whole of chapters 17 and 18.

In chapter 16, which offers a survey of the cardinal virtues, i.e., wisdom (*ḥikma*) (74,8–76,5), courage (*shajā'a*) (76,6–78,9), moderation (*'iffa*) (78,10–80,13) and justice (*'adāla*) (80,14–81,10), several additions, compared to Ibn Sīnā's *New Tract* become apparent:

- The part on wisdom is introduced by the evocation of the two faces of the soul, a doctrine inspired by Plato that predominates the psychological doctrine of Ibn Sīnā.[19] It may be worthwhile to mention that al-Ghazālī formulates the object of theoretical wisdom in a similar way as Ibn Sīnā does in the first chapter of the *Ilāhiyyāt* of the *Shifā'*, however by using *kalāmic* instead of philosophical terms (but al-Ghazālī explicitly endorses the law of non-contradiction on the logical level).
- Regarding courage, one finds a consideration on the necessity to acquire a praiseworthy character in searching the right path, specified in an Aristotelian manner as the equilibrium between two extremes.
- As to moderation, it is stressed that man must keep control over his appetites: the desire for food or sexual intercourse is destined to safeguard one's own life and that of one's species. In a similar vein,

[17] Pseudo-Nicolaus, *al-Madkhal ilā 'ilm al-akhlāq*, in Aristotle, *Al-akhlāq*, ed. 'A. Badawī (Kuweit, 1979), pp. 394–430, 413–14.

[18] The sign ~ signifies that the correspondence on the literal level is very small, but that one nevertheless has to do with a text that clearly inspired al-Ghazālī; the sign = means that there is a certain literal similarity.

[19] This is a well-known fact. For a recent study on this topic, see D. De Smet, 'La doctrine avicennienne des deux faces de l'âme et ses racines ismaéliennes,' *SI* 93 (2001), pp. 77–89.

> Ibn Sīnā, in his *Fī ʿilm al-akhlāq* (155,5–6),[20] insists that the experience of pleasures has no other goal than to maintain the existence of the individual and the species (he adds: or the community). Such idea has perhaps its ultimate source in Galen.[21]

With respect to justice, the notion of 'political justice' is briefly introduced. Herein, one may detect an Aristotelian influence, but, as I have stated elsewhere, al-Ghazālī's wording shows many affinities with Plato's discussion in the *Republic* and in addition to that seems to be indebted to Galen, al-Fārābī and the anonymous *Sirr al-asrār*, or, more probably, an earlier version of the latter.[22]

Despite their being embedded in a religious framework, these additions certainly reveal a philosophical background.

In chapter 19, one discovers a discussion (87,9–88,4) of the notion of shame (*ḥayāʾ*), including the notion of timidity (*khajal*), which corresponds largely to *Kitāb al-dharīʿa*, iii, 1 (146,7–147,11). The final part of the chapter (91,3–19) offers a general conclusion to the survey of the doctrine of the virtues and the vices. It insists in a genuine Aristotelian way on the golden mean, and declares that man can either become like an angel or like a beast.[23]

Chapter 20 starts with the distinction between three kinds of human beings according to the way in which they are interested in worldly goods (92–93). It corresponds very closely to *Kitāb al-dharīʿa*, i, 31 (64,17–65,18), except for a difference in wording in the last five lines. The second part of the present chapter (94–96) might have been inspired by chapter 32 of the *Kitāb al-dharīʿa*, insofar as it deals with two weaknesses in human behaviour, i.e., incapability (*quṣūr*) and deficiency (*taqṣīr*). But, whereas the former is presented in very similar terms, the discussion of the latter is quite different. Al-Ghazālī stresses the phenomenon of ignorance (*jahl*) in particular, which is either due to negligence or to the conscious rejection of the life in the hereafter, while claiming that it has no foundation whatsoever. Al-Ghazālī insists that a simple belief in the latter's existence is not sufficient: just as for this life, adequate

[20] Ibn Sīnā, *Fī ʿilm al-akhlāq*, in idem, *Tisʿ rasāʾil*, pp. 152–6.

[21] Galen (Jālīnūs), *Mukhtaṣar min Kitāb al-akhlāq*, pp. 190–211, especially ch. 2.

[22] See my 'Al-Ghazzālī's Political Thought: Elements of Greek Philosophical Influence,' *MUSJ* 57 (2004), pp. 393–410, pp. 395–8.

[23] This idea is already present in Galen (Jālīnūs), *Mukhtaṣar min kitāb al-akhlāq*, p. 201.

deeds are required. In this respect, the acquisition of a good moral conduct is absolutely necessary. In the present context, this emphasis on the life in the hereafter as developed here, is completely missing in al-Iṣfahānī.

In chapter 21, different types of happiness are first presented (97,2–14) in terms almost identical to *Kitāb al-dharīʿa*, i, 24 (48,3–18). Hereafter (97,14–21) five kinds of blessings are distinguished, i.e., otherworldly, psychic, corporeal, social and (by divine guidance) success-related ones (*tawfīqiyya*) and their mutual interdependence is underlined. *Kitāb al-dharīʿa*, i, 25 now constitutes the direct source. In a similar vein, the two next chapters of this latter work, i.e., chapters 26, 27 and also chapter 29 have been copied (with only a few minor modifications) in a more detailed examination of the social, corporeal and (by divine guidance) success-related 'good things' that constitutes the remaining part of the twenty-first chapter of the *Mīzān*.

Chapter 22 offers once more a distinction, but this time a fourfold one regarding the right means in order to obtain happiness. The discussion of the first three of them (104–105,14) is almost identical with *Kitāb al-dharīʿa*, i, 24 (48,19–50,7; 51,13–16). The fourth (105,15–107,5), which is related to the faculties of the soul, has been derived from the same work, although from a quite different place, i.e., iii, 9 (157,14–158,13). In a final remark, the small value of sensual pleasures is pointed out, especially in comparison to intellectual pleasures. In these lines, al-Ghazālī seems to have been inspired once more by a part of chapter i, 24 of the *Kitāb al-dharīʿa* (50,8–17).

Chapter 23 deals with three basic necessities for the human being: food (108–111,3), sexual intercourse (111,4–113,14) and anger (113,15–118). Regarding each of them, it is said what is praiseworthy, what reprehensible and what forbidden. The part on food has much in common with *Kitāb al-dharīʿa*, iii, 11, although the order of the text, as well as part of its wording has been changed. As to the section on sexual appetite, it is almost identical with the next chapter of the *Kitāb al-dharīʿa*. But the proper discussion of anger, including a brief evocation of its causes (the explanation given is almost identical with that presented in the *Problemata Physica*, ix, 20),[24] appears not to have its source in al-Iṣfahānī's work. So far, I have looked without success

---

[24] L.S. Filius, *The Problemata Physica Attributed to Aristotle. The Arabic Version of Ḥunain ibn Isḥāq and the Hebrew Version of Moses ibn Tibbon* (Leiden, 1999), p. 395.

for a possible source; at best, I can mention a very vague source of inspiration in *Kitāb al-dharīʿa*, iv, 3 (169,13–170,1). But, when al-Ghazālī, at the end of the chapter, discusses several terms related to anger, i.e., courage (*shajāʿa*), patience (*ṣabr*), exultation (*ghibṭa*), rivalry (*munāfasa*), envy (*ḥasad*) and moderation (*ʿiffa*), he once more relies on the *Kitāb al-dharīʿa*, more precisely iv, 3 (169,6–8); iv, 2; iv, 12 (somewhat modified) and iii, 13 (164,15–165,2).

In order to show the nobility of the activity of teaching, the twenty-fourth chapter starts (119,6–120,14) with the enumeration of three kinds of occupation of which governance is the highest, although one may distinguish within the latter four different levels.[25] This part is almost identical with the *Kitāb al-dharīʿa*, vi, 7. Also the second part of the chapter (120,15–123,8), which shows the nobility of intellection from all points of view—religion, reason and sensation—, is largely based on the latter work, more precisely ii, 1 (slightly reworked), ii, 2 (75,1–4, corresponding to 123,5–6 and 122,9–12); ii, 4 (78,20–79,4, corresponding to 122,7–9 and 123, 3–4) and ii, 5 (79,19–80,6, constituting the source of 122,12–123,2, except for some important modifications).

Chapter 25 insists that wisdom, including the belief in God, belongs to human nature (*fiṭra*). It copies by and large *Kitāb al-dharīʿa*, ii, 17.

The next chapter first presents the distinction between innate (*gharīzī*) and acquired (*muktasab*) intelligence (126,1–127,14), reproducing in a slightly modified way *Kitāb al-dharīʿa*, ii, 2. In both texts the soul and the body are compared in a Platonic fashion to a horseman and his horse. Once more based on *Kitāb al-dharīʿa*, but this time ii, 12 (99,12–15), the relation between rational and religious sciences is described as being complementary, the former functioning like drugs, the latter as regular food (127,15–21). Then, one finds a passage (127,21–128,5) that might be original to al-Ghazālī. It strongly blames those who make themselves guilty of blind imitation, *taqlīd*—an attitude so characteristic of al-Ghazālī! Finally (128,6–129), it is emphasized that one needs to possess knowledge of the realities of both the terrestrial realm and the life in the hereafter. This part extensively reproduces *Kitāb al-dharīʿa*, ii, 3 (although the final conclusion has to be qualified as a reformulation rather than as a copy).

---

[25] For a detailed analysis of this passage, and its possible sources, see my 'Al-Ghazzālī's Political Thought: Elements of Greek Philosophical Influence,' pp. 398–403.

In chapter 27, the duties of the disciple and the master are enumerated in great detail; ten regarding the former, eight with respect to the latter. Roughly speaking, al-Ghazālī bases his discussion here on *Kitāb al-dharīʿa*, ii, 18–29. But not every passage seems to be covered by these chapters of the latter work, and although one sometimes detects great resemblances, important modifications become apparent on other occasions. Hence, the list of similarities may be described as follows:

M. 130–133,15 (duties 1–4 of disciple) =~ Dh. ii, 24: the description of the first duty, i.e., the purification of one's soul, is more elaborate in the *Mīzān* and includes the discussion of two objections claiming the possible combination of bad moral conduct and the possession of real science; the presentation of the second and third duties, which consists in avoiding a strong linkage with this world as well as being proud of oneself, is very similar to that of the *Kitāb al-dharīʿa*; the discussion of the fourth duty consists in not paying attention to doctrinal differences or spurious opinions, it entails important textual modifications;

M. 133,16–134,3 = Dh. ii, 20 (112,15–113,4) and M. 134,3–6 = Dh. ii (113,13–17, modified). This part deals with the fifth duty of the disciple, i.e., his obligation to open his mind to all the sciences.

M. 134,7–12 = Dh. ii, 22 (115,3–7) and M. 134,12–19 = Dh. ii, 19 (12,3–12, somewhat modified). To learn the sciences in the correct order is the sixth duty of a student. It is also noted that divergent views among specialists do not constitute a proof of the imperfectness of a science.

M. 134,20–136,3 =~ Dh. ii, 22 (115,8–116,3, but the order of discussion has changed and there are more or less important modifications). This part presents the seventh duty of the disciple, i.e., to deal only in a general way with all the sciences, but to concentrate above all on the sole knowledge that really matters, i.e., the knowledge of God.

M. 136,4–12 = Dh. ii, 19 (111,15–112,4). To recognize the hierarchy of the sciences forms the eight duty of the disciple.

M. 136,18–138,7 = ~Dh. ii, 18 (110,9–111,3, amplified). A disciple has to know the different kinds of sciences according to whether they are dealing with pure terms, meanings related to terms, or pure meaning: this forms the ninth duty of the disciple. Then (138,8–140,17), al-Ghazālī insists that one has to progress in the acquisition of the sciences in such a manner that one attains the essential, ultimate knowledge. A comparison is made with the different preparations and stages of the pilgrimage to Mecca (inspired by Dh. 113,14?). As far as I can see, this part has been elaborated by al-Ghazālī, which also seems to apply to the next part (140,18–143,2). This part entails a request to examine seriously not only jurisprudence, but also the way of Sufism for the full understanding of the right practical life, presented thereafter in response to a triple interrogation.

M. 143,3–6 might have been inspired by Dh. ii, 22 (114,14 and M. 143,7–11 = Dh. ii, 21 (113,13–18, modified). The final part (144,1–145,3) of the presentation of the tenth duty of the disciple, which consists in coming as close as possible to God, stresses once more the value of the wisdom of Sufism and has probably been added by al-Ghazālī.

Regarding the eight obligations of the teacher—i.e., to consider his pupil as his child; to be guided by the Law-giver; not to withhold any advice; not to forbid openly, but by suggestion; not to disparage the value of other sciences; not to express matters that exceed the comprehension of his student; not to mention the existence of a truth that is beyond the grasp of the student; to act according to one's teaching, one may point out the following similarities:

M. 145,4–14 = Dh. 22 (slightly modified)—this passage constitutes a general introduction to the issue of the duties of the teacher. It compares four different degrees of having science to similar degrees of possessing goods;

M. 145,15–146,5 = Dh. ii, 25 (119,3–6,16–19);

M. 146,6–11 = Dh. ii, 25 (120,14–21). Al-Ghazālī adds (146,11–16) in what seems to be a personal statement that many students are often misled by their teachers' insane desire for glory;

M. 146,17–147,15: I have not found a direct source, but the passage might have been inspired by Dh. ii, 25, especially 119,20–120,1;

M. 147,16–148,3 =~ Dh. ii, 25 (120,1–14—the order of the discussion has been changed);

M. 148,4–9: maybe inspired by Dh. ii, 27;

M. 148,10–149,9 = Dh. ii, 26 (121,2–122,15)

M. 149,10–150,14 = Dh. ii, 26 (122,17–123,12)

M. 150,14–151,12 =~ Dh. ii, 29 (125,19–126,15—a few important modifications)

From what precedes, it is obvious that al-Ghazālī, in the elaboration of this chapter, has been inspired by al-Rāghib al-Iṣfahānī. But he does not hesitate to modify elements of his source whenever he judges it necessary. He is in no way guilty of any kind of plagiarism.

Chapter 28 deals with the acquisition of wealth, especially the existence of five obligations related to it. After a brief introduction (152,2–7), indicating the ambiguous character of wealth in what might be an elaboration of *Kitāb al-dharīʿa*, vi, 10 (205,10–11) or vi, 11 (207,3–4), the first obligation is defined as the knowledge of its rank (152,10–154) according to *Kitāb al-dharīʿa*, vi, 10 (205,11–206, but with changes in the order and somewhat amplified). The second obligation, i.e., to keep an equilibrium between income and expenses (155,1–19), starts (155,1–3) with a basic distinction between acquisition and fortune. This part is based on *Kitāb al-dharīʿa*, vi, 11 (207,20–208,2). Then (155,4–19) the text specifies what is allowed and what is forbidden in the acquisition of goods. If I am not mistaken, such wording is completely absent in the *Kitāb al-dharīʿa*. However, al-Ghazālī might have found inspiration for it in chapter 14 of part six of the latter work. The third obligation (157,1–158,5) concerns the three human basic needs, i.e., habitation, nourishment and clothing, specifying three kinds of quantity with respect to each of them, i.e., minimal, middle and maximal. I looked in vain for a possible source. I only found a small passage (157,9–14) that is clearly inspired by *Kitāb al-dharīʿa*, vi, 14 (210,20–211,1 and 211,6–8); it deals with three ways of keeping supplies of food. The fourth obligation consists in the right way of dealing with income and expenses. Its basic affirmation (158,6–13) is almost identical with *Kitāb al-dharīʿa*, vi, 19 (215,16–21). Then, some remarks about three categories of human beings that exist with regard to wealth (158,14–160,4) follow almost *verbatim Kitāb al-dharīʿa*, vi, 16. In a concluding remark (160,5–10) al-Ghazālī warns against the dangers involved in searching for wealth, quoting a qurʾānic verse (28:77), which is also quoted in *Kitāb al-dharīʿa*, p. 211,10. Finally, the fifth obligation, i.e., the possession of a right intention when giving or receiving, offers a basic statement (160,11–16) followed by an examination of the idea of asceticism (160,16–161). It is essentially based on *Kitāb al-dharīʿa*, vi, 17 (213,19–214,12, somewhat modified), including (160,13–16) a few lines corresponding to vi, 14 (211,3–6). It has to be added that the final remark[26] of the *Mīzān* (161,6–20) lacks, at least insofar as I can see, in the *Kitāb al-dharīʿa*.

In chapter 29, al-Ghazālī emphasises the necessity to avoid grief in this world. Its major part (162–164,6) is based on *Kitāb al-dharīʿa*, iv, 5 (171,14–173,10—but with one modification in the order). A final observation (164,7–165) on the need of a 'gradual' liberation from the links of attachment to this world has perhaps been added by al-Ghazālī.

Chapter 30 deals with the elimination of the fear of death. First (166–167,6) it is stressed that the remembrance of death is of great utility in this life in a somewhat modified version of *Kitāb al-dharīʿa*, iv, 6 (176,4–14). Then (167,7–168,5), based on *Kitāb al-dharīʿa*, iv, 5 (173,11–22), it is shown that fear of death, which may be manifested in four different ways, is completely deprived of sense. Finally (168,6–170), a distinction between three categories of men is presented according to the way in which they die. Once more, the source is *Kitāb al-dharīʿa*, iv, 6 (174–176,4—however, in the following order: 174,3–9; 176,1–4; 174,18–175,3; 174,9–18 and 175,17–176,1).

Regarding the last two chapters of the *Mīzān*, no source has been detected. They seem to form a general conclusion to the work and therefore may be the result of a very personal formulation of al-Ghazālī. The former of the two offers two indications (*ʿalāmatayn*) for those who are on their way to God, i.e., to weight one's acts on the balance of the *sharʿ* and to have God always present in one's heart. As to the last chapter, it insists that it is not so important to which school a thought belongs, but a genuine reflection is always needed. In this last idea, I see again a strong rejection of any form of *taqlīd*—a, if not *the* major crux of al-Ghazālī's thought.

As is the case with most of al-Ghazālī's writings, the *Mīzān* is to a large extent a combination of (somewhat modified) passages taken from works of his predecessors. In this present case, the major source is undoubtedly al-Rāghib al-Iṣfahānī's *Kitāb al-dharīʿa ilā makārim al-sharīʿa*—more than the half of it is traceable in the *Mīzān*. Also Ibn Sīnā's *New Ethical Tract* is almost completely reproduced, as well as some passages of his *Aḥwāl al-nafs*. But the obvious presence of all these kinds of modifications makes it very clear that there is much more involved than just

[26] In this remark, al-Ghazālī refers to king Solomon and evokes the stages of life in terms of travel, indicating that the present life may be compared to the stay in a hospice (*ribāṭ*).

blind copying. To examine the real scope of this work of reworking and further elaboration clearly exceeds the limits of the present study, but one may hope that one day it will receive a correct evaluation.[27] In order to do so, one will have to take into account all of al-Ghazālī's works that entail parallel passages with the *Mīzān*, as e.g., *Maʿārij al-quds fī madārij maʿrifat al-nafs*,[28] *Iḥyāʾ ʿulūm al-dīn*,[29] *al-Mustaẓhirī*[30] and *Tahāfut al-falāsifa*.[31] Especially with the former two, there are many similarities. Hence, they might offer important indications regarding an evolution in al-Ghazālī's thought and thereby help to better understand this great Islamic thinker.

[27] I have in mind the detailed comparison one finds for one fragment in M. Yasien, 'The Ethical Philosophy of Al-Ghārib Al-Iṣfahānī,' *JIS* 6 (1995), pp. 51–75, 71–3; see also idem, *The Path of Virtue. On the Ethics of al-Ghārib al-Iṣfahānī*, PhD diss. (University of Frankfurt, 1999), pp. 16–22 (I thank H. Daiber for having provided me with a copy of these pages).

[28] For the parts common with this work, see my '*Le Maʿārij al-quds fī madārij maʿrifat al-nafs*: un élément-clé pour le dossier Ghazzālī-Ibn Sinā?,' *AHDLMA* 60 (1993), pp. 27–55, especially 39–43.

[29] In a very provisory way, the following similarities between chapters (or, at least, parts of them) of the *Iḥyāʾ* and the *Mīzān* can be shown:

i,1,1–119–120	i,1,5–130–151
i,1,7,1–120–123	i,1,7,2–124–125
iii,1,3–53–56	iii,1,4–28–34
iii,1,7–126–129	iii,1,8–40–42
iii,1,9–42–43	iii,2,2–82–89
iii,2,3–61–63	iii,2,4–64–69
iii,2,5–70–72	iii,3,5–113
iii,5,2–113–115	iii,7,2–152–154
iii,7,11–155–161	iv,2,2,1–97–106
iv,2,3–116	

Let me insist once more that the present list is only a very rough one. A more detailed examination is needed in order to enumerate all the precise similarities. Let me add that I do not believe that the *Iḥyāʾ* offers a summarized version of the *Mīzān*, as claimed by M. Fakhry, *Ethical Theories in Islam* (Leiden, 1994), p. 193.

[30] One may compare the paragraphs 439–42 of this text to several fragments of *Mīzān*, pp. 53–7, i.e., 53–54, 2; 55, 9–56 and 57, 4–14.

[31] The tenth question, § 1–10 has much in common with *Mīzān*, pp. 23–7.

MS. MAHDAWI 514. AN ANONYMOUS COMMENTARY OF IBN MATTAWAYH'S *KITĀB AL-TADHKIRA*

Sabine Schmidtke

During the fourth/tenth and early fifth/eleventh centuries, Muʿtazilī *kalām* flourished under the Shīʿī reign of the Būyids in Iraq and western Persia. The vizier al-Ṣāḥib ibn ʿAbbād (d. 385/995) in particular favoured and promoted Muʿtazilī teachings. In 367/977 he appointed ʿAbd al-Jabbār ibn Aḥmad al-Hamadhānī al-Asadābādī as chief judge (*qāḍī 'l-quḍāt*) of Rayy, where the latter taught even after his dismissal from office in 385/995 until his death in 415/1025. Representing the Bahshamī school of the Muʿtazila,[1] named thus after its founder Abū Hāshim al-Jubbāʾī (d. 321/933), Qāḍī ʿAbd al-Jabbār became the undisputed head of the Muʿtazila after the death of his teacher Abū ʿAbd Allāh al-Baṣrī in 369/979.[2] ʿAbd al-Jabbār was commonly recognized as the most prominent *kalām* theologian of his time and attracted numerous students, among them Abū Rashīd al-Nīsābūrī, who became the leader of the school upon his teacher's death;[3] Abū Muḥammad Ḥasan

[1] For a detailed account of the Bahshamī school tradition, see M.T. Heemskerk, *Suffering in the Muʿtazilite Theology. ʿAbd al-Jabbār's Teaching on Pain and Divine Justice* (Leiden, 2000), pp. 13–71.

[2] For a detailed study on the life and work of Qāḍī ʿAbd al-Jabbār, see G.S. Reynolds, *A Muslim Theologian in the Sectarian Milieu. ʿAbd al-Jabbār and the* Critique of Christian Origins (Leiden, 2004).

[3] On him, see R.M. Frank, 'Abū Rashīd al-Nīsābūrī,' in *EI*², Suppl., pp. 31–2. Of his writings the following have been published: *Fī 'l-tawḥīd*, ed. M.ʿA. Abū Rīda (Cairo, 1969); and *Masāʾil al-khilāf bayna al-baṣriyyīn wa'l-baghdādiyyīn*, ed. M. Ziyāda and R. al-Sayyid (Beirut, 1979). An earlier partial edition of the latter work containing the first portion on substances and accidents was published by Arthur Biram under the title *Die atomische Substanzenlehre aus dem Buch der Streitfragen zwischen Basrensern und Bagdadensern* (Berlin, 1902), and a free translation of the entire text was produced by Max Horten (*Die Philosophie des Abu Rashid* [Bonn, 1910]).—For the identification of the text *Fī 'l-tawḥīd* as being part of Nīsābūrī's *Ziyādāt al-sharḥ* (and not of his *Dīwān al-uṣūl* as argued by the editor Abū Rīda), see R.C. Martin, 'The Identification of Two Muʿtazilite MSS,' *JAOS* 98 (1978), pp. 389–93. See also idem, *A Muʿtazilite Treatise on Prophethood and Miracles, being probably the* bāb ʿalā 'l-nubuwwah *from the* Ziyādāt al-sharḥ *by Abū Rashīd al-Nīsābūrī (Died First Half of the Fifth Century A.H.)*. Edited in Arabic with an English Introduction, Historical and Theological Commentaries, PhD thesis (New York University, 1975). Daniel Gimaret contested Martin's identification and suggested that the *Ziyādāt* as extant in the edition of Abū Rīda and in ms. British Library 8613

ibn Aḥmad ibn Mattawayh;[4] Abū 'l-Ḥusayn al-Baṣrī (d. 436/1044);[5] Abū 'l-Qāsim al-Bustī;[6] the Zaydī *sharīf* Abū 'l-Ḥusayn Aḥmad ibn Abū Hāshim Muḥammad al-Ḥusaynī al-Qazwīnī, known as Mānkdīm Shashdīw (d. 425/1034);[7] the Twelver Shī'ī al-Sharīf al-Murtaḍā (d. 436/1044);[8] the two Caspian Zaydī imams and brothers Abū Ṭālib Yaḥyā ibn al-Ḥusayn al-Buṭḥānī al-Nāṭiq bi'l-Ḥaqq (d. *c.* 424/1033),[9] and Aḥmad ibn al-Ḥusayn al-Mu'ayyad bi'llāh (d. 411/1020).[10] At the same time, Bahshamī doctrine was widely adopted in Jewish, and particularly Karaite circles.[11]

originated with a later author; see his 'Les uṣūl al-ḫamsa du Qāḍī 'Abd al-Ǧabbār et leurs commentaries,' *AI* 15 (1979), p. 73.

[4] On him, see W. Madelung, 'Ibn Mattawayh,' in *EI*[2], Suppl., p. 393; M. McDermott, 'Ebn Mattawayh,' in *EIr*; Ṣamad Muwaḥḥid, 'Ibn Mattawayh,' in K.M. Bujnurdī (ed.), *Dā'irat-i ma'ārif-i buzurg-i islāmī* (Tehran 1374/1996–), iv, p. 580.

[5] On him, see W. Madelung, 'Abū l-Ḥusayn al-Baṣrī,' in *EI*[2], Suppl., p. 25; M.J. Muqaddam, 'Abū l-Ḥusayn al-Baṣrī,' in Bujnurdī (ed.), *Dā'irat-i ma'ārif-i buzurg-i islāmī*, v, p. 368; D. Gimaret, 'Abu 'l-Ḥosayn al-Baṣrī,' in *EIr*. See also S.M. Stern, 'Ibn al-Samḥ,' *JRAS* (1956), pp. 31–44; E. Giannakis, 'The Structure of Abū l-Ḥusayn al-Baṣrī's Copy of Aristotle's *Physics*,' *ZGAIW* 8 (1993), pp. 251–8; M. Cook, *Commanding Right and Forbidding Wrong in Islamic Thought* (Cambridge, 2000), p. 218ff.

[6] On him, see the introduction to Abū 'l-Qāsim al-Bustī, *Kitāb al-baḥth 'an adillat al-takfīr wa'l-tafsīq* (*Investigation of the Evidence for Charging with* kufr *and* fisq), ed. and introd. W. Madelung and S. Schmidtke (Tehran, 1382/2004).

[7] He is the author of a commentary (*ta'līq*) on the lost *Sharḥ al-uṣūl al-khamsa* composed by 'Abd al-Jabbār. The work was published wrongly as a work by 'Abd al-Jabbār (ed. 'A. 'Uthmān [Cairo, 1384/1965; numerous reprints]). On him, see Gimaret, 'Les uṣūl al-ḫamsam,' p. 57ff.

[8] For al-Murtaḍā's theological views, see W. Madelung, 'Imāmism and Mu'tazilite Theology,' in T. Fahd (ed.), *Shī'isme imāmite* (Paris, 1970), pp. 13–29; M.J. McDermott, *The Theology of al-Shaikh al-Mufīd (d. 413/1022)* (Beirut, 1978). His most extensive theological works are his *Kitāb al-dhakhīra fī 'ilm al-kalām*, ed. A. al-Ḥusaynī (Qum, 1411/[1990–91]) and his *Kitāb al-mulakhkhaṣ fī uṣūl al-dīn*, ed. M.R. Anṣārī Qummī (Tehran, 1381/[2002]). On the *Kitāb al-dhakhīra*, see also S. Schmidtke, 'Nuskha-ī kuhan az *Kitāb al-dhakhīra-i* Sharīf Murtaḍā,' *Ma'ārif* 20/2 (1382/2003), pp. 68–84.

[9] On him, see W. Madelung, *Der Imam al-Qāsim ibn Ibrāhīm und die Glaubenslehre der Zaiditen* (Berlin, 1965), pp. 178–82; idem, 'Zu einigen Werken des Imams Abū Ṭālib an-Nāṭiq bi l-Ḥaqq,' *Der Islam* 63 (1986), pp. 5–10. An edition of his *Ziyādāt* to a commentary by Qāḍī 'Abd al-Jabbār on the *Kitāb al-uṣūl* by Ibn Khallād that are extant in what seems to be a unique manuscript (ms. Leiden 2949) is currently being prepared by Camilla Adang and Sabine Schmidtke.

[10] On him, see Madelung, *al-Qāsim ibn Ibrāhīm*, pp. 177–8.

[11] For the reception of Muslim *kalām* within Jewish circles, see H. Ben-Shammai, 'Kalām in Medieval Jewish philosophy,' in D.H. Frank and O. Leaman (eds.), *History of Jewish Philosophy* (London, 1997), pp. 115–48; D. Sklare, *Samuel Ben Ḥofni and his Cultural World. Texts and Studies* (Leiden, 1996); W. Madelung and S. Schmidtke, *Rational Theology in Interfaith Communication. Abu l-Ḥusayn al-Baṣrī's Mu'tazilī Theology among the Karaites in the Fāṭimid Age* (Leiden, 2006).

ʿAbd al-Jabbār and his pupils and later followers composed the authoritative works of the later Muʿtazila, among them *al-Mughnī fī abwāb al-tawḥīd wa'l-ʿadl*, a comprehensive exposition of Muʿtazilī doctrine that ʿAbd al-Jabbār dictated over a period of twenty years (360/970–380/990),[12] and his *al-Kitāb al-muḥīṭ fī 'l-taklīf* of which Ibn Mattawayh made an independent, explicative, and at times critical paraphrase entitled *al-Majmūʿ fī 'l-Muḥīṭ bi'l-taklīf.*[13] Manuscripts of most of these works have been preserved until today in Zaydī and Karaite Jewish repositories. While most of the published Muʿtazilī works are based on manuscripts that were discovered in the early 1950s by an expedition of Egyptian scholars in the Great Mosque of Sana'a,[14] the material housed by the many private and smaller public libraries in Yemen and the various European and Russian libraries where the bulk of Muʿtazilī writings of Karaite Jewish provenance is preserved, still needs to be explored in full.[15]

---

12 ʿAbd al-Jabbār al-Hamadhānī, *Kitāb al-mughnī fī abwāb al-tawḥīd wa'l-ʿadl*, ed. M. ʿIlmī et al. (Cairo, 1961–65).

13 Ibn Mattawayh, *Kitāb al-majmūʿ fī 'l-muḥīṭ bi'l-taklīf*, i, ed. J.J. Houben (Beirut, 1965); ii, ed. J.J. Houben and D. Gimaret (Beirut, 1981); iii, ed. J. Peters (Beirut, 1999). Margaretha T. Heemskerk is currently preparing a critical edition of the fourth volume. Whereas the original *al-Kitāb al-muḥīṭ* seems to be lost in the Islamic world, extensive fragments of the work are extant in the various collections housing material from the Karaite Genizah in Cairo. See H. Ben-Shammai, 'A Note on Some Karaite Copies of Muʿtazilite Writings,' *BSOAS* 37 (1974), pp. 295–304.

14 For the expedition and its results, see the report by Kh.Y. Nāmī, *al-Baḥtha al-miṣriyya li-taṣwīr al-makhṭūṭāt al-ʿarabiyya fī bilād al-Yaman* (Cairo, 1952); *Qā'ima bi'l-makhṭūṭāt al-ʿarabiyya al-muṣawwara bi'l-mīkrūfilm min al-Jumhūriyya al-ʿArabiyya al-Yamaniyya* (Cairo, 1387/1967). For further references, see G. Roper, *World Survey of Islamic Manuscripts*, iii (London, 1994), p. 645ff.

15 For recent efforts to search and catalogue the holdings of Yemeni public and private libraries, see ʿA. al-Ḥabshī, *Fihris makhṭūṭāt baʿḍ al-maktabāt al-khāṣṣa fī 'l-Yaman* (London, 1994); M. Wafādār Murādī, 'Fihrist-i ālifbā'ī-yi mīkrūfīlm-hā-yi tahiyya shude az Kitābkhāna-yi Jāmiʿ Ṣanʿā',' in *Kitābdārī wa-iṭṭilāʿ rasānī. Faṣlnāma-yi Kitābkhāna-yi Markazī wa-Astān-i Quds Raḍawī*, ii/3 (1378/1999), pp. 127–73; Ḥ. Anṣārī Qummī, 'Guzārishī az nuskha-hā-yi khaṭṭī-yi Yaman,' *Āyina-yi mīrāth* 3/4 (Spring 1380/2001), pp. 105–11; ʿA. ʿAbbās al-Wajīh, *Maṣādir al-turāth fī 'l-maktabāt al-khāṣṣa fī 'l-Yaman*, 2 vols. (Sana'a, 1422/2002). For Muʿtazilī materials preserved through the Karaites, see A. Jakovlevič Borisov, 'Muʿtazilitskie rukopisi Gosudarstvennoj Publičnoj Biblioteki v Leningrade,' *Bibliografija Vostoka* 8–9 (1935), pp. 69–95; see also idem, 'Ob otkrytych v Leningrade muʿtazilitskich rukopisjach i jich značenii dlja istorii musulmanskoj mysli,' *Akademija Nauk SSSR. Trudy pervoj sessii arabistov* 14–17 ijunka 1935 g. (Trudy Instituta Vostokovedenija 24), pp. 113–25. The two articles were reprinted in *Pravoslavniy Palestinskiy Sbornik* 99 (36) (2002), pp. 219–49 and in *The Teachings of the Muʿtazila. Texts and Studies I–II*, Selected and repr. F. Sezgin in coll. with M. Amawi, C. Ehrig-Eggert, E. Neubauer (Frankfurt, 2000), ii, pp. 17–57; D. Sklare (in coop. with H. Ben-Shammai), *Judaeo-Arabic Manuscripts in the Firkovitch Collections. The Works of Yusuf al-Basir. A Sample*

By contrast, next to nothing has been preserved of the Muʿtazilī literature prior to ʿAbd al-Jabbār, nor of rivals of the Bahshamiyya, such as the Baghdādī school or the Ikhshīdiyya. The same applies to the *œuvre* of ʿAbd al-Jabbār's student Abū 'l-Ḥusayn al-Baṣrī who deviated from the doctrines of his teacher and later became known as the founder of the last innovative school within Muʿtazilism. None of his theological writings has come down to us, with the exception of three extensive fragments of his *Taṣaffuḥ al-adilla* that have been preserved in the Abraham Firkovitch Collection (Russian National Library, St. Petersburg),[16] and the writings of later adherents to his doctrine are indispensable for a systematic reconstruction of his thought. The most important among these are Rukn al-Dīn ibn al-Malāḥimī al-Khwārazmī (d. 538/1144), the author of an extensive yet incompletely preserved *Kitāb al-muʿtamad fī uṣūl al-dīn* and of the shorter but completely extant *Kitāb al-fāʾiq fī uṣūl al-dīn*,[17] and a certain Taqī al-Dīn al-Najrānī (or: al-Baḥrānī) al-ʿAjalī (end of sixth/twelfth or early seventh/thirteenth century), the author of the *Kitāb al-kāmil fī 'l-istiqṣāʾ fīmā balaghanā min kalām al-qudamāʾ* containing a systematic comparison of the teachings of Abū 'l-Ḥusayn al-Baṣrī and the Bahshamiyya.[18]

Catalogue. Texts and Studies (Jerusalem, 1997); S. Schmidtke, 'The Karaites' Encounter with the Thought of Abū l-Ḥusayn al-Baṣrī (d. 436/1044). A Survey of the Relevant Materials in the Firkovitch-Collection, St. Petersburg,' *Arabica* 53 (2006), pp. 108–42; eadem, 'Muʿtazilī Manuscripts in the Abraham Firkovitch Collection, St. Peterburg,' in C. Adang, S. Schmidtke and D. Sklare (eds.), *A Common Rationality. Muʿtazilism in Islam and Judaism* (Würzburg, 2007), pp. 377–462; O. Hamdan and S. Schmidtke, 'Qāḍī ʿAbd al-Jabbār al-Hamadhānī (d. 415/1025) on the Promise and Threat. An Edition of a Fragment of his *Kitāb al-Mughnī fī abwāb al-tawḥīd wa-l-ʿadl* Preserved in the Firkovitch-Collection, St. Petersburg (II Firk. Arab. 105, fol. 14–92),' *MIDEO* 27 (2008) [in press].

[16] Abū 'l-Ḥusayn al-Baṣrī, *Taṣaffuḥ al-adilla. The Extant Parts*, introd. and ed. W. Madelung and S. Schmidtke (Wiesbaden, 2006).—Extant and edited is his *opus magnum* on legal methodology, *al-Muʿtamad fī uṣūl al-fiqh*, 2 vols., ed. M. Hamidullah, A. Bekir, and H. Hanafi (Damascus, 1964–65). The edition also comprises a brief text on juridical dialectic, entitled *Kitāb al-qiyās al-sharʿī*. For an analysis of the latter tract, see W.B. Hallaq, 'A Tenth-Eleventh Century Treatise on Juridical Dialectic,' *MW* 77 (1987), pp. 197–228.

[17] Ibn al-Malāḥimī, *Kitāb al-muʿtamad fī uṣūl al-dīn*, ed. W. Madelung and M.D. McDermott (London, 1991); idem, *Kitāb al-fāʾiq*, ed. W. Madelung and M.D. McDermott (Tehran, 1386/2007).—In addition, Ibn al-Malāḥimī's *Tuḥfat al-mutakallimīn fī 'l-radd ʿalā 'l-falāsifa* has been discovered some years ago. See Ḥ. Anṣārī, 'Kitāb-ī tāzihyāb dar Naqd-i falsafa. Paidā shudan-i Kitāb-i "Tuḥfat al-mutakallimīn-i" Malāḥimī,' *Nashr-i dānish* 18/3 (2001), pp. 31–2. Ḥasan Anṣārī, together with Wilferd Madelung and Sabine Schmidtke, is currently preparing an edition of this work.

[18] Taqī al-Dīn al-Baḥrānī (or: al-Najrānī) al-ʿAjālī, *al-Kāmil fī 'l-istiqṣāʾ fīmā balaghanā min kalām al-qudamāʾ*, ed. al-S.M. al-Shāhid (Cairo, 1420/1999). See also E. Elshahed,

Ms. Mahdawī 514 belongs to the literary legacy of the Bahshamiyya. It is a commentary on a work by Ibn Mattawayh on natural philosophy which is known under the titles *al-Tadhkira fī aḥkām al-jawāhir wa'l-aʿrāḍ* and *Kitāb al-tadhkira fī laṭīf [ʿilm] al-kalām*. The commentary, which is not given a title of its own (the colophon rather says: *hādhā ākhir Kitāb al-tadhkira*) and whose author is nowhere identified in the text as such, is preserved in an apparently unique manuscript copy housed at the Aṣghar Mahdawi Library in Tehran and is now available in a facsimile publication.[19] The entire codex consists of 191 leaves (19 × 25 cm) with 40/45 lines to a page, of which the commentary covers 188 folios.[20]

Das Problem der transzendentalen sinnlichen Wahrnehmung in der spätmuʿtazilitischen Erkenntnistheorie nach der Darstellung des Taqīaddīn al-Naǧrānī (Berlin, 1983).

[19] *An Anonymous Commentary on* Kitāb al-Tadhkira *by Ibn Mattawayh. Facsimile Edition of Mahdavi Codex 514 (6th/12th Century)*, introd. and indices S. Schmidtke (Tehran, 2006)—For a brief description of the codex, see M. Taqī Dānishpazūh, 'Fihrist-i nuskha-hā-yi khaṭṭī kitābkhāna-yi khuṣūṣī-yi Aṣghar Mahdawī,' *Nashriyya-yi Kitābkhāna-yi Markazī-yi Dānishgāh-i Tihrān* 2 (1341/1962), pp. 156–7.—Ibn Mattawayh's original work is preserved in five manuscripts: ms. Ambrosiana (Milan), Arabic C 104; mss. Sana'a, Maktabat al-Jāmiʿ al-kabīr (Sharqiyya) nos. 901, 560, 562, 561. For details, see G. Schwarb, S. Schmidtke, D. Sklare (eds.), *A Handbook of Muʿtazilite Manuscripts and Work*, ii. Repertory of Muʿtazilite authors, works and manuscripts (Leiden, forthcoming).—In 1975, a partial edition was published: Al-Ḥasan ibn Mattawayh al-Najrānī al-Muʿtazilī, *al-Tadhkira fī aḥkām al-jawāhir wa'l-aʿrāḍ*, ed. S. Naṣr Luṭf and F. Badīr ʿAwn (Cairo, 1975). Daniel Gimaret has completed a new critical edition of the entire text which will be published shortly (Cairo: IDEO).—Ibn Mattawayh's *Tadhkira* as well as the commentary presented here constituted two of the key texts for the study by Alnoor Dhanani: *The Physical Theory of Kalām. Atoms, Space, and Void in Basrian Muʿtazilī Cosmology* (Leiden, 1994). Dhanani gives the shelfmark of the Mahdawi codex erroneously as ms. Teheran Dānishgāh [sic] 514.

[20] At the end of the codex there are two heavily damaged fragments from two different *kalām* works, each written by a different hand. The first fragment (fol. 188b–189a) is clearly from a Muʿtazilī *kalām* work that evidently also belonged to the Bahshamī tradition, as it starts with a reference to Abū Hāshim al-Jubbāʾī and his *Kitāb al-jāmiʿ* and is again concerned with natural philosophy. The second fragment (fol. 189b–191b) begins with the remark *naqaltuhu min Kitāb al-nihāya fī 'l-kalām*, and contains a discussion of various aspects related to accidents (*aʿrāḍ*). It is written by a so far unidentified Ashʿarī author who refers to the Muʿtazila, whom he defines as *wa-hum Abū Hāshim wa-man tābaʿahu* (fol. 190a, l. 10), and to the philosophers as 'the opponents'. Apart from Abū Hāshim, the anonymous author of this fragment refers to Abū 'l-Hudhayl al-ʿAllāf (d. between 226/840 and 236/850), Abū 'l-Ḥusayn al-Baṣrī, Abū 'l-Qāsim al-Kaʿbī al-Balkhī and Hishām [ibn ʿAmr al-Fuwaṭī] (d. *c.* 230/844) (fol. 191a, l. 28), Abū Isḥāq al-Naṣībīnī (or al-Naṣībī) (fol. 191a, l. 29), Ibn ʿAyyāsh (fol. 191a, l. 29 and 191b, l. 3), and Abū Yaʿqūb al-Shaḥḥām (d. 257/871?) (fol. 191b, l. 1) among the Muʿtazilīs, and to the Imāmī theologian Hishām ibn al-Ḥakam (fol. 191b, l. 5–6). Among the Ashʿariyya (*aṣḥābunā*) the anonymous author refers to the *qāḍī* Abū Bakr [al-Bāqillānī] (d. 403/1013) (fol. 190a, l. 11, 12, 14, 16), the Imām al-Ḥaramayn [al-Juwaynī] (d. 478/1085) (fol. 190a, l. 11), the *ustādh* Abū Isḥāq (fol. 190a, l. 26, 28), by which Abū Isḥāq al-Isfarāyīnī (d. 418/1027) is meant, and to the not further identified *shaykhunā*

Both Ibn Mattawayh's *Tadhkira* and the commentary contain a detailed chapter on substances (*al-qawl fī 'l-jawāhir*), followed by two major sections devoted to physics (*al-kalām fī ithbāt al-juz' wa-furū'ihi*) and to 'biology' (*al-kalām fī 'l-ḥayāt*). These are further subdivided into several *aqwāl*, each consisting of numerous *fuṣūl*. The section on physics contains comprehensive discussions of the annihilation and restoration of substances (*fanā' al-jawāhir wa-i'ādatuhā*) and of atoms and bodies as well as their various properties, viz. colours (*alwān*), tastes (*ṭu'ūm*) and odours (*rawā'iḥ*), heat and cold (*ḥarāra wa-burūda*), pains and pleasures (*ālām wa-ladhdhāt*), sounds, speech and language (*aṣwāt wa-kalām*), spatial states (*akwān*), composition (*ta'līf*), pressure (*i'timād*), and moistness and dryness (*ruṭūbāt wa-yabas*). The section on biology deals in detail with desire and distaste (*shahwa wa-nifār*), capability (*qudra*), will and aversion (*irāda wa-karāha*), belief, knowledge, and assumption (*i'tiqādāt wa-'ulūm wa-ẓunūn*), rational investigation (*naẓar*), and perception (*idrāk*).

The style of the commentary is characteristic of the bulk of Mu'tazilī literature of the Bahshamī tradition that is amply available to us. It is an explicative paraphrase of Ibn Mattawayh's work that closely follows the original in argumentation and doctrinal outlook and can be read independently of the original *Tadhkira*.[21] The commentator differs from Ibn Mattawayh mainly in his wording and in the fact that he often subdivides the text further than is the case in the original *Tadhkira*. Moreover, the commentary by far exceeds the original in length.[22]

The commentary is replete with references to earlier Mu'tazilī authorities and contains numerous references to and quotations from Mu'tazilī writings that are lost, although in most cases the commentator seems to have relied on Ibn Mattawayh's *Tadhkira* for the earlier literature. The two authorities mentioned most often are Abū 'Alī and his son Abū Hāshim al-Jubbā'ī. Among the generation of Bahshamī Mu'tazilīs following Abū Hāshim, frequent mention is made of Abū 'Alī Muḥammad ibn Khallād,[23] Abū 'Abd Allāh al-Baṣrī (d. 369/979)

(fol. 190a, l. 31). The fragment may possibly be part of Fakhr al-Dīn al-Rāzī's *Nihāyat al-'uqūl fī dirāyat al-uṣūl* that is extant in manuscript.

21 For a discussion of the literary genre of commentary literature among the Bahshamīs, see Gimaret, 'Les uṣūl al-ḫamsa,' pp. 48–78.

22 Roughly estimated, ms. Mahdawī 514 consists of 262,260 words, whereas ms. Ambrosiana Arabic C 104, containing a complete copy of Ibn Mattawayh's *Tadhkira*, consists of 104,850 words.

23 On him, see D. Gimaret, 'Ebn K̲allād,' in *EIr*.

and Abū Isḥāq ibn ʿAyyāsh, and of the next generation the most prominent theologian referred to throughout the text is the Chief Judge ʿAbd al-Jabbār. Among the generation of ʿAbd al-Jabbār's students, the author repeatedly mentions Abū Rashīd al-Nīsābūrī and, of course, Ibn Mattawayh himself. In addition to the Bahshamī school tradition, the commentary contains comprehensive accounts of the views of opponents from within and outside the Muʿtazila. The views of Ibn al-Rāwandī (d. 245/860 or 298/912?) and of Ibn Zakariyyā al-Rāzī (d. 313/925 or 323/935) and of the philosophers in general (*al-awāʾil/al-falāsifa*) are often mentioned and refuted.[24] The views of rivals within the Muʿtazila, such as Muḥammad ibn ʿUmar al-Ṣaymarī (d. 315/927) and his disciples Abū Bakr ibn al-Ikhshīd (d. 326/938)[25] and Abū Aḥmad ibn Salama as representatives of the Ikhshīdiyya, and Abū 'l-Qāsim al-Kaʿbī al-Balkhī (d. 319/913), the founder of the school of the Baghdād in its scholastic phase, are also discussed in detail.

Throughout the entire commentary, different patterns of style can be observed at the beginning of the various *fuṣūl*. Numerous chapters begin with brief quotations from the original *Tadhkira*, introduced by formulas such as *qāla raḥimahu 'llāh*, followed by the opening words of the respective chapters *ilā ākhirihi/ilā ākhir al-faṣl* (fol. 1b–26b, 48a–91a, 101a–146b, 165a–). At times, the formula *qāla raḥimahu 'llāh* is missing and the chapter begins immediately with the opening words *ilā ākhirihi/ilā ākhir al-faṣl* (fol. 1b, 27b–35b, 52a, 59a, 60a, 65b, 80b, 81b, 83b, 114b, 181a–188a). This is followed by the commentary, introduced by formulas such as *qāla [al-shaykh] ayyadahu 'llāh/adāma 'llāh ʿulūwahu*. These eulogies suggest that the commentary was dictated by the *shaykh* and that the present manuscript was written down by a student attending his lectures. That the commentary was dictated in a teaching context is further supported by the fact that it neither quotes from nor comments upon Ibn Mattawayh's brief introduction, and that it lacks an introduction of its own (see Appendix). On fol. 157a, l. 40, the view of the Caspian Zaydī Imām Aḥmad ibn al-Ḥusayn al-Muʾayyad bi'llāh (d. 411/1020) is reported, which may indicate that the commentator himself was a Zaydī. Moreover, at one instance, the text says (fol. 100b,

---

[24] On both, see S. Stroumsa, *Freethinkers of Medieval Islam. Ibn al-Rāwandī, Abū Bakr al-Rāzī, and their Impact on Islamic Thought* (Leiden, 1999).
[25] On him, see D. Gimaret, 'Ebn al-Ek̲s̲īd,' in *EIr*.

l. 41) *qāla al-Shaykh Abū Jaʿfar Muḥammad ibn ʿAlī Mazdak adāma ʾllāh tawfīqahu*. According to an *isnād* to be found on the title page of a copy of *Taʿlīq Sharḥ al-uṣūl al-khamsa* by Ismāʿīl ibn ʿAlī al-Farrazādī, this Ibn Mazdak was a student of Ibn Mattawayh.[26] Since the wording of the passage deviates from the usual pattern of *qāla [al-shaykh] ayyadahu ʾllāh/ adāma ʾllāh ʿulūwahu*, Ḥasan Anṣārī has suggested Ibn Mazdak as the author of the present commentary.[27] The author, possibly Ibn Mazdak, changed his convention of introducing new chapters throughout the text at various occasions. Another frequent pattern he employs is to open a new *faṣl* with an explanatory note introduced as *al-gharaḍ bi-hādhā ʾl-faṣl/bihi al-kalām fī*... (fol. 35b–37b, 38a, 38b, 39a, 40a, 41a, 41a–42b, 43a–45b, 47a, 92a–100a, 147a–159a, 160b–164b), or to summarize what has been explained so far and to give the gist of what will now follow (*thumma lamma bayyana raḥimahu ʾllāh al-kalām fī*... *bayyana al-kalām fī*...) (fol. 38a, 38b, 39a, 39b, 40b, 41a, 42b, 164a). At times he simply indicates what the chapter will be about by the expression *faṣl fī*... (fol. 160a, 160b). In all these cases, the commentator evidently dispenses with quoting from the original *Tadhkira*. There are some few occasions where a chapter starts immediately with the commentary, introduced as a rule by the imperative *iʿlam* (fol. 95a, 95b).

The commentary is concluded by a colophon (fol. 188a, l. 16–18) stating that the end of the *Tadhkira* has been reached (*hādhā ākhir Kitāb al-tadhkira* [sic]) and that it (clearly referring to the commentary) was completed on 26 Shawwāl 570/19 May 1175. If indeed the commentary was dictated by Ibn Mazdak and taken down by a student of his,[28] the date as given in the colophon leaves no doubt that the present copy is not the *mustamlī*'s autograph.

A full study of the commentary described here still needs to be undertaken. In order demonstrate the character of the work in juxtaposition to Ibn Mattawayh's *Tadhkira*, this contribution will conclude with editions of the beginning and of the second part of *al-qawl fī fanāʾ al-jawāhir wa-iʿādatihā* devoted to restoration of substances, as they appear in both texts (ms. Mahdawī 514, fol. 1b, l. 1–13, 38a,

[26] See Mānakdīm, *[Taʿlīq] Sharḥ al-uṣūl al-khamsa*, p. 24, n. 1; Gimaret, 'Les uṣūl al-ḫamsa,' p. 60ff.

[27] Ḥ. Anṣārī, 'Kitābī az maktab-i mutakkilimān-i muʿtazilī Rayy,' *Kitāb-i māh dīn* 104–106 (1385/2006), pp. 68–75.

[28] Ḥasan Anṣārī (see previous note) suggests on the basis of the same *isnād* Ismāʿīl ibn ʿAlī al-Farrazādī to have been the one who took down the commentary.

l. 12–39a, l. 15). In the case of Ibn Mattawayh's *Tadhkira*, the edition prepared by Sāmī Naṣr Luṭf and Fayṣal Badīr ʿAwn (Cairo, 1975, pp. 33–34, 237–247) has been used. Obvious errors in the printed edition have been silently corrected.

شرح التذكرة (اب: 1–13)

قال الشيخ السعيد أبو محمد بن الحسن بن متويه رضي الله عنه: فصل اعلم أن المعلومات أجمع لا تخرج عن قسمة تتردد بين النفي والإثبات إلى آخر الفصل

قال الشيخ أيده الله: اعلم أنه رحمه الله: قصد بهذا الفصل حصر أجناس المعلومات وبيان أنواعها وقد حصرها قاضي القضاة عماد الدين رضي الله عنه في بعض المواضع على [...] [كلمة مطموسة في الأصل] فقال هي القديم والجواهر والأعراض، والأولى من القسمة ما ذكره صاحب الكتاب لترددها بين النفي والإثبات وإيرادها على وجه لا يمكن التشغيب فيهما وهي أن المعلومات لا تخلو إما أن تكون لها صفة الوجود أو لا تكون لها صفة الوجود، ما ليس لها صفة الوجود فهي المعدومات وما له صفة الوجود لا يخلو إما أن لا يكون لوجودها أول وإما أن يكون لوجودها أول فترددت هذه أيضاً كالأولى ما لا أول لوجوده ليس إلا القديم تبارك وتعالى والكلام فيه وفي صفاته يفرد بالذكر به ويميز عن الكلام في غيره من المعلومات تعظيماً له تبارك وتعالى. فالذي لوجوده أول إما أن تكون متحيزة عند وجودها وإما أن لا تكون متحيزة إذا وجدت فترددت هذه أيضاً كالأولة والثانية. فما يتحيز عند الوجود ليس إلا الجواهر وهي كلها جنس واحد على ما نبينه فلا نحتاج إلى تكلف القسمة فيها كما احتجنا إليها في الأعراض لوقوعها على أجناس وأنواع، وما لا يتحيز عند الوجود فهي الأعراض وإن لم يكن هذا حالها لأنا نجدها بأنها ما يعرض في الوجود ولا يجب لبثها كلبث غيرها من الجواهر والأجسام ونتحرز بقولنا: ولا يجب لبثها عن الأعراض الباقية كالألوان وغيرها فإنها وإن كانت باقية فإن بقاءها لا يجب كبقاء الجواهر وإذا ثبت هذا فجملة ما ثبت بالدليل أنه عرض اثنان وعشرون ذاتاً وهي الألوان والطعوم والروائح والحرارة والبرودة والرطوبة واليبوسة والآلام والأصوات والتأليف والاعتماد والأكوان والحيوة والقدرة والشهوة والنفار والاعتقاد والظن والنظر والفناء والإرادة والكراهة، وإن كانت في هذه الجملة ما نعلمها ضرورة على طريق الجملة، وأما ما عدا ذلك من العجز والسهو والنسيان والموت والإدراك فمما لا دلالة على إثباتها على ما نفصل القول فيها في مواضعها إن شاء الله

كتاب التذكرة (ص 33–34)

الحمد لله وبه نستعين وعليه نتوكل. اللهم إنا نستهديك طريق الحق ونسألك التوفيق للصدق ونعوذ بك من ركوب الهوى وتسليط الشبه ونعتصم بك من الزلل في القول والعمل ونرغب إليك في أن تجعل ما نأتيه خالصاً لك ومطابقاً لرضاك وأن نصلى على نبيك المصطفى محمد خير الأنام وعلى آله البررة الكرام. هذه تذكرة تشتمل على بيان أحكام المعلومات وأوصافها وتجتمع إلى الإيجاز في القول استقصاء في الأدلة والأسئلة وبياناً للأصول والفروع وبالله نستعين وعليه نتوكل.

اعلم أن المعلومات أجمع لا تخرج عن قسمة تتردد بين النفي والإثبات فإما أن تكون لها صفة الوجود [وهو المعبر عنه بالموجود] وإما أن لا تكون لها صفة الوجود وهو المعبر بالمعدوم، والذي له صفة الوجود فإما أن تكون حاصلة له عن أول أو لا عن أول وهذه القسمة كالأولى. فالذي لا أول لوجوده ليس إلا القديم وحده عز وجل والكلام فيه وفي صفاته ينفرد عن الكلام في غيره من المعلومات فلا نجمع بينه وبينها في الذكر إعظاما له تعالى. والذي لوجوده أول هو المعبر عنه بالمحدث وهو ينقسم إلى ما يتحيز عند الوجود وإلى ما لا يتحيز عند وجوده فالأول هو الجوهر والثاني هو العرض، وإن لم يكن ما ذكرناه حداً له.

والجوهر (فهو) جنس واحد فلا يحتاج من ذكر أقسامه إلى ما نحتاج إليه في العرض فإنه يقع على أنواع وأجناس وجملة ما يثبت بالدليل أنه عرض هو الألوان والطعوم والروائح والحرارة والبرودة والرطوبة واليبوسة والأكوان والتأليف والاعتماد والألم والصوت والحياة والقدرة والشهوة والنفار والإرادة والكراهة والاعتقاد والظن والنظر والفناء ويدخل في هذه الأنواع التي عددناها من الأقسام ما يطول ويكثر ولكني أذكر منها ما ينظم الفائدة إن شاء الله

شرح التذكرة (38أ:12–39أ:15)

فصل: اعلم أنه رحمه الله لما بيّن الكلام في أحكام الفناء وما يتصل بذلك أورد الكلام بعد في الإعادة وما يتصل بها من الأبواب فبدأ من ذلك بالجوهر
اعلم أن الجوهر يجوز عليه الإعادة وهذا لا شبهة فيه، وذلك لأن الجوهر باق وهو من فعل الله تعالى يختص بالقدرة عليه يقع منه ابتداءا لا عن سبب، فوجب أن لا يختص بالوجود في وقت دون وقت لأن الأوقات في هذا الباب على سواء فمتى خرج من أن يكون مقدوراً في بعض الأوقات بأن يصير موجوداً فإنه يصح أن يكون مقدوراً في وقت آخر بأن يعدم ويعاد في ذلك الوقت،

ولأن أفعال الله تعالى إذا كان يجوز فيها التقديم والتأخير فلو قدرنا أن القديم تعالى أخّر خلق الجواهر إلى وقت الإعادة فإنه يصح منه إيجادها في ذلك الوقت ابتداءً فكذلك وجب أن يصح منه إيجادها في ذلك الوقت على وجه الإعادة لأن الإعادة ضرب من الإيجاد.

فصل: الغرض بهذا الفصل الكلام في بيان الخلاف بين مشايخنا في كيفية الإعادة
وجملة القول في ذلك أنه لا خلاف بينهم أن الذات إذا كان مما يجوز عليه البقاء وكان مما يختص القديم تعالى بالقدرة عليه ابتداءً لا عن سبب فإنه يصح إعادته وإنما الخلاف بينهم فيما يدخل جنسه تحت مقدور العباد هل يصح إعادته أو لا؟

فذهب الشيخ أبو علي إلى أنه لا تصح إعادته وذهب الشيخ أبو هاشم إلى أنه يصح إعادته،

وأما مسبب السبب أنه هل يصح إعادته أو لا يصح فلأبي هاشم مذهبان، أحدهما أنه يصح إعادته ابتداءً والثاني أنه لا يصح إعادته لا ابتداءً ولا متولداً عن سبب،

كتاب التذكرة (ص 237–247)

فصل [في إمكان إعادة الجواهر بعد إفنائها]
اعلم أن الجواهر تصح إعادتها بعد إفنائها، وليس في ذلك خلاف بين شيوخنا. والدليل عليه أن الجوهر قد صح كونه مقدوراً لله جل وعز مبتدءً وصح أنه باق وأنه لا ينتهي في الوجود إلى حد لا يجوز وجوده من بعد. فإذا صحت هذه الجملة وجب أن لا تختص في صحة وجودها من جهة القادر لنفسه بحال دون حال وأن يكون إذا امتنع إيجاده وهو موجود فهو استحالة ذلك. فإذا عدمت الجواهر فقد عادت إلى ما كانت عليه في الأول وزال المانع. فكما وجب في الابتداء صحة أن يوجدها تعالى فكذلك من بعد لأن الإعادة ليست بأزيد من إيجاد مخصوص.
وبعد، فلا بد من صحة التقديم والتأخير على الجوهر لأنه باق غير متولد عن سبب ولا يدخل تحت القدرة. وإذا صح التقديم والتأخير عليه فلو قدرنا أن الله تعالى أخّر إيجاده إلى الوقت الذي جعلناه وقتاً للإعادة لكان لا بد من القول بصحة وجودها ابتداءً فكذلك إذا عدمت الجواهر الموجودة تجب صحة إعادتها لأن الإعادة من ضرب من تأخير إيجاد المعاد.

فصل [في جواز إعادة الباقيات]
اعلم أن شيوخنا رحمهم الله لم يختلفوا في جواز إعادة الباقيات إذا اختص القديم جل وعز بالقدرة عليها وكانت مبتدأة، وهذا سبيل الأجناس التي لا تدخل تحت القدر مما تختص بصحة البقاء والطريقة في صحة إعادتها ما تقدم وإنما اختلفوا في الباقي إذا استوى في القدرة عليه جميع القادرين كالتأليف وغيره.
فقال الشيخ أبو علي: إن الإعادة في هذا النوع وأمثاله مستحيل وإنما تصح فيما لا تتناول القدرة جنسه إذا كان باقياً.
قال أبو هاشم: بل يكفي كونه باقياً وأن يكون القديم جل وعز هو الفاعل له وإن كان جنسه داخلا تحت القدرة. هذا هو الذي قاله أبو هاشم
وله مذهبان في أن الواقع بسبب من جهة الله تعالى: هل يصح منه أن يوجده بعينه ابتداء أم لا؟ قال في الجامع الكبير: إنه يصح وجوده لا عن ذلك السبب بل مبتدءً فعلى هذا القول يكفي في شروط صحة الإعادة ما تقدم. وقال في الأبواب: بل لا يصح في المتولد عن سبب أن يوجد إلا عنه. فعلى هذا القول، وهو الصحيح من قوليه، يجب أن يزاد فيقال: وأن لا يكون متولداً عن سبب لا يبقى لأنه إذا لم يبق سببه وكانت إعادته لا تكون إلا بإعادة سببه أدى إلى أن لا يصح فيه الإعادة.

وهذا هو الصحيح وهو الذي اختاره قاضي القضاة. وإذا كان كذلك فشروط الإعادة عنده على المذهب الأول هو أن يكون الفعل مما يجوز عليه البقاء وكان من فعل الله جلّ وعزّ سواء كان متولداً عن سبب أو مبتدءً بعد أن تكون الإعادة على سبيل الابتداء، وعلى المذهب الثاني يعتبر في شروط الإعادة أن يكون من فعل الله تعالى وكان مما يجوز عليه البقاء ويكون مبتدءً لا متولداً عن سبب. وإنما قلنا: إن ما كان متولداً عن سبب فإنه لا يصح إعادته لأنه سببه لا يخلو إما أن يكون مما يجوز عليه البقاء أو لا يجوز عليه البقاء. فإن كان لا يجوز عليه البقاء لم يصح الإعادة في المسبب لأن سببه لا تصح إعادته من حيث لا يجوز عليه البقاء ولا يجوز أن يعاد بسبب آخر، ولا على وجه الابتداء لأن ذلك يؤدي إلى أن يكون له وجهان في الحدوث، أحدهما على وجه الابتداء والثاني على وجه التوليد. فإذا جاز أن يحصل على الوجهين في الوقتين جاز أن يحصل عليهما في وقت واحد إذ لا منافاة بينهما فيؤدي ذلك إلى صحة مقدور واحد بين قادرين. ولا يجوز أن يعاد أيضاً بسبب آخر لأن ذلك يؤدي إلى أن يكون المسبب الواحد له سببان يشتركان في توليده وذلك لا يجوز لأن فيه صحة مقدور واحد بين قادرين وإن كان سببه مما يجوز عليه البقاء فإذا أعيد المسبب يجب أن يعيد سببه أيضاً. فإذا أعيد سببه فإنه كما يتعلق بتوليد هذا المسبب على وجه الإعادة وجب أن يكون له مسبب آخر على وجه الابتداء كما يكون له على وجه الإعادة فيؤدي ذلك إلى أن يكون له مسببات ما لا تتناهي على وجه الابتداء وعلى وجه الإعادة، وذلك لأنه إذا تعدى في التوليد عن مسبب واحد إلى ما زاد ولا حاصر وجب أن يتعدى إلى ما لا نهاية له من المسببات وهذا كما نقول في القدرة أنها لو تعدت في التعلق عن وجه واحد إلى ما زاد عليه ولا حاصر، وجب أن تتعدى في التعلق إلى سائر الوجوه. وبعد، فإن السبب لو صح إعادة مسببه بإعادته أدى إلى أن يكون له مسببات ما لا تتناهي إذ لا اختصاص له ببعض المسببات دون بعض.

فإن قيل: ما أنكرتم أن هذه القضية تجب في السبب الذي تكون حالتا حدوثه وبقائه واحداً سواء في التوليد، وهلا فصلتم بين هذا السبب والمسبب الذي يختص بالتوليد حالة الحدوث كالمجاورة مع التأليف فإنها تختص بالتوليد حالة الحدوث، وقلتم: إن السبب الذي تكون حالة حدوثه وحالة بقائه سواء فإنه لو أعيد به مسببه أدى إلى أن يكون له مسببات ما لا يتناهي. وأما السبب الذي يختص بالتوليد حالة الحدوث فإنه إذا أعيد به مسببه لا يقتضي أن تكون له مسببات ما لا تتناهي كما فصلتم ما يبقى من الأسباب وبين ما لا يبقى الفصل الذي ذكر تم.

قيل له: هذا لا يصح لأن السبب الذي يختص بالتوليد حالة الحدوث فحالتا حدوثه أعني حالة الحدوث على الابتداء وحالة الحدوث على وجه الإعادة في باب التوليد لحالة الحدوث وحالة البقاء فيما يولد في حال البقاء فكما أن ما يولد في حال البقاء لو قيل أن مسببه يعاد به أدى إلى أن تكون له مسببات ما لا تتناهي فكذلك هذا

والصحيح على ما ذكره قاضي القضاة أن نجمع إلى الشروط التي تقدمت أن لا يكون متولداً أصلا سواء كان سببه باقياً أو غير باق. أما إذا لم يبق فالحال ما تقدم وإذا بقى فمن حقه أن يكون له في كل حال مسبب غير ما تقدم، كما يجب في القدرة أن يكون مقدورها في كل حال غير مقدورها في الحالة الأخرى. وعلى هذا ثبت في اللازم من الاعتماد أن الحاصل عنه في كل حال من الحركات غير ما تقدم.

فلو جوزنا الإعادة على مسببه لكان إنما يصح بإعادة سببه ومن حقه أن يوجد على وجه ابتداء جزءاً وعلى وجه الإعادة جزءاً آخر فيتعدى إلى ما لا يتناهى كما نقوله في القدرة.

وبعد، فإذا كان لهذا السبب مسببات فلو صحت إعادتها لم تكن بأن يعود بعضها أولى من بعض لأنه مع وجود السبب وزوال العوارض لا يقف وجود مسببه على الاختيار.

فإن قال: هذه القضية إنما تستمر في السبب الباقي الذي يستوى حال حدوثه وحال بقائه في التوليد كما قلتم في الاعتماد فأما إذا كان الكلام في الكون المولد للتأليف فهو إنما يولده في حال الحدوث فقط فقد صار ليس له إلا مسبب إعادة هذا الجزء الواحد من التأليف بعينه فقط فلا يقتضي التعدي الذي ذكرتم فهلا إذ سويتم بين المسبب الذي يبقى سببه والذي لا يبقى فضلتم هذا الضرب من التفضيل في الأسباب؟

قيل له: يجب في المجاورة، إن قلنا فيها بما ذكرته، أن يكون المتولد عنها حال الحدوث ثانياً غير ما يتولد عنها أولاً لأن حالتي الحدوث على هذه المجاورة تتنزل منزلة حالة البقاء مما يستوي في باب التوليد فيه حال الحدوث وحال البقاء فيلزم ما قلناه من قبل ويحل الكون محل الاعتماد في هذا الوجه، وإن كان الحال في الاعتماد أظهر.

فصل الغرض بهذا الفصل الكلام في أن ما لا يبقى من الذوات لا تجوز عليه الإعادة
لا خلاف بين مشايخنا في أن ما لا يبقى من الذوات لا يجوز عليه الإعادة. قال الفقيه: وأظن أن الخلاف في ذلك مع هؤلاء المجبرة من النجارية والأشعرية وأما مقدورات العباد فإنه لا يصح عليها الإعادة عندنا، [38ب] وذهب البغداديون إلى أن فيها ما يصح الإعادة.

أما الكلام في أن ما لا يبقى لا يصح عليه الإعادة هو أن القول بصحة الإعادة عليه قول بصحة وجوده في وقتين مع تخلل العدم فلو جاز وجوده في وقتين مع تخلل العدم فكذلك وجب أن يجوز وجوده وقتين من غير تخلل العدم إذ ليس هاهنا ما يمكن إن يشار إليه، فيقال أنه يمنع من ذلك إلا وجوده والوجود لا يصح أن يكون مانعاً من ذلك لأن الشيء وإن ثبت أنه يمنع ضده فإنه لا يمنع مثله بالاتفاق، وقد علمنا أن صفة الوجود في الذوات صفة واحدة. فإذا صح وجوده في الوقت الثاني من غير تخلل العدم فكذلك في الثالث والرابع وما والاه من الأوقات إذ لا اختصاص له ببعض الأوقات دون بعض فيؤدي إلى أن يصح وجوده في كل وقت ففي هذا إلحاق له بالباقيات، وهذا يؤدي إلى انقلاب ذاته لأن ما لا يبقى من الذوات لو صح عليه الإعادة لوجب أن يصح فيها التقديم والتأخير فإنه ما من وقت إلا ويصح فيه وجوده فسواء فيما قبل أو فيما بعد فإذا صح وجوده في كل وقت فلا يجب انتفاؤه بل يجب أن يكون باقياً أبداً، وهذا يؤدي إلى قلب ذاته.

وأما الكلام في أن مقدورات القدر لا يصح عليها الإعادة:

اعلم أن مقدورات القدر على ضربين، منها ما لا يجوز عليها البقاء فلا يصح عليها الإعادة ومنها (واما، الأصل) ما يجوز عليه البقاء. فلو قلنا: إنه يجوز عليه الإعادة فلا يخلو القول في ذلك إما أن يقال أن القديم تعالى يعيده أو يقال أن الواحد منا يعيده. لا يصح أن يقال أن القديم تعالى يعيده لأن ذلك يؤدي إلى أن يصح مقدوراتنا أن تكون مقدورات لله تعالى وذلك لا يجوز لاستحالة مقدور واحد بين قادرين، ولا يجوز أن يقال أن الواحد منا يعيده لأنه لا يخلو إما أن يعيد بتلك القدرة أو بغيرها من القدر. لا يصح أن يقال أنه يعيد بتلك القدرة لأن تلك القدرة لو صح أن يعاد بها مقدورها لأدى إلى أن تكون لها مقدوران في حالة واحدة على وجه الابتداء وعلى وجه الإعادة، وذلك لا يجوز كما بينا في السبب. وإن أعاد بغيرها من القدر لم يجز أيضاً لأن ذلك يؤدي إلى تعلق القدرتين بمقدور واحد وتعلق القدرتين بمقدور واحد يقتضي صحة مقدور واحد بين قادرين لأن هاتين القدرتين لو حصلت كل واحدة منهما لقادر واحد لكان المقدور الواحد مقدوراً لهما، وقد علمنا خلاف ذلك.

فصل [في استحالة إعادة ما لا يبقى]

فأما ما لا يبقى فالإعادة مستحيلة عليه وما يبقى إذا اختص في حدوثه بوقت فالإعادة غير جائزة عليه وذلك نحو مقدورات القدر. وقد ذهب الأشعري إلى جواز إعادة ما لا يبقى وفي المجبرة من جوز إعادة مقدورات القدر وهو محكي عن بعض البغداديين وغيرهم من أهل العدل.

والذي يبطل القول بصحة الإعادة فيما لا يبقى أن تجويزها يقتضي قلب جنسه من حيث تقتضي فيه جواز البقاء عليه لأن إعادته تقتضي صحة وجوده في وقتين، ثم لا فرق بين وجوده فيهما على وجه التوالى أو على وجه يتخلل بين الوقتين ثالث، لأن الصفة لا تحيل نفسها وإن أحالت غيرها. والوجود في الحالين صفة واحدة ولو وجد في وقتين أو صح ذلك فيه على ما ذكرنا لألحقه بالباقيات. وبعد، فإذا جوزت إعادته فقد قيل بجواز التقديم والتأخير عليه، لأنه إذا عدم صح أن يعاد في الثاني وصح تأخيره إلى الثالث والرابع، ثم كذلك في كل وقت فيجب إذا أن لا يشار إلى وقت إلا ويصح أن يوجد فيه معاداً، وهذا يقتضي أن الحالة التي تلي الحالة الأولى في صحة وجوده فيها كالحالة الثانية التي قد تخللها حال انقطع فيها الوجود وهذا يوجب صحة البقاء عليه واتصال الوجودية فيجب أن تمتنع عليه الإعادة وأن يكون كما اختص حدوثه بوقت أن يختص وجوده بوقت واحد وإلا لزم ما ذكرناه.

وبمثل هذه الطريقة نبطل جواز الإعادة على مقدور القدر لأنه وإن كان فيها ما لا تختص في الوجود بوقت فجميعها تختص في الحدوث بوقت فتلحق بما لا يصح البقاء عليه. والأصل فيه أن من حكم القدرة الواحدة أنها لا تتعلق والوقت والجنس والمحل واحد من الفعل فلو جوزنا الإعادة على مقدورها لبطل هذا الحكم لأنه لا بد من أن يعاد بهذه القدرة بعينها، فإن القدرة الأخرى لا تتعلق بهذا المقدور ولا تتعلق به قادر آخر. وإذا وجب ذلك، وقد صح أن مقدورها في كل حال غير مقدورها في حال أخرى، فيجب لو صحت إعادة مقدورها أن تتعلق القدرة الواحدة بجزء من الفعل على وجه الابتداء وبجزء آخر على وجه الإعادة فتخرج عن الحكم الذي ذكرناه.

فإن قيل: ما أنكرتم أن هذه القدرة تتعلق بمقدورها على وجه الإعادة ولا تتعلق بمقدور آخر على وجه الابتداء، قيل له: هذا لا يصح لأنه لو كان كذلك لوجب أن لو قدرنا بقاء القدرة والمقدور إلى وقت الإعادة أن لا يكون لها مقدور بعد ذلك، وهذا يؤدي إلى قلب جنسه وذاته فلا يجوز.

فإن قيل: لم لا يجوز أن يقال أنها تتعلق بكل واحد منهما على سبيل البدل بأن تتعلق بأحد منها على وجه الإعادة بدلاً من تعلقها بالآخر على وجه الابتداء ولا يلزم ما قلتم؟ قيل له: هذا لا يصح لأن هذا إنما يقال في الضدين، وكلامنا إنما وقع في المتماثل والمتجانس فثبت بهذه الجملة أن مقدورات القدر لا يصح عليها الإعادة، وإنما لا يصح عليها الإعادة لأمرٍ يرجع إلى القدر. فإذا ثبت أن انتفاء صحة الإعادة عليها لأمرٍ يرجع إلى القدر صح القول بصحة الإعادة فيما كان من فعل الله تعالى مما يدخل جنسه تحت مقدورات العباد خلافاً لما ذهب إليه أبو علي.

فصل لما بيّن رحمه الله الكلام في الإعادة وما يصح فيه الإعادة وما لا يصح، بيّن الكلام في أن الإعادة لا يجوز أن تكون لمعنى

فقد ذهب عباد بن سلمان الصيمري وهشام بن عمرو وأبو بكر الزبيري إلى أن الإعادة تكون لمعنى، وهكذا ذهبوا في الحدوث أنه لمعنى إلا أبو بكر الزبيري فإنه لم يساعدهم في الحدوث.

وذهب أبو الهذيل إلى أن الإعادة تتعلق بقول صادر من قبل الله تعالى كما قال في الإيجاد وكذلك الإفناء.

والدليل على أن الإعادة لا يجوز أن تكون لمعنى ما قد ثبت أن المعاد ليس له بكونه معاداً حالة وصفة أكثر من أنه وجد بعد عدم تقدمه وجود هذا هو المعنى في الإعادة، إلا أن أهل اللغة استطالوا هذه العبارة فأجروا لفظ المعاد. فإذا لم يكن له بكونه معاداً حالة وصفة فلا يصح أن يقال أنه لمعنى أو لنفسه، لأن استحقاق الصفة لمعنى أو للنفس فرع على ثبوت الصفة في الأصل.

والدليل على أن المعاد ليس له بكونه معاداً حالة وصفة هو الدليل بأن الباقي ليس له بكونه باقياً حالة وصفة لأن المعاد لو كان لمعنى لوجب لو أن الله تعالى أخر الإيجاد ابتداءً إلى وقت الإعادة لم يكن ذلك المعنى بأن يؤثر في وجوده على وجه الابتداء أولى من أن يؤثر

ولا يمكن أن يقال: إن في الثاني لا يصح أن يوجد بها \غير ذلك المعاد، لأن هذا يقتضي أن لو بقي الفعل الذي هو مقدور بهذه القدرة ولم ينتف ولم يعدم وكانت القدرة أيضاً باقية أن لا يكون لها مقدور في الوقت الثاني أصلاً وهذا يقتضي أن تعلقها بالمقدور هو موقوف على فناء هذا المعاد دون أن يرجع إلى ذاتها وقد عرفنا خلافه.

وليس يمكن أن يقال: إنها تتعلق بالجزء الواحد معاداً بدلاً من كونه مبتدءً بدلاً من كونه معاداً. فإذا وجد الفعل على أحد هذين الوجهين بطل صحة وجود الفعل الثاني على الوجه الآخر فلا نسلم تعلقها بالجزئين على هذين الوجهين، وذلك لأن الطريقة إنما تصح لو كان الفعلان ضدين فيقال بدخول البدل فيهما ونحن نصور الكلام في الإلزام الذي أوردناه في مثلين يصح وجودهما وإن اختلف الوجه الذي عليه يوجدان. وإذا صح أن المانع من جواز إعادة مقدورات القدر أمر يرجع إلى القدر وما يجب لها من الأحكام، فيجب أن يمتنع من القديم جل وعز أن يعيد الباقيات إذا فعلها ابتداءً، وإن كانت من جنس مقدور القدرة لأنه قادر لنفسه، فالمانع فيه غير حاصل.

فصل [في أن المعاد لا يعاد بإعادة]

والمعاد لا يكون معاداً بإعادة. وقد حكي ذلك عن العباد وهشام بن عمرو وأبي بكر الزبيري. وكذلك قالوا في المحدث أنه محدث بإحداث إلا أبو بكر الزبيري فإنه ما ساعدهما على ذلك في المحدث وأثبت المعاد معاداً لمعنى لما ظن أن له صفة زائدة على الحدوث فأجراها مجرى تحرك الجوهر في الحاجة إلى معنى.

وأما أبو الهذيل فإنه جرى على طريقته في الإحداث والإفناء والإعادة فجعل هذه الأمور متعلقة بقول وإرادة فيقول جل وعز «عد» أو «افن» أو «ابق» وقد تقدم إفساد ذلك في غير موضع.

وإنما منعنا أن يكون معاداً لمعنى لأنه ليس يفيد أكثر من وجود بعد عدم تقدمه وجود. فإذا عاد إلى الحالة الأولى قالوا: هذا معاد. واستطال أهل اللغة إيراد هذه الجملة فاقتصروا على ما ذكرناه من قولهم معاد فإذا صح أنه لا يكون موجوداً لعلة، فكذلك لا يكون معاداً لعلة.

وبعد، فإذا كان يصح من الله تعالى أن يوجد الجوهر ابتداء في الوقت الذي أوجده معاداً، فيجب أن لا يكون بد من وجود هذا المعنى فيه، فإذا أوجده على وجه الإعادة فيجب أن لا يصير هذا المعنى بأن يكون مقتضياً لكونه معاداً دون أن يكون مبتدءً إلا لمعنى آخر، ثم يتصل بمعان لا تتناهي وليس يجب ثبوت المعنى لأجل اختلاف الاسم عليه كما لم يجب ذلك في كونه باقياً. وكذلك فلا يجب لأجل أنه جاز وجوده وجاز أن يبقى معدوماً، أن يثبت

في وجوده على وجه الإعادة إلا لمعنى، ثم الكلام في ذلك المعنى كالكلام في المعنى الأول فيؤدي إلى ما لا يتناهي من المعاني.
فإن قيل: إنه إذا أخر إيجاده إلى وقت الإعادة فإنه لا يكون معاداً بل يكون مبتداءً فلا يجب أن يكون لمعنى، قيل له: هذا لا يصح لأن الإعادة ليس بأكثر من تأخير الإيجاد إلى وقت مخصوص، فمتى أخّر الإيجاد إلى ذلك الوقت كان معاداً وإن لم يسم معاداً فتغير الاسم لا تغير المعنى، وفي مسألتنا إذا أخّر الإيجاد إلى وقت الإعادة لزم ما ذكرناه من أنه لا يكون ذلك المعنى بأن يؤثر في الوجود على وجه الابتداء أولى من أن يؤثر على وجه الإعادة إلا بمعنى آخر، ثم يلزم إثبات ما لا يتناهى من المعاني.
فإن قيل: إنه حصل معاداً بعد أن لم يكن معاداً فيجب أن يكون ذلك لمعنى، قيل له: هذا لا يكون مقتضياً أن تكون الإعادة لمعنى إذ لو كان كذلك لوجب أن يكون الموجود موجوداً لمعنى لأنه حصل موجوداً بعد أن لم يكن موجوداً فوجب أن يكون ذلك لمعنى، فإن ارتكب ذلك فقد ثبت فيما تقدم فساد ذلك.

فصل الغرض بهذا الفصل الكلام بأن المعاد إذا أعيد فصفته في حال الإعادة هي صفته في حال الوجود ابتداءً لا مثالها ولا خلافها
وإنما قلنا ذلك لأنه لو صح أن يحصل على صفتين مثلين في حالتين لصح أن يحصل عليهما في حالة واحدة لأن الشيء لا يمنع مثله كما لا يمنع نفسه وإنما يمنع ضده، وهذا يؤدي إلى حصول التزايد في هذه الصفة وقد بينا أن هذه الصفة لا تقبل التزايد.

فصل ثم لما بيّن رحمه الله الكلام في الإعادة وما يصح إعادته وما لا يصح، بيّن الكلام من بعد في من يجب إعادته
وجملة القول في ذلك أنا لو خُلِّينَا والعقل لكُنَّا نقول بأن القديم تعالى لا تجب عليه إلا إعادة من يجب له عليه تعالى حق إما الثواب وإما العوض، فأما من لم يجب له عليه حق وإنما يجب لله عليه حق فإنه لا تجب إعادته إلا أن السمع ورد بأن الحيوانات كلها معادة، قال تعالى (وَإِذَا الْوُحُوشُ حُشِرَتْ) (سورة التكوير (81): 5)

والغرض بهذا الفصل الكلام في [+ أن ما، الأصل] من يعيده الله تعالى ما يجب عليه أن يعيده منه؟ فذهب أبو علي إلى أنه يجب أن يعيد كل جزء من الحي حتى الأقطع وجب عليه أن يعيد اليد المبانة منه. وهذا المذهب حكاه عنه أبو هاشم إلا أنه استبعد هذا المذهب منه وذهب الشيخ أبو هاشم إلى أنه إنما يجب أن يعيد الأجزاء التي لا يكون الحي حياً إلا

المعنى الذي قالوه، لأنه لو وجب ذلك في المعاد لوجب في الحادث، لأن هذه العلة موجودة فيه ومع هذا لم نقل بحدوثه لأجل معنى.

فصل [في أن صفة المعاد لا تتغير]
وكل ما يعاد فصفته بالوجود هي التي كانت أولاً لا أنه يحصل مثلها فإنه لو صح كونه على حالين بالوجود في وقتين لصح كونه عليها والوقت واحد. وهذا يقتضي صحة وقوع التزايد في هذه الصفة، وذلك مما قد أبطلناه من قبل.

فصل [في بيان وجوب الإعادة على الله]
اعلم أن كل من تجب إعادته فهو كل من له حق على الله جل وعز من ثواب وعوض. وإذا كان لا يمكن توفيره إلا بالإعادة وجبت، لأن الواجب ليس يتم دونها، ثم السمع قد دلنا على أنه تعالى يعيد من عليه حق كالعقاب ويعيد سائر الحيوانات ويتفضل عليها.

فأما القدر الواجب في إعادة من له أو عليه حق فهو القدر الذي لا بد منه في كونه حياً على ما قاله أبو هاشم دون ما قاله أبو علي في كتاب الإنسان من وجوب إعادة سائر الأبعاض حتى قال في من قطع يده أنه يجب إعادتها بعينها، وقد استبعد أبو هاشم هذه الحكاية عنه. وكما أوجب أبو هاشم ما ذكرناه فقد أوجب في المعاد من هذه الأجزاء أن يعاد التأليف الذي كان فيها معها. ثم حكى الشيخ أبو عبد الله رجوعه عن هذا القول إلى وجوب إعادة الحياة بعينها وبه قال هو. وذكر من بعد في جواب مسألة الصاحب أنه لا تجب إعادة الحياة بعينها وإنما تجب إعادة ما لا تكون حياة إلا له سواء كانت هذه بعينها أو غيرها. وهذا التحقيق في الخلاف بينهم قد حكاه قاضي القضاة في الخلاف بين الشيخين، ودل ذلك على اتفاقهم على وجوب إعادة ما لا يكون الحي حياً إلا معه من الأجزاء وإنما اختلفوا في

بها وأما ما عدا ذلك فلا يجب. قال: ويجب أيضاً أن يعيد التأليف بعينه، ثم رجع فقال: لا يجب إعادة التأليف بعينه وإنما يجب مثله، والذي يجب إعادته بعينه إنما هو الحياة، وهذا هو الذي ذهب إليه أبو عبد الله البصري، ثم أن أبا عبد الله ذهب في جواب مسائل الصاحب إلى أن الحياة أيضاً لا تجب إعادتها بعينها وإنما تجب إعادة مثلها، وهذا الخلاف على الحد الذي ذكرناه ذكره قاضي القضاة في الخلاف بين الشيخين، وبهذه الحكاية [39أ] نبّه على أنهم متفقون أن الذي يجب عليه تعالى إعادته إنما هي الأجزاء التي لا تكون الحي حياً إلا بها، وإنما وقع الخلاف بينهم في ما عدا هذه الأجزاء هل تجب إعادته بعينه أم لا؟ يبين ذلك أن من مذهبهم أن عرضاً من الأعراض إذا وجد في محل [لا يصح، الأصل] فكان لا يصح وجوده إلا في ذلك المحل، ومنهم من قال أنه تجب إعادة التأليف بعينه وكذلك تجب إعادة الحياة بعينها، ومنهم من قال: تجب إعادة الأجزاء كلها بأعيانها حتى الأقطع يجب أن تعاد يده المقطوعة، واختلافهم شتى على اتفاقهم أنه يجب على الله تعالى إيجاد الأجزاء التي لا تكون الحي حياً إلا بها، ويسقط بهذا تشنيع من شنع علينا بأن مشايخكم مختلفون في الإعادة، فإنهم لم يختلفوا في الإعادة وإنما اختلفوا في تفاصيلها.

وجملة القول أن الذي تجب إعادته إنما هي الأجزاء التي لا يكون الحي حياً إلا بها على ما ذهب إليه أبو هاشم وقرره أبو عبد الله، فأما ما عدا ذلك من التأليف والحياة فإنه لا تجب إعادته بعينه وإنما تجب إعادة ما لا يكون من التأليف إلا له، وكذلك إعادة الحياة التي لا تكون حياة إلا له. والدليل على صحة هذا القول أن الذي تجب إعادته إنما هو العاصي والمطيع، والمطيع والعاصي إنما هو هذه الجملة التي يكون الحي حياً بها وإليها يتوجه استحقاق المدح والذم والثواب والعقاب،

يبين ذلك [أن] حال الإيجاد مشبه بحال البقاء، وقد ثبت أن في حال البقاء إنما يتوجه المدح والذم والثواب والعقاب إلى هذه الأجزاء التي لا يكون الحي حياً إلا بها دون ما زاد عليها بدليل أنه لو تعدى إلى زيادة أو نقصان فإنه لا يتغير ما يستحقه من المدح والذم والثواب والعقاب. يزيد ما قلناه وضوحاً أنا لو قدرنا تجدد التأليف حالاً بعد حال وكذلك تجدد الحياة حالاً بعد حال فإنه لا يتغير ما يستحقه من المدح والذم والثواب والعقاب.

فأما الشيخ أبو هاشم حيث ذهب المذهب الأول في أن التأليف تجب إعادته بعينه قال: قد ثبت أن التمييز إنما يقع فيما يرجع إلى التخطيط والهيئة والتركيب والصورة، وذلك راجع إلى التأليف فتجب إعادته بعينه،

إلا أن الشيخ أبا عبد الله اعترض على هذا فقال: التأليف حكمه مقصور على المحل وما كان حكمه مقصوراً على المحل لا يقع به التمييز وإنما يقع التمييز فيما يرجع حكمه إلى

هل يجب ضم غيره إليه حتى يجري مجراه في وجوب الإعادة أم لا؟ ويبين ذلك ما قد تقرر من مذهبهم أن الموجود في محل لا يصح وجوده إلا فيه فلا بد لمن أوجب إعادة ذلك التأليف بعينه أو تلك الحياة بعينها أن يوجب وجودها في تلك الأجزاء لا غير. وعلى هذا قال الشيخ أبو علي بوجوب إعادة اليد المقطوعة، ولولا اتفاقهم على وجوب إعادة أجزاء الأصل لم يكن لذكر ذلك فائدة فيبطل تشنيع من يشنع عليهم في هذا الباب.

والصحيح أنه إنما تجب إعادة الأجزاء التي لا بد منها في كونه حياً لأنها هي المطيعة والعاصية والمؤلمة. فأما الأبعاض التي قد يبقى حياً من دونها فهي زوائد يصح فيها التبدل وكذلك الحال في المعاني.

يبين هذا أن حال الإعادة مشبهة بحال البقاء، ومعلوم أنه يبقى حياً مع فقد هذه الزوائد وهكذا فلو قدرنا أن الحياة والتأليف مما لا يصح البقاء عليهما وأنهما يحدثان حالاً بعد حال لما أثر ذلك في كون الحي هو الأول. فيجب إذا كان الذي يصح أن يحيا به زيد غير منحصر بعدد، أن تجوز إعادة الحياة التي كان حياً بها بعينها وأن يصح بدلا من ذلك إيجاد الحياة التي عرف من حالها ما ذكرناه.

ولما قال الشيخ أبو هاشم بالقول الأول جعل العلة في وجوب إعادة التأليف أن هذه الجملة به تبين من غيرها، لأن التميز يقع بين زيد وعمرو بالصورة، وإلا فالأجزاء من جنس واحد فلهذا وجبت إعادته كما وجبت إعادة الأجزاء التي لا بد منها في كونه حياً.

واعترض الشيخ أبو عبد الله ذلك فقال: كيف تقع البينونية بالتأليف مع أن حكمه يختص بالمحل، فالواجب وقوعها بما يرجع حكمه إلى الجملة وهو الحياة، وقد بينا أنه لا معتبر بواحد منهما وأن ما تجب إعادته هو القدر الذي لا بد منه في كونه حياً وما بعد ذلك فالعلم به موقوف على السمع، وقد دل على أن المثاب يعاد على أحسن صورة وأكملها، وأن المعاقب يعاد على صورة شوهاء تنفر الطباع، نسأل الله الكريم السلامة من عقابه.

الجملة وإن كان ولا بد فإنما تجب إعادة الحياة بعينها لأن حكمها يرجع إلى الجملة على أنا قد بينا أن الحياة لا تجب إعادته بعينها، وهذا القدر هو الذي يعتمد في العقل، فأما السمع فإنه ورد بأن الله جلّ وعزّ يعيد المثابين على أحسن الصورة وأبين خلق ويعيد المعاقبين على أقبح صورة وأشوه هيئة، اعاذنا الله من النار وجعلنا من الفائزين برحمته.

SUHRAWARDĪ ON MODAL SYLLOGISMS

Tony Street

On first encounter, Suhrawardī's modal syllogistic in the *Philosophy of Illumination* looks to be quite different from Avicenna's. Closer inspection, however, reveals that Suhrawardī's system is deeply Avicennan, more so than, for example, the system in Najm al-Dīn al-Kātibī's *Shamsiyya*. In this paper I examine what Suhrawardī achieves by reading all modal propositions as embedded in the phrase 'by necessity' (*bi'l-ḍarūra*), what help it can offer the historian of Arabic logic, and what it means for assessing Suhrawardī's larger philosophical project.

1. The Problem

Scholars who have looked at Suhrawardī's treatment of modal syllogisms to date have tended to take it to be one of the areas of Peripatetic philosophy which he attacks and reformulates. Prompting these assessments is the way Suhrawardī lays out the syntax of the propositions in his syllogistic. Here is how John Walbridge and Hossein Ziai describe his procedure:

> In the second discourse... Suhrawardī attacks the complex structure of the Peripatetic modal syllogistic. He argues that for any scientific purpose the modal proposition can be reduced to a universal necessary affirmative proposition with modality, quantification, and negation incorporated into the terms of the proposition. For example, 'It is possible that any man is literate' may be reformulated as 'It is necessary that all men are contingently literate.'... By means of this, all syllogisms can be reduced to one, a modalized iterated form of Barbara: 'Necessarily all A are B, and necessarily all B are C; therefore, necessarily all A are C.'[1]

Elsewhere, Ziai names the 'Peripatetic' from whom Suhrawardī is distancing himself, saying, 'This is among the important areas of formal

---

[1] Suhrawardī, *The Philosophy of Illumination*, ed. and trans. J. Walbridge and H. Ziai (Provo, 1999), introduction, pp. xxiv–xxv. I take all quotes from their translation of the *Philosophy of Illumination*, though I have written with one eye on the earlier edition by H. Corbin, *Oeuvres philosophiques et mystiques de... Sohrawardi*, ii (Tehran, 1952).

logic where Suhrawardī has departed from the Peripatetics (here specifically Avicenna)...'[2] Ziai further remarks that Dānesh-Pazūh regards this attempt to reduce all propositions to one type as ultimately Aristotelian, but Ziai clearly doubts this claim.[3] Overall, however, Ziai looks on these departures from Avicenna's logic as technical matters which should be left 'for a more specialized study of Suhrawardī's logic.'[4]

More recently, Walbridge has dealt with Suhrawardī's syllogistic in his *Leaven of the Ancients*. Walbridge believes that Suhrawardī has made important metaphysical decisions and rejected assumptions that motivate Aristotle's term logic.

> One large motivation to use term logic vanished with Suhrawardī's rejection of Aristotelian essentialism and essential definition [...]. Term logic, as Aristotle formulates it, is particularly useful for unpacking statements of essences and thus requires essentialist definition [...]. Suhrawardī rejected Aristotelian essentialism and held that essences could be known, if at all, only through direct acquaintance. It was in principle impossible to create definitions that were both complete and actually conveyed new knowledge. Detached from its scientific context, term logic loses importance in Suhrawardī's thought and can be reduced to his 'few simple rules'.[5]

I think—and I'm not sure I follow Walbridge's argument—this means that the new modal syntax is occasioned by a rejection of Aristotelian essentialism. Later on, while treating the propositional logic, Walbridge claims:

> Having rejected key features of Aristotelian essentialism as it applies to logic, Suhrawardī is thus pushed towards giving greater importance to propositional logic.[6]

To be fair, Walbridge admits that he is proceeding speculatively, and that more precise appraisal of Suhrawardī's logic will have to await editions of more of his works, and that, in any event, it's not 'always clear that his distinctive logical doctrines are exploited in the rest of his philosophy.'[7] Still, for Walbridge, the most promising line of speculation assumes an anti-essentialist motivation and function for the new logic and its distinctive form.

---

[2] H. Ziai, *Knowledge and Illumination* (Atlanta, 1990), p. 69.
[3] Ibid., p. 69, n. 3.
[4] Ibid., p. 75.
[5] J. Walbridge, *Leaven of the Ancients* (Albany, 2000), pp. 149–50.
[6] Ibid., p. 155.
[7] Ibid., p. 155.

This then, is the problem I want to consider in this study: Why does Suhrawardī reduce all his modal propositions in this way? Is it a departure from the Peripatetics and—more precisely—Avicenna? Is it an anti-essentialist move? I hope to show that, contrary to the claims of Ziai and Walbridge, we find in Suhrawardī an insightful twelfth-century defender and expositor of Avicenna's modal logic, a logic fitted out for essentialist metaphysics.

2. Preliminary Considerations

Suhrawardī's 'distinctive logical doctrines' appear only in the *Philosophy of Illumination*, a book which begins its treatment of logic speaking of other treatments Suhrawardī has written.

> Before I wrote this book and during the times when interruptions prevented me from working on it, I wrote other books in which I have summarized for you the principles of the Peripatetics according to their method. Among these books is the short work known as *Intimations of the Tablet and the Throne*. Many principles are summarized in it, despite its brevity [...]. But the present work has another method and provides a shorter path to knowledge than their method does. It is more orderly and precise, less painful to study [...].
> [...]
> We have reduced the famous 'tool' [*al-āla*] that guards thought from error to a small number of very useful rules. These are sufficient for the intelligent and for those who seek illumination. Whoever wishes to learn the details of this science—which is merely a tool—should consult the more detailed works.[8]

I take it that this entitles us to look to the *Logic of Intimations* to work out what is going on in the logic of the *Philosophy of Illumination*. In short, we can regard the two books as different expositions of the same logic.

What do we know about the *Logic of Intimations*?[9] Ziai's treatment of the book characterizes its method and substance as 'Peripatetic'.[10] This is a term so loose as to be useless. But if we examine the logic, we find that it is exactly the same modal logic that Avicenna presents. So—to take some telling examples—second-figure syllogisms with two absolute premises are sterile, first-figure syllogisms with possibility propositions as

[8] Suhrawardī, *Philosophy*, pp. 2 and 4.
[9] Suhrawardī, *Manṭiq al-talwīḥāt*, ed. A.A. Fayyāḍ (Tehran, 1955).
[10] Ziai, *Knowledge and Illumination*, pp. 41–5, 51–6, and 73–5.

minor premises (henceforth, M-minor syllogisms)[11] produce, and there is an exception to Aristotle's two-Barbara rule that uses a proposition of the form *every B is A while B* (which is a *waṣfī* proposition).[12] *Intimations* sets out Avicenna's account of the modal syllogisms without modification.

It is important to realise that Suhrawardī did not simply adopt the only logic with which he was familiar. Ziai stresses how important Abū 'l-Barakāt al-Baghdādī is for Suhrawardī, noting that he is one of the very few philosophers Suhrawardī quotes by name, albeit in insulting terms.

> It seems thus evident to me that Baghdādī should be regarded as an important immediate source for many of Suhrawardī's non-Peripatetic approaches to problems of philosophy [...]. Baghdādī and Suhrawardī both make serious attempts at reformulating many Avicennan philosophical principles, something no other philosopher does in such systematic manner.[13]

Turning to Abū 'l-Barakāt's logic in the *Evidential*, we find that the logic described differs from that presented by Suhrawardī and Avicenna (and, for that matter, from what is presented by Aristotle, which makes it interesting in its own right).[14] Abū 'l-Barakāt was not the source of Suhrawardī's logic. This means that Suhrwardī *consciously chose* to adopt Avicenna's approach because he had an alternative before him presented by a philosopher he regularly—if rudely—followed.

3. The Definitely Necessary Proposition

In the *Philosophy of Illumination*, Suhrawardī cuts a narrow swathe through the topics normally covered in texts on the syllogism. At the outset it

[11] See Appendix for an explanation of the mnemonics and symbolism used in this paper.

[12] See Suhrawardī, *Manṭiq al-talwīḥāt*: for [1] rejection of second-figure syllogisms with two absolutes, see p. 56, l. 14–18 and more straightforwardly p. 50, l. 9–10; [2] Barbara *MMM*, see p. 53, l. 6; Barbara *XMM*, see p. 53, l. 8–11; Barbara *LML*, see p. 53, l. 16, p. 54, l. 1; [3] Barbara *XLL* valid with *waṣfī* minor, see p. 53, l. 11–13. For Avicenna's account of each of these aspects, see T. Street, 'An Outline of Avicenna's Modal Syllogistic,' *Archiv für Geschichte der Philosophie* 84 (2002), pp. 129–60; for [1] see pp. 146–7, for [2] pp. 149–53, for [3] p. 153.

[13] Ziai, *Knowledge and Illumination*, p. 19.

[14] Abū 'l-Barakāt al-Baghdādī, *Kitāb al-muʿtabar* (Hyderabad, 1357 A.H.). Baghdādī accepts I-*MMM* (p. 148, pu), but rejects *LML* (p. 151, l. 24 et seq.); and he takes II-*XX* and *XM* to be productive; there is no treatment of the inferential contribution of the *waṣfī*. Against Aristotle, he rejects e.g. III–*LLL* (cf. p. 148, l. 19 et seq. and p. 151, l. 24 et seq.).

is worth noting two points. The first is that he is only interested in modalities in the *dhātī* reading:

> Further, when you say, 'All things that move necessarily change,' you should know that each and every thing described as moving is not necessarily changing because of its own essence, but because it is moving. Thus, its necessity depends on a condition and it is contingent in itself. By 'necessary', we mean only that which it has by virtue of its own essence (*wa-lā naʿnī bi'l-ḍarūrī illā mā yakūnu li-dhātihi fa-ḥasbu*).[15]

Secondly, Suhrawardī reads all modalities as belonging to the copula (*jihatu 'l-rabṭ*), though he will reformulate the syntax which best underlines that reading.[16]

It is perhaps easiest to say what these two points mean in negative terms. The first means that Suhrawardī is not interested in extensions to the logic explored by Avicenna using propositions of the form *every J is B while J*. The second means that he is not interested in whether propositions like *every man is a possible writer* may be inferred from a proposition like *possibly, every man is writing*. Neither point makes much sense apart from Avicenna's modal logic, nor do the technical terms that Suhrawardī is using. In Avicenna's logic, these restrictions mean the uses of the propositions will be straightforwardly essentialist (more on which in section five below). This sharp focus characterizes the whole of the logic section of *Philosophy of Illumination*.

Here is the famous passage in which Suhrawardī sets out his definitely necessary propositions:

> Since the contingency of the contingent, the impossibility of the impossible, and the necessity of the necessary are all necessary, it is better to make the modes of necessity, contingency, and impossibility parts of the predicate so that the proposition will become necessary in all circumstances. You would thus say, 'Necessarily all humans are contingently literate, necessarily animals, or impossibly stones.' Such a proposition is called the 'definitely necessary' [*al-ḍarūriyya al-batāta*]. In the sciences we investigate the contingency or impossibility of things as part of what we are investigating. We can make no definitive and final judgment except concerning that which we know necessarily. Even for that which is only true sometimes, we use the definitely necessary proposition. In the cases of 'breathing at some time', it would be correct to say, 'All men necessarily breathe at some time.' That men necessarily breathe at some time

[15] Suhrawardī, *Philosophy*, p. 17, l. 11, cf. ed. Corbin, p. 28, apu-pu.

[16] *Wa-idhā kānati 'l-qaḍiyyatu ḍarūriyyatan kafānā jihatu 'l-rabṭ fa-ḥasbu*; Suhrawardī, *Philosophy*, p. 18, l. 5, cf. ed. Corbin, p. 29, apu-pu.

> is always an attribute of man. That they necessarily do not breathe at some time is also a necessary attribute of a man at all times, even at the time that he is breathing. However, this is different from literacy. While literacy is necessarily contingent, it is not necessary that it be actualized [*laysat ḍarūriyyata 'l-wuqūʿ*] at some time.[17]

Which is to say, I think, that modal propositions that are fitted out to serve in the sciences must all have a certain necessity; if we talk of possibilities, they must be potentialities that belong necessarily to the thing, if we talk of actualized possibilities, they must be potentialities that are necessarily actualized for the thing. On the face of it, Suhrawardī is stipulating the following syntax for his necessity propositions (*al-qaḍāyā 'l-wājiba*):

> Every man necessarily must be an animal (*kullu insānin bi'l-ḍarūrati huwa yajibu an yakūna ḥayawānan*), or:
> necessarily, every J is a necessary B;

for his possibility propositions (*al-qaḍāyā 'l-mumkina*):

> Every man necessarily may be a writer (*kullu insānin bi'l-ḍarūrati huwa mumkinun an yakūna kātiban*), or:
> necessarily, every J is a possible B;

and, for the propositions recording necessarily actualised possibilities (what for Avicenna is the absolute, *al-muṭlaqa*, never named in *Philosophy of Illumination*):

> Every man necessarily must breathe at some time (*kullu insānin bi'l-ḍarūrati huwa mutanaffisun waqtan mā*), or:
> necessarily, every J is at one time an actual B.

Notice that in rendering the propositional form in English shorthand, I have made two decisions. First, I have embedded the whole proposition in the necessity operator (the 'necessarily' at the beginning of each rendition), and secondly, I have transformed what looks in Arabic to be a modification of the verb *kāna* in each proposition in such a way to show that the second modality belongs to the predicate. I hope this second decision will be accepted as simply respecting what Suhrawardī seems to me to be at pains to stipulate when he says that 'it is better that the modalities of necessity and its two counterparts be made parts of the predicate'. But some may need more persuading that I

[17] Suhrawardī, *Philosophy*, pp. 17–18; cf. ed. Corbin, p. 29, l. 1–12.

am entitled to render the first *bi'l-ḍarūra* as I have, since it sits in the middle of the Arabic examples Suhrawardī gives. Evidence in support of the rendition is provided in the fourth rule, on contradiction, where Suhrawardī sets out the contradictory of *every J is necessarily possibly B* as *not necessarily every J is possibly B* (*kullu fulānin bi'l-ḍarūrati huwa mumkinun an yakūna bahmānan naqīḍuhu laysa bi'l-ḍarūrati kullu fulānin huwa mumkinun an yakūna bahmānan*);[18] the scope of *bi'l-ḍarūra* is the whole proposition.

This second point is extremely important. Ziai talks of Suhrawardī's iterated modalities, but the iteration is only an artifact of the surface properties of Arabic syntax, and has no logical significance.[19] Walbridge compounds this error, and says this of the way Suhrawardī takes 'it is necessary that all humans are contingently literate':

> Although he is working here with terms rather than propositions, the result is not greatly different from those contemporary interpretations of modal logic that allow the inference 'it is possible that φ implies that it is necessary that it is possible that φ' ($\Diamond\varphi\supset\Box\Diamond\varphi$, a characteristic axiom of the S5 system of modal logic).[20]

As it happens, I agree that Suhrawardī needs modal intuitions as strong as S5, but it's clear from his account of the proposition and its contradiction that one of the modalities (be it necessity, possibility or 'at one time') belongs to the predicate, while the second belongs to the proposition as a whole. In short, the proposition with its modalized predicate is embedded in the first 'necessarily'. He does iterate modalities in the *Logic of Intimations*,[21] but not in the way intended by Walbridge and Ziai.

4. Suhrawardī's Syllogisms

With these truth-conditions stipulated for his propositions, Suhrawardī claims to be entitled to Barbara *LML* and *MMM* (respectively, every J is possibly B, every B is necessarily A, therefore every J is necessarily

[18] Suhrawardī, *Philosophy*, p. 19; cf. ed. Corbin, p. 31, l. 2–3. Notice that the Walbridge & Ziai edition iterates a modality at p. 19, l. 3 whereas Corbin at p. 31, l. 3 does not; I think Corbin's edition is right.

[19] Ziai, *Knowledge and Illumination*, p. 70.

[20] Walbridge, *Leaven of the Ancients*, p. 149.

[21] Suhrawardī, *Manṭiq al-talwīḥāt*, p. 53, l. 6: *mā yumkinu an yumkina yaḥkumu 'l-ʿaqlu bi-imkānihi*; this is only the S4 principle.

A; and every J is possibly B, every B is possibly A, therefore every J is possibly A).[22]

> Further, when the last term leads to the first term by means of the middle, the modes in the definite necessary proposition are made part of the predicate in one or both the premises, thus leading to the major. For example, 'All men are necessarily contingently literate, and all contingently literate beings are necessarily animals by necessity (or contingently walkers), therefore, all men are necessarily animals by necessity (or contingently walkers).'[23]

Suhrawardī only deals with necessity and possibility in *Philosophy of Illumination*. Given, however, that he talks of 'the possible which occurs in respect of every one at a certain time like breathing,' and that he seems to intend to include it as a contraction on the truth-conditions of the possibility proposition in his treatment of productive syllogisms, he would then also be defending Barbara *XMM* (every J is possibly B, every B is A, therefore every J is possibly A). Turning to *Logic of Intimations*, as we are recommended to do in the opening pages of *Philosophy of Illumination*, we find:

> Know that the conclusion in first-figure syllogisms follows the major in the mixed-premise syllogisms, except when the minor is possible and the major is existential (*wujūdiyya*). If we say 'Possibly every J is B' and 'Actually (*bi'l-wujūd*) every B is A', it is known from the nature of possibility that it may never actually occur (*jawāzu 'l-lā-wuqūʿ abadan*); so if the J is never described as B, it does not follow that the A comes to it actually, but only potentially, so it's possible.[24]

All of which means that Suhrawardī takes as productive the syllogisms that make Avicennan logic differ from, on the one hand, Aristotle's, and, on the other, post-Avicennan treatments like Najm al-Dīn al-Kātibī's. (Kātibī rejects Barbara *LML*, *MMM* and *XMM* because he insists on the sterility of all M-minor syllogisms.) And of course the first-figure base ramifies through the rest of the figures by virtue of—among other things—the principles of indirect reduction, which Suhrawardī accepts.[25]

---

[22] See Appendix for more information on these matters.

[23] Suhrawardī, *Philosophy*, pp. 22–3; cf. ed. Corbin, pp. 35–6.

[24] Suhrawardī, *Manṭiq al-talwīḥāt*, p. 53, l. 8–11.

[25] Suhrawardī, *Philosophy*, p. 26; ed. Corbin, p. 40, l. 2–9. It is worth noting this when reading Walbridge's claim that 'If there were only one kind of proposition, there would be only one kind of valid syllogism' (Walbridge, *Leaven of the Ancients*, p. 149).

We have to turn to *Logic of Intimations* for what Suhrawardī has to say about third-figure syllogisms. On purely modal second-figure moods, however, *Philosophy of Illumination* is thoroughly Avicennan. Suhrawardī puts forward Avicenna's argument from natures for a defence of his unAristotelian conclusions. Here is a late and well-phrased statement of the argument from Avicenna's *Book of Salvation*:

> ...because when J and A differ such that one of them is a subject of B always of necessity, or not at all at any time a subject of it; and the other is not always a subject of B, or not always[26] not a subject of it; then between the two natures there is an essential difference (*khilāfun dhātī*)...[27]

For example, take Camestres with a possibility proposition as major premise and a necessity proposition as minor, which Aristotle claims to conclude with a possibility proposition (*Prior Analytics* 38a25–26). Avicenna invites us to consider the two premises, *no J is possibly B*, and *every A is possibly B*. The second (major) premise tells us that B is possible for A's nature, whereas the first tells us that B is essentially excluded from J's nature; so A and J are essentially different natures, and the conclusion must be the necessity proposition, *no J is possibly A*.

Suhrawardī puts forward what is virtually the same argument more compendiously and in somewhat different terms, and then justifies it (and I give the same example, Camestres):

> For each of the propositions, what is possible for the subject of one is impossible for the subject of the other. Their two subjects are necessarily incompatible, yielding the conclusion that these two statements are propositions whose subjects are necessarily different.[28]

5. Interpretations of Avicenna's Logic

In this section, I deal with technical issues. Those without the stomach for a discussion of the history and substance of recent interpretations

---

I'm sure the only problem here is loose phrasing, but it's worth pointing out that while Suhrawardī may be saying there's only one perfect syllogism—though I doubt he is—, he recognizes that there are a number of valid, or more precisely, productive syllogisms.

[26] Reading *lā dā'iman* for *dā'iman*, following Huntingdon 534 at the Bodleian (Oxford), at folio 129b, pu.

[27] Ibn Sīnā, *Kitāb al-najāt*, ed. M. Kurdī (Cairo, 1331 A.H.), p. 58, l. 10–13.

[28] Suhrawardī, *Philosophy*, pp. 23–4; cf. ed. Corbin, pp. 36–7.

of Avicenna's logic can go straight to the last two paragraphs of the section for a summary of what is important.

Avicenna came to the major features of his modal logic at latest by his Jurjān period. All his works accept first-figure M-minor syllogisms (namely, Barbara *LML*, *MMM* and *XMM*) as productive though imperfect, and second-figure syllogisms with absolute premises as sterile.[29] This causes his account to differ from that in Aristotle's *Prior Analytics*, not only in the first figure, but in the second and third figures too.

My own doomed attempt at a semantics for Avicenna's logic was a modification of Rescher's semantics for Kātibī's *Shamsiyya*. Rescher gave a rendition of each of the propositions that corresponds to those deployed in Avicenna's account of the syllogism as follows (all are a-propositions, and one needs to be aware that ∃ before a predicate-letter means 'at one time'):[30]

> X: There is something that is at one time J, and whatever is at one time J is at one time B,
> $(\exists\chi)\exists\mathcal{J}\chi \wedge (\forall\chi)\ (\exists\mathcal{J}_\chi \supset \exists B_\chi)$
>
> L: There is something that is at one time J, and whatever is at one time J is a necessary B,
> M: There is something that is at one time J, and whatever is at one time J is a possible B.

These I modified—because of the M-minor syllogisms—so that the subject term was under a possibility operator, thus:

> X: There is something that is possibly J, and whatever is possibly J is at one time B,
>
> $(\exists\chi)\Diamond\mathcal{J}_\chi \wedge (\forall\chi)\ (\Diamond\mathcal{J}_\chi \supset \exists \mathrm{B}_\chi)$

and so on for the other propositions.

This attempt resulted in a semantics that gave all of Avicenna's syllogisms, but (1) didn't yield all of his conversions, and (2) made syllogisms he didn't explicitly accept valid. Further, (3) the semantics didn't make sense of Avicenna's proofs by upgrading, and (4) committed him to a strong principle of plenitude, a principle which he neither explicitly

[29] See Appendix for an introductory exposition of these matters.

[30] N. Rescher and A. van der Nat, 'The Theory of Modal Syllogistic in Medieval Arabic Philosophy,' in N. Rescher, R. Manor et al. (eds.), *Studies in Modality* (Oxford, 1974).

formulates nor endorses.[31] A similar model was adopted by Paul Thom in his *Medieval Modal Systems*, though Thom didn't commit Avicenna to a principle of plenitude but rather portrayed him as failing to distinguish between alethic and temporal modalities.[32] The semantics has since been referred to by Thom as the simple *de re* interpretation.

I've mentioned four problems with this interpretation (the conversions, the extra syllogisms, the proofs by upgrading, the principle of plenitude), but there turned out to be a fifth problem to which I hadn't adverted. Avicenna wants his necessity propositions (to give just one example) to be able to state metaphysical truths such as *every human is necessarily corporeal*.

> The metaphysical statement that humans are necessarily corporeal is before all else a statement of *de dicto* necessity. It is supposed to hold under all imaginable circumstances. True, it has a *de re* predicate: each possible human is supposed to have a necessary property, that of being corporeal. But the logical form of the whole statement is different from that of an accidental *de re* predication such as 'All (actually existing) possible animals are (as it happens) necessarily human,' which is true merely under the supposition that for a time no other animals exist.[33]

Thom proposes a set of propositions for his splendid revised interpretation—a mixed *de dicto/de re* interpretation—that gives plausible truth-conditions for the examples, and solves all five problems. The propositions are of this form (and I use Rescher's symbolism rather than Thom's only because I think Rescher's will be more familiar to Arabists):

X: Necessarily, whatever is J is at one time B,

$\Box[(\forall\chi)(\mathcal{J}_\chi \supset \exists B_\chi)]$

L: Necessarily, whatever is J is a necessary B
M: Necessarily, whatever is J is a possible B.

---

[31] For mention of the problems of (1) and (2), see Street, 'An Outline,' pp. 154–5; for (3) ibid. 141–2; for (4) see T. Street, 'Fa͟hraddīn ar-Rāzī's Critique of Avicennan Logic,' in U. Rudolph and D. Perler (eds.), *Logik und Theologie. Das Organon im arabischen und im lateinischen Mittelalter* (Leiden, 2005), pp. 99–116.

[32] P. Thom, *Medieval Modal Systems* (Ashgate, 2003), ch. 4, and esp. p. 66.

[33] Idem, 'Logic and Metaphysics in Avicenna's Modal Syllogistic,' forthcoming in S. Rahman, T. Street and H. Tahiri (eds.), *The Unity of Science in the Arabic Tradition. Science, Logic and Epistemology and Their Interactions* (Dordrecht, 2007).

That *de dicto* operator in which the whole proposition is nested is the key to solving the problems.

In short, to model Avicenna's modal propositions, we need to use what in modern logic are presented as different kinds of modal operators: a *de dicto* operator in which the proposition is embedded, and a *de re* operator which modifies the predicate. In 'necessarily, every man is rational', 'necessarily' is *de dicto* and makes the proposition true irrespective of whether or not any men actually exist—it makes the proposition stipulate truths about the nature of man. In 'necessarily, every man is necessarily rational', the second, *de re* necessity changes the proposition to mean that rationality is a necessary or essential property of the nature man. 'Necessarily, every man is possibly a writer' means that whatever has the nature of man necessarily has the potentiality to write. And 'necessarily, every man at one time breathes' means that whatever has the nature of man necessarily breathes at one or more points in time.

Arabic logicians did not have a pair of terms to make the distinction between *de re* and *de dicto*. What Suhrawardī does in *Philosophy of Illumination* is find a way to make the distinction without the requisite technical terms. In other words, his 'other method and shorter path to knowledge' involve a statement of the truth-conditions of the modals in a way that exactly parallels the only successful modern model that has been proposed for Avicenna's modal logic. In logic, Suhrawardī is a purist Avicennan with a flair for innovative and helpful exposition.

6. Concluding Remarks

I return to the questions which I set out in the first section of this paper. Suhrawardī reduces all his modal propositions to the appropriate definitely necessary version so that he can make explicit the truth-conditions for the propositions such that they work for the modal syllogistic he has chosen to defend. That syllogistic is Avicenna's. Since Suhrawardī has picked it out against at least one competing alternative, it is not a chance decision to defend Avicenna's account. And the propositions of this modal logic are fitted out to express the truths of essentialism, that is, truths about natures and the properties they must have essentially, or potentially, or occasionally.

I conclude by considering what this means for the way historians of philosophy should approach Suhrawardī's philosophical system.[34] Throughout their treatments of Suhrawardī's logic, both Walbridge and Ziai tend towards taking his logic as indicative of how he wants to shape his metaphysical project; as Walbridge puts it:

> What [Suhrawardī] does include in his logic is significant, for it provides a logical foundation and a parallel to his metaphysics and ontology later in the same work.[35]

But in fact, given the structure of Suhrawardī's modal logic, there are two ways ahead for the study of his philosophical system. Either we can reassess the place that logic has in it, decide that it is not important for him, and conclude that the essentialism implicit in the logic is irrelevant to his larger project; or we can reassess claims to find anti-essentialism in his metaphysics. Whatever, the current claim that his logic prefigures in an important way his anti-essentialism cannot stand, and must be revised.

---

[34] These remarks are in the spirit of some general remarks made in D. Gutas, 'Suhrawardī and Greek Philosophy,' *ArScPhil* 13 (2003), pp. 303–9; a review article of Walbridge, *Leaven of the Ancients*.

[35] Walbridge, *Leaven of the Ancients*, p. 143.

APPENDIX

INTRODUCTORY NOTES ON SYLLOGISTIC

I offer here a few references for those who want to find out something about about medieval modal logic, and specifically the mnemonics worked out in the medieval Latin west (among other things, the famous Barbara Celarent). I go on to explain the few mnemonics and symbols used in this paper.

The easiest introduction to traditional logic ('that solemn humbug', as Russell used to refer to it) is A.A. Luce's *Logic* (London, 1958). The little rhyme with the mood names is given on page 97. A much more solid introduction to the same material, along with a treatment of the modal logic, is given in Robin Smith's *Aristotle: Prior Analytics* (Indianapolis, 1989) pp. xiii–xxxi and in summary form pp. 229–35 (note that the letters Smith uses to represent the modalities differ from the ones I use). The ultimate references are Paul Thom's *The Syllogism* (Munich, 1981), *The Logic of Essentialism: an Interpretation of Aristotle's Modal Syllogistic* (Dordrecht, 1996), *Medieval Modal Systems* (Aldershot, 2003).

The first mood of the first figure is Barbara, given by Aristotle in the form:

> A belongs to all B (major premise), B belongs to all C (minor premise), therefore A belongs to all C (conclusion).

The major premise is so-called because it provides the predicate of the conclusion, whereas the minor premise provides the subject. Arabic logicians stated Barbara differently in two respects. First, they put the subject of the premise before the predicate, and secondly, they put the minor premise before the major:

> every C is B, every B is A, therefore every C is A,

which is just as obvious—or perfect—an inference as it is when stated in Aristotle's fashion.

This will, however, cause some problems for representing the Arabic logicians. The three occurrences of the letter 'A' in 'Barbara' mean that the major and minor premises and the conclusion are all A-propositions, that is, of the form 'every C is B'. The second mood, Celarent, has different vowels to show that the major premise and the conclusion are E-propositions, that is, of the form 'no C is B'. But now the order of the premises as stated in the Arabic will be out of step with the vowels in the Latin mood name:

> every C is B (a-proposition), no B is A (e-proposition), therefore no C is A (e-proposition).

Still, we should refer to this as Celarent because we can then compare it easily to the analyses of the same inference by Aristotle and the medieval Latin authors.

This will, however, cause one more problem of fit we need to take into account. The propositions in syllogisms often have modal operators: every C is possibly B (a possibility proposition or, in my rendition, an *M* proposition), or every C is necessarily B (a necessity proposition, or an *L* proposition). If the propositions that have no modal operators (in Arabic, the *muṭlaqa*) are called *X* propositions, we can then refer to various inferences that involve modal propositions by their Latin mood name followed by three letters, *M* or *L* or *X*, to designate what modality the conclusion will be from given modalizations of the premises. But—and this is important—the order of the letters designating the modal operators belonging to each proposition will follow the order of the Latin mood name (major minor conclusion) and not the order of the propositions as given in the Arabic. So the M-minor syllogisms:

> every C is possibly B, every B is possibly A, therefore every C is possibly A,

will be Barbara *MMM*;

> every C is possibly B, every B is A, therefore every C is possibly A,

will be Barbara *XMM*; and

> every C is possibly B, every B is necessarily A, therefore every C is necessarily A,

will be Barbara *LML*.

Avicenna differs from the account of syllogisms and modal syllogisms in the *Prior Analytics* in two fundamental respects. First, he rejects second-figure syllogisms with unmodalized (absolute, or *muṭlaqa*) premises;

and secondly, he claims first-figure M-minor syllogisms are productive though imperfect. The reasons behind both differences are deep, and derive from his essentialism. But note—as I show in this article—that Suhrawardī accepts the Avicennan account.

HISTORY OF SCIENCE

THE RELIGIOUS APPROACH TO NATURAL SCIENCES: THE CASE OF MINERALOGY IN THE IKHWĀN AL-ṢAFĀʾ AND IN ḤAMĪD AL-DĪN AL-KIRMĀNĪ

Carmela Baffioni

Among the various attempts to find a legitimate place for science in Islam, the procedure of Ismāʿīlī propagandists (*duʿāt*) is particularly original compared with those of the *falāsifa*, the thinkers inspired by Greek doctrines. For the Ismāʿīlīs, philosophy and religion are both necessary to achieve salvation; hence philosophy must be approached from a religious standpoint and its principles linked to prophetic knowledge. This is the case for *Rāḥat al-ʿaql*,[1] written by the Ismāʿīlī *dāʿī* Ḥamīd al-Dīn al-Kirmānī (d. after 1020). It is arranged as a 'city': each section is called a *sūr*, 'rampart' and has seven 'pathways' (*mashāriʿ*, pl. of *mashraʿa*), with the exception of the last one, which has fourteen. The domain of each of the natural sciences is evaluated according to the 'scale of religion' (*mīzān al-dīn*) and parallels are drawn between the structure, aspects and qualities of the Ismāʿīlī *daʿwa* and those of the natural world.

Minerals appealed deeply and variously to Muslim thinkers and scientists, both in themselves and in their relationship with the celestial world. Such relationships form the basis of alchemical speculations, which took as their starting point the reciprocal continuity established among all created beings. It was, of course, the same continuity that guaranteed salvation as attained through wisdom and scientific knowledge—and hence gave rise to encyclopaedias of science.

The Pure Brethren (Ikhwān al-ṣafāʾ) of Basra, the authors of the most famous of these encyclopaedias, the last edition of which is placed in the second half of the tenth century, devote a whole treatise to minerals, namely Epistle 19 (*On the Origin of Minerals*),[2] the fifth in the second section on natural sciences. They often deal with minerals in their *Rasāʾil*, generally in line with the contents of this Epistle.

[1] Ḥamīd al-Dīn al-Kirmānī, *Rāḥat al-ʿaql*, ed. M. Ghālib (Beirut, ²1983) [= K].

[2] Cf. *Rasāʾil Ikhwān al-ṣafāʾ wa-khullān al-wafāʾ*, ed. M. Ghālib, 4 vols. (Beirut, 1957), ii, pp. 87–131 [= IS].

Ḥamīd al-Dīn al-Kirmānī also speaks of minerals in various places in his *Rāḥat al-ʿaql*, though the main treatment of the subject is contained in the third and fourth pathways of the seventh rampart. Their titles are: 'On the three realms (*mawālīd*), which are minerals, plants and animals, firstly on mineral in that it is a body' (401, 2–3)[3] and 'On minerals in that they are a natural soul, endowed with actions and science. What is that science? What is that action?' (414, 2–3).

These chapters and their counterparts in the treatise *On Minerals* of the Ikhwān al-ṣafāʾ are the object of this article, with a focus on the theological meanings related to them—their origin, qualities and acts. The Ikhwān al-ṣafāʾ also take the opportunity to support their religious convictions through mineralogy.

The possible Ismāʿīlism of these authors has been widely considered; Daniel De Smet has even compared their doctrines—mainly their onto-cosmology—with those of Ḥamīd al-Dīn al-Kirmānī.[4] With regard to mineralogy, the technical details of the expositions of the Ikhwān al-ṣafāʾ and Kirmānī have recently been discussed by the present writer.[5]

It also seems worthwhile to compare the religious implications of both visions to focus more closely on the possible Ismāʿīlī commitment of the Ikhwān al-ṣafāʾ, or at least on the way in which Ismāʿīlī thinkers reworked their ideas, especially the religious foundation of their scientific theories. At the same time, I hope to enable better knowledge of a difficult and complex work such as *Rāḥat al-ʿaql*, which deserves further research in addition to the remarkable works by De Smet and Walker.[6]

---

[3] The references are to the page(s) and line(s) of Ghālib's editions quoted above (notes 1 and 2).

[4] Cf. D. De Smet, *La quiétude de l'intellect. Néoplatonisme et gnose ismaélienne dans l'œuvre de Ḥamîd al-Dîn al-Kirmânî (Xe/XIe s.)* (Leuven, 1995).

[5] Cf. C. Baffioni, 'L'influence des Ikhwān al-Ṣafāʾ sur la minéralogie de Ḥamīd al-Dīn Kirmānī,' paper presented at the Symposium 'Une lumière venue d'ailleurs', Louvain-la-Neuve, May 21–22, 2005, forthcoming in the Proceedings.

[6] Cf. D. De Smet, *La quiétude de l'intellect*; P.E. Walker, *Ḥamīd al-Dīn al-Kirmānī. Ismaili Thought in the Age of al-Ḥākim* (London, 1999). A Russian translation is available, cf. Ḥamīd al-Din al-Kirmānī, *Uspokoenie razuma*, trans. A.V. Smirnov (Moscow, 1995). For further bibliography cf. F. Daftary, *Ismaili Literature* (London, 2004), pp. 127–8. A doctoral thesis in "Studi sul Vicino Oriente e Magreb: specificità culturali e relazioni interculturali" with the title *La Filosofia della Natura nella* Rāḥat al-ʿaql *di Ḥamīd al-Dīn al-Kirmānī: Introduzione, Traduzione e Commento* has been presented in 2007 by my pupil Francesca Cicero (Università degli Studi di Napoli "L'Orientale", Dipartimento di Studi e Ricerche su Africa e Paesi Arabi).

The Religious Framework in the Ikhwān al-Ṣafā'

Each Ikhwānian treatise is, as we know, characterized by a *ḥaqīqa*, which is the 'true meaning' of the topics dealt with. The *ḥaqīqa* usually 'frames' the various sciences; in contrast, the various sciences are 'visible images' of that *ḥaqīqa*. From the religious point of view, the aim of Epistle 19 is to confirm that the world was created by God, as already stated in Epistles 42 and 32–3;[7] secondly, the Ikhwān introduce minerals as instances of divine justice.

It is no wonder that the Ikhwān al-ṣafā' accept creationism, because it is the official Muslim doctrine on the origin of the world, even though they give a well-known philosophical representation of it inspired by Neoplatonism. It is, however, particularly relevant that they see in minerals, an extremely technical field, a good occasion for demonstrating theodicy—that is, for its use with particular religious aims, probably also inspired by Muʿtazilism.

When they state the aim of the treatise—'to mention part of that which comes to be formed in the sub-lunar world in the course of time'—the Ikhwān introduce it as an aspect proper to the world of generation and corruption, in line with the treatment of eschatological realities in Epistle 38.[8] This correspondence is explained by the fact that if God wanted to annihilate it the spheres would come to a stop and the world would end, even though the celestial world is eternal and not subject to corruption.[9]

The *ḥaqīqa* of Epistle 19 is first stated in chapter 12, where technical discussion is temporarily set aside. The Ikhwān state that minerals were created to the advantage of animals, especially man, so that he can realize that the world was created from nothing by a wise God, Who subordinated the whole creation to him.[10]

The Ikhwān also consider the confusion of learned men, who need induction (*istidlāl bi'l-shāhid ʿalā 'l-ghā'ib* and *qiyās al-juz' ʿalā 'l-kull*) to be persuaded that the universe has an origin (*muḥdath*) and to be urged to investigate the ultimate cause of creation.[11] Many of these learned men, however, did not even recognize such an ultimate cause and believed

---

[7] Cf. IS, 87, 8–12.
[8] Cf. ibid., 87, 12–16.
[9] Cf. ibid., 87, 16–88, 14
[10] Cf. ibid., 114, 11–115, 1.
[11] Cf. ibid., 115, 1–5.

instead in the eternity of the world.[12] But God foreknew the doubts that would occur to them, so, He offered to their experience things that they could not doubt and that might then be a model for things that they were not able to see.[13] Generation and corruption of beings would convince them that the world is created and that such a perfect work necessarily presupposes an omnipotent Creator (here: *ṣāniʿ*). The fact that the four causes explained in Epistles 32–3 rule the whole also demonstrates that the universe is created from nothing.[14]

The function of mineralogy is not only to prove the existence of the Creator: the Ikhwān mention the properties of some minerals to introduce the topic they are mainly concerned with—that Nature is the agent that rules these substances.[15] Nature is a faculty of the heavenly Universal Soul, which permeates the whole terrestrial world; it moves all natural substances, using as its instruments beings (*ajsām*) that religious law calls 'angels'.[16] God is not in direct touch with beings: He only creates[17] them from nothing and charges them to fulfil His commandments;[18] it is proper to God to act without instruments.[19] In chapter 14, minerals[20] will be considered again, as instruments of Nature related to the motion of the heavens.[21]

The question of evil is repeatedly debated in this treatise. False opinions of it are related to 'dialecticians', people who try to attain truth without accepting God's message because they are convinced that science—that is, salvation—can be reached independent of divine help.[22] Some of them linked evil to God; others had recourse to two opposite principles, or to habit, or to the intervention of demons,[23] or

---

[12] Cf. ibid., 115, 6–13.

[13] Here, as often in the encyclopaedia, the old logical/theological principle of inference of the 'hidden' from the 'evident' is echoed.

[14] Cf. IS, 115, 13–116, 2.

[15] Cf. ibid., 116, 3–123, 18.

[16] Cf. ibid., 123, 18–124, 6. In 126, 4–11 the Ikhwān identify in Nature the 'influences' (*ta'thīrāt*) of the Universal Soul.

[17] The idea is given by the terms *abdaʿa/ibdāʿ* and *ikhtaraʿ/ikhtirāʿ*.

[18] Cf. IS, 126, 11–16.

[19] Cf. ibid., 127, 17–23. This is a well known *leitmotiv* of the Ikhwānian encyclopaedia. Cf. e.g. Ep. 8, i, 278, 3–12; Ep. 40, iii, 346, 20–23 and 359, 14–18.

[20] These are said, as in chapter 1, to be formed by the four elements—the 'mothers'; their agent is Nature and their goal the advantage of human beings and animals, cf. IS, 127, 2–7.

[21] Cf. ibid., 127, 8–16.

[22] This interpretation can be deduced by contexts such as those of Ep. 42, iii, 440, 19–441, 12.

[23] Cf. IS, 128, 1–11.

even to *tawallud*.[24] In another passage, we learn that they explained evil by chance or fate, by the influence of the stars, and even in terms of divine reward for accomplished deeds.[25] But as the Ikhwān have already stated, the truth is that the Creator is not in touch with bodies, just as kings are not in touch with the enterprises related to them;[26] through His servants He accomplishes deeds such as those described in Qurʾān 7:17; 2:17; 56:58–9 and 63–4. But even though God created the beings charged to fulfil His will, responsibility for their actions is theirs, not His.[27]

As in Epistle 42, deeds are linked to particular souls, who are considered as faculties of the Universal Soul, or even as instruments of Nature.[28] If they act in favour of religion and the world, they gain high esteem in the eyes of God.[29] Particular souls are rewarded for the good and punished for the evil they have accomplished.[30]

Kirmānī's Approach: Rethinking of 'Technical' Contexts

In Epistle 19, minerals are the *conditio sine qua non* for the announcement of the *ḥaqīqa*, but the scientific approach is stated as an 'aside'. What mainly interests the Ikhwān is once again the general aim of the encyclopaedia, which is to teach the knowledge—particularly religious knowledge—necessary to guarantee salvation. They take the opportunity to preach that the learner's soul is also a faculty of the Universal Soul, just as his body is one of the parts of the world, and hence that he must be careful of his deeds, from which divine reward will come.[31] In conclusion, the Ikhwān's religious approach to mineralogy consists

---

[24] The term indicates the generation of plants and animals directly from inanimate matter, cause and effect without intermediate point of volition, cf. R. Kruk, 'Tawallud,' in *EI*[2].

[25] Cf. IS, 130, 2–10.

[26] Cf. ibid., 128, 12–20.

[27] Cf. ibid., 128, 21–129, 17.

[28] Cf. ibid., chapter 13.

[29] Cf. ibid., 124, 10–125, 1. The case of Barzawayh, the physician of *Kalīla wa-Dimna*, is recalled to emphasize that nobody escapes his destiny. Qurʾān 51:56–7 is quoted in support.

[30] Cf. ibid., 130, 2–18.

[31] Cf. ibid., 130, 19–131, 2, with a quotation of Qurʾān 52:39–40.

in a foundation *ab extrinseco* of scientific research as a way towards salvation.[32]

Kirmānī, on the other hand, mingles religion and science in the stream of Ismāʿīlī *daʿwa*. So, in *Rāḥat al-ʿaql* the task of salvation proper to an encyclopaedia is directly fulfilled by the technical content. He does not have in mind here the beginners' salvation, but another equally 'technical' matter, the structure of Ismāʿīlī *daʿwa*. Therefore, every time he considers minerals—their origin, nature, properties, effects and so on—he compares them with aspects of the *daʿwa*.

This gives us the opportunity to list some points in which Kirmānī considers technical aspects of minerals in order to explain religious tenets. Needless to say, this procedure sharply differentiates him from the Ikhwān al-ṣafāʾ.

First, Kirmānī sees differences in minerals as an instance of the various divine acts through prophets and *imāms*:

> The fact that the divine matters (*al-mawādd*), while emanating and acting, confer distinction upon the Prophets and the *imāms* (*aṣḥāb al-adwār*) in that they are, among humans, within their [*scil.*, of the divine matters] purview (*fī ufqihā*) and joined to them, implies of necessity that the powers emanating from the heavenly bodies on all the parts of the earth confer distinction upon (some) minerals in that they are within the purview (*fī ufq*) of those powers at the highest (degree) compared with others (which are) at a lesser and at the least (degree), such as gold and mineral bodies (412, 10–15).

Second, different degrees of resistance to fire correspond to the different degrees of purity of members of the *daʿwa*:

> The fact that ranks of *ḥudūd*[33] in infallibility (*ʿiṣma*) and fault (*khaṭāʾ*) differ from one another (*mutafāwita*)—in fact infallibility and firmness (*thibāt*) in religion are proper to some of them and their judgements are like the those of the *nuṭaqāʾ* and of those who take their place in the preservation of the traditions and judgements of the religious community, while they are not proper to others inferior to them (in rank)—implies of necessity that the degrees of minerals with respect to resistance (*thibāt*) to fire, in enduring capacity and in decreasing must differ one from another (*mutafāwita*): in fact among them there are those that resist fire and last for a very long time (*ʿalā wajh al-dahr*) without anything decreasing them,

[32] In addition, discontinuity can be remarked in the explanation of the *ḥaqīqa*; it might even indicate textual corruption. Cf. C. Baffioni, *Appunti per un' epistemologia profetica. L'Epistola degli Ihwan al-Ṣafāʾ «Sulle cause e gli effetti»* (Naples, 2006).

[33] The degree of the *daʿwa* closest to the *imām*.

> such as gold, and those that decrease and do not resist fire, such as the other bodies (412, 15–20).

Third, according to Kirmānī, insofar as they are bodies, minerals are divided into 'that which can be forged (*yanṭariqu*) and that which cannot be forged', in other words what is malleable and that which is not. The text continues with the following division:

> Those that cannot be forged and are the oldest in being, because they are the most similar to earth, and the closest to it are divided into sulphurs, arsenics, salts, vitriols, alums, quicksilver, talc and others such as marcasite, antimonies, magnesia, lime, etc., and those that can be forged are divided into lead (*ānak*), tin (*raṣāṣ*),[34] iron, cupper, silver and gold (406, 3–8).

Six minerals are indicated because there are six divine Laws from Moses to Muhammad composed of *nāsikh* and *mansūkh* parts.[35] The fact that the sixth *sharīʿa* will not be abrogated and will last 'until resurrection' implies that gold lasts forever,[36] while innovations introduced into the Law by false prophets implies of necessity that mineral bodies must be corrupted when such bodies enter them, such as sulphur into silver or lead (*ibār*) into gold and diamond.[37]

The 'Motion' and 'Rest' of Intellect: An Occasion for Comparison

From the foregoing we should conclude that religious understanding of mineralogy according to the Ikhwān al-ṣafāʾ can hardly have been a model for Kirmānī. We know, indeed, that the *Rasāʾil* indicate a 'way' to be followed, whereas Kirmānī's work, which may also be considered an encyclopaedia of philosophical sciences, is an acquired patrimony where intellect enjoys calm and quietness.

But let me emphasize some points that might bring these works closer together. First, Kirmānī's 'dualistic' approach—that is, parallel

[34] I follow here the Russian translation (cf. pp. 287–8: 'svinec, olovo'). Cf. on the same line Freytag (G.W. Freytag, *Lexicon Arabico-latinum* [Halle 1830–37; repr. Beirut, 1975]), s. vv.). Kirmānī uses four terms to indicate 'lead': *raṣāṣ* is the generic word; *ānak* means 'pure lead', both black and white (cf. E.W. Lane, *Arabic-English Lexicon* [1863–93; repr. Cambridge, 1984], s.v.); *ibār* and *usrub* mean 'black lead' and *qalʿī* 'white lead' (that is, lead and tin respectively).

[35] K, 413, 1–4; on the meaning of the terms, see J. Burton, 'Naskh,' in *EI*².

[36] Cf. ibid., 413, 4–6.

[37] Cf. ibid., 413, 6–9.

examination of the minerals and the *daʿwa* structure as well—seems to be foreshadowed by the Ikhwān al-ṣafāʾ when they say that *ʿibāda* ("worship" in general; in Islamic Law, "acts of devotion") consists not only in prayer and fast, but also in enhancing prosperity (*ʿimāra*), both of religion and the world; God wants prosperity for both of them, because He is their master.[38] All people are His servants, and a lord loves the good servants and hates the bad ones, as stated in Qurʾān, 5:33–4 and 52:39.

It should be remembered that *ʿibāda*—in its double aspect exoteric (*ẓāhira*) and esoteric (*bāṭina*)—is one of the main starting points of *Rāḥat al-ʿaql*.[39] In Epistle 49 the Ikhwān speak of different *ʿibādāt*, each proper to one of the natural realms; with regard to minerals they say:

> The movement of mineral substances towards *ʿibāda* and the acknowledgement of the Creator (*al-mubdiʿ*)—praise be to Him—is their reception of figure and form: this is their act of devotion, and their obedience, submission and submissiveness, though if among them there are those that take pleasure in obedience and yearn it, those that are quicker in reception, better in form, greater and major in power and others inferior to this, and those that, disregarding [*ʿibāda*] do not receive any form nor are melted by fire, do not have radiance or limpidity and are of no use, such as firm massive rocks, dry stones (*al-ṣuwwat*), stones and salt soils (*al-araḍīn al-sibākh*) (iv, 210, 24–211, 5).

We might also say that in his encyclopaedia Kirmānī treats minerals as a support for and an introduction to the understanding of the Ismāʿīlī *daʿwa*; in other words that he takes information regarding the 'hidden' from the world that can be sensed—from what is 'evident', just like the Ikhwān al-ṣafāʾ.

With regard to the general approach to mineralogy, we know that for the Ikhwān the origin of minerals is related to the movements and aspects of heavenly bodies, which determine transformations in terrestrial soils.[40] On the other hand, Kirmānī lists minerals, together with plants and animals, as the second kind of existing beings after atmospheric phenomena—which are the first kind, as inferred from the foregoing twofold division of elements in *Rāḥat al-ʿaql*.[41] Yet the three

---

[38] Cf. IS, 125, 2–9.
[39] To confine ourselves to the third pathway, cf., e.g., 405, 10.
[40] Cf. IS, 88, 15–89, 5; 92, 5–93, 3.
[41] Cf. K, 401, 4–14.

natural realms in fact depend on 'meteorologica', to use Aristotle's term,[42] which means that although the Ikhwān al-ṣafā' look back to a 'remote cause', Kirmānī indicates the 'close cause' of the link between 'heaven' and 'earth'. Both of them recognize, however, the strict relationship between the super-lunar and sub-lunar worlds.

We know that differences among minerals are explained in both these works, though in different ways, by the different soils in which minerals are born.[43] Here I recall only that in addition to technical details Kirmānī says:

> The fact that the question (*amr*) of the *ḥudūd* finds its perfection in the foundation (*iqāma*) of religion and the existence of its products (*mawālīd*) after many longer and shorter cycles implies of necessity that the being of mineral bodies in their formation in mines must reach perfection after a very long period (*ba'd dahr*) and the passing of many centuries (*aḥqāb*) and years (412, 21–413, 1).

The Ikhwān al-ṣafā' also show a 'religious approach' to the question when they emphasize divine providence in describing places with sulphurous waters or with cold or mild winds.[44]

Stronger similarities between our texts can be remarked, however. The Ikhwān al-ṣafā' might have influenced Kirmānī directly in his presentation of gold. According to the Ikhwān, the 'perfect proportion' is realized in it: they speak of 'pure gold' (*al-dhahab al-ibrīz*) and say that it is formed from pure sulphur and quicksilver and from balanced heat.[45] For Kirmānī, gold precedes all minerals in nobility and essence;[46] it is thus like the prophets and the *ḥudūd* who take their place.[47] The discussion continues from the religious point of view. Plants and animals are compared to those who are charged with the 'external cult', command and prohibition, and to those who are charged with the 'external' and 'internal' cults.[48] Here we meet the word *'ibāda* once again.

[42] Cf. ibid., 401, 14–15.

[43] Cf. C. Baffioni, 'L'influence des Ikhwān al-Ṣafā' sur la minéralogie de Ḥamīd al-Dīn Kirmānī,' and 'La science des pierres précieuses dans l'Epître des Ikhwān-as-Ṣafā': entre les catalogues encyclopédiques et le commentaire philosophique,' paper presented at the Colloque International 'Aux origines de la géologie de l'antiquité à l'Age classique' (Paris, March 10–12, 2005), forthcoming in the Proceedings.

[44] Cf. IS, 99, 10–100, 10.

[45] Cf. ibid., 106, 14–18.

[46] Cf. K, 405, 6–7.

[47] Cf. ibid., 405, 3–6.

[48] Cf. ibid., 405, 7–11.

According to the Ikhwān, gold is, with ruby (*yāqūt*), the noblest substance.[49] Kirmānī says:

> The fact that the disposition of virtues by which perfection in the Laws is attained culminate in a Law that embraces them all, implies of necessity that the disposition of minerals must end in a body that embraces and contains them all and (is) nobler than the others, such as gold, which owing to its nobility is superior to every mineral (substance) (411, 11–14).

The Ikhwān describe gold in detail:[50] to its other qualities they add that it is yellow, stressing its link with the sun—*nāriyya*, the text says.[51] According to Kirmānī:

> The fact that there is a cause for the existence of the Prophet's Law—God bless him—and of that which he brought—a Law perfect in its rank ahead of the other traditions and Laws—namely the effusion of the divine power in the Prophet's soul—God bless him and his family—and the fact that (this effusion) completely receives the emanation of the First Intellect, in such a way that it becomes, thanks to it, a place of the emanation and an end to which it comes, like it and similar to it in nobility and perfection, implies of necessity that the fact that gold is gold results from its pre-eminence among minerals similar to it and the capacity of its matter to receive the power of the sun at the highest (degree), thanks to which the most complete emanation is emanated from it and the powers of the sun penetrate it in such a way that they make it similar to (the sun) in colour and beauty (411, 14–22).

In the fourth 'pathway', Kirmānī says that minerals, like every other existing being, have another aspect beyond the aspect for which they are bodies on which their life depends.[52] Minerals have a kind of 'soul', to which science and actions are linked, that rules their actions and protects them from that which might damage them; it brings them to perfection in that it helps them to reach their end.

On this occasion Kirmānī deals with the question of the reciprocal influences of minerals. He speaks of:

> The expansion (*inbisāt*) of some (minerals) towards others, the closure (*inqibāḍ*, 'contraction') of some from others, the affection (*ta'alluq*, adhesion) of some for others, the aversion (*munāfara*) of some for others, such as the aversion existing between iron and quicksilver—in fact it does

---

49 Cf. IS, 116, 4–5.
50 Cf. ibid., 116, 5–21.
51 Cf. ibid., 116, 11–12.
52 Cf. K, 406, 1–2.

> not adhere or expand to it, nor does iron leave it to penetrate into itself or to adhere to itself as it adheres to gold and silver unless after treatment (*bi-ʿilāj*); and like the enmity (*ʿadāwa*) existing between gold and lead (*ibār*)—in fact nothing is damaged by gold like lead because of its aversion (*nikāba*) for it; like the enmity existing between lead and other bodies such as silver and copper—in fact it corrupts and destroys them and for this (reason) is used in purification (*khalāṣ*) of bodies; and like the enmity existing between diamond and lead—nothing damages or breaks diamond but lead; and like the love (*maḥabba*) existing between gold and quicksilver, or between sulphur and gold, or between arsenic and silver, or between iron and copper—in fact, when they are mixed together they do not get separated; and like the love which is found between a magnet and iron—iron is attracted towards it and it is not extended toward other bodies...(414, 19–415, 12).

The Ikhwān al-ṣafāʾ have already spoken of 'spiritual' properties of mineral substances: these have a 'sensitive power' (*shuʿūr khafī* and *ḥass laṭīf*)[53] because of which:

> One nature has affection (*taʾlafu*) for another nature, one nature is related (*tunāsibu*) to another, one nature adheres (*talṣaqu*) to another, one nature is friend (*taʾnasu*) of another, one nature subjugates (*taqharu*) another, one nature consolidates (*taqwā ʿalā*) another, one nature is enfeebled by (*tuḍʿafu ʿan*) another, one nature ignites (*tulahhibu*) another, one nature loves (*tuḥibbu*) another, one nature is agreeable (*taṭību*) for another, one nature corrupts (*tafsudu*) another, one nature whitens (*tabyaḍḍu*) another, one nature reddens (*taḥmarru*) another, one nature flees from (*tahrubu min*) another, one nature hates (*tubghiḍu*) another, one nature is mixed (*tumāziju*) to another (110, 13–19).

Diamond and gold are quoted as instances of *taʾalluf* between minerals,[54] then magnets and iron,[55] and finally the stones that attract flesh, hair, nails and straw.[56]

Kirmānī also indicates the religious implications of these influences: the Book and the Law given to the Prophet must be linked to the *aʾimma* or they would decay, so minerals are preserved thanks to powers spread by God.[57] As in the *sharīʿa* some parts came to replace others, being mostly in contrast with them,[58] so the powers of minerals annihilate

[53] Cf. IS, 110, 10.
[54] Cf. ibid., 110, 20–22.
[55] Cf. ibid., 110, 22–111, 3.
[56] Cf. ibid., 111, 4–5.
[57] Cf. K, 415, 22–416, 3.
[58] Kirmānī refers here to the problem of *nāsikh* and *mansūkh*.

one another and are contrary to one another; and as in the Law there are parts that replace in accordance with what is replaced, so some of the powers of minerals are in accordance with others.[59] But the possession of 'a soul and a power' is also related to the well known division of minerals into 'that which can be forged and that which cannot'.[60] It should be recalled that when the Ikhwān al-ṣafā᾽ state that God does not need instruments to act, they mention the 'intelligent' who consider the actions of minerals, and of the magnet in particular: they realize that the agent and the mover are distinct from minerals 'because the body, in that it is a body, has no power of action'.[61]

For Kirmānī, the multiplicity of divine signs and traditions implies of necessity that minerals that 'have power and on which the advantage of the world depends' are many more than the six mentioned in correspondence to the six first Laws,[62] that is, minerals that can be forged. More particularly, the variety of beings belonging to the realm of religion implies that among minerals many are of great advantage when used and others bring only small benefits, but that there is a general usefulness in them as a whole.[63]

Consequently, the 'ignorant' in the world of religion are compared to salts, sulphurs, arsenics, the various kinds of vitriols and minerals: even if they can be used in some way, they are all poisonous and eating them hurts natural life.[64] The remoteness of the 'ignorant' from balance and the fact that they receive what fits them from the stars and the natural powers, implies of necessity that:

> The cause of the fact that these kinds of vitriols, arsenics, sulphurs, etc. are according to their natures and forms (is) the fact that the influences of the stars join them, and this from a balanced (situation) to a (situation) far (from balance), so that the way in which th(ose) influences act on them is like the way of fire, that acts on something and coagulates it, (while it acts) on another and melts it, according to the nature of the thing acted upon (411, 7–11).

---

[59] Cf. K, 416, 3–8.
[60] Cf. ibid., 408, 18.
[61] Cf. IS, 126, 6–7.
[62] Cf. K, 416, 8–10.
[63] Cf. ibid., 410, 18–20.
[64] Cf. ibid., 411, 1–3.

The Ikhwān al-ṣafā᾽ remark that the influence of stones is similar to that of medicines, and is due to divine wisdom and goodness; immediately after, we find the 'definition of God' given by Moses and Aaron to Pharaoh,[65] from which a personal idea of God is deduced that is here considered a medicine for the soul and elsewhere as the result of a reworking of Aristotle's categories.[66]

Finally, it is especially remarkable that the Ikhwān speak of a kind of 'evolution' from the mineral to the vegetal realm: when they distinguish three kinds of minerals according to the soils in which they are born, the second kind is

> That which is produced in the cave of the mountains and in the deep of the seas, whose making (*nadj*) is not accomplished but in one year or more, such as pearl and corals: one of them is vegetable, and it is corals, and the other animal, and it is pearl (91, 7–8).

In another passage, the Ikhwān exalt pearl, silk and honey,[67] explicitly stating that they are animal products, excellent even if very small, that come from the shell, the silkworm and the bee.[68]

Kirmānī's classification also starts from the bottom with gypsum, 'the most similar to earth', and finishes to the highest, corals, which owing to their perfection also encroach on the vegetable realm.[69]

The link between super-lunar and sub-lunar worlds is here developed into two hierarchical visions of the universe. The idea of a hierarchical structure of universe is one of the main tenets of Ismāʿīlī science and thought, and it is obviously found in Kirmānī.[70] But it is also easily traceable in the Ikhwānian encyclopaedia,[71] and in Epistle 19 as well.

---

[65] Cf. IS, 111, 15–21; Qur᾽ān 20:49–50 is quoted. Cf. also Qur᾽ān 23:23–6.

[66] The Ikhwān al-ṣafā᾽ substitute nine 'philosophical questions' for the ten Aristotelian categories (cf. e.g. Cat. 1b 26–7), and, in particular, position, possession, action and passion are replaced by the questions: 'Is it?' 'Why is it?' and 'Who is it?'. With regard to God, however, the basic questions are only: 'Is He?' and 'Who is he?' Cf. Ep. 42, iii, 513, 22–514, 11, commented on in C. Baffioni, *L'Epistola degli Ikhwān al-Ṣafā᾽ «Sulle opinioni e le religioni»* (Naples, 1989), pp. 60 and 193 and in eadem, *Appunti per un' epistemologia profetica*, p. 42. Here, Qur᾽ān 26:23ff, similar in content, is quoted.

[67] Cf. IS, 114, 2–9.

[68] Pearl is the best example of the properties of minerals, cf. ibid., 113, 2–22.

[69] Cf. K, 405, 11–21; 410, 3–5 and 409, 8–10.

[70] Cf. ibid., 403, 17–405, 3.

[71] From the political point of view, for example, when the Ikhwān develop well-known Platonic ideas. Cf. C. Baffioni, 'The "General Policy" of the Ikhwān al-Ṣafā᾽:

In conclusion, our comparisons show that *Rāḥat al-ʿaql* was partially influenced by the Ikhwān in a special field of the natural sciences such as mineralogy and also in some aspects of its theological approach; such similarities, however, cannot yet demonstrate the Ismāʿīlī militancy of the Ikhwān al-ṣafāʾ.

Plato and Aristotle Restated,' in A. Arnzen and J. Thielmann (eds.), *Words, Texts and Concepts Cruising the Mediterranean Sea* (Leuven, 2004), pp. 575–92.

THE CREATION AND INNOVATION OF MEDIEVAL HEBREW MEDICAL TERMINOLOGY: SHEM TOV BEN ISAAC, *SEFER HA-SHIMMUSH**

Gerrit Bos

Introduction

During the Middle Ages when Linnaeus had not yet been born and there was no uniform binary system for identifying plants and herbs, the risk of a doctor administering the wrong drug was certainly very real. Such a risk would be especially acute at a time when a doctor would move to and settle in a different country, in a different linguistic environment. Jewish doctors were especially confronted with this problem when several of them emigrated in the wake of the Berber invasions of the Almoravids and Almohads into southern Spain in the eleventh and twelfth centuries, to the Christian northern part of Spain and to southern France, from a society, where Jews used and understood Arabic next to Hebrew and Romance, to a society where they lost their knowledge of Arabic. Because of this shift in languages an

* Note: In this article, which is an adapted version of part of the introduction to the edition of Shem Tov Ben Isaac's first glossary of medico-botanical terms, I present the reader with some of the results of the research into the Hebrew-Arabic linguistic component featuring in this glossary. The research carried out in the years 2001–2006 is part of an interdisciplinary project running at the Martin-Buber Institute for Jewish Studies of the University of Cologne and at the Department of Romance Philology of the Free University Berlin. The project is dedicated to publishing the edition and the analysis of various unedited scientific texts written in Hebrew characters that belong to the area of medico-botanical literature. Within this project the Cologne group, consisting of Gerrit Bos and Martina Hussein, is responsible for the Hebrew-Arabic linguistic material, while the Berlin group, consisting of Guido Mensching and Frank Savelsberg, is in charge of the Latin-Romance material. This article follows upon a first article published in 2001 ('Shem Tov Ben Isaac, *Glossary of Botanical Terms, nrs 1–18*,' *JQR* 92 [2001], pp. 1–20), in which Bos and Mensching presented the glossary for the first time, followed by a detailed analysis of the first eighteen entries of the list. I thank Gad Freudenthal, Tzvi Langermann and Guido Mensching for their comments to an earlier draft of this article. For the Romance material, see, among others G. Mensching and F. Savelsberg, 'Reconstrucció de la terminologia mèdica occitano-catalana del segle XIII a través de llistats de sinònims en lletres hebrees. Edició i anàlisi del vint-i-novè llibre del *Sèfer ha-Ximmuix* de Xem Tov ben Isaac de Tortosa,' in *Actas del I congrès de l'estudi dels jueus en territori de llengua catalana* (Barcelona, 2004), pp. 69–81.

urgent need arose for 'lexica or glossaries in which technical-medical expressions have been listed alphabetically, especially the names of simple medicines.'[1]

However, this particular literary genre has been sorely neglected by modern research in spite of the fundamental bibliographical surveys composed by Moritz Steinschneider in the nineteenth century, in which he pointed to the importance of this particular genre for the deciphering of individual plant names in pharmacological fragments, and especially recommended the editions of the glossaries composed by Shem Tov Ben Isaac, and another one extant in ms. Florence, Mediceo Laurenziana Or. 17.[2] The only notable exceptions are a recent concise survey of Hebrew medical glossaries in manuscript, composed by J.P. Rothschild as an appendix to an article on the manuscript tradition of the Hebrew-Italian glossary on Maimonides' *Guide of the Perplexed*, prepared by Moses of Salerno,[3] and a list with Romance elements edited by Magdalena Nom de Déu.[4] The apparent lack of interest on the part of scholars in this particular field stands in sharp contrast to that in the field of Arabic studies which can boast of a recent bibliographical survey by Ullmann,[5] while one of its best known glossaries, namely that compiled

[1] M. Steinschneider, 'Zur Literatur der "Synonyma",' in J.L. Pagel (ed.), *Die Chirurgie des Heinrich von Mondeville* (Berlin, 1892), pp. 582–95 (English translation: Bos). Some of the following issues have been discussed in G. Bos, 'Hebrew Synonym Literature, Some Notes on a Virtual Forgotten Genre' (Forthcoming in the proceedings of the 'Internationale Fachtagung 'Mittelalterliche medizinisch-botanische Fachtexte: Hebräisch und Romanisch im Kontext des Lateinischen und Arabischen,' Berlin, May 2–4, 2002); G. Bos and G. Mensching, 'Shem Tov Ben Isaac, Glossary of Botanical Terms'; G. Bos and G. Mensching, 'Hebrew Medical Synonym Literature: Romance and Latin Terms and their Identification,' *Aleph* 5 (2005), pp. 169–211.

[2] Cf. M. Steinschneider, 'Donnolo. Farmakologische Fragmente aus dem 10. Jahrhundert,' *Virchows Archiv* 39 (1867), pp. 304–17; idem, 'Zur Literatur der "Synonyma"'; idem, 'Glossar zu den Synonymen Cap. IX des Antidotarius,' in *Die Chirurgie des Heinrich von Mondeville*, pp. 596–625; idem, *Die hebräischen Übersetzungen des Mittelalters und die Juden als Dolmetscher* (Berlin, 1893; repr. Graz, 1956), pp. 838–40.

[3] J.P. Rothschild, 'Remarques sur la tradition manuscrite du glossaire hébreu-italien du Commentaire de Moïse de Salerne au Guide des égarés (en appendice, note sur les glossaires médicaux hébreux; liste de manuscrits hébreux contenant des glossaires),' in J. Hamesse and D. Jacquart (eds.), *Lexiques bilingues dans les domaines philosophique et scientifique (Moyen âge—Renaissance)* (Turnhout, 2001), pp. 49–88.

[4] J.R. Magdalena Nom de Déu (ed.), *Un glosario hebraico aljamiado trilingüe y doce 'aqrabadin' de origen catalán* (Siglo XV) (Barcelona, 1993).

[5] Cf. the recent bibliographical survey in M. Ullmann, *Die Medizin im Islam* (Leiden, 1970), pp. 288–92. For an extensive discussion of this particular subject cf. Bos, 'Hebrew Synonym Literature, Some Notes on a Virtual Forgotten Genre'.

by Maimonides under the title *Sharḥ asmāʾ al-ʿuqqār*, is available both in a critical edition and in French,[6] Hebrew,[7] and English translations.[8]

Shem Tov Ben Isaac

One of the most prominent composers of such synonym lists was Shem Tov Ben Isaac, who was born in 1198 in Tortosa (Catalonia). He is known to have traveled in the Near East for business. After 1229 he began to study in Barcelona under R. Isaac Ben Meshullam at the age of thirty; subsequently he spent some time in Montpellier, and was then active as a physician and translator in Marseilles.[9] While Marseille was formerly an independent and sovereign city, it was forced in the year 1257 to recognize the sovereignty of Charles of Anjou who in return gave the inhabitants a constitution (*Les Statuts de Marseille*) which considered Jews and Christians as equal citizens.[10] He protected and intervened on behalf of the Jews when necessary, as in 1276 when he issued a special decree against the inquisitors who tried to introduce even more stringent measures against the Jews than those decided at the fourth Lateran Council in 1215 and in subsequent Councils.[11] The Jewish doctors in Marseilles found themselves in an especially privileged position, not only with regard to their colleagues in the rest of the Provence, but also with regard to their fellow-citizens, as they enjoyed special rights. As Marseilles suffered from frequent epidemics and there were only few Christian physicians, they were by sheer necessity tolerated in the city and sometimes even employed by the municipality,

---

[6] Maimonides, *Sharḥ asmāʾ al-ʿuqqār* (*L'explication des noms des drogues*), ed. M. Meyerhof (Cairo, 1940).

[7] S. Muntner, *Beʾur shemot ha-refuʾot. Teshuvot refuʾiyot* (Jerusalem, 1969).

[8] *Moses Maimonides' Glossary of Drug Names*, trans. F. Rosner (Haifa, 1995).

[9] On Shem Tov Ben Isaac, his life and literary activity, see E. Renan, *Les Rabbins français du commencement du quatorzième siècle* (Paris, 1877; repr. Farnborough, 1969), p. 592; Steinschneider, *Die hebräischen Übersetzungen des Mittelalters und die Juden als Dolmetscher*, pp. 741–5; H. Gross, *Gallia Judaica. Dictionnaire géographique de la France d'après les sources Rabbiniques* (Paris, 1897), pp. 375–6; S. Muntner, 'R. Shem Tov Ben Isaac of Tortosa about the Life of the European Jewish Doctor and his Ethics,' *Sinai Jubilee Volume* (Jerusalem, 1957), pp. 321–7; G. Sarton, *Introduction to the History of Science*, ii/2 (New York, 1975), pp. 845–6; J. Shatzmiller, *Jews, Medicine and Medieval Society* (Berkeley, 1994), pp. 44–5.

[10] Cf. A. Crémieux, 'Les Juifs de Marseille au Moyen Age,' *REJ* 46 (1903), pp. 1–47, 246–68, pp. 3–4.

[11] Cf. Gross, *Gallia Judaica*, p. 368.

contrary to the Jewish doctors in the rest of the Provence.[12] It was in the city of Marseilles that Shem Tov translated into Hebrew the medical encyclopaedia entitled *Kitāb al-taṣrīf li-man ʿajiza ʿan al-taʾlīf* (*The Arrangement of Medical Knowledge for Him Who is Not Able to Compile a Book for Himself*), composed in the tenth century by the Andalusian physician Abū ʾl-Qāsim Khalaf ibn ʿAbbās al-Zahrāwī, known in the western world as Abulcasis.[13] Calling it *Sefer ha-Shimmush*, Shem Tov started his translation in the year 1254,[14] completed it in 1258,[15] and then possibly wrote a revision in which he was engaged in the year 1261.[16] Next to the *Kitāb al-taṣrīf* Shem Tov translated Abū Walīd Muḥammad ibn Rushd's Middle Commentary on Aristotle's *De anima*,[17] Abū Bakr Muḥammad ibn Zakariyyāʾ al-Rāzī's medical encyclopaedia *Kitāb al-Manṣūrī*,[18] and Hippocrates' *Aphorisms* with Palladius' commentary.[19]

Shem Tov's translation of al-Zahrāwī's *Kitāb al-taṣrīf* is especially important as it shows the attempt to create a new Hebrew medical terminology based on the terminology of the Bible, Mishnah and Talmud, as well as the medieval commentaries and translations. And in some cases he uses the method of loan-translation or semantic borrowing for terms not attested in any Hebrew source.[20] Shem Tov's

---

[12] Cf. Crémieux, 'Les Juifs de Marseille,' p. 37; I. Alteras, 'Jewish Physicians in Southern France during the 13th and 14th Centuries,' *JQR* 68 (1977–78), pp. 209–23, p. 215.

[13] On the *Kitāb al-taṣrīf* see D. Jacquart and F. Micheau, *La médecine arabe et l'occident mediéval* (Paris, 1990), pp. 139–41 and passim.

[14] Cf. Introduction, ms. Oxford, Hunt. Don. 1 (= cat. Neubauer 2118), fol. 23b.

[15] Cf. Colophon, ms. Paris, BN héb. 1163, fol. 239a.

[16] The assumption that Shem Tov revised his translation is based on his account of the medical incident occurring in Marseilles in the year 1261 in the introduction to his translation featuring below. See as well Renan, *Les Rabbins français du commencement du quartorzième siècle*, p. 592; Sarton, *Introduction to the History of Science*, p. 846, and Shatzmiller, *Jews, Medicine and Medieval Society*, p. 45. Steinschneider, *Die hebräischen Übersetzungen des Mittelalters*, p. 741, remarks that he completed the translation between 1261–1264 (= Muntner, 'R. Shem Tov Isaac of Tortosa,' p. 322).

[17] Cf. Steinschneider, *Die hebräischen Übersetzungen des Mittelalters und die Juden als Dolmetscher*, p. 148; Ibn Rushd, *Middle Commentary on Aristotle's* De Anima, ed. and trans. A.L. Ivry (Provo, 2002), pp. xxviii–xxix, 150, n. 69.

[18] Cf. Steinschneider, *Die hebräischen Übersetzungen des Mittelalters und die Juden als Dolmetscher*, pp. 725–6.

[19] His commentary is no longer extant in Greek, but it has recently been rediscovered by Hans Hinrich Biesterfeldt and Y. Tzvi Langermann, who hope to publish soon a preliminary study of Palladius' commentary, to be followed by a full edition and analysis.

[20] S. Muntner (art. 'medicine,' in *Encyclopaedia Judaica* 11, col. 1189) remarks that the translation of al-Zahrāwī's *Kitāb al-taṣrīf* by Abraham Shem Tov (sic) is 'of particular

main reason for his translation of the *Kitāb al-taṣrīf* was that the Jews would have easier access to medical knowledge and not be dependent on non-Jewish doctors anymore.[21] As to his method of translation, he remarks that for diseases the names of which he knows from the Bible or Rabbinic literature or from Romance, he uses either of those names. And that for diseases, organs, drugs, wild and domestic animals, insects and vermin, the names of which he does not know from these sources, he uses the Arabic term, because nowadays knowledge of Hebrew has been lost, due to the length of the Exile.[22]

In order to stress the intention of his translation once more, Shem Tov extols its qualities, remarking that it is useful for individuals and crowds, wise and foolish, kings and poor people in all times and all places, and emphatically warns physicians against swapping similar remedies or buying a certain remedy they do not know from a pharmacist who then gives them something else instead of it, on purpose or by mistake.[23] Shem Tov adds that many people, who were first mistaken themselves, then misled others concerning the names of herbs, seeds and plants in some countries.[24] As an example he gives the case of a plant that has four species, one of which was called by the Persian doctors 'fanjangusht' (chaste-tree, *Vitex agnus castus* L.),[25] meaning 'five leaves', while the Christians called it 'pentaphyllon' (cinquefoil-*Potentilla*

importance because he introduced a new Hebrew terminology based mainly on terms used in the Talmud.'

[21] Cf. Introduction to *Sefer ha-shimmush* edited by Muntner, 'R. Shem Tov Ben Isaac of Tortosa,' pp. 324–5, §2–3; E. Feliu and J. Arrizabalaga, 'El pròleg de Semtov ben Issac, el Tortosí, a la seva traducció hebrea del *Taṣrīf* d'Abū al-Zahrāwī,' *Tamid* 3 (2000–2001; Barcelona, 2002), pp. 66–95, p. 80, §2–3; Steinschneider, *Die hebräischen Übersetzungen des Mittelalters*, p. 742.

[22] Cf. Muntner, 'R. Shem Tov Ben Isaac of Tortosa,' p. 326, §9; Feliu and Arrizabalaga, 'El pròleg de Semtov ben Issac, el Tortosí,' p. 82, §9; Steinschneider, *Die hebräischen Übersetzungen des Mittelalters*, p. 743. summarizes the text as follows: 'Alle Krankheiten, Glieder, Mittel etc., für welche er einen Ausdruck in der Bibel, in der Sprache der Weisen oder in der Landessprache gefunden, habe er danach benannt, sonst den arabischen Terminus beibehalten, da sich durch das Exil die Kenntnis der hebr. Sprache vermindert habe.'

[23] Cf. Muntner, 'R. Shem Tov Ben Isaac of Tortosa,' p. 327, 13, 15–16; Feliu and Arrizabalaga, 'El pròleg de Semtov ben Issac, el Tortosí,' p. 84, 13, 15–16.

[24] Cf. Muntner, 'R. Shem Tov Ben Isaac of Tortosa,' p. 327, 17; Feliu and Arrizabalaga, 'El pròleg de Semtov ben Issac, el Tortosí,' pp. 84–5, 17.

[25] Cf. Maimonides, *Sharḥ asmāʾ al-ʿuqqār*, ed. Meyerhof, no. 308; Dietrich (ed.), *Dioscurides Triumphans*, i, 72. The actual meaning of the Persian term is not 'five leaves' but 'five fingers' (*panj angusht*); cf. Vullers, *Lexicon Persico-Latinum Etymologicum*, i, p. 374: 'quinque digiti'.

reptans L.),[26] which also means 'five leaves'.[27] This misled, says Shem Tov, many physicians. Some of them think that 'fanjangusht' is a large fragrant tree that is as large as a man or even so large that a man can hide underneath it, and that its fruits are called 'agnus castus', while 'pentaphyllon' is a plant that has no stem but only five leaves which grow from the earth. Others say, Shem Tov continues, that 'fanjangusht' is not the mentioned tree, but a plant which has the effect of completely annulling sexual lust,[28] and that 'pentaphyllon' is a plant that has a stem.[29]

Another example of a possible mix-up of remedies, is taken from Maimonides, who warns in the case of the 'black nightshade' (*Solanum nigrum* L.) of prescribing the wrong species.[30] For many physicians prescribe one of its species for internal diseases, and another similar species, which can be easily distinguished from the previous one when it has been dried and its fruits become black, when it is still fresh, for external diseases, as it is poisonous. When this last species is mistaken for the first one and its juice is taken internally, it causes severe suffocation, hiccups and nausea accompanied by yawning and vomiting of blood.[31] And how the place where one species is grown can influence its form

[26] Cf. Maimonides, *Sharḥ asmāʾ al-ʿuqqār; Moses Maimonides' Glossary of Drug Names* (Haifa, 1995), no. 263; Dietrich (ed.), *Dioscurides Triumphans*, iv, 37.

[27] Cf. Muntner, 'R. Shem Tov Ben Isaac of Tortosa,' p. 327, 17; Feliu and Arrizabalaga, 'El pròleg de Semtov ben Issac, el Tortosí,' p. 85, 17.

[28] For the effect of the 'chaste tree' as an anti-aphrodisiac cf. *Ibn al-Jazzār on Sexual Diseases and their Treatment*, ed. and trans. G. Bos (London, 1997), p. 250: 'Galen maintained that the women of the inhabitants of Athens used to spread chaste-tree and then sleep upon it during their high festivals so that the lust for coitus would leave them' (= Galen, *De simpl. med. temp. et fac.*, VI, 1; ed. C.G. Kühn, *Claudii Galeni Opera Omnia* [Leipzig, 1821–33; repr. Hildesheim, 1967], xi, p. 808).

[29] Cf. Muntner, 'R. Shem Tov Ben Isaac of Tortosa,' p. 327, 18; Feliu and Arrizabalaga, 'El pròleg de Semtov ben Issac, el Tortosí,' p. 85, 18.

[30] Cf. Muntner, 'R. Shem Tov Ben Isaac of Tortosa,' pp. 327–8, 19; Feliu and Arrizabalaga, 'El pròleg de Semtov ben Issac, el Tortosí,' p. 85, 19.

[31] Shem Tov's quotation is taken from Maimonides' *On Poisons* (forthcoming ed. and trans. Bos, ch. 91): 'To the [substances] taken by mistake belongs the soporific type of black nightshade (*Solanum nigrum* and var.), for we often prescribe black nightshade juice among the ingredients to be taken for diseases of the internal organs. One of its varieties that has black seeds and that is soporific is [sometimes taken] by mistake when [the seeds] are [still] green before they turn black. Upon drinking, it immediately causes severe dryness, hiccups and vomiting of blood. Its treatment: hasten to let him vomit by means of the general emetics which have been described before. Then let him vomit for the last time by means of water and honey. Then let him drink a large quantity of water and honey. When he has digested something of it, he should take another drink of water and honey. He should do so for a day and a night. And then he should feed himself as usual...' Instead of 'dryness' Shem Tov has 'suffocation' which

is stressed by Shem Tov by quoting from Maimonides' statement in the *Mishneh Torah, Sefer Zera'im* that one species can assume many [different] forms depending on the place and on the cultivation of the soil, and that these two forms of one species can be so different that they look like belonging to two different species altogether.[32] Again in the name of Maimonides, Shem Tov adds that the opposite can also happen, namely that two different species are so similar to one another that their appearance seems to be the same. And quoting from the same source Shem Tov gives a wide variety of examples, both from plants and trees, such as garden lettuce (*Lactuca sativa*) and wild lettuce (*Lactuca scariola*), chicory (*Cichorium intybus*) and wild chicory (*Cichorium pumilum*), garden leek (*Allium porrum*) and field leek (*Allium ampeloprasum*), garden coriander (*Coriandrum sativum*) and wild coriander (*Smyrnium conatum* Boiss.).[33] Shem Tov concludes his extensive quotation from Maimonides by exclaiming that if in the case of crops the Torah is very strict and makes it obligatory to know the different forms of plants and trees and their fruits because of the prohibition of *kil'ayim* (mixing together of different species), how much more so should this be the case for remedies when human life is at stake.[34] That the wrong administration of a drug can have fatal consequences is illustrated by Shem Tov by relating an incident that happened around the year 1260 in his home town Marseille, in which two Christian doctors administered a patient a purgative of half a drachm of white hellebore as a result of which the patient died of suffocation.[35] The following incident, related by Shem Tov, not only endangered the life of the patient, but that of the whole Jewish community:

> An ignorant, foolish man, a fellow Jew, arrived in the city of Marseilles, claiming to be a physician, although he was [in fact] estranged from that

is similar to the variant reading in ms. P (Paris, BN, héb. 1211): خناق and Moses ibn Tibbon's Hebrew translation: אסכרה (angina).

[32] Muntner, 'R. Shem Tov Ben Isaac of Tortosa,' p. 328, 19; Feliu and Arrizabalaga, 'El pròleg de Semtov ben Issac, el Tortosí,' p. 85, 19; Maimonides, *Mishneh Torah, Sefer Zera'im, Hilkhot Kil'ayim* III, 1.

[33] Muntner, 'R. Shem Tov Ben Isaac of Tortosa,' p. 328, 20–21; Feliu and Arrizabalaga, 'El pròleg de Semtov ben Issac, el Tortosí,' pp. 85–6, 20–21; Maimonides, *Mishneh Torah, Sefer Zera'im, Hilkhot Kil'ayim* III, 2–3.

[34] Muntner, 'R. Shem Tov Ben Isaac of Tortosa,' p. 328, 21; Feliu and Arrizabalaga, 'El pròleg de Semtov ben Issac, el Tortosí,' p. 86, 21.

[35] Muntner, 'R. Shem Tov Ben Isaac of Tortosa,' pp. 328–9, 23; Feliu and Arrizabalaga, 'El pròleg de Semtov ben Issac, el Tortosí,' pp. 87–8, 23; Shatzmiller, *Jews, Medicine and Medieval Society*, p. 79.

science. Worse, he really had no medical knowledge at all. His patient had been bedridden for a long time, suffering from arthritis, *retet* in Hebrew. In his ignorance [the 'doctor'] ordered the root of a certain herb to be boiled and that [the patient] should drink the liquid from it. This ignoramus knew neither the strength nor the properties of this herb, nor that it was potentially harmful. As a result of this treatment, the sick man fell into a coma, lost his memory, and [eventually] lost his mind. His face and eyes turned red, his tongue dry, and his throat parched. Unable to speak or breathe, his body turned cold. Saliva and moisture came out of his mouth and he was unable to control it. That night he died suddenly. The ignoramus rose early the next morning to visit his patient, as doctors do. He found him sleeping deeply, in the torpor of death. Nevertheless, he assured [the patient's] relatives and neighbors that this was the way in which the herb worked and that [the patient] was merely asleep and would eventually wake up. Still [the doctor] decided to go into hiding. When evening came, the bailiff, [*shoter*] ordered him to present himself before him, otherwise he would be hit with a fine of one hundred marcae. [The 'doctor'] left the city, escaping on the second day of the feast of Passover of the year five thousand and twenty one (March 18, 1261). Had it not been for the fact that the Christian happened to be a foreigner, we all would have been in great danger on his account.[36]

That Jewish doctors had to be particularly careful and think twice before treating a Christian patient is borne out by the following statement featuring in *Sefer ha-yosher*, a medical treatise composed at the end of the Middle Ages:

> We Jewish doctors in the Diaspora have to possess extraordinary knowledge, for the Christian doctors envy us and challenge us, so that at times we have to provide explanations about our procedures [lit., 'science']. And if they discover any ignorance on our part they say, 'He kills gentiles.' This is the reason I advise each and every Jew not to [even] touch a gentile if he is not able to answer [the questions of those Christian doctors] in natural sciences.[37]

[36] Muntner, 'R. Shem Tov Ben Isaac of Tortosa,' p. 328, 22; Feliu and Arrizabalaga, 'El pròleg de Semtov ben Issac, el Tortosí,' pp. 86–7, 22; trans. Shatzmiller, *Jews, Medicine and Medieval Society*, pp. 83–4; but see as well Crémieux, 'Les Juifs de Marseille,' p. 254, who states that never, following the misdemeanor of an individual, a complaint was filed against the Jewish community of Marseille as a whole, because of the protection the Jews enjoyed on the basis of the 'Statuts de Marseille' and because of the spirit of tolerance predominant in the city.

[37] English trans.: Shatzmiller, *Jews, Medicine and Medieval Society*, p. 85; Hebrew text and German trans.: M. Güdemann, *Geschichte des Erziehungswesens und der Cultur der Juden in Italien während des Mittelalters* (Vienna, 1884; repr. Amsterdam, 1966), p. 237, German trans., p. 337, n. XVII; Hebrew text.

The Synonym Lists

To minimise the risk of mistaking medicinal herbs and drugs, because of ignorance or because of the lack of proper terminology in the actual vernacular, and to provide both Jewish doctors and patients with the proper medico-botanical terminology, Shem Tov composed 'a list of roots and herbs in the language of the Bible and of the Sages, of blessed memory, according to the interpretation most commentators agreed upon, in the vernacular language and in Arabic, alphabetically arranged.'[38] This list, which consists of two separate lists of synonyms and is part of book twenty-nine of his translation of *Kitāb al-taṣrīf*, is extant in the following manuscripts:

1. Ms. Paris, BN héb. 1163. This manuscript contains books eighteen to thirty and was copied in a Sephardic script in the fourteenth century.[39] The two lists feature on fols. 191a–198a.
2. Ms. Oxford, Hunt Donat 2 (Neubauer 2119). This manuscript contains books seven to twenty-nine and was copied in the year 1369 in a Sephardic cursive script by Asher ben Abraham ha-Kohen in the city of Trets (Provence).[40]
3. Ms. Vatican Ebr. 550. This manuscript dating from the fourteenth, or, possibly fifteenth century, only contains the glossary.[41]

Of the five chapters of the original Arabic text of book twenty-nine, Shem Tov modified the first two chapters to serve his purpose, since, as the author states, the Jews neither need nor profit from a translation of the first two chapters dealing with foreign and different names of plants

---

[38] Muntner, 'R. Shem Tov Ben Isaac of Tortosa,' p. 327, 12:
ואכתוב לכם עוד במאמר תשעה ועשרים שמות העקרין והעשבין בלשון מקרא ובלשון חכמים ז"ל כפי מה שפרשו בהם רוב המפרשים והסכימו עליהם ובלשון לועזים ובלשון הגרי על פי הא"ב
Feliu and Arrizabalaga, 'El pròleg de Semtov ben Issac, el Tortosí,' p. 84, 12.

[39] H. Zotenberg (ed.), *Catalogues des Manuscrits Hébreux et Samaritains de la Bibliothèque Nationale* (Paris, 1866).

[40] See A. Neubauer, *Catalogue of the Hebrew Manuscripts in the Bodleian Library* (Oxford, 1886; repr. 1994) and *Supplement of Addenda and Corrigenda*, compiled under the direction of M. Beit-Arié, ed. R.A. May (Oxford, 1994). For Trets cf. Gross, *Gallia Judaica*, pp. 244–5, and B. Richler, *Hebrew Manuscripts in the Bibliotheca Palatina in Parma. Catalogue*, Palaeographical and codicological descriptions M. Beit-Arié (Jerusalem, 2001), no. 1526 (= Parma 1953; De Rossi 1053).

[41] Cf. N. Allony and D.S. Loewinger, *List of Photocopies in the Institute*. Part III: *Hebrew Manuscripts in the Vatican*, p. 71, no. 550.

in Greek, Syriac and Persian.[42] In the introduction to the first list the author specifies this list in the sense that it not only covers roots and herbs, but also instruments, bodily parts and other items.[43] As to the second list he remarks: 'I have also composed an explanation of the drugs and herbs in the vernacular and Arabic so that someone who goes on a distant journey will know their names in both languages. And I have arranged them alphabetically.'[44]

Thus, Shem Tov created two independent lists which he added to the *Sefer ha-shimmush*, one starting with the Hebrew or Aramaic term, followed by the Arabic synonym, and then in about seventy percent of the entries the vernacular term which usually is Middle Latin or Old Occitan. This list was intended to help the reader with the identification and clarification of the Hebrew terminology used by the author in his translation of the *Kitāb al-taṣrīf*. The second list starts with the vernacular term, followed by its Arabic synonym and sometimes supplemented with its Hebrew or Aramaic aequivalent and was to be used and consulted independently from the *Sefer ha-shimmush*. As a source for the proper biblical Hebrew synonym for the Arabic term, Shem Tov consulted Saʿadya ben Yosef al-Fayyūmī, better known as Saʿadya Gaon[45] (882–942) and Abū 'l-Walīd ibn Marwān, i.e. Jonah ibn Janāḥ (died after 1040).[46] Research has shown that Ibn Janāḥ in turn relied heavily upon Saʿadya's biblical translations and commentaries for iden-

---

[42] Ms. Paris, BN héb. 1163, fol. 191a:
ובראותי שאין לנו צורך ולא תועלת בהעתקת שני השערים הראשונים בשמות הזרות והשונות שיש לעשבים והסמים בלשון יון וארמית ופרס

[43] Ms. Paris, BN héb. 1163, ibid.:
עיקרין ועשבין וכלים ואיברין וזולתם.

[44] Ms. Paris, BN héb. 1163, ibid.:
ועוד כתבתי באור הסמים והעשבים בלשון לעז ובלשון הגרי למען דעת המרחיק נדוד שמותם בשתי הלשונות .וסדרתי כל אחד ואחד על דרך אלף בית.

[45] For Saʿadya Gaʾon, philosopher and exegete, poet and polemicist, legist and communal leader see H. Malter, *Saadia Gaon. His Life and Works* (Philadelphia, 1942). For a fundamental study of the language comparisons in his linguistic works and for his Bible translations which served as a source for subsequent scholars see Maman, *Comparative Semitic Philology in the Middle Ages*, esp. pp. 162–79. The diffusion of Saʿadya's works in the Provence is otherwise known from the *Sefer Doreš rešumot* which quotes from Saʿadya's long commentary on Genesis in Hebrew (cf. Y.T. Langermann, 'A Citation from Saadia's Long Commentary to Genesis in Hebrew Translation,' *Aleph* 4 [2004], pp. 293–7).

[46] For Jonah ibn Janāḥ, the undisputed master of Sephardic linguistics, who lived in the first half of the eleventh century, see *Encyclopaedia Judaica*, viii, cols. 1181–6, s.v. 'Ibn Janāḥ, Jonah' (D. Tenne). For the language comparisons in his works see Maman, *Comparative Semitic Philology in the Middle Ages*, pp. 299–370.

tifying Biblical botanical terms.[47] Shem Tov remarks that he especially chose these two authors because he agreed to their identification of the relevant Hebrew and Arabic terminology. In the case of Saʿadya Gaon, his Arabic translations and/or commentaries to the Torah, Isaiah, Psalms, Proverbs, and Job often provided Shem Tov Ben Isaac with the required Hebrew equivalent to a specific Arabic term.[48]

An example of an identification preserved in Saʿadya's writings is the following entry (SeSh1—Alef 1):[49]

אהלות או אהלים ב"ה ענבר

ʾHLWT or *ʾHLYM*, Arab. *ʿNBR*

Hebrew *ʾHLWT* or *ʾHLYM* means 'aloe wood', *Aloexyllon Agallochum* and *Aquilaria Agallocha*, both from India, and features in the Bible, e.g. Ps 45:9 (KB 19; CD 1: 146; LF 3:411–414).

Arabic *ʿanbar* can mean: 1. 'ambergris',[50] 2. 'Saffron',[51] 3. 'Wars'.[52] 4. *al-ḥajar al-ʿanbarī*: 'Ambrastein' (RS 56).

For the identification of *ʾHLYM* as *ʿanbar*, cf. Saʿadya Prov 7:17 (SM 69):

נפתי משכבי מור אהלים וקנמון

[47] I. Löw, *Die Flora der Juden* (Vienna, 1928–34; repr. Hildesheim, 1967), iv, p. 178; Tenne, 'Ibn Janāḥ, Jonah,' col. 1185. For a detailed account of his sources see Maman, *Comparative Semitic Philology in the Middle Ages*, pp. 299–370.

[48] Saadya allegedly also composed a translation of the Five Scrolls and of Ezra. Of the edition of the Five Scrolls with Sa'adya's translation by Kafiḥ (Jerusalem, 1962) only Esther is considered to be authentic. For the question of Saadya's commentaries and/or translations to the Bible see Y. Ratzaby, *Tafsir Yeshaʿyahu le-Rav Saʿadya. Saadya's Translation and Commentary on Isaiah* (Kiriat Ono, 1993), pp. 7–8; M. Polliack, *The Karaite Tradition of Arabic Bible Translation. A Linguistic and Exegetical Study of Karaite Translations of the Pentateuch from the Tenth and Eleventh Centuries C.E.* (Leiden, 1997), pp. 77–80.

[49] SeSh1—Alef 1 refers to our forthcoming edition of the first list of synonyms from Shem Tov Ben Isaac, *Sefer ha-Shimmush*, bk. 29. In keeping with the purposes of this article I have omitted the Romance material.

[50] Lane, *Arabic-English Lexicon*, L 2168, cites various origins for *ʿanbar* as ambergris as given in the earlier literature: 'It issues from a source in the sea: a fish, marine beast, or a vegetable in the bottom of the sea'. It is now generally conceded to be a morbid secretion of the sperm whale intestine, which is fragrant when heated.

[51] Lane, ibid.

[52] Yellow powder from Memecylon tinctorium WILLD. or Flemingia rhodocarpa BAK. (SP 798).

(I have sprinkled my bed with myrrh, aloes, and cinnamon):[53]

וקד בכ׳רת צ׳ג׳עי במסך וענבר ועוד טיב

Ibn Janāḥ (IJ 24): فسّر فيه الصندل (this [term] is explained as 'sandal-wood'), and the gloss in ms. Rouen n. 18: وقيل عنبر (according to others it is: *ʿanbar*).[54]

As for Jonah ibn Janāḥ, Shem Tov certainly used the *Kitāb al-uṣūl*, a dictionary of Biblical Hebrew,[55] as in the following entry (SeSh 1—Samekh 22):

סמדר ב״ה פקאח

SMDR, Arab. *FQ'Ḥ*

Hebrew *SMDR* means 'flower buds of the vine' or '(berry) in the building stage' (KB 759; JD 998; LF 1:72f) and features in the Bible (Song 2:13) and in Rabbinic literature, e.g. in mOrl 1.7.

Arabic *fuqqāḥ* means 'a flower, or blossom, of any plant' (L 2424).

For the identification, cf. Ibn Janāḥ (IJ 495) on the mentioned Bible verse:

هو فقاح الكرم

and Maimonides on mOrl 1.7 (MK 1:401). In his translation of the same verse Saʿadya (SH 53) uses the Arabic aequivalent סמנדר for Hebrew סמדר. See as well MCS 457:1439.

Next to the *Kitāb al-uṣūl* Shem Tov probably consulted Ibn Janāḥ's *Kitāb al-talkhīṣ*, a book on simple drugs, measures and weights in which he gives synonyms of the drugs in a.o. Arabic, Syriac, Hebrew, Greek, Berber, Spanish, and Latin.[56] This work has unfortunately been lost

[53] All the English translations from the Bible are from *The New JPS Translation according to the Traditional Hebrew Text* (Philadelphia, 1985).

[54] For the glosses in ms. Rouen, cf. the extensive discussion in Maman, *Comparative Semitic Philology in the Middle* Ages, pp. 300–316.

[55] *Kitāb al-uṣūl*, ed. Neubauer (= IJ).

[56] See Ullmann, *Die Medizin im Islam*, p. 272; Tenne, 'Ibn Janāḥ, Jonah,' col. 1182; Z. Amar and Y. Sari, 'Liqquṭim mi-milon shemot ha-refu'ot shel R. Jonah ibn Janaḥ,' *Leshonenu* 63 (2000–2001), pp. 279–91.

but survives in a number of quotations by subsequent authors, as, for instance al-Idrīsī (d. 1166), who was active at the court of king Roger II of Sicily and composed the *Kitāb al-jāmiʿ li-ṣifāt ashtāt al-nabāt wa-ḍurūb anwāʿ al-mufradāt* (*Compendium of the Properties of Diverse Plants and Various Kinds of Simple Drugs*).[57] This work survives in an incomplete manuscript in Istanbul (Fatih Library, no. 2610) and in a complete manuscript in Teheran (Kitābkhāna-i Majlis-i Sanā, 18120). These manuscripts actually represent two different redactions, the Istanbul manuscript preserving the synonyms to the names of plants and drugs, while the Teheran manuscript has omitted them.[58] For an example of a quotation from Ibn Janāḥ's *Kitāb al-talkhīṣ* preserved by al-Idrīsī see the entry SeSh 1—Alef 14 in our edition. A second author preserving material from Ibn Janāḥ's *Kitāb al-talkhīṣ* is Seʿadyah ibn Danān from Granada (fifteenth century) who composed the *Sefer ha-shorashim*, a dictionary of biblical Hebrew in which he drew heavily on Ibn Janāḥ and quotes him explicitly not less than 77 times.[59] For an example of a quotation see the entry SeSh 1—Alef 14.

With regard to the sources consulted by Shem Tov for the identification of the Arabic synonyms with terminology in Rabbinic Hebrew or Aramaic, we only have the author's general reference to 'medieval commentators'. To identify these medieval commentators has proven to be especially problematic. One prominent medieval commentator Shem Tov probably had recourse to, was Saʿadya Gaʾon who not only was an important Bible commentator and translator, but also a prominent linguist who dealt extensively with the explanation of difficult terms in the Mishnah in a genre known as *Alfāẓ al-Mishnah*, of which several examples exist in the Genizah and which consists, as Brody remarks, of 'a series of short glosses in Arabic on Hebrew words and expressions, according to the order of the Mishnaic text.'[60] Unfortunately, these lexical explanations only survive partially and are, moreover, for the

---

[57] Facsimile edition in 3 vols. by F. Sezgin (Frankfurt, 1995); see Amar and Sari, 'Liqquṭim mi-millon shemot ha-refuʾot shel R. Jonah ibn Janaḥ'.

[58] Cf. facs. ed. Sezgin, pp. VII–VIII.

[59] Seʿadyah ibn Danān, *Sefer ha-Šorašim*, ed. M. Jiménez Sánchez (Granada, 1996) (= SID).

[60] R. Brody, *The Geonim of Babylonia and the Shaping of Medieval Jewish Culture* (New Haven, 1998), pp. 268–9. While Allony attributed these lexical explanations to Saʿadya Gaʾon, Abramson ('Millon ha-Mishnah le-Rav Saʿadya Gaʾon,' *Leshonenu* 18 [1954], pp. 49–50) and subsequently Maman (*Comparative Semitic Philology in the Middle Ages*, p. 169, n. 12) argued that they are not Saʿadya's at all. According to Brody (*The Geonim of Babylonia*, p. 269), Allony's identification was correct, as confirmed by 'further manuscript

major part still in manuscript.[61] An example of a derivation possibly going back to Saʿadya's *Alfāẓ al-Mishnah* is the following entry (SeSh 1—Gimmel 3):

גופנן ב״ה סבסתאן

GWPNN, Arab. *SBST'N*

The Hebrew term *GWPNN* features in Rabbinic literature (e.g. mDemai 1.1) and means 1) 'fennel', Foeniculum vulgare MILL. (FM 27) and 2) Cordia Myxa L. (LF 1:296–7; 3:462; AEY 168; DAS 2:295; FZ 241–2).

Arab. *sibistān* or *sabistān* is the Arabic form of the Persian *sag-pistān* (VL 1:360) and is identified with Cordia Myxa L., the fruit of the sebesten-tree (M 202; ID 57:20).

For the identification of the two terms, cf. Saʿadya (SAM 184:78); see as well Maimonides on mDemai 1.1:

גופנן: נוע מן אלבקול ישבה אלשבת וקיל אלסבסתאן

(a plant similar to aneth and according to others to sebesten) (MK 1:132).

A second medieval commentator Shem Tov consulted was Maimonides whose commentary on the Mishnah contains a wealth of medico-botanical synonym terminology,[62] and who in turn relied on earlier sources, possibly Saʿadya's explanatory lists and certainly Ibn Janāḥ's works as he states explicitly in his introduction to his *Glossary of Drug Names*.[63] As we have seen above, Shem Tov was familiar with Maimonides' commentary, had access to it and consulted it. In a few cases such a derivation from Maimonides is beyond any doubt as they are literal

discoveries in the Genizah, along with a comparison of citations in Seʿadyah's name and interpretations contained in his other works'.

[61] Three of the fragments were published by N. Allony in *Meḥkarei Lashon we-Sifrut*, i: *Pirqei Rav Saʿadya Gaʾon* (Jerusalem, 1986), pp. 137–98. A large Geonic fragment covering the commentary of nearly half the Mishnah is being prepared for publication by the Institute for the Complete Israeli Talmud but has not been published so far.

[62] Ed. J. Kafiḥ, *Mishnah ʿim Perush Rabbeinu Moshe Ben Maimon. Makor we-Targum*, 6 parts in 7 vols. (Jerusalem, 1963–68). For this commentary see I.M. Ta-Shma, *Ha-Sifrut ha-Parshanit la-Talmud*, 2nd rev. ed. (Jerusalem, 2000–2004), i, p. 185ff.

[63] Maimonides, *Glossary of Drug Names* (trans. Rosner, p. 4).

quotations from his commentary on the Mishnah, as in the following entry (SeSh 1—Shin 45):[64]

שוע הו אלתלביד לאנה ידלך וג׳ה אלשי אלמלבד ותרגום וטח הבית וישוע ית ביתא

ŠWʿ, i.e. mending, covering, because he rubs the surface of something [to be] mended and Targum Onkelos has for **וטח הבית** (the house shall be scraped; Lev 14:42): **וישוע ית ביתא**

Hebrew *ŠWʿ* means 'smoothing, plastering' (JD 1538; LW 4:522; KA 8:119; DAS 5:16, 106; KT 1:142, 545f) and features in Rabbinic literature, e.g. in mKil 9.8.

Shem Tov's explanation is a literal quotation from Maimonides' commentary on the mentioned Mishnah (MK 1:217):

הו אלתלביד לאנה ידלך וג׳ה ד׳לך אלשי אלמלבד ותרגום וטח הבית וישוע ית ביתא.

Arabic *talbīd* means 1) 'forming, pressing, felting, making one's hair stick together; 2) lining, covering, mending (WKAS 1:109–10).

For the remaining terms the medieval commentary and responsa literature composed by Geonim—that is the heads of the Jewish academies in Babylonia—next to Saʿadya, proved to be an important source consulted by Shem Tov. These commentaries and responsa contain a wealth of botanical material and are a valuable source of information for the scientific, technical terminology in Arabic and Hebrew.[65] However, with a few exceptions,[66] the consultation of these particular sources is problematic, as some texts have been edited in the past but

[64] Other examples are Ayin 32; Quf 27,48; Shin 45; Tav 12.

[65] For instance, the responsa composed by Sherira and Hai Gaon contain as S.W. Baron (*A Social and Religious History of the Jews*, 2nd rev. and enl. ed. [New York, 1952–85], viii, p. 229) remarks 'so many attempts at correct identification of names [of plants] and their relation to the previously known species.'

[66] A major exception is the *Ozar ha-Ge'onim. Thesaurus of the Gaonic Responsa and Commentaries* (Haifa, 1928–43), the monumental collection of Geonic responsa and commentaries in the order of the Talmud tractates (to Bava Meziʿa), edited by Benjamin Lewin.

without proper indices, other texts are still in manuscript and yet others have to be considered lost.[67]

An example of a quotation from Geonic sources is (SeSh 1—Alef 22):

אקקיא ב"ה אקאקיה

ʾQQYʾ, Arab. *ʾQʾQYH*

Aramaic *ʾQQYʾ* from Greek ἀκακία (KG 2:125; LS 46) is the 'Arabic gum' won from the acacia tree (Acacia nilotica or Acacia senegal) and features in Rabbinic literature, e.g. in bGitt 69b as *ʾQQYʾ* (variants: *ʾQWQYʾ*, *ʾQQYʾ*), where it is stated: 'for anal worms one make a dressing of acacia, aloe juice, white lead, silver dross, an amulet-full of phyllon and the excrement of doves and apply it to the affected part' (cf. JD 113; LW 1:156; SDA 161; LF 2:388). The regular Hebrew term for the acacia tree is the biblical שטה (see LF 2:377; FO 98; FEB 236–242).

For the Arabic equivalent *aqāqiyā* (DT 1:70; M 12) to the Aramaic *ʾQQYʾ*, cf. the Geonic Responsum to the Talmudic passage quoted:

אקקיא כדי יסמי באל עראק

(*ʾQQYʾ*: thus it is called in Babylonia); see LO *Teshuvot* on bGitt 69b, p. 157. The Arabic equivalent for the Hebrew שטה is *sanṭ* (cf. Saʿadya's commentary on Isaiah 41:19 [DS 62]).

In some cases we could only retrieve a Geonic source indirectly, namely through its quotation in the *Arukh*, a lexicon on the Talmud and Midrashim composed by Nathan Ben Jeḥiel of Rome and completed in 1101. Most of the Arabic plant names quoted in the *Arukh* are derived from Geonic sources, as Löw pointed out in his monumental *Flora und Fauna der Juden*.[68] An example of such a quotation from the *Arukh* is

---

[67] An example of a text still largely in manuscript is the *Kitāb al-ḥāwī*, a dictionary on Biblical and Rabbinic Hebrew written in Judaeo-Arabic and composed by Hai Gaon, of which substantial portions have survived in the Genizah; cf. Brody, *The Geonim of Babylonia*, pp. 330–31, and Maman, *Comparative Semitic Philology in the Middle Ages*, pp. 371–4.

[68] LF 4: 158: 'Die arabischen Pflanzennamen des Aruch stammen mittelbar oder unmittelbar aus gaonäischen Quellen'; see as well idem, entry 'Plants,' *Jewish Encyclopaedia* (New York, 1907), x, pp. 81–2.

SeSh 1—Dalet 8 in our edition. A final author who preserved material from earlier sources which are otherwise lost and which were consulted by Shem Tov is Tanḥum Ben Joseph Ha-Yerushalmi, who was active in Jerusalem in the thirteenth century and composed a dictionary of difficult terms featuring in Maimonides' *Mishneh Torah*, entitled *Kitāb al-murshid*.[69] For his lexicographic explanations of these terms Tanḥum consulted a variety of sources, amongst them different Geonic authors such as Saʿadya, Sherira, Hai, but above all Nathan's *Arukh*, Ibn Janāḥ's works and Maimonides' commentary on the Mishnah.

However, some of the terms featuring in the list composed by Shem Tov cannot be retrieved neither in biblical nor in rabbinic literature, nor in the standard lexica, nor in the medical and botanical literature. An example of such a botanical term is the following (SeSh 1—Aleph 16):

ארי הארץ ב"ה מאזריון

'RY H'RṢ, Arab. *M'ZRYWN*

Hebrew *'RY H'RṢ*, lit., 'lion of the earth' designates Daphne mezereum und Var. The Hebrew term is not attested in secondary literature (cf. BM 387, n. 2) and was possibly coined by Shem Tov as a loan-translation of the Arabic أسد الأرض which in turn is a loan-translation (via Syriac?, cf. LA 33) of the Greek χαμαιλέων (LS 1975; DT 4: 135, esp. n. 4; cf. al-Idrīsī (IJS 1:272): مازريون... بالعربية أسد الأرض).

Māzaryūn is the standard Arabic term for the same plant (cf. DT 4:135; M 237). It features, for instance, in Maimonides' *Medical Aphorisms* (XXI, 85) and is translated by N as: לבריאולה and by Z as: מאזריון הוא לריאוֹלַה.

These terms not only belong to the field of botany and pharmaceutics but also to that of pathology such as in the following example (SeSh 1—He 11):

---

[69] The *Kitāb al-murshid* was edited by B. Toledano, letters Alef-Kaf (Tel Aviv, 1961), by J. Dana, letter Tav (MA thesis [Jerusalem, 1969]), and by Hadassa Shy, letters Lamed-Tav (doctoral thesis, [Jerusalem, 1975]). For a study of this dictionary and excerpts from it see W. Bacher, *Aus dem Wörterbuche Tanchum Jeruschalmi's* (Strasbourg, 1903). An anonymous summary of this work can be found in *Qiẓẓur al-Kafi*, ms. Berlin 153. A new edition and study of this work prepared by Hadassa Shy has been published recently: *Al-Murshid al-Kâfi. The Lexicon of Tanḥūm ben Yosef Hayerushalmi to Mishné Tora of Maimonides*, with a translation from Judaeo-Arabic into Hebrew and references (Jerusalem, 2005).

המעדת המעים ב"ה זלק אלאמעא
HM'DT HM'YM, Arab. *ZLQ 'L'M'*

Hebrew *HM'DT HM'YM* which is not attested in secondary literature is possibly coined by Shem Tov as a Hebrew loan-translation of the Arabic *zalaq al-am'ā'* 'Dysenteria spuria' (SN 111; KZ 37; cf. BM 1117). The Arabic term features in Maimonides' *Medical Aphorisms* (XXII, 36; XXIII, 80, 90, 93, 94), and is translated by N as:

המעדת המעים/מעידת המעים

and by Z as:

המעדת המעים/חלקות או המעדת המעים

Moses ibn Tibbon translates the Arabic as: הגרת המעים in Maimonides' *On the Regimen on Health*; cf. BMR IV, 22.

Next to the method of loan-translation Shem Tov possibly employed the method of semantic borrowing in order to create a Hebrew medical terminology. An example is the following (SeSh 1—Ḥet 19):

חוג ב"ה דואר
ḤWG, Arab. *DW'R*

Hebrew ḤWG means 'circle' or 'to make a circle' (LW 2:21; BM 1460; KB 283) and features in the Bible (e.g. Job 26:10) and Rabbinic literature (bḤag 10b).

Arabic *duwār* means 'circle', and as a medical term 'vertigo, giddiness in the head' (L 931; SN 255). As a medical term it features in Maimonides' *Medical Aphorisms* (XV, 14: الدوار والسدر) and is translated by Nathan ha-Me'ati as

סבוב הראש ומבוכתו הנקרא אשקוטומיא ווטגיני

and by Z as הסתימות והסקוטומיאה.

The source of the identification could not be retrieved. As far as we know Shem Tov probably introduced the Hebrew term in the sense of 'vertigo' by semantic borrowing from the Arabic.

Especially striking is the creation of a new special terminology in the field of different kinds of inflammations or tumors all of them in combination with the Hebrew term צמח which already features in Rabbinic literature in the special sense of 'morbid growth, swelling, ulcer'. Thus we find צמח בין הצלעות (SeSh 1—Zadeh 20) for Arabic *dhāt aljanb* (pleurisy); צמח בטרפשא (SeSh 1—Zadeh 21) for Arabic *shawṣa* (pleurisy); צמח המוח (SeSh 1—Zadeh 22) for Arabic *sirsām* (phrenitis); צמח בלובן העין (SeSh 1—Zadeh 23) for Arabic *zurqa* (glaucoma); צמח אדמדם (SeSh 1—Zadeh 24) for Arabic *falghamūnī* (inflamed tumor); צמח גדול ועמוק (SeSh 1—Zadeh 25) for Arabic *makhba'* (an ulcer affecting the flesh, not the bones or tendons); צמח בצפורן האצבע (SeSh 1—Zadeh 26) for Arabic *dāḥis* (whitlow); צמח מתחת הלשון for Arabic *ḍafdaʿ* (ranula), and finally: צמחים דקים בכל העור (SeSh 1—Zadeh 27) for Arabic *buthūr* ('pimples, or small pustules').

However, these terms did not leave any trace in subsequent literature as far as we know now, possibly because the term צמח—ambiguous at it was—was not a very felicitous one as a general term for inflammation or tumor. Instead of it, the term מורסה became the common one in medieval Hebrew medical literature, representing the Arabic *waram* and Latin *tumor*.

Outlook

The occurrence of these terms in the list raises the question of their originality. Are they the product of the translation activity of earlier or contemporary Jewish translators or medical authors whose works Shem Tov had access to, or was he the first one to coin these new terms, in order to fill the gap of a missing Hebrew technical lexicon? In order to answer this question we have to see which Hebrew translators were active before or simultaneous with Shem Tov, so that he might have been able to consult their translations of medical works, and especially pharmaceutical ones, for the medical-pharmaceutical terminology.[70]

[70] We can disregard those works in which pharmaceutics and pharmaceutical therapy played no role, as in Samuel ibn Tibbon's translation of ʿAlī ibn Riḍwān's commentary to Galen's *Ars parva* (Cf. Ullmann, *Die Medizin im Islam*, pp. 158–9) which he translated in Béziers in the year 1199 under the title *Perush Melakhah qeṭannah.* See Steinschneider, *Die hebräischen Übersetzungen des Mittelalters*, pp. 733–4; *Jewish Encyclopaedia*, vi, pp. 549–50 (Max Schloessinger); G. Freudenthal, 'Les sciences dans les communautés juives médiévales de Provence,' *REJ* 152 (1993), pp. 29–136, p. 49; D. Romano, 'La

The only major translator who qualifies is Moses Ben Samuel ibn Tibbon who was active between 1240 and 1283, possibly first in Naples and then in Montpellier from 1254 on. Thus, he translated, previous to Shem Tov, Maimonides' *Fī tadbīr al-ṣiḥḥa* in 1244 under the title *Al Hanhagat ha-beri'ut*.[71] And contemporary to Shem Tov he translated in 1257 al-Rāzī's antidotarium *Kitāb al-aqrābādhīn al-kabīr*[72] under the title *Aqrabadin*,[73] in the same year or in 1267 Maimonides' *Commentary on Hippocrates' Aphorisms*,[74] in 1259 Ibn al-Jazzār's medical encyclopaedia *Zād al-musāfir*[75] under the title *Zedat ha-derakhim*,[76] and in 1260 Averroes' commentary on Ibn Sīnā's poetical summary of the *Qānūn*,[77] entitled *al-Urjūza fī 'l-ṭibb*.[78] Of unknown date are his translations of Maimonides' treatises *On Poisons*,[79] and *On Hemorrhoids*.[80] As none of these works, with the exception of the antidotarium, is a pharmaceutical handbook that could have been consulted easily by Shem Tov and as most of these translations are contemporary to the time in which Shem Tov translated the *Kitāb al-taṣrīf*, it seems unlikely that he consulted Moses ibn Tibbon's translations. A final verdict in this matter has to wait until these translations have been analyzed with regard to their Hebrew medical-botanical terminology. Two other major thirteenth-century translators of medical works from the Arabic into Hebrew, namely Nathan

transmission des sciences Arabes par les Juifs en Languedoc,' in M.H. Vicaire and B. Blumenkranz (eds.), *Juifs et judaïsme de Languedoc* (Toulouse, 1977), pp. 369–70.

[71] A critical edition of the Arabic text and Hebrew translations by Gerrit Bos is forthcoming.

[72] Cf. Ullmann, *Die Medizin im Islam*, p. 303.

[73] Cf. Steinschneider, *Die hebräischen Übersetzungen des Mittelalters*, p. 730; Romano, 'La transmission des sciences Arabes,' p. 372; *Jewish Encyclopaedia*, vi, p. 548 (Max Schloessinger).

[74] Cf. Steinschneider, *Die hebräischen Übersetzungen des Mittelalters*, p. 769.

[75] For this encyclopaedia see *Ibn al-Jazzār on Sexual Diseases*, ed. and trans. Bos, pp. 5–11.

[76] Cf. Steinschneider, *Die hebräischen Übersetzungen des Mittelalters*, p. 704.

[77] Cf. Ullmann, *Die Medizin im Islam*, p. 155.

[78] Cf. Steinschneider, *Die hebräischen Übersetzungen des Mittelalters*, p. 699.

[79] See Steinschneider, *Die hebräischen Übersetzungen des Mittelalters*, p. 764. A critical edition of the Arabic text and Hebrew translations by Gerrit Bos is forthcoming.

[80] See Steinschneider, *Die hebräischen Übersetzungen des Mittelalters*, p. 763. While Steinschneider rejected the ascription to Moses ibn Tibbon, there is new evidence confirming his authorship; cf. the forthcoming critical edition of the Arabic text and Hebrew translations.

ha-Me'ati[81] and Zeraḥyah Ben Isaac Ben She'altiel Ḥen,[82] mentioned above, were both active in Rome subsequently to Shem Tov, as Nathan was active between 1279 and 1283 and Zeraḥyah between 1279 and 1291. Therefore, the occurrence of a similar botanical terminology in their translations of Maimonides' *Medical Aphorisms* (see below) is to be ascribed to the influence of Shem Tov on these translators, unless they drew on another hitherto unknown source. The question whether Shem Tov had access to other synonym lists composed by earlier authors, is open as most of these lists are anonymous, undated and research into them is still in its infancy.[83] However, a first selective enquiry into some of these lists provided a negative answer to this question. The same holds good for Arabic compendia which contain alphabetical lists of plants with synonyms in different languages, amongst them Hebrew, such as al-Idrīsī's *Kitāb al-jāmiʿ li-ṣifāt ashtāt al-nabāt*. Research into this compendium showed that the novel Hebrew terminology used by Shem Tov does not feature in this medical compendium. A final source Shem Tov might have consulted is original medical compositions in Hebrew with synonym terminology composed at an earlier date than his glossary. The only surviving examples are the *Sefer Asaph* also called *Sefer refu'ot*, a book that was known in southern Italy in the tenth century and was reedited (or possibly even composed) by the southern Italian doctor Shabbetai Donnolo (913–982),[84] and the *Sefer ha-yakar* also called

[81] For Nathan ha-Me'ati (of Cento), see H. Vogelstein and P. Rieger, *Geschichte der Juden in Rom* (Berlin, 1895–96), i, pp. 398–400, Steinschneider, *Die hebräischen Übersetzungen des Mittelalters*, p. 766; Freudenthal, 'Les sciences dans les communautés juives médiévales de Provence,' pp. 69–70.

[82] On Zeraḥyah see Vogelstein and Rieger, *Geschichte der Juden in Rom*, i, pp. 271–5, 409–18; Steinschneider, *Die hebräischen Übersetzungen des Mittelalters*, p. 766; A. Ravitzky, *Mishnato shel R. Zeraḥyah b. Isaac b. She'altiel Ḥen*, doctoral thesis (Jerusalem, 1977), pp. 69–75; *Aristotle's* De Anima, Translated into Hebrew by Zeraḥyah ben Isaac ben She'altiel Ḥen, ed. G. Bos (Leiden, 1994), pp. 1–4; Freudenthal, 'Les sciences dans les communautés juives médiévales de Provence,' pp. 67–9.

[83] For some first results see Bos and Mensching, 'Hebrew Medical Synonym Literature.'

[84] Cf. Shatzmiller, *Jews, Medicine and Medieval Society*, p. 11. For a recent discussion of scholarship of this book see E. Lieber, 'Asaf's Book of Medicines: a Hebrew Encyclopaedia of Greek and Jewish Medicine, possibly compiled in Byzantium on an Indian model,' *Dumbarton Oaks Papers* 38 (1984), pp. 233–49. For the materia medica in the *Sefer Asaph* see especially L. Venetianer, 'Asaf Judaeus. Der aelteste medizinische Schriftsteller in hebraeischer Sprache,' *Jahresbericht der Landes-Rabbinerschule in Budapest* 38–40 (1915–17).

Sefer merkaḥot likewise composed by Shabbetai Donnolo.[85] However, the consultation of these works during the editing of Shem Tov's first list has shown us that the terminology featuring in these works is very different from that used by Shem Tov.

To summarize, Shem Tov Ben Isaac applied two procedures in order to create a Hebrew medical terminology. On the one side he consulted the works of Saʿadya Gaʾon, Jonah ibn Janāḥ and the Geonim in order to find the Hebrew-Aramaic equivalent to the Arabic term at hand, on the other side he saw himself forced to fall back on terminology created through the method of loan-translation and semantic borrowing because of a gap in the existing Hebrew medical-botanical lexicon. As far as we know at the moment several of these loan-translations are attested for the first time in Shem Tov's list, were coined by him after the Arabic and were then adopted by other translators such as Nathan ha-Meʾati and Zeraḥyah Ben Isaac Ben Sheʾaltiel Ḥen.[86]

List of Abbreviations

AEY P. Auerbach and M. Ezrahi, '*Yalkut Zemahim*,' *Leshonenu* 1 (1929), pp. 161–395.

BM E. Ben Yehuda, *Millon ha-Lashon ha-Ivrit. Thesaurus Totius Hebraitatis et Veteris et Recentioris*, 17 vols. (Berlin, 1910–59; repr. Tel Aviv, 1948–59).

BMR Maimonides, *On the Regimen of Health*, Critical edition of the Arabic text and Hebrew translations G. Bos (forthcoming).

CD David J.A. Clines (ed.), *The Dictionary of Classical Hebrew*, 5 vols. (Alef-Nun) (Sheffield, 1993–2001).

DAS G. Dalman, *Arbeit und Sitte in Palästina* (Gütersloh, 1928–43; repr. Hildesheim, 1964–87, Berlin 2001).

DS J. and H. Derenbourg (eds.), *Targum Sefer Yeshaʿyahu ʿim Liqquṭim mi-Beʾuro be-Lashon ʿaravit le-Rabbeinu Saʿadyah Gaʾon Ben Yoseph ha-Fayyumi*, Version arabe d'Isaïe de R. Saadia ben Iosef al-Fayyoûmî (Paris, 1896; repr. Hildesheim, 1979).

DT A. Dietrich (ed.), *Dioscurides Triumphans. Ein anonymer arabischer Kommentar (Ende 12. Jahr. n. Chr.) zur Materia medica* (Göttingen, 1988).

FE J. Feliks, *Ezei Peri le-Mineihem. Fruit Trees in the Bible and Talmudic Literature* (Jerusalem, 1994).

FEB J. Feliks, *Ezei Besamim, Yaʿar we-Noy. Trees: Aromatic, Ornamental and of the Forest in the Bible and Rabbinic Literature* (Jerusalem, 1997).

FM J. Feliks, *Marʾot ha-Mishnah, Seder Zeraʿim* (Jerusalem, 1967).

[85] Cf. Shatzmiller, *Jews, Medicine and Medieval Society*; H.J. Zimmels, 'Science,' in C. Roth (ed.), *The Dark Ages. Jews in Christian Europe 711–1096* (Tel Aviv, 1966), pp. 297–301. The *Sefer ha-Merkaḥot* was edited by S. Muntner, in *R. Shabbetai Donnolo. Kitvei ha-Refuʾah* (Jerusalem, 1949), pp. 1–23. For a new edition cf. L. Ferre, 'Donnolo's *Sefer ha-yaqar*: New Edition with English Translation,' in G. Lacerenza (ed.), *Šabbeṯay Donnolo. Scienza e cultura ebraica nell' Italia del secolo X* (Naples, 2004), pp. 1–20.

[86] For a list of these terms see the introduction to our edition of the first list.

FO J. Feliks, *Olam ha-Zome'ah ha-Miqra'i* (Tel Aviv, 1957).
FZ U. Feldman, *Zimchei ha-Mishnah* (Tel Aviv, n.d.).
ID A. Issa, *Dictionnaire des noms des plantes en Latin, français, anglais et arabe* (Cairo, 1930).
IJ Abū 'l-Walīd Marwān ibn Janāḥ, *Kitāb al-uṣūl. The Book of Hebrew Roots*, ed. A. Neubauer (Oxford, 1875).
IJS al-Idrīsī, *Kitāb al-jāmiʿ li-ṣifāt ashtāt al-nabāt wa-ḍurūb anwāʿ al-mufradāt*, facs. ed. F. Sezgin (Frankfurt, 1995).
JD M. Jastrow, *A Dictionary of the Targumim, the Talmud Bavli and Yerushalmi, and the Midrashic Literature* (New York, 1950).
KA A. Kohut, *Arukh shalem. Aruch Completum*, and Krauss, *Tosefet he-Arukh*. Additamenta. repr. in 9 vols. (Tel Aviv, 1970).
KB L. Koehler and W. Baumgartner, *The Hebrew and Aramaic Lexicon of the Old Testament*, subsequently revised by W. Baumgartner and J.J. Stamm, with assistance from B. Hartmann, Z. Ben-Hayyim, E.Y. Kutscher, P. Reymond, trans. and ed. under the supervision of M.E.J. Richardson (Leiden, 1994–2000).
KG S. Krauss, *Griechische und Lateinische Lehnwörter im Talmud, Midrasch und Targum*, Mit Bemerkungen von Immanuel Löw (Berlin, 1898–99).
KT S. Krauss, *Talmudische Archäologie*, 3 vols. (Leipzig, 1910–12).
KZ H. Kroner, *Zur Terminologie der arabischen Medizin und zu ihrem zeitgenössischen hebräischen Ausdrucke*, an der Hand dreier medizinischer Abhandlungen des Maimonides (Berlin, 1921).
L E.W. Lane, *Arabic-English Lexicon* (London, 1863–79).
LA I. Löw, *Aramäische Pflanzennamen* (Leipzig, 1881).
LF I. Löw, *Die Flora der Juden*, 4 vols. (Vienna, 1928–34; repr. Hildesheim, 1967).
LO B. Lewin, *Oṣar ha-Geonim*. Thesaurus of the Geonic Responsa and Commentaries (Haifa, 1928–43).
LS H.G. Liddell and R. Scott, *A Greek English Lexicon*, Revised and augmented throughout by H.S. Jones a.o. (Oxford, 1925–40; repr. with supplement Oxford, 1989).
LW J. Levy, *Wörterbuch über die Talmudim und Midraschim*, Nebst Beiträgen von H. Leberecht Fleischer, Zweite Auflage mit Nachträgen und Berichtigungen von L. Goldschmidt (Berlin, 1924).
M Maimonides, *Sharḥ asmāʾ al-ʿuqqār*, ed. M. Meyerhof (Cairo, 1940); English translation from the Hebrew translation by S. Muntner by F. Rosner (Haifa, 1995).
MCS A. Maman, *Comparative Semitic Philology in the Middle Ages. From Saʿadiah to Ibn Barūn* (10th–12th C.) (Leiden, 2004).
MK *Mishnah ʿim Perush Rabbeinu Mosheh Ben Maimon. Makor we-Targum*, ed. J. Kafiḥ, 6 parts in 7 vols. (Jerusalem, 1963–69): MK 1 = Zera'im; MK 2 = Moʿed; MK 3 = Nashim; MK 4 = Nezikin; MK 5 = Kodashin; MK 6 = Toharot.
N Nathan ha-Meʾati
RS J. Ruska (ed. and trans.), *Das Steinbuch des Aristoteles. Mit literargeschichtlichen Untersuchungen nach der arabischen Handschrift der Bibliothèque Nationale* (Heidelberg, 1912).
S J. Derenbourg (ed.), *Targum ḥamishah ḥumshei Torah be-lashon ʿaravit le-Rabbeinu Saʿadyah Gaʾon Ben Yoseph ha-Fayyumi. Version arabe du Pentateuque de R. Saadia Ben Iosef al-Fayyoûmî* (Paris, 1893; repr. Hildesheim, 1979).
SAM Saʿadya Gaʾon, *Alfāẓ al-Mishnah*, ed. N. Allony in *Meḥkarei Lashon we-Sifrut*, i: *Pirqei Rav Saʿadya Gaʾon* (Jerusalem, 1986), pp. 137–98.
SDA M. Sokoloff, *A Dictionary of Jewish Babylonian Aramaic of the Talmudic and Geonic Periods* (Ramat Gan, 2002).
SeSh G. Bos and M. Hussein – G. Mensching and F. Savelsberg, Shem Tov Ben Isaac, *Sefer ha-Shimmush*, Bk. 29, Glossary one (forthcoming).

SH *Ḥamesh Megillot. Shir Ha-Shirim, Rut, Kohelet, Ester, Ekhah. ʿIm Perushim ʿattiqim ha-yoẓeʾim la-Or Paʿam Rishonah ʿal Pi Kitvei Yad be-ẓeruf Mevoʾot Heʿarot we-Heʾarot*, ed. J. Kafiḥ (Jerusalem, 1962).

SM *Mishlei ʿim Targum we-Pherush ha-Gaon Rabbeinu Saʿadya Ben Yoseph Fayyumi*, ed. J. Kafiḥ (Jerusalem, 1976).

SN D. Sontheimer, 'Nachricht von einer arabisch-medicinischen Handschrift, vermutlich des Ibn-Dschezla,' *Janus* 2 (1847), pp. 246–72, repr. in F. Sezgin (ed.), *Beiträge zur Geschichte der arabisch-islamischen Medizin. Aufsätze* i (1819–1869) (Frankfurt, 1987), pp. 92–128.

SP W. Schmucker, *Die pflanzliche und mineralische Materia Medica im* Firdaus al-ḥikma *des ʿAlī ibn Sahl Rabban aṭ-Ṭabarī*, doctoral thesis (University of Bonn, 1969).

VL I.A. Vullers, *Lexicon Persico-Latinum Etymologicum*, 2 vols. (Bonn, 1855–64; repr. Graz, 1962).

WKAS M. Ullmann et al., *Wörterbuch der klassischen arabischen Sprache* (Wiesbaden, 1957–).

Z Zeraḥyah Ben Isaac Ben Sheʾaltiel Ḥen

WEATHER FORECASTING, LUNAR MANSIONS AND A DISPUTED ATTRIBUTION: THE *TRACTATUS PLUVIARUM ET AERIS MUTATIONIS* AND *EPITOME TOTIUS ASTROLOGIAE* OF 'IOHANNES HISPALENSIS'*

Charles Burnett

We are now quite fully informed of the tradition of meteorology from Greek, through Arabic into Western European Latin culture, thanks to the work of Hans Daiber, Hidemi Takahashi, Paul Lettinck, Pieter Schoonheim, Resianne Fontaine, A.C. Smet, Gudrun Vuillemin-Diem and others.[1] Serious study of its sister discipline, weather forecasting, however, has only just begun.[2] We find both sciences referred to in

* I am very grateful to David Juste for supplementing information on manuscripts from his forthcoming volumes of *Catalogus codicum astrologicorum latinorum*, and for making several helpful suggestions, and to Dorian Greenbaum, Paul Kunitzsch, Shlomo Sela, and Renate Smithuis for several details.

[1] H. Daiber, *Ein Kompendium der Aristotelischen Meteorologie in der Fassung des Hunain ibn Ishāq* (Amsterdam, 1975); idem, 'The *Meteorology* of Theophrastus in Syriac and Arabic Translation,' in W.W. Fortenbaugh and D. Gutas (eds.), *Theophrastus: His Psychological, Doxographical, and Scientific Writings* (New Brunswick, 1992), pp. 166–293; H. Takahashi, *Aristotelian Meteorology in Syriac. Barhebraeus, Butyrum sapientiae, Books of Mineralogy and Meteorology* (Leiden, 2004); P. Lettinck, *Aristotle's Meteorology and its Reception in the Arab World, with an edition and translation of Ibn Suwār's "Treatise on Meteorological Phenomena" and Ibn Bājja's "Commentary on the Meteorology"* (Leiden, 1999); Alexander of Aphrodisias, *Commentary on the* Meteorologica *of Aristotle*, in William of Moerbeke's translation, ed. A.J. Smet (Leuven, 1968); *Otot ha-shamayim. Samuel Ibn Tibbon's Hebrew Version of Aristotle's Meteorology*, ed. R. Fontaine (Leiden, 1995); P. Schoonheim, *Aristotle's Meteorology in the Arabico-Latin Tradition* (Leiden, 2000); G. Vuillemin Diem, 'Zu Wilhelm von Moerbekes Übersetzung der aristotelischen *Meteorologie*. Drei Redaktionen, ihre griechischen Quellen und ihr Verhältnis zum Kommentar des Alexander von Aphrodisias,' in R. Beyers et al. (eds.), *Tradition et traduction. Les textes philosophiques et scientifiques grecs au Moyen Age latin* (Leuven, 1999), pp. 115–66, and eadem, *Aristoteles Latinus X 2.1 Meteorologica. Translatio Guillelmi de Morbeka. Praefatio* (Leiden, 2007).

[2] After the earlier forays of Gustav Hellmann (*Denkmäler Mittelalterliche Meteorologie* [Berlin, 1904] and 'Die Wettervorhersage im ausgehenden Mittelalter [XII. bis XIV. Jahrhundert],' *Beiträge zur Geschichte der Meteorologie*, II.8 [Berlin, 1917], pp. 169–229), the subject has been mapped out by the bibliographies of Arabic and Latin texts by Fuat Sezgin (*Geschichte des arabischen Schrifttums*, vii, Astrologie—Meteorologie und Verwandtes bis ca. 430 H [Leiden, 1979]) and Stuart Jenks ('Astrometeorology in the Middle Ages,' *Isis* 74 [1983], pp. 185–210 and 562), respectively, and the survey by Gerrit Bos and Charles Burnett: *Scientific Weather Forecasting in the Middle Ages: The Writings of al-Kindī* (London, 2000).

Arabic literature as *ʿilm al-āthār al-ʿulwiyya* ('the science of celestial effects'; in Latin: 'de impressionibus superioribus'), though the latter was also referred to as *ʿilm aḥdāth al-jaww* ('the science of the phenomena of the atmosphere'; 'de mutationibus aeris') and, more pedestrianly, *fī 'l-maṭar* or *al-amṭār* ('on rains'; *de imbribus* or *de pluviis*). Several traditions can be identified in ancient Greek and Arabic works relevant to weather forecasting:

1) Weather signs, as found in several ancient texts (Virgil, *Georgics*, Bk 1, Pliny, *Natural History*, Bk 18, Aratus, *Phaenomena*,[3] Ptolemy, *Tetrabiblos*, Bk 2, ch. 13, and the text specially devoted to it: the Pseudo-Aristotle, *De signis*).[4] Many of these signs are of popular origin, and, in the Arabic world, appear in poetry and rhymed prose (*sajʿ*).[5]
2) The Peripatetic tradition of Aristotle's *Meteorologica* with its commentaries and derivatives.[6]
3) Calendars of weather changes during the year, represented in Antiquity by the *parapegmata* (again Pliny, *Natural History*, Bk 18, Columella, *De re rustica*, Bk 11, Ptolemy, *Phaseis*, and Aetios of Amida, *Tetrabiblos*, III, ch. 164) and in Arabic, especially by the *anwāʾ* tradition, which dates to pre-Islamic times. The *anwāʾ* (singular *nawʾ*) are the twenty-eight stars or asterisms whose rising and setting throughout the year indicate specific changes in the weather. Their distinctive names are Arabic (some perhaps recall pre-Islamic local deities), and give a nomenclature that is completely different from that of the constellations of the zodiac of the Greek tradition.[7]

---

[3] E.g. in the translation of Germanicus, ed. A. Le Boeuffle (Paris, 1975), pp. 44–59.

[4] This text is extant in Greek and in a Latin translation made by Bartholomaeus of Messina in the mid thirteenth century, but extant in two variant versions in Oxford, Corpus Christi College, ms. 243, fol. 48vb–52ra and 52rb–53ra respectively.

[5] E. Robertson, 'Arab Weather Prognostics,' *JRAS* (1930), pp. 377–89 and W.J. Sersen, *Arab Meteorology from Pre-Islamic Times to the Thirteenth Century A.D.*, PhD thesis (University of London, 1976).

[6] See note 1 above. One should note that the Arabic version of *Meteorologica*, made by Yaḥyā ibn al-Biṭrīq (ed. C. Petraitis [Beirut, 1967]), is a free paraphrase of the Greek text, with many omissions and additions.

[7] *Le Calendrier de Cordoue*, ed. R. Dozy, revised by C. Pellat (Leiden, 1961); M. Forcada, 'L'expression du cycle lunaire dans l'ethnoastronomie arabe,' *Arabica* 47 (2000), pp. 37–77. Another set of fixed stars were thought to bring bad weather: see P. Kunitzsch, 'Zur Tradition der "Unwettersterne",' *ZDMG* 122 (1972), pp. 108–17 (repr. in idem, *The Arabs and the Stars* [Northampton, 1989], article XVI).

4) Weather-forecasting texts of purportedly Indian origin using the same 28 divisions (with a variant of 27 divisions), and the same nomenclature. Since the presence of the Moon in these divisions is significant, these were referred to as 'lunar mansions' (*manāzil al-qamar*).

All four of these traditions were brought together in two *Letters* (*rasāʾil*) by Yaʿqūb ibn Isḥāq al-Kindī, the 'philosopher of the Arabs' (d. after 866).[8] Substantial portions of these *Letters* share common sources with other texts, identified up to now only in Latin:

1) The section on the lunar mansions in al-Kindī, *Letter II*, sections 79–133 corresponds to Jafar Indus, *Liber imbrium* 44–91, 140–57 and the parallel sections in *Sapientes Indi*.[9]
2) Al-Kindī, *Letter II*, sections 134–57 corresponds intermittently with sentences in *Apertio portarum*[10] and the *Tractatus pluviarum* of 'Iohannes Hispalensis'.

In this article I would like to give the evidence we have for the last of these texts; namely, the *Tractatus pluviarum et aeris mutationis secundum magistrum Iohannem Yspalensem*, i.e. 'the treatise on rains and the change of the atmosphere according to Master John of Seville'. Two elements concerning this text will be highlighted: (1) the forms of the names of the lunar mansions, which are aberrant from those usually found in Latin texts since they are Latinised rather than given in transliteration; (2) the attribution to John of Seville, the prolific translator of works on the science of the stars from Arabic into Latin in the second quarter of the twelfth century.

---

[8] The two *Letters* are extant as such only in Hebrew, but a compendium deriving from the two *Letters* with the addition of some other material survives in Latin as the 'Liber Alkindii de mutatione temporum'; all these texts are edited in Bos and Burnett, *Scientific Weather Forecasting*. The date of al-Kindī's death is inferred from the juncture between real events and hypothesised events in his *Risāla fī mulk al-ʿarab wa-kammiyyatihi* ('Letter on the Rule of the Arabs and its Duration') which falls after the rebellion (*fitna*) of al-Mustaʿīn in 866 C.E.: see Abū Maʿshar, *On Historical Astrology*, ed. K. Yamamoto and C. Burnett, 2 vols. (Leiden, 2000), i, pp. 539 and 602.

[9] These two Latin texts, deriving from the same Arabic original, are edited in C. Burnett, 'Lunar Astrology: the Varieties of Texts Using Lunar Mansions with Emphasis on *Jafar Indus*,' *Micrologus* 12 (2004), pp. 43–133 (see pp. 60–124).

[10] See pp. 226–7 below.

The *Tractatus pluviarum* has been identified in the following manuscripts:

1. Cambridge, Trinity College, O II 40 (James no. 1144), s. xv, fol. 110r–v.
2. Cracow, Biblioteka Jagiellonska, 1843, s. xv, fol. 6082v (*sic*).
3. London, British Library, Sloane 636, s. xv, fol. 78r–v.
4. Munich, Bayerische Staatsbibliothek, Clm 27, s. xv, fol. 211r–213r.
5. Munich, Bayerische Staatsbibliothek, Clm 667, s. xv, fol. 13v.
6. Munich, Bayerische Staatsbibliothek, Clm 18940, A.D. 1527, fol. 98r–99v.
7. Oxford, Bodleian Library, Bodley 464, A.D. 1318, fol. 126r.
8. Oxford, Bodleian Library, Can. Misc. 396, s. xiii^{ex}, fol. 92rb–93va.
9. Paris, Bibliothèque nationale de France, 7316A, s. xiv, fol. 45r–47r.
10. Paris, Bibliothèque nationale de France, lat. 7328, s. xiv–xv, fol. 72vb–74rb.
11. Prague, Národní Knihovna České Republiky, Univ. VI.F.7 (1144), s. xv, 123r–v (1–22 only).
12. Vienna, Österreichische Nationalbibliothek, cvp. 2436, s. xiv, fol. 136v–138r.[11]

The *Tractatus pluviarum* is a portion of a larger work, the *Epitome totius astrologie* (*Ysagoge* and *Liber quadripartitus*), similarly attributed to John of Seville.[12] This is a popular work including an introductory book (*Ysagoge*) and four 'parts', covering the major branches of astrology. At various points in the text the 'present date' is stated as being 1142 (with slight variants),[13] a date confirmed by the values given to the longitudes of

[11] A text 'Iohannes Hispalensis de mutatione aeris' is listed in the table of contents of Paris, BNF, lat. 10268, fol. [1]. A reference to *Tractatus pluviarum*, sentence 32, occurs in the incipit of a work in Munich, Bayerische Staatsbibliothek, Clm 7633, fol. 216r–217r: 'De portarum aut valvarum apertionibus. De apertione magna portarum loquitur Hispalensis parte prima Quadripartiti…'

[12] That it is extracted from the larger work, rather than incorporated into it, is suggested by the reference within the *Tractatus pluviarum* to a previous section of the *Epitome* (see sentence 7 and note on p. 255 below).

[13] *Ysagoge*, ch. 1, ms. London, Royal 12.C.XI (= J), fol. 30r: 'Est in ipso unum de syderibus primi honoris est hodierno tempore anno .1142. 15 grad. 32 m.; *Secunda pars*, c. vii (De locis stellarum fixarum): ibid., fol. 41ra: 'Primum sidus Capitis Dyaboli cuius

the fixed stars mentioned by the author, which are advanced from the epoch date of those of Ptolemy by approximately 1° degree in 70 years, as the author himself comments.[14] The *Tractatus pluviarum* occupies most of the 'prima pars' of the *Epitome*, which deals with 'general astrology'. The *Epitome* exists, complete or partially, in numerous manuscripts:

1. Berlin, Staatsbibliothek Preussischer Kulturbesitz, lat. fol. 192, s. xv, fol. 85vb (*Quarta pars* only).
2. Boston, Medical Library, 4, s. xv^{ex}, fol. 1r–50v.
3. Catania, Biblioteca Universitaria, 87, s. xv, fol. 38–71v.
4. Edinburgh, Royal Observatory, Crawford 3.29, 1317 A.D., fol. 16–22.
5. Erfurt, Wissenschaftliche Bibliothek, Amplon. F 394, s. xiv^{in}, fol. 68r–79r.
6. Erfurt, Wissenschaftliche Bibliothek, Amplon. O 84, s. xiv^{1}, fol. 1r–37r.
7. Erfurt, Wissenschaftliche Bibliothek, Amplon. Q 223, s. xiv^{ex}, fol. 57v–85v.
8. Erfurt, Wissenschaftliche Bibliothek, Amplon. Q 377, s. xiv^{2}, fol. 7–11 (*Quarta pars* only).
9. Erfurt, Wissenschaftliche Bibliothek, Amplon. Q 379, s. xiv^{mid}, fol. 26r–37v. *Secunda, Tertia* and *Quarta pars*.
10. London, British Library, Royal 12.C.XI, s. xiv, fol. 30r–55v.
11. Lyons, Bibliothèque municipale, 329, s. xv^{ex}, fol. 270ra–285va (copied from Paris, BNF, lat. 7321).
12. Madrid, Biblioteca nacional, 10009, s.xiii, fol. 39r–46v (*Secunda pars* only).
13. Madrid, Biblioteca nacional, 10063, s. xiii, fol. 10v–11rb (*Quarta pars* only).
14. Oxford, Bodleian Library, Digby 38, s. xiv, fol. 84–86v (*Ysagoge* and part of *Prima pars*).

longitudo est in Tauro 14 gr. 46 min. anno .1142. alibi completo'. Another reference to the date occurs only in the printed *Epitome*, sig. T3v (at the end of *Quarta pars*): 'Est itaque exordium ipsarum mansionum secundum tabulas quae sunt factae in motu Solis a circulo recto a sedecim gradibus Arietis hoc tempore 1142 annorum Christi'.

[14] Paul Kunitzsch shows how the longitudes of the stars have been advanced in respect to Ptolemy's values by approximately 15°55', in his 'Abū Maʿšar, Johannes Hispalensis und Alkameluz,' *ZDMG* 120 (1970), pp. 103–25 (repr. in idem, *The Arabs and the Stars*, article XVII), at pp. 115 and 119. This implies a rate of precession of 1° in 70 years, which is explicitly stated in the printed *Epitome* sig. K1r.

15. Oxford, Bodleian Library, Digby 149, s. xiii, fol. 195r–200r (*Tertia* and *Quarta pars*).
16. Oxford, Bodleian Library, Canon. Misc. 517, s. xv, fol. 9ra–10ra (last chapter of *Quarta pars* only: *Electiones secundum Indos et Dorotheum*).
17. Paris, Bibliothèque nationale de France, lat. 7303, s. xv, fol. 104ra–117ra (*Ysagoge* omitted).
18. Paris, Bibliothèque nationale de France, lat. 7306, s. xv, fol. 64v–87r.
19. Paris, Bibliothèque nationale de France, lat. 7321, s. xv (A.D. 1448), fol. 122r–54v.
20. Paris, Bibliothèque nationale de France, lat. 7338, s. xv, fol. 1ra–42va.
21. Prague, Národní Knihovna České Republiky, Univ. VI.F.7 (1144), s. xiv, fol. 155r–162v (*Ysagoge* and *Prima pars*).
22. Vatican, Biblioteca apostolica Vaticana, lat. 4080, s. xv, fol. 1r–40v.
23. Vatican, Biblioteca apostolica Vaticana, lat. 4082, s. xv^{in}, fol. 121ra–138rb.
24. Vatican, Biblioteca apostolica Vaticana, lat. 4087, s. xiv, fol. 38v–60r.
25. Vatican, Biblioteca apostolica Vaticana, Reg. lat. 1452, s. xiv, fol. 58r–76v.
26. Venice, Biblioteca Nazionale Marciana, Lat. Z. 343 (1877; Valentinelli, Cl. XI, 102), s. xiv, fol. 107r–131r.
27. Venice, Biblioteca Nazionale Marciana, Lat. Z. 344 (1878; Valentinelli, Cl. XI, 104), s. xiii^{in}, fol. 1ra–30rb.
28. Vienna, Österreichische Nationalbibliothek, cvp. 2436, s. xiv, fol. 13–15 (last chapter of *Quarta pars* only) and fol. 206r–228v (full text).[15]
30. Vienna, Österreichische Nationalbibliothek, cvp. 5442, s. xv (after A.D. 1441), fol. 158r–179v.
31. Vienna, Österreichische Nationalbibliothek, cvp. 5463, s. xv, fol. 148–79.

[15] Note that the scribe indicates that there is another 'translation' of the text of the *Epitome*: 'aliam translacionem huius operis vide circa tabulas Alphonsi in pergameno, que est multo latior ista' (fol. 228rb).

The text was printed in Nuremberg in 1548, with the following title page: *Epitome totius astrologiae, conscripta a Ioanne Hispalensi Hispano Astrologo celeberrimo, ante annos quadringentos, ac nunc primum in lucem edita. Cum praefatione Ioachimi Helleri Leucopetraei, contra Astrologiae adversarios. Noribergae in officina Ioannis Montani et Ulrici Neuber, Anno Domini MDXLVIII.* The printed text differs considerably from that found in the manuscripts I have consulted. This may be due partially to the stylistic changes of the editor, Joachim Heller, and to the fact that he used more than one manuscript.[16]

The similarity between the separately occurring *Tractatus pluviarum* and the text within the *Epitome* is so great that they can be regarded, for editorial purposes, as manuscripts of a single text.

[16] The order of the text of *Tractatus pluviarum et aeris mutationis* in the printed edition is as follows (I give the headings in the printed version, and the sentence numbers of the edition below): sig. G1r *Caput VI. De dispositione aeris* (+ Scire volens generaliter an annus sit futurus pluviosus, considera in anni revolutione scilicet in hora coniunctionis vel oppositionis, an signum sit aquae et de mansione aquosa lunae et eius dominus sit aquosus vel in signo aquoso vel in mansione aquosa et quae sit natura gradus coniunctionis vel oppositionis inde et naturam planetarum existentium in praedictis locis vel ad ea respicientium) **1–6**; *Additio* (**7–17, 19–20**); *In vestustiori exemplari* (+ Lunae vero status consideratur cum incipit intrare in aliquam praedictarum portarum, nam secundum eius statum erit tempus, usque dum intret in aliam ianuam) **21–22, 8–9, 27, 32** (+ Verbi gratia, Luna separatur a Marte et coniungitur Veneri, quorum domus scilicet Aries et Libra sunt oppositae. Sic Luna existente in quarta s<cilic>et signo foeminino, si eam prospiciat retrogradus existens in signo foeminino, erit pluvia); *in alio codice* **32**; *Additio ex alio codice* **33**; *De dispositione aeris Caput VII. de eodem* (+ Iudicabis autem particulariter de aeris mutatione sic. In coniunctione Solis et Lunae, considera quod signum sit ascendens in ipsa et cuius sit naturae an sit mansio humida vel sicca et dominus signi an sit pluvialis naturae vel non, et signum in quo est ipse cuius sit naturae et qui planetae sint in signo ascendente, vel cum domino eius, vel in eorum aspectu similiter, qui sint coniuncti Lunae, vel aspiciant eam tunc, vel qua in domo, vel qua mansione sit ipsa Luna. Quod si plura testimonia sint planetarum et signorum praedictorum in pluvia illa media lunatio erit, scilicet usque ad tempus oppositionis pluviosa... Semper autem in oppositione consideramus gradum luminaris existentem supra terram) **23, 25–6** (+ Est etiam videndum an in mansione humida vel sicca finiatur eadem computatio, secundum quod iudicabis), **28, 30** (+ Summe autem notandum est quod tempore aestatis duo testimonia pluvae minus valent quam unum in hieme. Sic duo serenitatis minus faciunt in hieme quam unum in aestate. Et quod in una terra pluit, in alia non, potest esse vel quia hora conversionis habet aliud ascendens vel quia signum unius terrae in ipsa hora habet planetam pluvialem et alterius terrae signum habet serenum planetarum. Videamus ergo planetas cuiusque partis et signum.); *Caput VIII. De partibus mundi distributis planetis et signis* **35–55**; *Aphorismi in alio codice inventi* **56–65**; *Caput IX. de duodenariis signorum* **66–7** (+ Quod iudicium Latinis est insolitum); *Caput X. de fridariis planetarum* (additional); *Caput XI. de mansionibus Lunae* **69–101**. It is noticeable that some passages are repeated in different words, and that the lunar mansions are referred to several times before they are described.

Much of the material in the first third of the *Tractatus pluviarum* appears in almost the same wording, but in a rearranged order, as a text with the title *Apertio portarum*, 'the opening of the doors'.[17] This is found in the following manuscripts:

1. Berlin, Staatsbibliothek Preussischer Kulturbesitz, lat. 632 (formerly Boncompagni ms. 4), fol. 67v–68v.
2. Boston, Medical Library, 20, s. xiv, fol. 184r–185v.
3. Cambridge, Clare College 15, before A.D. 1280, fol. 6v.
4. Erfurt, Wissenschaftliche Bibliothek, Amplon. Q 363, s. xiii–xiv, fol. 75r–v.
5. Munich, Bayerische Staatsbibliothek, Clm 2841, s. xv, fol. 13r–v.
6. Munich, Bayerische Staatsbibliothek, Clm 11067, s. xv, fol. 93ra–94ra.
7. Munich, Bayerische Staatsbibliothek, Clm 25005, s. xv, fol. 49r–v.
8. Oxford, Bodleian Library, Bodley 464, A.D. 1318, fol. 127r–v (incomplete).
9. Paris, Bibliothèque nationale de France, lat. 7303, s. xv, fol. 120vb–121rb.
10. Paris, Bibliothèque nationale de France, lat. 7324, s. xv, fol. 67r.
11. Paris, Bibliothèque nationale de France, lat. 7445, s. xv, fol. 8va–vb.
12. Paris, Bibliothèque nationale de France, lat. 18504, s. xiv, fol. 187v–191r.
13. Prague, Národní Knihovna České Republiky, Univ. III.C.2 (433), s. xv, fol. 167r–169v.
14. Vienna, Österreichische Nationalbibliothek, cvp. 3162, s. xv, fol. 233r–234v.
15. Vienna, Österreichische Nationalbibliothek, cvp. 5438, A.D. 1430, fol. 117ra–rb.

The correspondence of the sentence numbers of *Apertio portarum* with those of *Tractatus pluviarum* is as follows (the related sections of al-Kindī, *Letter II*, are added):

[17] This text is edited, from mss. HP, in Bos and Burnett, *Scientific Weather Forecasting*, pp. 385–93.

Apertio portarum	*Tractatus pluviarum*
1 (definition of 'apertio portarum')	–
2–3	8
4	9
5	10
6–7	11
8–9	12
10	13
11	14
12	15
–	16
13–14	23a
	23b = al-Kindī, *Letter II*, 156
15–16	24
17	25
18	26
19–23	–
24–26, 28 (confused text)	1–3
27	4
29 (first half)	5
–	6
29 (second half)	7
30–32	21 = al-Kindī, *Letter II*, 148–50
33	22 = al-Kindī, *Letter II*, 146–7
34	27 = al-Kindī, *Letter II*, 137
35	28 = al-Kindī, *Letter II*, 134
–	29–31 = al-Kindī, *Letter II*
36	32
–	33
37	34
38 (another definition of 'apertio portarum')	–
–	35–101

The main addition in *Apertio portarum* is a section on delaying or hastening the day of rain (19–21), while *Tractatus pluviarum* adds some jejune philosophical reflections (**17–20**), and a reference to cities (**33**) which looks forward to the latter part of the treatise. In the shared passages there is some variation in terminology between *Apertio portarum* and *Tractatus pluviarum*, the former using 'vide' and 'principium' where the latter uses 'considera' and 'exordium'. In general the Latin of *Apertio portarum* is smoother than that of the *Tractatus pluviarum*, which retains

Arabic syntax.[18] The text in *Tractatus pluviarum* is closer to what is found in al-Kindī: e.g., *Tractatus pluviarum*, sentences 29–31, which are missing in *Apertio portarum*, occur in al-Kindī, *Letter II*, 134, and *Tractatus pluviarum* 23 (last part) occurs in al-Kindī, *Letter II*, 156. These facts suggest that the text of *Tractatus pluviarum* (and *a fortiori* of the *Epitome*) is closer to the common source than is that of *Apertio portarum*. A comparison with chapter 38 in al-Kindī's *Kitāb aḥkām al-nujūm*, which also shares material with *Apertio portarum* and *Tractatus pluviarum*, leads to the same conclusion.[19]

Finally, Prague, Národní Knihovna České Republiky, Univ. VI.F.7 (1144), as well as including copies of both the *Tractatus pluviarum* (partial) and the *Epitome*, provides a table on fol. 103r giving, in respective columns, the asterisms of the constellations of the lunar mansions; their serial number; the signs of the zodiac they are in; their names and degrees; and their elemental qualities. This table can also be found, without the asterisms, in the printed edition of the *Epitome* on sig. T4r, and after copies of *Sapientes Indi*, within a short text beginning 'Cornua Arietis est quedam stella...'[20] in mss. Cambridge, Clare College 15, fol. 6r–v and Parma, Biblioteca Palatina 720, s. xiv, fol. 438v. Another form of this material is found in British Library, Sloane 702, s. xv, fol. 72r–v, where a text beginning 'Sciendum quod 28 sunt mansiones Lune in zodiaco...' gives the 28 lunar mansions, with their names, asterisms and elemental qualities (given in the form of verbs: '...tenet medium (inter siccum et humidum),...desiccat,...humectat' etc.), followed by another table giving their degrees.

The nomenclature for the lunar mansions distinctive of *Tractatus pluviarum* also appears in

1) a text appended to the *Epitome* in Madrid, BN, 10063, 11rb, Vat. lat. 4082, fol. 138rb–140ra, Vat. lat. 4087, fol. 60r–62r, Venice, BNM, Lat. Z. 344 (1878), fol. 30v, and the printed edition, sig. S3r–T3v: 'De electionibus Indorum et Dorothii secundum mansiones Lunae viginti octo'.[21]

---

[18] See *Tractatus pluviarum* **14** (n. 360) below.

[19] For these parallel passages see the notes to the translation of the *Tractatus pluviarum*, sentences **1–14**.

[20] Edited in Burnett, 'Lunar Astrology,' pp. 124–5.

[21] See Burnett, 'Lunar Astrology,' p. 50.

2) in Paris, Bibliothèque nationale de France, lat. 7277, s. xiv, fol. 1v, where the lunar mansions (with their positions verified in Tournai in A.D. 1367) are listed after two columns of the Latin transliterations of the Arabic names of the lunar mansions as 'nomina earundem Latina'.[22]
3) Leopold of Austria's *Compilatio de astrorum scientia*, pr. Augsburg 1489, Tractatus I.[23]
4) Ms. Catania, Biblioteca Universitaria, 87, s. xv, fol. 317r, where they appear with a column of Arabic names and with coordinates for A.D. 1440.
5) Cornelius Agrippa, *De Occulta Philosophia Libri tres*, II, ch. 33, where they are given as translations after the Arabic names.[24]
6) the margin of the version of Pseudo-Aristotle, *De luna* printed in *Sacratissime astronomie Ptholemei liber diversarum rerum*, P. Liechtenstein, Venice, 1509, fol. 13r–v, from which they have been copied into ms. London, British Library, Add. 10775, s. xvi, fol. 329v–330v.[25]

The last two works extend the use of this nomenclature to the realm of talismanic magic.

It is clear, then, that the influence of 'John of Seville's' text (or at least of the nomenclature for the lunar mansions that it contains) was widespread. It remains to look more closely at the text itself, and to identify its sources and context.

The text, on the whole, is loosely arranged. Predictions are made for various periods of time: years, quarters and months (**1–7**), the beginning, middle or end of the year (**21**), and days (**23**). The first requires the observation of the conjunction or opposition of the Moon and Sun (i.e. the New Moon or Full Moon) that precedes the entry of the Sun

[22] L. Thorndike, 'Notes on Some Astronomical, Astrological and Mathematical Manuscripts of the Bibliothèque Nationale, Paris,' *Journal of the Warburg and Courtauld Institutes* 20 (1957), pp. 112–72, see p. 115. What is probably the same table is also found in Paris, BNF, lat. 7445, fol. 62v–63r and lat. 18504, fol. 113r–v, 170r and 193r.

[23] See P. Kunitzsch, 'Abū Maʿšar, Johannes Hispalensis und Alkameluz,' p. 120, n. 27.

[24] Cornelius Agrippa, *De Occulta Philosophia Libri tres*, II, ch. 33, ed. V. Perrone Compagni (Leiden, 1992), pp. 347–50. This nomenclature had already been reproduced from Agrippa by Moritz Steinschneider in 'Ueber die Mondstationen (Naxatra), und das Buch Arcandam,' *ZDMG* 18 (1864), pp. 118–201 (See Tabelle I).

[25] See C. Burnett, 'Arabic, Greek, and Latin Works on Astrological Magic Attributed to Aristotle,' in J. Kraye, W.F. Ryan and C.B. Schmitt (eds.), *Pseudo-Aristotle in the Middle Ages* (London, 1987), pp. 84–96, see p. 92.

into Aries—the beginning of the astronomical year. Within this section 'the opening of the doors' is described (**8**). Then follows a description of the 'lot of rain' (**25**), and the effect of the Moon entering different signs of the zodiac in the 'hours' of different planets (**26**), the rainy effect of the planets in Scorpio, Capricorn and Aquarius (**28–30**), the effect of a planet's retrograde movement (**31**), and the significance of the Moon in various positions in respect to planets (**32**), signs of the zodiac (**33**), and astrological places (**34**). The mention of the zodiacal sign of a state (**33**) prompts the author to list the regions assigned to each of the planets (**35–41**) and the states or cities belonging to each of the signs (**42–53**). An addendum to this (**54–5**) discusses the sign and degree of cities which are presumably of particular concern to the author: Jerusalem, Rome, Pisa, Lucca, Palermo and 'Africa' (i.e. the province of Ifrīqiya = present day Tunisia). The author returns to observations based on the New Moon or Full Moon immediately preceding the entry of the Sun into Aries (**56–8**), and the particular indications of Saturn, Mars, Jupiter and Mercury (**60–5**), but this time the prognostications concern matters other than rain: the abundance of the crops, drowning or thirst, wars, thieves, death and earthquakes.[26]

The author next turns to the doctrines of the Indians. The first doctrine attributed to them is that of the 'twelfth parts' of the signs of the zodiac (**66–7**) which are mentioned as having relevance to the state or city under that sign. The major part of this Indian lore, however, is that of the twenty-eight lunar mansions, which occupies the rest of the text (**68–101**). These are classified according to whether they are moist and indicate rain, dry, or temperate. A method for forecasting on which day of the month the rain will fall, using the lunar mansions, concludes the short treatise.

The text is rigorously practical in tone. The only statement of theory is that 'the Moon receives power from the planets and gives it to the earth—this power is determined by their nature and different conditions' (these are listed) (**17**). No examples are given, however, of how far these methods of forecasting work in practice.[27] For the most part the author is drawing from astrological traditions which can be recognized.

[26] This probably betrays the fact that the text has been excerpted from a treatise which deals with general astrology as a whole: in this case, the *Prima pars* of the *Epitome*.

[27] Personal experience is evoked only in respect to determining the sign and degree of Pisa (we shall return to this below). A tantalising excerpt from what would seem to be a

The observation of the conjunction immediately preceding the Sun's entry into Aries to predict the events of the coming year is found in other texts on 'the Revolutions of the Years'.[28] The concept of the 'opening of the doors' (*fatḥ al-abwāb, apertio portarum*), however, is specific to the prediction of rain. It is defined in al-Qabīṣī's *Introduction to Astrology*, 4 [22] as being the condition 'when an inferior planet applies to a superior planet and their houses are in opposition.'[29] Each planet has two signs of the zodiac assigned to it as its 'houses', while the Sun and the Moon have one house each. Mercury's houses are opposite those of Jupiter; Venus's to those of Mars; while one of Saturn's houses is opposite that of the Moon, the other of the Sun. In the pairs, Mercury and Jupiter, Venus and Mars, Moon and Saturn, Mercury, Venus and the Moon are inferior planets, whilst Jupiter, Mars and Saturn are superior planets. The Sun is neither inferior nor superior, but in the middle, and its status with Saturn, as we shall see, is different from that of the other pairs.

Al-Qabīṣī's definition is confirmed by that found in *Tractatus pluviarum*, **8**: 'The "openings of the doors" are the conjunctions of the Sun or the Moon with Saturn or their aspect to the planet; and the same with Jupiter with Mercury, and Venus with Mars.'[30] Later in *Tractatus pluviarum* (**32**) the additional factor of the Moon separating from one of the pair of planets and applying to the other is described as being the 'opening of the great door-leaves', and causing rain and winds. This latter kind of 'opening the door' is described by the eighth-century astrologer, ʿUmar ibn al-Farrukhān al-Ṭabarī, in his *Kitāb mukhtaṣar*

variant version of the text mentions 'experience' of the veracity of one prediction: Krakow, Biblioteka Jagiellonska, 1843, fol. 203v (within a text entitled 'De coniunctionibus planetarum: Nota de coniunctionibus planetarum: coniunctio enim tantum superiorum est triplex, scilicet minor, maior et media…'): Hispalensis dicit: Et ego expertus sum pluries cum ascendens coniunctionis est signum aquaticum et Luna ante oppositionem intrat illud signum in eodem die quando Luna intrat…' To the passage equivalent to *Tractatus pluviarum*, **34**, *Apertio portarum* has added 'ut semper experimur'.

[28] E.g. Abraham Ibn Ezra, *Liber coniunctionum planetarum et revolutionum annorum mundi qui dicitur de mundo vel seculo*, in *Abrahe Avenaris Iudei Astrologi peritissimi in re iudiciali opera*…, trans. Petrus de Abano (Venice, 1507), fol. lxxvi–lxxxv (see fol. lxxxii va).

[29] Al-Qabīṣī, *The Introduction to Astrology*, ed. C. Burnett, K. Yamamoto and M. Yano (London, 2004), pp. 137 and 346–7. Note that there is a qurʾānic precedent, in that rains are described as being poured through 'the doors of heaven' (*abwāb al-samāʾ*) in Q 54:11.

[30] In *Apertio portarum* this definition is preceded by a Latin phrase that could well be a translation of al-Qabīṣī's definition: 'Apertio portarum dicitur proprie cum coniungitur planeta inferior planete superiori et fuerint cum hoc eorum domus opposite'.

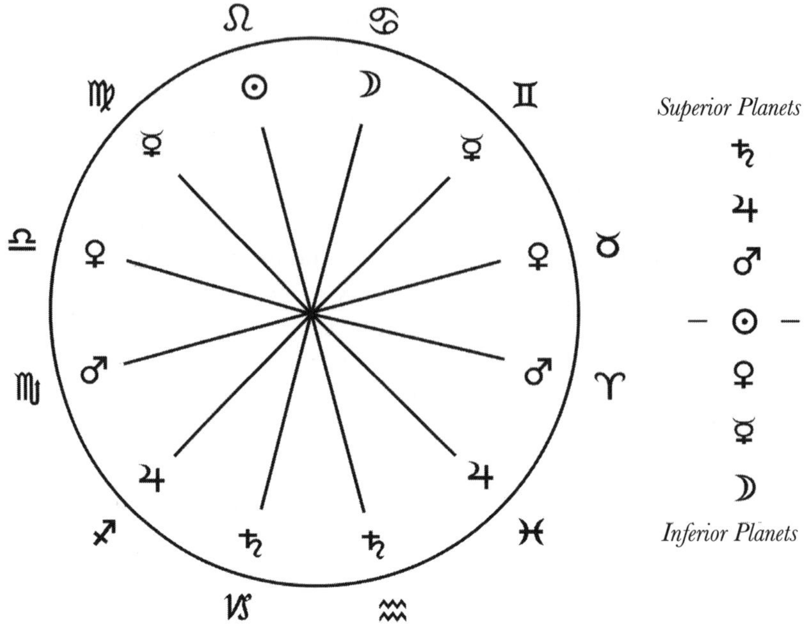

The zodiacal houses of the planets and their order of superiority

al-masāʾil, ch. 83: 'Look at the Moon, and when it separates from Venus and applies to Mars or separates from Mars and applies to Venus, there is an opening of a door. When it separates from Jupiter and applies to Mercury or separates from Mercury and applies to Jupiter, there is an opening of a door. When it applies to Saturn alone, whether there is separation or not, there is an opening of a door.'[31] Both kinds of 'opening of the doors' are described more obscurely in Abū Maʿshar's *Kitāb al-sirr* ('book of the secret'), in ms. Escorial, Biblioteca Real, 938, fol. 1–20. For the first, see fol. 2r: 'As for rain, it happens as a result of the opening of the doors, because each planet except the Sun, causes the wind and rain which comes from the direction of the higher planets, e.g. the Moon causes what comes from Saturn, because its sphere (*falak*) is opposite its (Saturn's) sphere, and similarly for the other planets. So, from these lower planets one knows what wind and rain will happen

[31] Bos and Burnett, *Scientific Weather Forecasting*, pp. 439–40 and 446–7. ʿUmar's text is incorporated into Ibn Abī 'l-Rijāl's *Kitāb al-bāriʿ* and summarised in a chapter 'on winds' (*fī 'l-riyāḥ*) attributed to al-Kindī in ms. Tehran, Majlis-i Shūrā-i Milli 6452 (see ibid., pp. 421–2, 426–7, and 432).

in the world....' For the second kind see fol. 9v: 'If you wish to know the opening of the doors of rain, look at the two planets which are following (?)[32] the sphere, and if the Moon separates from one of the two and applies to the other, there is an opening of a door, and when any one of them applies to its companion, there is an opening of the door, and the best opening of the door is when the two are looking at (*or* opposite) their houses.'

A different differentiation between greater and lesser openings of the door occurs in a short text attributed to al-Kindī, *Risāla fī aḥdāth al-jaww* ('Letter on the phenomena of the atmosphere'): 'The greater opening is the application of the Sun to Saturn—Saturn is like the male principle, the Sun is like the female principle... As for the smaller opening, it is the application of each one of the lower planets to the Lord of the opposite of its house' (the effects of Mercury's application to Jupiter, Venus's application to Mars and the Moon's application to Saturn are described).

Abraham Ibn Ezra, quoting a certain 'Abū Shāriq', one of the pair of planets is the 'key' opening the other one.[33] An additional factor to take into account is the position of the Moon in one of its twelve 'stations' (*marākiz*, 'phases') in respect to the Sun, which are confusingly also called 'doors' and 'keys'.[34] It is these latter 'doors' which are mentioned in sentences **7** and **24** of the *Tractatus pluviarum*.[35]

Also belonging to the astrological tradition is the 'lot of rain' (*sahm al-maṭar, pars pluviarum*). A lot is a point on the ecliptic circle which is determined by counting the number of degrees from one place on the circle to another, counting the same number of degrees from a third place (usually the ascendant) and making predictions from the degree

[32] This word is a correction of 'the doors'; the astrological implications of the word are not clear.

[33] Abraham Ibn Ezra, *Sefer ha-Olam*, ed. J.L. Fleischer, *Ozar ha-Hayyim* 13 (1937), p. 46 (Bos and Burnett, *Scientific Weather Forecasting*, p. 49).

[34] 'Doors': 'Umar ibn al-Farrukhān's chapters 83 and 84 (Bos and Burnett, *Scientific Weather Forecasting*, pp. 447–8), al-Kindī's *Letter I*, chapter 4 (ibid., pp. 343–4), and in al-Kindī's *Forty Chapters*, ch. 38 [675–8] (ibid., pp. 398–9 and 401–2); 'keys': Abraham Ibn Ezra, *Sefer ha-Olam* (ibid., p. 49).

[35] Another kind of 'opening of the doors' appears in the last sentence of *Apertio portarum*, which does not have any equivalent in *Tractatus pluviarum*: 'Additionally, the opening of the doors is spoken of when the Lord of the ascendant aspects or conjoins the Lord of the seventh house' (ibid., p. 389). This is paralleled in Abraham Ibn Ezra's *Sefer ha-Olam*, Paris, Bibliothèque nationale de France, héb. 1058, fol. 91a: 'if the planet is in aspect to the lord of the seventh house, then it is named "opening of the door"' (information from Shlomo Sela).

where the counting ends. A 'lot of rain according to Abū Maʿshar' is mentioned in several Arabic and Latin manuscripts of al-Qabīṣī's *Introduction to Astrology*, and in al-Bīrūnī's *Kitāb al-tafhīm*: 'measure by day from the Moon to Venus, by night the opposite, and cast out from the ascendant.'[36] A 'lot of rain' does not feature among the 97 lots described in Abū Maʿshar's *Great Introduction to Astrology*,[37] but there are two 'lots of rain' in the same author's *Kitāb al-sirr.* Neither of these, however, corresponds to the lot mentioned in al-Qabīṣī's *Introduction to Astrology*, and by al-Bīrūnī. The first one instructs one to 'measure by day from the degree of the conjunction (of the Sun and the Moon) to the house of the Moon, and by night the opposite, and cast the degrees from the ascendant of the conjunction, and where they arrive is the lot.'[38] The second lot (corresponding to the lot described in *Tractatus pluviarum* **25**) is: 'Another lot is known. You can tell from it what happens day by day. When the Sun rises each day, you measure <the degrees> from the Sun to Saturn and you cast them from what is collected (?) from the degrees of the Moon, and where they arrive, is the lot.'[39] Both lots are described in what is virtually a translation of this passage of the *Kitāb al-sirr* in Abraham Ibn Ezra's *Reshit Hokhmah.*[40] The second lot alone appears in the Latin text which combines al-Kindī's two *Letters on Weather Forecasting* (*De mutatione temporum*).[41]

The *Tractatus pluviarum* abbreviates the predictions of this lot, as given by Abū Maʿshar (which correspond to those in Ibn Ezra and al-Kindī): 'If it falls in a house of Saturn, there is intense coldness on that day, if it falls in a house of Jupiter, there will be strong wind, if it falls in a

[36] See al-Qabīṣī, *Introduction to Astrology*, p. 155 and 361 and al-Bīrūnī, *The Book of Instruction in the Elements of the Art of Astrology*, facsimile and trans. by R. Ramsay Wright (London, 1934), section 479, no. 13 (p. 293).

[37] Abū Maʿshar, *Liber Introducotrii maioris ad scientiam judiciorum astrorum*, 9 vols. (Naples, 1995–96); see iii, pp. 613–60, v, pp. 319–86 and viii, pp. 149–69.

[38] Ms. Escorial 938, fol. 17r.

[39] Ibid.

[40] Abraham Ibn Ezra, *The Beginning of Wisdom*, ed. F. Cantera and R. Levy (Baltimore, 1939), pp. lxxii–lxxiii, 121 and 231 and *Principium sapientie* in *Abrahe Avenaris Iudei Astrologi peritissimi in re iudiciali opera…*, trans. Petrus de Abano, fol. xxx ra–b. Ibn Ezra states that the first lot is to be cast out 'from the ascendant in the morning or evening'; he does not mention Abū Maʿshar, but attributes the second lot to Enoch (= Hermes). The second lot also appears in Ibn Ezra's *Sefer ha-Olam*, Paris, Bibliothèque nationale de France, héb. 1058, fol. 91b. Both lots occur, without attribution, also in Sahl ibn Bishr's *Fatidica*, as translated by Hermann of Carinthia: S.M. Low-Beer, *Hermann of Carinthia, the "Liber imbrium", the "Fatridica" and the "De indagatione cordis"*, PhD thesis (City University of New York, 1979), p. 225.

[41] Bos and Burnett, *Scientific Weather Forecasting*, p. 298, L2.

house of Mars, there is heat and hot winds (simooms) on that day, if it falls in a house of Venus, rain will come on that day, if it falls in the house of the Sun, there is the same, if it falls in a house of Mercury, there is a strong wind on that day, and if it falls in the house of the Moon, rain and coldness come on that day.'[42]

The division of each sign of the zodiac into twelve parts (**66–7**), of which the first has the nature of the sign, and the subsequent parts have the nature of each of the following signs respectively, is attributed in the *Tractatus pluviarum* to the Indians. Although the Indians used twelfth parts,[43] this doctrine belongs to the ancient tradition of Babylonian and Greek astronomy, and is not attributed to the Indians either by Abū Maʿshar, in his *Great Introduction* V, 18 (a doctrine of 'the Ancients') nor by Ibn Ezra where the twelfth-parts are attributed (1) to each of the planets respectively; (2) to each of the signs respectively, and the second of these doctrines is attributed to Enoch (Hermes).[44]

The assignment of regions and cities to planets and signs of the zodiac (35–53) is close to that found (for the regions) in Abū Maʿshar, *Kitāb al-milal wa'l-duwal*, II ch. 4,[45] and for the states or cities the same author's *Great Introduction*, VI, ch. 9, together with the list in an addition to *Kitāb al-milal wa'l-duwal* found in Escorial, Biblioteca real, 937, fol. 33a–b.[46]

The list of twenty-eight lunar mansions and their classification into humid, dry and moderate mansions is in broad agreement with the information in al-Kindī's *Letter II*, Jafar Indus's *Liber imbrium*, *Sapientes Indi*, the Calendar of Cordova, and al-Farghānī's *Thirty Chapters on the*

---

[42] Ms. Escorial 938, fol. 17r.

[43] *The Yavanajātaka of Sphujidhvaja*, ed. D. Pingree, 2 vols. (Cambridge, 1968), ii, p. 210.

[44] Abraham ibn Ezra, *The Beginning of Wisdom*, ch. 2, pp. xi, 40 and 231 and *Principium sapientie* in *Abrahe Avenaris Iudei Astrologi peritissimi in re iudiciali opera*..., trans. Petrus de Abano, fol. i vb. Ibn Ezra refers back to this passage in *Sefer ha-Olam*, Paris, Bibliothèque nationale de France, héb. 1058, fol. 89b.

[45] Abū Maʿshar's classification is followed by ʿAlī ibn Abī 'l-Rijāl's *Kitāb al-bāriʿ*, VIII, 36.

[46] Abū Maʿshar, *On Historical Astrology*, ed. Yamamoto and Burnett, i, pp. 515–17. These lists have parallels in ʿAlī ibn Abī 'l-Rijāl's *Kitāb al-bāriʿ*, VIII, 35, al-Qabīṣī's *Introduction to Astrology*, 1 [25]–[36], and Abraham ibn Ezra, *Beginning of Wisdom*, ch. 2 (ed. Cantera and Levy, pp. viii–xxxix, 36–77, 156–89 and *Abrahe Avenaris*... *opera*, fol. iiii recto–xv verso).

Science of the Stars,[47] except that the eighteenth mansion is dry rather than moist, and the twenty-first mansion is temperate rather than moist.

To his sources the author of *Tractatus pluviarum* has added his own distinctive stamp. First of all, in respect to the lunar mansions, instead of simply transliterating them, as did the compilers of the *Alchandreana* in the late tenth century,[48] the translators of the other extant texts on weather forecasting, and John of Seville and Gerard of Cremona in their versions of al-Farghānī's *Thirty Chapters*, he tried to find Latin translations for them all.[49] In some cases, he was evidently translating Arabic names which are not the ones most commonly used for the lunar mansions; especially in respect to a group of names which convey a quaint pastoral imagery for the asterisms: a shepherd leading his calves down to a place where they can drink, while a bucket full of water, suspended from a beam, is being drawn up from a well.[50]

In assigning the planets to regions and the signs to states, he has inserted familiar places into the Arabic lists: England (**35, 53**), 'Barbary' (**45**), Italy (**46**), Athens (**47**), Sardinia (**50**) and Macedonia (**51**). Most significant is the list of ascendants of cities that follows (**55**), which does not follow any Arabic text,[51] and includes the only instance of a correction of received knowledge in the light of the author's own experience.

Who then is this author? The manuscripts of *Tractatus pluviarum* and of the *Epitome* are unanimous in ascribing the texts to John of Seville.[52]

---

[47] The terminology for the 28 mansions and the complexions of the mansions in all these sources, except al-Farghānī, are set out in parallel columns in Bos and Burnett, *Scientific Weather Forecasting*, pp. 576–81. For al-Farghānī see *Mohammedis filii Ketiri Ferganensis*...*Elementa Astronomica, Arabice et Latine*, ed. J. Golius (Amsterdam, 1669), pp. 77–9 and *Alfragani astronomorum peritissimi compendium id omne quod ad Astronomica rudimenta spectat complectens, Ioanne Hispalense interprete* (Paris, 1546), pp. 76–9.

[48] See D. Juste, *Les Alchandreana primitifs. Étude sur les plus anciens traités astrologiques latins d'origine arabe (Xe siècle)* (Leiden, 2007).

[49] Paul Kunitzsch already had already noted this characteristic, which is shared by the author of a work on the construction of an astrolabe beginning 'Dixit Iohannes: Cum volueris facere astrolabium accipe auricalcum optimum', which includes a list of the fixed stars added to the *rete* of an astrolabe: see his 'Abū Maʿšar, Johannes Hispalensis und Alkameluz,' pp. 119–20.

[50] See the comments to the translation of the terms for mansions 20, 22, 26, 27, and 28 in the translation of *Tractatus pluviarum* below (p. 264).

[51] Note, however, that ascendants of a completely different set of cities, without degrees, immediately follows the sign-state assignation in the Escorial addendum to Abū Maʿshar, *Kitāb al-milal wa'l-duwal*, ed. Yamamoto and Burnett, i, p. 519.

[52] The attribution of the *Ysagoge* and the first three *Partes* of the *Epitome* to Johannes Hispalensis is also found in the The *Speculum astronomiae*, a very popular critical survey

However, there are several problems with this ascription. The terminology differs considerably from that of John of Seville's translations: e.g. 'honor' is used instead of 'exaltatio' for 'exaltation', 'lapsum/domus lapsa' instead of 'cadens' for 'cadent place', 'dominus vigoris' instead of 'almubtaz' for dominant planet, and so on. In the *Epitome* Arabic authorities are rarely referred to, and Arabic transliterations are avoided, as we have already observed in regard to the lunar mansions. The translator, John of Seville, however, shows no hesitation in retaining transliterations of Arabic terms in his Latin versions. One can directly compare John of Seville's translation of the lunar mansions in al-Farghānī's *Thirty Chapters*, chapter 20, with the equivalent passages in the *Epitome*. For example, in its description of the eighth mansion (*al-nathra* = the tip of the nose) the *Thirty Chapters* gives 'Octava vocatur Anathera, et nominatur Os Leonis, et est quale pitacium nubis modicum inter duas stellas parvas' for the *Epitome*'s 'Octavam dicunt Nebulosam, suntque sydera duo cum nube media'; for the ninth mansion (*al-ṭarf*, 'the eye') the *Thirty Chapters* gives 'Atarphe, quod interpretatur summitas (apparently reading *al-ṭaraf* = 'outermost point'),' for the *Epitome*'s 'Oculum'; and for the the twenty-first mansion (*al-balda*, 'the place') the *Thirty Chapters* gives 'Albeldah, et est quoddam pitacium celi modicum in quo nulla est stella' for the *Epitome*'s 'Desertum, et est locus sine syderibus'.

Also striking is the focus on Italian cities in *Tractatus pluviarum*, sentences **54–5**, when John of Seville's translation activity is associated with 'Limia' (most probably the valley of the Lima/Limia in northern Portugal) and Toledo. The mention of a personal experience of the ascendant of Pisa, points rather to another scholar: namely, Abraham ibn Ezra. Shlomo Sela has remarked that in the second Hebrew version of the *Sefer ha-Olam*, extant in ms. Vat. Ebr. 477, fol. 89v, Ibn Ezra gives a list of 22 cities, each accompanied by its zodiacal sign and degree; but for Lucca and Pisa he has additionally made personal observations:

---

of the astrological literature available in the mid-thirteenth century: P. Zambelli, *The* Speculum Astronomiae *and its Enigma* (Dordrecht, 1992), pp. 226, 230, 234 and 236. These are the only texts mentioned in the *Speculum* that are not translations from Arabic.

> Pisa: some say that its sign is Piscis, but, according to my own observations, its sign is Aquarius 3°; Lucca, according to my own observations on two occasions, its sign is Cancer in the term of Jupiter.[53]

This passage is remarkably similar to *Tractatus pluviarum* **55**:

> They also said that Pisa's sign is Pisces; but by experience it is Aquarius 2°. The sign of Lucca is Cancer.

The difference between 2° and 3° can be explained by a common confusion of two similar Hindu-Arabic numerals, while in the case of Lucca's ascendant, the *Epitome* abbreviates the information. Abraham ibn Ezra (*c.* 1092–1165) was in Lucca as well as in several other cities in Italy in the early 1140s (when the *Epitome* was written), before moving to Béziers, Narbonne, Rouen, and finally London.[54] He wrote at least two collections of astrological works, which correspond closely in arrangement and nature of subject matter to the *Epitome* with its 'Introduction' and 'Four Parts', and initial comparison between his texts and that of the *Epitome* suggest that the author of the *Epitome* belonged to the circle of Ibn Ezra.[55] Abraham, like the author of the *Epitome*, eschews transliterations of Arabic words, and tries to forge an astrological terminology in Hebrew. In this case the *Epitome* could be put alongside several other Latin works on arithmetic and geometry, astrology, astronomy and the astrolabe, that are attributed to Ibn Ezra or written 'on his dictation'.[56] The reasons why such a text should be ascribed to John of Seville remain to be explained.

---

[53] S. Sela, 'Abraham ibn Ezra's Scientific Corpus: Basic Constituents and General Characterization,' *ArScPhil* 11 (2001), pp. 91–149, at p. 102, n. 34 (translation slightly altered).

[54] See S. Sela and G. Freudenthal, 'Abraham Ibn Ezra's Scholarly Writings: a Chronological Listing,' *Aleph* 6 (2006), pp. 13–55.

[55] See R. Smithuis, *Abraham ibn Ezra the Astrologer and the Transmission of Arabic Science to the Christian West*, PhD thesis (University of Manchester, 2004), ch. 3: 'The authorship of the *Ysagoge* and *Liber quadripartitus*,' pp. 169–99 and tables on pp. 358–83.

[56] These works are discussed in R. Smithuis, 'Science in Normandy and England under the Angevins: the Creation of Avraham ibn Ezra's Latin Works on Astronomy and Astrology,' in G. Busi (ed.), *Hebrew to Latin: Latin to Hebrew* (Berlin, 2006), pp. 23–59.

APPENDIX

EDITION AND TRANSLATION OF *TRACTATUS PLUVIARUM ET AERIS MUTATIONIS SECUNDUM MAGISTRUM IOHANNEM YSPALENSEM*

The following edition is based on ms. R, which has been collated with ms. S of the *Tractatus pluviarum*, and mss. JV of the *Epitome*, while the departures from this text in mss. HP of *Apertio portarum* are noted in the apparatus. The section on the lunar mansions additionally uses the tables in mss. HQ. Small changes in word order, and the presence or absence of 'in' and 'et' are not normally noted. Editorial deletions are indicated by square brackets.

H	=	Cambridge, Clare College 15
J	=	London, British Library, Royal 12.C.XI
P	=	Paris, Bibliothèque nationale de France, lat. 18504
Q	=	Parma, Biblioteca Palatina 720
R	=	Paris, Bibliothèque nationale de France, 7316A
S	=	Vienna, Österreichische Nationalbibliothek, cvp. 2436
V	=	Vatican, Biblioteca apostolica Vaticana, Reg. lat. 1452
pr_M	=	*Epitome totius astrologiae*, Nürnberg, 1548, sig. T4r

Tractatus Pluviarum et Aeris Mutationis Secundum Magistrum Iohannem Yspalensem[57]

1 Volens aeris scire naturam in annis singulis, considera[58] coniunctionem[59] Solis et Lune vel[60] oppositionem que est antequam Sol in Arietem intret,[61] in[62] quo signo sit Saturnus tunc. **2** Et si fuerit[63] in

[57] Capitulum ad sciendum naturam aeris V
[58] R omits
[59] in coniunctione HP
[60] HP add 'considera'
[61] RJV add 'considera'
[62] R omits
[63] sit JV

igneo signo, annus erit calidus; si in frigido, talis; si in humido vel sicco, huiusmodi.[64] **3** Et misceas cum natura signi naturam termini in quo est Saturnus, et[65] natura cuiusque termini est ut sui domini,[66] misceasque[67] cum Saturno naturas planetarum qui sibi associantur vel eum[68] aspiciunt. **4** Sic iudicabitur[69] de Iove cum associatur[70] Luna Soli[71] vel sunt in oppositione[72] antequam Sol intret[73] aliquod[74] signorum mobilium que caput[75] quarte circuli sunt.[76] **5** Sic est iudicium Veneris et Mercurii in unoquoque[77] mense in tempore coniunctionis vel[78] oppositionis Solis et Lune. **6** Sic etiam[79] iudicium mensis; si in coniunctione sit ascendens firmum, mensem signabit integrum.[80] **7** Sic est iudicium Lune in quacumque[81] predictarum portarum.

8 Et portarum apertiones sunt coniunctiones[82] Solis vel Lune cum Saturno vel eorum aspectus ad[83] Saturnum, Iovis[84] eodem modo cum Mercurio, Veneris[85] cum Marte. **9** Et[86] Venus significat humiditatem, Mercurius ventos, Saturnus nubila et frigus, Mars ventos a dextro et calorem,[87] Iupiter temperiem aeris ventosque a sinistro, quod[88] in

[64] talis JSV
[65] quia S
[66] Et si fuerit in igneo signo...sui domini] Et in suo termino minuitur de caliditate. Et si fuerit in termino Martis, augebitur caliditas anni et sic de aliis predictorum ducum (?) secundum naturam eorum HP
[67] Misceas quoque HP
[68] associantur vel eum] sociantur vel qui eum HP
[69] iudica HP, iudicabis V
[70] sociatur HP
[71] Sole R
[72] vel sunt in oppositione] vel opponitur S, et sit in oppositione HP
[73] intrat V
[74] R adds 'aliorum'
[75] capite R, que quidem signa mobilia caput HP
[76] HP add 'Misceas quoque naturam cum natura termini in quo est Saturnus, et natura cuiuslibet termini ut natura sui domini' = 3 above.
[77] quoque R
[78] et S
[79] sic etiam] in J, etiam R
[80] mense significabit integro JV
[81] unaquaque HP
[82] R omits, coniunctio JV
[83] HP add 'ipsum'
[84] Saturni R
[85] Venus JS
[86] Sed HP
[87] a dextro et calorem] sed a dextro calorem HP
[88] qui R, quia S

Cancro maxime[89] significat, sicut Mars in Capricorno; et Sol calidus et siccus.[90]

10 Si apertio portarum fuerit cum Mercurio, ventus cum[91] pluvia erit.[92] **11** Si Mars, considera iuncture[93] Solis et Lune vel oppositionis dominum;[94] erunt lampades et fulmina et tonitrua; auge huic vel deme[95] nature Martis naturam[96] signi et[97] termini in quo ipse est. **12** Quod si Saturnus aspicit in loco Martis et est in domo terre,[98] significat domorum destructionem; si in domo aque, monstrat frigus, nubila[99] et tenebras. **13** Quod si planeta retrogradus ibi fuerit, significat pluviam magnam. **14** Luna in pluvia posse habet.[100] **15** Et Sol nunc pluviam nunc serenitatem portendit;[101] nam cum est in loco inferiori circuli puncti,[102] pluviam, cum[103] in superiori, serenitatem significat.[104] **16** Sic est iudicium omnium planetarum.

17 Et Luna accipit virtutem a planetis datque[105] terre, prout eorum natura existit[106] et ut planete permutantur[107] a dextro ad sinistrum vel[108] econverso vel a velocitate ad tarditatem vel econverso[109] vel a directione

---

89 magis HJP
90 HP omit 'et Sol calidus et siccus'
91 vel V
92 Si apertio…erit] Mercurius itaque si aspexerit dominos apertionis portarum, vel fuerit cum eis, et coniunctus in coniunctione Solis et Lune vel in oppositione, erit ventus et pluvia HP
93 iunctures V
94 domum J. Si Mars…dominum] Quod si portarum apertio fuerit cum Marte ut dictum est de Mercurio HP
95 minue P
96 nature Martis naturam] secundum Martis naturam et HP
97 vel JV
98 JV add 'Saturnus'
99 monstrat frigus nubila] monstrat frigus nivem J, significat frigora imbres HP
100 Luna vero significat pluviam HP
101 significat JSV. HP omit 'et Sol nunc pluviam nunc serenitatem portendit'
102 predicti JV, S omits. Sol autem cum in infimo circulo simul et cum inferioribus planetis (?) HP
103 cum] significat JSV
104 JSV omit.
105 dans JSV
106 R adds 'accipit'
107 permittantur S
108 et V
109 V omits 'vel a velocitate…econverso'

ad retrogradationem vel econverso, vel ab oriente ad occidentem[110] vel econverso vel ingredi unum angulorum vel egredi, sic aeris permutatio erit. **18** Nam si immobilis[111] est, movetur et econverso motu ventorum. **19** Sic facit coniunctio planetarum vel aspectus. **20** Sic significat aspectus astrorum firmorum.[112]

21 Volens scire[113] an[114] in exordio[115] anni vel medio vel[116] fine[117] pluvia multa sit futura,[118] considera[119] locum Veneris cum intrat Sol in[120] Libram; que si est sub luce Solis sitque occidentalis, pluvia erit multa in exordio,[121] in fine autem pauca; si orientalis[122] et visibilis, erit pluvia econverso. **22** Signa multe pluvie sunt [signa aque et][123] Aquarius, finis Capricorni, Aries, Taurus et Leo; his fortiora Piscis,[124] Aquarius, Scorpio.[125]

23 Volens[126] scire diem pluvie, considera tempus[127] coniunctionis Solis et Lune vel[128] oppositionis, et considera[129] quot gradus sunt[130] inter Lunam et proximum planetam qui pluviam significat;[131] pro gradu quoque[132] diem accipe; [133] mobilium signorum significatio velox, duum[134]

---

[110] a directione...occidentem] ab ascensu inferioris partis circuli puncti (predicti V) ad superius JSV

[111] immobili R

[112] fixorum JSV. S adds 'Capitulum secundum'. V adds 'Ad sciendum quando pluvia sit ventura'

[113] Scire autem volens J, Scire volens V, Item scire volens HP

[114] si HP, J omits

[115] principio HP

[116] in exordio...vel] in medio vel in V

[117] J adds 'si sit'

[118] multa sit futura] si sit pluvia ventura multa J, sit ventura multa S

[119] vide HP

[120] JHPV omit

[121] exordio] JV add 'anni', principio anni illius qui incipit a Libra et HP

[122] occidentalis autem V

[123] signa aque et] aquatica HP, significativa J, signativa aque V

[124] Pisces J

[125] Scorpius V

[126] HP add 'itaque'

[127] diem HP

[128] et V

[129] vel oppositionis et considera] et vide HP

[130] sint HP

[131] qui significat pluviam] pluvialem HP

[132] gradu quoque] quolibet gradu JV, unoquoque gradu HP

[133] diem accipe] accipe secundum signorum velocitatem HP

[134] duum] et duum JS, et duorum V

corporum mediocris, firmorum tarda. **24** Et considera[135] portas Lune, et considera[136] quis planeta aspicit[137] Lunam vel ei est iunctus; tunc et ex hoc poteris scire[138] diem initii[139] pluvie et finem.[140]

25 Et quot sint[141] gradus equales inter Solem et Saturnum quoque mane[142] considera[143] et totidem gradus computa a loco[144] Lune; quo pertingunt,[145] ibi pars pluvie erit. **26** Que computatio si erit[146] in domo Lune vel Veneris, significat pluviam, si in domo Mercurii vel Iovis, significat[147] ventos, si Martis vel Solis, serenitatem, si Saturni, significat nubila.

27 Cum[148] intrat Luna Cancrum in hora Solis, vel[149] Virginem in hora Veneris, vel Sagittarium hora Lune, vel Geminos hora Mercurii, vel Taurum hora Martis, vel Libram hora Iovis, significat pluviam vel ventos ad modum nature signi in[150] quo est dominus[151] hore.

28 Cum est in Scorpione Venus vel in Capricorno vel Aquario, coniungiturque Lune vel eam[152] aspicit, significat pluviam multam. **29** Sic[153] Mercurius in statione[154] cum aliquo trium predictorum signorum. **30** Sic fit tribus planetis in Aquario iunctis.[155]

---

[135] vide HP
[136] JVHP omit
[137] aspiciet HP
[138] ei est iunctus... scire] quis iunctus est quod (?) tunc ex hoc scies HP
[139] diem initii] initium JV
[140] finis HP
[141] sunt JHP
[142] quoque mane] unoquoque die JSV, HP omit
[143] considera] cum scire R, vide HP
[144] gradu HP
[145] quo pertingunt] quo pertingit et R, quo pertingerit V, et quo pervenerit HP
[146] computatio si erit] quidem si fuerit HP
[147] HP omit 'significat', V omits 'pluviam... significat'
[148] Item cum HP
[149] R adds 'in'
[150] V omits
[151] JSV add 'signi vel'
[152] V omits
[153] Similiter JS, Similiter et V
[154] V adds 'vel'
[155] V omits

31 Planeta cum est retrogradus a prima statione ad[156] oppositionem Solis plus notat pluviam quam cum est ab oppositione Solis[157] ad secundam stationem.

32 Et cum Luna seiungitur[158] a coniunctione vel aspectu alicuius planete coniungiturque alii[159] planete[160] coniunctione[161] vel aspectu et sunt domus illorum planetarum sibi contrarie,[162] hoc est[163] apertio magnarum valvarum,[164] in qua necesse est evenire[165] ventos vel pluviam ut est natura[166] illorum planetarum.

33 Quod si Luna sit in signo[167] civitatis vel in aliquo suorum angulorum, quod significat erit forte[168] in illa urbe; si[169] non est in angulis sed in signo considerante signum urbis, significatio minor; si in signo non considerante, significatio[170] cassa.

34 Et[171] cum Luna in domo[172] femina fuerit, ut[173] orientalis et[174] occidentalis quarta[175] et signo feminino et considerat eam[176] planeta retrogradus in signo feminino, necessario pluvia erit.[177]

---

[156] usque ad V
[157] JSV omit
[158] separatur HP
[159] coniungiturque alii] et coniungitur alteri HP
[160] V adds 'cum'
[161] corpore HP
[162] HP add 'utputa si separetur a Venere et iungatur Marti'
[163] hoc est] hec tunc dicitur HP
[164] J adds 'sive portarum'
[165] venire HP
[166] ut est natura] secundum naturam HP
[167] JSV add 'alicuius'
[168] erit forte] forte dicitur JSV
[169] JSV add 'autem'
[170] V omits 'minor...significatio'
[171] HP add 'iterum'
[172] quarta HP
[173] sit JS, vel V
[174] vel JSV
[175] HP omits 'ut orientalis...quarta'
[176] et considerat eam] considera cum V, et aspiciat eam HP
[177] necessario pluvia erit] denecessario habet pluere ut semper experimur HP; V adds 'De partibus quas habent planete in terra'

35 Antiqui[178] astrologi dixerunt quod partes quas Luna habet in terra sunt occidentales et Yspania,[179] Anglia,[180] Armenia, Francia; habet posse in parte sexta populorum.[181]

36 Sol autem in parte[182] secunda eius orientalis terra,[183] terra Azin[184] et Zarozin.[185]

37 Saturnus in parte prima populorum[186] et India et Ethiopia.

38 Iuppiter[187] in quarta parte, Persia, Bebil et Baldech,[188] Alirach.[189]

39 Mars in parte septima, in Trachis[190] ultra mare.

40 Venus in parte quinta, Arabia.

41 Mercurius in parte tertia, in Grecia, Dalibē.[191]

42 Civitates signorum: urbs Arietis: Bebil, Persia, Palestina, Esmosal, Eschasar.[192]

43 Tauri: Asia Minor, Cabros, Speen.

44 Geminorum: Iurgen, Tabrasten, Armenia[193] maior, Adargen, Babilonia, Barcha.

---

[178] V omits. *Only a small selection of the variants in the spelling of the place names in* **35–53** *has been given in the following notes.*
[179] Yspania] in gena id est in V. JS add 'et'
[180] S adds 'et', V adds 'et Hyspania'
[181] sexta plarum J, plarum 6 V
[182] Atque J, V omits
[183] orientalis terra] orientalis J (*cf.* orientalis plagae *in the printed edition*)
[184] Azym J, Arin V
[185] Zozam J, Abin, al. Ozz et zozo V
[186] plarum JV (*cf.* plagarum *in the printed edition*)
[187] JV add 'autem'
[188] Bedil et Bedel (Beldel J) et JV
[189] et Alirac J
[190] intrans J, in Turchis V
[191] in Grecia Dalibē] in Grecia Cloalem J, Grecia Dialem V
[192] Elmosal Elcasar J, Elmosal Elasal V
[193] Hermenia JR

45 Cancri: Barbaria, Affrica, Ermedia, Roma, Ladechia,[194] Armenia[195] minor.

46 Leonis: Trachorum[196] terra, Italia, Damascus, Alcufa, Thos,[197] Ieguth Meguth, Albazara.[198]

47 Virginis: Masimar,[199] Nibeth, Eclites, Athene.

48 Libre autem Nara,[200] Carmena, terra maxima Ethiopum.

49 Scorpionis: Igesi, Tarasus, Mecche.

50 Sagittarii: Alieman, Phylistim, Sicilia, Sardinia.

51 Capricornii: Zinda,[201] Amen, Macedonia.

52 Aquarii: Surri,[202] Fergana.[203]

53 Piscis: Allexandria, Anglia.

54 Astrologi dixerunt[204] quod Iherusalem est Cancri; dicunt[205] alii quod Leo .19.[206] gradibus; secundum veritatem autem[207] Aquarius est eius signum, Leo signum Rome. **55** Dixerunt etiam[208] Pise[209] signum esse Piscem,[210] experimento autem .2. gradus Aquarii, Lucce vero

---

194 Lathechia V
195 Hermenia JR
196 Turchorum V
197 Alcufa Thos] Alcufatos J, Alcufathos V
198 Ieuit Megut Helbaizara J
199 Marsuuiar J, Nicimar V
200 autem Nara] vel Nata J, diana V
201 Zuida JV
202 Sugii J, Frigii V
203 Fergadia J
204 Astrologi dixerunt] Astronomi dicunt V
205 Cancri dicunt] Cancri J, Cancri signum V
206 .9. R
207 vero JV
208 R omits 'etiam'
209 Pisce V
210 Pisces J, Piscis V

Cancrum, signum Palermi primum gradum[211] Leonis, signum Affrice .4. gradus Leonis.

56 Semper[212] considera cum Luna seiungitur[213] a coniunctione Solis vel aspectu oppositionis[214] secundum quod quis earum[215] propior[216] fuerit Soli in ingressu in capite[217] Arietis, cum quo planeta coniungatur[218] vel prospiciat[219] aspectu pleno. **57** Et si[220] est de planetis fortune,[221] erit annus bonus et salus corporis,[222] et pluet in suo tempore et messis erit[223] multa, fructus et pecudes cum similibus.[224] **58** Et si planeta ille dat vim planete infortunato vel[225] dat vim stanti[226] in angulo et ille planeta dat vim stanti[227] in domo lapsa, finis anni malus. **59** Et quisque planeta[228] monstrat ut sua est[229] natura. **60** Et natura Saturni est sitis,[230] suffocation[is] in aqua; cum est Saturnus in aliquo signorum que dicuntur pars Solis, significat submersionem in aqua; si[231] in his que pars Lune dicuntur, sitim.[232]

---

211 primus gradus JSV
212 Semperque JS
213 seiunguntur V
214 aspectu oppositionis] oppositione JSV
215 earum *scripsi*, eorum JRSV
216 proprior V
217 in capite] capitis JSV
218 iungatur SV
219 prospiciat] eam (eum V) aspiciat JSV
220 sic V
221 fortunis JSV
222 salus corporum J
223 erunt V
224 cum similibus] consimilibus SV
225 JSV add 'si Luna'
226 stando R
227 et ille planeta dat vim stanti] planete stanti R. J omits 'in angulo... stanti'
228 JSV omit
229 sua est] est eius JSV
230 sitis] fons JSV
231 V adds 'sit Mars'
232 sit JS, V omits

61 Mars[233] si est in signo calido, facit calidum[234] et siccum;[235] si est[236] in signo humane similitudinis,[237] significat bella, latronum multitudinem,[238] eritque[239] bellum validum ingresso Marte sub luce Solis vel in[240] domum sui dedecoris. **62** Sic significat Iuppiter ingressus in lucem Solis. **63** Considera an Mars sit orientalis vel occidentalis vel a dextro vel a sinistro; in quorum parte erit Mars, superabunt. **64** Et si fuerit[241] Mercurius cum Marte in exordio anni vel aliqua alia de quartis, significabit[242] mortalitatem in terris que sunt signi[243] in quo iunguntur. **65** Si Mercurius cum Saturno erit fueritque[244] in signo terree nature, significat terremotum.[245]

66 Indi gnari[246] considerant partem que dicitur duodecima, hoc est ut[247] dones ab exordio[248] signi 2 gradus et semis ipsi signo; [contingunt[249] cuique gradui[250] duorum et semis[251] .12. gradus equales signi[252]] sic attribuas cuique sequenti[253] signo, scilicet .2. gradus et semis; sic usque in finem signorum facies.[254] **67** Ex hoc potest sciri signum ad quod pertingit[255] .12a. pars[256] planete fortunati vel infortunati, et notatur sic bonum vel malum quod habebunt civitates illius signi.[257]

---

[233] V omits
[234] facit calidum], significat calorem JV, facit calorem S
[235] siccitatem JSV
[236] si est] et cum JS
[237] signo humane similitudinis] signis humane figure V
[238] JSV omit
[239] et erit JSV
[240] R omits
[241] JSV omit
[242] significat JSV
[243] que sunt signi] quas significat signum JSV
[244] sicutque J, suntque V
[245] motum V
[246] Indi gnari] Ignari V
[247] V omits 'est ut'
[248] ab exordio] a principio V
[249] continguntque J, tinguntque V
[250] graduum J, domorum V
[251] V omits 'duorum et semis'
[252] V omits
[253] equali V
[254] V omits 'Sic...facies'
[255] pertingunt RSV
[256] .12. partes V
[257] V adds title 'De .28. mansionibus'.

68 Indi distribuerunt circulum in .28. partes quas mansiones Lune appellaverunt;[258] queque .12. gradus et sex septimas comprehendit.[259]

69 <1>[260] Et initium illarum est a duobus syderibus magnis que Arietis Cornua appellantur;[261] harum[262] similitudo hec est:[263]  t<emperata>[264]

70 <2> Pars secunda Venter dicitur que est tria sydera, hoc modo:[265]  s<icca>

71 <3> Tertia Caput Tauri dicitur,[266] que est sex sydera perexigua,[267] et sunt qui nominant Caudam Arietis; Arabes autem[268] pluvialem.[269] 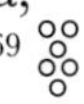 t<emperata>

72 <4> Quarta Sydus magnum et rubeum et ante illud[270] sydera parva. Maius[271] vocatur Oculus sinister[272] Tauri.[273]  h<umida>

[258] appellamur J, vocaverunt V, vocant S

[259] queque .12. gradus et sex septimas (septime R) comprehendit] quarum queque .13. gradus et tertiam partem continet J, quarum quelibet .13. et partem terciam continet V

[260] *R numbers the mansions in the margin.*

[261] dicuntur V

[262] huius J, cuius V

[263] V omits 'est'. Et initium... hec est] Prima mansio vocatur cornu Arietis et compositus a tribus stellis sic S

[264] *The qualities are only in R and the printed edition (sig. H1v–H2v).*

[265] que est tria sydera hoc modo] que est tria sidera huius similitudinis J, que sunt tria sydera V; Pars secunda... hoc modo] Secunda vocatur Venter Arietis et constat ex duabus stellis sic S

[266] JV omit

[267] que est sex sydera perexigua] sex sydera per signa vel sic V

[268] Arabes autem] strabos vero J

[269] Tertia... pluvialem] Tertia dicitur Caput Tauri seu Cauda Arietis Arabes vero vocant eam pluvialem et ex sex stellis sic componitur S

[270] et ante illud] ante illa JV

[271] magnus V

[272] HQ omit

[273] Quarta... Tauri] Quarta ex .6. stellis et una magna et rubea que dicitur Oculus Tauri S

73 <5> Quinta est[274] Caput canis validi,[275] et sunt .3. sydera parva[276] inter duo magna, hoc modo.[277] Hec quidam tres digitos appellant.[278] s<icca>

74 <6> Sexta est quam Indi dicunt[279] Sydus parvum[280] lucis magne.[281] t<emperata>

75 <7> Septimam vocant[282] Brachium Leonis, suntque duo sydera clara.[283] h<umida>

76 <8>[284] Octavam[285] dicunt Nebulosam,[286] suntque sydera duo cum nube media.[287] t<emperata>

77 <9> Nonam autem[288] Oculum[289] dicunt, sunt etiam sydera tria.[290] s<icca>

78 <10> Decima Frons[291] dicitur; sunt sydera quatuor.[292] h<umida>

---

[274] dicitur S

[275] canis validi] cancruandi R

[276] JSV omit

[277] duo magna hoc modo] duo parva hoco modo V vel sic, duas parvas sic componitur S

[278] Hec quidam tres digitos appellant] R omits, Hec quidem digitos appellant J, quam quidam tres digitos appellant S, Hec quidem .3. digitos appellant V

[279] Sexta est quam Indi dicunt] Sextam Indi dicunt J, Sexta vocatur ab Indis S, Sexta Indi dicunt quod est V

[280] stella parva HQ; JSV add 'et'

[281] S adds 'que ex una sit composita'

[282] Septima vocant R, Septima dicitur S

[283] suntque duo sydera clara] suntque duo sidera multum lucencia J, suntque duo sydera V, et ex duabus stellis perlucidis sic componitur S

[284] V omits <8> and <9>

[285] J adds 'vero'

[286] Nebula HQ

[287] Octavam...media] Octava dicitur nebulosa et ex duabus stellis cum nube media componitur S

[288] vero J

[289] Oculus Leonis HQ

[290] Nonam...tria] Nona dicitur Oculus Leonis et ex tribus stellis sic componitur S

[291] Frons J *after correction*, Fors R, Sors SV, Frons Nebule HQ, Frons Leonis *printed edition*

[292] sunt sydera quatuor] et ex quatuor componitur S, sunt quatuor V

79 <11> Undecima dicitur Capillus; sunt sydera duo.[293] oo t<emperata>

80 <12> Duodecima Cauda Leonis dicitur; est sydus maximum.[294] o humida[295]

81 <13>[296] Tertiadecima Canis dicitur; sunt sydera quatuor ooo o t<emperata>

82 <14> Quartadecima Spica dicitur; sydus est[297] maximum. o t<emperata>

83 <15> Quintadecima Cohoperta dicitur;[298] sunt[299] tria sydera parva.[300] ooo h<umida>

84 <16> Sextadecima sunt sydera magna duo seiuncta que Cornua Scorpionis dicuntur.[301] oo h<umida>

85 <17> Decima septima dicitur Corona; sunt tria sydera super caput Scorpionis[302] lineatim posita magne lucis; prope ea in eadem linea[303] duo alia, hoc modo:[304] ooo oo h<umida>

---

[293] sunt sydera duo] et sunt sidera duo JV, et ex duabus componitur sic S

[294] est sydus maximum] et est sydus maximum V, que est stella magna S

[295] humida *printed edition*, R omits

[296] S omits <13> and <14>

[297] sydus est] et est sydus JV

[298] Cooperta dicitur JS, dicitur Cooperta V, Velamen HQ

[299] et sunt pr; J adds 'autem'

[300] cohoperta…parva] dicitur Cooperta et ex tribus parvis stellis sic componitur S

[301] sunt sydera…dicuntur] sunt duo magna sydera que cornua Scorp. dicuntur J, sunt sydera magna duo (duo magna V) que Cornua dicuntur Scorpii V, dicitur Cornua Scorp. et ex duabus stellis sic componitur S

[302] Scorpii V

[303] J omits 'in eadem linea'

[304] duo alia hoc modo] duo alia posita J, duo alia V; sydera…hoc modo] stelle linealiter super caput Scorpionis et alie parve in eadem lina sic posite S

86 <18> Decima octava[305] Cor Scorpionis,[306] estque sydus grande[307] rubeum inter duo parva.[308] ooo s<icca>

87 <19> Decima nona est[309] Cauda Scorpionis;[310] sunt duo lucentia sydera; in eadem linea novem alia sydera.[311] oo.ooooooooo h<umida>

88 <20> Vicesima dicitur Trabs;[312] sunt octo sydera lucida,[313] quatuor in cinctura, alia extra.[314] oo oo oo oo h<umida>

89 <21> Vicesima prima dicitur Desertum, et est locus sine syderibus. t<emperata>

90 <22> Vicesima secunda dicitur Pastor;[315] sunt duo sydera parva, unum erectum in sinistro, alterum[316] in dextro descendens.[317] Arabes autem[318] vocant sanitatis[319] occisorem, quod habetur[320] ante illud[321] sydus aliud dictum Aries.[322] oo o h<umida>

91 <23> Vicesima tertia dicitur Gluciens,[323] et sunt duo sydera parva.[324] o o t<emperata>

---

[305] S adds 'dicitur'

[306] Scorpii Q

[307] grande] magnum et J

[308] Cor Scorpionis . . . parva] est sydus magnum (sydus magnum est V) rubeum inter duo parva et dicitur Cor Scorpionis V, S adds 'sidera sic positum'

[309] dicitur S

[310] Cauda Scorpionis] Cor Scorpii V, R adds 'et'

[311] novem alia sydera] JV omit, sic S

[312] turbis J (alias trabs J *in the margin*), trabis V, trabes HQ

[313] lucentia V

[314] quatuor in cinctura alia extra J] quatuor in cinctura R, .4. inter .4. extra V; sunt octo . . . extra] et ex octo stellis quarum quatuor sunt in cinctura zodiaci et alie quatuor extra sic posite S

[315] Pistor HQ

[316] aliud V

[317] descendens *scripsi*, descendente MSS

[318] Arabes autem] Ambos JV

[319] R omits

[320] quod habetur] et est qui J, qui JV

[321] aliud V

[322] sunt duo sydera . . . Aries] seu occisor sanitatis et apud quosdam Aries et ex tribus stellis sic componitur S

[323] Gluciens] Degluciens HQ

[324] sydera parva] parva sydera sic S, sydera parva ut V

92 <24> Vicesima quarta dicitur sydus Fortune,[325] et sunt duo sydera, unum magis altero lucens.[326] °o t<emperata>

93 <25> Vicesima quinta dicitur Papilio;[327] sunt duo sydera retro ante sydera quatuor.[328] oo oo oo s<icca>

94 <26> Vicesima sexta dicitur primus Hauriens[329] aquam, et sunt duo sydera lucida.[330] o o s<icca>

95 <27> Vicesima septima dicitur secundus Hauriens[331] aquam, et sunt duo sydera lucida.[332] o o h<umida>

96 <28> Vicesima octava dens dicitur Piscis.[333] ooooooo ooooooo t<emperata>

97 Indi dicunt quod mansiones que pluviam significant sunt .11., scilicet[334] Oculus Tauri, Brachium Leonis, Frons,[335] Cauda Leonis, Cohoperta, Cornu[336] Scorpionis, Corona sui capitis, eiusdem Cauda, Trabs,[337] Pastor, et primus Hauriens,[338] scilicet 4, 7, 10, 12, 15, 16, 17, 19, 20, 22, 27.[339] **98** Et sex mansiones sunt sicce, scilicet Venter Arietis, Caput Canis Validi, Oculus, Cor, Papillio, secundus Hauriens,[340] scilicet

---

[325] sydus Fortune] Fortunata HQ, sydus Fortuna JV
[326] unum magis altero lucens] unum altero magis lucet JV, quorum unum altero magis lucet sic S
[327] Papira HQ, Papillio R, Lapacia V
[328] retro ante sydera quatuor] JV omit, sic S
[329] primus Hauriens] primus Auriens RV, Degluciens primus HQ
[330] et sunt duo sydera lucida] etc. V; J adds 'sic'
[331] secundus Hauriens] Hauriens secundus HQ, secundus Auriens V
[332] V omits; J adds 'sic', S adds 'sic composita'
[333] Dens dicitur Piscis] Dens HQ, dicitur Piscis J, Piscis dicitur V. VS add 'et est ex .14. stellis sic'
[334] qui sunt J
[335] Fors JR
[336] Cor J, Cornua V
[337] Turbis J, Trabis V
[338] Hauriens secundus R
[339] *V omits* scilicet *and the numbers. J adds the numbers above the names of the mansions rather than listing them here.*
[340] Hauriens primus R

2, 5, 9, 18, 25, 26.[341] **99** Undecim[342] alie sunt temperate, scilicet 1, 3, 6, 8, 11, 13, 14, 21, 23, 24, 28.[343]

100 Debes quoque[344] mense considerare horam coniunctionis Solis et Lune et eius oppositionis et eius quadrati a dextro et a[345] sinistro, et in quibus mansionibus Lune, humidis scilicet[346] vel siccis, sint[347] quinque planete, et in qua sit mansione ipsa Luna, et que mansio eius[348] Lune sit ascendens ab oriente, sicque cognoscere poteris in quo die mensis[349] pluvia sit futura. **101** Et semper debes considerare ipsius terre signum et suos angulos suumque dominum omnemque suum[350] aspectum.

The Treatise on Rain and the Change of the Atmosphere, according to Master John of Seville

1 If you wish to know the nature of the atmosphere each year, consider which sign Saturn is in when the Sun and Moon are in conjunction or opposition before the Sun enters Aries. **2** If it is in a fiery sign the year will be hot, and likewise for a cold, moist or dry sign—the year will have the same quality.[351] **3** You should mix with the nature of the sign that of the term in which Saturn is—the nature of each term is according to the nature of its Lord—and you should mix with Saturn the natures of the planets which are in conjunction with it, or which aspect it. **4** One will judge in the same way with Jupiter when the Moon

341 *V omits* scilicet *and the numbers. J adds the numbers above the names of the mansions rather than listing them here.*

342 Et .11. V

343 JV *omit* scilicet *and the numbers.*

344 unoquoque JV

345 JV omit

346 JV omit

347 sunt V

348 JV omit

349 die mensis] demon (!) V

350 omnemque suum aspectum] et eius aspectum J, suumque aspectum V. V adds 'Explicit prima pars huius artis'

351 Cf. al-Kindī, *Forty Chapters*, ch. 38, [669]: 'If you wish to know the phenomena of the atmosphere (*aḥdāth al-jaww*), establish the ascendant of the conjunction before the Sun enters Aries. If Saturn is in one of the signs when you establish the ascendant, the year will be according to the nature of that sign: when it is hot, the year will be hot; when it is cold, it will be cold; when it is dry, it will be dry; and when it is moist, it will be moist.'

is in conjunction[352] with the Sun or they are in opposition, before the Sun enters any of the tropical signs,[353] which are the beginnings of the quarters of the circle. **5** One judges in this way with Venus and Mercury in every month when the Sun and Moon are in conjunction or opposition. **6** One judges in this way for the month; if in the conjunction the ascendant is a fixed (sign), the judgement will hold true for the whole month. **7** One judges in this way with the Moon in each of the aforementioned 'doors'.[354]

8 The 'openings of the doors' are the conjunctions of the Sun or the Moon with Saturn or their aspect to it; and the same with Jupiter with Mercury, and Venus with Mars. **9** Venus indicates moisture, Mercury winds, Saturn clouds and cold, Mars winds from the right and heat, Jupiter moderation of the atmosphere and winds from the left (especially when it is in Cancer, just as Mars's indication is especially in Capricorn), and the Sun indicates warmth and dryness.[355]

10 If the opening of the doors happens with Mercury, there will be wind with rain. **11** If Mars, look at the Lord of the conjunction or opposition of the Sun and the Moon: there will be 'torches', lightning and thunder; add to this or take away from the nature of Mars the nature of the sign and term in which it is. **12** But if Saturn is aspecting rather than Mars, and is in the house of earth,[356] it indicates the

---

[352] Note that the author uses different terms for the conjunction or syzygy of the Moon and Sun: the nouns 'coniunctio' (**1, 5, 6, 8, 23, 56, 101**) and 'iunctura' (**11**), and the verb 'associare' (as here).

[353] Cf. al-Kindī, *Forty Chapters*, ch. 38, [670] (continuation of citation in n. 351 above): 'When the Sun enters any cardine you should look in which of the signs Jupiter is. Judge from it in the same way as you judged from Saturn, but for that quarter of the year, and mix (*umzuj*) the indications of the quarter with those of the year, and pass judgement according to that.'

[354] 'Aforementioned' refers to an earlier passage in the *Prima pars* of the *Epitome*, ms. V, fol. 62ra, *Epitome*, sig. F1v: 'Ianuae Lunae sunt .12. ...' The 'doors' here are the phases of the Moon in respect to the Sun, elsewhere described as 'stations' (*marākiz*); see Bos and Burnett, *Scientific Weather Forecasting*, pp. 342–4 and 394.

[355] al-Kindī, *Forty Chapters*, ch. 38, [674]: 'Know that Venus indicates moistures (*ruṭūbāt*), Mercury winds, Saturn clouds, darkness of the atmosphere (*ẓulmat al-jaww*) and cold, Mars the blowing of southern winds and heat, especially when it is in Capricorn or its cardines and most strongly in Capricorn, Jupiter northern winds, especially when it is in Cancer or in its shares (*ḥuẓūẓ*) in the houses and most strongly in Cancer, and the Sun indicates heat and dryness.' Capricorn and Cancer are respectively the signs of the exaltation of Mars and Jupiter.

[356] I.e. the earthy house of its two houses, namely Capricorn.

destruction of houses; if in the house of water,[357] it shows cold, clouds and darkness.[358] **13** But if a retrograde planet is there, it indicates great rain.[359] **14** The Moon has power over rain.[360] **15** But the Sun portends now rain, now a clear sky; for when it is in the lower point of its circle, it indicates rain, when in the higher, clear sky. **16** One judges in this way for all the planets.

17 The Moon receives power from the planets and gives it to the earth—this power is determined by their nature. For, as the planets change from right to left or *vice versa*, or from swiftness to slowness or *vice versa*, or from direct movement to retrogression or *vice versa*, or from East to West[361] or *vice versa*, or enter or leave one of the cardines, so the atmosphere will be changed. **18** For if it is stationary, it moves or doesn't move by the movement of the winds. **19** Such is the effect of the conjunction or aspect of the planets. **20** Such is the indication of the fixed stars.

21 If you wish to know whether there will be much rain at the beginning, in the middle or at the end of the year, look at the position of Venus when the Sun enters Libra; if it is under the rays of the Sun and in the west, there will be much rain at the beginning, but little at the end. If it is in the east and visible, the rain will be in the oppo-

---

[357] I.e. Aquarius, which although being aery is particularly associated with moisture.

[358] Cf. al-Kindī, *Forty Chapters*, ch. 38, [676–7]: 'When the Moon arrives at one of these stations (*marākiz*: i.e. those refered to in sentence **7** above) and a lower planet applies from opposition (*or* aspect) to one of the higher planets, it brings forth its nature...And if this opening (*fatḥ*) occurs to Mercury, it indicates winds with rain, if Venus, abundant rain and cold, if Saturn, cold and continuous darkness of the atmosphere and continuous destruction and clouds piled on clouds...When Mars aspects the indicators and especially the Moon, it indicates the occurrence of thunder and lightning.'

[359] Cf. al-Kindī, *Forty Chapters*, ch. 38, [679]: 'Also call to witness the retrogression of the indicators, for these increase the abundance of water.'

[360] Cf. Abū Maʿshar, *Kitāb al-sirr*, ms. Escorial 938, fol. 14v: 'kānat al-quwwa li'l-qamar fī ḥāl al-maṭar'. *Tractatus pluviarum* retains the Arabic syntax whereas *Apertio portarum* has changed the text into something more Latinate: 'Luna vero significat pluviam'.

[361] Instead of 'from direct movement to retrogression or *vice versa*, or from East to West' mss. JSV have 'from the ascent from the lower part of the point of the circle to the higher part'. The unusual terminology 'inferius (partis) circuli punctum' also occurs in **15**.

site situation.[362] **22** The signs of much rain are Aquarius, the end of Capricorn, Aries, Taurus and Leo; the strongest of these are Pisces, Aquarius and Scorpio.[363]

23 If you wish to know the day of the rain, look at the time of the conjunction or opposition of the Sun and Moon, and look at the number of degrees there are between the Moon and the next planet that indicates rain; take a day for each degree; the indication of tropical signs is swift, that of bicorporeal, medium, that of fixed, slow.[364] **24** Look at the doors[365] of the Moon, and see which planet aspects the Moon or is joined to it; then from this too you can know the day of the beginning and the end of the rain.

25 See how many equal degrees there are between the Sun and Saturn each morning and count the same number of degrees from the position of the Moon; where they arrive will be the lot of rain.[366] **26** If the number arrives in a house of the Moon or Venus, it indicates rain, if in a house of Mercury or Jupiter, rain, if of Mars or the Sun, clear sky, if of Saturn, clouds.[367]

27 When the Moon enters Cancer in the Sun's hour, Virgo in Venus's hour, Sagittarius in the Moon's hour, Gemini in Mercury's hour, Taurus in Mars's hour, or Libra in Jupiter's hour, it indicates rain or

---

[362] Cf. al-Kindī, *Letter II*, 148–50 (Bos and Burnett, *Scientific Weather Forecasting*, pp. 225, 260 and 308), in which the first part is the same, but the prognosis then varies: 'If you wish to know whether there will be more rain at the beginning, in the middle or at the end of the year…If Venus is under the Sun, west of it, it indicates much rain at the beginning of the year and at the end. But when it is west of it about one and half months later on it indicates much rain in the middle of the year.'

[363] This list is closest to that in al-Kindi, *Letter II*, 146–7 (ed. Bos and Burnett, pp. 224–5, 260 and 308). This includes the inconsistencies of this text but reverses the last three signs: 'Pisces, Aquarius and Scorpio'.

[364] Cf. al-Kindī, *Letter II*, 156 (ed. Bos and Burnett, pp. 225, 261 and 309): 'Tropical signs indicate the fastest time, bicorporeal signs medium, and fixed signs a more distant (?) time.'

[365] See n. 354 above.

[366] This lot of rain is the second of the two lots described in Abū Ma'shar, *Kitāb al-sirr*, ms. Escorial 938, fol. 17r (see p. 234 above).

[367] The predictions from the lot in the following sentence abbreviate considerably the predictions given by Abū Ma'shar (which correspond to those in Ibn Ezra and al-Kindī): see pp. 234–5 above.

winds in accordance with the nature of the sign in which the Lord of the hour is located.[368]

28 When Venus is in Scorpio, Capricorn or Aquarius, and conjoins or aspects the Moon, it indicates much rain. **29** Similarly, Mercury in its station with any of these signs. **30** Similarly, when three planets are conjoined in Aquarius.[369]

31 When a planet is retrograde from the first station to being opposite the Sun, it is more indicative of rain than when it is between being opposite the Sun and its second station.

32 When the Moon separates from conjunction or aspect with any planet and conjoins another planet, either by conjunction or by aspect, and the two houses of those planets are opposite to each other, this is the opening of the great door-leaves,[370] in which winds or rain according to the nature of those planets necessarily occur.

33 But if the Moon is in the sign of a state or in any of its cardines, it will have a strong indication for that city; if it is not in the cardines, but in a sign 'looking towards' the sign of the city, its indication will be less; if it is in a sign that does not look towards it, its indication is empty.

34 When the Moon is in a feminine house,[371] such as the eastern and western quarters,[372] and in a feminine sign and a retrograde planet 'looks towards' it from a feminine sign, rain will necessarily fall.

---

[368] Cf. al-Kindī, *Letter II*, 137 (this is exactly the same until the prediction): 'When the Moon enters Cancer in the Sun's hour... it indicates rain in its proper times and wind according to the <Lord of the> sign and according to what will be of the substance of the Moon itself and what will be of the substance of the signs.'

[369] Cf. al-Kindī, *Letter II*, 134–6: 'When Venus is in Scorpio, Capricorn or Aquarius, there will be much rain. Similarly, Mercury in Scorpio, Capricorn or Aquarius indicates rain. When Venus and Mercury are in these signs which we mentioned and the Moon is in opposition, conjunction or quartile aspect with them, rain will arise.'

[370] Cf. 'Umar ibn al-Farrukhān, *Kitāb mukhtaṣar al-masāʾil*, ch. 83 (see pp. 231–2 above).

[371] *Apertio portarum* has 'quarter'.

[372] The eastern and western quarters are generally classified as being masculine.

35 The ancient astrologers have said that the regions which the Moon has on the earth are western and Spain, England, Armenia, and France;[373] it has power over the sixth part of the peoples.

36 The Sun has the second part: the land of the Orient, the land of Arin (Ujjayn) and Zarozin (?).

37 Saturn has the first part of the peoples, and *India* and Ethiopia.

38 Iupiter has the fourth part: Persia, *Babylon* and Baghdad, Iraq.

39 Mars has the seventh part: *the Turks* beyond the sea.

40 Venus has the fifth part: *Arabia*.

41 Mercury has the third part: Greece and Daylam (?).

42 The states of the signs:[374] the city of Aries: *Babylon,* ***Persia, Palestine***, Esmosal, Eschasar.[375]

43 Taurus: **Asia Minor, Cyprus, Isfahan**.

44 Gemini: ***Jurjān,*** **Tabaristan, Greater** ***Armenia***, *Adharbayjān*, Babylonia, *Barqa*.

45 Cancer: the land of the Berbers, ***Africa***, *Numidia* (?), **Rome**, Ladhakia, *Lesser Armenia*.

46 Leo: **the land of** ***the Turks***, Italy, **Damascus, Kufa**, *Ṭūs*, Gog and Magog, Basra.

[373] The regions shared with Abū Maʿshar's *Kitāb al-milal wa'l-duwal*, bk 2, ch. 8 [4] (see ed. Yamamoto and Burnett, i, p. 606) are italicised. In the case of the Moon, ʿAlī ibn Abī 'l-Rijāl, *Kitāb al-bāriʿ*, which otherwise follows Abū Maʿshar, substitutes 'the Maghreb, al-Andalus and France (*Ifranj*)' for 'China' (ibid., i, p. 513).

[374] The states shared with Abū Maʿshar, *Great Introduction* (ed. Lemay, *Liber Introductorii maioris ad scientiam judiciorum astrorum*), vi, ch. 9 are italicised; those shared with the Escorial addendum to Abū Maʿshar's *Kitāb al-milal wa'l-duwal* (ed. Yamamoto and Burnett, i, pp. 514–19), in bold.

[375] The last two places are equivalent to Adharbayjān in Abū Maʿshar.

47 Virgo: Nishapur (?), Tibet (?), Eclites (Crete?), Athens.

48 Libra: Nara (?),[376] **Karmān** and the immense land of the ***Ethiopians***.

49 Scorpio: ***al-Hijāz***, Tarsus (?), Mecca.

50 Sagittarius: Yemen, Palestine, **Sicily**, Sardinia.

51 Capricorn: ***Sind, Oman***, Macedonia.

52 Aquarius: Surri (?), Farghānā.

53 Pisces: ***Alexandria***, England.

54 The astrologers have said that Jerusalem belongs to Cancer; others say that Leo 19° <is its sign>; the truth is that Aquarius is its sign, while Leo is the sign of Rome.[377] **55** They also said that Pisa's sign is Pisces; but by experience it is Aquarius 2°. The sign of Lucca is Cancer;[378] that of Palermo is Leo 1°, that of Africa, Leo 4°.

56 Always observe, when the Moon separates from conjunction or opposition to the Sun (whichever is closer to the Sun entering the beginning of Aries), which planet the Moon conjoins or looks at with a full aspect. **57** If it is one of the planets of good fortune, the year will be good and bodies will be sound, and it will rain at the right time and there will be plenty of crops, fruit, animals etc. **58** If that planet pushes power on a malefic planet or pushes power on a planet standing in a cardine and that planet pushes power on one standing in a cadent place,[379] the end of the year will be bad. **59** The indication of each planet is in accordance with its nature. **60** The nature of Saturn is thirst, and drowning in water; when Saturn is in one of the signs which

---

[376] Cf. 'Hara (Hare)' in the Latin version of the Escorial addendum to Abū Maʿshar's *Kitāb al-milal wa'l-duwal* (= Bukhārā, read as 'bi-Hara'): ed. Yamamoto and Burnett, ii, p. 142.

[377] Jerusalem is assigned to Cancer and Leo to Rome in the Escorial addendum to Abū Maʿshar's *Kitāb al-milal wa'l-duwal* (ibid., i, p. 519).

[378] Cf. Abraham ibn Ezra, *Sefer ha-Olam* (see n. 44 above).

[379] For this astrological terminology see al-Qabīṣī, *Introduction to Astrology*, 3 [18], pp. 99 and 304.

are called 'the domain of the Sun',[380] it indicates drowning in water; if it is in those which are called 'the domain of the Moon', thirst.

61 If Mars is in a warm sign, it causes warmth and dryness; if it is in a sign of something like a man,[381] it indicates wars and a multitude of thieves, and the war will be fierce when Mars enters under the rays of the Sun or into the house of its fall. **62** Jupiter indicates the same when it has entered the rays of the Sun. **63** See whether Mars is east or west or right or left; in whichever of these directions Mars is situated, they (the protagonists) will win. **64** If Mercury is with Mars at the beginning of the year or in any of the quarters, it will indicate death in the lands which belong to the sign in which they conjoin. **65** If Mercury is with Saturn and in a sign of an earthy nature, it indicates an earthquake.

66 The wise Indians take into account the division which is called the 'twelfth': this is that you give, from the beginning of a sign, 2½ degrees to that sign, and you distribute the same—namely 2½ degrees—to each sign that follows, and do this until the end of the signs. **67** From this can be known whether the sign to which the twelfth-part arrives belongs to a benefic or malefic planet,[382] and in this way the good or evil befalling the citizens of that sign will be known.[383]

68 The Indians have divided the circle into 28 parts which they have called 'the mansions of the Moon'; each comprises twelve degrees and six sevenths of a degree.[384]

69 The starting-point of these mansions is from two large stars which are called 'the Horns of the Ram'. They look like this:  temperate.

[380] I.e. the Sun's half of the zodiac, from Leo to Capricorn.

[381] These are the zodiac signs that take the shape of a human being, fully or partially, namely Gemini, Virgo, Sagittarius and Aquarius.

[382] The planets here would be the Lords of the signs to which the individual twelfth-parts are assigned.

[383] Cf. Abraham ibn Ezra, *Sefer ha-Olam*, Paris, Bibliothèque nationale de France, héb. 1058, fol. 89b: 'Know in which sign the twelfth falls, as I have explained in *Reshit Hokhma*, and from that you may know any good or evil that befalls any city' (information from Shlomo Sela).

[384] The readings of mss. JV assume a division of the zodiac into 27 mansions of 13 and one third degrees each, rather than into 28 mansions (see Bos and Burnett, *Scientific Weather Forecasting*, pp. 366–72).

70 The second part is called 'the Belly', which consists of three stars, like this:  dry.

71 The third is called 'the Head of the Bull', which consists of six very small stars, and some people call them 'the Tail of the Ram', but the Arabs call the mansion 'the Rainy One':[385]  temperate.

72 The fourth: a large, red star,[386] before which is a small star. The larger one is called 'the Left Eye of Bull': 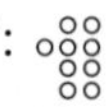 moist.

73 The fifth is 'the Head of the Strong Dog';[387] it consists of three small stars between two large ones, like this: 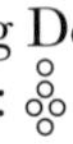 Certain people call these 'the Three Fingers':[388] dry.

74 The sixth is what the Indians call 'the Small Star of Great Light': temperate.

75 They call the seventh 'the Foreleg of the Lion'; there are two bright stars:  moist.

76 They call the eighth 'the Cloudy One';[389] there are two stars with a cloud in the middle: 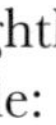 temperate.

---

[385] The 'head of the Bull' describes the position in the zodiac; 'rainy' is how the Bedouin Arabs regard *al-thurayyā*, which is the Arabic name for the Pleiades: see P. Kunitzsch and M. Ullmann, *Die Plejaden in den Vergleichen der arabischen Dichtung* (Sitzungsberichte der Bayerische Akademie der Wissenschaften. Philosophisch-historische Klasse, Jahrg. 1992, Heft 4), p. 32.

[386] Aldebaran (a proper name) is aptly described as a conspicuous red star; Ibn Ezra, *Keli ha-Nehoshet* (first two versions) also describes it as a 'red star' (Smithuis, 'Abraham Ibn Ezra,' p. 366).

[387] This name for the head of Orion also occurs amongst the fixed stars in the *Epitome* (Kunitzsch, 'Abū Maʿšar, Johannes Hispalensis und Alkameluz,' p. 19, Anmerkung v). It has a parallel in Ibn Ezra's use of the term *ha-kelev ha-gibbor* (translated by Petrus de Abano as 'canis validus') for Orion in his *Reshit Hokhmah*, chapter 1: see Abraham Ibn Ezra, *The Beginning of Wisdom*, ed. R. Levy and F. Cantera (Baltimore, 1939), pp. vi and 153.

[388] On these three stars see P. Kunitzsch, *Untersuchungen zur Sternnomenklatur der Araber* (Wiesbaden, 1961), no. 19. The translation here implies *al-aṣābiʿ* ('fingers') where the Arabic *anwāʾ* tradition has *al-athāfī* ('hearthstones').

[389] For the description of constellations as 'nubes' ('clouds') in 'Iohannes Hispalensis'' list of fixed stars see Kunitzsch, 'Abū Maʿšar, Johannes Hispalensis und Alkameluz,' p. 117, Anmerkung l.

77 They call the ninth 'the Eye'; there are also three stars: ∘∘∘ dry.

78 The tenth is called 'the Forehead'; there are four stars: ∘∘∘∘ moist.

79 The eleventh is called 'the Hair'; there are two stars: ∘∘ temperate.

80 The twelfth is called 'the Tail of the Lion'; it is a very large star: ∘ moist.

81 The thirteenth is called 'the Dog';[390] there are four stars: ∘∘∘∘ temperate.

82 The fourteenth is called 'the Ear of Corn';[391] it is a very large star: ∘ temperate.

83 The fifteenth is called 'the Covering';[392] there are three small stars: ∘∘∘ moist.

84 The sixteenth is two large stars separated from each other, which are called 'the Claws of the Scorpion': ∘∘ moist.

85 The seventeenth is called 'the Crown'; there are three stars of great light placed in a line over the head of the Scorpion, with two others in the same line close to them, like this: ∘∘∘ ∘∘ moist.

86 The eighteenth is called 'the Heart of the Scorpion'; it is a large red star between two small ones: ∘∘∘ dry.[393]

87 The ninteenth is 'the Tail of the Scorpion'; there are two shining stars; in the same line there are nine further stars: ∘∘.∘∘∘∘∘∘∘∘∘ moist.

[390] This is a literal translation of the Arabic *al-ʿawwāʾ* ('the barking one' = Boötes).

[391] This is a literal translation of *al-sunbula*, the usual Arabic name for Virgo as well as of the star α Virginis.

[392] While the usual form of the Arabic name—*al-ghafr*—means 'a coat of mail', the translation 'cohoperta' suggests *ghufra* = 'cover'. Mss. HQ give a synonym for 'cohoperta': *velamen*.

[393] Most other weather-forecasting treatises make this mansion moist; the printed version makes it temperate.

88 The twentieth is called 'the Beam';[394] there are eight bright stars, four on the girdle, the other outside:  moist.

89 The twenty-first is called 'the Desert';[395] it is a place without stars: temperate.[396]

90 The twenty-second is called 'the Shepherd';[397] there are two small stars, one rising on the left, the other descending on the right. The Arabs call it 'the Slayer of Health'; it is held that, before that star is another one called 'the Ram':  moist.

91 The twenty-third is called 'the Swallower'; there are two small stars:  moist.

92 The twenty-fourth is called 'the Star of Fortune'; there are two stars, one shining more than the other: 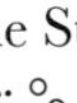 temperate.

93 The twenty-fifth is called 'the Tent'; there are two stars behind four stars:  dry.

94 The twenty-sixth is called 'the First Pourer of Water'; there are two bright stars:  dry.

394 The name of the 20th mansion, *al-naʿāʾim*, is usually understood as 'ostriches'. However, according to Lane (s. v. zurnūq) *al-naʿāma* can also mean the 'beam' over an Arabic well, to which is attached the 'well rope' (*al-rishāʾ*, an alternative name for mansion no. 28) with which the 'bucket' (*dalw*, the Arabic for Aquarius = mansions 26–27) is lowered into the well. I owe this explanation to Paul Kunitzsch. Note that, in the treatise on the astrolabe (*Keli ha-Nehoshet 1*) Ibn Ezra writes: 'Al-naʿāʾim: its translation is 'the beams' (Smithuis, 'Abraham Ibn Ezra,' p. 185). See Steinschneider, 'Ueber die Mondstationen' (n. 24 above) p. 152. This may suggest that Latin mss. HQ's 'trabes' (in the plural) may be more authentic than 'trabs' in the singular.

395 This seems to be a translation of the Arabic description of the asterism as a *ruqʿa*—an area or place—without stars. Information from Paul Kunitzsch.

396 Most other weather-forecasting treatises make this mansion moist.

397 'Pastor' is also used in the 'Construction of the Astrolabe' by 'Iohannes' for α Oph (in Sgr) = *al-rāʿī* (P. Kunitzsch, *Typen von Sternverzeichnissen in astronomischen Handschriften des zehnten bis vierzehnten Jahrhunderts* [Wiesbaden, 1966], p. 32), but is out of place here. On the other hand the name, *al-rāʿī*, is sometimes given to the star λ Sag. which stands between the two quadrangles of the eight stars of no. 20, where it is regarded as the 'shepherd' leading the '(small) cattle' (*naʿam*, rather than ostriches *al-naʿāʾim*) to the drinking place in the Milky Way: see Kunitzsch, *Untersuchungen*, nos 238a–b. The 'slayer of health' retains a literal translation of *al-dhābiḥ* ('the slaughterer'). The 'shepherd' may be here simply out of association with the ram and the slaughterer.

95 The twenty-seventh is called 'the Second Pourer of Water'; there are two bright stars: ○○ moist.

96 The twenty-eighth is called 'the Tooth of the Fish': ○○○○○○○ ○○○○○○○○ temperate.

97 The Indians say that the mansions that indicate rain are 11 in number; namely, the Eye of the Bull, the Foreleg of the Lion, the Forehead, the Tail of the Lion, the Covering, the Horn of the Scorpion, the Crown on its head, its Tail, the Beam, the Shepherd and the First Pourer, i.e., numbers 4, 7, 10, 12, 15, 16, 17, 19, 20, 22, 27. **98** Six mansions are dry; namely the Belly of the Ram, the Head of the Strong Dog, the Eye, the Heart, the Tent and the Second Pourer, i.e. numbers 2, 5, 9, 18, 25, 26. **99** The other eleven are temperate, i.e. numbers 1, 3, 6, 8, 11, 13, 14, 21, 23, 24, and 28.

100 You ought also to look every month at the hour of the conjunction and opposition of the Sun and the Moon and its right and left quadrant, and in which mansions of the Moon—i.e. in the moist or dry ones—the five planets are located, in which mansion the Moon itself is, and which mansion of that Moon is ascending in the east. Thus you will be able to know on which day of the month the rain will fall. **101** You should always look at the sign of the land itself and its cardines and its Lord and every aspect in it.

THE *BOOK OF ANIMALS* BY ARISTOTLE

Lou Filius

Within Greek literature a long-standing controversy about the *Book of Animals* exists concerning whether all ten books are written by Aristotle or not. Books I–VI have always been considered authentic, but VII–IX presented problems and this is even more true of book X.[1] Nowadays it is generally believed that book X is not part of the *Historia animalium* (*HA*), but a separate text with the title ὑπὲρ τοῦ μὴ γεννᾶν.[2] According to the list of Ptolemaeus, and the Appendix of the Anonymus Menagii,[3] this book has been added to the *HA*. It is not strange that they supposed it to belong to the *HA*, since other books in the *HA* also dealt with reproduction (if the order of Balme and Gotthelf is to be followed). In that case one can easily understand that this book, ὑπὲρ τοῦ μὴ γεννᾶν, about the impossibility of reproduction was accepted as book X of the *Book of Animals* after Book IX, which discusses reproduction. The controversy about the authorship of this book, i.e. ὑπὲρ τοῦ μὴ γεννᾶν, seems to have been decided in favour of Aristotle. Yet, a certain degree of unanimity about the total number of books arose, when the Arabs became interested in the Greek manuscripts.

Another controversial issue is the order of the books. The latest edition of Balme and Gotthelf follows the old list before Theodorus Gaza, i.e. the order I–VI, VII (Bekker VIII), VIII (Bekker IX), IX (Bekker VII), X.[4] The order of books in the Arabic manuscripts, however, corresponds to the order of Ptolemaeus' list. Moreover, in the Arabic world,

---

[1] Cf. Aristotle, *History of Animals*, books VII–X, ed. D.M. Balme, prepared for publication by A. Gotthelf (Cambridge, Mass., 1991), pp. 1–13 and *Histoire des Animaux*, Tome I, livres I–IV, ed. and trans. P. Louis (Paris, 1964), pp. vii–xxxii. Other books have been quoted in the notes of these books.

[2] Cf. also P.J. van der Eijk, 'On Sterility ("HA X"), a Medical Work by Aristotle,' *Classical Quarterly* 49 (1999), pp. 490–502.

[3] P. Moraux, *Les listes anciennes des ouvrages d'Aristote* (Louvain, 1951), pp. 272, 278 and 297. Also the *Vita Hesychii* mentions ten books, cf. I. Düring, *Aristotle in the Ancient Biographical Tradition* (Göteborg, 1957), p. 87, n. 155. The list of Diogenes Laertius mentions nine books, see ibid., p. 47, n. 102. for the Arabic tradition of the list of Ptolemaeus, see ibid., p. 225, nn. 49–51. The *Fihrist* mentions simply ten books, ibid., p. 194.

[4] P. Louis followed the order of Theodorus Gaza and Bekker.

two other books formed part of the *Book of Animals: The Generation of Animals*[5] and *De partibus animalium*[6] in addition to the *Historia animalium.*[7] In Arabic literature, this set of books (*HA* 1–10, *PA* 11–14, *GA* 15–19) is referred to as *The Book of Animals, Kitāb al-ḥayawān*. No discussion existed in the Arabic world about book X of the *HA*. They apparently followed the list which was known from late Antiquity.

A new edition of the Arabic translation of the *Book of Animals* is currently being prepared at the University of Leiden,[8] based not only on the Tehran manuscript used by ʿA. Badawī,[9] but also on a London manuscript, which was clearly the Vorlage of the Tehran manuscript.[10] Therefore, the London manuscript is the basis of the new edition, although only the beginning of the text (books I–III, till 514b 16) and the last part (IX–X from 582a 32) are extant. The other books are lost. The second part of the fifth book (from 550a 8) has not come down to us in Arabic. The translation of Scotus[11] from Arabic into Latin lacks also this part of the fifth book.

It is often stated that al-Biṭrīq or his son Yaḥyā translated the text from Syriac,[12] but this is not mentioned in the *Fihrist*.[13] The *Fihrist* only records the existence of a Syriac translation which was supposed to be better than the Arabic version. Even though many Syriacisms can be identified in the Arabic translation,[14] one might still ask whether these

[5] Aristotle, *Generation of Animals. The Arabic Translation commonly ascribed to Yaḥyâ ibn al-Biṭrîq*, ed. J. Brugman and H.J. Drossaart Lulofs (Leiden, 1971).

[6] *The Arabic Version of Aristotle*'s Parts of Animals, *Book XI–XIV of the* Kitāb al-Ḥayawān, ed. R. Kruk (Amsterdam, 1979).

[7] A new edition is being prepared by J. Mattock (his papers are used), R. Kruk, J. den Heijer and L.S. Filius. There is a previous edition of the *HA* by ʿA. Badawī, *Kitāb ṭibāʿ al-ḥayawān* (Kuwait, 1977).

[8] By R. Kruk, J. den Heijer and L.S. Filius.

[9] Aristotle, *Kitāb ṭibāʿ al-ḥayawān li-Arisṭūṭālīs*, ed. ʿA. Badawī (Kuwait, 1977).

[10] Cf. L.S. Filius, 'The Arabic Transmission of the *Historia Animalium* of Aristotle,' in A. Vrolijk and J.P. Hogendijk (eds.), *O Ye Gentlemen. Arabic Studies on Science and Literary Culture in Honour of Remke Kruk* (2007), p. 27.

[11] The Latin translation by Scotus has been edited by Aafke M.I. van Oppenraaij, *Generation of Animals* (Leiden, 1992) and *Parts of Animals* (Leiden, 1998). The *Historia Animalium* will follow.

[12] Cf. ed. Brugman and Drossaart Lulofs, pp. 1–17 and G. Endress, *Die arabischen Übersetzungen von Aristoteles' Schrift* De Caelo, doctoral thesis (Frankfurt am Main, 1966), pp. 113–15.

[13] Ibn al-Nadīm, *Kitāb al-Fihrist*, ed. G. Flügel (Leipzig, 1871–72), p. 251 and ed. R. Tajaddud (Tehran, 1391/1971), p. 312. Cf. B. Dodge, *The* Fihrist *of al-Nadīm* (New York, 1970), ii, p. 605.

[14] Cf. ed. Brugman and Drossaart Lulofs, pp. 4–17.

are due to a Syriac intermediary translation. R. Kruk has already expressed some doubts about such an intermediary:

> Whether there was any relation between this Syriac translation, of which nothing further is known, and the Arabic one which has come down to us, cannot be ascertained. There is not even any proof that the Arabic text was translated from the Syriac, either from this old translation or from a translation specially made by the translator as a preparation for the translation in Arabic (as was often done). Several Syriacisms may be pointed out in the translation, but these only prove that the translator's language was strongly influenced by Syriac.[15]

An answer to this problem requires a careful examination of the texts. Expressions like فهو هو, وايضا, etc. appear frequently in the whole *Book of Animals*.[16] Expressions like شيء من (500a 10), to render 'a quantity of something,' seem to be a Syriac feature. Furthermore, there are some Syriac words like تنور for θώραξ and اذان for βράγχια (e.g. 511a 5; 533b 4), and some names of animals might have a Syriac origin. Yet, these examples do not suggest the definite existence of a Syriac intermediary, since Christians like al-Biṭrīq or his son Yaḥyā were native speakers of Syriac and therefore we can assume that they used Syriacisms and technical terms of the Syriac scientific tradition, even when writing Arabic.

Let us consider some further evidence. First of all, Ibn al-Nadīm judged al-Biṭrīq's translation to be poor. The Syriac translation was better. Unfortunately, such a Syriac translation is not preserved. Until the end of the Umayyad period and even somewhat later, Greek and Arabic were often used in social and commercial relations, judging from the considerable amount of bilingual Greek and Arabic papyri of deeds and contracts in Egypt and elsewhere in the Middle East, but this type of translation depended nearly always on accidental circumstances.[17] Although Ibn al-Nadīm did not know Greek himself, he had good informants about the different subjects of his books, e.g. Yaḥyā ibn ʿAdī for philosophical matters.[18] This second-hand information as well as his knowledge of Greek manuscripts allowed him to compare

[15] Cf. R. Kruk remarks in her edition of *PA*, p. 22, cf. also p. 14.

[16] Cf. ed. Brugman and Drossaart Lulofs, pp. 4–10.

[17] Cf. D. Gutas, *Greek Thought, Arabic Culture. The Graeco-Arabic Translation Movement in Baghdad and Early ʿAbbāsid Society (2nd–4th/8th–10th Centuries)* (London, 1998), pp. 23–4.

[18] Gutas, *Greek Thought, Arabic Culture*, p. 72.

the Arabic translation with the Syriac one and judge this translation of the *HA*. It seems probable that he assessed the Arabic translation on the basis of the Syriac translation or even on the original Greek manuscript(s), with or without competent informant(s). Perhaps he saw the numerous Syriacisms in the Arabic translation and based his judgment on this criterion. The use of Syriacisms, however, was common in the old translations, since the translators of the Greek books were almost exclusively Christians, who used Syriac as an intermediary for their translation into Arabic. Thus, for Ibn al-Nadīm, this was probably not the main point of objection.

But if the Syriac translation has a better quality, it must have been one without strange ways of rendering the original and without the many reductions which make the preserved Arabic version at times unclear. In I 494a 13–16 τὸ δὲ σαρκῶδες κάτωδεν στῆθος is translated ما تحت طرف القدم يسمى صدر القدم, but the Greek στῆθος can also be translated into Arabic as كف القدم. It is unlikely that this Greek word was translated into Syriac with the equivalent of the Arabic word صدر. This example gives an idea of why Ibn al-Nadīm thought the Arabic version a translation of inferior quality.

Another strange feature is the text of VIII 615a 17–18. Perhaps the translator was ignorant of the meaning of the Greek words, where adjectives, belonging to the recently mentioned *trochilos* (the wren),[19] have become different names of birds, namely the so-called δυσάλωτος, δραπέτης τὸ ἦθος.

A word in VIII 615b 24, θρυλεῖται, seems to have been unknown to the translator, since he translated "there exists doubt about it" instead of "it is commonly reported".

Many words are simply not translated at all, perhaps because they were not commonly used and therefore unknown to the translator, e.g. VIII 613a 23 and 613a 28 παλευτριῶν "the decoys". In Scotus' translation, this word is also missing in both places.

If the translator did know the meaning of the different Greek words, Ibn al-Nadīm might have criticised the translation for its inaccuracy, e.g. II 502a 14: δόρατα is translated with "whips" سياط. Scotus rendered it as *sotulares*, transcribing the Arabic word in an incorrect way.

VIII 616b 34 δειπνοφόρος καὶ ἔπαγρος "it brings meals home and looks for prey" was rendered in Arabic as يأكل الحيوان الصغير "it eats

[19] See *HA*, ed. Balme-Gotthelf, p. 271, n. 2.

little animals", VIII 619b 13 ὠτίδες "bustards" is translated into Arabic العقبان "eagles", VIII 624a 6–7 καὶ κάτω συνυφεῖς (Balme and Gotthelf: συνυφὲς) ποιοῦσιν (B&G + τε) ἕως τοῦ ἐδάφους ἱστοὺς πολλοὺς "and woven continuously below they make many webs down to the floor" became in Arabic ويصير الزوايا على اوتاد البناء "They make corners on the pins of the construction". Another example is VIII 624a 17–18 ἡ δὲ συνεχὴς ἀλοιφὴ τούτῳ πισσόκηρος, ἀμβλύτερον καὶ ἧττον τῆς μίτυος "The smearing that is next to this is pissoceros, a blander and less medicinal stuff than the mitys" which was translated as وان خلط به موم وزفت يكون دواء اقوى واكثر منفعة "and if the wax has been mixed with it and smeared with pitch, a medicine is stronger and more beneficial". The last example was perhaps also difficult to translate because of the peculiar words.

Another reason Ibn al-Nadīm criticized the Arabic translation were the many omissions, e.g. all from Book VIII: 609a 14–15, 609b 16 (which exists in Scotus' translation), 609a 26, 610a 11, 610a 32–3 etc.

How can these shortcomings in the Arabic translation be explained? The only explanation is that the translator translated the text directly without any intermediary from Greek into Arabic. He was not used to working in this way and therefore made many mistakes, sometimes shortening the text because of its length, rightly or wrongly. These characteristics make it a good example of how texts were first translated directly from Greek into Arabic. When compared with later translations by Ḥunayn ibn Isḥāq, this text also shows the enormous improvement of translations from Greek and the development of scientific language.[20]

On the other hand, a closer look at the failures might be rewarding. The different readings in particular show what type of manuscripts the translator used for his translation. He did not study the manuscripts thoroughly enough, and sometimes incorrect readings are due to a faulty reading of the Greek manuscripts.

Sometimes the translation reflects the wording of a specific Greek manuscript tradition. For example, in VII 599a 15, the Arabic translation, مرسلا, corresponds with the reading of mss. AC ἀπολύτων. Badawī followed here Bekker and other editions (ἀναπολύτων) and changed the Arabic word to ملصق.

[20] Cf. M. Ullmann, *Wörterbuch zu den griechisch-arabischen Übersetzungen des 9. Jahrhunderts* (Wiesbaden, 2002), pp. 47–8.

In VII 606b 9, ἄπλατον was translated as ليس عرض له, as if it was derived from τὸ πλάτος "breadth" and, as ἀπλάτης, meant "without breadth". The true meaning, however, has to be derived from πελάζω, "unapproachable", and the word has to be translated as "terrible" or "monstrous".

VII 605b 15: ἀνάπλεων, but mss. AC have ἀναπνέων, in Arabic تنفست.

VII 602b 31: τῷ πλόμῳ, but Arabic follows AC τῷ πάγῳ and has في اوان شدة الشتاء وزمهرير, in Latin: *quando glacies accidit*. A very strange translation.

VIII 616b 10: τὸν δὲ πόδα χωλός ἐστιν, but ms. P reads χλωρός instead of χωλός. On the basis of this ms., the Arabic text reads اخضر الرجل, "with green feet" through the following word χλωρίων (the oriole).

Stranger still seems VIII 632a 7, where διαφθείρεται is translated into Arabic as اختلاف, i.e. "difference" = διαφέρει. This might have already been a mistake in Greek manuscripts (albeit in none of those which are preserved), which was adopted by the translator. An error of the translator is more likely.

The spelling of the technical terms was always the traditional one, usually the Syriac transliteration. Some examples:

The Greek word σελάχη is normally spelled سلاشي, but strangely, in VII 591a 10 and VII 598a 12 it is سلاخي, following the Greek spelling. In 540b 17 σελάχη is rendered as سلاخيا and 511a 5 σέλαχος is rendered as سلاخوس. This supports the assumption that the translator was working directly with the Greek text. Cf. also χρύσοφρυς in V543b 3, VII 598a 10, 598a 21 and 599b 33 اخرسفيد and ἄχαρνος in VII 591b 1 اخرنوس, etc.

In short, we must conclude that the translation was done directly from the Greek text into Arabic without a Syriac intermediary. The translation was made at the beginning of the period of translation, but the identity of the translator is by no means certain. The analysis of individual translation techniques has not progressed far enough[21] to allow us to identify ours as the translation Ibn al-Biṭrīq, whether Yuḥannā or

[21] H. Daiber, *Aetius Arabus. Die Vorsokratiker in arabischer Überlieferung* (Wiesbaden, 1980), pp. 5–74 is an excellent example of analysis of the translation of the *Placita philosophorum* by Qosṭā ibn Lūqā. Further R. Arnzen, *Aristoteles'* De Anima. *Eine verlorene spätantike Paraphrase in arabischer und persischer Überlieferung* (Leiden, 1998), pp. 145–74 and G. Endress, *Die arabischen Übersetzungen von Aristoteles' Schrift* De Caelo. The continua-

Yaḥyā,[22] or Usṭāth.[23] M. Ullmann has brought together some points to identify the translator, but, as he stated, 'die Aristoteles-Versionen sind durch eine eigene Diktion geprägt, die nicht mit der Ausdrucksweise der Galen-Version SM übereinstimmt.'[24] The search for the translator of such a book so important for the early history of translations of scientific literature in the Arab world will be a difficult one.

tion of the *Greek & Arabic Lexicon*, edited by G. Endress and D. Gutas, might make it possible to discern the different translators.

[22] D.M. Dunlop, 'The Translations of al-Bitriq and Yahya (Yuhanna) b. al-Bitriq,' *JRAS* (1959), pp. 140–50. For a commentary on it, Endress, *Die arabischen Übersetzungen von Aristoteles' Schrift* De Caelo, p. 90.

[23] Endress, *Die arabischen Übersetzungen von Aristoteles' Schrift* De Caelo, pp. 113–15.

[24] Ullmann, *Wörterbuch zu den griechisch-arabischen Übersetzungen*, p. 58.

THE EARLIEST KNOWN SCHEMES OF ISLAMIC SACRED GEOGRAPHY

Mónica Herrera-Casais and Petra G. Schmidl[1]

فالكعبة قبلة لأهل المسجد والمسجد قبلة لأهل الحرم والحرم قبلة لأهل الآفاق

The Ka'ba is the qibla *for people in the Sacred Mosque. The Sacred Mosque is the* qibla *for people in the sacred area around it. The sacred area is the* qibla *for people in all regions of the world.*

Ibn Raḥīq (Yemen, eleventh century)[2]

1. Introduction

Following injunctions in the Qurʾān and much discussed in the *sunna*, the *qibla*, or sacred direction of Islam, is highly significant in daily Muslim life.[3] Not only the five daily ritual prayers have to be performed towards the Kaʿba in Mecca, but also other religious duties such as the recitation of the Qurʾān, announcing the call to prayer, ritual slaughter of animals, and the burial of the dead. Islamic sacred geography is the notion of the world being centred on the Kaʿba, and those who followed it proposed facing the *qibla* by means of simple

---

[1] This paper arose out of a seminar on Arabic scientific manuscripts at the Institute for the History of Science at Frankfurt University. We thank David King for his patience, encouragement and critical suggestions. We are also grateful to Doris Nicholson at the Bodleian Library in Oxford, and Hars Kurio at the Staatsbibliothek Preussischer Kulturbesitz in Berlin for sending free copies of manuscript folios. The authors are responsible for all inaccuracies and mistakes. [Abbreviations: KHU = Ibn Khurradādhbih; MUQ = al-Muqaddasī].

[2] In his treatise on folk astronomy studied by P.G. Schmidl, *Volkstümliche astronomische Abhandlungen aus dem mittelalterlichen arabisch-islamischen Kulturraum. Zur Bestimmung der Gebetszeiten und der Qibla bei al-Aṣbaḥī, Ibn Raḥīq und al-Fārisī*, doctoral thesis (University of Frankfurt, 2005, published under the title *Volkstümliche Astronomie im islamischen Mittelalter. Zur Bestimmung der Gebetszeiten und der Qibla bei al-Aṣbaḥī, Ibn Raḥīq und al-Fārisī* [Leiden, 2007]), B, 22a, 10f.

[3] For a general introduction to the *qibla*, see A.J. Wensinck, 'Ḳibla i. Ritual and Legal Aspects,' in *EI*[2].

folk astronomical methods, independent of mathematical geography.[4] Islamic sacred geography involves three main principles:

- The axes of the rectangular base of the Kaʿba are astronomically oriented while the four corners of the building point roughly to the four cardinal directions. The major axis is oriented towards the rising point of Canopus and the setting points of the Handle of the Plough; the minor axis is aligned to the rising point of the sun at the summer solstice and its setting point at the winter solstice.[5] In early Islamic folklore, the walls are also associated with the winds, their limits being defined by similar astronomical phenomena.[6]
- The world around the Kaʿba is divided into sectors, each one being associated with a segment of the perimeter of the building.[7]
- *Qibla* directions are intended to face the segment of the perimeter of the Kaʿba that is associated with one's locality, as if one where standing directly in front of that part of the building and facing at the same astronomical indicators[8] or *qibla* stars.[9]

[4] On determining the qibla in folk astronomy, see D.A. King, 'Makka 4. As the Centre of the World,' in *EI*²; Schmidl, *Volkstümliche Astronomie*, esp. pp. 99ff and 3f, n. 1 for the term 'folk astronomy'. Both folk and mathematical aspects are mentioned in D.A. King, *In Synchrony with the Heavens. Studies in Astronomical Timekeeping and Instrumentation in Medieval Islamic Civilization* (= *SATMI*), i *The Call of the Muezzin* (*Studies i–ix*) (Leiden, 2004), viia. Mathematical aspects of the *qibla* are covered by idem, 'Ḳibla ii. Astronomical Aspects,' in *EI*²; and idem, *World-Maps for Finding the Direction and Distance to Mecca. Innovation and Tradition in Islamic Science* (Leiden, 1999).

[5] On the astronomical orientation of the Kaʿba, see G.S. Hawkins and D.A. King, 'On the Orientation of the Kaʿba,' *Journal for the History of Astronomy* 13 (1982), p. 102ff [repr. in King, *Astronomy in the Service of Islam* (London 1993), xii]; and Schmidl, *Volkstümliche Astronomie*, M, 29b,14–30a,15.

[6] P.G. Schmidl, 'Zur Bestimmung der Qibla mittels der Winde,' in P. Eisenhardt, F. Linhard and K. Petanides (eds.), *Der Weg der Wahrheit. Aufsätze zur Einheit der Wissenschaftsgeschichte. Festgabe zum 60. Geburtstag von Walter G. Saltzer* (Hildesheim, 1999), p. 135ff; A. Heinen, *Islamic Cosmology. A Study of as-Suyūṭī's al-Hayʾa as-sanīya fī al-hayʾa al-sunnīya with Critical Edition, Translation, and Commentary* (Beirut, 1982), p. 157f; D.A. King, 'Maṭlaʿ,' in *EI*².

[7] King, 'Makka,' p. 180f; Schmidl, *Volkstümliche Astronomie*, O, 43a, 20f.

[8] An example of this general practice is described by al-Dimyāṭī (Egypt, *fl.* twelfth century): cf. D.A. King, 'Architecture and Astronomy. The Ventilators of Medieval Cairo and Their Secrets,' *JAOS* 104 (1984), p. 115 [revised in idem, *SATMI*, viib, p. 806f].

[9] The term '*qibla* star' is attested in the work of Ibn al-Qāṣṣ (Tabaristan, tenth century): Arabic text in D.A. King, *A Catalogue of the Scientific Manuscripts in the Egyptian National Library (in Arabic)*, ii (Cairo, 1986), p. 52 (4.4.1); see also J.-C. Ducène, 'Le Kitāb Dalāʾil al-qibla d'Ibn al-Qāṣṣ: Analyse des trois manuscrits et des emprunts d'Abū Ḥāmid al-Gharnāṭī,' *ZGAIW* 14 (2001), p. 169ff, esp. p. 182.

The beginning of the tradition is unknown, but it most probably originated in Mecca, by the time of the Prophet, with the naming of the four corners of the Kaʿba after the regions towards which they point and which the Arabs knew from their trading activities.[10] David King has identified about twenty schemes of Islamic sacred geography in thirty sources from the ninth to the nineteenth century, ranging from texts on folk and mathematical astronomy, geography, and sacred law, to encyclopaedias.[11] The oldest known schemes appear in the geographical works of Ibn Khurradādhbih (second half of ninth century) and al-Muqaddasī (second half of tenth century). The first is a textual scheme based on the concept of counter-*qibla* or direction from the perspective of the Kaʿba. The second is a graphical scheme of *qibla* directions towards the sacred building. Ibn Khurradādhbih's treatise is one of the sources for that of al-Muqaddasī, and their schemes share common features, but no definite connection can be established between them.

2. The Scheme of Ibn Khurradādhbih

Abū 'l-Qāsim ʿUbayd Allāh ibn Khurradādhbih (also spelled Khurdādhbih) is one of the earliest Arab geographers. He was born in Khurasan in the first quarter of the ninth century, worked in al-Jibāl (NW Iran) and the province of al-ʿIrāq, and died about 300/910.[12] Ibn Khurradādhbih wrote several works, among them one on folk astronomy

[10] This idea needs to be documented but has been proposed already by King, 'Makka,' p. 181. The names of three corners are mentioned in the major *ḥadīth* collections: the Yemeni corner (*al-rukn al-yamānī* in al-Bukhārī, *Ḥajj* 80 *et al.*); the Western corner (*al-rukn al-gharbī* in Aḥmad ibn Ḥanbal, *Musnad* i, 45/313 and *passim*); and the Eastern corner named as the Black Stone (*al-ḥajar*, *al-ḥajar al-aswad*, *al-rukn al-aswad* in al-Bukhārī, *Ḥajj* 56) or *rukn al-ḥajar* in Aḥmad ibn Ḥanbal, *Musnad* i, 246; the northern corner does not appear to be attested.

[11] D.A. King's unpublished survey *The Sacred Geography of Islam*; see the summaries in idem, 'Makka,' and idem, 'The Sacred Geography of Islam,' in T. Koetsier and L. Bergmans (eds.), *Mathematics and the Divine. A Historical Study* (Amsterdam, 2005), p. 161ff. Some examples are in K. Miller, *Mappae Arabicae. Arabische Welt- und Länderkarten des 9.–13. Jahrhunderts in arabischer Urschrift, lateinischer Transkription und Übertragung in neuzeitliche Kartenskizzen*, v (Stuttgart, 1931), p. 149ff [facsimile repr. in F. Sezgin (ed.), *Islamic Geography* 241 (Frankfurt, 1994)].

[12] M. Hadj-Sadok, 'Ibn Khurradādhbih,' in *EI*²; C. van Arendonk, 'Ibn Khordādhbeh,' in *EI*¹; *GAL* I, p. 258, and *S* I, p. 404; B.A. Rosenfeld and E. İhsanoğlu, *Mathematicians, Astronomers and Other Scholars of Islamic Civilisation and Their Works (7th–19th c.)* (Istanbul, 2003), p. 60 (no. 120); S. Maqbul Ahmad, 'Ibn Khurdādhbih,' in H. Selin (ed.), *Encyclopaedia of the History of Science, Technology, and Medicine in Non-Western Cultures* (Dordrecht, 1997), p. 423; A. Miquel, *La géographie humaine du monde musulman jusqu'au milieu du 11ᵉ siècle*, 4 vols. (Paris, 1967–88), i, p. 87ff (see also index).

(*Kitāb al-anwāʾ*), now lost.[13] But he is best known for his *Geography*, usually called *Kitāb al-masālik wa'l-mamālik* (*The Book of Itineraries and Kingdoms*) that was compiled about 232/846 and gradually increased by additions until about 272/885. Two surviving manuscripts of eastern provenance preserve different versions of the book:

- Vienna (Österreichische Nationalbibliothek): ms. 2403 (Mixt. 783),[14] datable before the mid-twelfth century.
- Oxford (Bodleian Library): ms. Uri 993 (Hunt. 433),[15] copied in 630/ 1232.

The determination of the *qibla* is mentioned in the extended title of the *Geography* that appears immediately after the pious introduction: *Hādhā kitāb fīhi ṣifat al-arḍ wa-binyat al-khalq ʿalayhā wa-qiblat ahl kull balad wa'l-mamālik wa'l-masālik ilā nawāḥī al-arḍ* ('This is a book containing the description of the earth, the distribution of its peoples, the *qibla* for the inhabitants of all localities and countries, and the itineraries to the furthest regions'). The title reads slightly awkwardly as if the reference to the *qibla* had been inserted in the wrong place. Nevertheless, we feel comfortable in attributing the scheme of sacred geography to Ibn Khurradādhbih rather than considering it to be a later interpolation.

The chapter with the scheme follows a short description of the features of the earth at the beginning of the book. The heading reads: *Qiblat ahl kull balad* (*The* qibla *for the inhabitants of all localities*) in the Vienna manuscript, and *Dhikr al-qibla fī 'l-nawāḥī* (*On the* qibla *in [all] the regions*) in the Oxford manuscript.

Two editions with French translations of the *Geography* have been published. The first by Barbier de Meynard (1865) is based on the Oxford manuscript, which was the only one available at the time, but it is a corrupted and abridged version of the work. It was later revised by De Goeje (1889) in light of the Vienna manuscript as the main source; it is the closest to the original though still incomplete.[16] The

---

[13] Mentioned in *GAS*, vii, p. 348.

[14] H. Loebenstein, *Katalog der arabischen Handschriften der Österreichischen Nationalbibliothek. Neuerwerbungen 1868–1968, Teil 1* (*Codices Mixti ab nr. 744*) (Vienna, 1970), p. 197.

[15] J. Uri, *Bibliothecae Bodleianae codicum manuscriptorum orientalium catalogus. Pars 1* (Oxford, 1787), p. 216.

[16] KHU (C. Barbier de Meynard), 'Le livre des routes et des provinces' (Arab. text and French trans.), JA, 6ème série, 5 (1865), pp. 5ff, 227ff, 446ff [repr. in *Islamic Geography* 29 (1992), p. 1ff]: the editor mentions another manuscript in Istanbul of which nothing is known (p. 7f); and KHU (M.J. De Goeje), *Liber viarum et regnorum* (Arab. text

scheme has been briefly described with a different interpretation in several works of David King.[17]

2.1 *Arabic Text and Translation of the Scheme*

The Vienna manuscript is the main source for the edition of the Arabic text which is located in folios 2v–3r. The scheme appears in the Oxford manuscript in folios 2v–3r, with several words omitted; variants are noted in the critical apparatus.[18]

[a] قبلة أهل كلّ بلد [a]

[S1] فقبلة [b] أهل أرمينية وآذربيجان وبغداد وواسط [c] والكوفة والمدائن والبصرة وحلوان والدينور ونهاوند وهمذان وإصبهان والريّ وطبرستان وخراسان كلّها وبلاد الخزر وقشمير الهند إلى حائط الكعبة الّذي فيه [d] بابها [e] وهي من القطب الشماليّ [e] عن يساره إلى وسط المشرق . [S2] وأمّا التبّت وبلاد الترك والصين والمنصورة فخلف وسط المشرق [بـ] أجزاء (؟) [f] لقرب [g] قبلتهم من الحجر الأسود . [S3] وأمّا قبلة أهل اليمن فصلاتهم إلى الركن اليمانيّ [h] ووجوههم إلى وجوه أهل أرمينية إذا صلّوا . [S4] [i] وأمّا قبلة [i] أهل المغرب وإفريقية ومصر والشام والجزيرة [j] فـ[من] وسط [j] المغرب [إلى القطب الشماليّ] وصلاتهم [من الركن الغربيّ (؟)] إلى الركن الشاميّ ووجوههم [k] إذا صلّوا [k] إلى وجوه أهل المنصورة إذا صلّوا . [l] فهذه قبل القوم والنحو الّذي يصلّون إليه . [l]

[a–a] ms. Oxford: ذكر القبلة في النواحي. [b] ms. Oxford: قبلة . [c] Omitted in ms. Oxford. [d] ms. Vienna: فيها crossed out, فيه above the line. [e–e] ms. Vienna: وهو من القطب; ms. Oxford: والقطب.

[f] ms. Vienna: بثمانية أجزاء (with the word بثمانية in the margin) and so edited by De Goeje; ms. Oxford: بستّة أجزاء . These values of eight and six degrees (of due east) respectively are copyists' mistakes.

[g] ms. Oxford: تُعرف. [h] Omitted in ms. Oxford. [i–i] ms. Oxford; MS Vienna: وقبلة

[j] ms. Oxford: ووسط . [k-k] Omitted in ms. Oxford. [l–l] This last sentence is missing in ms. Oxford.

---

and French trans.) (Leiden, 1889) [repr. in *Islamic Geography* 39 (1992)]: reference to the lost Istanbul manuscript in p. xiii, n. 3, and dates for the compilation in p. xx.

[17] King, *Sacred Geography*, 2.1, 3.1; King, 'Makka,' p. 183, fig. 2; idem and R. Lorch, 'Qibla Charts, Qibla Maps, and Related Instruments,' in J.B. Harley and D. Woodward (eds.), *The History of Cartography*, ii/1 *Cartography in the Traditional Islamic and South Asian Societies* (Chicago, 1992), p. 190f and fig. 9.1.

[18] Use of conventional signs: [ ] = missing in the Arabic text or correction; ( ) = inserted to assist the flow of the text. The sectors are here identified as S1–S4. See KHU (De Goeje), p. 3f (trans.), 5 (ed.); KHU (Barbier de Meynard), pp. 27f (ed.), 230f (trans.).

The qibla *for the inhabitants of all countries*

[Sector 1] The *qibla* for the inhabitants of Armenia, Azerbaijan, Baghdad, Wasit, Kufa, al-Madāʾin, Basra, Hulwan, Dinawar, Nihawand, Hamadhan, Isfahan, Rayy, Tabaristan, all of Khurasan, the land of the Khazars and Kashmir in India is to the wall of the Kaʿba where the door is located. It is between the North Pole to the left of (the wall) as far as due east.

[Sector 2] (The *qibla* for) Tibet, the land of the Turks, China and Mansura is a section (*bi-ajzāʾ*) (?) behind due east, because of the proximity of the *qibla* to the Black Stone.

[Sector 3] The *qibla* for the peoples in the Yemen: they pray towards the Yemeni corner, with their faces opposite to the faces of the people of Armenia when they (also) pray.

[Sector 4] The *qibla* for the inhabitants of the Maghrib, Ifrīqiya, Egypt, Syria and al-Jazīra is from due west [to the North Pole]. They pray [from the western corner(?)] to the Syrian corner, with the faces (of the Maghribis) opposite to the faces of the inhabitants of Mansura when they (also) pray.

These are the *qibla*s for (all) the peoples and the directions in which they have to pray.

2.2 *Commentary*

The scheme of Ibn Khurradādhbih is based on counter-*qibla* directions from the perspective of the Kaʿba looking out to other regions. The concept surely derives from the earliest geographical divisions of the world around the Kaʿba that were implied in the naming of the corners of the building. It is diametrically opposite to the *qibla* or local sacred direction towards the Kaʿba. The *qibla* is defined here as opposite directions for the places at cardinal points which are the Yemen–Armenia and the Maghrib–Mansura. The associated sections of the perimeter of the Kaʿba can be interpreted both as counter-*qibla* or *qibla* depending on the point of view of the observer, respectively, from or towards the Kaʿba.

Similar schemes constructed with counter-*qibla*s might have been in circulation even before Ibn Khurradādhbih's time, but none of them have apparently survived. Almost all later ones, from the following century onwards, concentrate on and fully integrate the qibla as a cartographical element, most probably for practical reasons, to avoid the conversion of counter-*qibla* into *qibla* directions. However, two late examples of counter-qibla schemes are to be found in the nautical atlases of the sixteenth-century Tunisian scholar ʿAlī al-Sharafī.[19]

In terms of astronomical orientation, Ibn Khurradādhbih defines the counter-*qibla* by the location of the North Pole (*al-quṭb al-shamālī*). The first sector gives rudimentary prescriptions for finding its position in Mecca, which is to the left when facing north east with the main wall of the Kaʿba behind. In sacred geography, both the North Pole and the Pole Star are used alternatively as *qibla* indicators, as attested in the treatises of the thirteenth-century Yemeni scholars al-Aṣbaḥī and al-Fārisī.[20] The astronomer al-Bīrūnī (Ghazna, *fl.* eleventh century) declares Polaris the nearest bright star to the pole, and hence to serve as substitute for it.[21] Most sources, such as al-Muqaddasī and especially another Egyptian legal scholar called al-Dimyāṭī (*fl.* twelfth century), give instructions for standing with the pole at different parts of the body (the back, shoulders or the head) so that one can be correctly oriented to the *qibla*. The method is rather approximate but very easy to use.

The regions of the world are organized in four sectors that basically irradiate from the Kaʿba towards the cardinal points. The sectors are associated with unequal segments of the perimeter of the building including the corners and two of the walls. The extant version of the scheme is corrupted in the second and fourth sectors. Each sector provides with:

- geographical data (regions and localities),
- associated section of the Kaʿba,
- counter-*qibla* and *qibla* directions.

---

[19] See the forthcoming thesis on the atlases by Mónica Herrera-Casais. For a reproduction of al-Sharafī's schemes, see King, *World-Maps for Finding the Direction and Distance to Mecca*, p. 55, fig. 2.3.4.

[20] Studied by Schmidl, *Volkstümliche Astronomie*, p. 388f; see also M. Rius, *La Alquibla en al-Andalus y al-Maghrib al-Aqṣà* (Barcelona, 2000), p. 249ff.

[21] R.R. Wright, *The Book of Instruction in the Elements of the Art of Astrology of al-Bīrūnī* (trans. and facsimile of *Kitāb al-Tafhīm li-awāʾil ṣināʿat al-tanjīm*) (London, 1934), p. 77 [repr. Baghdad, n.d.; and in F. Sezgin (ed.), *Islamic Mathematics and Astronomy* 29 (Frankfurt, 1998)]; reference in P. Kunitzsch, 'al-Ḳuṭb,' in *EI*², p. 543.

The *first sector* includes the region extending from the Caucasus, Mesopotamia and central Asia (Iran, the Caspian and Khurasan) as far as Kashmir. It is associated with the front of the Kaʿba, which is the north eastern wall where the door is located. The *qibla* varies from due south for Armenia as far as due west for Kashmir.

The *second sector* refers to the region of central Asia (the land of the Turks or Turkistan), Tibet, China and Mansura (in modern south Pakistan). It is associated with the Black Stone on the eastern corner and a small section of the adjacent wall. The expression *fa-khalf wasṭ al-mashriq* (lit. meaning *behind due east*) should here indicate turning to east north east that would be the appropriate counter-*qibla*. However, in a scheme of the Yemeni legal scholar Ibn Surāqa (*fl. c.* 1000), Tibet (which is mentioned twice as Tubbat and al-Tubbatān) appears to be associated with the same section of both the north eastern and the south eastern walls.[22] In the manuscripts of Ibn Khurradādhbih's text, the description of that section as a certain number of degrees from due east is possibly a corruption, as the *qibla* is normally defined by means of elements of the structure of the Kaʿba.[23] Those degrees might have originated by copyists' misreading of expressions such as *tayāmun*^{an} *qalīl*^{an} (turning a little to the right) that later sources attest for giving directions.[24] Our suggestion—*tayāmun*^{an} *tayāsur*^{an} (turning a little to the right or left)—would explain the wrong reading *bi-thamāniya* and even *bi-sittat ajzā'*. The associated section of the north eastern wall should correspond to the *multazam*, which is the space between the Black Stone and the door against which pilgrims press their breasts in the course of circumambulation. The *qibla* varies from due west for Mansura that is opposite to the *qibla* for the Maghrib (sector 4), and slightly south west for the other places (including continental China). Only the coasts of Southeast Asia (Indochina), if they are here con-

[22] The scheme is transmitted by Ibn Raḥīq. Tubbat is registered together with Basra, Ahwaz, Fars, Isfahan, Kirman, Sijistan and China; while al-Tubbatān is included with Kabul and Kandahar. The first goes to the north eastern wall, while the second to the south eastern wall: see Schmidl, *Volkstümliche Astronomie*, B, 23a–25b.

[23] Nevertheless, Ibn Surāqa and al-Fārisī describe sections of walls by measuring cubits from the corners: see Schmidl, *Volkstümliche Astronomie*, B, 24b,11–25a,5 and M, 30a,16–30b,7; and summary in King, 'Makka,' p. 184, tab. 1.

[24] King, *SATMI*, p. 803 (comm.), 823 (text).

sidered, are problematic as their *qibla* should be more to the west and even slightly north west.[25]

The *third sector* applies exclusively to the Yemen that is associated with the southern corner of the Kaʿba, also called the Yemeni corner. The *qibla* is defined as the opposite direction to the *qibla* for the people of Armenia (sector 1), and therefore it is due north.

The *fourth sector* stretches from the Maghrib as far as Syria and Upper Mesopotamia (al-Jazīra), and is most probably associated with the north western wall of the Kaʿba. The Maghribis should be the ones facing the western corner that is opposite to the eastern corner for Mansura (sector 2). The others should face the so called Syrian corner that many schemes, such as that of al-Muqaddasī, identify with the northern corner.[26] The *qibla* will then vary from due east for the Maghrib to due south for Syria.

1.1 *Summary of Data and Reconstruction*

Two striking features become obvious in the reconstruction of the scheme. The first and second sectors overlap in the north eastern quadrant, while the southern quadrants remain empty, except for the Yemen. No places are considered for the Indian Ocean area to the south east and for Ethiopia and beyond to the interior of Africa to the south west.

The overlapping sectors, the omission of southern regions and, especially, no associated sector for the south eastern wall of the Kaʿba, reappear in the scheme attributed to al-Muqaddasī.

[25] This is the mathematical *qibla* given by C. Schoy, 'Die Mekka- und Qiblakarte,' *Kartographische und schulgeographische Zeitschrift* 6 (1917), p. 184f (map) [repr. F. Sezgin et al. (eds.), *Beiträge zur arabisch-islamischen Mathematik und Astronomie*, 2 vols. (Frankfurt, 1988), i, p. 157ff; also in *Islamic Geography* 18 (Frankfurt, 1992), p. 106ff]. Malaysia and the islands around still belong to India, according to KHU (De Goeje), p. 65ff; the borders of China are described as follows (p. 69): 'Ce pays est borné par la mer, le Tibet, le pays des Turcs, et, à l'occident, l'Inde. A l'orient de la Chine sont les pays d'al-Wākwāk (le Japon)….'

[26] Apparently, only a few sources apply 'Syrian corner' to the western corner, see survey in Schmidl, *Volkstümliche Astronomie*, p. 369ff.

Tab. 1

	Segment of the Kaʿba	Counter-*qibla*	Regions	*Qibla*	Description of the *qibla*
1	main wall (from N corner to E corner)	N → E	from Armenia to Kashmir	S → W S (Armenia) W (Kashmir)	opposite the Yemen for people in Armenia
2	Black Stone and section of [NE] wall [up to the door]	E → [ENE]	Turkestan, China, Tibet and Mansura	W → [WSW] W (Mansura)	opposite [the Maghrib] for Mansura
3	Yemeni corner (= S corner)	S	Yemen	N	opposite the people in Armenia
4	[NW wall (from W corner to Syrian corner)]	W → [N]	from the Maghrib to Syria and Upper Mesopotamia	E → [S] [E (Maghrib)] [S (Syria)]	opposite Mansura [for the Maghribis]

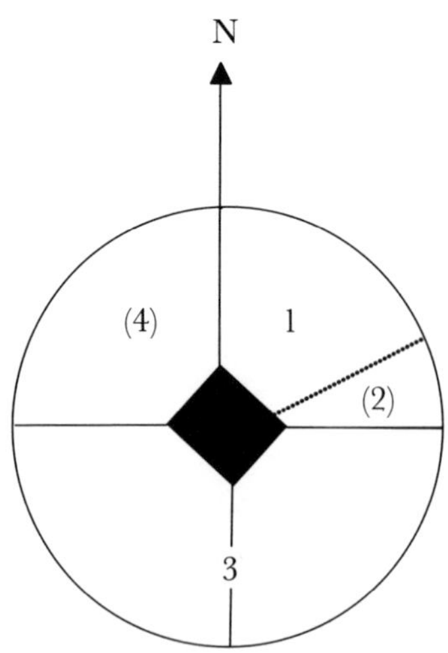

Fig. 1. [The associated segment of the Kaʿba is uncertain and subject to interpretation for the sectors in brackets]

3. The Scheme Associated with al-Muqaddasī

Shams al-Dīn al-Muqaddasī (also called al-Maqdisī) was born in Jerusalem about 334/945 and lived at least until about 380/990.[27] He belongs to the Balkhī School of geography that produced the so-called *Atlas of Islam*,[28] though he is the only representative who features sacred geography. His *Aḥsan al-taqāsīm fī maʿrifat al-aqālīm* (*The Best Divisions for the Knowledge of the Regions*) was compiled about 375/985, and is extant in two versions, though several others with summarized chapters might have also been in circulation. The scheme is only registered in the second or Berlin version of the book, of which two manuscripts survive:[29]

- Berlin (Staatsbibliothek Preussischer Kulturbesitz): ms. Ahlwardt 6034 (Sprenger 5),[30] datable *c.* 1500, and copied by a distinctive eastern *naskhī* hand.
- Berlin (Staatsbibliothek Preussischer Kulturbesitz): ms. Ahlwardt 6033 (Sprenger 6),[31] a modern copy of Sprenger 5, datable *c.* mid-nineteenth century.

In addition to the textual problem of the form of the *Geography*, the attribution of the scheme to al-Muqaddasī should be taken with caution. The scheme is included in the chapter on the climates that is missing all together from the first (or Istanbul) version of the book.

---

[27] Miquel, 'al-Muḳaddasī,' in *EI*²; J.H. Kramers, 'al-Muḳaddasī,' in *EI*¹; *GAL* I, p. 264, and *S* I, p. 410f; Rosenfeld and İhsanoğlu, p. 87 (no. 215); B. Rosenfeld, 'A Supplement to *Mathematicians, Astronomers and other Scholars of Islamic Civilisation and their Works (7th–19th c.)*,' *Suhayl* 4 (2004), p. 92; R. Mushtaqur, 'al-Maqdisī,' in Selin (ed.), *Encyclopaedia of the History of Science, Technology, and Medicine in Non-Western Cultures*, p. 753; Miquel, *La géographie humaine*, i, esp. p. 313ff (see also index). For the name, see A. Fischer, 'Al-Maqdisī and al-Muqaddasī,' *ZDMG* 60 (1906), p. 404ff [repr. in *Islamic Geography* 32 (1992), p. 293ff].

[28] See G.R. Tibbetts, 'The Balkhī School of Geographers,' in Harley and Woodward, *History of Cartography*, esp. pp. 110, 111, 122f.

[29] MUQ (A. Miquel), *La meilleure répartition pour la connaissance des provinces* (part. French trans.) (Damascus, 1963), p. xxiv ff (introd.). The first version of the *Geography* is preserved in Istanbul (Süleymaniye Kütüphanesi): ms. Ayasofya 2971bis, and dated in 658/1260. A late copy of it (dated 1256/1840) is preserved in Leiden (Universiteitsbibliotheek): Cod. Or. 2063 (ms. Leiden = ms. C in De Goeje's edition). These manuscripts have not been consulted.

[30] W. Ahlwardt, *Die Handschriften-Verzeichnisse der Königlichen Bibliothek zu Berlin. 17. Band: Verzeichniss der arabischen Handschriften*, v (Berlin, 1893), p. 363. The manuscript was brought from India by the German Orientalist Aloys Sprenger.

[31] Ahlwardt, *Handschriften-Verzeichnisse*, p. 362f.

The reference to the *qibla* in the title of the chapter could be both, original or a later addition: *Dhikr al-aqālīm al-ʿālam wa-markaz al-qibla* (*On the Climates of the World and the Direction of the* Qibla). The expression *markaz al-qibla* is used here in the sense of focus, and it is not attested in the sources that usually render *dhikr* or *maʿrifat* (*jihāt* or *ḥudūd*) *al-qibla*, or *dāʾirat al-qibla*. The author's general concern on the *qibla* is exposed at the beginning of the chapter, but there is no allusion to the scheme that appears abruptly at the end. According to internal evidence, the latter has been constructed at different stages, perhaps because of an addition or copyists' negligence.

The Arabic text of al-Muqaddasī's *Geography* has been edited and established by De Goeje (Leiden, 1877; 2nd rev. ed. 1906) using for the second edition Sprenger 5 and Sprenger 6, as well as the Leiden manuscript with a late copy of the Istanbul version.[32] He omitted not only the scheme of sacred geography but also all the maps. The existence of the scheme is simply indicated in a note that reads: *Hic sequitur in B* (= Sprenger 5) (*media pag. 34*) *mappa qua diversarum regionum kibla exponitur*.[33] Traditionally, scholars have dedicated little attention to such schemes, cartographical devices and illustrations in Islamic scientific manuscripts, which have often been omitted from the eventual publication.[34] A good example is an index diagram of the regions of the world that also slipped from Reinaud's edition and translation (1848) of the *Geography* of Abū 'l-Fidāʾ (672–732/1273–1331). The diagram was eventually on display at the Bibliotheca Palatina Exhibition of manuscripts in Heidelberg in 1986.[35]

---

[32] MUQ (M.J. De Goeje), *Descriptio imperii moslemici* (Arabic text) (Leiden, 1906), p. v (introd.) [repr. in *Islamic Geography* 36 (1992)]. Concerning the manuscripts used by De Goeje, see MUQ (B. Collins), *The Best Divisions for the Knowledge of the Regions* (English trans.) (Reading, 1994; repr. 2001), p. xv (preface); a different opinion is in the introduction to reprinted edition of MUQ (G.S.A. Ranking and R.F. Azoo), *Aḥsanu-t-taqāsīm fī maʿrifati-l-aqālīm* (part. English trans.) (Calcutta, 1897–1910) [repr. in F. Sezgin (ed.), *Geography* 8 (Frankfurt, 1989), p. vi].

[33] MUQ (De Goeje), p. 61, note m.

[34] See D.A. King, 'Some Illustrations in Islamic Scientific Manuscripts and their Secrets,' in G.N. Atiyeh (ed.), *The Book in the Islamic World. The Written Word and Communication in the Middle East* (New York, 1995), esp. pp. 162, 164, and figs. 9.10, 9.12–16, 9.20.

[35] This is mentioned by King, 'Some Illustrations in Islamic Scientific Manuscripts and their Secrets,' p. 174, fig. 9.23. The place-names were transposed instead to the general index, see footnotes in J.-T. Reinaud and M.G. de Slane, *Géographie d'Aboulféda* (Arabic text of *Taqwīm al-buldān*) (Paris, 1840), p. 75; and J.-T. Reinaud, *Géographie d'Aboulféda* (French trans.) (Paris, 1848), 1, p. 98 [repr. in *Islamic Geography* 13 (1992) and 278 (1998)]. See also E. Mittler et al. (eds.), *Bibliotheca Palatina. Katalog zur Ausstellung*

Konrad Miller's *Mappae Arabicae* (1931) provide with a first transcription and translation of the scheme of al-Muqaddasī.[36] Of the several partial translations of the *Aḥsan al-taqāsīm*, all based in De Goeje's edition, the chapter on the climates is rendered by Ranking and Azoo (1897) and Miquel (1963), who do not refer at all to the scheme.[37] Collin's modern and complete, but nontechnical translation (1994), gives an English version with reproduction.[38] A preliminary study is to be found in King's unpublished book on *Sacred Geography*.[39]

3.1 *Text and Translation*

We have edited al-Muqaddasī's remarks on the *qibla* using Sprenger 5 (p. 31). The scheme appears only in this manuscript (p. 34), and both the chapter on the climates and the sacred geography are written by the same hand. The scheme was deliberately left out from the modern copy (= Sprenger 6), in which place the following note was added (p. 60): *Wafī hādhā 'l-maḥall ṣūrat Makka wa-dā'irat al-bilād ḥawlahā* (In this place there is a picture of Mecca and a circular scheme of the localities around).[40] The sectors have been numbered starting from the north point in the picture of the Kaʿba.

[*Fragment on the* qibla]

وعلى هذا صنّف من ذكرنا كتبهم في هذا الباب ونحن ننقل [a] منها وعمّن لقينا من كبراء المنجّمين هذا الباب لأنّه علم يحتاج إليه في سمت القبلة ومعرفة مواضع الأقاليم منها . فإنّي رأيت خلقا قد اختلفوا في القبلة وحوّلها وتماروا فيها ولو عرفوا الوجه في ذلك ما اختلفوا فيها ولا غيّروا ما وضعه الأوائل .

[a] Difficult to read in Sprenger 5; De Goeje: ننقل ; Sprenger 6: فنقول .

Those (authors) that we mention in this chapter have written their books according to these (principles). In this chapter we quote from

vom 8. Juli bis 2. November 1986 Heiliggeistkirche Heidelberg. Text- und Bildband (Heidelberg, 1986): general description of the manuscript without photos on p. 95.

[36] Miller, *Mappae Arabicae*, v, p. 149ff (Kibla-Karten 1).

[37] MUQ (Miquel), p. 125ff; and MUQ (Ranking and Azoo), fasc. 1, p. 98ff.

[38] MUQ (Collins), pp. 54 and 407 (map III).

[39] King, *Sacred Geography*, 2.2, 3.3.

[40] Between round parentheses and possibly in the hand of Sprenger, as he ordered the copy of the manuscript and added several marginal notes.

them, and from the eminent astronomers that we have encountered, because it is knowledge required for (determining) the direction of the *qibla* (*samt al-qibla*), and the relative location of the climates.[41] I have seen people with differing (opinions) on the *qibla*, (who) changed it (from one direction to another), and argued vehemently about it. If they had known the (correct) procedures they would not have differed on that, nor changed what the predecessors had established.

[*The scheme of sacred geography*]

[S1] هذه قبلة الجار إلى وليلة [a] وطنجة : ما بين النسر الطائر والواقع فإذا طلعا فاستقبل فيما بينهما الكعبة يصحّ لك .
[S2] هذه قبلة السند وغز نين : ما بين الجديّ والعيّوق فإذا [b] طلع العيّوق [b] فاستقبل بينهما .
[S3] هذه قبلة العراق إلى سمرقند وفارس وكرمان : بين المقام إلى الباب فإذا طلعت الثريّا فاستدبرها .
[S4] هذه قبلة [...] [c] ثلاثة آلاف فر سخ ونواحيها : تجعل بنات نعش إذا طلعت على عينك اليمنى والجديّ على اليسرى تكون القبلة أمامك .
[S5] هذا حدّ بلد الروم وهو أربعة آلاف فرسخ .
[S6] هذه قبلة النبيّ صلّى الله عليه وسلّم : إذا طلع سهيل فاجعله على عينك اليمنى و النسر الطائر على عينك اليسرى بين الركن الشاميّ والميزاب .
[Sector 7 (empty)]
[S8] هذه قبلة الأردنّ [أ]رديل وميّافارقين و برذعة و أر مينية : بين الر كن الشاميّ إلى مقام إبراهيم [d] فإذا طلع العيّوق فاجعله على نقرة قفاك اليمنى وتستقبل القبلة .

[a] ms.: ويلة and always written with spelling mistakes in the *Geography* (cf. MUQ [De Goeje], *Descriptio*, 1906, p. 57 n.: "B ليله ", and p. 220 n.: "B hic et supra ليله . C ليه ").
[b–b] ms.: بلغ الوقت . [c] The place names are missing. [d] ms.: إبرهيم .

[Sector 1] This is the *qibla* for al-Jār[42] as far as Walīla (Volubilis)[43] and Tangier: (The *qibla* stars are) Altair and Vega. Face the Kaʿba in between them when they both rise. (This is) correct.

[41] In this context, the climates refer to the regions or sectors of the world.
[42] Al-Jār was the port of Medina on the Red Sea until almost the end of the Middle Ages: A. Dietrich, 'al-Djār,' in *EI*[2]. See MUQ (De Goeje), pp. 12, 53, 69, 83, 97, 107, 110.
[43] Ancient Volubilis, nowadays a Moroccan archaeological site. It is uncertain whether it served as capital for the Roman province of Mauritania Tingitana. It became

[Sector 2] This is the *qibla* for al-Sind and Ghaznīn:[44] (The *qibla* stars are) the Pole Star and Capella. Turn your face in between them when [Capella rises].

[Sector 3] This is the *qibla* for al-ʿIrāq as far as Samarqand, Fārs and Kirmān: (it is associated with the section) between the Station of Abraham and the door (of the Kaʿba). Turn your back to the Pleiades when they rise.

[Sector 4] This is the *qibla* for [...], which (covers) three thousand *farsakh*s, and its neighbouring territories:[45] stand so that the (stars of the) Handle of the Plough rise at your right eye, and the Pole Star is at your left (eye). The *qibla* will be in front of you.

[Sector 5] This is the sector for the land of the Byzantines, which (extends) four thousand *farsakh*s.

[Sector 6] This is the *qibla* of the Prophet—may God bless Him and grant Him salvation—: Stand so that Canopus rises at your right eye, and Altair (rises) at your left eye. (The *qibla* is associated with the section) between the Syrian Corner and the waterspout.

[Sector 7 (empty)]

[Sector 8] This is the *qibla* for Jordan, [A]rdabīl, Mayyāfāriqīn, Bardhaʿa and Armenia: [it is associated with the section] between the Syrian Corner and the Station of Abraham. [Stand] so that Capella rises to the right of your neck furrow, and you will be facing the *qibla*.

---

the centre of the Idrīsid kingdom by the end of the eighth century, before the founding of the new capital in Fez. The place was already in ruins in the twelfth century, and contemporary sources mention it under the Berber name of Tizra (meaning *stones*): M. El Mansour, 'Walīlī,' in *EI*². See MUQ (De Goeje), pp. 57 and 220.

44 C.E. Bosworth, 'Ghazna,' in *EI*²: Ghaznīn is the spelling to be found in MUQ and the anonymous *Ḥudūd al-ʿālam* (end of tenth century) for Ghazna. Yāqūt (Syria, 1179–1229) considers it the correct, learned form: see *Jacut's geographisches Wörterbuch*, ed. F. Wüstenfeld, 6 vols. (Leipzig, 1866–70), iii, p. 798. See MUQ (De Goeje) 1906, pp. 50, 295ff, 303ff, 349, 482, 486. According to G. Cornu, *Atlas du monde arabo-islamique a l'époque classique (IXᵉ–Xᵉ siècles)* (Leiden, 1985), p. 148: 'Circonscription (*kūra*) du H̱urāsān et sa capitale (*qaṣaba*) pour Muqaddasī, cité du Siğistān pour les autres géographes'.

45 *Nāḥiya* (pl. *nawāḥin*) is a territory belonging to a province (*iqlīm*) but with certain original features of physical or human geography or economy: MUQ (Miquel), p. 23 (§19), n. 51.

3.2 *The Illustration of the Kaʿba*

At the centre of the scheme, there is a picture of the Kaʿba with the label of 'The Holy House of God' (*bayt Allāh al-ḥarām*). It is inscribed in concentric circles that are subdivided into the eight sectors. Their limits irradiate from the four corners and the middle-points of the four walls of the building. The orientation of the Kaʿba does not fit with the arrangement of the sectors around. This reveals that illustration and sectors were constructed independently, without considering the basic association of the latter with the perimeter of the Kaʿba (see tab. 3).

The elements of the structure of the Kaʿba and the sacred precinct are labelled correctly in relation to one another. Only the door that should be located at the north eastern wall appears to be misplaced at the south eastern wall. The sides of the rectangular base of the building are drawn at an approximate ratio of 8:7.

1.2 *Commentary*

The literal reading of the scheme is presented in the graphic. Critical interpretation leads to a differing and more complete reconstruction that is given below. It appears that the limits for the sectors result naturally from elements of the perimeter of the Kaʿba (meaning the Kaʿba in Mecca and not the illustration). The Syrian or northern corner (S6 and S8), the door (S3), the Station of Abraham (S3 and S8) and the waterspout (S6) are used as markers in the text.

Table 2

Names of the corners of the Kaʿba

الحجر	the (Black) Stone	= Eastern corner
اليماني	Yemeni (corner)	= Southern corner
الغربي	Western (corner)	= Western corner
الشامي	Syrian (corner)	= Northern corner

Elements of the Kaʿba and the sacred precinct

الباب	the door	(next to E corner, should be in NE wall)
ميزاب	waterspout (*mīzāb*)	(in NW wall)
الحطيم	*al-ḥaṭīm*	(in front of NW wall)
مقام إبرهيم	Station of Abraham (*Maqām Ibrāhīm*)	(in front of NE wall)
زمزم	*Zamzam*	(in front of NE wall)

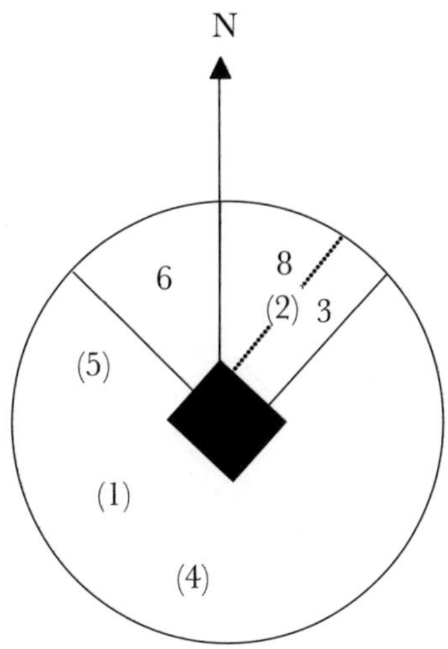

Fig. 2. [The associated segment of the Kaʿba is uncertain and subject to interpretation for the sectors in brackets]

The sectors provide with the following data:

- geographical data (place names and distance measurements),
- associated section of the Kaʿba,
- *qibla*-stars,
- and body orientation.

Most corrections apply to the position of the *qibla*-stars (such as rising or setting), and prescriptions for placing them on parts of the body (facing or turning the back to the *qibla*-stars, etc.). Such instructions are often confused or even omitted in the sources, because of corruption in the transmission of the schemes or perhaps confusion with counter-*qibla* descriptions in earlier ones. The latter might be the case in the first and second sectors.

The *first sector* stretches from the Red Sea (al-Jār) as far as the western limit of the Maghrib (Walīla [Volubilis] and Tangier). The east north east direction indicated by facing the rising of the *qibla*-stars, Altair and Vega, is problematic. One would stand in the more appropriate east south east *qibla* by turning the *back* to the *setting* of the *qibla*-stars (fig. 3). The region should be associated with the north western wall of the Kaʿba.

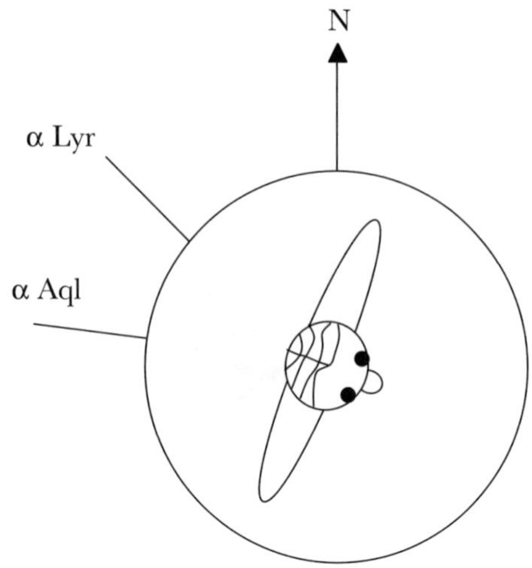

Fig. 3. *qibla* for sector 1

The *second sector* applies to medieval Pakistan (west of the Indus) and Ghazna (in modern east Afghanistan). The *qibla* here should be (west) south west, but neither facing the rising nor the setting of Capella indicates this direction. On the contrary, turning the back in between the Pole Star and the rising of Capella will indicate such *qibla*. The region should be associated with the eastern part of the Kaʿba.

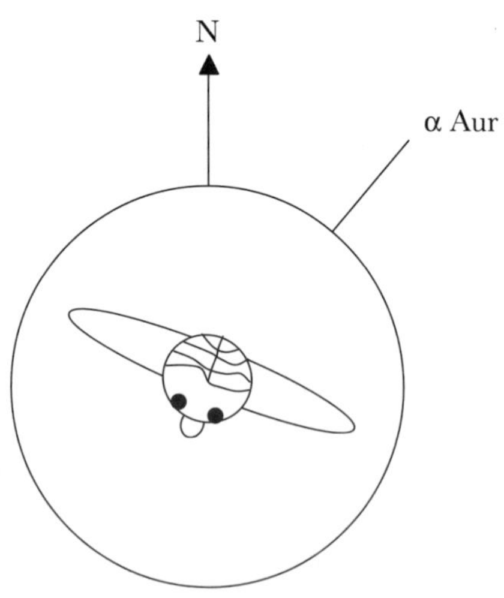

Fig. 4. *qibla* for sector 2

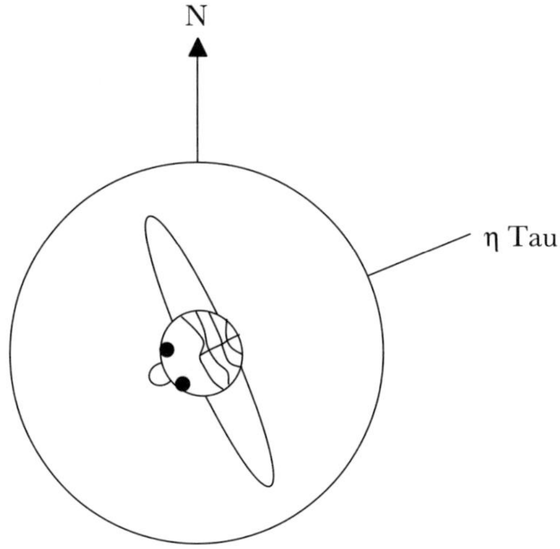

Fig. 5. *qibla* for sector 3

The *third sector* includes medieval Iraq and part of central Asia (from Fārs and Kirmān as far as Samarqand in the Transoxania). The astronomical instructions (turning the back to the rising of the Pleiades) give a roughly correct south west direction for the *qibla*. The scheme associates the region with the middle part of the north eastern wall of the Kaʿba, between the Station of Abraham and the door.

Place names are missing in the *fourth sector* that refers to a region of 3000 *farsakh*s. If a surface measurement is intended, a second value, either the width or length, is missing. The astronomical instructions (rising of the Handle of the Plough at the right eye, while the Pole Star is at the left eye) give a north north east direction for the *qibla*. This would be appropriate for the east and interior of Africa, and the sector should be associated with a section of the south western wall of the Kaʿba.

According to the *Murūj al-dhahab* of al-Masʿūdī (Baghdad and Cairo, tenth century), areas in square *farsakh*s were compiled by the Iraqi astronomer al-Fazārī already in the eighth century.[46] *The administration of Rūmiya of the Romans* (Rome) is the only place that al-Fazārī lists

[46] D. Pingree, 'The Fragments of the Works of al-Fazārī,' *JNES* 29 (1970), p. 116f.

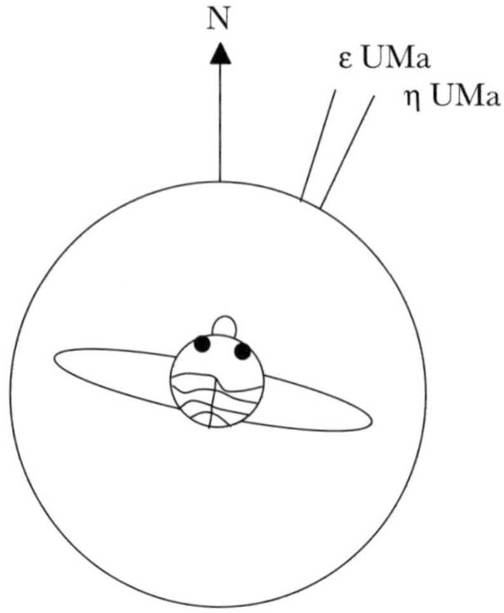

Fig. 6. *qibla* for sector 4

covering 3000 by 700 farsakhs,[47] but it should have a south east *qibla*, and would be very unusual for an early scheme of sacred geography. Nubia in Africa covers only 1500 by 500 *farsakhs* in al-Fazārī's list.

The *fifth sector* applies to Byzantium (the land of al-Rūm) extending 4000 *farsakhs*. Accordingly, the sector should be associated with a section of the north western wall of the Kaʿba. Specifications for the associated horizon phenomena are missing. The data differs from al-Fazārī's values both for the administration of the Romans in Constantinople (5000 by 420 *farsakhs*), and for the western Romans (3000 by 700 *farsakhs*).[48] On the other hand, al-Muqaddasī's chapter on the climates ends with a statement on the circumference of the earth, where a value of 8000 *farsakhs* for al-Rūm is attributed to a certain Abū 'l-Jald.[49] Similar measurements for the climates and some of the regions of the world that

[47] Ibid., p. 116.

[48] Ibid., p. 116.

[49] MUQ (De Goeje), p. 62; MUQ (Miquel), p. 136. De Goeje has pointed out that Abū 'l-Jald is the *kunya* of the traditionist Jīlān ibn Farwa (see n. a). The statement might have been taken from Ibn Faqīh (M.J. de Goeje, *Compendium Libri Kitâb al-Boldân* [Leiden, 1885], p. 4 [repr. in *Islamic Geography* 38 (1992)]) and appears with some alterations in Yāqūt who attributes it to a certain ʿUmar ibn Jīlān and elsewhere to the succesor Qatāda ibn Diʿāma (Basra, about 680 to about 735) (Yāqūt [ed. Wüstenfeld], i, pp. 16 and 19). The first might be the same as Abū 'l-Jald (cf. W. Jwaideh, *The Introductory Chapters of Yāqūt's Muʿjam al-Buldān* [Leiden, 1959], p. 24, n. 4). Both traditions might have a common origin.

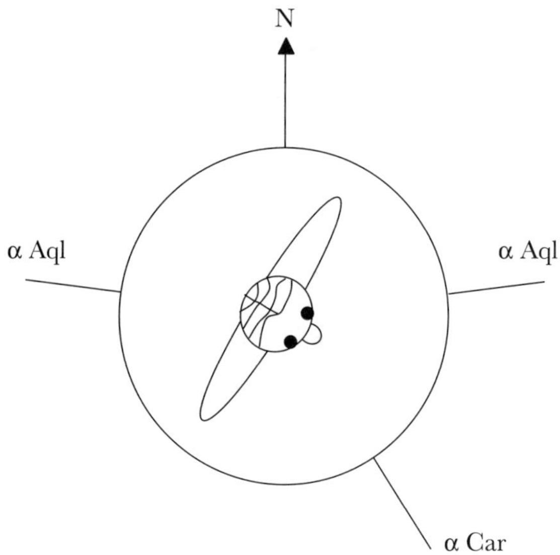

Fig. 7. *qibla* for sector 6

are included in the chapter (before and after the scheme) are unrelated to the sacred geography.

There are no place names for the *sixth sector*. With the data at hand, it could apply again to the Maghrib, as well as al-Andalus, which is missing from the scheme. The instructions place the rising of Canopus at the right eye and of Altair at the left eye. Though not explicit in the text, the setting of Altair at the left eye would be physically impossible for bodily orientation. The *qibla*-stars indicate a southeast direction that is suitable for the associated segment of the Kaʿba. It is the section of the north western wall between the Syrian corner and the waterspout that is named here as the *qibla* of the Prophet. Similarly, Ibn Raḥīq describes the qibla of the Prophet opposite to the waterspout.[50]

The *seventh sector* is empty, and perhaps it is simply drawn for graphical symmetry. As no other sector is associated with the south eastern wall of the Kaʿba, it should be considered that the Indian Ocean has been excluded, as in the scheme of Ibn Khurradādhbih. But it is problematic that the eastern regions of the Arabian Peninsula, which should be

[50] Schmidl, *Volkstümliche astronomische Abhandlungen*, B, 22a, 9; U. Rubin, 'The Kaʿba: Aspects of its Ritual Functions and Position in Pre-Islamic and Early Islamic Times,' *JSAI* 8 (1986), p. 104.

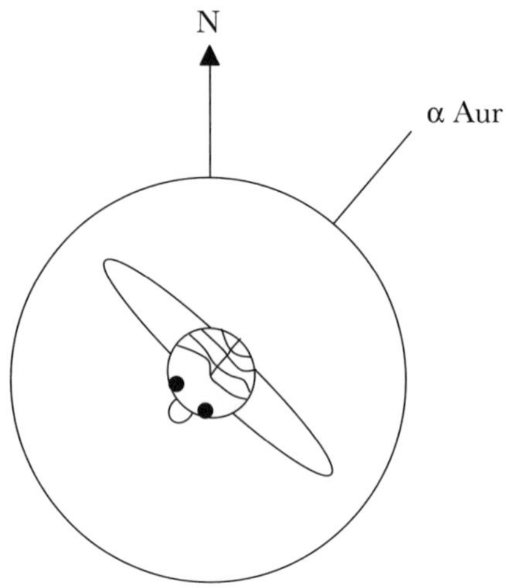

Fig. 8. *qibla* for sector 8

associated with that wall, have also been left out. The textual information might have also been lost in the process of copying.

The *eighth sector* includes the Levant up to Armenia and the Caspian (Jordan, Ardabīl, Mayyāfāriqīn, Bardhaʿa and Armenia). Turning one's back to the rising of Capella, as stated in the text, indicates a roughly south western *qibla*. But the *qibla* should vary from south south west to south south east for the places involved. The sector is associated with the section of the north eastern wall of the Kaʿba between the Syrian corner and the Station of Abraham, overlapping part of the third sector.

If one turns one's back instead to the setting of Capella, one gets a south east direction which is also appropriate for a part of the region. But then, the associated sector of the Kaʿba would shift to the north western wall (perhaps between the corner and the waterspout).

3.3 *Summary of Data and Reconstruction*

Our analysis reveals that the sectors are in disorder. The majority concentrates and even overlaps (S3 and S8) in the northern quadrants, which were better known at the time and cover most of the Islamic world. None appears to be associated with the south eastern wall. The location of sectors in the south western wall is uncertain. The elements

Table 3

	Section of the Kaʿba (picture)	Section of the Kaʿba (text)	Place names	*Qibla* by the stars	Distances
1	N NW	[W NW]	al-Jār, Walīla and Tangier	ENE [should be: ESE]	—
2	N NE	[NE E]	al-Sind and Ghazna	ENE [should be WSW]	—
3	NE E	NE (between Station of Abraham and the door)	al-ʿIrāq, Samarqand, Fārs and Kirmān	SW	—
4	E SE	[S SW]	—	NNE	3000 *farsakhs*
5	SE S	[NNW?]	Byzantium (al-Rūm)	—	4000 *farsakhs*
6	S SW	NW W (between Syrian corner and waterspout)	—	[ESE?]	—
7	SW W	[E S(?)]	—	—	—
8	W NW	N NE (between Syrian corner and Station of Abraham)	Jordan, Ardabīl, Mayyāfāriqīn, Bardhaʿa and Armenia	SW [could be: SE]	—

of the perimeter of the Kaʿba, such as the waterspout or the door, that also gather in the north western and specially the north eastern wall, favour that subdivision into sectors.[51] Apart from the corners and the blocked door, no other markers can be established for the southern walls.

3.4 *Final Remarks*

The scheme associated with al-Muqaddasī could have been planned either with seven or eight sectors. In geography, seven is an appealing number that matches with the seven climate division of the earth from

[51] For other examples of subdivisions, see King, 'Makka,' fig. 4, 5, 6, and 7; Schmidl, *Volkstümliche Astronomie*, M, 29b,14–33a,1, M, 36b, M, 37a.

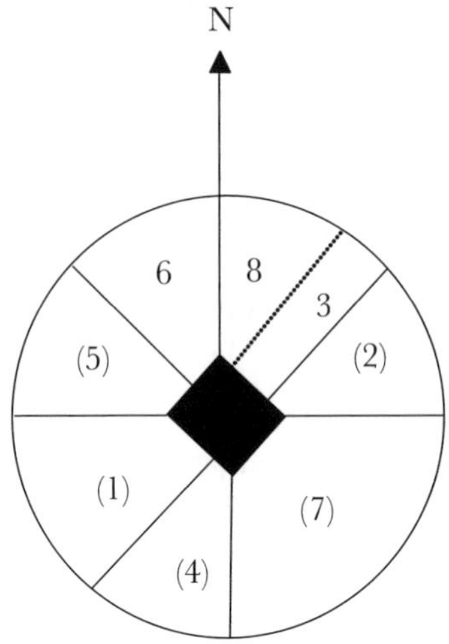

Fig. 9.

Greek Antiquity, the Persian system of the seven *keshwars* (kingdoms of the world), and the seven heavens and seven earths of Islamic cosmology.[52] As the original scheme might be different from the extant graphical representation, the eighth sector might be a later addition. But in case the instructions were lost in the transmission process, the sector would have possibly been associated with the south eastern wall of the Kaʿba, so that the picture of the inhabited world would then be completed.

Significant regions are missing from al-Muqaddasī's scheme, such as Egypt, al-Andalus and the Yemen. The latter is normally associated with the southern corner of the Kaʿba, and could be here integrated in the fourth sector. Recurrent place names in sacred geography, such as, Jerusalem which, in addition, is al-Muqaddasī's hometown, Damascus, Baghdad or Kabul, have also been left out. On the other hand, unusual localities are listed, such as al-Jār or Walīla that, however, are well documented in descriptive geography, and both in the treatises of Ibn Khurradādhbih and al-Muqaddasī. Considering the corrupted state of the only available copy of the scheme, it is not possible to determine whether or not those regions and localities were ever included, or were simply dropped by copyists.

[52] A. Miquel, 'Iḳlīm,' in *EI*²; A. Heinen, 'Samāʾ,' in *EI*².

Similar confusion reigns with the remaining *farsakh* distances in the fourth and fifth sectors that perhaps were originally mentioned for each sector, unless the sectors derive from different sources.

4. Conclusion

Ibn Khurradādhbih and al-Muqaddasī provide the earliest known schemes of Islamic sacred geography, and both hint at the first stages of development of this tradition. The schemes are constructed from different perspectives that also involve differing conceptions of the *qibla*: i.e. counter-*qibla* from the Kaʿba or *qibla* towards the Kaʿba. The counter-*qibla* structure in Ibn Khurradādhbih reveals an early and primitive cartographical approach to the schemes that perhaps were initially conceived as simple representations of a Kaʿba-centred world. The *qibla* could have developed as a cartographical element at a second stage, and mainly for practical reasons, to avoid the conversion of counter-*qibla* into *qibla* directions.

The interpretation of both schemes is problematic, and requires significant corrections which are sometimes uncertain. Most of the later schemes are not so corrupted, but in the case of al-Muqaddasī, repeated copying has made the instructions not only contradictory but useless. Both schemes seem to share significant features which are the overlapping sectors in the northern sections of the Kaʿba, and the south eastern wall of the building with no associated region.

The schemes of Ibn Khurradādhbih and al-Muqaddasī attest the gradual subdivision process of the sacred space of Islam from the most natural four-sector segmentation into a multiplicity of sectors. The schemes of the Yemeni legal scholar Ibn Surāqa, a main authority in the field, evolve up to twelve sectors about the year 1000.[53] The tradition he started set the subject on a firmer basis.

[53] One of Ibn Surāqa's schemes is discussed in Schmidl, *Volkstümliche Astronomie*, B, 23a,5–25b,12; on all three see King, 'Makka,' p. 183ff.

APPENDIX

QIBLA *STARS*

The morning rising and evening setting amplitudes are based on parameters calculated for the latitude of Mecca in the year 800 C.E.

Table 4

Star Names			Morning rising or evening setting amplitude	MUQ
Altair	*al-nasr al-ṭāʾir*	α Aql	$\psi(_{\alpha\ \mathrm{Aql}}) = +7°$	1, 6
Canopus	*suhayl*	α Car	$\psi(_{\alpha\ \mathrm{Car}}) = -58°$	6
Capella	*al-ʿayyūq*	α Aur	$\psi(_{\alpha\ \mathrm{Aur}}) = +48°$	2, 8
the Handle of the Plough	*banāt naʿsh*	εζη UMa	$\psi(_{\eta\ \mathrm{UMa}}) = +63°$ and $\psi(_{\varepsilon\ \mathrm{UMa}}) = +73°$	4
Pleiades	*al-thurayyā*	–	$\psi(_{\eta\ \mathrm{Tau}}) = +21°$	3
the Pole Star	*al-judayy*	α UMi		4
Vega	(*al-nasr*) *al-wāqiʿ*	α Lyr	$\psi(_{\alpha\ \mathrm{Lyr}}) = +42°$	1

'UNMASKING THE CRAFT': 'ABD AL-LAṬĪF AL-BAGHDĀDĪ'S VIEWS ON ALCHEMY AND ALCHEMISTS

N. Peter Joosse

innamā naḥnu ṣabbāghūn
'Wir sind ja bloss Färber'
'We are only dyers'
Mary the Copt[1]

There is not much known about the two treatises on alchemy which 'Abd al-Laṭīf al-Baghdādī wrote during the second part of his lifetime, and as a logical consequence there is of course hardly any information available about his views on alchemy and its practitioners.

Only in 1962, with Samuel Miklos Stern's extensive description of the so-called Bursa manuscript,[2] the scholarly world became acquainted with the existence of ten thus far unknown treatises on a variety of subjects by 'Abd al-Laṭīf and with one treatise (no. 9) by Alexander of Aphrodisias on the 'differentia specifica'. Two of these treatises (no. 6 and no. 7) deal with the 'art' or 'craft' of alchemy.

A good two years later, the German scholar Albert Dietrich released an edition and a German translation of Bursa no. 9: the *Risāla li'l-Iskandar fī 'l-faṣl khāṣṣatan wa-mā huwa* by Alexander of Aphrodisias, which was accompanied by a description of the entire Bursa manuscript.[3] In the year 1972 Paul Ghalioungui and Said Abdou presented us with a detailed description of the Bursa manuscript in the Arabic language,[4]

[1] In 'Abd al-Laṭīf al-Baghdādī, *Risāla fī mujādalat al-ḥakīmayn al-kīmiyā'ī wa'l-naẓarī* (ms. Bursa, Hüseyin Çelebi 823, fol. 111v). Mary the Copt is a legendary authority in the field of alchemy, cf. *GAS* iv, pp. 70–73.

[2] S.M. Stern, 'A Collection of Treatises by 'Abd al-Laṭīf al-Baghdādī,' *Islamic Studies* (Karachi) 1 (1962), pp. 53–70. Reprinted in idem, *Medieval Arabic and Hebrew Thought* (London, 1983) No. XVIII.

[3] A. Dietrich, *Die arabische Version einer unbekannten Schrift des Alexander von Aphrodisias über die Differentia specifica* (Nachrichten der Akademie der Wissenschaften in Göttingen: I. Philologisch-historische Klasse, Jahrgang 1964, Nr. 2; Göttingen, 1964), pp. 85–148.

[4] 'Abd al-Laṭīf al-Baghdādī, *Maqālatān fī 'l-ḥawāss wa-masā'il ṭabī'iyya/Risāla li'l-Iskandar fī 'l-faṣl/Risāla fī 'l-maraḍ al-musammā diābīṭis by Abd Al-Latif Al-Baghdadi*, ed. P. Ghalioungui and S. Abdou (Kuwait, 1392/1972), pp. 27–35.

and in 1993 Seyfullah Sevim published a short essay on ʿAbd al-Laṭīf's views on alchemy and chemistry in Turkish.[5] But the by far most elaborate publication on the subject came to us in 1988 through the unpublished PhD dissertation of the Swiss scholar Franz Allemann.[6] Allemann prepared a critical edition of Bursa no. 6, to which he added a German translation and an extensive commentary apart from some very useful appendices. Unfortunately, Allemann could not hold the limelight and accordingly his dissertation did not receive the attention it so well deserved. In fact, it became almost forgotten, which caused a rather undesirable and unjust situation, because his dissertation gives us an excellent insight into the author ʿAbd al-Laṭīf al-Baghdādī, his material, his sources and his social *Umfeld*.

The author Muwaffaq al-Dīn ʿAbd al-Laṭīf ibn Yūsuf al-Baghdādī (557–12 Muḥarram 629/1162 9 November 1231) dealt with the subject of alchemy in two tractates. The first tractate (no. 6), which bears the title *Risāla fī mujādalat al-ḥakīmayn al-kīmiyāʾī wa'l-naẓarī* (a dispute between an adherent of alchemy and an adherent of theoretical philosophy) and has been preserved in the unique manuscript Bursa, Hüseyin Çelebi 823, fol. 100b–123b, has aroused my interest to a large extent. The treatise has been written during ʿAbd al-Laṭīf's first visit to Aleppo (613–617/1216–1220), and was most probably revised in the city of Erzincan in the year 622 by the author himself. The dispute is most likely fiction although it must be partly based on discussions between ʿAbd al-Laṭīf and his former advisors and mentors like for example Ibn Nāʾilī and Yāsīn al-Sīmiyāʾī (Yāsīn, the Magician, or rather: Yāsīn, the Illusionist).[7]

The definition of the genre presents us with a serious problem. In discussions preceding the publication of this article, it was often suggested to me that the anecdotes, which form the major part of the tractate were merely composed to serve as amusement or jest. It was

[5] Seyfeddin (Seyfullah) Sevim, 'Abdüllâtif Bağdadi'nin kimya-simya hakkındaki görüşleri,' in A.H. Köker (ed.), *Abdüllâtif Bağdâdi*. Gevher Nesibe Sultan Anısına Düzenlenen Abdüllâtif Bağdâdi Kongresi tebliğleri, 14 Mart 1992 Kayseri (Erciyes Üniversitesi Gevher Nesibe Tıp Tarihi Enstitüsü, Yayın 14; Kayseri: Erciyes Üniversitesi Matbaası, 1993), pp. 57–67.

[6] F. Allemann, *ʿAbdallaṭīf al-Baġdādī: Ris. fī Muġādalat al-ḥakīmain al-kīmiyāʾī wan-naẓarī („Das Streitgespräch zwischen dem Alchemisten und dem theoretischen Philosophen"). Eine textkritische Bearbeitung der Handschrift: Bursa, Hüseyin Çelebi 823, fol. 100–123 mit Übersetzung und Kommentar*, PhD diss (University of Bern, 1988).

[7] Ibid., pp. 62–3.

stated that these anecdotes were to correspond qua contents and style with the communications of al-Jawbarī on quacks and charlatans and that they could be interpreted in two different ways. However, this is far removed from the truth in view of the fact that amusement has the connotation of entertainment. These forbidding and often sinister anecdotes are hardly entertaining and qua atmosphere they on no account show similarities to the rather light-hearted stories of al-Jawbarī.[8] Moreover, there is no question of ambiguity here, so that they can only be interpreted in one way. They have to be considered a grave and stern warning against the concealment of so-called 'unspeakable things' and against the dangerous increase of utter simplicity and gullibility among the masses in the author's time. Allemann proposed to classify the tractate under the heading 'Die Kontroversliteratur über die Wahrheit der Alchemie'.[9] But because in this specific type of literature the discussion takes place on the basis of premises, which can neither be proved nor disproved, a solution or a step forward in the discussion can never be realized.[10] At the first glance, the treatise in question indeed seems to focus on the discussion of the truth, the value and the scientific character of alchemy. However, it soon becomes clear that its actual goal is to brand alchemy as a complete fraud and to show its evil and immoral nature. Therefore, it is not too hard to predict that in the given dialogue the alchemist is defeated by the philosopher, who in the course of his speech quotes a number of anecdotes of contemporary cases of fraudulent alchemists, con-artists and foolish dupes.

I am also in favour of a classification of our tractate under the header 'Kontroversliteratur', but believe that this classification is not sufficient to give a comprehensive overview of the genre. In order to give a more detailed and well-considered picture of the genre, a subdivision should be introduced. Thus it goes without saying that our tractate should be classified under a new header, which may be defined as parenetic literature (*paränetische Literatur*), a type of literature in which the author intends to warn against the pitfalls of the ideologies, aims and objectives of false alchemists/physicians, charlatans and quacks as

---

[8] E. Wiedemann, 'Über Charlatane bei den Muslimen nach al-Gaubarī,' *Sitzungsberichte der Physikalisch-medizinischen Sozietät zu Erlangen* 43 (1911), pp. 206–32; repr. in idem, *Aufsätze zur arabischen Wissenschaftsgeschichte*, i (Collectanea vi/1), ed. W. Fischer (Hildesheim, 1970), pp. 749–75.

[9] Allemann, *'Abdallaṭīf al-Baġdādī*, pp. 1, 64; cf. also M. Ullmann, *Die Natur- und Geheimwissenschaften im Islam* (Leiden, 1972), pp. 249–55.

[10] Cf. Ullmann, *Die Natur- und Geheimwissenschaften*, p. 249.

well as to provide a holistic understanding that should he not warn the people, the communities would suffer untold harm. In short, should his warnings not be taken to heart, the masses would not be protected against the false ideologies, motives and pretences propounded by the deceivers of this world.

The main goal of these false alchemists was always the production of the elixir, which was also called by the name 'Philosophers' stone' or 'sublime stone'. They believed in the substantial transmutation of metals and thought that the 'differentia specifica' of metals could be produced during an artificial process, which in the end would always lead to the transformation of lead and other base metals into the precious metals gold and silver.[11] The true alchemists, or rather chemists, were more interested in the colouring or dying of metals and in producing alloys. They rejected the substantial transmutation of metals vehemently.[12]

Before entering at some length into the stories itself, it might perhaps be fitting here to call the attention once more to the author Muwaffaq al-Dīn ʿAbd al-Laṭīf ibn Yūsuf al-Baghdādī, who is also known as Ibn al-Labbād, the son of the felt-maker, by briefly introducing him in 'what lies beneath'.

In his *Paradise of Wisdom*, the ninth-century Arab physician ʿAlī ibn Rabbān al-Ṭabarī laconically observes that one should not live in a country in which four things do not occur: a sound government, running water, useful medicine and a cultivated and skilful physician. The educated physician, the *ḥakīm*, became one of the idealized pictures of medieval Arab society.[13] Having had no proper education from a famous teacher caused one's honour to be at stake, was considered a very shameful shortcoming and could bring much harm to one's future career.[14] ʿAbd al-Laṭīf's father Yūsuf, a devoted scholar in the field of the religious law and by no means a felt-maker, must have been extremely aware of this problem. He must have encouraged his son from

[11] That this idea is wide-spread shows its frequent occurrence in works of fiction; cf. for instance M. Toonder, 'De loodhervormer,' in *Een groot denkraam* (Amsterdam, 1972), pp. 185–245 (separate ed.: Amsterdam, 2005); cf. also J.K. Rowling, *Harry Potter and the Philosopher's Stone* (London, 1997).

[12] Allemann, *ʿAbdallaṭīf al-Baġdādī*, pp. 42–4; Ullmann, *Die Natur- und Geheimwissenschaften*, pp. 257–61; E.J. Holmyard, *Alchemy. The Story of the Fascination of Gold and the Attempts of Chemists, Mystics, and Charlatans to Find the Philosophers' Stone* (Harmondsworth, 1957), pp. 13–14.

[13] H. Schipperges, 'Zum Bildungsweg eines arabischen Arztes,' *Orvostörténeti Közlemények* 60–61 (1971), pp. 13–31, especially p. 13.

[14] Schipperges, *Zum Bildungsweg*, p. 16.

a young age to learn from a multitude of teachers. ʿAbd al-Laṭīf was exceptionally eager to learn, but at the same time extremely impatient and often bored to tears by his teachers. Because the biographical and autobiographical *curriculum vitae* of ʿAbd al-Laṭīf is rather extensive and readily available to many, it does not make good sense to treat it here in full detail. Instead of it, I shall present in a more or less chronological order, some examples of the development of ʿAbd al-Laṭīf's attitude towards his teachers and mentors and towards the ones intended by him to become his teachers and mentors.[15]

At an early age ʿAbd al-Laṭīf was put under the care of Kamāl al-Dīn ʿAbd al-Raḥmān al-Anbārī: ʿAbd al-Laṭīf could not understand any of this scholar's continuous and considerable jabber, although he states that Kamāl al-Dīn's other students seemed pleased enough with it. However, Kamāl al-Dīn did not like to teach children and sent the young ʿAbd al-Laṭīf to his pupil Abū Bakr al-Wajīh al-Wāsiṭī of whom ʿAbd al-Laṭīf afterwards would say that he outstripped him in powers of memory and understanding. ʿAbd al-Laṭīf used al-Wajīh to get ahead. He surpassed him and enjoyed the double advantage of al-Wajīh's and Kamāl al-Dīn's initiatory company. Hibatallāh ibn al-Tilmīdh taught ʿAbd al-Laṭīf in medicine. ʿAbd al-Laṭīf calls Ibn al-Tilmīdh in his *Kitāb al-naṣīḥatayn* the only person who, in the true sense of the word, is worthy enough to be called a physician.[16] Ibn Abī Uṣaybiʿa, however, states in his *ʿUyūn* that Hibatallāh was not of such high merit and that ʿAbd al-Laṭīf only praised him so highly because of his extreme partiality for the Iraqis. Apart from Ibn al-Tilmīdh there is only one other physician to receive the same amount of credits from ʿAbd al-Laṭīf. This person is Abū Jaʿfar Aḥmad ibn Abī 'l-Ashʿath (d. 360/970). ʿAbd al-Laṭīf puts him on a par with Hippocrates and Galen and calls him the last physician in the Islamic period, worthy to be counted among these 'Ancients'.[17] It is reported that ʿAbd al-Laṭīf wrote a compendium of Ibn Abī 'l-Ashʿath's *Kitāb al-ḥayawān* (*The Book of Animals*), and of his

---

[15] Cf. among others: S.M. Toorawa, 'The Educational Background of ʿAbd al-Laṭīf al-Baghdādī,' *Muslim Education Quarterly* 13 (1996), pp. 35–53; G. Makdisi, *The Rise of Colleges. Institutions of Learning in Islam and the West* (Edinburgh, 1981), pp. 84–91; Ibn Abī Uṣaybiʿa, *Kitāb ʿuyūn al-anbāʾ fī ṭabaqāt al-aṭibbāʾ*, ed. Imruʾu 'l-Qays ibn al-Ṭaḥḥān (August Müller), 2 vols. (Cairo, 1299/1882), ii, pp. 201–13.

[16] Cf. Stern, 'A Collection of Treatises,' p. 64; Allemann, *ʿAbdallaṭīf al-Baġdādī*, p. 13; ʿAbd al-Laṭīf al-Baghdādī, ms. Bursa, fol. 89v, l. 17–89r, l. 1.

[17] Cf. Stern, 'A Collection of Treatises,' p. 62; ʿAbd al-Laṭīf al-Baghdādī, ms. Bursa, fol. 73v, l. 17.

Kitāb al-qūlanj (*The Book of Colic*). His compendium of the *Kitāb al-ḥayawān* is expected to be the source of the many Ibn Abī 'l-Ashʿath quotations found in later works.[18]

ʿAbdallāh ibn Nāʾilī (who is sometimes also called Ibn al-Tātalī or al-Bābilī) initially filled ʿAbd al-Laṭīf's heart with a yearning for all knowledge, but in retrospect he calls him a dabbler who attached value to 'procedures' thought contemptible and trivial. One day ʿAbd al-Laṭīf addressed him in the following way: 'If you had devoted the time you have wasted in the pursuit of the Craft to some of the Islamic or rational sciences you would today be without equal, waited on hand and foot. This alchemy nonsense simply does not have the answers you seek'. In search of a new master ʿAbd al-Laṭīf went from Baghdad to Mosul, but the man whom he approached, proved disappointing. Al-Kamāl ibn Yūnus was only partially learned and much misguided. His love of alchemy and his work in connection therewith, so absorbed his mind and his time that he thought little of anything else. Then, he heard of the fame of Shihāb al-Dīn al-Suhrawardī (al-Maqtūl) and wanted to visit him in Diyarbakır, but fortunately he read some of Suhrawardī's books before commencing his journey. Suhrawardī proved to be a deluded fool. ʿAbd al-Laṭīf considered his own marginal and supplementary notes with which he was not satisfied better than the arguments of this sot, and he found in his works a clear proof for the ignorance of his contemporaries. Suhrawardī afterwards became the personal astrologer and alchemist of al-Malik al-Ghāzī, one of the sons of Ṣalāḥ al-Dīn. Suhrawardī appeared to have lost his senses by then. He was recognized as a wandering Sufi and became a self-declared prophet. It is said that he was dirty in appearance and never cut his nails or hair. He was a walking flea-circus and lice crept over his face and clothes. Whosoever saw him tried to outflank him or ran away from him! Ṣalāḥ al-Dīn ordered his death, apparently under the pressure of the local *ʿulamāʾ*. However, Suhrawardī isolated himself and died from hunger at the age of barely thirty-six.[19] Other sources, however, state that Suhrawardī was executed publicly.[20]

---

[18] R. Kruk, 'Ibn abī 'l-Ashʿath's *Kitāb al-Ḥayawān*: a Scientific Approach to Anthropology, Dietetics and Zoological Systematics,' *ZGAIW* 14 (2001), pp. 119–68, especially pp. 162–3.

[19] S. Jadon, 'The Physicians of Syria during the Reign of Ṣalāḥ Al-Dīn 570–589 A.H. 1174–1193 A.D.,' *Journal of the History of Medicine and Allied Sciences* 25 (1970), pp. 323–40, 338–9.

[20] H. Ziai, 'Al-Suhrawardī,' in *EI*².

ʿAbd al-Laṭīf left for Damascus in 586/1190 where he studied under al-Kindī al-Baghdādī al-Naḥwī. ʿAbd al-Laṭīf describes him as a fine-looking *shaykh* with a keen wit, but very self-satisfied and troublesome and offensive to his associates. ʿAbd al-Laṭīf surpassed him on many topics and soon left his side. After that, ʿAbd al-Laṭīf set out for Jerusalem and Acre (ʿAkkā) in order to join up with Ṣalāḥ al-Dīn's camp. There he asked al-Q̲ādī al-Fāḍil, Ṣalāḥ al-Dīn's chief counsellor and director of his chancellery, to be sent to Cairo. He obtained permission to go and a letter of introduction was given to him. In Cairo ʿAbd al-Laṭīf sought for three persons. The first one he met was Yāsīn al-Sīmiyāʾī. ʿAbd al-Laṭīf found him utterly absurd, a liar, a conjuring cheat and a charlatan. It was said of him that he could do things even the prophet Moses was unable to do, that he could produce minted gold whenever he wished, and of any quantity he wished, and that he could turn the waters of the river Nile into a tent in which he would then sit with his friends. The Jewish scholar Moses Maimonides was the second person ʿAbd al-Laṭīf met in Cairo. He was a man of the highest merit, but overcome with the love of leadership and a courtier of those in high station. He wrote a book with the title *Kitāb al-dalāla* (*The Guide for the Perplexed*) and cursed anyone who transcribed it into anything but Hebrew script. ʿAbd al-Laṭīf looked through it and found it to be an evil book that corrupted the articles of faith and law with elements the author thought would reform them. One of his works was on medicine, based on the sixteen books by Galen and on five books by others. He took it upon himself not to alter a single word unless it was an 'and' or a 'so', and, as a matter of fact, copied sections in their entirety. Finally, ʿAbd al-Laṭīf encountered Abū 'l-Q̲āsim al-Shāriʿī. It is only in this person that ʿAbd al-Laṭīf finds the fulfilment of his desire: 'It is you I seek!' he cries out when meeting Abū 'l-Q̲āsim for the very first time in Cairo. Abū 'l-Q̲āsim introduced ʿAbd al-Laṭīf, among others, to the works of Alexander of Aphrodisias and Abū Naṣr al-Fārābī. ʿAbd al-Laṭīf later declares that he had no faith in these authors, because he used to think that the whole of philosophy had been comprehended by Ibn Sīnā and commented upon in his works. Abū 'l-Q̲āsim tamed ʿAbd al-Laṭīf's defiance, and wore down his natural intractability, until he inclined to his side, putting one foot forward and the other back. He would surpass ʿAbd al-Laṭīf in producing proofs and in the strength of his argument, whereas ʿAbd al-Laṭīf would surpass him in disputation and use of language. They remained constant companions and were inseparable morning till night. Their harmonious relationship only ended with the master's death. After Abū 'l-Qāsim only masters *extraordinaires* would do, namely rulers: Ṣalāḥ al-Dīn, al-Malik al-Afḍal, al-Malik al-ʿAzīz, al-Malik al-Ghāzī, ʿAlāʾ al-Dīn Dāʾūd ibn Bahrām and Shihāb al-Dīn Ṭughril Atabeg.

A new ʿAbd al-Laṭīf, cleansed and purified, arises. Because he is rejuvenated after the fruitful and mutually complimentary relationship with Abū 'l-Qāsim, ʿAbd al-Laṭīf gets more and more convinced of the superiority of the books of the 'Ancients' and of the works of al-Fārābī over the writings of Ibn Sīnā. Although ʿAbd al-Laṭīf from now on mainly focuses on the science of logic, there is still that voice in his head, which tells him to rebel against the deceiving craft of alchemy and to deal with it in a final way. After all, ʿAbd al-Laṭīf was in his younger days a convinced adherent of alchemy. He studied books by Jābir ibn Ḥayyān and Ibn Waḥshiyya on alchemical transformation and experimentation. In the *ʿUyūn* it is told that he started to practice the illusory art and made frivolous experiments of error.

The most potent of the influences that led him astray was, however, that of Ibn Sīnā, by his *Book on the Art* (*of Alchemy*), in which his philosophy appears to have attained completion. According to ʿAbd al-Laṭīf, this completion added nothing to philosophy, but rather derogated from it. Therefore, he considered Ibn Sīnā a false philosopher with a bad moral character and ended up feeling only contempt for the man and his *œuvre* since Ibn Sīnā drank wine, kept the company of prostitutes and composed his work under the influence of alcohol and other stimulants. But Ibn Sīnā was as a matter of fact unfavourably disposed towards alchemy and alchemists. He rejected the substantial transmutation of metals and merely wrote about the possibility of accidental transformation of metals.[21]

Ibn Sīnā's *Book on the Art* (*of Alchemy*) is not known to us within the Arabic tradition. It is a falsification,[22] and does not—as Allemann has suggested—form part of a series of Shiʿite-mystical tractates known collectively as *ḥikma mashriqiyya/mushriqiyya* (or: Eastern wisdom/radiant wisdom). Allemann also lacks proof in stating that ʿAbd al-Laṭīf gradually became a notorious opponent of the so-called *ḥikmat al-ishrāq* (or:

[21] Allemann, *ʿAbdallaṭīf al-Baġdādī*, pp. 35–51; Holmyard, *Alchemy*, pp. 88–95.

[22] It is most likely the *De anima in artis chemicae principes, Avicenna atque Geber* (Basel, 1572), which is a Latin version of an Arabic forgery attributed to Avicenna that should not be confused with the genuine *De anima* or *Liber sextus de naturalibus*.

Philosophy of Illumination).[23] It is most significant that al-Fārābī, who was honoured greatly by ʿAbd al-Laṭīf, actually believed in the possibility of the substantial transmutation of metals, but only explained this possibility in theory. He did not present us with records of experiments or recipes and hardly ever mentioned the elixir in his writings.[24]

However, the father of the false science of alchemy is, according to ʿAbd al-Laṭīf, Jābir ibn Ḥayyān, a scoundrel of the first water, who led astray many of the great scholars of subsequent generations like for instance Abū Bakr al-Rāzī (lat: Rhazes) and Ibn Miskawayh. 'All the blood, that was shed because of this art, all the money that vanished, every sin that was committed, all the great minds that went to ruin, and every person who because of alchemy dwelled from the straight path, is chargeable to Jābir. It is his sin,' recapitulates ʿAbd al-Laṭīf bitterly. He goes on by saying that if a thousand persons like him spent their whole lives to try and efface the traces of alchemy they could not erase the slightest part of it, for Jābir wrote four thousand books, that these books are spread in the entire world and stuff bookcases, that a single purge cannot cleanse the impurities of the world and that one farmer cannot uproot the weeds of a thousand gardens, unless he were a prophet. Jābir's words are pretences, misrepresentations and fabrications and are just as dangerous to the people as the teeth and claws of barking jackals and wild boars who rout up the earth.

Overcome by a deep-rooted hatred now for the false art, ʿAbd al-Laṭīf began to express his feelings in two treatises, the aforementioned *Risāla fī mujādalat al-ḥakīmayn* and another one, the *Risāla fī 'l-maʿādin wa-ibṭāl al-kīmiyāʾ*. Below we shall present some of the anecdotes which ʿAbd al-Laṭīf collected for his first treatise. The second treatise mainly deals with the formation of minerals and distinguishes between genuine and useful chemical operations and the procedures of the false alchemists.

Let us first explore the scene and the scenery a bit: ʿAbd al-Laṭīf sketches a desolate, chaotic picture of his social and geographical environment—briefly that means the triangle Aleppo-Baghdad-Damascus. However, it is not the Near East as we like to picture it. ʿAbd al-Laṭīf pictures an

[23] Allemann, *ʿAbdallaṭīf al-Baġdādī*, pp. 16 and 45–6.
[24] Ibid., pp. 52–7.

unholy no man's land which on the outside might resemble a landscape as painted by Pieter Brueghel the Elder.

In it, grotesque gruesomeness and sinister situations are coupled with an almost complete absence of upholding the law. The atmospheric description roughly approximates that of a gold-rush *avant la lettre*, a gold fever without gold and is similar qua monstrosity and atrocity to ʿAbd al-Laṭīf's description of the famine in Egypt during the years 597–598/1200–1202, in which we encounter such realistic, or rather surrealistic, tales of cannibalism. The image of roasted or boiled little children is vividly present in the mind's eye and the anecdote of the fat lady who was killed, cut and sliced, and converted into a human flesh pudding with vinegar, the *sikbāj*, still makes the flesh creep![25] Medical doctors were also greatly in demand, especially general practitioners. Not because they tasted so good, but chiefly because they were considered easy prey. The hunters just laid in wait until the house-calls were made and then they pounced.

Blacksmiths, bakers and millers desert their work abruptly, notable citizens (judges, legal experts, notaries) neglect municipal administration, only to devote themselves fully to the production of the elixir, a schoolmaster leaves his pupils in the lurch just to execute gross experiments, judges let themselves be cheated and deceived by sly and pretentious swindlers, people who belong to the nobility and are elected for the office of vizier wholly run aground, are obliged to sell all their assets and finally end up in jail, merely for the fact that they became the victims of these false alchemists. However, we should pay attention to the anecdotes! Are they exaggerated? Most probably! Is there a germ of truth in them? Without a doubt! Much of the action in the majority of these anecdotes takes place during the Crusades. Poverty and famine reigns everywhere and Ṣalāḥ al-Dīn is, during this period, entangled in fatiguing attempts to consolidate his newly acquired power in Egypt, while at the same time, he is fighting against

[25] ʿAbd al-Laṭīf al-Baghdādī, *The Eastern Key: Kitāb al-ifādah wa'l-iʿtibār*, trans. K.H. Zand and J.A. and I.E. Videan (Cairo, 1204/1964), esp. pp. 223–55. It is still a difficult matter to decide on whether these horror stories are an eyewitness account by ʿAbd al-Laṭīf al-Baghdādī or represent a topos, a literary stereotype. The fifteenth-century author al-Maqrīzī quotes the story of the fat lady in his *Kitāb al-mawāʿiẓ wa'l-iʿtibār bi-dhikr al-khiṭaṭ wa'l-āthār*. Apparently he does not seem to refer to the Egyptian famine of 1200–1202 C.E., but to the famine during the 'great crisis': *al-shidda al-ʿuẓmā*: 1066–1073 C.E. Many thanks to Dr. Johannes den Heijer (TCMO, Leiden University, The Netherlands) for supplying me this very interesting information.

the crusaders for the possession of Jerusalem. There was chaos (*fitna*) and confusion all over and in all ranks: among the rural population and among the city-dwellers; among the rich and the poor; among the dignitaries; among the military; and certainly also among the scholars in- and outside the institutes of learning, the *madāris*.[26] A situation like that must have been a fertile soil for the coming into existence of such grim and gruesome anecdotes. Alchemy in all its weird and wonderful manifestations may apparently have provided a favourable outlet for angst, chaos, confusion, dissension, ignorance and poverty; although one of course cannot deny that a passion for gold, a love for gain, evidently was the principal motive for the common people and the upper crust to start practicing the Craft.

Many people held the conviction that the 'sublime stone' was the blood. ʿAbd al-Laṭīf mentions his meeting with the judge (*qāḍī*) of Baalbek who buried ten Damascene *raṭl* (= 18 kg and 50 pounds) of blood in secret places. When ʿAbd al-Laṭīf asked him if he fulfilled the acts necessary for the processes of distillation, purification, coagulation, solution and fixation, the judge replied: 'Do not talk nonsense, man. That is useless stuff. I can do without it very well!' The judge showed ʿAbd al-Laṭīf a silver dirham a couple of days later, which supposedly was created out of blood through a secret procedure, which the judge preferred to keep to himself. Of course, the dirham was genuine and made of pure silver, but the judge did not want to look like a fool in front of ʿAbd al-Laṭīf and pretended that he produced the coin himself.[27] This example clearly illustrates how far people were willing to go in their self-deception and in the deception of others. The same judge was once approached by an Egyptian man, who made him believe that urine was the 'Philosophers' stone'. They co-operated and obtained the urine from the eunuchs, who often held the position of bath-superintendents and in this way collected seven hundred earthenware jugs of the specific liquid for the special price of five jugs for one dirham. They placed the urine in large containers and left these in the sun to bake until the urine became thick. Because of the ghastly stench, the neighbours came out of their houses and scolded the judge and his new friend badly.

[26] N.A. Faris, 'Arab Culture in the Twelfth Century,' in N.P. Zacour and H.W. Hazard (eds.), *A History of the Crusades* (general editor: K. Meyer Setton), v: *The Impact of the Crusades on the Near East* (Madison, Wis., 1985), ch. I, 6; M. Chamberlain, *Knowledge and Social Practice in Medieval Damascus, 1190–1350* (Cambridge, 1994), pp. 91–107.

[27] Allemann, *ʿAbdallaṭīf al-Baġdādī*, p. 85; ʿAbd al-Laṭīf al-Baghdādī, ms. Bursa, fol. 107v.

Therefore, the Egyptian proposed to bring the 'concentrated' urine to his house, but the judge would not hear of this plan, being afraid that the Egyptian would take off without giving him his share of the gold. They haggled about it for weeks and finally split up, being completely estranged from each other.[28] Unfortunately, ʿAbd al-Laṭīf does not clear up what was done with the urine afterwards.

A similar event took place (at a deserted site) on the outskirts of Baghdad. A man, who endeavoured to produce the elixir, created a long-necked bottle out of glass and filled it with a mixture of body-juices (most likely: blood, urine and sperm). He possibly also stirred in some watery stool and buried the lot for a period of seven days in decaying manure. Then he dug it out again and opened the bottle. But, out of sheer greed, he opened it too hastily and a pillar of fire rose up, which made him fall to the ground. The bottle fell on the floor and broke into pieces so that the stench was able to permeate the room. Somebody found him lying there on the ground and brought him to town. The people recognized him and carried him home. He was ill for eight months, but did not make a good recovery, for the accident caused him to become a mentally disturbed person.[29]

Something that happened in the *al-Khātūniyya* area of Baghdad is that a teacher sold his property, sent his pupils home, and from then on, occupied himself only with the execution of alchemistical experiments. He collected blood and egg yolks and deliberately caused them to rot. After many difficult years filled with countless setbacks, he finally got lucky and by means of putrefaction, worms and maggots were generated spontaneously. Unfortunately, the creatures started eating each other up and in the end only one worm survived. It became mighty big because the false alchemist fed it twice daily with freshly cupped blood. When the alchemist ultimately reached the point where he could feed the worm with the amalgam, a metal mixed with quicksilver (mercury), which had to change into an elixir in the worm's stomach, he had to expose the worm to intense heat. Hardly had he started to place the animal on a leaf in the burning sun, when a pussycat grabbed it and quickly ran away with it. The alchemist jumped up like a mother whose baby had just been snatched from under her eyes and went after the

[28] Allemann, *ʿAbdallaṭīf al-Baġdādī*, pp. 85–6; ʿAbd al-Laṭīf al-Baghdādī, fol. 107v–107r.

[29] Allemann, *ʿAbdallaṭīf al-Baġdādī*, p. 86; ʿAbd al-Laṭīf al-Baghdādī, fol. 107r.

naughty cat, panic-stricken, like a headless chicken. He climbed from roof to roof, falling down every time and hurting himself, but the pussycat disappeared without leaving a single trace![30] These experiences must have had the effect of a nightmare on the unfortunate alchemist. Perhaps the pussycat's name was *qābūs*!

Another story from Baghdad is the following titillating tale: in the middle of summer, twenty town dignitaries had gathered in an unoccupied house in a dead-end street. They had met there to execute a specific chemical process called *bāb ḥabs al-ziʾbaq* (= the process of the secret—lit. 'holding back'—of the quicksilver). They double locked the door of the house, dug a deep pit, filled it up with dung and buried in it a bulbous iron container, a kind of *qumqum*, which was filled with quicksilver and sealed several times with the so-called *ṭīn al-ḥikma*, 'the clay of wisdom', which is fermented clay mixed with a little dung, cut animal-hair and salt. They lighted the fire under the container, but this threw out such an intense heat that they had to take off their clothes and sit around the fire naked. Telling stories and having a lot of fun, the good gentlemen forgot to observe the fire so that all of a sudden the container started to make a terrible noise and exploded in their faces. They landed on the roof of the house, fell on the floor, jumped up again and hurried for the outer-door. Twenty naked and barefooted old and young men with baldheads and long beards came out of the house all at once so that the women, who were spinning in the oriel rooms, believed that the *jinn*, the demons, had come to the surface in broad daylight. Soon the men were chased by the whole town and even the police were hot on their trail and eager to throw them in jail. (The Arabic sentence reads here *bāb ḥabs al-luṣūṣ* = the door to the cell of thieves.)[31]

Because alchemy—or as may be preferred—experimenting with chemicals was in popular demand at every level of society, many impostors saw their way clear to make big money out of it. And thus, we should also mention Jaʿfar, the alchemist here. He liked to play games with people with a high social status and loved to ridicule them. One fine day he conceived a plan to do an experiment with human excrements. So, he asked three well-off citizens, who were not averse

---

[30] Allemann, *ʿAbdallaṭīf al-Baġdādī*, pp. 86–7; ʿAbd al-Laṭīf al-Baghdādī, fol. 107r.

[31] Allemann, *ʿAbdallaṭīf al-Baġdādī*, pp. 87–8; ʿAbd al-Laṭīf al-Baghdādī, fol. 108v–108r.

to gold, to join him at his house for an extended period. While they stayed with him, he only fed them on honey, eggs and unleavened bread and denied them the pleasure of all exquisite foods. Moreover, he ordered them to collect their excrements for a period of forty days and commanded them to distil, filter, contract, liquefy and harden it again. In every phase he let them taste and smell their own excrements and most certainly also that of the others and asked them after their sensory perception. This experiment took six months to conduct and during this period the three men presented Jaʿfar with the finest wines [*sic*!] and also helped him to indulge in wild excesses, although they were known to bear an irreproachable character.[32] Of course, Jaʿfar was 'full of shit' and unfortunately he had an unusual preference for offering his 'victims' shit!

In Mosul a Moroccan alchemist, a damned devil and a godforsaken rebel, made the streets unsafe. He used to sweet-talk and butter up the municipal elite and the town dignitaries, so that they left their wives, sons and daughters in his care. He committed outrageous and shameful acts and among others practiced anal sex with barefaced boys and told them that the 'sublime stone' was his sperm. He abused them for three long years until finally his deception and lies came to light. Then, he was executed in public in a particularly nasty and humiliating way.[33]

Worthwhile to relate may also be the story of one of the companions of al-Suhrawardī (al-Maqtūl), an emir of the Seljuqs of Anatolia, known as the Seljuqs of Rūm, whose opinion it was that the eyes were the 'Philosophers' stone'. For a while, he took great pains to collect the eyes of sheep and goats, but when once 12,000 Franks were butchered near Acre, he and a few of his friends headed for the battlefield and cut out the eyes of the dead soldiers, whereas other practitioners of the Craft took away their gall-bladders. ʿAbd al-Laṭīf refers here to the chronicler and personal secretary to Ṣalāḥ al-Dīn, ʿImād al-Dīn al-Kātib al-Iṣfahānī, who appeared to have been an eyewitness to these atrocities. From the ambiguous and vague way in which ʿAbd al-Laṭīf tells the story, one gets an eerie feeling as if more is meant than meets the eye. Did al-Kātib al-Iṣfahānī also contribute to the 'stealing' of the eyes or was he just an innocent bystander? Some false alchemists took the view that the perfect elixir could be produced only with the eyes of a

[32] Allemann, *ʿAbdallaṭīf al-Baġdādī*, pp. 88–9; ʿAbd al-Laṭīf al-Baghdādī, fol. 108r.
[33] Allemann, *ʿAbdallaṭīf al-Baġdādī*, p. 111; ʿAbd al-Laṭīf al-Baghdādī, fol. 117v.

strong, healthy human being: blond, blue-eyed and of fair complexion. Occasionally, they bought a slave for that purpose, but now and then they also set a trap to catch people who were fully unaware of the danger they found themselves in. The alchemists tied their hands and feet and pulled out their eyes with a type of fishhook, but they did it in such a violent manner that they often teared off a piece of flesh, which was attached to the brain.[34]

ʿAbd al-Laṭīf informs us that Suhrawardī's followers mainly consisted of common people and riff-raff. They were either singers and flute-players or owners of public houses and inns in which vulgar amusement ran rampant. A highwayman was once hanged on the gallows in Aleppo; a few charlatans, supporters of Suhrawardī, agreed on stealing his eyes. They promised one from within their midst a dinar if he could get hold of the hanged person's eyes. The chosen one came at midnight, gripped the dead man's neck and cut the rope which was bound around his throat. Because the blood and the wind in his abdomen were pressed together, strange croaking sounds escaped from his throat. It was as if the dead man had said something very nasty or pronounced a curse. The false alchemist became frightened out of his wits and passed out on the spot.[35]

There must have flourished a lucrative, yet illicit and illegal, trade in body parts in these days and in that area during a relatively short period if we take it as a fact that alchemy was widespread and practiced by many. However, reports of interference by the local governments (or the central government) or any specific jurisprudence (*fiqh*) with regard to a prohibition of these evil and objectionable practices seem to be fully absent, which is perhaps caused by the fact that the whole area was in such a gigantic turmoil during the Crusades. The rulers could not cope with the situation and as a consequence public security all but collapsed. Decisions on any matter whatsoever were put on a back burner or postponed indefinitely during this period of chaos and confusion. Thus by the time the peace had been restored, the boom of alchemy with all its undesirable side effects past its peak and declined, so that the officials did not feel the necessity to amend or to change the laws.

---

[34] Allemann, *ʿAbdallaṭīf al-Baġdādī*, pp. 109–10; ʿAbd al-Laṭīf al-Baghdādī, fol. 116r.

[35] Allemann, *ʿAbdallaṭīf al-Baġdādī*, p. 110; ʿAbd al-Laṭīf al-Baghdādī, fol. 116r.

Because traditions have developed independently of each other and history has the propensity to reinvent itself most of the time in a more or less cyclical mode, it is definitely not strange that the trade in body parts, with the aim of using them as an ingredient in an elixir, a medicine, a (magic) potion or an ointment, is not restricted to a particular geographical area or a specific period in time. Even now, in modern-day South Africa, there occurs the phenomenon of murder for the use of human body parts in the making of potions, the so-called *muti* murder. The *muti* traders often attack young and healthy children. Recently, a 6-year-old boy and a 10-year-old boy were killed for their body parts. In the first case, the boy's head was cut off and kept in a fridge in expectation of prospective customers to come. In the latter case, the boy died in hospital more than a week after attackers had hacked off his hands, ears and penis.[36] Some clients are willing to pay the trader huge amounts of money for 'strong' *muti*. The trade in strong *muti* has led to some people raiding mortuaries and stealing body parts from corpses and selling them to *inyangas* (diviners, traditional healers). It has to be emphasized here strongly that *muti* has nothing in common with that which is sometimes erroneously called bush medicine, in which plant matter and occasionally parts of animals are being used for potions and ointments etcetera. *Muti* murder or medicine murder is also distinctly different from ritual murder, when a victim is sacrificed in public for the 'benefit' of the whole community. *Muti* murder is committed by individuals and for selfish motives. It is a criminal act strongly tied to the beliefs in the power and benefits of human body parts and to certain beliefs in witchcraft. It is highly remarkable that also in this situation any specific jurisprudence with regard to these dreadful offences is lacking. The South African government (by means of The Ralushai Commission) proposed and drafted a new Act: The Witchcraft Control Act. In it, related issues such as *muti* murder shall almost certainly be addressed, because *muti* murder is ever-increasing and not yet covered by any existing Act.[37]

At the end of ʿAbd al-Laṭīf's second treatise on alchemy (no. 7) we find an addition. In it, he advises us that he was just re-reading his

---

[36] Report from the South African newspaper the 'Cape Argus' of Saturday 21–08–2004.

[37] A. Minnaar, 'Legislative and Legal Challenges to Combating Witch Purging and *Muti* Murder in South Africa,' in J. Hund (ed.), *Witchcraft Violence and the Law in South Africa* (Pretoria, 2003), pp. 73–92.

work in Erzincan in the year 622 A.H., when by a curious coincidence a visitor was announced who turned out to be an alchemist and who asked for an introduction from ʿAbd al-Laṭīf to al-Malik ʿAlāʾ al-Dīn Dāʾūd ibn Bahrām, the ruler of the city, so that he might use the body of a slave or a condemned criminal for the preparation of his elixir![38] Having learned a little bit more about ʿAbd al-Laṭīf's flamboyant and perhaps somewhat warped personality, his use of often abrasive and excessive language, his sometimes unwarranted prejudice and his slight but undeniable inclination towards exaggeration, one can only begin to imagine what ʿAbd al-Laṭīf's reply to the poor alchemist must have been like. Still I gain the impression that all his display is justifiable and evidently not without a reason. It merely functioned as a means to an end. ʿAbd al-Laṭīf al-Baghdādī was a rebellious intellectual, but unmistakably a rebel with a cause.

[38] Stern, 'A Collection of Treatises,' p. 67; Allemann, *ʿAbdallaṭīf al-Baġdādī*, p. 26; ʿAbd al-Laṭīf al-Baghdādī, fol. 131v, l. 8–132v, l. 7.

MATHEMATICAL GEOGRAPHY IN FIFTEENTH-CENTURY EGYPT: AN EPISODE IN THE DECLINE OF ISLAMIC SCIENCE

David A. King

1. Introductory Remarks

Islamic mathematical geography deals with lists of longitudes and latitudes for numerous localities and the associated world-maps fitted with proper longitude and latitude grids and with localities properly marked according to their coordinates. The history of this subject is now, thanks mainly to the groundbreaking research of Ted and Mary Helen Kennedy and, more recently, Fuat Sezgin, a recognized discipline within the history of Islamic science.[1] Whilst some of the earliest tables of coordinates from the ninth and tenth centuries have survived, no such early world-maps are preserved for us.[2]

---

[1] E.S. Kennedy and M.H. Kennedy, *Geographical Coordinates of Localities from Islamic Sources* (Frankfurt, 1987). See also E.S. Kennedy, 'Mathematical Geography,' in R. Rashed and R. Morelon (eds.), *Encyclopedia of the History of Arabic Science*, 3 vols. (London, 1996), i, pp. 185–201. The most recent volumes of F. Sezgin, *Geschichte des arabischen Schrifttums*, 12 vols. to date (Leiden, 1967 onwards, since 2000 Frankfurt am Main), are x–xii: *Mathematische Geographie und Kartographie im Islam und ihr Fortleben im Abendland* (2000). The forthcoming volumes of Sezgin's monumental work dealing with the sources for mathematical and descriptive geography are awaited with anticipation. The standard bio-bibliographical sources for Muslim astronomers and mathematicians after the period covered by Sezgin (v: *Mathematik*, 1974, vi: *Astronomie*, 1978, vii: *Astrologie, Meteorologie und Verwandtes*, 1979) are H. Suter, 'Die Mathematiker und Astronomen der Araber und ihre Werke,' *Abhandlungen zur Geschichte der mathematischen Wissenschaften* 10 (1900), and 'Nachträge und Berichtigungen,' ibid. 14 (1902), pp. 157–85, repr. Amsterdam, 1982, and again in idem, *Beiträge zur Geschichte der Mathematik und Astronomie im Islam*, ed. F. Sezgin et al., 2 vols. (Frankfurt, 1986), i, pp. 1–285 and 286–314; C. Brockelmann, *Geschichte der arabischen Litteratur*, 2 vols. (Leiden, ²1943–49), and Supplementbände, 3 vols. (Leiden, 1937–42); C.A. Storey, *Persian Literature. A Bio-bibliographical Survey*, ii/1: A—Mathematics, B—Weights and Measures, C—Astronomy and Astrology, D—Geography (London, 1972); D.A. King, *A Survey of the Scientific Manuscripts in the Egyptian National Library* (Winona Lake, Ind., 1986); B.A. Rosenfeld and E. İhsanoğlu, *Mathematicians, Astronomers and Other Scholars of Islamic Civilisation and their Works (7th–19th C.)* (Istanbul, 2003), with a supplement in *Suhayl. Journal for the History of the Exact and Natural Sciences in the Islamic Civilisation* (Barcelona) 4 (2004), pp. 87–139.

[2] Kennedy, 'Suhrāb and the World-Map of Ma'mūn,' in J.L. Berggren and B.R. Goldstein (eds.), *From Ancient Omens to Statistical Mechanics. Essays on the Exact Sciences*

a) *Mathematical Geography in Iran and Central Asia*

In some recent publications I have drawn attention to the two most influential traditions of mathematical geography in the Islamic East—Greater Iran and Central Asia—that lasted from the eleventh century at least until the seventeenth century.[3] A driving force behind the activity in these two traditions, often overlooked in certain modern writings, was the determination of the *qibla* for all localities in the world.[4] These two different traditions of geographical coordinates and the associated world-maps, alas mainly lost,[5] and *qibla*-directions are the following:

First we have the tradition of al-Bīrūnī (Central Asia, *fl. c.* 1050), followed by ʿAbd al-Raḥmān al-Khāzinī (Marw, *fl. c.* 1125), Sanjar

---

Presented to Asger Aaboe = *Acta Historica Scientiarum Naturalium et Medicinalium* 39 (1987), pp. 113–19. This important study has been overlooked in some modern writings on ninth-century cartography.

[3] See D. King, *World-Maps for Finding the Direction and Distance to Mecca. Innovation and Tradition in Islamic Science* (Leiden, 1999), and idem, *In Synchrony with the Heavens. Studies in Astronomical Timekeeping and Instrumentation in Islamic Civilization*, 2 vols., vol. 1: *The Call of the Muezzin. Studies I–IX* (Leiden, 2004), and vol. 2: *Instruments of Mass Calculation. Studies X–XVIII* (Leiden, 2005), especially VIIc.

[4] On the *qibla* see the article 'Ḳibla. i. Ritual and legal aspects' by A.J. Wensinck in *EI*². My article 'Ḳibla, ii. Astronomical aspects' in *EI*² (1979) is long outdated. In it I expressed my surprise that in spite of all of the activity amongst Muslim scientists to determine the *qibla* on the basis of geographical coordinates and accurate trigonometric formulae, the majority of medieval mosques from one end of the Islamic world to the other are oriented in directions that do not correspond to the computed values. By the time the article 'Makka, iv. As centre of the world' appeared in 1987, the basic documentation of medieval techniques for finding the *qibla* by the legal scholars, based on folk astronomy and using astronomical horizon phenomena to face an astronomically-aligned Kaʿba, had been achieved. For a new survey of medieval *qibla* determinations see King, *Synchrony*, i, pp. 741–71; on the consequences for Cairo see ibid., pp. 773–823. For the situation in al-Andalus and the Maghrib, as described in texts on the sacred law and folk astronomy, see M. Rius, *La Alquibla en al-Andalus y al-Magrib al-Aqṣà* (Barcelona, 2000). For a survey of over 20 different medieval schemes of sacred geography—the notion of the world divided into sectors around the Kaʿba and the associated non-mathematical *qibla* directions—see D. King, *The Sacred Geography of Islam*, forthcoming, with a summary in the *EI*² article 'Makka. iv'. For a more detailed analysis of some Yemeni schemes of sacred geography see P. Schmidl, *Volkstümliche astronomische Abhandlungen aus dem mittelalterlichen arabisch-islamischen Kulturraum*..., 2 vols., (Leiden, 2007), and for the earliest schemes see the contribution of Mónica Herrera Casais and Petra Schmidl to this volume.

[5] Much credit is due to Joachim Lelewel for his reconstruction of various world-maps. See his *La géographie du moyen âge*, with an *Atlas composé de cinquante planches* (Brussels, 1850–57; repr. in F. Sezgin et al. [eds.], *Islamic Geography*, vols. 129–33 [Frankfurt, 1993]), especially the atlas volume: figs. II (Ibn Yūnus), V (al-Bīrūnī), XIX (al-Ṭūsī), XX (*Kitāb al-aṭwāl*), XXI (Ibn Saʿīd al-Maghribī), XXII (al-Marrākushī), and XXIII (Abū 'l-Fidā'), also King, *World-Maps*, pp. 32–3, for his map based on the coordinates of al-Ṭūsī.

al-Kamālī (Yazd, *fl. c.* 1300), Ibn al-Shāṭir (Damascus, *fl. c.* 1350), and some lesser authorities. al-Bīrūnī surely prepared several world-maps, and I have shown that one of them was used by al-Khāzinī.[6] Alas, none of these maps has survived. A world-map based on the coordinates of al-Khāzinī, themselves derived from a world-map by al-Bīrūnī, is preserved in a treatise on folk astronomy by Sirāj al-Dīn al-Sajāwandī (1210): this is a singularly wretched copy that could serve no practical purpose.[7] It is possible that some of the numerous surviving maps of Greater Iran are based on one or other of these eastern Islamic traditions.[8]

Second we have the tradition of the anonymous and enigmatic *Kitāb al-aṭwāl wa'l-'urūḍ li'l Furs* (Isfahan? Eleventh century?), Naṣīr al-Dīn al-Ṭūsī (Maragha, *fl. c.* 1260), Ghiyāth al-Dīn al-Kāshī and Ulugh Beg (both Samarqand, *fl. c.* 1425), and some anonymous scholars who compiled an enormous table that included *qibla* directions and distances to Mecca (Kish near Samarqand, *fl. c.* 1450).[9] For only one of these various sources have any traces of a serious world-map survived: the coordinates in the anonymous tables from Kish underlie three remarkable world-maps engraved on brass that were made in Isfahan in the late seventeenth century. These display the direction and distance to Mecca at the centre, and Jan Hogendijk has found the evidence to confirm my hypothesis that the mathematics underlying them was known in tenth-century Baghdad and eleventh-century Isfahan.[10] It is perhaps too much to hope that a precursor to the Safavid world-maps will show up. The geographical data was used in gazetteers on Iranian astrolabes until the early eighteenth century.[11]

---

[6] King, *World-Maps*, pp. 41–2 and 71–5, as well as App. D on pp. 564–85.

[7] See King, 'A World-Map in the Tradition of al-Bīrūnī (*ca.* 1040) and al-Khāzinī (*ca.* 1120) presented by Sirāj al-Dīn al-Sajāwandī (1210),' to appear in the *Festschrift* for Professor Hossam Elkhadem in a special issue of *Archives et bibliothèques de Belgique* in 2006. This appears to be the only evidence that al-Sajāwandī (see Suter, 'Die Mathematiker und Astronomen der Araber,' p. 192; Brockelmann, *GAL*, I, pp. 470–71, and *S* I, pp. 650–51; King, *Survey of the Scientific Manuscripts*, nos. C4 = G11, Rosenfeld and İhsanoğlu, *Mathematicians, Astronomers and Other Scholars*, no. 581), well known in Ḥanafī legal circles, worked in Egypt, though this is still questionable. Apparently no biographical information is available on him.

[8] For some examples see King, *World-Maps*, pp. 141–8.

[9] King, *World-Maps*, pp. 42–3 and 149–68, and King, *Synchrony*, i, p. 846.

[10] On two of these maps see King, *World-Maps*, pp. 195–364, as well as App. A on pp. 456–77 and App. C on pp. 552–63, and on the third, as well as on Hogendijk's discoveries, see King, *Synchrony*, i, pp. 825–46.

[11] King, *World-Maps*, pp. 175–86.

b) *Mathematical Geography in Egypt and Syria*

In my study of mathematical geography in the Islamic East, I remarked that the magnificent world-map on silk prepared in Fustat in the year 964, at a reported cost of 22,000 *dīnārs*, which has not survived, had—as far as we know—no successor in Egypt, or, for that matter, in Syria either.[12] Although the celebrated astronomer Ibn Yūnus (Cairo-Fustat, *fl. c.* 990) in his geographical tables relied heavily on those of al-Khwārizmī, and in the thirteenth century the tables of the Andalusī Ibn al-Zayyāt (d. 1058) were clearly available in Egypt, there was, with one notable exception, no serious work done on coordinates and no serious world-maps prepared, this in spite of a colourful tradition of mathematical astronomy and instrumentation in Mamlūk Egypt and Syria.[13] The only exception was the tables of al-Marrākushī (Cairo, *fl. c.* 1280), who presented his own measurements of the latitudes of various localities between the Maghrib and Cairo.[14] I have investigated several Egyptian and Syrian tables of geographical coordinates, showing them to be mainly of little initiative and, where calculation was involved for *qibla*-values, rather carelessly computed.[15] A world-map with a corrupt longitude scale (but no latitude scale) and numerous localities marked in defiance of any reconstruction of the scale in a thirteenth century copy of an anonymous geographical work, *Kitāb gharāʾib al-funūn wa-mulaḥ al-ʿuyūn*, probably compiled in Egypt in the late eleventh century.[16] A world-map presented by the historian-encyclopaedist Ibn Faḍlallāh in his encyclopedic *Masālik al-abṣār c.* 1340 (see Fig. 2) bears a sophisticated looking grid, but this is completely out of place on a map of the simple *mappa mundi* tradition, and not a single locality indicated on the map is positioned according to any geographical tables.[17]

Two Syrians, however, stand out in their activities relating to geographical coordinates. The first is the scholar-prince Abū 'l-Fidā (Hama,

[12] King, *World-Maps*, p. 35.

[13] For the context see the already outdated study King, 'The Astronomy of the Mamluks,' *ISIS* 74 (1983), pp. 531–55, repr. in idem, *Islamic Mathematical Astronomy* (London, 1986; 2nd rev. edn., Aldershot, 1993), III, and the new insights in F. Charette, *Mathematical Instrumentation in Fourteenth-Century Egypt and Syria. The Illustrated Treatise of Najm al-Dīn al-Miṣrī* (Leiden, 2003), pp. 5–31.

[14] See Kennedy and Kennedy, *Geographical Coordinates*, p. xxv *sub* MAR.

[15] King, *World-Maps*, pp. 76–86 and Tables F1–F8 on pp. 600–22.

[16] See J. Johns and E. Savage-Smith, '*The Book of Curiosities*: a Newly Discovered Series of Islamic Maps,' *Imago Mundi* 55 (2003), pp. 7–24, especially pl. 1 and pp. 11–13.

[17] King, *World-Maps*, pp. 34–7.

fl. c. 1325). He documented over 1200 pairs of coordinates from several earlier sources, some of which are no longer extant. His list is critical in the sense that he carefully indicated his sources, but uncritical in the sense that it is devoid of comment. Nevertheless, his heart was in the right place. His list has been particularly useful in reconstructing part of the lost *Kitāb al-aṭwāl wa'l-ʿurūḍ*, which underlies the main Eastern Islamic tradition mentioned above.[18] The second Syrian of merit is Shams al-Dīn al-Khalīlī, who in *c.* 1360 calculated the *qiblas* of 40-odd localities in Syria and Palestine, mainly accurately to the nearest minute. Even this achievement pales in comparison with his table of almost 3,000 entries giving the *qibla* to minutes, usually accurately, for each degree of latitude and each degree of longitude difference from Mecca in the entire Muslim world.[19] But his *qibla* table was not widely used after his time, not least because it was simply too complicated for practical purposes.

An Egyptian scholar, some 150 years after Abū 'l-Fidā, tried to carry out the same exercise, and his endeavour is the topic of the present study. But first we need to introduce the notion of the climates, which was of fundamental importance in Islamic mathematical geography.

c) *The Fate of the Climates of Antiquity in Medieval Islam*

The Greeks divided the inhabited earth into seven latitudinal 'climates' (κλίματα) whose centres and boundaries are defined in terms of the length of longest daylight—see Fig. 1. The climates were of paramount importance not only in ancient and medieval mathematical geography, but also in mathematical astronomy and astronomical instrumentation.[20] The problem with the climates defined by these criteria is that the latitudes of their centres and boundaries change by a few minutes

[18] Kennedy & Kennedy, *Geographical Coordinates*, pp. xix–xx *sub* FID.

[19] King, 'Al-Khalīlī's Qibla Table,' *JNES* 34 (1975), pp. 81–122, repr. in King, *Islamic Mathematical Astronomy*, XIII; also King, *Synchrony*, i, pp. 386–93.

[20] The standard work on the climates in Antiquity is E. Honigmann, *Die sieben Klimata und die πόλεις ἐπίσημοις. Eine Untersuchung zur Geschichte der Geographie und Astrologie im Altertum und Mittelalter* (Heidelberg, 1929). Already in the Hellenistic world, new but unhappy definitions of the climates started to appear: see B. Stautz, 'Die früheste bekannte Formgebung der Astrolabien,' in A. von Gotstedter (ed.), *Ad radices. Festband zum fünfzigjährigen Bestehen des Instituts für Geschichte der Naturwissenschaften Frankfurt am Main* (Stuttgart, 1994), pp. 315–28, especially p. 318. For an introduction to the climates in Islamic descriptive geography see André Miquel's article 'Iḳlīm' in *EI*[2]. On the importance of the climates in medieval Islamic and European science see King, *Synchrony*, i, pp. 688–9, etc., and ii, pp. 925–32.

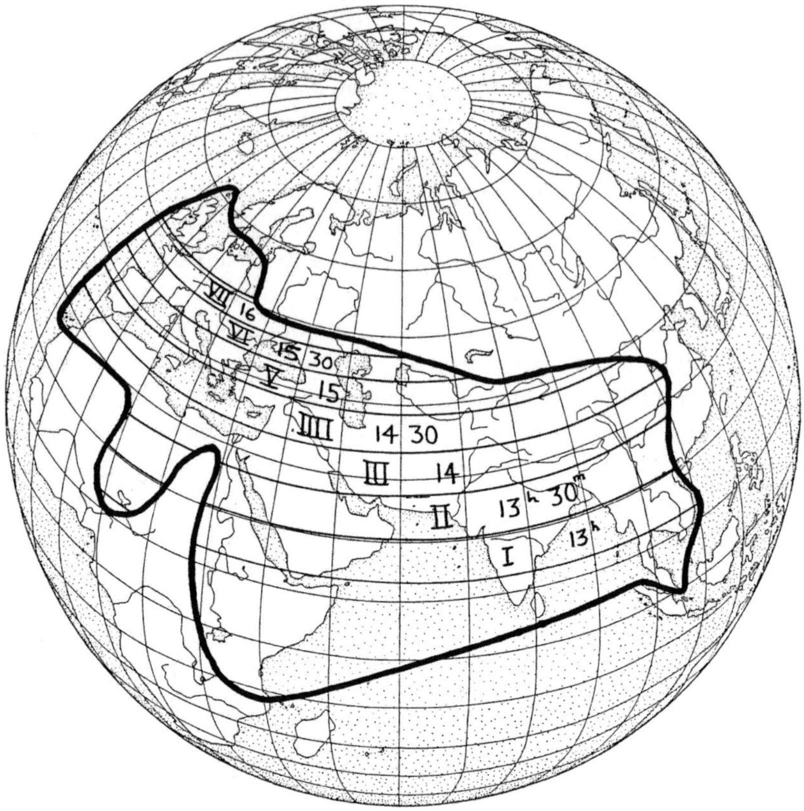

Fig. 1: The climates of Antiquity and the world known to Ptolemy. The world of Ibn al-ʿAṭṭār was the same, but he used a different definition of the climates. (From D. King, *World-Maps for Finding the Direction and Distance to Mecca: Innovation and Tradition in Islamic Science* [Leiden, 1999], p. 24.)

over the centuries because the obliquity of the ecliptic changes slowly with time.[21]

In this study we use the sexagesimal notation standard in the history of science; in the Islamic sources numbers are expressed in alphanumerical (*abjad*) notation.[22] Ptolemy had derived 23;51,20° for the obliquity

---

[21] See the *EI*² articles 'Mayl' [= declination and obliquity] by D.A. King and 'Minṭaḳat al-burūdj' [= ecliptic and obliquity] by P. Kunitzsch. The necessary technical background is provided in King, *Synchrony*, i, pp. 27–38.

[22] Sexagesimal numbers are written in the form p;q,r, representing p + q/60 + r/3600. On the Arabic *abjad* notation see R.A.K. Irani, 'Arabic Numeral Forms,' *Centaurus* 4 (1955), pp. 1–12, repr. in D.A. King and M.H. Kennedy (eds.), *E.S. Kennedy,*

and his latitudes for the climates are based on that parameter. Muslim astronomers, already in the early ninth century onwards, determined better values of this parameter, 23;33° and 23;35°, that of Ptolemy being long out of date. The maximum length of daylight D in equinoctial degrees or hours (where 360° = 24 hours, so that 1 hour = 15°) is a function of the terrestrial latitude (φ) and the obliquity of the ecliptic (ε):

$$D(\varphi,\varepsilon) = 2 \{ 90^\circ + \sin^{-1} (\tan \varepsilon / \tan \varphi) \} \text{ in equinoctial degrees, or}$$
$$2 \{ 90^\circ + \sin^{-1} (\tan \varepsilon / \tan \varphi) \} / 15 \text{ in equinoctial hours}$$

Hence, for a specific climate C_n ($n = 1, 2, \ldots, 7$) defined by the length of longest daylight D_n, if the obliquity changes, then so does the latitude of the climate, $\varphi_n(D_n,\varepsilon)$. The latitudes of the climates corresponding to the most widely used values of the obliquity are as follows:

Table 1—The latitudes of the climates

	C_1	C_2	C_3	C_4	C_5	C_6	C_7
ε \ max D	13^h	13;30	14	14;30	15	15;30	16
23;51°	16;27°	23;49	30;21	36;1	40;53	45;1	48;32
23;35°	16;39	24;5	30;40	36;22	41;14	45;22	48;53

The climates presented no problem to most Muslim scientists. Thus, for example, al-Bīrūnī dealt with them in detail presenting a surplus of numerical information (based on ε = 23;35°) even on their areas.[23] In Greater Iran they seem to have remained as Ptolemy had defined them into the early modern period,[24] although different values based on different values of the obliquity were in circulation. However, the changing latitudes of the climates confused some scholars who were not completely in control of their subject or who were bound by traditionalism.[25] The standard value of the obliquity used by Mamlūk

Colleagues and Former Students. Studies in the Islamic Exact Sciences (Beirut, 1983), pp. 710–21.

[23] See A. Dallal, 'Al-Bīrūnī on Climates,' *Archives internationales d'histoire des sciences* 34 (1984), pp. 3–18.

[24] King, *World-Maps*, pp. 230–34.

[25] This is one reason why, for example, many astrolabe-makers over the centuries preferred to keep on using, either explicitly or implicitly, Ptolemy's outdated value of the obliquity 23;51° in their presentations of the latitudes of the climates or the length of daylight for different latitudes. See King, *Synchrony*, ii, pp. 948–58.

astronomers was 23;35°, derived by Ibn Yūnus *c.* 990, but already used by his predecessors Ḥabash al-Ḥāsib and al-Battānī,[26] as well as al-Bīrūnī and Naṣīr al-Dīn al-Ṭūsī (Iran, *c.* 1250), before he measured it properly. In the following table we present the latitudes of the climates as presented by Ptolemy, and in two Islamic works, the *Tadhkira fī ʿilm al-hayʾa* by al-Ṭūsī,[27] and the popular summary *al-Mulakhkhaṣ fī ʾl-hayʾa* by Maḥmūd ibn Muḥammad ibn ʿUmar al-Jaghmīnī (Khwarazm, *fl. c.* 1221/22),[28] whose complicated history has now been to some extent unravelled by Jamil Ragep.[29] In the following table, adopted from Ragep, the latitudes of the middles of the climates are presented for Ptolemy (PT), based on ε = 23;51°, and al-Ṭūsī (TS), based on ε = 23;35°, and three different versions of al-Jaghmīnī: the 1246 Laleli copy (LL), the German translation (RH), and a commentary by Qāḍī Zāde al-Rūmī (QZ).[30]

In Egypt, some scholars proposed a different definition of the climates, to make them begin at the equator and embrace all latitudes up to the Arctic Circle. We find such aberrant definitions already in the geographical tables of Najm al-Dīn al-Miṣrī *c.* 1325.[31] They are also implicit in the world-map of Ibn Faḍlallāh—see Fig. 2[32]—in which, however, the

---

[26] See King, *Synchrony*, i, pp. 44, 164 and 247.

[27] F.J. Ragep, *Naṣīr al-Dīn al-Ṭūsī's Memoir on Astronomy* (al-Tadhkira fī ʿilm al-hayʾa), 2 vols. (New York, 1993), i, pp. 250–53, and ii, pp. 469–71. On his values for the climates see also King, *World-Maps*, p. 233.

[28] Translated in G. Rudloff and A. Hochheim, 'Die Astronomie des Maḥmūd ibn Muḥammad ibn ʾOmar al-Gagmīnī,' *ZDMG* 47 (1893), pp. 213–75, (repr. in F. Sezgin [ed.], *Islamic Mathematics and Astronomy*, lxxvii [Frankfurt, 1998]); see especially pp. 260–61. On the author see Suter, 'Die Mathematiker und Astronomen der Araber,' no. 403; C.A. Storey, *Persian Literature. A Bio-bibliographical Survey*, ii/1 (London, 1972), no. C88; King, *Survey*, no. G17; and Rosenfeld and İhsanoğlu, *Mathematicians, Astronomers and Other Scholars*, no. 547. See also the next note and the text to n. 45 below.

[29] See now J. Ragep, 'On Dating Jaghmīnī and his Mulakhkhaṣ,' in M. Kaçar and Z. Durukal (eds.), *Essays in Honour of Ekmeleddin İhsanoğlu*, 2 vols. (Istanbul, 2006). I am grateful to Jamil Ragep for a preprint of this paper.

[30] On Qāḍī Zāde see Suter, 'Die Mathematiker und Astronomen der Araber,' no. 430; Storey, *Persian Literature*, nos. A16 and C118; King, *Survey*, no. G50; and Rosenfeld and İhsanoğlu, *Mathematicians, Astronomers and Other Scholars*, no. 808. Ragep cites an Iranian edition from *c.* 1880 of his commentary on al-Jagmīnī.

[31] On his use of the climates see King, *World-Maps*, p. 79, where he states that the fifth climate runs from 31° to 36° (!!).

[32] See King, *World-Maps*, pp. 34–7. This map has nothing to do with the world-map of the ʿAbbāsid Caliph al-Maʾmūn, contrary to various claims in the modern literature. That map, which has not survived, is known to have had an orthogonal grid: see Kennedy, 'Suhrāb'. Besides, the sophisticated grid on Ibn Faḍlallāh's map is entirely superfluous, not least because no cities are marked on the map.

Table 2—The latitudes of the climates according to Ptolemy (ε = 23;51°), al-Ṭūsī (ε = 23;35°), and three versions of al-Jaghmīnī (after Ragep)

C	D max	PT	TS	LL	RH	QZ
C_1	13^h	16;27° [0]	16;37,30° [a]	16;27	15;37 [!]	16;37
C_2	13;30	23;51 [+2]	24;5 [0]/24;40 [!][b]	23;51	24;40 [!]	24;40 [!]
C_3	14	30;22 [+1]	30;40 [0]	30;22	30;40	30;40
C_4	14;30	36;0 [−1]	36;22 [0]	36;0	35;22 [!]	36;22
C_5	15	40;56 [+3]	41;15 [+1]	40;56	41;15	41;15
C_6	15;30	45;1 [0]	45;21 [−1]	45;1	45;21	45;21
C_7	16	48;32 [+1]	48;52,30 [c]	48;32	48;52	48;52
C_7 end	16;15	50;4 [0]	50;20 [−5]	50;25 [!]	50;20	50;20

[a] Accurately to seconds 16;38,48 (!!).[b] Here the 40 minutes have probably been taken from the entry for C_3. [c] Accurately to seconds 48;52,35°.

scale for the climates shows them to be of the same width. The centres of the Ptolemaic climates C_1, C_2, …, C_6 become the beginnings of the Egyptian climates E_2, E_3, …, E_7; also, E_1 begins at the equator, and E_7 ends at the Arctic Circle. It is perhaps significant that the historian Ibn Khaldūn *c.* 1377 attributes this arrangement to Ptolemy himself, rather than recognizing it as a distortion of Ptolemy's scheme.[33]

2. The Geography of Ibn al-ʿAṭṭār

Our source is a single manuscript, which shows how precarious and arbitrary our chances are of reconstructing history. Ms. HL 2520 (formerly 2469) in the Khoda Bakhsh Library in Patna is a collection of mainly astronomical treatises by a single author. The manuscript was catalogued in a masterful fashion in 1937 by Maulavi Abdul Hamid.[34] The author is the fifteenth-century Egyptian astronomer Muḥibb al-Dīn (Abū ʿAbdallāh) Muḥammad ibn Muḥammad and known as Ibn al-ʿAṭṭār, whose name is not new to the standard bio-bibliographical sources for late Islamic science.[35] Some of the treatises contained in it

[33] Ibn Khaldūn, *The Muqaddimah. An Introduction to History*, trans. F. Rosenthal, 3 vols. (Princeton, N.J., 1958, ²1967), i, pp. 112–13.

[34] M. Abdul Hamid, *Catalogue of the Arabic and Persian Manuscripts in the Oriental Public Library at Bankipore*, xxii (Arabic MSS): Science (Bihar, Patna, 1937), pp. 93–101 (no. 2469), especially pp. 94–5. These pages are full of information on Ibn al-ʿAṭṭār.

[35] The first mention of Ibn al-ʿAṭṭār in the modern literature is in Suter, 'Die Mathematiker und Astronomen der Araber,' no. 431, listing his treatise on different types of quadrants and a set of astronomical tables in Oxford (see n. 37 below). *GAL* II, pp.

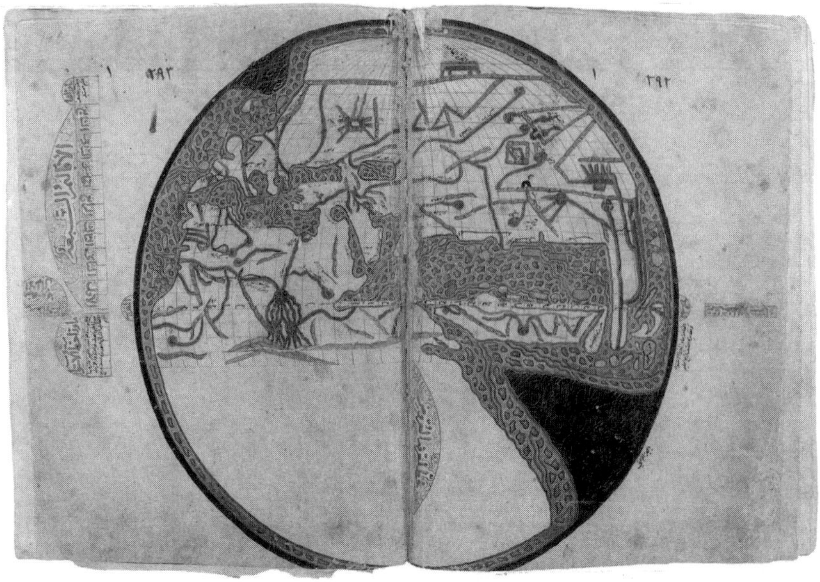

Fig. 2: The climates in the world-map presented by Ibn Faḍlallāh are defined in a different fashion from those in Fig. 1. The unhappy scale to the left of the map divides the climates more or less equally between the equator and the Arctic Circle. (From King, *World-Maps*, p. 34.)

are known from other copies. However, fol. 11v–25r preserve for us a unique copy of Ibn al-ʿAṭṭār's geographical work which has no title, but begins: *hādhihi ʿurūḍ al-bilād wa-aṭwāluhā wa-inḥirāfuhā ʿan al-jihāt fī 'l-aqālīm al-sabʿa wa-ghayruhā*, 'These are the latitudes and longitudes of localities in the seven climates and their *qibla* directions with respect to the cardinal directions'. See Fig. 3. This text was copied in 874 A.H. [= 1469/70]—see the Appendix for the colophons. Ibn al-ʿAṭṭār has already attracted our attention as the author of a treatise entitled *Kashf al-qināʿ fī rasm al-arbāʿ*, which deals with different kinds of quadrants, and reveals that he was in touch with the impressive developments in earlier Mamlūk instrumentation; indeed, he had studied with some of

157–8, and *S* II, p. 158, mentions all of the relevant treatises in the Patna manuscript. King, *Survey*, no. C66, mentions his treatise on quadrants, his treatise on the construction of curves on astrolabe plates, and a lost treatise on astronomical timekeeping entitled *Jawharat al-yawāqīt*, albeit associated with Shams al-Dīn Muḥammad ibn al-ʿAṭṭār al-Bakrī al-Shāfiʿī (see further ibid., no. C124 *ad* 3.1.28). Maybe this is the father of our author? Rosenfeld and İhsanoğlu, *Mathematicians, Astronomers and Other Scholars*, no. 813, has no 'new' materials, but conveniently lists all of the 'old' ones.

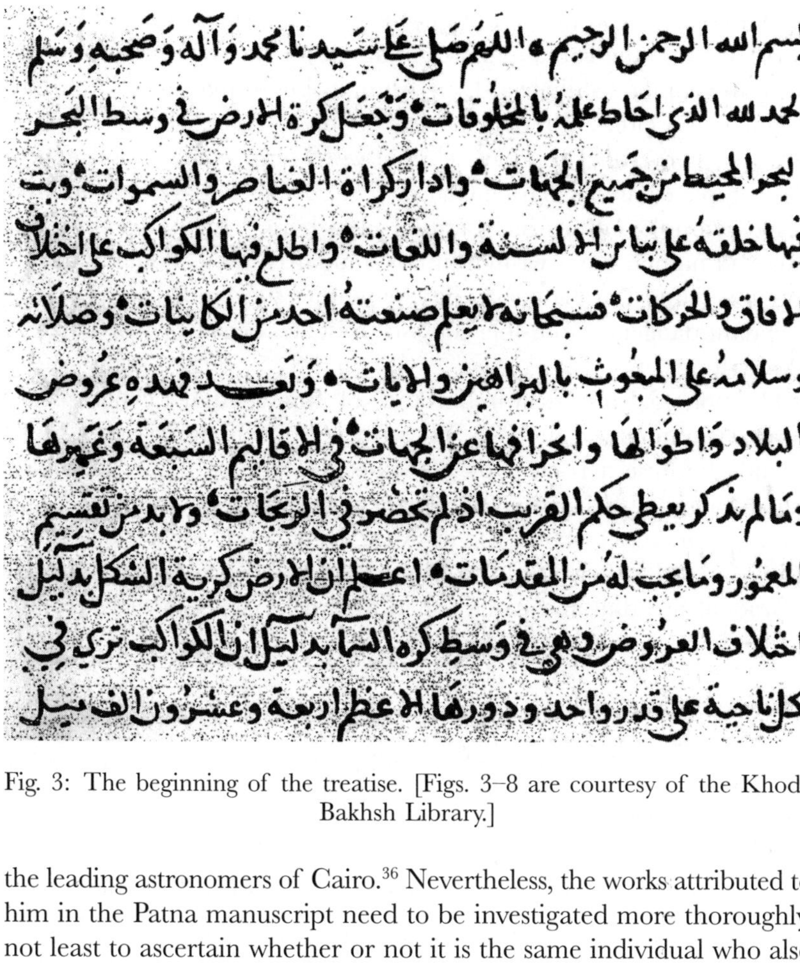
بسم الله الرحمن الرحيم اللهم صلى على سيدنا محمد وآله وصحبه وسلم
الحمد لله الذي احاط علمه بالمخلوقات وجعل كرة الارض في وسط البحر
البحر المحيط من جميع الجهات وادار كرة العناصر والسموات وبث
فيها خلقه على تباين الالسنة واللغات واطلع فيها الكواكب على اختلاف
الافاق والحركات فسبحانه لا يعلم صنعته احد من الكاينات وصلاته
وسلامه على المبعوث بالبراهين والايات وبعد فهذه عروض
البلاد واطوالها وانحرافها عن الجهات في الاقاليم السبعة وغيرها
وما لم يذكر يعطى حكم القريب اذ لم تحصر في الزيجات ولا بد من تقسيم
المعمور وما يجب له من المقدمات اعلم ان الارض كرية الشكل بدليل
اختلاف العروض وهي في وسط كرة السما بدليل ان الكواكب ترى في
كل ناحية على قدر واحد ودورها الاعظم اربعة وعشرون الف ميل

Fig. 3: The beginning of the treatise. [Figs. 3–8 are courtesy of the Khoda Bakhsh Library.]

the leading astronomers of Cairo.[36] Nevertheless, the works attributed to him in the Patna manuscript need to be investigated more thoroughly, not least to ascertain whether or not it is the same individual who also wrote treatises in Damascus.[37]

---

[36] On his activities within the context of Mamlūk astronomy see Charette, *Mathematical Instrumentation*, pp. 7, 19–20, etc. On his treatise on quadrants see idem and King, *The Universal Astrolabe of Ibn al-Sarrāj*, forthcoming. Charette has located a detailed commentary on this treatise by Ibn al-ʿAṭṭār himself, which remains to be studied: see his forthcoming paper 'The Locales of Islamic Astronomical Instrumentation,' *History of Science* 44 (2006), pp. 123–38.

[37] Ms. Oxford Bodleian Laud. Or. 221 (Uri I.974) contains an extensive set of planetary tables, stated to be for the longitude of Damascus, said to be based on those of the celebrated fourteenth-century astronomer Ibn al-Shāṭir and attributed to (and copied by?) Muḥammad ibn al-ʿAṭṭār. The date mentioned in the instructions on the first page is 930 A.H. The mean-motion tables begin with the date 870 A.H. and have entries for each 30 years. For each of the five planets there are double-argument equation tables. Dr. Benno van Dalen kindly showed me a microfilm of the Oxford manuscript.

The copy of Ibn al-ʿAṭṭār's *Geography*, and various copies of other works of his in the Patna manuscript are in the hand of another Egyptian astronomer, (Shihāb al-Dīn) Aḥmad ibn Aḥmad ibn Timurbāy al-Ḥanafī al-Qādirī, a student (*tilmīdh*) of Ibn al-ʿAṭṭār, who read it back to his teacher.[38] Ibn Timurbāy's elegant *naskhī* script sometimes degenerates completely in his marginalia (see Fig. 8), and we have encountered worse examples of his handwriting elsewhere.[39] In the treatise and tables all numerals are written in the standard Arabic alphanumerical notation,[40] and Ibn Timurbāy has sometimes succumbed to all of the attendant dangers of possible copyists' mistakes because of his careless handwriting.[41]

The potential historical importance of Ibn al-ʿAṭṭār's geographical work was announced in 1999.[42] It attracted our attention again more recently because it is cited by the sixteenth-century Tunisian scholar ʿAlī al-Sharafī al-Ṣafāqusī in the text accompanying his sea atlas.[43] The work does little to enhance Ibn al-ʿAṭṭār's reputation, for it is naïve and ill informed. Furthermore, Ibn Timurbāy distinguishes himself by his occasional dreadful and careless handwriting. What concerns us here

---

[38] King, *Survey*, no. C94, and Rosenfeld and İhsanoğlu, *Mathematicians, Astronomers and Other Scholars*, no. 908, mention *al-Barq al-sāṭiʿ fī mukhtaṣar al-Bārīʾ*, a commentary on the treatise of Ibn Abī 'l-Rijāl (also *GAS* VII, p. 187); a second astrological treatise *Risāla fī maʿrifat al-ḥawādith al-sufliyya min dalālāt al-ashkhāṣ al-ʿulwiyya*; an astrological treatise copied in 894 A.H. in his own hand; and some solar and lunar tables in his hand. See also the next note on another manuscript copied by him. Perhaps our man was a grandson of the Mamlūk *amīr* Tīmūrbughā Mintāsh. See also the text to note 71 below.

The vowelling of the Turkic name T-m-r-Bāy or T-m-r-Bughā has caused some problems. In the *Patna Catalogue* (n. 34) our author becomes 'Aḥmad Tamirtāshī' (p. 95) and 'Aḥmad ibn Muḥammad al-Tamirtāshī' (p. 96). In *GAS* VII, p. 187, he becomes 'Aḥmad ibn Tīmūrbughā (or Tamrīta)'. In Rosenfeld and İhsanoğlu, *Mathematicians, Astronomers and Other Scholars*, no. 908, he becomes Aḥmad ibn Tamirbuqā. In the *Encyclopedia of Islam* we encounter the Mamlūk sultan (al-Ẓāhir) Temirboghā (*reg.* 1467–68) (IV, col. 462b), and the Mamlūk *amīr* Tīmūrbughā Mintāsh *c.* 1390 (VI, cols. 548a and 580b), but in the Index these become Tīmūrbughā. It seems that in Arabic the long vowels of Tīmūr (meaning 'iron or steel' in Turkish) could be shortened when followed by such appendages, as in Timurtāsh and even Timurtash. In modern Turkish the orthography Temür is preferred.

[39] See King, *Survey*, pl. LXXXIII on p. 303, illustrating a fragment of an astrological history, as yet unstudied, of the campaigns between the Mamlūks and Ottomans around Adana during the period 1470–73.

[40] See n. 22.

[41] On copyists' errors in geographical tables see King *World-Maps*, pp. 161–3.

[42] King *World-Maps*, p. 83.

[43] This is the topic of a doctoral dissertation by Mónica Herrera Casais at the Institute of History of Science in Frankfurt and the Universidad de La Laguna (Tenerife).

are, firstly, Ibn al-ʿAṭṭār's statements about the latitudes of the climates and Ibn Timurbāy's marginalia, which may also be due to the author, since Ibn Timurbāy read the text back to his teacher, and secondly, their joint presentation of a substantial corpus of geographical data.

3. Ibn al-ʿAṭṭār on the Climates

Ibn al-ʿAṭṭār uses the climates as defined by, say, Ibn Khaldūn.[44] When he writes 'the nth climate: its latitude is from the latitude of the previous climate to φ_n degrees', the values φ_n are in fact *the middles* of each of the Ptolemaic climates. Also he writes 'the latitude of the first climate is from the equator to φ_1 degrees', and 'the limit of (the seventh climate) is latitude 66;25°'. So the climates have been stretched to fill all the space between the equator and the Arctic Circle. In addition, since the Arctic Circle is at latitude 66;25° = 90°–23;35°, we might expect that our author would use obliquity 23;35° throughout.

In any case, the latitudes associated with each of the climates are given as follows, with some marginal additions labelled *Mg* below. Other marginal insertions, which are labelled in the text *(al-)Mulakhkhaṣ*, here abbreviated *Mu*, indicate that our astronomers have compared the values with those in the very popular summary of Ptolemaic astronomy entitled *al-Mulakhkhaṣ fī 'l-hayʾa* by al-Jaghmīnī.[45] We here distinguish between the Laleli copy, *LL*, the German translation, *RH*, and Qāḍī Zāde's commentary, *QZ*. We sometimes refer to Ptolemy's values, *PT*, based on ε = 23;51°, which are essentially as found in *LL*, whereas al-Ṭūsī's values, TS, are based on ε = 23;35°, and recur in *RH* and *QZ* (see Table 2 above). In some cases, values for the climates are given in the geographical tables that follow (§4), here abbreviated *GT*. In the accompanying notes, the reader should bear in mind that the phrase 'has been computed for a certain value of the parameter ε' means 'lifted from a source in which the latitudes were based on that value for ε':

---

[44] See the text to n. 33 above.
[45] See n. 28 above.

Table 3—The latitudes of the climates given by Ibn al-ʿAṭṭār

E_1 begins at latitude 0°	This confirms that we are dealing with the Egyptian definition of the climates.
Beginning of E_1 (middle of C_1): 16;25°, confirmed in GT, Mg: (16);35°, Mu: 16;37°	16;25° has been computed for ε = 23;51° (accurately 16;27°, as in PT/LL) or is a copyist's error for 16;35°; 16;35° has been computed for ε = 23;35° (accurately 16;39°). 16;37°, probably originally rounded from TS 16;37,30, agrees with QZ
Beginning of E_2 (middle of C_2): 24;40°, Mg: 23;50°, Mu: 24;40° or 23;50° (unclear)	24;40° is an aberrant error—perhaps resulting from some copyist combining the degrees of 24;5° of C_2 in TS and the minutes of 30;44 for C_3—which is found in copies of TS and recurs in RH/QZ. 23;50° is probably rounded from 23;51°, which is found in PT/LL.
Beginning of E_3 (middle of C_3): 30;40°, Mu: 30;40°	30;40° has been accurately computed for ε = 23;35°, and is found in TS/RH/QZ
Beginning of E_4 (middle of C_4): 36;22°, Mu: 36;22°; GT: 36;0°	36;22° agrees with TS/QZ and has been accurately computed for ε = 23;35°. 36;0° is computed for ε = 23;51° (accurately 36;1°) and agrees with PT/LL.
Beginning of E_5 (middle of C_5): 40;15°, Mu: 41;15°, GT: 40;55°	40;15° is probably an error for 40;55°, perhaps rounded from the 40;56° in PT/LL; 41;15° is computed for ε = 23;35° and is found in TS/RH/QZ.
Beginning of E_6 (middle of C_6): 45;0°, confirmed in GT, Mg: (45);21°; Mu: 45;21°	45;0° has been computed for ε = 23;51° or rounded (the accurate value is 45;1°, as in PT/LL); 45;21°, as in TS/RH/QZ, is computed for ε = 23;35° (accurately 45;22°).
Beginning of E_7 (middle of C_7): 48;30°, confirmed in GT, Mu: (48);52°, Mg: 50;25°	48;30° has been computed for ε = 23;51° (accurately 48;31°) or miscopied from 48;32° as in PT/LL. 48;52°, as in TS/RH/QZ, was computed for ε = 23;35° (accurately 48;53°). 50;25° is the upper limit of C_7 according to the Ptolemaic definition and is accurately computed for ε = 23;35°: TS/RH/QZ have 50;20°, but LL has 50;25°.
E_7 ends at latitude 66;25°	The end of E_7 at 66;25° implies an ε of 23;35°, still using the Egyptian definition of the climates.

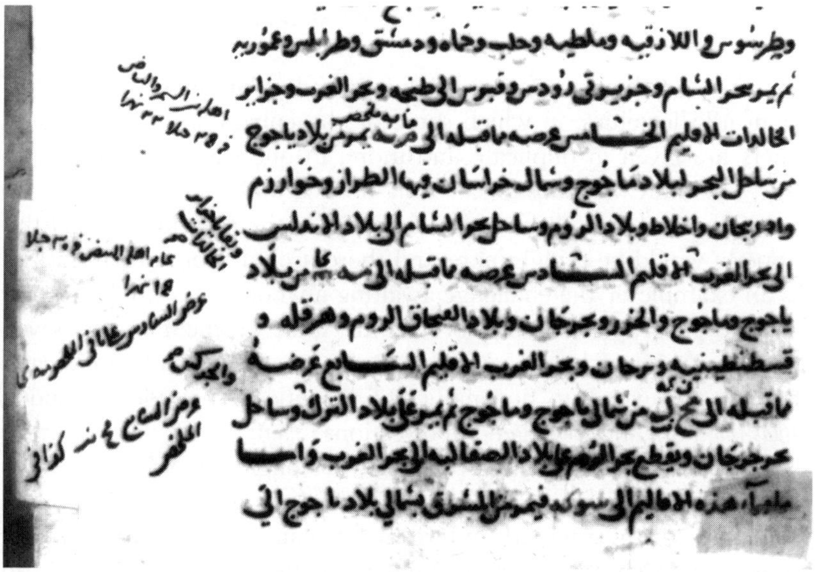

Fig. 4: The information on the 5th, 6th and 7th climates.

In other words, the information presented here is a hopeless jumble (by the higher standards of Islamic mathematical geography in contemporaneous Iran and Central Asia). The latitudes in the text and in the marginalia have been taken from at least two separate sources and have been shuffled a few times.

The numerical information on the climates is followed by lists of localities in each one—see Fig. 4; these will not concern us here, save to note that they show no advance over the previous 600 years.[46] All this does not reflect well on either the author or the copyist, and it does not augur well for the geographical tables that follow.

[46] We present one set of data here, namely, the localities in the 'third climate' on fol. 12v: 'Northern China and India, including Qandahar, Sind, then Kabul, Kirman, Sijistan, Mahdiyya, coast of al-Basra, including Istakhr (Persepolis), Siraf, Shiraz, Ahwaz, Sapur, then al-ʿIrāq, including Wasit, Anbar, south of Kufa and Basra, Baghdad, Syria or Damascus (*al-Shām*), Acre, Jerusalem, Gaza, then Madyan [= Madyan Shuʿayb], Qulzum, Cairo, and the south of Alexandria, Barqa, south of Ifriqiya, Kairouan, to the Maghrib and the encompassing Ocean (*al-baḥr al-muḥīṭ*).'

4. Ibn al-ʿAṭṭār's Geographical Data

The text is followed by a table of longitudes, latitudes and *qiblas* of about 425 localities, in which some of the entries are left blank and others are given in duplicate and more, in one case, even quintuplicate. There are 12 folios of carefully-framed tables with 30 entries on each folio, and the other entries are given in the margins. To edit this properly would be a nightmare and there are far more urgent tasks.[47] As an example of the challenge awaiting anybody who would venture into the geographical tables of Ibn al-ʿAṭṭār, we mention just seven sets of entries: see Figs. 5–6 for some of these. The heading indicates that we can expect to find names of localities, longitudes, latitudes, *qiblas* (*inḥirāf*), and their general directions (*jiha*), that is, the quadrants between the cardinal points. We use L for longitude, φ for latitude as before, and q for *qibla*, measured from the local meridian. The values for Mecca are particularly interesting because for the former their sources are identified, if cryptically, with the four labels *Qānūn*, Ibn Saʿīd, Kūshyār and *Rasm*. These, and other relevant sources, here in chronological order, are abbreviated as follows, in the tradition of the corpus of Islamic geographical data compiled by Ted and Mary Helen Kennedy (hereafter K&K), which provides the key to any investigations of Islamic coordinates.[48]

PTO	Ptolemy (*c.* 140)
RSM FID	*Rasm al-rubʿ al-maʿmūr* (*c.* 820), cited by Abū 'l-Fidāʾ (FID) (K&K: RES)
KHU	the *Kitāb ṣūrat al-arḍ* by al-Khwārizmī (*fl. c.* 830)
KHW	other sources for al-Khwārizmī
KHZ	a related list in an Istanbul manuscript
SUH	Suhrāb (*fl. c.* 930)
YUN	Ibn Yūnus (*fl. c.* 990)
KUS	Kūshyār ibn Labbān (*fl. c.* 1000)
BIR	*al-Qānūn al-Masʿūdī* of al-Bīrūnī (*c.* 1025)
BIR FID	values attributed to BIR by Abū 'l-Fidāʾ (FID)

[47] Most other late medieval Egyptian geographical tables are full of errors of one sort or another: see King *World-Maps*, pp. 76–84, and various tables edited on pp. 600–622.

[48] Kennedy & Kennedy, *Geographical Coordinates*. The abbreviations are on pp. xv–xxxvii.

ATW	The anonymous *Kitāb al-aṭwāl wa'l-ʿurūḍ li'l-Furs*, compiled in Isfahan (?) in the eleventh (?) century (K&K: ATH)
ATW FID	values attributed to ATW by Abū 'l-Fidāʾ (FID)
SNJ	the reconstructed values in the *Sanjarī Zīj* of ʿAbd al-Raḥmān al-Khāzinī (see King, *World-Maps*, pp. 71–5 and 564–85)
ZAY	the *Kitāb aḥkām al-marjān*... by Isḥāq ibn al-Ḥasan al-Zayyāt (*fl. c.* 1040)
SAA	the *Kitāb basṭ al-arḍ fī 'l-ṭūl wa'l-arḍ* of Ibn Saʿīd al-Maghribī (d. 1286)
SAA FID	values attributed to SAA by Abū 'l-Fidāʾ (FID)
MAR	al-Marrākushī (*fl. c.* 1280)
BNA	Ibn al-Bannāʾ (*fl. c.* 1300)
LYD	a fourteenth-century Egyptian list, being essentially a selection from ZAY, but with a list of the Egyptian pilgrim-stations appended
FID	Abū 'l-Fidāʾ (*c.* 1325), cites RSM, FID, ATW, BIR, SAA, amongst others
SHA	Ibn al-Shāṭir (*fl. c.* 1350), derived from SNJ
AIN	the *Āʾīn-i Akbarī* of Abū 'l-Faḍl ʿAllāmī (India, *fl. c.* 1580), derived from ATW and its derivatives

Table 4—The latitudes of various localities given by Ibn al-ʿAṭṭār

Locality, L, φ, q, stated source				*Comments*
Mecca[49]				
67; 0°	21;20°	N[!]SE[!]	BIR	as in BIR, confirmed by BIR FID, also in SNJ and SHA. On the entry for the qibla see below.
67;31	21;30	–	SAA	SAA & SAA FID have 61;31°/21;31°, which is clearly a double copyist's error for 61;0°/21;0°.[50] The latter coordinates are as found in several other sources, especially the derivative SAA & BNA. It is the value 21;30° that surprises us here.

[49] Kennedy & Kennedy, *Geographical Coordinates*, pp. 225–6.

[50] The symbols for '0,' of Greek origin have been mistasken for the vaguely similar ligature for *lām-alif*: see Irani, 'Arabic Numeral Forms,' pp. 11–12.

Table 4 (*cont.*)

Locality, L,	*φ,*	*q,*	*stated source*	*Comments*
67;13	21;40	–	KUS	KUS has 67;10°/21;40°, which is repeated in some dependent sources. φ = 21;40° is an early value, also found in ATW: see below.
67; 0	21; 0	–	RSM	as in RSM FID, KHU, SUH, YUN, also ZAY & BNA. These are the coordinates used by Ibn Yūnus to compute the qibla for Cairo-Fustat at 55°/30°—see below.

Fig. 5–6: The entries for Mecca and Saba, together with Dongola and Baṭn Marr, and Cairo and Ḥāmma.

The values 21°, 21;20°, 21;30° and 21;40° for the latitude of Mecca were apparently all derived in the ninth century, by whom, alas, we do not know.[51] For the *qibla* at Mecca there is written the word *shamāl*, 'north', which is absurd,[52] and this is associated with the quadrant of south-east (*j-q* for *janūbī-sharqī*), which is even more absurd. The same quadrant is associated with the next entry, for Taif, which has *zawāl*, 'meridian', for the *qibla*.

Cairo/Fustat[53]

53; 5°	30;10°	13; 0° SE	–	written by the name *Miṣr*; no other source has these values.
[6]4;50	29;15	–	–	MAR has 64;50° / 29;55°; no other sources have longitudes with 50 minutes; KHZ has φf = 29;55° and SUH has 29;15°. On φ = 29;15° and 29;55° see below.
–;30	30;0	–	–	no other sources have longitudes with 30 minutes; φ = 30° common after YUN.
34;40	[30;0°]	–	–	no φ is given; KHZ has L = 54;40°, which explains the absurd copyists' error; several later sources have 54;40°, with φ = 30;0°.
55; 0	30;0	37;0° SE	?	YUN, BNA, and various Yemeni sources, have these coordinates. On the *qibla* value see below. The 'source' here indicated by '?' may simply be the word *Miṣr* repeated.

The value 29;15° for the latitude of Cairo seems to a careless mistake for the 29;55° used by al-Marrākushī.[54] Furthermore, two different values for the *qibla* at Cairo are presented without comment: they are 13°, a careless scribal error for 53°, and 37°, the former being measured from the meridian and the latter from the east west line. The single direction is that calculated by Ibn Yūnus *c.* 990 (for coordinates 55°/30° for Cairo and 67°/21° for Mecca), which was generally favoured by the Mamlūk astronomers.[55]

---

[51] See D. King, 'Too Many Cooks...—A Newly-Rediscovered Account of the First Muslim Geodetic Measurements,' *Suhayl* 1 (2000), pp. 207–41, especially pp. 225–6.

[52] One would expect a phrase such as *qiblat al-ʿālam*, '*qibla* of the whole world'.

[53] Kennedy & Kennedy, *Geographical Coordinates*, pp. 111–12.

[54] Although it may be that the 29;15° is an older value: see King, *Synchrony*, ii, p. 932, n. 25.

[55] King, *Synchrony*, i, pp. 268–70 and 804.

Damascus[56]				
60; 0	33;30	31;10 / 29; 9	–	BIR has these coordinates, as well as various late sources. On the two *qibla* values see below.

The *qibla* value of 29;9° appears to be a scribal error for 29;4° (*ṭā'* for *dāl*!), the value of al-Khalīlī for these coordinates and 67;0°/21;30° for Mecca.[57] The other value does not correspond to any known medieval values.

Two lesser localities are featured in the extract in Fig. 5, namely, Dongola in Nubia and Baṭn Marr near Mecca. Also, as was the case with some earlier Mamlūk Egyptian tables, all manner of irrelevant localities—places that had been destroyed or abandoned centuries before—are listed. Modest amounts of information are given for two such localities of this kind—Saba' in the Yemen (Sheba, modern Marib) and Madīnat Ḥamma in Southern Ifrīqiya—featured in Figs. 5–6, but the details are of considerable historical interest for other reasons, not least for identifying the source of the data:

Dongola[58]				
53;40°	14;30°	–	–	as in SUH and probably ATW FID (K&K give L = 43;40°); BIR has 53;40°/14;0°
53;40	17; 0	81;30 SW[!]	–	f as in ZAY and MAR; ZAY has L = 53;0°. On the *qibla* value see below.
Baṭn Marr[59]				
66;30	21;55	5; 0 SE	–	only as in ATW FID, as in AIN, a late Indian source relying on ATW (now with 77;0°/21;55° and a meridian 10° to the west). On the *qibla* value see below.
Saba'[60]				
64°	17;10°		–	as in KHU & RSM FID; PTO has 73;40°/17;10°.

[56] Kennedy & Kennedy, *Geographical Coordinates*, pp. 99–100.

[57] King, *World-Maps*, entries F4–93 on p. 615 and F7–14 on p. 620, and *Synchrony*, i, entry 14 on p. 392.

[58] Kennedy & Kennedy, *Geographical Coordinates*, pp. 109–10 (two entries).

[59] Kennedy & Kennedy, *Geographical Coordinates*, p. 87.

[60] Kennedy & Kennedy, *Geographical Coordinates*, pp. 289–90.

68; 0	17;30	12; 0 SE[!]	–	no sources have φ = 17;30°. On the *qibla* value see below.
68;0	14; 0		–	no value given for L; ATW FID & BIR have 68;0°/14;0°.
Ḥāmma[61]				
10;20°	28;30°	–	–	this locality occurs only in PTO (33;40°/28;15°), KHU (34;30°/28;30°) and SUH (34;20°/28;30°).

The *qibla*-value for Baṭn Marr is attested in an anonymous fourteenth-century list of stations on the pilgrim road from Cairo to Mecca (LYD), where it is associated with other coordinates (to which it does not correspond).[62] The *qibla* of Dongola is between 70° and 75° in other Egyptian sources.[63] The *qibla* of Sabaʾ is absurd, and bears no relation to medieval Egyptian values.[64] The longitude given for Ḥāmma shows that our two astronomers had no idea where Ptolemy intended it to be.

Finally, we consider the entries at the bottom of the last page of tables: see Fig. 8. These are:

Hiraqla (Ereğli)[65]				
57;20°	46;35°	29;20°		ATW has 57;22°/46;30°. No other early sources have L between 57° and 58°. φ = 46;35° attested in KHU, KHZ, SUH, BIR, *etc*. The *qibla* value is lifted from SNJ (or one of its derivatives): there, however, it is associated with coordinates 53;25°/46;35°.[66]
Yājūj *al-dākhila* ('Inner Gog')[67]				
172;30°	63;0°	–	–	KHU has these coordinates for Yājūj *al-dākhila*, which locality does not occur elsewhere. SUH and various later sources have these coordinates for Yājūj and Mājūj.

[61] Kennedy & Kennedy, *Geographical Coordinates*, p. 133 (confused?).
[62] King, *World-Maps*, entry F4–M11 on p. 618.
[63] King, *World-Maps*, entries F2–51, F3–84, and F4–20, on pp. 604, 609 and 613.
[64] King, *World-Maps*, entry F2–52 on p. 604.
[65] Kennedy & Kennedy, *Geographical Coordinates*, pp. 138–9.
[66] King, *World-Maps*, entry 36 on p. 567.
[67] Kennedy & Kennedy, *Geographical Coordinates*, p. 377.

In brief, the information presented is confused, and more of it would need to be evaluated in order to identify all of the various sources that have been used. However, this would probably be an exercise with ever diminishing returns. With but one exception, Mecca, no sources are named, but these are not the sources used for other localities. The fact that there is not even a clear statement of the *qibla* at Cairo, and the fact that al-Khalīlī's *qibla* value for Damascus has been miscopied, are poor reflections on Ibn al-ʿAṭṭār and his student.

5. Ibn Timurbāy on Distances between Cities

By his colophon on fol. 20v, Ibn Timurbāy has penned the following remark:

> If you multiply the number of degrees and fractions thereof between two localities by fifty six and two thirds, the result will be the number of miles between them.

This conversion factor was attributed by al-Farghānī (Baghdad, *fl. c.* 850) to the earliest Muslim geodetic measurements in the early ninth century. Al-Bīrūnī regarded it with some suspicion since it is slightly different from the values in the actual reports of these measurements.[68] It would be interesting to know Ibn Timurbāy's source for the value. Inevitably, he does not present the complicated formula for finding the distance in degrees between two localities, which had been known for centuries.

6. Concluding Remarks

Almost 500 years had passed since al-Bīrūnī had written on the divergences between the coordinates in different sets of geographical tables that were in circulation in his time. Alas, his writings on this topic have not survived.[69] Almost 150 years had passed since the scholar-prince Abū 'l-Fidāʾ had noted different sets of coordinates in his *Geography*, properly identifying his sources. But one cannot conduct such research seriously if one is using and / or producing manuscripts in which the numbers are not uniquely identifiable or already hopelessly corrupt anyway. Scholars like Ibn al-ʿAṭṭār and Ibn Timurbāy were amongst

[68] King, 'Too Many Cooks,' pp. 216 and 219.

[69] They are listed in King, *World-Maps*, p. 344, n. 65 (after Boilot).

the leading scholars of their time, but they fell victim to the manuscript tradition. Certainly, their hearts were in the right place, but circumstances beyond their control rendered their efforts little more than an exercise in futility. And although they surely did not realize it, they were contributing to the decline of Islamic science. In this case, we are witness to an uncritical handling of a superfluity of uncontrolled data, without regard to the reliability of the sources, which could not be checked anyway. Not even the two inconsistent values for the *qibla* at Cairo received a comment.

Mathematical geography in medieval Europe was not much better, because there the coordinates in the *Toledan Tables*, a careless mix of values from al-Khwārizmī and al-Battānī, reigned supreme until the recovery of Ptolemy's original coordinates in the fifteenth century.[70] These then had to be corrected, which was a long and complicated procedure that sometimes involved recourse to Islamic lists. In the mid eighteenth century the German scientist Tobias Mayer drew maps of Europe based on the different sets of coordinates available to him, graphically demonstrating the need for new geodetic measurements. Such important undertakings were very much in the spirit of al-Bīrūnī and the compilers of the enigmatic *Kitāb al-aṭwāl wa'l-ʿurūḍ li'l-Furs* in the eleventh century, as well as poor Ibn al-ʿAṭṭār in the fifteenth.

[70] See, for example, N. Swerdlow, 'The Recovery of the Exact Sciences of Antiquity: Mathematics, Astronomy, Geography,' in A. Grafton (ed.), *Rome Reborn. The Vatican Library and Renaissance Culture* (Washington, D.C., 1993), pp. 125–68.

APPENDIX

MORE INFORMATION ON IBN AL-ʿAṬṬĀR

The colophons of the text of the treatise and the geographical tables (fol. 17v and 20v—see Figs. 7–8) are of some historical interest. They read:

> The copyist completed the reading of this text before its author (*balagha kātibuhu qirāʾa[-tan? for -tahu] ʿalā muʾallifihi*), the *shaykh*, *imām*, and scholar (*ʿālim*) and luminary (*ʿallāma*) Muḥibb al-Dīn Muḥammad ibn Muḥammad ibn Aḥmad ibn Aḥmad ibn Muḥammad ibn Mushrif, whose noble ancestor was known as Ibn al-ʿAṭṭār—may God treat him with kindness and with the blessing of his chosen Prophet and (all) the Muslims—in the months of the year 874 A.H. [= 1469/70 C.E.]. This was said and written (? *qālahu wa-katabahu* ?) by Aḥmad ibn Aḥmad ibn Timurbāy—may light clothe his sleep (? *w-l-b-s* ? *nawmahu al-nūr* ?)[71]—al-Ḥanafī al-Qādirī—may God forgive him and his parents and the best of the prophets and his family (?? *wa-li-khayr al-[a]nbiyā[ʾ] wa-li-ālihi* ??). God knows best.
>
> Completed by the hand of the needy servant of God, Aḥmad ibn Aḥmad ibn Timurbāy al-Ḥanafī al-Qādirī, student of the scholar Muḥibb al-Dīn Ibn al-ʿAṭṭār, may God treat both me and him with His hidden (? *khafī*) kindness, [and] His favour and generosity, in Rajab *al-fard* of the year 874 (A.H.) [= January, 1470].

Elsewhere in the Patna manuscript, at the end of a work labelled *al-Fuṣūl* on the coordinates for marking astrolabes and quadrants, Ibn Timurbāy writes:[72]

> This copy was checked against the manuscript of the author, then its coyist read it back to the author, who is the *shaykh*, the *imām*, the scholar and luminary Muḥibb al-Dīn Muḥammad ibn Muḥammad ibn Aḥmad ibn Aḥmad Ibn Muḥammad ibn Mushrif, whose ancestor was known as Ibn al-ʿAṭṭār al-Bakrī by lineage, al-Shāfiʿī by rite, al-Wafāʾī by (Ṣūfī) order.… Written by Aḥmad ibn Aḥmad al-Timurbāy al-Ḥanafī al-Qādirī.

---

[71] On the problems of identification of our copyist's grandfather see n. 38 above. The phrase used here is curious because Timurbāy must have been about 100 years old or deceased.

[72] *Patna Catalogue*, p. 96.

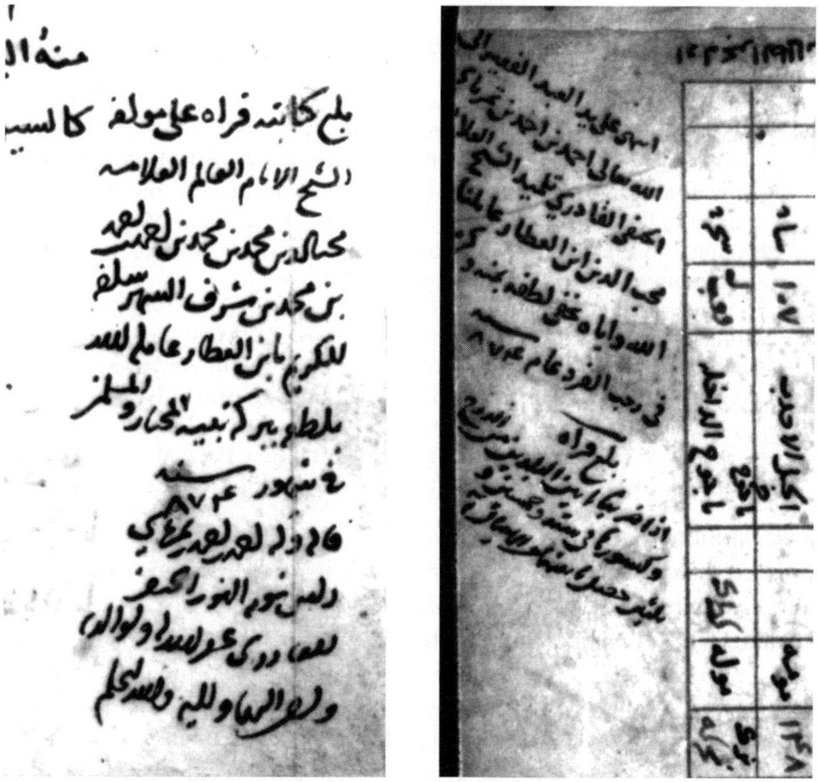

Figs. 7–8: The colophons on fol. 17v and 20v.

On fol. 78r of the Patna manuscript, Ibn al-ʿAṭṭār informs us:[73]

> I spent about thirty years working on this subject (of astronomy) from eight hundred thirty A.H. [= 1426/27] to the year 859 [= 1454/55] then until this date, which is the year 871 [= 1466/67].

At the end of some copies of his treatise on quadrants, we find the following note by Ibn al-ʿAṭṭār:[74]

> All of this is from the useful information (*fawāʾid*) of our *shaykh*, the *ustādh*, the luminary (*ʿallāma*) Nūr al-Dīn al-Naqqāsh ibn ʿAbd al-Qādir in the year eight hundred and thirty (A.H.) [= 1426/27]. He, through

[73] *Patna Catalogue*, p. 94. The date is given as 849 in the reproduction of the Arabic text but 859 in the English translation.

[74] Recorded with numerous errors in King, *Fihris al-makhṭūṭāt al-ʿilmiyya al-maḥfūẓa bi-Dār al-Kutub al-Miṣriyya*, 2 vols. (Cairo, 1981 and 1986), ii, pp. 368–9.

> his modesty and humanity, when he saw that I excelled over my peers, he took me and passed me on to the unique teacher (*ustādh*) in the(se) difficult (*ʿazīza*) subjects, our *shaykh*, the teacher, the luminary Shihāb al-Dīn al-Majdī al-Shāfiʿī, may God make his knowledge useful to us. I studied carefully under him (*ḥaqqaqnā ʿalayhi*) the principles of this science, and I read under him about the different kinds of sundials, horizontal and vertical, and cosmology, and other subjects relating to the sciences of religion in the sacred law and other aspects, at the Azhar Mosque and at his home in the next to (the Mosque) and at the Madrasa of Jānī Beg. I accompanied him for close to twenty years. In God is success and from Him we ask for help. This was reported by the luminary Muḥibb al-Dīn Muḥammad ibn Muḥammad known as Ibn al-ʿAṭṭār ibn Aḥmad ibn Aḥmad ibn Muḥammad ibn Mushrif ibn Badr al-Dīn ʿAbd al-Khāliq ibn al-Ḥasan ibn Dāʾūd ibn Idrīs ibn ʿAbd al-Jalīl ibn ʿAbd al-Kāfī ibn Ṭalḥa ibn ʿAbdallāh ibn ʿAbd al-Raḥmān ibn Abī Bakr al-Ṣiddīq, may God be pleased with him and make him content, and may He make Paradise his resting place. Amen. Praise be to God, Lord of the Worlds.

This cannot be a complete genealogy back to the Caliph Abū Bakr, because it does not span six centuries. It is of course naïve to expect that it would. His teachers Ibn al-Majdī and Nūr al-Dīn (ʿAlī ibn Muḥammad) al-Naqqāsh are well known.[75]

* * * * *

Acknowledgements: In 1998 Dr. Sonja Brentjes kindly furnished me with a set of incomplete and poor-quality photocopies of parts of the Patna manuscript. During the winter semester of 2004–2005, I read the text of Ibn al-ʿAṭṭār's *Geography* with Petra Schmidl and Mónica Herrera Casais in my Arabic Scientific Manuscripts seminar. In December, 2005, I received a microfilm of the complete text of the *Geography* thanks to the kindness of Dr. Imtiaz Ahmad, director of the Khoda Bakhsh Library. In January, 2006, Prof. Jamil Ragep kindly sent me a preprint of his paper on al-Jaghmīnī, which proved to be of considerable relevance to the present study and saved me not least from the error of thinking that all was well with the climates in the Islamic East.

---

[75] See King, *Survey*, nos. C62 and C74, also Rosenfeld and İhsanoğlu, *Mathematicians, Astronomers and Other Scholars*, nos. 815 and 847, as well as Charette, *Mathematical Instrumentation*, pp. 19–20.

ʿABD AL-LAṬĪF AL-BAGHDĀDĪ'S *KITĀB AL-ḤAYAWĀN*: A CHIMAERA?

Remke Kruk

ʿAbd al-Laṭīf al-Baghdādī and his Zoological Interest

Medieval Islamic physicians were by the very nature of their work interested in zoology, and the colorful and learned scholar and physician ʿAbd al-Laṭīf al-Baghdādī (d. 629/1231) is no exception. His interest, however, went further than that of his average colleague. Best known in this respect is his small but noteworthy section on the animals of Egypt in his *Kitāb al-ifāda wa'l-iʿtibār*.[1]

Included in this section are descriptions of the chicken, remarkable for the extensive description of the consummate craft of artificial hatching; of the donkeys, the cows, and the horses particular to Egypt; of the crocodile, the skink, the hippopotamus, the electric ray[2] and two other types of fish used for consumption, plus the tortoise and a kind of shellfish, the tellin. The descriptions are quite lively, describing what the animals look like, how they live, how they are caught and to what uses they are put, including some medical uses.

ʿAbd al-Laṭīf's name repeatedly turns up in the zoological sections of post-thirteenth-century encyclopedic works, sometimes in connection with descriptions of animals living in Egypt. None of these references are traceable to the *Kitāb al-ifāda*, which demonstrates that ʿAbd al-Laṭīf's interest in zoology also came up in other works of his. In some of the quotations, explicit mention is made of a *Book on Animals*, or *On the Natures of Animals*, of ʿAbd al-Laṭīf's hand. No such title is included in Ibn Abī Uṣaybiʿa's bibliography of ʿAbd al-Laṭīf's work.[3] Several of the titles in this bibliography, however, indicate a zoological interest:

---

[1] ʿAbd al-Laṭīf al-Baghdādī, *The Eastern Key/Kitāb al-ifāda wa'l-iʿtibār of ʿAbd al-Laṭīf al-Baghdādi*, trans. K.H. Zand and J.A. and I.E. Videan (Cairo, 1204/1964 [London, 1965]), pp. 79–106.

[2] On this animal, see S. de Sacy, *Relation de l'Égypte par Abd-Allatif, médecin arabe de Baghdad, 1162–1231 A.D.* (Paris, 1810), p. 167, n. 53.

[3] Ibn Abī Uṣaybiʿa, *ʿUyūn al-akhbār fī ṭabaqāt al-aṭibbāʾ*, ed. A. Müller, 2 vols. (Cairo, 1882–84). Quoted in the article as IaU.

- *Maqāla fī 'l-siqanqūr* (IaU ii, p. 212, l. 3). No mss. of this text or references to it are known.
- *Kitāb al-mudhish fī akhbār al-ḥayawān* (IaU ii, p. 213, l. 13ff). It is not quite clear in Ibn Abī Uṣaybiʿa's text where the title stops, and the undated modern Beirut edition of Nizār Riḍā (published by Dār Maktabat al-Ḥayāt) includes the sequel of the text, *al-mutawwaj bi-ṣifāt nabiyyinā ʿalayhi afḍal al-ṣalāt wa'l-salām*, in this title instead of presenting it as a separate item. If this should be correct, which I do not consider likely, it would invalidate Ullmann's suggestion (*Naturwissenschaft*, 31) that this title is identical with the *Kitāb (ṭabāʾiʿ) al-ḥayawān*.
- *Maqāla fī jawāb masʾala suʾila ʿanhā fī dhabḥ al-ḥayawān wa-qatlihi wa-hal dhālika sāʾigh fī 'l-ṭabʿ wa-fī 'l-ʿaql kamā huwa sāʾigh fī 'l-sharʿ* (IaU ii, p. 213, l. 23). A treatise about the question whether the slaughtering and killing of animals is acceptable according to natural instinct and reason as well as according to law. No mss. or references to this text are known.

Ibn Abī Uṣaybiʿa also reports three compendia made by ʿAbd al-Laṭīf of earlier zoological works:

- *Ikhtiṣār Kitāb al-ḥayawān li-Arisṭūṭālīs* (IaU ii, p. 211, l. 19)
- *Ikhtiṣār Kitāb al-ḥayawān li'l-Jāḥiẓ* (IaU ii, p. 211, l. 11; HKh iii, p. 122, l. 4, no. 4662)
- *Ikhtiṣār Kitāb al-ḥayawān li'bn Abī 'l-Ashʿath* (d. 360/970; IaU ii, p. 213, l. 7f; Hkh iii, p. 122, l. 4, no. 4662.)

As to the second of these compendia: Ḥājjī Khalīfa (*Kashf*, ed. Flügel, iii, p. 122) remarks that ʿAbd al-Laṭīf made a *mukhtaṣar* of Jāḥiẓ' *Kitāb al-ḥayawān*, as was also done by Abū Qāsim Hibatallāh ibn al-Qāḍī al-Rashīd Jaʿfar (d. 608/1211 according to Ḥājjī Khalīfa). I have not so far been able to discover any further information about this Hibatallāh and his book on animals.

None of these compendia have been preserved, and apart from Ḥājjī Khalīfa, no references to their titles have so far been found in later books. On the other hand, a number of sayings about animals ascribed to ʿAbd al-Laṭīf occur in post-thirteenth-century encyclopedic works, as was said above, and if there is a book title mentioned in connection with the references, it is given either as *Kitāb ṭabāʾiʿ al-ḥayawān* or simply as *Kitāb al-ḥayawān*.

As to the nature and existence of this *Book on (the Natures) of Animals* the matter is further confused by Ghalioungi, who says that according to Ḥājjī Khalīfa, ʿAbd al-Laṭīf's book on animals is just a compendium of the animal books of Hibat Allāh and Ibn Abī 'l-Ashʿath, a statement for which I have not found any support.[4] So, the question of ʿAbd al-Laṭīf's zoological writings is something of a mystery. Did there exist a *Book on (the Natures of) Animals* written by ʿAbd al-Laṭīf, and if so, what did it contain? And did he indeed write three compendia of older animal books, viz. of Aristotle, Jāḥiẓ and Ibn Abī 'l-Ashʿath? Or was there maybe just one compendium in which he included material from all three authors, and did this start to circulate as a 'book on animals' or 'book on the natures of animals', or even as a ' book of amazing things of what is told about animals', *Kitāb al-mudhish fī akhbār al-ḥayawān*?

A crucial point in the discussion is ʿAbd al-Laṭīf's supposed compendium of Ibn Abī 'l-Ashʿath's *Kitāb al-ḥayawān*, written in the tenth century. This was not a widely known work. Only a limited set of quotations, all, with one exception, traceable to the thirteenth-century *Kitāb badāʾiʿ al-akwān* of Ibn Abī 'l-Ḥawāfir (see below), can be found in post-thirteenth century literature.[5] Ibn Abī 'l-Ḥawāfir may of course personally have had access to Ibn Abī 'l-Ashʿath's work, but there is also a possibility that he took his quotations from a zoological compilation by ʿAbd al-Laṭīf, with whom his family had a professional connection: his grandfather Jamāl al-Dīn ʿUthmān ibn Abī 'l-Ḥawāfir (d. 595/1198), a well-known physician, had at some time practiced in the same hospital as ʿAbd al-Laṭīf. These later compilers, as is well known, were usually more inclined to use second-hand material than to consult the sources themselves, and they rarely mentioned their intermediate source.

When we see, for instance, that an author such as Ibn al-Akfānī (d. 749/1348) says that he based himself in his *Asās al-riyāsa fī ʿilm al-firāsa*

---

[4] ʿAbd al-Laṭīf al-Baghdādī, *Maqālatān fī 'l-ḥawāss wa-masāʾil ṭabiʿiyya; Risāla li'l-Iskandar fī 'l-faṣl; Risāla fī 'l-maraḍ al-musammā diyābīṭis*, ed. P. Ghalioungui and S. Abdou (Kuwait, 1392/1972), p. 166.—What Ghalioungi says is presented as a literal quotation from Ḥājjī Khalīfa's *Kashf al-ẓunūn* (Istanbul, 1941), i, 696: *inna kitāb al-ḥayawān li'l-Baghdādī mā huwa illā mukhtaṣar kitāb al-ḥayawān li-Abī 'l-Qāsim Hibatallāh ibn al-Qāḍī al-Rashīd wa-kitāb al-ḥayawān li'bn Abī 'l-Ashʿath*. The edition used by Ghalioungui was not available to me and I have not succeeded in locating the quotation in Flügel's edition (London, 1842) of Ḥajjī Khalīfa's *Kashf al-ẓunūn*.

[5] On Ibn Abī 'l-Ashʿath's work, see R. Kruk, 'Ibn abi-l-Ashʿath's *Kitāb al-ḥayawān*: a Scientific Approach to Anthropology, Dietetics and Zoological Systematics,' *ZGAIW* 14 (2001), pp. 119–68.

for zoological information on 'al-Jāḥiẓ, Aristotle and Ibn al-Ashʿath (*sic*)',[6] we can of course not exclude the possibility that he personally consulted Jāḥiẓ' *Kitāb al-ḥayawān* and the Arabic translation of Aristotle's zoological works, which were still current in his day, but it is much more likely that he simply used quotations handed down by others and conveniently left out the name of the compiler. This applies even more to the little known *Kitāb al-ḥayawān* of Ibn Abī 'l-Ashʿath. Thus we may very well have an echo here of a zoological compendium, or compendia, providing quotations from Jāḥiẓ, Aristotle and Ibn Abī 'l-Ashʿath, and possibly even of such a compendium written by ʿAbd al-Laṭīf.

In the following we will try to analyze the available information from post-thirteenth-century sources on ʿAbd al-Laṭīf's zoological writings and see whether this leads to substantial new results.

Textual Evidence

Our starting point are the references to ʿAbd al-Laṭīf's *Book on Animals* given by Ullmann,[7] which include those of De Somogyi.[8] Given the scanty amount of work that has been done on the post-thirteenth-century encyclopedic tradition, especially on unpublished works, there is always a hope that more material may come to light, even though it is not likely that this may yield so much material that one might actually try to reconstruct ʿAbd al-Laṭīf's *Kitāb al-ḥayawān* from it, as Hamidullah e. a. attempted to do for the lost *Book on Plants* of al-Dinawarī (of which, by the way, ʿAbd al-Laṭīf is also reported to have made a compendium [IaU ii, p. 211, l. 15]). A scrutiny of sources likely to contain useful material yielded the following results.

A. *Sources*

1. Ibn Abī 'l-Ḥawāfir (d. 701/1301); Ibn al-Athīr (*fl.* probably in the first part of the fourteenth century); al-Waṭwāṭ (d. 718/1318)

[6] J.J. Witkam, *De Egyptische arts Ibn al-Akfānī (gest. 749/1348) en zijn indeling van de wetenschappen*, doctoral thesis (University of Leiden, 1989), p. 74. Witkam cites ms. Paris, BN Arab. 2762, fol. 28a.

[7] M. Ullmann, *Die Natur- und Geheimwissenschaften im Islam* (Leiden, 1972) p. 31, n. 1.

[8] J. de Somogyi, 'Index des sources de la *Ḥayāt al-ḥayawān* de Ad-Damīrī,' *JA* (Juillet–Septembre 1928), pp. 5–128, 73–4.

A number of the references given by Ullmann come from Jamāl al-Dīn ʿUthmān ibn Abī 'l-Ḥawāfir's *Kitāb badāʾiʿ al-akwān fī manāfiʿ al-ḥayawān*[9] and Ibn al-Athīr's *Tuḥfat al-ʿajāʾib wa-ṭurfat al-gharāʾib*.[10] Unlike Ibn Abī 'l-Ḥawāfir, Ibn al-Athīr quotes Qazwīnī, e.g. on fol. 261b. As Ullmann already noticed, substantial parts of the zoological section of Ibn al-Athīr's *Tuḥfa* are very closely related to—possibly directly copied from—Ibn Abī 'l-Ḥawāfir's *Kitāb al-badāʾiʿ*.[11] This applies in particular to the *ṭabāʾiʿ* part of the animal descriptions, not to the *khawāṣṣ wa-manāfiʿ*.

As to the ʿAbd al-Laṭīf quotations in Ibn al-Athīr's work, the disappointing result of a comparison is that all except one of these quotations can be traced to Ibn Abī 'l-Ḥawāfir's book,[12] and very likely originated from it.

The result is the same when we look at another encyclopedic work with a large section on animals, not included among Ullmann's source references for ʿAbd al-Laṭīf's *Kitāb al-ḥayawān*. This is the *Kitāb mabāhij al-fikar wa-manāhij al-ʿibar* of Yaḥyā al-Waṭwāṭ al-Kutubī.[13] Waṭwāṭ, in true *adab* style, has replaced the *khawāṣṣ* part of his entries by poetry fragments referring to the animal in question.

The *ṭabāʾiʿ* parts, as it turns out, are practically identical to Ibn Abī 'l-Ḥawāfir's, just as most of Ibn al-Athīr's. It is unlikely, however, that the texts of Ibn al-Athīr and al-Waṭwāṭ are mutually dependent, because they have completely different entries on the giraffe, an animal not included by Ibn Abī 'l-Ḥawāfir.[14]

---

[9] Ullmann, *Natur- und Geheimwissenschaften*, pp. 33–4. Ms. used in the article: Dublin, Chester Beatty 4352.

[10] This Ibn al-Athīr is another than the well-known historian, one of three brothers, of that name, and for that reason Ullmann refers to him as 'pseudo-Ibn al-Athīr'. We will, however, simply call him Ibn al-Athīr. References in this article are to the Berlin ms., Ahlwardt Ar. 6163.

[11] Ullmann, *Natur- und Geheimwissenschaften*, p. 38.

[12] For the present purpose I consulted only two of the existing mss., Dublin Chester Beatty Ar. 4352 and Berlin, Ar. 6167 (= Sprenger 1923). Quotations are to the Chester Beatty ms. A more extensive collation of mss. may yet change the picture.

[13] Ullmann, *Natur- und Geheimwissenschaften*, p. 34; a facsimile edition of this work was published by Sezgin: Jamāl al-Dīn al-Waṭwāṭ, *Manāhij al-fikar wa-mabāhij al-ʿibar; Encyclopaedia of Four Natural Sciences*, 2 vols. (Frankfurt, 1990), and a critical, but not quite satisfactory edition appeared in Beirut in 2000: al-Waṭwāṭ, *Mabāhij al-fikar wa manāhij al-ʿibar*, ed. ʿA.A. al-Ḥarbī, (Beirut, 1420/2000). This edition gives a number of textual parallels, but has no indices, no usable table of contents, and the references to mss. are muddled because the editor mixes up the recto and verso pages in the photographic material of the mss. which he used. References in the article are to both editions.

[14] See R. Kruk, 'Elusive Giraffes. Ibn abī l-Ḥawāfir's *Badāʾiʿ al-akwān* and Other Animal Books,' in A. Contadini (ed.), *Proceedings of the Arab Painting Conference* (London, September 2004; forthcoming).

2. Other encyclopedic works
The following encyclopedic works contain references to ʿAbd al-Laṭīf's zoological work, although not a large number:

- Damīrī (d. 808/1405–6), *Ḥayāt al-ḥayawān al-kubrā*;[15]
- Qalqashandī (d. 821/1418), *Ṣubḥ al-aʿshā*.[16]

The nature of the references found in these works will be discussed below.

3. Encyclopedic works yielding no results
No explicit references to a zoological work by ʿAbd al-Laṭīf have been found in Qazwīnī's *ʿAjāʾib al-makhlūqāt*; Dimashqī's *Nukhbat al-dahr*; ʿUmarī's *Masālik al-abṣār*; Nuwayrī's *Nihāyat al-adab*; Jildakī's *Durrat al-ghawwāṣ*; Ghuzūlī's *Kitāb al-maṭāliʿ al-budūr*.

B. *The quotations; strands of transmission*

The zoological references to ʿAbd al-Laṭīf found in the works mentioned above under 1) and 2) are here arranged under the names of the animals which occur in them. Explicit references to books ascribed to ʿAbd al-Laṭīf in the quotations are highlighted in italics.

The basic questions to be answered are:

a) How many strands of transmission can be discerned in the quotations? Can all the quotations be traced to Ibn Abī 'l-Ḥawāfir, or are there also independent strands?
b) How many of the quotations are definitely, or very likely, connected to a zoological work of ʿAbd al-Laṭīf?
c) Do the quotations tell us more about ʿAbd al-Laṭīf's zoological writings than was already known?

1. *Ḍabb* and *waral* cluster. Sources: Ibn Abī 'l-Ḥawāfir/Ibn al-Athīr/Waṭwāṭ cluster; Damīrī.

---

[15] Al-Damīrī, *Ḥayāt al-ḥayawān al-kubrā*, 2 vols. (Cairo, 1383/1963).
[16] Al-Qalqashandī, *Ṣubḥ al-aʿshā*, 4 vols. (Cairo, 1331/1913–1338/1919).

NB: It is difficult to decide exactly which species were meant by the Arabic words. According to Ghaleb,[17] who mostly presents modern usage, the *waral* is the monitor, the *ḍabb* the uromastix, *ḥirbāʾ* the chameleon, and *ḥirdhawn* the agama. He gives no equivalent for the *shaḥmat al-arḍ*, which here indicates a kind of lizard. The name *shaḥmat al-arḍ* is also used for other animals, such as the millipede (Kazimirski), and they are also said to be the same as the *kharāṭīn*, earthworms. *Wazagh* is the gecko (Kazimirski). In the translation I have maintained the Arabic names.

Ibn Abī'-Ḥawāfir, fol. 141b, l. 4 (idem, ms. Berlin 6167, fol. 137a), entry on the *waral*: the entry starts with:

الورل وهو الحرذون.

There is no mention of ʿAbd al-Laṭīf, unlike in Ibn al-Athīr and Waṭwāṭ, which start with a reference to him:

Ibn al-Athīr (fol. 261a13ff–b), entry on the *waral*:

قال عبد اللطيف البغدادي فى كتاب مسمى بالطبائع الحيوان الورل والضب والحرباء وشحمة الأرض والوزغ كلها متناسبة فى الخلق فأما الورل فهو الحرذون.

"Abd al-Laṭīf al-Baghdādī says in a book called *The Natures of the Animals*: "The *waral*, the *ḍabb*, the *hirbāʾ*, the *shaḥmat al-arḍ* and the *wazagh* are all of a (Waṭwāṭ adds: closely, *mutaqāriba*) related nature. As to the *waral*, it is the same as the *ḥirdhawn*." (Continues): "Those who speak about the natures of animals say that there is no animal that copulates more than he (i.e. the *waral*), and none that is slower in it" etc.' (Whether this last quotation was included by ʿAbd al-Laṭīf cannot be ascertained)

Waṭwāṭ (2000, p. 364): practically the same text as Ibn al-Athīr. The most significant textual difference is in the title of ʿAbd al-Laṭīf's book: "Abd al-Laṭīf says in *a book on animals by him* (*kitāb al-ḥayawān lahu*)'.

[17] E. Ghaleb, *Dictionnaire des sciences de la nature*, 3 vols. (Beirut, 1965–6).

Damīrī (ii, p. 78, l. 8ff), entry on the *ḍabb*:

قال عبد اللطيف البغدادي الورل والضب والحرباء وشحمة الارض والوزغ كلها متناسبة فى الخلق وللضب ذكران وللانثى فرجان كالورل والحرذون.

"Abd al-Laṭīf says that the *waral*, the *ḍabb*, the *ḥirbāʾ*, the *shaḥmat al-arḍ* and the *wazagh* are all of a related nature. The *ḍabb* has two penises and the female two vulvae, just like the *waral* and the *ḥirdhawn*.'[18]

Damīrī (ii, p. 396, l. 28ff), entry on the *waral*:

قال عبد اللطيف البغدادي الورل والضب والحرباء وشحمة الأرض والوزغ كلها متناسبة فى الخلق فأما الورل فهو الحرذون.

"Abd al-Laṭīf said that the *waral*, the *ḍabb*, the *hirbāʾ*, the *shaḥmat al-arḍ* and the *wazagh* are all of a related nature. As to the *waral*, it is the *ḥirdhawn*.' (Continues): 'There is no animal that copulates more than he'. Damīrī's text, where the introduction to these last words, *fa-zaʿama aṣḥāb al-ḥayawān*, has disappeared, clearly belongs to a later stage than the text of the Ibn Abī 'l-Ḥawāfir-cluster.

A reaction to the opinion of ʿAbd al-Laṭīf is found in Damīrī (i, p. 290, l. 2):

والحق أنه غير الورل خلافاً لعبد اللطيف البغدادي.

'Contrary to what ʿAbd al-Laṭīf says, the *waral* is not in fact identical with the *ḥirdhawn*.'

Conclusion: within the *waral/ḍabb* cluster of quotations, the only words that are ascribed to ʿAbd al-Laṭīf are: 'The *waral*, the *ḍabb*, the *hirbāʾ*, the *shaḥmat al-arḍ* and the *wazagh* are all related by nature. As to the *waral*, it is the same as the *ḥirdhawn*.' An intriguing point is that this quotation is not found in the Ibn Abī 'l-Ḥawāfir mss. that I consulted, a matter which complicates the matter of the relationship between those

[18] For this last bit of information, cf. also Ibn Abī 'l-Ḥawāfir, fol. 93b, about the *ḍabb*, without mention of ʿAbd al-Laṭīf: *yuqālu inna lahu nazkayn wa-li'l-ḍabba farjayn kamā li'l-waral.*

texts. At the present stage of research, however, nothing conclusive can be said about this question.

2. The *dalaq-ibn ʿirs-sammūr-nims* cluster. Sources: Ibn Abī 'l-Ḥawāfir/Ibn al-Athīr/Waṭwāṭ cluster; Damīrī; Qalqashandī.

NB: the information about these four animals is mixed up in the sources, and clearly all three words were occasionally used for weasel, maybe also for the ferret, probably also the ermine, and various species of marten. *Dalaq* nowadays is the marten, *ibn ʿirs* the weasel, and *sammūr* the sable marten, while *nims* is used for the ichneumon. In the translation, I have simply retained the Arabic words.

Ibn Abī 'l-Ḥawāfir (fol. 17b, l. 2ff), entry on the *ibn ʿirs*:

ويوجد فى منازل أهل مصر وقال عبد اللطيف البغدادي وأظنه الحيوان المسمى بالدلق وإنما يختلفٍ وبره ولونه بحسب البلاد وفى طبعه أنه يسرق ما يوجد من الذهب والفضة وإن وجد حبوباً خلطها وهو عدو الفأر ويصيده ويقتله ويقال إن عداوته له أشد من عدواة السنور وخوف الفأر من السنور أشد كما يخاف الدجاج من ابن عوى أشد من خوفه من الثعلب.

(About the *ibn ʿirs*) 'It is found in the houses of the people in Egypt. ʿAbd al-Laṭīf al-Baghdādī says: "I think that it is the animal that is called *dalaq*. Its pelt and color, however, differ according to country. It is in its nature to steal any gold and silver there is (Ibn al-Athīr and Damīrī: that it finds), and if it finds beads, it mixes them (idem, ms. Berlin 6167, fol. 22a and Waṭwāṭ [2000, p. 375]; Ibn al-Athīr, fol. 264a: *khāṭahā*, strings them). It is the enemy of the mouse, and he catches and kills it. One says that its enmity towards it is stronger than that of the cat, while the mouse is more afraid of the cat, just as the chicken is more afraid of the jackal than of the fox."' (Continues): story about a trapped mouse that is ransomed by its mother.

Ibn al-Athīr (fol. 264a, l. 9ff), Waṭwāṭ (2000, p. 375), entry on the *ibn ʿirs*, the weasel: the same as above, with a few insignificant textual variants.

Damīrī (ii, p. 170, l. 21ff), entry on the *ibn ʿirs*: the same as above, with slight variants. The most important of them is the omission of the passage on the beads:

وقال عبد اللطيف البغدادي وأظنه الحيوان المسمى بالدلق وإنما يختلف لونه ووبره بحسب البلاد قال وفي طبعه أنه يسرق ما وجد من فضة وذهب كما يفعل الفأر وربما عاد الفأر فقتله ولكن خوف الفأر من السنور أشد من خوفه منه قال وهو كثير الوجود فى منازل أهل مصر.

(About the *ibn ʿirs*) "ʿAbd al-Laṭīf al-Baghdādī says: "I think that it is the animal that is called *dalaq*. Its pelt and color, however, differ according to country." He says: "It is in its nature to steal any gold and silver that it finds, just like the mouse does. Sometimes it shows enmity towards the mouse and kills it, but the mouse is more afraid of the cat. It is frequently found in the houses of the people in Egypt".' (Continues): story about a trapped mouse that is ransomed by its mother.

Ibn Abī 'l-Ḥawāfir (fol. 130b–131a, l. 13ff), entry on the *nims*, as part of what is said about the *sammūr*:

ويقرب منه حيوان يسمى الدلق وهو أيضاً يفترس ويكرع الدم.

'An animal called *dalaq* closely resembles it. It also kills its prey and sips its blood.'

Ibn al-Athīr (fol. 252a paen.), Waṭwāṭ (2000, p. 233), entry on the *nims*, as part of what is said about the *sammūr*: the same text. Ibn al-Athīr omits *ayḍan*.

Damīrī and Qalqashandī have a slightly different text and both include ʿAbd al-Laṭīf's name:

Damīrī (i, p. 338 paen.ff), entry on the *dalaq*:

قال عبد اللطيف البغدادي وهو يفترس فى بعض الأحايين ويكرع الدم.

"ʿAbd al-Laṭīf says: "It sometimes kills his prey and sips the blood".'

Qalqashandī (ii, p. 50, l. 2f), entry on the *dalaq*:

قال عبد اللطيف البغدادي وهو يفترس فى بعض الأحايين ويكرع فى الدم وذكر ابن فارس أنه النمس.

"Abd al-Laṭīf says: "It sometimes kills his prey and sips the blood".' Ibn al-Fāris says that it is the *nims*.

Conclusion: the ʿAbd al-Laṭīf information on the *ibn ʿirs/dalaq* transmitted here by Ibn Abī 'l-Ḥawāfir, Ibn al-Athīr, Damīrī and Qalqashandī is basically identical, but Damīrī and Qalqashandī show evidence of a slightly independent strand of transmission.

Sammūr. Included, with some additional information on the *dalaq*, in the entry on the *nims*. Sources: Ibn Abī 'l-Hawāfir/Ibn al-Athīr/Waṭwāṭ cluster—Damīrī-Qalqashandī.

Ibn Abī 'l-Ḥawāfir (fol. 130b, l. 13ff), entry on the *nims*:

وحكا عبد اللطيف البغدادي في كتاب الحيوان عن ابن ماسويه أن السمور حيواناً جريء ليس فى الحيوان أجرأ منه على الإنسان سريع الحضر لا يؤخذ إلا بالحيل فيحتال فى صيده إن تدفن له جيفة فيغتال بها ولحمه حلوٍ والترك يأكلونها وجلده لا يدبغ كسائر جلود الحيوان ويقرب منه حيوان يسمى الدلق وهو أيضا يفرس ويكرع الدم وذكر ابن فارس فى كتاب المجمل فى اللغة أنه النمس أيضا.

"Abd al-Laṭīf al-Baghdādī has told in his *Book on Animals* on the authority of Ibn Māsawayh[19] (Ibn al-Athīr, fol. 252a paen.–b, Waṭwāṭ 2000, p. 233: Ibn Māssa) that the *sammūr* is a courageous animal, there is no animal that is more courageous against man than this. It quickly alarmed (Ibn Abī 'l-Ḥawāfir: *ḥiḍr*, Ibn al-Athīr and Waṭwāṭ: *ḥidhr*). It can only be caught by tricks. A trick that is used to trap it is to bury a dead body and then grab it with it. Its meat is nice, and the Turks eat it. Its pelt cannot be dyed like other pelts of animals. An animal called *dalaq* closely resembles it; it also kills its prey and sips the blood.'

Ibn al-Athīr (fol. 252a–b), Waṭwāṭ (2000, p. 233), entry on the *nims*: the same, with slight textual variations. Waṭwāṭ adds at the end: *wa-dhakara ibn Fāris fī kitāb al-mujammal fī 'l-lugha annahu al-nims ayḍan*, a passage that does not quite fit in here. It is what Qalqashandī (see above) said about the *dalaq*.

[19] About a *Kitāb al-ḥayawān* by Ibn Māsawayh, see Ullmann, *Natur- und Geheimwissenschaften*, p. 19.

Damīrī (ii, p. 34, l. 16ff), entry on the *sammūr*:

وقال عبد اللطيف البغدادي إنه حيوان ليس فى الحيوان أجرأ منه على الإنسان لا يؤخذ إلا بالحيل وذلك بأن تدفن له جيفة فيغتال بها ولحمه حار والترك يأكلونها وجلده لا يدبغ كسائر جلود الحيوان.

'Abd al-Laṭīf says that there is no animal that is more courageous against man than this. It can only be caught by tricks, namely by burying for it a dead body and then grabbing it with it. Its meat is hot, and the Turks eat it. Its pelt cannot be dyed like other animal pelts.' (Continues): something about Nawawī.

Qalqashandī (ii, p. 49, l. 7f):

قال عبد اللطيف البغدادي إنه حيوان ليس فى الحيوان أجرأ منه على الإنسان لا يؤخذ إلا بالحيل.

'Abd al-Laṭīf al-Baghdādī says: "There is no animal more courageous against man than this. It can only be caught by tricks".'

Conclusion: it is difficult to decide which part of the quotations, apart from the information from the *Book of Animals* about the courage of the *sammūr*, is definitely connected to ʿAbd al-Laṭīf. Yet the information clearly all belongs to the same strand of transmission.

3. *Dajāj*. Source: Qalqashandī.

A quotation found only in Qalqashandī (ii, p. 70, l. 6):

قال عبد اللطيف البغدادي أمر كل قوم من الكسب بحسب مقدورتهم.

'Abd al-Laṭīf al-Baghdādī said: "All people have been ordered to gain their living according to what has been allotted to them".'

It is unlikely that this quotation comes from a book on animals. I have not been able to trace it to a work by ʿAbd al-Laṭīf.

Conclusion: there is no indication that this quotation comes from a *Book on Animals*.

4. *Fīl*: Source: Damīrī.

Damīrī (ii, pp. 227–8), entry on the *fīl*:

وقال عبد اللطيف البغدادي إنها تحمل سبع سنين ولا ينزو إلا على فيلة واحدة وله عليها غيرة شديدة فإذا أتم حملها وأرادت دخلت النهر حتى تضع ولدها لأنها لا تلد إلا وهى قائمة ولا فواصل لقوائمها وتلد والذكر عند ذلك يحرسها وولدها من الحيات.

"ʿAbd al-Laṭīf al-Baghdādī says that she is seven years pregnant and only mates with one female, about which he is extremely jealous. When her pregnancy is completed and she wants to bring forth she enters the river in order to have her baby there, because she can only give birth while standing, her limbs having no joints. She then gives birth while the male guards her and her and her child at that time against snakes.' (text continues with a story about an elephant killing his caretaker)

Conclusion: no book title is given, but the information typically is the sort of material that one would expect in a book on animals. The quote cannot be traced to Ibn Abī 'l-Ḥawāfir. This suggests the possibility of an independent strand going back to ʿAbd al-Laṭif, possibly to his *Kitāb al-ḥayawān*.

5. *Kabsh*. Source: Damīrī.

Damīrī (ii, p. 271, l. 22f), entry on the *kabsh*:

قال عبد اللطيف البغدادي هذه المعالجة تصلح للأعراب الذين يعرض لهم هذا المرض من يبس.

"ʿAbd al-Laṭīf al-Baghdādī says: "This treatment is right for Bedouins who are afflicted by this illness because of dryness".'

Conclusion: there is no indication that this quotation comes from a *Book on Animals*. It may come from a medical work of ʿAbd al-Laṭīf. So far, I have not been able to trace it. It is not found in his *Kitāb fī uṣūl mufradāt al-ṭibb wa-kayfiyyāt ṭabāʾihā*.[20]

[20] Ms. used: Paris, BN Ar. 2670,2.

6. *Samandal*. Sources: Damīrī-Qalqashandī.

Damīrī (ii, p. 34, l. 5ff), entry on the *samandal*:

قال ورأيت بخط شيخنا العلامة عبد اللطيف البغدادي أنه قال قدم للملك الظاهر بن الملك الناصر صلاح الدين صاحب الحلب قطعة سمندل عرض ذراع فى طول ذرعين فصاروا يغنسونها فى الزيت ويوقدونها حتى يفنى الزيت وترجع بيضاء كما كانت.

'He (sc. Ibn Khallikān, d. 681/1282–3) says: "I have seen in the handwriting of our learned shaykh ʿAbd al-Laṭīf al-Baghdādi that he said: 'To al-Malik al-Ẓāhir, the son of al-Malik al-Nāṣir Ṣalāḥ al-Dīn the ruler of Aleppo a piece of *samandal* was brought of one cubit wide and two cubits long. They drenched it in oil and put it on fire until the oil had disappeared and it became just as white again as before'."

Qalqashandī (ii, p. 78, l. 6f):

قال ورأيت بخط عبد اللطيف بن يوسف البغدادي أنه أهدى للظاهر بن السلطان صلاح الدين صاحب الحلب قطعة منه عرض ذراع فى طول ذراعين فغمست فى الزيت وقربت من النار فاستعلت حتى فنا الزيت ثم عادت بيضاء كما كانت وبعضهم يقول إنه وحش كالثعلب وإن ذلك يعمل من وبره.

'He (sc. Ibn Khallikān, d. 681/1282–3) says: "I have seen in the handwriting of our learned shaykh ʿAbd al-Laṭīf al-Baghdādi that a piece of it of one cubit long and two cubits broad was brought to al-Ẓāhir, the son of sultan Ṣalāḥ al-Dīn, the ruler of Aleppo. They drenched it in oil and brought it close to the fire so that it burned until the oil had disappeared. Then it became just as white again as before. Some of them say that it is a wild animal like the fox, and that that is made from its pelt".'[21]

Conclusion: it is very unlikely that this stems from a book on animals. Dr. N.P.G. Joosse, who is preparing a monograph on ʿAbd al-Laṭīf, suspects that it was part of ʿAbd al-Laṭīf's third autobiography, written late

[21] Cf. Ibn Khallikān, *Wafāyāt al-aʿyān*, ed. I. ʿAbbās (Beirut, 1978), vii, p. 44 (biography of the poet Yaʿqūb ibn al-Ṣābir al-Manjanīqī), trans. MacGuckin de Slane (Paris, 1842–51), iv, p. 376.

in his life and dedicated to his son.—The salamander is not mentioned by Ibn Abī 'l-Ḥawāfir.

7. *Thaʿlab.* Source: Ibn Abī 'l-Hawāfir/Ibn al-Athīr/Waṭwāṭ cluster.

Ibn Abī 'l-Ḥawāfir, fol. 38b, l. 10–12 (entry on the *thaʿlab*):

وفرو الثعلب أفضل الأوبار ومنه الأسود والأبيض والخلنجي وأدونه الأعرابى لقلة وبره وخشن لونه وهذا قول عبد اللطيف البغدادي في كتاب الحيوان الذي صنفه.

'The fox has the best pelt. There are among them black, white, and *khalanjī*[22] ones. The worst is the *aʿrābī* ("the Bedouin one") because of its thin pelt and its rough color. This is what ʿAbd al-Laṭīf al-Baghdādī says in *The Book on Animals* that he has written (Ibn al-Athīr: a book on animals by him, *kitāb al-ḥayawān lahu*).'

Ibn al-Athīr (fol. 251b, l. 2), Waṭwāṭ (2000, p. 230): the same text, apart from the variant mentioned above.

وهذا قول عبد اللطيف البغدادي في كتاب الحيوان الذي صنفه.

Conclusion: this bit of information clearly is connected to a *Book on Animals* by ʿAbd al-Laṭīf.

Conclusion

We must conclude that at most three strands of transmission are discernible in the existing quotations from ʿAbd al-Laṭīf's *Book on (the Natures of) Animals*. The first strand consists of the quotations from the Ibn Abī 'l-Ḥawāfir/Athīr/al-Waṭwāṭ cluster of texts. Some of these quotations are also found in Damīrī and Qalqashandī. In the quotations found in both groups of texts, the material offered by Damīrī and Qalqashandī sometimes overlaps partly with that of the other texts, but they (jointly) also contain some exclusive material. This may possibly belong to a

[22] According to Lane, the *khalanj* is a kind of tree from the wood of which vessels are made. Nowadays *khalanj* is the botanical name for heather and its various species. A brownish color?

second strand of transmission. The third strand is formed by the quotations exclusively found in Damīrī.

So, does this textual evidence throw any new light on the nature and existence of ʿAbd al-Laṭīf's zoological compendia and his book on animals?

As to the compendia:

- Aristotle and Jāḥiẓ: no. So far, there is no indication that any quotations from Aristotle's zoological works found in post-ʿAbd al-Laṭīf literature were taken from a work by ʿAbd al-Laṭīf. The same can be said about the quotations from Jāḥiẓ' *Kitāb al-ḥayawān.*
- Ibn Abī 'l-Ashʿath: none of the Ibn Abī 'l-Ashʿath quotations found in these later works refer to ʿAbd al-Laṭīf either. All of them can be traced back to Ibn Abī 'l-Ḥawāfir, who mentions (Ibn Abī 'l-Ḥawāfir fol. 143b, colophon) that he has made use of 'what has been said by the scholars in this field, namely Jāḥiẓ, Ariṣṭū and Ibn Abī 'l-Ashʿath, who have spoken about animals, and [...]' Here he mentions a number of others, without bringing in ʿAbd al-Laṭīf.

So there is nothing in these quotations to support the idea that ʿAbd al-Laṭīf wrote a compendium of Ibn Abī 'l-Ashʿath's *Kitāb al-ḥayawān* or a general zoological compendium including quotations from his work. Yet there is a difference between the Ibn Abī 'l-Ashʿath quotations and those from Aristotle and Jāḥiẓ: Ibn Abī 'l-Ashʿath references are exclusively found in these later works, while Aristotle and Jāḥiẓ quotations are widely present from the ninth/tenth century on, and could have been taken from any number of works.

This is disappointing, because there is evidence in ʿAbd al-Laṭīf's work that he appreciated Ibn Abī 'l-Ashʿath's scholarly approach. In his *Kitāb al-naṣīḥatayn*, ʿAbd al-Laṭīf puts him forward as the scholar who in Islamic times updated (*aḥyā wa-yujaddidu*) the medical knowledge of the ancient Greeks, thus fulfilling the same role that had earlier been fulfilled by Hippocrates and Galen.[23] Moreover, Ibn Abī Uṣaybiʿa tells us

[23] *Kitāb al-naṣīḥatayn*, ms. Bursa Hüseyin Çelebi 823, 5, fol. 73b, l. 14–17 (about medical science): 'God, praised be He, sent people to bring its traces to life again and to renew its obliterated outlines and destroyed cornerstones, just as happened in the time of Hippocrates, who gave new life to the medical science of his great forebear Asclepius, and then in the time of Galen, who gave new life to Hippocrates' medical science, and the last one whom I have known in the Islamic community was Abū Jaʿfar Aḥmad ibn Muḥammad ibn Abī 'l-Ashʿath.' Ghalioungui's rendering of this passage

that ʿAbd al-Laṭīf made a compendium (*ikhtiṣār*) of Ibn Abī 'l-Ashʿath's book on colic (*qūlanj*) (IaU ii, p. 213, l. 7). His next reference, to a compendium by ʿAbd al-Laṭīf of Ibn Abī 'l-Ashʿath's very interesting *Book on Animals*, is all the more intriguing because of the emphasis which Ibn Abī 'l-Ashʿath puts in this work on personal observation. It is tempting to think that this would have appealed to ʿAbd al-Laṭīf, who was a great advocate of personal verification. But, as shown above, nothing has so far turned up to support this idea: the source material as it stands does not provide additional evidence for the existence and nature of any zoological compendium by ʿAbd al-Laṭīf.

As to his *Book on (the Natures of) Animals*:
Explicit mentioning of a *Kitāb (ṭabāʾiʿ) al-ḥayawān* written by ʿAbd al-Laṭīf only occurs in the Ibn Abī 'l-Ḥawāfir/Ibn al-Athīr/Watwāṭ cluster, so, with the exception of the passage on the *dalaq*, basically only in Ibn Abī 'l-Ḥawāfir's *Badāʾiʿ al-ḥayawān*. The entries in which it is found are those on the *ḍabb-waral*, the *dalaq-ibn ʿirs-sammūr-nims*, and the *thaʿlab*. In the entry on the *fīl*, only found in Damīrī, ʿAbd al-Laṭīf is mentioned as the source without a book title, but the information is typically of the kind that is found in zoological literature. It does not go back to Ibn Abī 'l-Ḥawāfir.

The impression given by these few explicit quotations is that a *Kitāb (ṭabāʾiʿ) al-ḥayawān* of ʿAbd al-Laṭīf indeed existed, and that it was a collection of fairly standard bits of information on animals, material that could also aptly have been included under the title *al-Mudhish fī akhbār al-ḥayawān* (IaU ii, p. 213, l. 13ff), suggested by Ullmann to be an alternative title for the *Kitāb al-ḥayawān*. It is not, at first sight, the type of work that one would expect from ʿAbd al-Laṭīf, who certainly was not a compiler of *adab*-type material. On the other hand, some of the zoological entries in his *Kitāb al-ifāda*, such as that on the crocodile, are also of a conventional *adab* nature.

As to the other quotations explicitly referred to ʿAbd al-Laṭīf: it is very unlikely that the story about the *samandal*, found in Damīrī and Qalqashandī, was taken from a zoological text. As to the *dajāj* (Qalqashandī) and the *kabsh* (Damīrī): the former could, for instance,

(cf. n. 4, p. 30) is not very accurate: he says that ʿAbd al-Laṭīf characterizes Ibn Abī 'l-Ashʿath as 'the last doctor worthy of that name'.

have occurred in a philosophical work, while the latter most likely stems from a medical context. None of these three entries can be connected to Ibn Abī 'l-Ḥawāfir.

Concluding, we can say that a *Kitāb (ṭabʿaʾiʿ) al-ḥayawān* circulated under the name of ʿAbd al-Laṭīf, but that very little is known about it, although at first sight there appear to exist a substantial number of quotations. On closer analysis, this number can be reduced to basically three, all belonging to the same strand of transmission, plus two more from a different strand that are of a doubtful nature.

A TREATISE ON METEOROLOGY BY MUḤAMMAD IBN MŪSĀ AL-ṬĀLISHĪ (MS. DAIBER COLLECTION II, 82)*

Hidemi Takahashi

1. Introduction

Besides the immeasurable debt I personally owe Prof. Daiber as the ever-patient and ever-resourceful supervisor of my doctoral dissertation, as one of those engaged in Islamic and germane studies in Japan, I count myself among those who are indebted to him for the fact that a wondrous collection of some 520 Arabic, Persian and Turkish manuscripts has found its home in Japan at the Institute of Oriental Culture in the University of Tokyo, increasing the number of such manuscripts found on Japanese soil by a significant proportion and providing (mostly) young Japanese scholars with valuable material to work with.[1] It seemed

---

* I am indebted to Prof. Modjtaba Sadria, a neighbour, at the time of writing, at Chuo University, for his help in deciphering the Persian quotations which occur in the manuscript discussed here, and to my friend Mr. Jean Fathi-Chelhod (Riyadh) for saving me from a good number of blunders in the edition of the Arabic text.

[1] The manuscripts are described in: H. Daiber, *Catalogue of the Arabic Manuscripts in the Daiber Collection. Institute of Oriental Culture. University of Tokyo* (Tokyo, 1988); and idem, *Catalogue of the Arabic Manuscripts in the Daiber Collection ii. Institute of Oriental Culture. University of Tokyo* (ibid., 1996).—Digital images of all the manuscripts in the collection have recently been made accessible online in the database section of the website of the Institute of Oriental Culture at: http://ricasdb.ioc.u-tokyo.ac.jp/daiber/db_index.html. At the time of going to press, the instructions for the use of these pages are available only in Japanese, but Prof. Shigeru Kamada of the Institute of Oriental Culture informs me that preparations are under way for the provision of English instructions. Prof. Kamada has further had the kindness of conveying to me the names of the following publications by Japanese and non-Japanese scholars which are based to a greater or lesser extent on the study of manuscripts in the collection: Yasushi Tonaga, 'Controversies on the Orthodoxy of *Waḥdah al-Wujūd* in the Late-Mamlūk Period' (in Japanese), *The World of Islam* (Association for Islamic Studies in Japan) 33/34 (1990), pp. 51–72 (Suyūṭī, *Tanbīh al-ghabī fī tanzīh Ibn al-ʿArabī*, ms. Daiber [i], 156); Hiroshi Tarui, 'Treatise on Sufi Love: Hüdâ'î's "Seed of Love"—edition, translation and commentary' (in Japanese), *Annals of Japan Association for Middle East Studies* 9 (1994), pp. 273–307 (*Ḥabbat al-maḥabba*, ms. Daiber ii, 142, fol. 239v–246v, and Bologna, Marsigli 240_8); Shigeru Kamada, 'Manuscripts of the Classics (5): Islamic Studies' (in Japanese), *Reconstruction of the Study of the Classics* 5 (Jan. 2000), front and inside cover (with photograph and Japanese translation of a page out of ms. Daiber [i], 23 [Mawlānāzādah, *Sharḥ Hidāyat al-ḥikma*]; the Japanese text of the article, but not the

therefore appropriate—almost imperative—that a contribution from Japan to this *Festschrift* should consist of an edition of a text from that collection. The choice of the text to be edited was determined by two factors, namely my own inability to handle Arabic texts on other topics with any competence (not that I have been able to deal with this one with much competence, as will be seen) and the fact that the tradition of Aristotelian meteorology in Semitic languages is one of the areas Prof. Daiber has had a particularly strong interest in at least since the time of composition of his first published monograph.[2]

The treatise edited below is found in ms. Daiber Collection ii, 82, a manuscript consisting of six folios and written in a *naskhī* hand judged by Prof. Daiber to be of the seventeenth century.[3] The manuscript gives no specific title for the treatise, but the text of the treatise ends on

photograph, also accessible at www.classics.jp/Contents/Assets/publication/NLpdf/NL05SHA.pdf); Tatsuya Kikuchi, 'The Principle of the Evil in the Early Druze Thought' (in Japanese), *Annals of Japan Association for Middle East Studies* 16 (2001), pp. 139–53 (the 'Druze Canon' [*al-Ḥikma al-sharīfa/Rasāʾil al-ḥikma*], ms. Daiber ii, 64 & ii, 153); idem, '*Taqīya* and Metempsychosis in the Early Thought of al-Durūz' (in Japanese), *The World of Islam* 57 (2001), pp. 1–20; Kazuo Morimoto, *Sayyids, Genealogists, Naqibs: a Study on the Genealogical Literature of Sayyid/Sharifs from Late 10th through Early 15th Century* (in Japanese), PhD dissertation (Graduate School of Humanities and Sociology, University of Tokyo, 2004); Biancamaria Scarcia Amoretti, 'Women's Names in Early Islamic Pro-Shiʿite texts on the Genealogy of the Talibiyyin,' *Medieval Prosopography* 23 (2002), pp. 141–65; Yaḥyā ibn al-Ḥasan al-ʿAqīqī, *Kitāb al-muʿqibīn min wuld al-Imām Amīr al-Muʾminīn*, ed. M. al-Kāẓim (Maktabat Āyat-Allāh al-ʿUẓmā al-Marʿashī al-Najafī, 1422 A.H.q/1380 A.H.s/2001 C.E.) (photographs and edition of ms. Daiber [i], 127, published without permission).—I have had occasion myself to make a small addition to the information given by Daiber on ms. Daiber ii, 43, namely on the identity of the copyist Gregory Michael b. Barṣawmo of ʿUrbīsh, Syrian Orthodox metropolitan of Gargar 1591–1618 (H. Takahashi, *Barhebraeus: a Bio-Bibliography* [Piscataway, 2005], p. 210f).

[2] *Ein Kompendium der aristotelischen Meteorologie in der Fassung des Ḥunain ibn Isḥâq* (Amsterdam, 1975). His other major contributions to this field include the edition of the Syriac and Arabic versions of Theophrastus' *Meteorology* ('The *Meteorology* of Theophrastus in Syriac and Arabic Translation,' in W.W. Fortenbaugh and D. Gutas [eds.], *Theophrastus. His Psychological, Doxographical, and Scientific Writing* [New Brunswick, 1992], pp. 166–293), as well as the publication under his auspices of: R. Fontaine, *Otot ha-Shamayim. Samuel Ibn Tibbon's Hebrew Version of Aristotle's* Meteorology (Leiden, 1995); P. Lettinck, *Aristotle's* Meteorology *and its Reception in the Arab World. With an Edition and Translation of Ibn Suwār's* Treatise on Meteorological Phenomena *and Ibn Bājja's* Commentary on the Meteorology (Leiden, 1999); P.L. Schoonheim, *Aristotle's* Meteorology *in the Arabico-Latin Tradition* (Leiden, 1999); H. Takahashi, *Aristotelian Meteorology in Syriac. Barhebraeus,* Butyrum sapientiae, *Books of Mineralogy and Meteorology* (Leiden, 2004).—Another text I might have made an attempt at editing is a treatise on the preconditions for the appearance of a rainbow found in ms. Daiber Collection ii, 125, fol. 35 (Daiber, *Catalogue*, ii, p. 174, cf. ibid., Intro., p. xv).

[3] Daiber, *Catalogue*, ii, p. 109f.

fol. 6r with the words, *tammat al-nuskha al-Ṭālishīya*, while in the margin of fol. 1r. we find the words, *risāla-yi Ṭālishī, īn ast az ʿilm ḥikmat al-ʿayn*. Prof. Daiber identified the author as Muḥammad ibn Mūsā al-Ṭālishī, about whom little seems to be known beyond the fact that he hailed from the region of Ṭālish (Talesh/Talysh) on the southwestern coast of the Caspian,[4] and that he wrote, at the command of the Aq-Qoyunlu Sultan Yaʿqūb (1478–90) in 884/1479, a commentary on *Kitāb ḥikmat al-ʿayn*, a work on metaphysics and physics composed by Najm al-Dīn ʿAlī ibn ʿUmar al-Qazwīnī al-Kātibī (d. 1276 or 1294), a disciple of Naṣīr al-Dīn al-Ṭūsī.[5] The identity is confirmed by a passage in the text of the treatise (§4.0.1.3 in the text below) where the author refers to an earlier work of his called *Sharḥ Ḥikmat al-ʿayn*, a reference which also indicates that the treatise here is not a part of *Sharḥ Ḥikmat al-ʿayn*, but an independent treatise written at a later date.

The treatise consists of four sections (*uṣūl*) dealing with 1) 'vapour' and its products such as cloud and rain, 2) 'smoke' and its products such as thunder and lightning, 3) winds, and 4) rainbow and halo, along with an introductory section (*tamhīd*) dealing with the definition of the 'body' and transformation of the four elements, and a concluding section (*khātima*) dealing with the formation of water sources and earthquakes. The content, or at least the subject matter, of the treatise is largely Aristotelian-Avicennian, and the only source the author explicitly

[4] It may be noted in this connection that the one place-name mentioned in the text of the treatise is Ardabīl (§2.4), which is in the close vicinity of Ṭālish.

[5] *GAL* i, p. 467 (mentioning a manuscript in what is now the National Library in Cairo, catal. Khidīwīya vii, 97); ʿU.R. Kaḥḥāla, *Muʿjam al-muʾallifīn. Tarājim muṣannifī al-kutub al-ʿarab* (Damascus 1957–61), xii, p. 64 (mentioning that the commentary was written in 884/1479 'for [*bi-sm*] Sultan Yaʿqūb who died in 896'); King Faisal Center for Research and Islamic Studies, *Khizānat al-turāth* [CD Rom] (Riyadh, sine dato), no. 35758 (mentioning, besides the Cairo manuscript, a further manuscript in Kuwait, Maktabat Kullīyat al-Ādāb wa'l-Makhṭūṭāt, 724/3; and giving the name of the author as Muḥammad Amīn ibn Mūsā).—The information accessible at the time of writing does not allow me to determine whether Nicholas Rescher is correct in identifying our Muḥammad ibn Mūsā al-Ṭālishī with Muḥyī al-Dīn al-Tālijī (aliter al-Tālishī), the glossator of Ḥusām al-Dīn al-Ḥasan al-Kātī's (*c.* 1290–1359) commentary on Athīr al-Dīn al-Abharī's (d. 1264) *Kitāb al-Īsāghūjī* (N. Rescher, *The Development of Arabic Logic* [Pittsburgh, 1964], p. 239; cf. *GAL* i, p. 465, *S* i, p. 842; Daiber, *Catalogue*, p. 45f, ms. 55; Daiber, *Catalogue*, ii, p. 111, ms. 83; *Khizānat al-turāth*, no. 79739 [mentioning manuscripts in al-Azhar Library in Cairo and the National Library in Abu Dhabi]).—*Khizānat al-turāth* (no. 81322), which likewise identifies al-Ṭālishī, al-Tālishī and al-Tālijī, names a third work by this author entitled *Sharḥ Jihat al-waḥda*, with manuscripts in Kuwait (137/1) and St. Petersburg (Institute of Oriental Studies, no. 1252; with copy in Jumʿa al-Mājid Cultural and Heritage Centre, Dubai, no. 3931).

mentions in the treatise is Ibn Sīnā's *Kitāb al-shifā'* (in §§1.5, 2.4, 4.3, 4.6, 4.9). This does not mean, however, that our author is always in agreement with Ibn Sīnā, or that he does not deal with items not dealt with in the *Shifā'*. Even a cursory look at the text reveals that a number of technical terms which do not occur in the *Shifā'* but are known to occur in other Arabic works on meteorology are used there, such as the term *ḥarīq* ('conflagration', §2.6), known to occur, for example, in Fakhr al-Dīn al-Rāzī's *Mabāḥith al-mashriqiyya*,[6] and the term *zamharīr*, used in §1.7 to designate a kind of snow formed in the air as in Kindī,[7] and in the form *kurat al-zamharīr* in the sense of the 'sphere of extreme cold' (§1.3, 2.2), a usage found in the *Rasā'il Ikhwān al-ṣafā'*, Zakarīyā' ibn Muḥammad al-Qazwīnī's *'Ajā'ib al-makhlūqāt* and Abū 'l-Barakāt al-Baghdādī's *Kitāb al-mu'tabar*.[8] The account of how wind is generated differs significantly from the account in the *Shifā'* in defining wind as 'moving air' (as opposed to 'smoke', §3.1)[9] and devoting much space to the generation of wind through expansion and contraction of air (§3.2), an idea which in this form probably goes back to al-Kindī but is not given much attention by Ibn Sīnā.[10] The discussion on the shape of the rainbow (§4.3–5) also goes some way beyond what we find in the *Shifā'* and no doubt reflects some of the developments in the discussion in the period subsequent to that of Ibn Sīnā.[11]

The treatise presented here is one that stands near the end of a long tradition of discussions on Aristotelian meteorology in Arabic, and a proper understanding of the treatise requires a close comparison of the material with the literature on the subject especially from the period subsequent to that of Ibn Sīnā. Limitations of space and time (compounded by my sloth), however, as well as the difficulty of access-

---

[6] See note ad loc. below.

[7] See Lettinck, *Aristotle's* Meteorology, p. 110. The word *zamharīr* itself is one that is found in the Qur'ān (76:13).

[8] See Lettinck, *Aristotle's* Meteorology, pp. 53, 115. On the relationship between the *Rasā'il Ikhwān al-ṣafā'* and Qazwīnī's *'Ajā'ib*, see Takahashi, *Aristotelian Meteorology in Syriac*, p. 41f, n. 28

[9] On the significance of this, see Lettinck, *Aristotle's* Meteorology, p. 161, Takahashi, *Aristotelian Meteorology in Syriac*, p. 489f.

[10] See Lettinck, *Aristotle's* Meteorology, pp. 107f, 176, 177f.

[11] On the discussion of this topic in the period following that of Ibn Sīnā, see Lettinck, *Aristotle's* Meteorology, pp. 283–300. It was apparently Ibn al-Haytham (965–1040), a contemporary of Ibn Sīnā, who first applied to the discussion of the shapes of the halo and the rainbow the 'law of equal angles' of incidence and reflexion (Lettinck, *Aristotle's* Meteorology, p. 286), which is referred to here in §§4.0.1.1, 4.4 and 4.5.

ing some of the relevant materials, have meant that it has not been possible to provide any detailed analysis of the material in the notes to the translation below. Although, as I have just indicated, the content of our treatise is not always in agreement with the *Shifāʾ*, in the notes to the translation I have limited myself largely to referring to the passages of the *Shifāʾ* where the corresponding discussion is found,[12] since that is—as it evidently was in Ṭālishī's time—the standard work against which subsequent works in the field need to be compared. To these have been added the pages of Lettinck's study where summaries of those passages of the *Shifāʾ* may be found,[13] since in that work the reader has the advantage of being able to find summaries of what other Arabic works on Aristotelian meteorology have to say on a particular subject by flicking to and fro on either side of the pages where the summary of the *Shifāʾ* is given. I have also frequently added references to Qazwīnī's *ʿAjāʾib al-makhlūqāt*, a relatively late work, with which our treatise shares a number of features, although the similarities are not such as to indicate direct use of this work by the author of our treatise.[14]

In editing the text, I have generally tried to remain faithful to the readings of the single manuscript from which I was working, retaining the orthography of the manuscript except where it was judged that this might be misleading or confusing for the reader.[15] The paragraph division and numbering inserted for ease of reference, along with all the punctuation, are mine. In the text and translation chevrons (<>) indicate instances where I suspect lacunae in the manuscript. Obeli (†) in the text and asterisks (***) in the translation indicate passages where it has not been possible to make sense of the Arabic text. Besides the Persian note on fol. 1r mentioned above, the manuscript has a number of other marginal additions in Arabic and Persian. The Arabic notes,

---

[12] The page and line references are to the volume in the 'Cairo edition' of the *Shifāʾ* covering the *fann* on mineralogy and meteorology: Ibn Sīnā, *Al-Shifāʾ. Al-Ṭabīʿīyāt. 5. Al-Maʿādin wa'l-āthār al-ʿulwīya*, ed. ʿA. Muntaṣir et al. (Cairo, 1385/1965).

[13] Lettinck, *Aristotle's* Meteorology.

[14] The references are to the edition by F. Wüstenfeld (*Zakarija Ben Muhammed Ben Mahmud el-Cazwini's Kosmographie*, Erster Theil. Die Wunder der Schöpfung [Göttingen, 1849; repr. Frankfurt, 1994]) and the translation by H. Ethé (*Zakarija Ben Muhammed Ben Mahmûd el-Kazwîni's Kosmographie*, Die Wunder der Schöpfung. Erster Halbband [Leipzig, 1868; repr. Frankfurt, 1994]).

[15] I have, for example, followed the manuscript in retaining the *madda* in such cases as هوآء, الخآتمة; in the orthography of the *hamza* in such instances as اضائة (as opposed to اضاءة; but منشاؤه is changed to منشؤه); and in neglecting the concord between the verb and the following feminine or plural noun.

which all seem to have some relation to the content of the text, are given in the apparatus below. The Persian additions are quotations from poetry with no relevance that I can figure out to the content of the text and have not been reproduced.[16] Two quotations in Persian verse are also found copied in the body of the text, and, although these are probably interpolations by our copyist, they have been left where they are in the text (§§0.2.8, 0.2.9).

Finally, it needs to be stated that it is with a great deal of apprehension that I present the text and translation below, since what is presented here is clearly far from the level of perfection scholars worldwide associate with the one honoured in this *Festschrift*, as well as the level my former supervisor would have expected from his former student. As will be seen, a good number of difficulties remain unsolved in the text and translation given below. For all these problems to be solved, however, and for this *Festschrift* nevertheless to appear on time, one would need, like Joshua bar Nun, to stay the sun and the moon. I beg the indulgence, therefore, of my *Doktorvater* and of other readers for the eyesores that remain.

2. Muḥammad al-Ṭālishī, *Treatise on Meteorology* (Text)

[0.0] (1r)[(1)] بسم الله الرحمن الرحيم وبه نستعين. الحمد لوليه والصلوة على نبيّه محمد وآله الطيّبين الطاهرين وصحبه اجمعين. اما بعد فهذه رسالة مرتبة على تمهيد واصول اربعة وخاتمة. التمهيد مشتمل على مقدمتين.

[0.1] الاولى فى ماهية الجسم واقسامه على الاجمال. فالجسم جوهر قابل للابعاد الثلثة المتقاطعة على الزوايا القائمة، اعنى الطول والعرض والعمق. وهوعلى قسمين بسيط ومركب. فالمركب ما يلتأم من اجسام مختلفة الحقائق، اما تام، اذا كان له صورة من شانها ان تحفظ زمان يعتد به كالمواليد الثلثة، اعنى المعادن والنباتات والحيوانات، واما غير تام، اذا لم يكن له صورة كذلك كالبخار والسحاب. والبسّيط ما لا يكون ملتأما منها، وهو اما علوّى وهو الافلاك بما فيها من الكواكب، واما سفلى وهو عناصر الاربعة التى هى النار والهواء والارض والماء. فالنار حارّ يابس، والهواء حار رطب، والماء بارد رطب، والارض بارد يابس.

[16] Some of the quotations have been cut away with the edge of the page and are no longer decipherable. Besides these marginal additions, the last page of the manuscript (fol. 6v) has quotations in Persian verse attributed to شيخ نظامى [i.e. Niẓāmī Ganjawī] and عبيد زاقى, along with lists of honorific titles for a *shaykh, ʿulamāʾ, ḥukamāʾ wa-aṭibbāʾ*, an *amīr* etc.

[0.2.1] المقدمة الثانية فى انقلاب العناصر بعضها ببعض. قد عرفت ان لكل من العناصر كيفيتين. فنقول ان لكل منها حدّ او رتبة، اذا جاوزت عنها فى احَد جانبى الافراط والتفريط، لم يبق استعدادُ مادة ذلك العنصر لصورته، ويحصل استعداد صورة اخرى يناسبها تلك الكيفية، فتخلع مادة الصورة الاولى، وتلتبس الثانية.

[0.2.2] فالنار اذا انكسر صورة حرارتها باحاطة الهواء وغيرها، صارت حرارتها لحرارة الهواء انسب، لم يبق استعداد مادتها لصورتها، ويحصل[2] لها استعداد صورة الهواء، فتخلع صورتها وتلبس صورته. وهكذا فيما بين سائر العناصر.

[0.2.3] ولنورد امثلة جزئية لانقلاب بعضها ببعض تشحيذا للاذهان وتسكينا للخواطر. اما انقلاب النار بالهواء، فكانقلاب شعلة السّراج بالهواء بسبب احاطة الهواء بها، فانها اذا ارتفعت عن مبدئها، اعنى الفتيلة، اثر الهواء المحيط بها فيما يجاوره، فينقلب شيئا فشيئا اليه، فيصير على هيئة المخروط الى ان ينطفئ[3] بالكلية. وانطفاء السراج بالريح انقلابها الهواء بسرعة لشدة تاثيره فيها ونفوذه فى اعماقها.

[0.2.4] واما انقلاب الهواء بالنار، فكما يشاهد فى كور الحدادين اشد منافذها والح فى النفخ، فان الهواء الكائن ثمه ينقلب نارا. ومن هذا القبيل انه اذا اطفئ[4] سراج وجعل فوق سراج[5] آخر، فانه يشتعل باستيلاء الحرارة على الهواء المجاور للسّراج المنطفئ[6] مع شدة استعداده للاشتعال. ومن هذا القبيل ايضا حصول النار عن قرع مثل الزند على الحجر المخصوص، والنار الحاصلة ثمه ليست كائنة فى موضع تماس[7] احدهما، والا لاحس بها ولانطفأت[8]† بغورهما فى الماء مدة مديدة، بل لان احدهما اذا ضرب على الآخر بشدة تسخن الهواء الكائن عنه مع ما بين الزند والحجر المخصوص من مناسبة مخصوصة لا يعرفها، فينقلب نارا.

[0.2.5] واما انقلاب الهواء ماء، فكما يشاهد من القطرات الماء النازلة من سطح(1v) الحمام، فان هوائه المستعد لتاثير البرودة بتسخنه ولطافته، اذا لاقى سقفه البارد بالنسبة اليه، تاثر منه وانقلب ماء، فينزل. ومن هذا القبيل تراكم القطرات على ظاهر كوز فيه الجمد وفوقه طاس كب عليه، واجزاءٍ رشّية بعلوّ سطح حديد او حجر او خشب اذا نفخ[9] عليه قريبا من الفم وهو مفتوح.

[0.2.6] واما انقلاب الماء هواء، فكما يشاهد من انقلابه اليه عند غليانه او مروره مدة مديدة عليه. ومن هذا القبيل جفاف ثوب مبلول اذا اثر فيه شيء من المسخنات، والهواء شيئا† اذا كان حارّا بتلطيفه، وجعله مستعدا لأن ينقلب هواء.

[0.2.7] واما انقلاب الماء بالارض وبالعكس، فكما يشاهد من تكون حجر الرخام وغيره من الماء، وسيلان بعض الاحجار مياها بحيل اصحاب الاكسير.

[0.2.8] ومن فوائد هذا البحث والاطلاع على انقلاب بعض العناصر تجويز امر الطوفان وبعض معجزات الانبياء، عليهم السلام، كانفجار المياه من الاحجار وغيرها على قواعد الحكماء ايضه[10]، ومنع استحالة ما اشتهر فى السنة العوام من صيرورة الحيوانات احجارا، اذ ليس بعد الحيوان من الحجر اكثر من بعد الماء منه، وهو ظاهر، والاطلاع على ان العناصرَ كلها حقيقة واحدة تطورت[11] بالاطوار وماهية بسيطة[12] تعددت بالاعتبار اذ الانقلابَ

للحقائق بالاغيار† تامّل تدر ما هو المط بالاخيار† –شعر آن كسى كه ز شهر آشنايیست داند كه متاع ما كجاييست.

[0.2.9] والتنبيه. لمّا سئل بين العناصر المتضادة بالكيفيات المتعاندة باعتبار الصور من انقلاب البعض الى البعض بالمناسبة التى تحصل لها بالمجاورة والمخالطة، فتارة تخرج النار من كمال حرارتها ولطافتها الى كمال كثافة الارضية وبرودتها، وبالعكس. †فعليك النظر والاعتبار، ومجانبة الاشرار الى مجاورة الاخيار† - شعر آن چه بصد عمر كسى يافته همنفسى در نفسى تافته. داغ بلندان طلب اى هوشمند تا شوى از داغ بلندان بلند.

[1.1] الاصل الاوّل فى البخار. وهو ما يحصل منه فى الجوّ اذا تسخن المياه والاراضى الرطبة من اشعة الكواكب وغيرها، انقلب بعض الاجزاء المائية هواء، ومال لطيفة الى مكانه الطبيعىّ مستصحبا لبعض آخر من الاجزاء المائية بحيث لا يميز الحسّ بينهما، ويسمّى ذلك المجموع بخارا. وقلما يخلو بخار عن اجزاء ارضية، على ما يدل عليه مشاهدة تلك الاجزاء فى مياه الامطار والثلوج اذا جعلت فى اناء حتى يصفو. وقد يحصل البخار من تاثير البرودة فى الهواء، وقلب بعض اجزائه ماء مختلطا ببعض آخر كذلك.

[1.2] ثم اذا كان الهواء حارّا ولم يكن ريح، تحلل[13] الاجزاء المائية وانعدم البخار بالكلية. وان لم يكن كذلك، بل حصل فى البخار غلظ وكثافة فى الجملة، بسبب اصابة ادنى برودة، او تراكم بعض اجزائها على بعض بانضمام بخار اليه بريح، او غيره، يسمّى غيما. والغيم انً كان نازلا من العلو الى السفل، كان قرينة صحو الهواء وعدم المطر، سيّما اذا كان عقيب المطر. واذا كان صاعدا، كان من علامات السحاب والمطر.

[1.3] واذا اشتد تكاثف البخار باحد الاسباب المذكورة، يسمّى سحابا، سواء وصل(2r) الى كرة الزمهرير او لا.

[1.4] واذا لاقى بعض الاجزاء المائية الكائنة فيه بعضا، واذا انقلب بعض اجزائه ماء ببرودة الهوآء او بالبرد المحتبس فيه فى احاطة الهواء الحار به،[14] ونزل بميله الطبيعى الى حيزه[15]، يسمى مطرا. واذا اعرض فى مقابلة الريح الموجبة لتراكم الابخرة والمطر جبل او ريح اخرى، اشتد تركم الابخرة، وكثر[16] الامطار، ومن ثم كثرة الامطار فى جبال بعض البلاد الحارّة نحو الحبشة[17].

[1.5] واذا كان البرودة اشد مما ذكر، او كان استعداد الاجزاء المائية لتاثير البرودة الحاصلة من بعد[18] الطرفين اشتد بتسخنها ولطافتها، انجمد بردا. و<ذلك> اذا اشتد انجمادها وكان محيطاتها ملساء[19] وترى عليها بلة. والا، كان ثلجا غليظا كثيفا، وذلك [20]اذا كان اشتداد البرودة واستعداد الاجزآء المذكورة اكثر. و<حبات> البرد قد تكون كبيرة وقد تكون[21] صغيرة، لكون القطرات المائية التى تكون هى منها كذلك، او لتفاوت[22] المسافة، فمتى كانت طويلة، تحلل اجزاؤها[23] بالحركة وصارت صغيرة، واذا كان قصيرة، بقيت على ما هى عليها، ولا يتحلل كثيرة. وايضا حبات البرد قد تكون مستديرة لاصابة البرودة اياها بعد الاجتماع التام وصيرورتها بالطبع كريا، او لتحلل زواياها لطول المسافة، وقد تكون[24] مضلعة لانتفاء الامرين معا. وهو فى الربيع والخريف اكثر مما هو فى الصيف والشتاء.

وذلك لان حصوله من السحب الرطبة، وهى فى الفصلين الاولين، سيما الاول، اكثر مما فى الاخيرين، اما فى الصيف فلكثرة تحليل المادة واستيلاء اليبوسة على الهواء، واما فى الشتاء فعلى ما قال الشيخ فى الشفاء من انه اذا اشتد البرودة فيه، حصل الثلج، والا، لم يحصل شىء. تامل.

[1.6] واذا اشتد البرودة او كان الاجزاء المائية، بل السحابة الرطبة قبل الانجماد، انجمدت قبل ان يحصل منها انضمام تام وتلاق كامل، ونزلت على هيئة القطن المندوف، يسمى ذلك ثلجا. وانما ذكرنا قيد زيادة الاستعداد حيث ذكرنا، لانا نشاهد ان الهواء المتسخن ينقلب ماء ببرودة لا ينقلب غيره اليه باضعاف تلك البرودة، كما يرى فيما مر من هوآء الحمام. وان الماء الحارّ ينجمد قبل الماء البارد، ومن هذا الانجماد<...> . الابخرة الخارجة عن الفم فى حواليه مع هوائها اقل بردا مما هو ابعد منها.

[1.7] وقد يشاهد نزول اجزاء صغيرة متجمدة، تسمى فى بعض السنة زمهريرا، وغير متجمدة عن هواء صاف، وذلك لاستيلاء برد شديد او قليل، وحصول غلبة تكاثف وثقل فيما يكون فيه من الاجزاء المائية او فيما ينقلب من اجزائه ماء، حتى نزل.

[1.8] واذا استولى برد بدخول الليل على هوآء رطب او ابخرة[25] لطيفة رطبة فى بعض المواضع، انقلب ذلك الهواء او البخار المجاورين لوجه الارض ماء، او اجتمع الاجزاء المائية الكائنة فيه او المنقلبة منه ماء، وثقل، حتى نزل، ولكن لا يحس بها لصغرها حتى اجتمع على وجه الارض، ويسمى طلا ان لم ينجمد، وصقيعا ان انجمد.

[2.1] الاصل الثانى. فى الدخان. و<هو>[26] ما يحصل فى الجوّ من تسخن الاراضى اليابسة والمواضع الكبيرتية باشعّة الكواكب وغيرها. انقلب اجزاء الهواء الكائن فيها نارا مختلطة باجزآء (2v) هوائية وارضية بحيث لا يميز الحس بينها، ويسمى ذلك المجموع دخانا. وقلّما يخلو دخآن عن بخار وبالعكس، ولكن ايّما كان غالبا يسمّى باسمه.

[2.2] ثم اذا ارتفع الدخان بتسخنه، انقطع عن الارض. فقد يكون اذا وصل الى كرة الزمهرير، احتبس فى السحاب الذى كان ثمه او يكون من الذى كان معه. ثم ان بقى على حرارته، خرج من السحاب الى العلو خارقا اياه.

[2.3] واذا نزل لثقل[27] حصل فيه بانتفاء الاجزاء النارية عنه واستيلاء الاجزاء الارضية عليه او لاسباب اخرى كاقتضاء الفاعل وغيره، فلبقآء حرارة فيه دون اجزاء نارية محرقة، يؤثر فى اى شىء تصادفه من غير احراق، كما قيل انه يصادف صرة ويذيب الذهب من غير ان يحرق الكيس. وذلك التاثير مختلف، فاذا كان غليظا شديد الحرارة، كان تاثيره اشد ويسود موضع اصابته، واذا كان لطيفا، لا يوثر كذلك.

[2.4] وربما اشتدّ استيلاء البرودة عليه بحيث ينجمد حجرا او غيره بحسب الاجزاء الكائنة، بل يمكن ان يتولد من امتزاج الابخرة والادخنة مزاج يفيض عليه صورة ونفس حيوانية، كما اشتهر فى لسان العوام من نزول الاحجار والحيوانات فى الامطار. ولعل الحجر الذى كان الآن فى جامع اردبيل واشتهر بين عامته انه نزل من السماء من هذا القبيل. وقد حكى الشيخ فى الشفاء باراء وروله† احجارا كبيرة غريبة، لا نطول الكلام بذكرها.

[2.5] وعلى اىّ تقدير يحصل من خرقة اجزآء[28] السحاب صوت شديد، يسمى رعدا. ويمكن ايضا من انخراق السحاب بريح عاصفة او خروج ادخنة محتبسة بانضمام اجزاء السحاب بعضها الى بعض او ريح او غيرهما، ومن حصول الاصطكاك بين اجزاء السحاب بريح او غيرها، ومن حركة الهوآء فيما بين اجزآء السحاب. الا يرى ان الهوآء الصافى اذا اشتد حركته حصل صوت قوىّ، فما ظنك بين اجزاء السحاب الكثيفة.

[2.6] وان لم ينقطع الدخان المرتفع عن الارض، بل بقى احد طرفيه متصلا بها، فربما اشتعل طرفه المرتفع لقربه من النار او بامر آخر، فينزل الاشتعال جزءً فجزءً حتى يصل الى الارض، فترى على شكل عمود نازل من العلو. يسمى ذلك حريقا. ثم ان كان مادته غليظة، احرق كل ما يصادفه، والا فلا، كما لا يحرق شعلة القطن المندوف شيئا.

[2.7] وان غلب على الدخان اللطيف الكائن فى السحاب اجزاء دهنية، واشتعل بشدة الحركة والاصطكاك او لسبب[29] آخر، و<حرك> حركة سريعة غير مستقيمة اما لاختلاف قوام المادة او لاعوجاج المميز بين السحب، يسمى[30] برقا. ويجعل ايضا اعوجاجه وتحيل حركته لوقوع مادة دخانية لطيفة فى تلك المسافة واشتعال احد طرفيه بسبب من الاسباب ساريا منه الى الطرف الآخر منطفئا[31] بسرعة للطافة المادة، هذا لكن لا يخلو عن بعد.

[2.8] واعلم ان الرعد كثيرا ما يكون بدون البرق، واما عكسه فقليل، حتى قيل انه لا يكون البرق بدون الرعد، حيث لا يخلو[32] البرق عن اصطكاك وخرق[33] موجب للرعد، لكنه قد يكون خفيفا[34] بحيث لا يسمع. وايضا البرق اذا اشتعل من اصطكاك اجزاء السحابة، كان تابعا للرعد. واذا فصل الاصطكاك او الخرق[35] من اشتعاله لحرارة محبوسة فيما بين اجزاء السحاب او لغيره، وحركته نحو جهة، كان الامر بالعكس، لكن على التقديرين يرى البرق قبل سمآع الرعد، اما فى الاول فظاهر[36]، واما فى الثانى (3r) فلعدم توقف رؤية الاشياء المقابلة على زمان بخلاف سماع الصوت <... الصوت> الحاصل[37] من ضرب الفاس بعد مشاهدة ضربه بزمان.

[2.9] واذا جاوز الدخان الذى غلبت عليه الدهنية الطبقة[38] الزمهريرية، حتى وصل الى النار اما بقرب منها، واشتعل بحرارة النار او بحركته او بحرارة محتبسة فيه، وحركته على الاستقامة لعدم الاختلاف فى مادته وممرّه، وينطفئ[39] بسرعة ابطأ[40] من البرق بقليل لكون لطافة مادته، يسمى شهابا. ويحتمل ان يكون استقامته وتحيل حركته لما ذكرناه فى البرق، بل يحتمل ايضا ان كلا منهما ريح دخانية يتحرك فى الهواء المتشابه[41] على الاستقامة وفى السحاب المختلف القوام على الاعوجاج.

[2.10] وان كان الدخان[42] المذكور غليظا واشتعل بلا تحرك، تصوّر[43] على اشكال مختلفة، فتارة يكون على صورة حيوان ذى قرن وهيئة ذوابة وذنب[44] وكوكب ذى ذنب الى غير ذلك مما يتفق على هيئة المادة. وتختلف مرة بقائها بحسب غلظ المادة ورقتها. ويتحرك نحو المغرب متابعة للحركة اليومية. قال الاستاد روح الله روحه انى قد رصدت بعض تلك الاثار مدة ستة اشهر فوجدته متحركا حركة بطيئة من المغرب الى المشرق ومائلا عن الشمال الى الجنوب.

[2.11] وان لم يشتعل الدخان المذكور او كان اشتعاله(45) قليلا لعدم الدهنية او قلتها،(46) يرى على هيئة ظلمانية او حمرة، اما للاشتعال القليل، واما لوقوع شعاع الشمس عليه، كما ترى السحب قبل طلوع الشمس وبعد الغروب احمر.

[2.12] ويمكن ان يصير تلك الصّور اسبابا لكسوف الشمس وخسوف القمر واحتجاب(47) بعض الكواكب. ولعل الكسوف والخسوف الذين(48) يجتمعان مع القيد كما يفهم من الكتب الفقهية يكون من هذا القبيل. واجتماع الكسوف الذى يكون عند مفارقة الغيرين والخسوف الذى يكون عند تقابلهما مع احد القيدين ممتنع.

[2.13] وقد يرتفع فى المواضع الكبريتية دخان لطيف، اذا دخل الليل، استولى عليه الرطوبة، سيّما اذا كان بعد المطر، ويحصل فيه بذلك سيلان دهنى قابل للاشتعال. فقد يشتعل باشعة الكواكب وغيره، ولكن لا يحرق كنار(49) بخار خمر طرح فيه الملح والنوشادر وشعلة نار† تحرق الخمر†. ويحتمل ان لا يختص اشتعاله بالليل، ولكن لا يرى فى النهار لغلبة ضوء الشمس على ضوئه.

[2.14] تذنيب. اعلم ان الاثار المذكورة المشتعلة ان كانت تحت كرة(50) الاثير، كان بقاء اشتعالها مدة على سبيل تجدد الشعل وتعاقبها، او ينتفى بعضها لميله(51) الى حيّزه الطبيعى. وانطفاؤه ان كانت المسافة بينه وبين الكرة المذكورة طويلة ومتعلقا ببعض آخر بمادته الى ان يبقى قابلية المادة لذلك الامر، فالشعل(52) المتعلقة بالاجسام التى عندنا كالسراج من هذا القبيل. وان كانت <كذلك>، فبقاؤها(53) يكون باعتبار بقاء نار مخصوصة، وايضا انتفاء(54) تلك الاثار اما بانتفاء نيرانها لانتفاء مادة قابلة تثبت(55) هى بها، او بلطافتها بحيث لا يرى باعتبار انتفاء كثافة المادة الارضية.

[2.15] وحدوثها دليل على حدوث الرياح وقلة الامطار وفساد الجوّ العالى ويبوسة الهواء اوحرارته وعلى طريان †كثرة الامراض الحارة اليابسة المهلكة.

[3.1] الاصل الثالث (3v) فى الريح وكيفية حصولها. وهى عبارة عن الهوآء المتحرك لا بالعرض. وسبب حركته قد يكون اندفاع دخان او سحاب او هوآء آخر الى السفل لثقل حركة †حصل فيها باستيلاء البرودة او بتكاثف اجزائها وتركب بعضها على بعض، او لدفع(56) الفلك اياها بحركته(57) الدورية او غيره الى السفل.

[3.2] وقد تحقق فى موضعه ان تداخل المقادير محال، وان الاجسام العنصرية قابلة للتخلخل والتكاثف، وان الخلاء محال.(58) فاذا تحرك هواء مخصوص بسبب من الاسباب، تحرك ما قدامه لاستحالة التداخل، ولكن لما كان قابلا للتكاثف وكان فيما يقابله ممانعة ما، يحصل فيه تكاثف ما، وهكذا الى ان ينتهى الى ما يتكاثف ولا يحرك ما يلاقيه، على مثال ما اذا طرح حصاة(59) فى حوض، فانه تحدث فيه دوائر متعاقبة متفاوتة فى الحركة الى ان ينتهى الى دائرة لا يحدث بعدها دائرة اخرى. وقد يحدث فى الهوآء المتحرك لتحركه لطافة وتخلخل يوجب زيادة الدفع، فيتدرج(60) الريح من البطوء الى السرعة، الى ان ينعكس الامر ويسكن الريح. فاذا وصل الهوآء المتحرك من العلو الى قريب من الارض، ولا يقوى على تحريك الهواء

المجاور اياه فى سمت حركته لممانعته اياه، يحصل له حركة بالضرورة لامتناع التداخل نحو جهة اخرى. وقد يتحرك المتحرك الى غير سمت حركة المحرك[61] اياه لممانعة امر عن سمته او زيادة مناسبة الى غيرها او غير ذلك. وقد يحصل الريح من تخلخل الهواء او تكاثفه، اما الاول فلانه اذا زاد مقداره بالتخلخل، دفع ما يجاوره الى ما يليه، وهكذا الى ان ينتهى الى ما لا يندفع، كما قلنا آنفا. واما الثانى فلانه اذا نقص مقداره فى التكاثف، انجذب اليه ما يجاوره لاستحالة الخلآء وتلازم سطوح الاجسام، وهكذا الى ان ينتهى الى سفل مكان المجتذب بالتخلخل، ولا ينجذب اليه بكليته، فيسكن الريح.

[3.3] ولا يذهب عليك ان بعض الاسباب يقتضى ان يحدث من موضع واحد رياح متعدد الى جهات مختلفة الا يمنع مانع وبعض اخر يقتضى عكس ذلك. وظهر[62] انه قد يحصل رياح جزئية لاسباب صناعية.

[3.4] وقد يحدث ريح على هيئة الاستدارة لتصادم الريحين الآتيين من جهتين مختلفتين. وقيل، يمكن ان يكون الاستدارة لاعوجاج المادة الريحية، كما يلتوى الشعر لاعوجاج منبته. وبقاء هذه الرياح على هيئتها زمانا يكون لاسباب خارجية او لكثافة مادتها[63] الرطبة اللزوجة. وقد تشتد هى بحيث تقلع الاشجار وترفع الاحجار.

[3.5] <...> على الاراضى الحارة اليابسة الكبريتية، ويحدث فيها حرارة منضخمة موجبة للتعفن، فتعفن ايّ حيوان تصادفه بحيث يضمحل بالكلية، وتسمى هذه الرياح سموما.

[3.6] تذنيب. اسماء الريّاح المشهورة عند العرب اربعة. فان كان هبوبها من جانب قطب الشمالى فشمال، وقطب الجنوب فجنوب، ومن المشرق صبا،[64] ومن المغرب دبور، وغير هذه الاربعة تسمى نكباء اى المعوج.

[3.7] وليعلم ان مبدأ[65] الريح وممرّها اذا كان مواضع رطبة باردة، كانت الريح ايضه كذلك، وان كانت حارة يابسة، كانت هى ايضه كذلك، اذا لم يمنع مانع. فلما كان منشأ[66] الشمال وممرّها جبال كثيرة المطر والثلج، كانت باردة رطبة على عكس الجنوب. ولما كان منشأ[67] الصبا[68] والدبور وممرّهما قريبين من الاعتدال، كانتا ايضا كذلك، ولكن لما (4r) كان مرور الصبا على البرارى والصحارى وحركتها على وفق حركة الشمس، كان اسخن من الدبور التى منشؤها[69] ومرورها على الجبال والبحار وحركتها على خلاف حركة الشمس. ومن البين ان كون الرياح على هذا النمط ليس كليا، اذ ربما يكون الشمال حارة بسبب الاوضاع والاقطار الفلكية ومرورها على بعض المواضع الحارة، سيّما اذا قربت من الجنوب، وعليه فقس الباقية.

[3.8] وهبوب الريح من ناحية فيها الشمس يكون اكثر لكثرة تبخير الاشياء الرطبة وتلطيف المواد الغليظة،[70] ومن هذا ظاهر[71] ريح الاسحاب، وقد يحصل على عكس هذا لاسباب اخرى.[72] والاكثر ان الريح تكون موجبة لدفع الابخرة وصحو[73] الجوّ فى مبدأ[74] الحدوث وسببا لاجتماع الابخرة وتراكمها فى الاخر. وقد ينعكس ذلك لكثرة الابخرة والجبال المانعة فى الابتدآء دون الانتهآء. وايضا قد يصير الريح لصحو[75] الهوآء بدفع الابخرة بالتحليل وتفريق اجزاء السحب، وقد يكون على العكس بتبخير بعض الاجسام الجامدة وجمع الاجزاء

البخارية والسحابية. وايضا المطر قد يكون موجبا لانتفاء الريح بتقليل الاجزاء البخارية والدخانية، وقد يكون سببا لترطيب الاجسام المعدّ لتبخيرها.

[4.0] الاصل الرابع. فى بيان قوس قزح والهالة. اما قوس قزح فهى قطعة دائرة ملونة بالوان مختلفة تحصل فى الجو من الاجزء الرشيّة فى مقابلة الشمس. وهى يكون نصف دائرة، وقد يكون اقل منه. واما الهالة فهى دائرة بيضاء تامة او غير تامة تحصل حول القمر او كوكب اخر. وبيانها موقوف على مقدمات.

[4.0.1.1] الاولى فى بيان كيفية رؤية الاشياء. واعلم ان المذاهب الذى يعتد به فيها اثنان. الاول مذهب اصحاب الشعاع، وهو ان تخرج من البصر شعاع تميل[76] الى ان يقع على سطح المرئى، فينكشف هو عنده به. ورؤية الاشياء فى مثل المرايا عند هذه الطائفة لانعكاس الاشعة الواقعة على صقيل الى اشياء يكون بنسبتها من المراة كنسبة الرائى اليها، اى يكونان فى جهة واحدة منها بحيث يكون زاويتا الشعاع والانعكاس متساويين فى سطح واحد مستو. وكون الزاويتين كذلك، وان لم يقم عليه برهان قطعيّ، لكنه مما يدل عليه التجربة عند وقوع شعاع الشمس الداخل من كوة على ماء وانعكاسه الى جدار يقابلها[77] مرتفعا بارتفاعه منخفضا بانخفاظه على قد واحد.

[4.0.1.2] والثانى مذهب اهل الانطباع، ورؤية الاشياء عندهم لانطباع صورتها فى الباصرة[78] بتوسط جرم الشفاف بينهما. وهم فى رؤية الاشياء فى مثل المرايا يفترقون فرقتين. الاولى يقول، كما ان الجرم الشفاف يؤدى صورة المرئى المقابل الى الباصرة ويوجب انطباعها فيها، لذلك الجرم الصقيل يؤدى صورة شيء يكون نسبته منه كنسبة الناظر اليه ويوجب انطباعها فيها. والثانية يقولون، تنطبع اولا صور الاشياء فى ذلك الصقيل، ثم تنطبع تلك الصورة المقابلة للباصرة فيها. وظهر[79] ان انطباع الصور فى المرايا، سيّما اذا كانت اعظم كنصف[80] الفلك، غير معقول، والا لما تبدل موضع المرئى منها عند تبدل مكان الناظر وبقاء المرئى على وضع واحد منها.

[4.0.1.3] وقد ابطل ايضا مذهب الشعاع بوجوه متعددة، وانى ابطلته ببرهان قطعى هندسى على ما ذكرته فى شرح (4v) حكمة العين. فتعيّن ان المذهب المتصوّر هو <عكس>[81] مذهب الفرقة الاولى من تينك الفرقتين،[82] ولكن لما لم يختلف[83] المقصود ههنا باعتبار المذاهب كثير اختلاف، وكان مذهب الشعاع اشهر، اجرينا الكلام على وفقه.

[4.0.1.4] ولتعلم ان رؤية الاشياء فى المرايا على المذهب المتصوّر ومذهب الشعاع من قبيل اغلاط الحس، ومنشؤه[84] ان النفس، لما كانت معتادة برؤية الاشياء بالمقابلة ولم تنتبه ههنا لانعكاس الاشعة وتأدية[85] الصقيل غير المقابل، تخيلت ان المرئى ههنا ايضا فى المقابلة.

[4.0.2] المقدمة الثانية. فى ان المراة اذا كانت صغيرة بحيث لا تحيط قاعدة المخروط الشعاعىّ المنعكس منها الى شيء بقدر محسوس، لا تؤدى شكله ولا لونه، نعم اذا اكثرت المرايا الصغيرة المتجاورة، تؤدّى لونه وضوئه فى الجملة.

[4.0.3] المقدمة الثالثة. فى ان المراة انما تحكى صورة المرئى لو لم ينفذ الشعاع عنها اما لكثافتها واما لوقوع جرم كثيف خلفها.

[4.0.4] المقدمة الرابعة. هى ان التجربة دالة على ان المرآة الملونة لا تؤدّى لون المرئى على ما هو عليه بل مخلوطا بلونها وبلون مجاورها.

[4.1] واذا تمهدت هذه المقدمات، فاعلم انه اذا اجتمعت اجزاء رشية مائية صقيلة حين ما كانت الشمس فوق الافق قريبة منه <فى> احد جانبى الشرق والغرب، وكان ورائها سحاب كثيف او جبل يمنع نفوذ الشعاع بتمامه فيها، فاذا نظر الناظر اليها مستدبرا[86] الى الشمس، انعكس الشعاع[87] البصرى منها اليها، ولكن لما كانت الاجزاء صغيرة جدّا لم تؤدّ جرم الشمس ولا لونها وضوئها كما هو، فنرى الوان مختلفة حاصلة من لون الجسم الكثيف المختلف القوام والكثافة مع الشمس بل ضوئها.

[4.2] وقد شاهدت مع جمع كثير قوس قزح[88] فى بيداء لم يكن ثمة جبل ولا سحاب اصلا. ويمكن ان يكون سبب ذلك تراكم الاجزاء وتكاثفها مختلطة باجزاء ارضية بحيث يمنع نفوذ شيء من الاشعة وتوجب انعكاسها.

[4.3] والقوس قد يكون نصف الدائرة وقد يكون اقل منه على ما قاله الشيخ في الشفاء من ان الناظر والشمس كلاهما على محور القوس، فان كانت على الافق كان مركز القوس ايضه عليه، فيكون هو نصف الدائرة، وان كانت فوقه، كان مركزها تحته، فكانت اقل من النصف. ومن ههنا يظهر ان الشمس اذا كانت فى نصف النهار او قريبة منه على سمت الرأس[89] او قريبة منها، لا يحصل قوس. وان بعدت عنها فى احد الاطراف بعدا صالحا، يرى القوس بحسب البعد، فكلما كان العكس اكثر، كان القوس اكبر. وايضه كلما كانت الاجزاء بعيدة عن الناظر، كانت القوس اوسع، وكلما كانت قريبة منه، كانت اضيق بحسب اتساع قاعدة المخروط الشعاعى المنعكس وضيقها.

[4.4] واستدارة القوس لما سبق من تساوى زوايتى الشعاع والانعكاس. فانه لا يمكن ان يقع على الشمس من الاشعة المنعكسة منها الا ما كانت من الاجزاء كانت نسبة الشمس والناظر اليها على السوآء على ما مر. ولا يكون تلك الاجزاء الا على هيئة الاستدارة، سوى ما يقتضيه قطر الشمس بحسب الرؤية، كما يشهد به الفطرة السليمة والرجوع الى البرهان الهندسى. لا يقال انعكاس الاشعة عن تلك الاجزاء الى الشمس يقتضى ان ينعطف الاشعة المنعكسة بحيث يتقاطع بعضها بعضا، لانا نقول (5r) وجد انها بعد الانعطاف الى اطراف الشمس لا يقتضى ان يكون انعطافها على هيئة التعاطف. وانما يلزم ان† لو كان جرم الشمس اصغر مما انعكس منه الخطوط، اعنى الاجزاء الرشية، وليس كذلك بحسب نفس الامر.

[4.5] فان قيل ما ذكرتم فى كيفية الانعكاس يقتضى ان يرى الشمس بالاشعة المنعكسة على الاجزاء الواقعة فى[90] جانب تنفر القوس ايضا، قلت يمكن ان لا ينعكس منها شعاع يحكى ضوء الشمس، بل السهم، وما يحويه لقربها ينفر فيما يقع عليه، ولا تنعكس الا الى ان ينتهى[91] فى البعد على السهم، الى ان تقوى على النفوذ التام، فلا ينعكس[92] الى الشمس. بل الاشبه ان يجعل القوس انما هو باعتبار رؤية جرم الشمس بتمامه عن جميع الاجزاء الكائنة على هيئة الاستدارة، بان يرى جرمها من جملة واقعة على تلك الهيئة، كما يدل عليه التجربة من رؤية شىء من مرآة حين ما استدبره اليه الناظر او رآها[93] على هيئة

الاستدارة من احد طرفى الشمال والجنوب فى الاخر، مع ادنى تغيير يقع المرآة بالنسبة الى وجه الناظر بالموازاة† وعدمها تغييرا†. يمكن ان يكون واقعا فى الاجزاء المختلفة بحسب الاوضاع وبذلك يندفع ما، اذا احتيج الخاطر ان رؤيتها من المواضع المختلفة الاوضاع لا يستقيم على قاعدة تساوى زوايتى الشعاع والانعكاس. فان قيل رؤية جرم الشمس بتمامه عن جميع الاجزاء على هذا الوجه يقتضى تلاقى الاشعة المنعكسة، وقد ثبت بطلانه، قلت انعكاس الاشعة من الاجزاء المذكورة ليس عن مرآة واحدة بل عن مرآة متعددة مختلفة بحسب الوضع، كما اشير اليه آنفا.

[4.6] واما تلونها بالوان مختلفة فانه، وان سبق اليه الاشارة، الان الحق ان طيته غير ظاهرة، على ما اعترف به الشيخ وكل ما وصل الينا فى الكتب المشهورة مما لا يلتفت اليه، كما يظهر على من يطالعها ويتامل فيها.

[4.7] وقد شاهدت حين ما كانت الشمس فوق الافق قريبة منه فى بعض المواضع والقى †نحو خشبة† الماء يظهر ظله تحت حمرة مائلة الى الصفرة فى احد طرفى الظل، وهو الذى يلى الشمس، وتعليقا† فى الطرف الآخر. فتجزم جزما حدسيا بان تلوّن قوس قزح ايضا يكون من هذا القبيل، وان لم يظهر حقيقية. ويعرف منه ايضا بطلان بعض ما ذكر فى تلونها.

[4.8] وقد لاح فى بعض الليالى، حين ما كان القمر فى كمال الاضائة، قوس بيضاء على شكل قوس قزح، ومليته† ما مرّ، الا ان عدم تلونه بالوان مختلفة يمكن ان يكون لقلة اضائة الليل، فيرى لون القمر على ما هو عليه، الا ترى[94] ان النار ترى فى الليل بيضاء شفافة، وفى النهار وعند استيلاء الشمس كثيفة حمراء.

[4.9] وقد حكى الشيخ انه قد رأى[95] قوسا على هيئة اصل قوس قزح والوانها فى الحمام حين وقوع الشمس على جاماته، وحكم بانها ليست على سبيل التخييل لعدم <انتقالها> بانتقال[96] الناظر. وحكى ايضا انه قد يتصور قوس كذلك[97] على حول سراج.

[4.10] واما حصول الهالة، فبان يقع بين الناظر والقمر او كوكب اخر غيم رقيق، فيه اجزاء رشية صقيلة، يتادى من اجزاء كانت نسبتها من الناظر والقمر على السواء لون القمر على ما سبق فى القوس، او بان يقع بينه[98] وبين القمر (5v) ايضا غيم رقيق يمنع نفوذ بعض الاشعة، فينعطف من اجزاء كانت على النسبة المذكورة الى القمر خاليا لونه.

[4.11] وعدم الاجسآم بالغيم لاختفائه تحت شعاع القمر. وقد سبق فى القوس ما يعلم منه عدم لزوم الغيم فى الهالة ايضا. ودائرة ترى حول الشمس بيضاء ومتلونة بالوان قوس قزح تامة او غير تامة تسمى طفاوة[99].

[4.12] والهالة قد يكون كبيرة وقد يكون صغيرة، وقد يكون متعددة، باعتبار قرب الاجزاء المذكورة من الناظر وبعدها عنه وتعدد طبقاتها. فمتى كانت قريبة منه ترى اكبر، وبعيدة ترى اصغر، لكون الخطوط المتقاطعة فى الاولى اطول وفى الثانية اقصر، وذلك يستلزم كون الهالة كون الاولى اوسع مما فى الثانية، على ما يشهد به التخيل الصحيحة والبرهان الهندسى ايضا. ومن هذا يبين ان الهالة اذا تعددت، تعددت طبقات الاجزاء كانت بعضها اكبر من بعض، ان الكبرى تحت الصغرى.

[4.13] وهى وان كانت كثيفة صحيحة الاستدارة، سيّما اذا كانت حول الشمس او تبقى الى ان يتكاثف الابخرة ويحدث السحاب، كانت من علامات المطر. وان لم يكن كذلك، تدل على عكس ذلك.

[5.1] خاتمة. فيما يتولد فى الارض من الابخرة والادخنة. اذا حدثت فى الارض لاشعة الكواكب وللاوضاع الفلكية او بغيرها ابخرة وادخنة، فان كانت فى وجه الارض رخاوة ومنافذ تصلح لخروجها عنها، خرجت بالتدريج. وان كان وجهها صلبا ولم يكن منافذ تصلح كذلك، احتبست فيها. ثم اذا اصابت الابخرة برودة، انقلبت ماء. فان كان قليلا جدا لم يحسن به او يحسن بنداوة. وان لم يكن قليلا كذلك ولكن يكون وجه الارض صلبا وكانت مرتفعة بحيث لا يقوى على الخروج، ترشح منها، او تحرك حركة بطيئة ان كان ضعيفا، وحركة سريعة ان كان فيه قوة. وايضا ان كانت الابخرة كثيرة وتعاقب انقلابها ماء، دام المآء، والا انقطع. وان ارتفع اطراف الارض بحيث لا يقوى الماء على الجريان عليها، حصل ماء راكد اكثر او قليل دائم او منقطع لما مرّ آنفا. وليعلم ان المياه[100] المذكورة، كما يمكن حصولها من الابخرة، يمكن ايضا من مياه الامطار وترشح الانهار والبحار. وقد يظهر مما ذكرنا لمية كثرة العيون فى الجبال والاراضى المرتفعة دون الصحارى والاراضى المطمئنة. وهى كثرة اسباب حدوث المياه وشرط جريانها، اعنى صلابة الوجه وكثرة الامطار والثلوج واحتباس الابخرة باستيلاء البرودة على الظاهر وارتفاع احد جانبى الارض وانخفاض الاخر فى الاول دون الثانى. وقد يكون بعض المياه المذكورة حارة بمجاورة معدن الزرنيخ والكبريت، ولذا يحس من بعضها رائحة الكبريت. وقد يكون سبب الحرارة مجاورة الادخنة الحارة، ومن ثم يكون مياه العيون فى الشتآء اسخن مما فى الصّيف لكثرة احتباس الابخرة والادخنة فى الشتآء تحت الارض دون الصيف.

[5.2] واذا كثرت الادخنة والابخرة بحيث لا يسعها تجاويف الارض، مالت بالضرورة نحو الخروج، ولما لم يكن فى وجهها مسامات[101] تخرج هى منها، تزلزلت وتحركت، الى ان (6r) ينقلب بعضها هوآء يخرج من المسامات[102] الضيقة، او يحدث فيها تجاويف يسعها او شقوق تخرج هى منها.

[5.3] واذا كانت الابخرة غالبة، واصابتها برودة الارض بعد ان حصلت فيها بالحركة سخونة مقدمة لتاثيرها، انقلب مياها، وحصلت عيون راكدة او جارية بحسب كثرة الابخرة وقلتها مع سائر الشروط. وقد يكون حدوث العيون بحدوث المنافذ التى يخرج منها مياه كانت محتبسة فى الارض. وقد يصير الزلزلة سببا لانعدام العيون القديمة لانتفاء الابخرة بالتخلخل او بخروجها من المسامات[103] الحادثة، <او> لحدوث تجاويف، وتجاويف مال اليها المآء.

[5.4] وقد يقع الزلزلة بحركة الهوآء المحتبس فى الارض حركة شديدة ليست من الاسباب. وقد يشاهد حين الزلزلة دخان وشعل نيران لغلبة الدخان وانقلاب بعضه لشدة الحركة نارا. وقد يسمع لشدة الحركة وتصادم الاجزاء البخارية والدخانية شدة صوت هائل.

[5.5] تذنيب. ولما كانت حركات اسباب الزلزلة على جهات متعددة، كان تحرك الارض ايضه كذلك، فتارة يرى انها ترتفع، واخرى مائلة الى جهة، اخرى يحسّ بحركة كحركة

الرعشة. ولما كان حدوث الزلزلة لاحتباس كثرة الابخرة والادخنة فى الارض، ففى اىّ وقت كان حدوثها فيه مع سبب الاحتباس اكثر، كانت فيه اكثر. ومن ثمه كانت فى الربيع والخريف والليل اكثر مما فى الصيف والشتاء والنهار. وقد يصير الكسوف الذى يوجب استيلاء البرودة على وجه الارض واحتباس الابخرة والادخنة فى تجاويفها دفعة سببا لها.

[5.6] تتمة. تنبيه. ضرر الزلزلة افناء[104] بعض الحيوانات وتخريب وحر وافساد وجه الارض واعدام العيون والابار والقنوات وحدوث الرياح المفرطة من خروج الابخرة وغيرها. ومنافعها حدوث العيون ورياح صالحة[105] وامطار نافعة من خروج الابخرة وقلع الجبال والاحجار المضرة وترهيب العاقل وارادة بعض الاشارات،[106] اذا زلزلت الارض زلزالها واخرجت الارض اثقالها.

[5.7] وتنبيه على الارض مع وثاقتها[107] والجبال مع صلابتها كيف تزلزلت عن رياح ضعيفة وابخرة نحيفة، فما بال النشئة الضعيفة الانسانية، بل الرتبة الاعتبارية الكبانية عند صدمات عاصفات القهر الالهى وصواعق نداء، اذا دكت الارض دكا دكا، وبست الجبال بسا، فكانت هبآء، منبشا لمن الملك اليوم، لله الواحد القهار. تمت النسخة الطالشية.

Textcritical apparatus

1 In margine sinistro fol. 1r, supra: رساله طالشى اينست از علم حكمة; et infra: والسموات واقعة (cf. 0.2.3); et شحذ السكين اذا حددت :العين ساكنة غير متحرك وبه قال المسلمون واهل الكتب خلافا للمنجمين والفلاسفة. 2 وتحصل cod. 3 ينطفى cod. 4 اطفى cod. 5 سرج cod. ازهار. 6 المنطفى cod. 7 تماسى cod. 8 ولانطفت cod. 9 اذا نفخ: او انفخ cod. 10 In marg.: اى كما من قواعد اهل السنة والجامعة 11 تطوت cod. 12 بسيط cod. 13 يحلل cod. 14 الحار به: الحارية cod. 15 حيز cod. 16 وكثير cod. 17 الحيثية cod. 18 احد cod. 19 ملسئا cod. 20 ذلك in marg. cod. 21 يكون cod. 22 لتفاوة cod. 23 تحلّل اجزائها cod. 24 الحار به: يكون cod. 25 ابخرت cod. 26 Cf. supra 1.1. 27 الثقل cod. 28 لاجزاء cod. 29 ليست cod. 30 ويسمى cod. 31 منطفيا cod. 32 يخ cod. 33 وحرق cod. 34 حفيفا cod. 35 الحرق cod. 36 فظ cod. 37 الحاصلة cod. 38 الدهنية طبقةالطبقة الدهنية: cod. 39 وينطفى cod. 40 ابطاء cod. 41 والمتشابه cod. 42 الدحان cod. 43 يتصوّر cod. 44 ذنب cod. 45 اشتعالة cod. 46 او قلتها : اقلتها cod. 47 احتراق cod. 48 اللذين cod. 49 نار cod. (cf. *Shifā'* 68.16f). 50 كر cod. 51 لمتله cod. (cf. supra, 1.4 بميله الطبيعى الى حيزه) 52 فالشعل: فى الشعل cod. 53 فبقائها cod. 54 Praebet افتضاء in textu, et انتفاء فى in marg. cod. 55 نبتشث ut vid. cod. 56 يدفع cod. 57 بحركة cod. 58 مح cod.

59 حصارّ cod. 60 فيندرج cod 61 المتحرك cod. 62 وظَ cod. 63 مادته cod. 64 صباء cod. 65 مبداء cod. 66 منشاء cod. 67 منشاء cod. 68 الصباء cod. 69 منشائها cod. 70 الغليظ cod. 71 ظهر cod. 72 آخر cod. 73 وصحب cod. 74 ميل cod. 75 لصحب cod. 76 Fort. تمتد legendum (cf. *Shifā'* 40.12 فيمتد), 77 تقابلها cod. 78 الباصر cod. 79 وظ cod. 80 كنصف: in marg. cod. 81 Cf. *Shifā'* 44.5. 82 الفرفتين cod. 83 تختلف cod. 84 ومنشاؤه cod. 85 ونارية cod. 86 مستديرا cod. 87 شعاع cod. 88 قوس وقزح cod. 89 الرائى cod. 90 الواقعة فى الواقعة فى cod. (dittogr.). 91 يتنهى cod. 92 فينعكس :فلا ينعكس cod. 93 راها cod. 94 يرى cod.1; ترى cod. corr. 95 رى cod. 96 انتقال cod. 97 ذلك cod. 98 بينها cod. 99 طفارة cod. 100 الماء cod. 101 مسانات cod. 102 مسانات cod. 103 المسانات cod. 104 اغناء cod. 105 صالحه cod. 106 الاشاراقت cod. 107 وثافتها cod.

3. Muḥammad al-Ṭālishī, *Treatise on Meteorology* (Translation)

[0.0] In the name of God the Merciful the Compassionate, and it is Him we ask for help. Praise be to his friend, and blessed be His prophet Muhammad and his noble and pure family and all his companions. This is a treatise consisting of an introduction, four sections and a conclusion. The introduction includes two preliminary matters.

[0.1] First [preliminary matter]: on what body [*jism*] is and its divisions in general. A body is a substance [*jawhar*] subject to the three dimensions which intersect at right angles, i.e. length, width and depth. The body is of two kinds, simple and compound. A compound body is what is produced through a combination of bodies differing in nature. It is either complete—as when it has a form which can be retained over a significant length of time, like the bodies in the three kingdoms of nature [*mawālīd*], i.e. minerals, plants and animals—or incomplete—as when they do not have such form, like vapour and clouds. A simple body is what is not produced through a combination of them. It either belongs to the upper regions, namely the spheres with the stars which are in them, or to the lower regions, namely the four elements, fire, air, earth and water. Fire is hot and dry, air hot and moist, water cold and moist, and earth cold and dry.[17]

[0.2.1] Second preliminary matter: on transformation of elements into one another. You have learned that each of the elements has two qualities. We say: each quality has its limit, or range. When it oversteps it in one of the two directions of excess or deficiency, the susceptibility of the material of that element for its form no longer remains and turns into susceptibility for another form which is associated with that quality, so that the material takes off the first form and puts on a second.

[0.2.2] When the form of the fire's heat is broken through encirclement by air etc., its heat becomes closer to the heat of air, the susceptibility of its material for its form no longer remains, and the material acquires a susceptibility for the form of air, so that it takes off the form of fire and puts on the form of air.—Likewise with the other elements.

[0.2.3] Let us present particular examples of transformation of certain elements into others to whet the intellect and to reassure the minds. An example of transformation of fire into air is the transformation of the flame of a lamp into air due to its encirclement by air. As it rises from its source, i.e. the wick, the air surrounding it acts upon the past nearest to it, so that the flame is gradually transformed into air, and acquires a conical shape, until it vanishes completely. The extinction of a lamp by wind involves a rapid transformation of fire into air due to the intensity of the air's action upon it and the air's penetration into its depths.

[0.2.4] An example of transformation of air into fire is observed in the furnace of blacksmiths, where the fire's penetration is most intense and the blowing is most violent, so that the air that is there is transformed into fire.[18] A similar case is that, when a lamp is extinguished and placed above another lamp, it catches fire because of the heat's prevalence over the air adjoining the extinguished lamp, along with the intensity of the lamp's susceptibility to inflammation.[19] Another similar case is

[17] Cf. Qazwīnī, *ʿAjāʾib*, p. 89, l. 3–9 (trans., p. 182).

[18] Cf. Qazwīnī, *ʿAjāʾib*, p. 89, l. 29f (trans., p. 184); Ibn Sīnā, *Dānish-nāma-yi ʿAlāʾī*, trans. M. Achena and H. Massé, *Avicenna. Le Livre de science* [Paris, 1955–58], ii, p. 40.

[19] This example of an extinguished lamp being lit again by another placed below it (which is also found in other works, such as Abū ʾl-Barakāt al-Baghdādī, *Kitāb al-muʿtabar* [Hyderabad, 1357–58 A.H.], ii, p. 222, l. 11–16, cf. Lettinck, *Aristotle's* Meteorology, p. 83, Takahashi, *Aristotelian Meteorology in Syriac*, p. 559) is used by Qazwīnī not in

the generation of fire from the stroke of a drill, or something similar, against a special stone. The fire which arises there is not there [from the beginning] at the point of contact of one of the two.—Otherwise, it would be perceptible, ** it has been extinguished through their immersion in water over a prolonged period [?].—[The fire arises], rather, because when one of them strikes the other with intensity, the air which is there is heated by the stroke, along with the special, unknown, relationship that exists between the drill and the special stone, and is transformed into fire.

[0.2.5] An example of transformation of air into water is observed in the waterdrops that fall from the ceiling of a bath. When the air of a bath, which is susceptible to the influence of coldness because of its being heated and its fineness, meets the ceiling which is cold in relation to it, it is influenced by it and is transformed into water, so that it falls. A similar case is the accumulation of waterdrops on the outside of a jug which has ice inside and a saucer laid over it,[20] and of spray-formed particles on a surface of iron, stone or wood, when one blows upon it from nearby with an open mouth.

[0.2.6] An example of transformation of water into air is observed when it is transformed into it by boiling and by air passing over it for a prolonged period. A similar case is the drying of damp clothes when something which has been heated acts upon them—and air is [such] a 'thing' when it is hot with its refinement—and makes it susceptible to transformation into air.

[0.2.7] Examples of transformation of water into earth and vice versa are observed in the formation of marble etc. from water, and in the liquefaction of certain stones into water through the art of the alchemists.[21]

[0.2.8] Among the uses of this kind of study and investigation about the transformation of certain elements is the recognition of the fact of the deluge and some of the miracles of the prophets—may peace be upon them—such as making water gush out of rocks, in terms of the precepts of the philosophers, and the refutation of the view that consid-

the discussion of transformation of elements, but in the discussion of shooting stars at *ʿAjāʾib*, p. 91, l. 22–24 (trans., p. 188).

[20] Cf. Qazwīnī, *ʿAjāʾib*, p. 89, l. 23–27 (trans., p. 184); Ibn Sīnā, *Dānish-nāma*, loc. cit.

[21] Cf. Qazwīnī, *ʿAjāʾib*, p. 90, l. 1–3 (trans., p. 184); Ibn Sīnā, *Dānish-nāma*, loc. cit.

ers impossible what is commonly known among the populace about the changing of animals into stones, since the difference between animals and stones is no greater than the difference between water and stones, and this is evident. The investigation about the fact that all the elements are of one nature which has been made to evolve and a simple quiddity which has multiplied through progression, since transformation into several natures through differentiation ***—As the poet says: 'He who is in the know from the city knows where our merchandise is.'

[0.2.9] Note: When someone asks [about transformation] between elements which are opposite of one another in their qualities and which are resistant to one another in terms of their forms, [we say that this occurs] by way of transformation of one [element] into another due to an affinity which they acquire through juxtaposition and mixture, so that fire sometimes goes out from the utmost level of its heat and fineness to an utmost level of the thickness and coldness of an earthy substance, and vice versa. It is for you to observe and consider, and [move away from] association with the wicked to the company of the good [?].—As the poet says: 'In a hundred lifetimes is found someone with whose breath one's every breath is intertwined.' 'Seek the deepest pain, wise one, so that you may gain depth through the deepest pain.'[22]

[1.1] First Section: on vapour. Vapour is something that is generated in the atmosphere when water and moist earth are heated by the rays of the stars etc., so that some of the watery particles are transformed into air and are drawn in its fine state towards its natural place, taking in its trail another group of watery particles in such a way that sense cannot distinguish between the two. That ensemble is called 'vapour'. Vapour is rarely free of earthy particles as is demonstrated through the observation of those [earthy] particles in water from rain or snow, when the water is placed in a vessel until it becomes clear. Sometimes vapour is generated by the influence of coldness on air, when some particles of air are transformed into water, mixed with other particles of that kind.[23]

[1.2] Then, when the air is hot and there is no wind, the watery particles are dissipated and vapour disappears in its entirety. If it is not

---

[22] Cf. Niẓāmī, *Makhzan al-asrār*, 'dānistān-i Farīdūn bā Āhū'.

[23] Cf. Qazwīnī, *'Ajā'ib*, p. 93, l. 28f (trans., p. 192) (not very close); cf. further Lettinck, *Aristotle's* Meteorology, pp. 54–8.

so, but the vapour as a whole acquires density and thickness due to being subjected to the least coldness, or due to the accumulation of its particles through vapour being added to it by wind, or due to some other cause, it is called 'mist' [*ghaym*]. If the mist is descending from above to below, it is an indication of clear weather and absence of rain, especially when it immediately follows rain. When it is ascending, it is a sign of cloudy weather and rain.[24]

[1.3] When the thickening of vapour is intensified by one of the reasons mentioned, it is called 'cloud' [*saḥāb*], whether it reaches the sphere of extreme cold [*kurat al-zamharīr*][25] or not.[26]

[1.4] When the watery particles which are in the cloud meet one another, and when some of its particles are transformed into water by the coldness of air or the cold confined in it through encirclement by hot air[27] and the water descends because of its natural inclination towards its place, it is called 'rain' [*maṭar*]. When a mountain or another wind stands as an obstacle in the face of the wind that causes the accumulation of vapours and the rain, the accumulation of the vapours intensifies and rain increases, and that is the reason for the abundance of rain in the mountains in some hot countries, such as Ethiopia.[28]

[1.5] When the coldness is more intense than has just been mentioned, or the susceptibility of the watery particles to the influence of the coldness arising from the distance[29] from the two [upper and lower] limits [of the stratum of air] is intensified by their being heated and by their fineness,[30] they become frozen as 'hail' [*barad*]. That happens when their freezing is intense, their peripheries are smooth and moisture is seen on them. Otherwise, they become dense and thick snow, and that

---

[24] With the second half of the paragraph, cf. *Shifā'*, p. 38, l. 7–9 (Lettinck, *Aristotle's* Meteorology, p. 113).

[25] Cf. Introduction above.

[26] Cf. Qazwīnī, *'Ajā'ib*, p. 93, l. 29–p. 94, l. 4 (trans., p. 192f).

[27] I.e. because of cold fleeing from heat by *antiperistasis* (see Aristotle, *Meteorology* 348b 2 etc.; cf. Lettinck, *Aristotle's* Meteorology, p. 112).

[28] Rain: cf. *Shifā'*, p. 35, l. 15–p. 36, l. 5 (Lettinck, *Aristotle's* Meteorology, p. 112); Qazwīnī, *'Ajā'ib*, p. 94, l. 4–6 (trans., p. 193).

[29] *Bu'd* here is my emendation for the manuscript's *aḥad*. The middle layer of the sphere of air is the coldest because of its distance from the sphere of fire above and from the layer near the earth which is warmed by the reflexion of rays from the ground (Aristotle, *Meteorology* 340a 26–9; Abū 'l-Barakāt, *Mu'tabar*, ii, p. 213, l. 13–16 [Lettinck, *Aristotle's* Meteorology, p. 114]; Qazwīnī, *'Ajā'ib*, p. 92, l. 28–p. 93, l. 2, trans., 190f. [Lettinck, *Aristotle's* Meteorology, p. 53]; Takahashi, *Aristotelian Meteorology in Syriac*, p. 421f).

[30] Cf. 0.2.5 above and 1.6 (with n. 33) below.

happens when the intensification of the coldness and the susceptibility of the above-mentioned particles are greater. Hailstones are sometimes larger and sometimes smaller, [either] because the water-drops from which they are formed are so, or because of the discrepancy in the distance [through which they travel]. For when the distance is long, their particles are dissolved by the movement and they become small, and when it is short, they remain as they are and they are not dissolved much. Furthermore, hailstones are sometimes round, because of coldness reaching them after their complete coalescence and their being naturally spherical, or because of the dissolution of their corners due to the long distance. Sometimes they are crooked because of the absence of the two factors together. Hail is more frequent in spring and autumn than in summer and winter. That is because it is formed from moist clouds, and these are more frequent in the first two seasons, especially in spring, than in the latter two. In summer [they are rare] because of the frequent dissipation of the material and prevalence of dryness over air, in winter for the reason given by the Shaykh [Ibn Sīnā] in the [*Book of*] *Healing*, namely that 'when coldness intensifies in it, it becomes snow, and otherwise, it becomes nothing'.[31] Consider.

[1.6] When the coldness is intense, or the watery particles—rather the moist cloud before it is frozen—become frozen before they are subjected to complete coalescence and perfect union, and fall in the shape of carded cotton, that is called 'snow' [*thalj*].[32] We have mentioned the requirement concerning the increase in susceptibility in its place, for we see that air which has been heated is transformed into water by coldness where unheated air is not transformed into it by coldness several

[31] On hail, cf. *Shifā'*, p. 36, l. 12–p. 38, l. 5 (Lettinck, *Aristotle's* Meteorology, p. 112f), and for the sentence quoted, *Shifā'*, p. 36, l. 13–15; cf. Qazwīnī, *'Ajā'ib*, p. 94, l. 15f (trans., p. 193).

[32] On snow, cf. *Shifā'*, p. 36, l. 10f (Lettinck, *Aristotle's* Meteorology, p. 112); Qazwīnī, *'Ajā'ib*, p. 94, l. 8–10 (trans., p. 193).—The comparison to 'carded cotton' is applied not to snow but to stratified clouds in the passage of Qazwīnī immediately following (*'Ajā'ib*, p. 94, l. 12), as well as in the *Rasā'il Ikhwān al-ṣafā'* (ed. B. Bustānī [Beirut, 1957], ii, p. 74, l. 14; trans. F. Dieterici, *Die Naturanschauung und Naturphilosophie der Araber im zehnten Jahrhundert. Aus den Schriften der lautern Brüder*, Die Philosophie bei den Arabern im X. Jahrhundert n. Chr. V [Berlin, 1861; repr. Hildesheim, 1969], p. 81). One of the places where it is found applied to snow as here is Ghazālī's *Maqāṣid al-falāsifa* (ed. S. Dunyā [Cairo, 1961], p. 340, l. 23–26; cf. H. Takahashi, 'Barhebraeus und seine islamischen Quellen. Têḡraṯ têḡrāṯā (Tractatus tractatuum) und Ġazālīs Maqāṣid al-falāsifa,' in M. Tamcke (ed.), *Syriaca. Zur Geschichte, Theologie, Liturgie und Gegenwartslage der syrischen Kirchen. 2. Deutsches Syrologen-Symposium (Juli 2000, Wittenberg)* [Münster, 2002], pp. 147–75, here p. 162, n. 37).

times as strong, as was seen above in the case of the air in a bath.[33] Hot air becomes frozen before cold air, and because of this freezing <…>. Vapours, which go out from the mouth into its surroundings with their air, are less cold than what is further away than them.
[1.7] One sometimes sees small particles descending from clear air, frozen—which is called *zamharīr* in some traditions[34]—and not frozen. That is due to the prevalence of coldness, be it intense or not, and the occurrence of an overwhelming condensation and heaviness in those watery particles which are in the air or in those of particles of air that are transformed into water, with the result that it descends.
[1.8] When coldness prevails over moist air or fine, moist vapours at nightfall in some places, that air or vapour near the surface of the earth is transformed into water, or the watery particles, which are in the air or which are transformed from air into water, gather together, and grow heavy, so that they descend, but one does not perceive them because of their small size, until they have gathered on the surface of the earth. This is called 'dew' [*ṭall*] if it not frozen, and 'frost' [*ṣaqīʿ*] if it is frozen.[35]

[2.1] Second Section: on smoke. Smoke is something that is generated in the atmosphere from the heating of dry grounds and sulphureous places by the rays of the stars and other causes. The particles of air which are there are transformed into fire, mixed with airy and earthy particles in such a way that sense cannot distinguish between them, and that ensemble is called 'smoke'. Smoke is rarely free of vapour, and vice verse, but the ensemble is called by the name of whichever [of the two] that predominates.[36]
[2.2] Then, as the smoke rises because of its heat, it becomes separated from the earth. Sometimes, when it reaches the sphere of extreme cold [*kurat al-zamharīr*], it becomes confined in the clouds which are there or [in the clouds] which are formed from [the vapour] that was

---

[33] Cf. 0.2.5 (also 1.5) above. The notion goes back to Aristotle (*Meteorology* 348b 30–349a 4); cf. *Shifāʾ*, p. 37, l. 2f (Lettinck, *Aristotle's* Meteorology, p. 112f).
[34] Cf. Introduction, p. 366 above.
[35] Cf. *Shifāʾ*, p. 36, l. 6–9 (Lettinck, *Aristotle's* Meteorology, p. 112); Qazwīnī, *ʿAjāʾib*, p. 94, l. 17f (trans., p. 194).
[36] Cf. *Shifāʾ*, p. 39, l. 1–6; p. 67, l. 4f; Qazwīnī, *ʿAjāʾib*, p. 97, l. 20–22 (trans., p. 200).

accompanying it. Then, if it retains its heat, it exits from the clouds upwards, tearing them [in the process].[37]

[2.3] When it descends—because of heaviness generated in it by the expulsion of the fiery particles from it and the prevalence of earthy particles over it, or because of some other cause, such as the exigency of the efficient cause—because of the retention of heat in it but not of the combustive fiery particles, it affects whatever object it strikes without burning it. It has been said, for example, that it strikes a purse and melts the gold in it without burning the bag. That effect varies. For, when it is thick and intensely hot, its effect is more intense and it blackens the spot it strikes, but when it is fine, it does not have the same effect.[38]

[2.4] Sometimes, prevalence of coldness over it intensifies in such way that it is frozen into stone and other things depending on the particles that occur. In fact, it is possible for a mixture to be born from the mixing of the vapours and smokes, into which the form and soul of an animal flow, as is made known by what people say in common parlance about 'stones and animals falling' in rain. Perhaps the stone which is now in the mosque of Ardabīl and is famed among its populace to have fallen from the sky is of this kind. The Shaykh [Ibn Sīnā] has recounted in the [*Book of*] *Healing* the opinions ** extraordinary large stones, but we shall not dwell on them here.[39]

[2.5] In any case, from the tearing of parts of clouds an intense sound called 'thunder' [*ra'd*] is generated.[40] It can also arise from the tearing of clouds by violent winds and the exit of smokes confined [in the clouds] through the coming together of parts of clouds with one another or [with] wind or [with] something else, or from the occurrence of collision between parts of clouds because of wind or something else, or from the movement of air between parts of clouds.—Does one not see that a strong sound occurs when the movement of clear air intensifies? What do you think then will happen between parts of thick clouds?[41]

[2.6] If the rising smoke is not separated from the earth, but one end of it remains attached to it, sometimes, its rising end is ignited because

[37] Cf. Qazwīnī, *'Ajā'ib*, p. 97, l. 22–4 (trans., p. 200).

[38] Cf. *Shifā'*, p. 70, l. 9–16 (Lettinck, *Aristotle's* Meteorology, p. 236); Qazwīnī, *'Ajā'ib*, p. 97, l. 24–8 (trans., p. 200).

[39] Cf. *Shifā'*, p. 71, l. 1f; p. 5, l. 15–p. 6, l. 11.

[40] Cf. Qazwīnī, *'Ajā'ib*, p. 97, l. 24f (trans., p. 200)

[41] Cf. *Shifā'*, p. 68, l. 1–4 (Lettinck, *Aristotle's* Meteorology, p. 234f).

of its proximity to fire or some other cause. Then, the inflammation gradually descends until it reaches the earth, so that it appears in the shape of a column descending from above. That is called 'conflagration' [*ḥarīq*]. Then, if its material is thick, it will burn everything it strikes, but, if not, then it will not [burn], just as the flame of carded cotton does not burn anything.[42]

[2.7] If oily particles predominate over the fine smoke that is present in clouds, and the smoke is inflamed by the intensity of the movement and collision or some other cause, and it moves quickly but not in a straight line, either because of the difference in the consistency of the material or the crookedness of the gaps between the clouds, that is called 'lightning' [*barq*]. Its crookedness and movement are also attributed to the occurrence of fine smoky material in that space and inflammation of one end of it by some cause, which [inflammation] travels from there to the other end and is extinguished quickly because of the fineness of the material. This, however, is not free of improbability.

[2.8] Know that thunder often occurs without lightning, but the reverse is rare, so that it is said that lightning does not occur without thunder, since lightning cannot occur without the collision and tearing which cause thunder, but the thunder is sometimes so faint that it is inaudible. Furthermore, when it is inflamed by the collision of parts of clouds, the lightning follows the thunder. When the collision or tearing occurs separately from its inflammation, [the latter being] due to the heat confined in between parts of clouds or due to some other cause, and its movement is towards one side, the reverse will be the case. Either way, however, the lightning is seen before the thunder is heard, in the first case for obvious reasons, and in the second because of the absence of a pause over time in visual perception of things which one faces, unlike with audition of sound, <as is proven by the fact that the sound> arising from the stroke of an axe <is heard> after its stroke is seen by an interval of time.[43]

[2.9] When the smoke in which oiliness predominates passes the stratum of extreme cold, so that it comes to the [sphere of] fire or near to

[42] Cf. Qazwīnī, *ʿAjāʾib* p. 91, 1.20–24 (trans., p. 188); also Fakhr al-Dīn al-Rāzī's *Mabāḥith al-mashriqiyya*, ed. Hyderabad (1343 h.), ii, p. 190, l. 6–12; ed. M.M. al-Baghdādī (Beirut (1410/1990), ii, p. 196, l. 9–16, where the term *ḥarīq* is applied to the phenomenon described here (cf. Lettinck, *Aristotle's* Meteorology, p. 238f; see further Takahashi, *Aristotelian Meteorology in Syriac*, pp. 556–9).

[43] Cf. *Shifāʾ*, p. 68, l. 19–p. 69, l. 8 (Lettinck, *Aristotle's* Meteorology, p. 235f); Qazwīnī, *ʿAjāʾib*, p. 97, l. 29–p. 98, l. 3 (trans., p. 201).

it, and is inflamed by the heat of the fire or by its own movement or by the heat confined in itself, while its movement is in a straight line because of the absence of variation in its material and its path, and it is extinguished at a speed a little slower than that of lightning because of the fineness of its material, it is called a shooting star [*shihāb*]. It may be that its being straight and its movement are due to the reasons mentioned about the lightning, but it may also be that both of them are smoky wind, and move in a straight line in air which is homogenous, but in a crooked way in clouds which differ in consistency.[44]

[2.10] If the smoke mentioned is thick and is inflamed without moving about, it takes on different shapes, so that it sometimes occurs in the form of a horned animal, and in the shape of a lock of hair, a tail and a tailed star, as well as in other forms that fit the shape of the material. The duration of their persistence varies according to the thickness or thinness of the material, and they move towards the west following the diurnal movement.[45] The Master—God rest his soul—said: 'I once observed some of those phenomena for a period of six months, and found them moving slowly from west to east and tending from north to south.'

[2.11] If the smoke mentioned is not inflamed or its inflammation is of little extent because of the absence or paucity of the oiliness, it has the appearance of darkness or redness, either because of its low-level inflammation or of the sun's rays falling on it, just as the clouds appear red before sunrise and after sunset.[46]

[2.12] Those appearances can be the causes of the eclipses of the sun and the moon and the occultation of some stars. Perhaps the solar and lunar eclipses, which occur in conjunction with the requisite condition as can be understood from the books of *fiqh*, are of this kind. The conjunction of a solar eclipse occurring in separation from the other two with a lunar eclipse occurring with the two in opposition and with one of the two conditions is impossible.

[2.13] Sometimes in sulphureous places fine smoke rises at nightfall, moisture prevails over it, especially if it is after rain, and an oily flow prone to inflammation is formed in it because of that. This is then sometimes ignited by the rays of the stars and other causes, but it does

[44] Cf. *Shifā'*, p. 71, l. 4–10 (Lettinck, *Aristotle's* Meteorology, p. 80).

[45] Cf. *Shifā'*, p. 71, l. 10f; p. 73, l. 10–p. 74, l. 2 (Lettinck, *Aristotle's* Meteorology, p. 80f); Qazwīnī, *'Ajā'ib*, p. 91, l. 27–30 (trans., p. 188).

[46] Cf. *Shifā'*, p. 74, l. 5–9 (Lettinck, *Aristotle's* Meteorology, p. 82).

not burn [other things], like the fire in the vapour of wine to which salt and sal ammoniac have been added and the flame of fire ***. It may be that its inflammation is not limited to night time, but it cannot be seen during the day because the light of the sun is stronger than its light.[47]
[2.14] Appendix: Know that if the above-mentioned inflammatory phenomena occur below the sphere of aether, the persistence of their inflammation over a period is [effected] through the renewal and succession of the flames, while some of them disappear because of their inclination towards their natural domain. Its extinction [occurs], if the distance between it and the above-mentioned sphere is long and it is dependent on another part for its material, in order for the capacity of the material for that condition to remain. For the flames dependent on bodies that are with us, like the lamp, are of this kind. That being the case, their persistence will be related to the persistence of a particular fire, while the disappearance of these phenomena occurs either with the disappearance of their fires due to the disappearance of the host material to which the fire is fixed, or with the refinement of the material, so that they are no longer visible as a result of the disappearance of the thickness of the earthy material.[48]
[2.15] The occurrence of these phenomena indicates the occurrence of winds, scarcity of rain, corruption of the upper atmosphere, dryness or hotness of air, and spread of hot, dry and deadly diseases.[49]

[3.1] Third Section: on wind and how it is generated. 'Wind' is the term for air which is moving [for a reason that is] not accidental. The cause of the air's movement may be smoke, clouds or other air being propelled downwards because of heaviness ** generated in them through the prevalence of coldness or through condensation of its particles and joining together of one part of them with another, or because of downward propulsion of them by the circular movement of the sphere or something else.[50]
[3.2] It has been established in its rightful place that it is impossible for masses to penetrate each other, that elemental bodies are subject to rarefaction and condensation, and that [the existence of] void is impossible.

---

47 Cf. *Shifā'*, p. 68, l. 13–18 (Lettinck, *Aristotle's* Meteorology, p. 235).
48 Cf. *Shifā'*, p. 71, l. 15–p. 72, l. 17 (Lettinck, *Aristotle's* Meteorology, p. 81).
49 Cf. *Shifā'*, p. 74, l. 14f.
50 Cf. *Shifā'*, p. 58, l. 6–9 (Lettinck, *Aristotle's* Meteorology, p. 177); Qazwīnī, *'Ajā'ib*, p. 94, l. 26–p. 95, l. 9 (trans., p. 194f).

When, then, particular air is made to move by some cause, what is in front of it [also] moves because of the impossibility of interpenetration. When, however, it is subject to condensation and there is resistance [to it] among what stands in its face, it undergoes condensation, and thus, until it ends up being condensed, and no longer causes what it meets to move, just in the same way as when pebbles are thrown into a basin, successive circles dissimilar in movement are formed, until it ends up as a circle after which no other circle is formed. Sometimes, fineness and rarefaction occur in moving air because of its movement, which cause an increase in propulsion, so that the wind gradually changes from being slow to fast, until the process is reversed and the wind subsides. When the moving air comes from above near to the earth, but is unable to move the neighbouring air in the path of its movement because of its resistance to it, it will, of necessity, because of the impossibility of interpenetration, acquire a movement towards one side. Sometimes, the moving [air] moves along a path different from that of the movement of what is moving it because of the resistance of something [diverting it] from its path, or of an increase in its attachment to something, or for some other reason. Wind is sometimes caused by rarefaction and condensation. [It is caused by] the first, because when its volume increases through rarefaction, it pushes neighbouring [air] towards what is next after it, and thus [the process continues] until it comes to something that will not be pushed, as we have said before. [It is caused by] the second, because when its mass decreases through condensation, the neighbouring [air] is pulled towards it because of the prohibition of void and the inseparability of the surfaces of bodies, and thus [the process continues] until it comes to the bottom of the place [occupied by the air] that is being pulled, which cannot be pulled towards it in its entirety, with the result that the wind subsides.[51]

[3.3] Do not forget that certain causes force several winds [blowing] in different directions to arise from one place unless there is some hindrance and other [causes] require the reverse. It is clear that particular winds arise from artificial causes.

[3.4] Winds sometimes occur in the form of a whirl because of the collision of two winds coming from two different directions. It has been said that the whirl can arise from the crookedness of the material of

[51] See Introduction, p. 366 above; cf. *Shifāʾ*, p. 59, l. 12–16 (Lettinck, *Aristotle's* Meteorology, p. 177f; also 176).

the wind, just as hair is made curly by the crookedness of its source. The persistence of these winds in their forms over time is due [either] to external reasons, or to the density of their moist and viscous material. These winds sometimes become so intense that they uproot trees and lift up stones.[52]

[3.5] <...> on hot, dry and sulphureous grounds, and enormous, putrefying heat arises in them, and they cause whatever animal they meet to putrefy, so that it disappears entirely. These winds are called *samūm*.[53]

[3.6] Appendix: The names of winds that are well-known among the Arabs are four. If it blows from the direction of the north pole it is called 'north wind' [*shimāl*], [if from] the south pole 'south wind' [*janūb*], [if] from the east 'east wind' [*ṣaban*], and [if] from the west 'west wind' [*dabūr*]. Those other than these four are called *nakbāʾ*, or 'crooked'.[54]

[3.7] It should be known that if the places where the wind begins and passes over are moist and cold places, the wind, too, will be so, and if hot and dry, it, too, will be so, provided there are no contrary conditions. Since the places where the north wind originates from and passes over are mountains with much rain and snow, it is cold and moist, in contrast to the south wind. Since the places where the east wind and west wind originate from and pass over are closer to moderation, they are also thus, but since the passage of the east wind is over steppes and deserts and its movement conforms to that of the sun, it is hotter than the west wind, whose place of origin and passage are over mountains and seas and whose movement is opposite that of the sun. It is evident that it is not always the case that winds are generated in this fashion, since the north wind is often made hot by reason of positional relationships, of the celestial quarters, and of its passage over some hot places, especially as it approaches the south—and similarly with the remaining winds.[55]

---

[52] Cf. *Shifāʾ*, p. 60, l. 16–p. 61, l. 7 (Lettinck, *Aristotle's* Meteorology, p. 178f); Qazwīnī, *ʿAjāʾib*, p. 95, l. 9–18 (trans., p. 195f).

[53] *Samūm*, though not mentioned by Ibn Sīnā in the *Shifāʾ* is mentioned by him in his *Kitāb al-najāt* (ed. M. Fakhrī [Beirut, 1405/1985], p. 193, l. 19; as a name for burning [*muḥtariq*] wind).

[54] Cf. *Shifāʾ*, p. 61, l. 17–p. 62, l. 1 (Lettinck, *Aristotle's* Meteorology, p. 179); Qazwīnī, *ʿAjāʾib*, p. 95, l. 19–30 (trans., p. 196).

[55] Cf. *Shifāʾ*, p. 62, l. 2–p. 63, l. 3 (Lettinck, *Aristotle's* Meteorology, p. 179f); Qazwīnī, *ʿAjāʾib*, p. 96, l. 1–8, p. 96, l. 12–15, p. 96, l. 29–p. 97, l. 3, p. 97, l. 6–9 (trans., pp. 197–9).

[3.8] Wind blows more from the direction in which the sun is because of the increase in vaporisation of moist objects and refinement of thick materials, and for this reason wind assists [the formation of] clouds, but sometimes the opposite happens for other reasons. Usually the wind brings about the expulsion of vapours and clearance of the atmosphere where it begins to occur, and causes gathering and accumulation of vapours at the end. Sometimes that is reversed because of the abundance of vapours and obstructing mountains at the start but not at the goal. Furthermore, wind sometimes causes clearance of the air by pushing away the vapours through dissipation and dispersing the particles of clouds, but sometimes the opposite happens through the vaporisation of frozen bodies and gathering of the particles of vapour and cloud. Also, rain sometimes brings about the disappearance of wind through the reduction of the vapoury and smoky particles, and sometimes becomes a cause for moistening of bodies, which helps their vaporisation.[56]

[4.0] Fourth Section: explanation of the rainbow and the halo. The rainbow is a segment of a multi-coloured circle that is generated in the atmosphere by spray-formed particles opposite the sun. It is a semi-circle, and sometimes smaller than that. The halo is a complete or incomplete white circle that is generated around the moon or another star.[57] The explanation of them rests on a number of preliminary matters.[58]

[4.0.1.1] The first preliminary matter concerns the explanation of how things are seen. Know that there are two opinions of import concerning vision. The first is the opinion of the adherents of the 'ray theory', which is that rays go out from the eye and proceed until they fall on

[56] Cf. *Shifā'*, p. 59, l. 17–p. 60, l. 12, p. 63, l. 4–12, p. 64, l. 12–p. 65, l. 8 (Lettinck, *Aristotle's* Meteorology, pp. 178, 180, 181).

[57] Cf. *Shifā'*, p. 47, l. 4.

[58] Qazwīnī also gives us four *muqaddimāt* at the beginning of his section on the halo, rainbow etc., although they are not quite the same as the four here in content and order (*'Ajā'ib* p. 98, l. 11–p. 100, l. 6), and attributes these to al-Qāḍī 'Umar ibn Sahlān al-Sāwī [al-Sāwajī] (*fl.* mid-twelfth century; see *GAL* i, p. 456, *S* i, pp. 763, 817, 819, 830), who was the author of a Persian treatise on meteorology (*al-Risāla al-Sanjarīya fī al-kā'ināt al-'unṣuriyya*, published in M.T. Dānishpazūh, *Dū risāla dar bārā-yi āthār-i 'ulwī* [Tehran, 1337/1958], pp. 1–56; unfortunately inaccessible to me at the time of writing). Fakhr al-Dīn al-Rāzī had organised the points made by Ibn Sīnā into seven *muqaddimāt* in his *Mabāḥith al-mashriqiyya* (ed. Hyderabad, ii, pp. 176–8; ed. Beirut, ii, pp. 181–3, cf. Lettinck, *Aristotle's* Meteorology, p. 283f; the second, third and fourth points here correspond to Rāzī's third, sixth and fourth).

the surface of the object seen, so that the object becomes manifest to the eye with the rays. According to this school, the visual perception of objects in the likes of mirrors is due to the reflexion of rays falling upon a smooth [surface] towards objects whose [positional] relationship to the mirror is like the relationship of the viewer to the mirror, that is to say, the two are on the same side of the mirror in such a way that the angles of [the incidence of] the rays and their reflexion are equal on a level surface.[59] The fact that the angles are so, even if there is no definitive proof for it, is among what is demonstrated by experiment when the sun's rays entering through a skylight fall on water and are reflected towards a wall facing it, rising and falling in the same shape.[60]

[4.0.1.2] The second is the opinion of the adherents of the 'imprinting theory'. According to them, visual perception of objects occurs because of imprinting of their forms in the eye through the mediation of a transparent body that is between them. Concerning the visual perception of objects in the likes of mirrors, they are divided into two groups. The first group says that, just as the transparent body conveys the form of the facing object of vision towards the eye and causes the form to be imprinted in the eye, that smooth body [i.e. mirror etc.] conveys the form of an object whose relationship to it is like the relationship of the viewer to it and causes it be imprinted in the eye. The second group say that the forms of the objects are first imprinted in that smooth body, and then that form facing the eye is imprinted in the eye. It is clear that [the notion that] forms are imprinted in mirrors, especially when they are very large like a half of the celestial sphere, is absurd. Otherwise, the [apparent] position of the object seen should not change in relation to the mirror, when the position of the viewer changes but the position of the object remains the same in relation to the mirror?[61]

[4.0.1.3] The 'ray' theory has been refuted in several ways. I myself have refuted it using a definitive geometrical proof in accordance with what I have said in the *Commentary on the Philosophy of the Eye* [*Sharḥ Ḥikmat al-ʿayn*]. You should realise that the 'image' theory[62] is the reverse of the

[59] On the application of the 'law of equal angles,' see Introduction, n. 11, above. The law is also used by Qazwīnī (*ʿAjāʾib*, p. 98, l. 17f, trans., p. 202).

[60] Cf. *Shifāʾ*, p. 40, l. 11–16 (Lettinck, *Aristotle's* Meteorology, p. 277f).

[61] Cf. *Shifāʾ*, p. 42, l. 6–10, also p. 40, l. 8–10 (Lettinck, *Aristotle's* Meteorology, p. 278).

[62] I.e. the theory described at *Shifāʾ*, p. 41, l. 1–p. 42, l. 5 (Lettinck, *Aristotle's* Meteorology, p. 278).

opinion of the first of these two groups [i.e. of the 'ray' theory]. Since, however, there is no great difference in the result between the opinions [for the purpose of our discussion] here and since the 'ray' theory is the best known, we shall conduct our discussion in accordance with it.[63]

[4.0.1.4] You should know that the visual perception of objects in mirrors according to the 'image' theory and 'ray' theory is a kind of error of sensation, the source of the error being the fact that the soul, since it is accustomed to seeing things in contraposition and is not aware of the reflexion of rays and the conveyance by smooth bodies of objects not in contraposition, imagines that the object seen is also in contraposition.

[4.0.2] Second preliminary point: on the fact that when the mirror is so small that the base of the cone formed by the rays reflected from it towards the object cannot encompass the outline of the object sensed, it will not convey its shape or colour. When, however, there are many small mirrors next to each other, they as a whole will convey its colour and light.[64]

[4.0.3] Third preliminary point: on the fact that the mirror only reproduces the shape of the object seen if the rays cannot pass through it, either because of its density or because of the presence of a dense body behind it.

[4.0.4] Fourth preliminary point is that experiment shows that a coloured mirror does not convey the colour of the object seen as it is, but mixed with its colour and in a colour close to it.[65]

[4.1] Now that these preliminary matters have been sorted out, know that if spray-formed, watery and smooth particles gather at a time when the sun is above the horizon but near it either in the east or the west and there is a dense cloud or mountain behind them preventing the complete penetration of the rays through them, then, when the observer looks at them with his back to the sun, the visual rays are reflected from the particles towards the sun, but, because the particles are very small, they will not convey the bulk, colour or light of the sun as they are, so that we see different colours arising from the dense body differing in consistency and density along with the sun, or rather, its light.[66]

---

[63] Cf. *Shifā'*, p. 43, l. 1–5 (Lettinck, *Aristotle's* Meteorology, p. 278).

[64] Cf. *Shifā'*, p. 44, l. 3–9 (Lettinck, *Aristotle's* Meteorology, p. 279); Qazwīnī, *'Ajā'ib*, p. 99, l. 21–3 (trans., p. 204).

[65] Cf. Qazwīnī, *'Ajā'ib*, p. 99, l. 24–6 (trans., p. 196).

[66] Cf. Qazwīnī, *'Ajā'ib*, p. 100, l. 14–20 (trans., p. 205f).

[4.2] I once saw, together with a large crowd, a rainbow in a desert where there was no mountain or cloud at all. The cause of that may be the accumulation and condensation of [watery] particles mixed with earthy particles in such a way as to prevent the penetration of any rays and bringing about their reflexion.[67]

[4.3] The rainbow is sometimes a semi-circle and sometimes smaller than that, in accordance with what the Shaykh [Ibn Sīnā] has said in the [*Book of*] *Healing* concerning the fact that both the observer and the sun are on the axis of the rainbow. If the sun is on the horizon, the centre of the rainbow will also be on the horizon, so that the rainbow will be a semi-circle, but if it is above it, the centre of the rainbow will be below it, so that it will be smaller than a semi-[circle]. From this it will be clear that a rainbow will not occur if the sun is at the meridian or near it, and at the zenith or near it. If the sun moves away from them by a substantial distance in one direction, the rainbow will appear [varying in size] according to that distance, and it will grow larger as the inversion increases. Also, the further away the particles are from the observer, the wider the rainbow will be, and the closer they are to him, the narrower it will be, in accordance with the width or narrowness of the base of the cone [formed by] the reflected rays.[68]

[4.4] The circularity of the rainbow is due to what has been said about the angles of [the incidence of] the rays and reflexion being equal. For, of the rays reflected from the particles, the only ones which can fall on the sun are those [reflected] from those particles to which the sun and the observer are in the same relationship, in accordance with what has been said. Those particles can only occur in the form of a circle, except for what the diameter of the sun requires depending on [its relationship to the source of] vision,[69] as is confirmed by sound intellect and reference to geometrical proof. Let it not be said that for the rays to be reflected from those particles towards the sun they must be deflected in such a way that they will intersect each other. For we say, it has been found that upon being deflected towards the edges of the sun, the rays need not be deflected so as to form converging lines. That would be necessary only if the body of the sun were smaller than

---

[67] Cf. *Shifā'*, p. 50, l. 8–p. 53, l. 6 (Lettinck, *Aristotle's* Meteorology, p. 281).

[68] Cf. *Shifā'*, p. 53, l. 9–p. 54, l. 4 (Lettinck, *Aristotle's* Meteorology, p. 282).

[69] This clause and what follows are obscurely worded, but I take the argument here to be about the fact that the circle formed is not a line but a band with a certain width.

what the lines are reflected from, i.e. the spray-formed particles, but that is not the case in accordance with the nature of the matter.

[4.5] If someone says, 'what you have said about how reflexion occurs requires that the sun be seen also in the rays reflected from the particles which are in the part where the rainbow fades away', I say, it may be that rays are not reflected from them which reproduce the light of the sun, but [only such rays as to reproduce] a pale colour, and what comprises that part, because of the sun's [?] proximity, appears faint although the rays fall on it; and the rays are not reflected, until they reach a part further away in a pallid state, [and this continues] until they are able to penetrate [through the particles] completely, so that they are no longer reflected towards the sun. More likely, however, is that the rainbow is only produced when the body of the sun can be seen in its fullness from all the particles that occur in the form of a circle, accompanied by the fact that the sun's body is seen in all [the particles] which fall in that form, as is demonstrated in the observation of an object in a mirror when the observer has his back to it or sees the mirror [?] in the form of a circle from one of the extremities of north and south at the other [?]. With the least movement, the mirror becomes parallel in relation to the face of the observer and ***. It may be that the rays fall on particles that are different in position and for that reason become dislocated, when the notion is postulated that their observation from places differing in position does not rest on the rule of the equality of the angles of incidence and reflection. If it is said that seeing the body of the sun in its fullness from all the particles in this manner requires the meeting of the reflected rays, this has already been refuted [when] I said the reflexion of the rays from the above-mentioned particles is not from a single mirror, but from many mirrors differing in position, as I have already indicated.[70]

[4.6] Concerning the rainbow being coloured with different colours, although I have indicated this before, the truth now is that the reason for it is not clear, as the Shaykh [Ibn Sīnā] has admitted, along with everything that has come down to us in the well-known books, which is not worthy of consideration, as will be clear to all who read and study them.[71]

---

[70] Like al-Shaykh al-Ra'īs in the following paragraph, I admit defeat in my attempts so far to draw satisfactory sense out of the text of this paragraph.

[71] Cf. *Shifā'*, p. 50, l. 14f, p. 54, l. 4–7, p. 56, l. 1f.

[4.7] I once observed, when the sun was above the horizon but near to it in a certain place and I came across *** water, whose shade appeared a kind of red tending towards yellow at one end of the shade, namely that nearer the sun, and ** at the other end.[72] You may draw a conclusion by conjecture that the coloration of the rainbow is also of this kind, even if the truth is not clear, and from this may be known the invalidity of some of the statements about it coloration.

[4.8] A white rainbow once appeared during the night when the moon was in its full brightness. *** has already been given, except that the absence of coloration with different colours may be due to the lack of illumination at night, so that the colour of the moon is seen as it is; [for] do you not see that the fire appears white and transparent at night but dense and red during the day and when the sun is strong.[73]

[4.9] The Shaykh has recounted that he once saw an arc in the shape and colours of a rainbow in a bath when the sun was falling on the cups [i.e. glass cups in the skylight], and judged that it was not by way of illusion because of the absence of its movement with the movement of the observer.[74] He also recounted that an arc like that was formed around a lamp.[75]

[4.10] The formation of the halo is [either] due to the fact that a thin cloud containing spray-formed, lucid particles occurs between the observer and the moon or another star, so that the colour of the moon is conveyed from those particles which stand in equal relationships to the observer and the moon, in accordance with what has been said concerning the rainbow, or due to the fact that, again, a thin cloud occurs between him and the moon, which prevents the penetration of some of the rays, so that they are reflected from those particles, which stand in the relationship mentioned, towards the moon, [conveying] its colour only.[76]

[4.11] The reason why the substances in the cloud itself are not seen is because it is concealed by the rays of the moon.[77] What has already been said about the rainbow tells us that the cloud is not a requisite in a halo either. A circle which is seen around the sun, [which may be]

---

[72] The sentence, which is evidently corrupt, seems to be about observation of rainbow colours produced by refraction of light due to its passage through water.

[73] Cf. *Shifāʾ*, p. 57, l. 4–16.

[74] Cf. *Shifāʾ*, p. 52, l. 6–10; Qazwīnī, *ʿAjāʾib*, p. 101, l. 3–5 (trans., p. 206f).

[75] Cf. *Shifāʾ*, p. 52, l. 3f.

[76] Cf. Qazwīnī, *ʿAjāʾib*, p. 100, l. 7–13 (trans., p. 205).

[77] Cf. *Shifāʾ*, p. 48, l. 11–14 (Lettinck, *Aristotle's* Meteorology, p. 280).

white or in rainbow colours, [and] complete or incomplete, is called *ṭufāwa*.[78]

[4.12] The halo may be large or small, or multiple, according to the nearness or farness of the particles mentioned from the observer and the multiplicity of their layers. For when they are near him, it appears larger, and when far away, smaller, because the intersecting lines are longer in the first case and shorter in the second, and that necessitates that the halo in the first case be larger than in the second, as is witnessed to by sound imagination, as well as geometrical proof. From this, it will be clear that when the haloes are multiple, the layers of the particles are multiple, some of them being larger than others, the larger below the smaller.[79]

[4.13] When the halo is thick and forms a perfect circle, especially when it is around the moon or persists until the vapours thicken and a cloud is formed, it is among the signs of rain. If it is not so, then it indicates the opposite of that.[80]

[5.1] Conclusion: on what is generated in the earth from vapours and smokes. When vapours and smokes are generated in the earth due to the rays of the stars and the positions of the spheres or other causes, if the surface of the earth is loose and has passages which allow them to flow out from it, they will gradually flow out. If its surface is solid and there are no passages which allow them to do so, they will be confined in it. Then, when the vapours meet coldness, they are transformed into water. If its [quantity] is very small, one [either] does not sense it, or senses the moisture. If it is not so small, but the surface of the earth is solid and the earth is elevated in such a way that it cannot flow out, it seeps out from it; or moves slowly, if it is weak; or quickly, if it has strength. Furthermore, if the vapours are abundant and their transformation into water follows in succession, the water will persist; if not, it will come to an end. If the edges of the earth are elevated in such a way that the water cannot flow over them, it will become standing water, varying in its quantity and its persistence for the reasons mentioned above. It should be known that just as the kinds of water under discussion can arise from vapours, they can also [arise] from rainwater and

[78] Cf. *Shifā'*, p. 49, l. 4–7, p. 49, l.13–p. 50, l. 6 (Lettinck, *Aristotle's* Meteorology, p. 280f).

[79] Cf. *Shifā'*, p. 49, l. 8–10 (Lettinck, *Aristotle's* Meteorology, p. 280).

[80] Cf. *Shifā'*, p. 49, l. 7f (Lettinck, *Aristotle's* Meteorology, p. 280).

seepage from rivers and seas. The reason for the abundance of springs in mountains and elevated areas, but not in deserts and low-lying areas, will be clear from what we have said. The reason is the abundance of the causes for the generation of waters and the precondition for their being able to flow, namely the solidity of the surface, abundance of rain and snow, confinement of the vapours with the prevalence of coldness on the exterior, and the elevation of one side of the ground and depression of the other. Some of the waters mentioned are made hot by the proximity of lodes of arsenic and sulphur, and for that reason sulphury smell is perceived with some of them. Sometimes the cause of the heat is the proximity of hot smokes, and hence spring water is warmer in winter than in summer because more vapours and smokes are confined under the earth in winter than in summer.[81]

[5.2] When the smokes and vapours increase to such an extent that the cavities in the earth cannot accommodate them, they will of necessity tend towards exit. If there are no pores through which they can escape on the earth's surface, they will shake and move about, until some part of them is transformed into air, which [then] escapes through the narrow pores, or [until] cavities which can accommodate them are formed, or a cleft through which they can escape.[82]

[5.3] When vapours predominate [over smoke], and the coldness of the earth strikes them after heat has been generated in them by the movement, making them more prone to the influence of coldness, they are transformed into water, and springs of water are formed, [which will be either] standing or running, depending on the quantity of the vapours along with the rest of the preconditions. Formation of springs sometimes occurs with the formation of passages through which the waters confined in the earth can escape.[83] Earthquakes sometimes become a cause of annihilation of old springs because of the disappearance of vapours through rarefaction or through their outflow through the pores which are formed, or because of the formation of cavities, [since] water will be drawn towards [such] cavities.

[5.4] Earthquakes sometimes occur with intense movement of air confined in the earth for no particular reason. During an earthquake smoke and inflammation of fires are sometimes observed because of the predominance of smoke and the transformation of part of it into

---

[81] Cf. *Shifā'*, p. 13, l. 4–p. 14, l. 10 (Lettinck, *Aristotle's* Meteorology, p. 142f).

[82] Cf. Qazwīnī, *'Ajā'ib*, p. 149, l. 4–13 (trans., p. 303f).

[83] Cf. *Shifā'*, p. 17, l. 8–10 (Lettinck, *Aristotle's* Meteorology, p. 219).

fire through the intensity of movement.[84] Sometimes a terrifying sound is heard because of the intensity of the movement and the intense collision of the particles of vapour and smoke.[85]

[5.5] Appendix: As the movements that cause the earthquake occur in multifarious ways, so also are the ways in which the earth moves [multifarious]. Sometimes the earth appears to rise, sometimes to incline sideways, and sometimes one feels a movement like that of a shiver.[86] Since the occurrence of the earthquake is due to the confinement of a large quantity of vapours and smokes in the earth, it increases at whatever time the generation of them increases together with the cause of confinement. For this reason, it is more frequent in spring and autumn and at night than in summer and winter and during the day.[87] The solar eclipse, which brings about a prevalence of coldness over the face of the earth and sudden confinement of the vapours and smokes in its cavities, sometimes becomes its cause.[88]

[5.6] Final remarks. Note: The harm of the earthquake includes the extinction of some animals, devastation, heat, corruption of the face of the earth, annihilation of springs, wells and water channels, and generation of excessive winds due to the outflow of vapours. Its benefits include formation of springs, [generation] of salutary winds and beneficial rains due to the outflow of vapours, clearance of obstructive mountains and rocks, terrifying of the intelligent and indication of warnings, 'when the earth quakes and shakes off her burdens' [Qurʾān 99:1–2].[89]

[5.7] Note: How the earth with its firmness and the mountains with their solidity are made to quake by feeble winds and thin vapours! How much more will this be the case with feeble humanity? Rather ***, in the violent commotion of divine Subduing and the thunderbolts is the summons, when 'the earth is made level' [Qurʾān 89:21] and 'the mountains are made to crumble' [ibid. 56:5], so that they become dust, revealing 'to whom the kingdom belongs on that day. To God, the One, the Subduer' [Qurʾān 40:16].

—Here ends the transcript of al-Ṭālishī.

[84] Cf. *Shifāʾ*, p. 17, l. 10–14 (Lettinck, *Aristotle's* Meteorology, p. 219).
[85] Cf. *Shifāʾ*, p. 17, l. 14–16, p. 19, l. 9–14 (Lettinck, *Aristotle's* Meteorology, pp. 219, 220).
[86] Cf. *Shifāʾ*, p. 19, l. 3–9 (Lettinck, *Aristotle's* Meteorology, p. 220).
[87] Cf. *Shifāʾ*, p. 18, l. 14–18 (Lettinck, *Aristotle's* Meteorology, p. 220).
[88] Cf. *Shifāʾ*, p. 18, l. 19–p. 19, l. 2 (Lettinck, *Aristotle's* Meteorology, p. 220).
[89] Benefits and harms of earthquakes, cf. *Shifāʾ*, p. 17, l. 8–14, p. 19, l. 14–15.

ZUR ÜBERLIEFERUNG VON ARISTOTELES *PA* IV 9. 684b 22 *ὥσπερ εἴ τις νοήσειεν ἐπ' εὐθείας*

Aafke van Oppenraay

In *De partibus animalium* iv. 9 vergleicht Aristoteles die Anatomie der Cephalopoda mit der anderer Tiere. In 684b 21–685a 3 beschreibt er die Unterschiede in der Position der inneren Organe und die Folgen dieser Position für die anatomischen Unterschiede. Er versucht dabei, die Position der Organe mit einer virtuellen Linie zu visualisieren, auf der die Organe als Punkte (A, B, C und D) gekennzeichnet sind. Der betreffende Text wird hier in drei Fassungen zitiert: das griechische Original von Aristoteles (mit Übersetzung), die anonyme arabische Übersetzung (*c.* 900 nach Christus) und die lateinische Übersetzung, die Michael Scotus um 1215 in Toledo auf Grundlage der arabischen Fassung anfertigte.

> [b21]Ἀμφοτέρων γὰρ τοῦτον [22]ἔχει τὸν τρόπον ἡ φύσις, *ὥσπερ εἴ τις νοήσειεν ἐπ' εὐθείας*, [23]καθάπερ συμβέβηκεν ἐπὶ τῶν τετραπόδων ζῴων καὶ τῶν [24]ἀνθρώπων, πρῶτον μὲν ἐπὶ τῷ ἄκρῳ τῷ ἄνω στόμα τι τῆς εὐθείας [25]κατὰ τὸ A, ἔπειτα τὸ B τὸν στόμαχον, τὸ δὲ Γ τὴν [26]κοιλίαν· ἀπὸ δὲ τοῦ ἐντέρου μέχρι τῆς διεξόδου τοῦ περιττώματος, [27]ᾗ τὸ Δ. Τοῦτον μὲν οὖν τὸν τρόπον ἔχει τοῖς ἐναίμοις [28]ζῴοις… [34]Τὰ δὲ μαλάκιά τε καὶ στρομβώδη τῶν ὀστρακοδέρμων [35]ἔχει αὑτοῖς μὲν παραπλησίως, τούτοις δ' ἀντεστραμμένως. [a1]Κέκαμπται γὰρ ἡ τελευτὴ πρὸς τὴν ἀρχήν, [2]ὥσπερ ἂν εἴ τις τὴν εὐθεῖαν ἐφ' ἧς τὸ E κάμψας προσαγάγοι [3]τὸ Δ πρὸς τὸ A.[1]
>
> Die Natur dieser beiden [d.h. Cephalopoden und Hartschaligen] verhält sich nämlich folgendermaßen: *Wenn man sich eine Gerade vorstellt*, wie es bei den vierfüßigen Tieren und bei den Menschen der Fall ist, so hat man zuerst am oberen Ende der Geraden den Mund, entsprechend dem Punkt A, dann B, die Speiseröhre, darauf Γ, den Magen, dann folgt der Darm bis zum Ausgang der Ausscheidungsprodukte, dieser ist Δ. So verhält es sich bei den „Bluttieren" <Wirbeltiere>… Die Cephalopoden und die Hartschaligen mit gewundenem Gehäuse <Schnecken> ähneln sich untereinander sehr, sind aber den anderen entgegengesetzt. Denn das

[1] Der griechische Text nach der Edition von I. Bekker (Berlin, 1831); b 24 wurde aber zitiert nach der Edition von P. Louis (Budé, Paris, 1957 [1990]): ἄνω στόμα τι: ἄνω στόματι *codd.*, Bekker: ἀνὰ στόμα τι Bussemaker (Paris, 1854).

Ende biegt sich zum Anfang hin, als wenn einer die Gerade E <AΔ> umbiegend den Punkt Δ nahe an den Punkt A bringt.[2]

فطباع هذين الصنفين بقدر النوع الذى ذكرنا ولذلك سيرها سير مستقيم مثل العرض الذى يعرض للحيوان الذى له اربعة ارجل وللناس ايضا . فاما الانسان فله فم فى رأسه اعنى فى الناحية العليا من جسده ثم له معدة وبعد المعدة بطن وبعد البطن معاء يتهالى موضع خروج الفضلة . فهذه الاعضاء فى الحيوان الدمى على مثل هذه الحال...وفى الحيوان الذى يسمى مالاقيا والذى يشبه الصنف الذى يسمى اسطرنبوس خلاف اذا قيس الى الحيوان الخزفى الجلد وفيه موافقة ومشابهة بالخلقة من قبل ان اواخر جسدها مثنية الى ما يلى اوائلها كما يفعل الذى يثنى الخط المستقيم اعنى الذى عليه الف وبآ يثنيه الى الموضع الذى عليه دال.

Natura ergo istorum duorum modorum est sicut diximus. *Et propter hoc ambulant uniformiter*, sicut accidit animalibus quadrupedibus, et hominibus etiam. Homo vero habet os in capite, scilicet in parte superiori corporis, deinde habet stomachum, deinde ventrem, et post ventrem intestinum perveniens ad locum exitus superfluitatis. Istae ergo res in animalibus sanguinem habentibus sunt secundum hanc dispositionem... Malakie autem et quod assimilatur satirinoz diversantur quando conferuntur ad animalia durae testae. Et assimilantur in creatione, quoniam posteriora corporis sunt incurvata versus principium, sicut si linea recta AB esset incurvata ad locum C.[3]

Die Überlieferung des griechischen Textes im Abschnitt *PA* 684b 22 ὥσπερ εἴ τις νοήσειεν ἐπ᾽ εὐθείας bis einschließlich b28 καὶ ὁ θώραξ καλούμενος,[4] wie auch der Ausdruck 685a2 ἐφ᾽ ἧς τὸ E, werden von einigen Herausgebern des Textes als korrupt betrachtet, obwohl der

[2] Übersetzung von L.N. Scharfenberg, *Die Cephalopoden des Aristoteles im Lichte der modernen Biologie* (Trier, 2001), S. 34–5. Frau Scharfenberg hat, auf Anregung von Herrn Prof. Wolfgang Kullmann (Freiburg im Breisgau), in ausgezeichneter Weise, sowohl als Philologin als auch als Biologin, die Beschreibungen von Aristoteles untersucht und die Wirkungsgeschichte dieses Textes verfolgt: „Die vorliegende Arbeit untersucht die Berichte des Aristoteles über die μαλάκια (Cephalopoda) und setzt sie zu den Erkenntnissen der modernen Biologie in Beziehung, um zu erhellen, ob und inwieweit Aristoteles' Angaben zutreffend sind und woher er sein Wissen bezogen haben könnte... Denn hier zeigen sich die Leistungen des antiken Naturwissenschaftlers, und es wird deutlich, dass die moderne Biologie nicht ohne die von Aristoteles geschaffenen Voraussetzungen denkbar ist, und dass sie in vielfältiger Weise mit ihm verbunden ist—trotz alles Trennenden." (S. 16). Es wäre zu begrüßen, wenn mehr Forscher mit einer solch glücklichen Kombination von Expertenwissen in ähnlich akribischer Weise die aristotelische Zoologie im Licht der heutigen Wissenschaft erforschen würden.

[3] Arabischer Text nach der Edition von R. Kruk, *The Arabic Version of Aristotle's Parts of Animals* (Amsterdam, 1979), S. 124–5. Arabisch-lateinische Übersetzung von Michael Scotus nach der Edition von A.M.I. van Oppenraay, *Aristotle, De animalibus. Michael Scot's Arabic-Latin Translation. Part two, Books XI–XIV: Parts of Animals* (Leiden, 1998), S. 181; Anm. S. 293–5.

[4] Der Kürze halber wurde hier nicht der vollständige griechische Text abgedruckt.

Text, so wie er in den Handschriften überliefert ist, mit einer kleinen Korrektur eine plausible Bedeutung ergibt.[5] A.L. Peck[6] fand eine Korrektur des Textes notwendig und ging dabei ziemlich weit: Er nahm an, dass es durch eine fälschlicherweise an dieser Stelle angedeutete Zeichnung zu einer Korruption gekommen war, und ergänzte den korrupten Text durch Angaben aus der arabischen Version des Textes in der lateinischen Übersetzung von Michael Scotus folgendermaßen:

> [ἀμφοτέρων γὰρ τοῦτον ἔχει τὸ τρόπον ἡ φύσις] *et propter hoc ambulant uniformiter* <ἀλλ' οὐ> καθάπερ συμβέβηκεν ἐπὶ τῶν τετραπόδων ζῴων καὶ τῶν ἀνθρώπων, *homo vero habet os in capite, scilicet in parte superiori corporis,* [ἔπειτα τὸν στόμαχον, ἔπειτα δὲ τὴν κοιλίαν]
>
> [Die Natur dieser beiden verhält sich nämlich folgendermaßen:] *und deshalb laufen sie einförmig,* <und nicht> so wie es den Vierfüßigen und den Menschen zukommt. *Der Mensch hat in der Tat den Mund im Kopf, d. h. im Teil des oberen Körpers,* [dann folgt die Speiseröhre, darauf der Magen].[7]

Peck hat sich von der Tatsache täuschen lassen, dass die beiden beschriebenen Diagramme[8] in der arabischen, also auch in der arabisch-

---

[5] Vgl. Anm. 1.

[6] Peck ändert den Text übrigens auch an anderen Stellen in diesem Abschnitt drastisch. Vgl. Aristotle, *Parts of Animals*, ed. A.L. Peck (London, 1937, 1983), S. 356 und 432. Pecks Bemerkung „One of the MSS. (Merton 278) of Michael Scot's version has an entirely different diagram; the three MSS of Scot at Cambridge have no diagram at all, nor has the Arabic MS B.M. Add. 7511" in seiner „Additional Note on 684b 21–29" (S. 432) trifft nicht zu. In der arabisch-lateinischen Tradition werden die Diagramm-Buchstaben in 684b 25–7 ausgelassen, nicht aber die inhaltliche Beschreibung. Abbildungen gibt es in den arabischen und lateinischen Handschriften nicht. In 685a 2–3 dagegen ist die Buchstabenreihe vereinfacht. Statt der Linie E, die aus den Punkten A–D besteht (wie in 684b 25–7 beschrieben) und die so gekrümmt wird, dass die Punkte A und D aufeinander zugehen, wird eine Linie AB beschrieben, deren Enden so gekrümmt werden, dass sie in einem Punkt C zusammenkommen (so Scotus; D Arabisch). Hier gibt der Korrektor des ms. Merton 278 eine Illustration zum Inhalt als Marginalie  (mit Buchstaben a-b-c) ähnlich wie im Abschnitt *GA* 742b 26–8, wo er am Rande 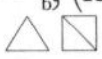 notiert beim Text *semper erunt tres anguli aequales duobus rectis* (sc. *angulis*), *et semper erit diameter incommensurabilis costae* („Denn daß die Winkel im Dreieck gleich sind zwei Rechten, findet immer statt, und daß der Diameter der Seite incommensurabel ist, ist etwas Ewiges." Übers. H. Aubert und F. Wimmer [Leipzig, 1860]).

[7] Zitat und Übersetzung nach Scharfenberg, S. 34, Anm. 59.

[8] Vgl. auch Anm. 7. Im griechischen Text (684b 23–7) ist das Diagramm für die Anatomie der Position der inneren Organe im Rumpf von Menschen und Vierfüßern eine gerade Linie, auf der die Buchstaben A (Mund), B (Speiseröhre), C (Magen) und D (Darmtrakt) verzeichnet sind. Dann wird in 685a 1–3 ein analoges Diagramm für Cephalopoda und Schnecken gegeben, mit den gleichen Buchstaben, jedoch auf einer gekrümmten Linie. Der arabische Übersetzer hat die Buchstaben in 684b 23–7 weggelassen. (Im Gegensatz zu meiner von Pecks Argumentation angeregten Annahme auf S. 295 meiner Edition glaube ich nicht, dass die Buchstaben ein späterer Einschub

lateinischen Tradition, nicht wortwörtlich, sondern in verstümmelter Form überliefert sind. Dies führte bei ihm zu der Annahme, sie könnten zu einem bestimmten Zeitpunkt in die griechische Tradition eingeschoben worden sein, da die arabische Übersetzung des Textes, die im 9. Jahrhundert nach Christus entstand, sich auf ältere griechische Handschriften stützt als die hier erörterte. Diese hätten ältere und ursprüngliche Lesungen von Aristoteles enthalten können.

Beide Buchstabenreihen sind aber gleich lautend in den erhaltenen griechischen Handschriften überliefert, wie auch in der graeco-lateinischen Tradition.[9] Deshalb scheint mir der griechische Text dieser Passage, sowohl was die Textüberlieferung als die Bedeutung angeht, über jeden Zweifel erhaben.

Wie Laila Scharfenberg bin ich der Meinung, dass der überlieferte griechische Text von Aristoteles so beibehalten werden muss (in der Lesung von P. Louis: ἄνω στόμα τι).[10] Pecks radikale Rekonstruktion des griechischen Textes scheint mir unnötig und auch nicht sinnvoll. Das heißt aber, dass der abweichende arabische Text einen korrumpierten griechischen Text zur Grundlage gehabt haben muss.

Während meiner Arbeit an der Edition der arabisch-lateinischen Fassung von Michael Scotus ergab sich ein Problem bei der Interpretation von b 22 ὥσπερ εἴ τις νοήσειεν ἐπ' εὐθείας. Es ist offensichtlich, dass die arabische und die arabisch-lateinische Übersetzung an dieser Stelle nicht mit der griechischen Quelle übereinstimmen. Über die Ursache des Unterschieds sind Scharfenberg und ich jedoch unterschiedlicher Meinung.[11]

Scharfenberg meint, Scotus habe die zweideutige arabische Stelle falsch interpretiert.[12] Meiner Meinung nach hat aber Scotus das

in den griechischen Text sind; sie sind nämlich in der ganzen griechischen Tradition anwesend.) Dadurch gerät er in 685a 1–3 bei der Übersetzung des zweiten Diagramms in Schwierigkeiten, außerdem sieht er nicht, dass τῶν ὀστρακοδέρμων eine nähere Bestimmung zu στρομβώδη ist. Er paraphrasiert es als gekrümmte Linie AB, deren Enden in *Punkt D* zusammenkommen. Scotus hält diese letztere Lösung für unlogisch und korrigiert sie zu *Punkt C*. Die Bedeutung der Diagramme aus dem griechischen Ausgangstext ist jedoch in der arabischen wie in der arabisch-lateinischen Übersetzung völlig verständlich. Vgl. Aristotle, *De animalibus*, Part Two, ed. van Oppenraay, Anm. S. 293–5.

[9] P.B. Rossi, 'La zoologia di Aristotele nel Medioevo: recenti edizioni,' *Aevum* (Mailand) 78 (2004), S. 580, Anm. 16.

[10] Vgl. Anm. 1.

[11] Aristotle, *De animalibus*, Part Two, ed. van Oppenraay, S. 293–95; Scharfenberg, *Cephalopoden*, S. 34–6.

[12] Scharfenberg, *Cephalopoden*, S. 35–6, Anm. 59.

Arabische hier äußerst plausibel interpretiert. Natürlich stimme ich Scharfenberg zu, wenn sie feststellt, dass das Element in den arabischen und arabisch-lateinischen Texten anschließend nicht mehr in den Kontext passt. Die freie Übersetzung, die Scotus an dieser Stelle gibt, kann darauf deuten, dass auch er sich dessen bewusst war. In meinem Kommentar zu den Abschnitten 684b 22–24 und 685a 2–3 musste ich mich seinerzeit kurz fassen. Ich konnte daher nur andeuten, dass meiner Meinung nach der arabische Übersetzer den Text seiner griechischen Vorlage aus irgendeinem Grund nicht gut hatte lesen können, und anschließend von einer Lesart ausgegangen ist (nämlich des Stammes πορεύ- ω/ομαι), die zu einer Verwirrung innerhalb des gelesenen Textes geführt haben könnte. Ich möchte diesen Vorschlag hier gern näher erläutern.[13]

Ich gehe davon aus, dass der arabische Text nach seiner logischsten und selbstverständlichsten Bedeutung interpretiert werden muss, in Übereinstimmung mit dem Sprachgebrauch des arabischen Übersetzers, der sich meistens nicht komplizierter, gekünstelter und verzerrter Konstruktionen bedient.

Scharfenberg weist auf die Möglichkeit einer anderen Interpretation des Arabischen ولذلك سيرها سير مستقيم hin: Außer der wörtlichen Übersetzung „und deshalb ist ihre Fortbewegung eine aufrechte (gerade?) Fortbewegung", welche sich auf den Gang bezieht, könne der arabische Text auch bedeuten „und deshalb ist ihr Verlauf (nämlich des Verdauungssystems) ein gerader Verlauf". Scharfenberg argumentiert, dass der Satzteil so aufgefasst besser in den Kontext zu passen scheint, der ja nicht von der Fortbewegung der Tiere handelt. Der Kontext handelt aber auch nicht vom Verdauungssystem, er beschreibt ausschließlich die Position der inneren Organe in den Körpern der verschiedenen Tierarten (Hartschalige, Weichschalige, Cephalopoda, Schnecken, Insekten)

[13] Es freut mich sehr, dass ich mit diesem Artikel an der Festschrift für Hans Daiber, meinen Doktorvater, Freund und Mentor, beitragen kann. Ich danke an dieser Stelle Herrn Prof. Wolfgang Kullmann (Freiburg i. B.) für den Hinweis auf die Notiz von Frau Scharfenberg und auf das hier beschriebene Übersetzungsproblem. Ich danke Herrn Kullmann, Herrn Prof. G.K.R. Endress (Bochum), Frau Prof. R. Kruk (Leiden) und meinem Institutskollegen Dr. H.J.M. Nellen (Den Haag) sehr für Anmerkungen, die im Beitrag verarbeitet wurden. Herrn Prof. C.J. Ruijgh† verdanke ich die ausführliche syntaktische und paläographische Analyse des griechischen Textes (vgl. Anm. 18). Ich danke meinem Kollegen Dr. J.A.W. Gielkens für die Hilfe bei der Übersetzung ins Deutsche und Herrn Dr. Wim Raven und Frau Dr. Anna Akasoy für die Möglichkeit, diesen Beitrag in dieser Festschrift zu veröffentlichen. Frau Akasoy verdanke ich außerdem Korrekturen stilistischer Art.

verglichen mit den inneren Organen beim Menschen und bei Vierfüßern. Außerdem stimmt der syntaktische und inhaltliche Anschluss im arabischen Text nicht, wie Scharfenberg selber ausführlich darlegt. Der Satzteil würde bei dieser Interpretation innerhalb des arabischen Textes vielleicht weniger fremd erscheinen, es bliebe aber unklar, von welchem griechischen Satzteil dies eine Übersetzung sein sollte.[14] Man könnte annehmen, dass der arabische Übersetzer ὥσπερ εἴ τις νοήσειεν ἐπ' εὐθείας von sich aus durch Wörter ersetzt hat, die seiner Meinung nach besser in den Kontext passten. Dies entspräche aber nicht seiner sonstigen Arbeitsweise. Wenn er in seiner Übersetzung dann und wann vom griechischen Text abweicht, dann immer, weil er den griechischen Text falsch liest oder falsch interpretiert, und er tut dies meistnachweisbar aus paläographischen oder syntaktischen Gründen.

Scharfenberg schließt die Möglichkeit aus, dass der arabische Übersetzer das Griechische an dieser Stelle nicht richtig gelesen hat und versucht darum dem Arabischen eine Bedeutung zu geben, die meines Erachtens große Schwierigkeiten bereitet. Meistens geht der arabische Übersetzer so vor, dass er einfach buchstäblich das übersetzt, was seinem Verständnis nach im Griechischen steht; Scotus macht das Gleiche mit dem arabischen Text. Es ist nicht sehr wahrscheinlich, dass der arabische Übersetzer sich an dieser Stelle, entgegen seiner normalen Praxis, auf einmal eine abstruse, schwer erklärliche Lösung ausgedacht haben sollte. Außerdem stimmt der Anschluss mit ولذلك (*et propter hoc*) nicht; dieser deutet schon auf einen anders gelesenen griechischen Text hin.

Meines Erachtens handelt es sich hier, wie so oft in diesem Text, um einen vom arabischen Übersetzer falsch gelesenen Satzteil, der von Scotus sinngemäß übersetzt wurde (*et propter hoc ambulant uniformiter*), weil er das anscheinend angemessener fand als die wörtliche Übersetzung *et propter hoc est ambulatio eorum ambulatio recta*. Es könnte z.B. sein, dass ihm *recta* an dieser Stelle zu doppeldeutig erschien, weil sowohl die Bedeutung „in einer geraden Vorwärtsbewegung" als auch „aufrecht, vertikal" möglich sind.[15] Man kann sich dabei natürlich fragen, ob *uniformiter* eine unmissverständliche Lösung ist.

---

[14] مستقيم wird in diesem Kontext übrigens natürlich schon mit dem Verdauungssystem assoziiert. So findet man z.B. الحيوان المستقيم المعاء (*Scot* animalia recti intestini) für εὐθυέντερα 675a 21 (vgl. 675b 9, b 27).

[15] Vgl. Aristotle, *De animalibus*, Part Two, ed. van Oppenraay, Index Latino-arabus s.v. rectus (S. 414).

Es geht mir in meinem Kommentar darum festzustellen, dass Scotus Formen von سار fast immer mit *ambulare*, *ambulatio* übersetzt, und dass man daher davon ausgehen kann, dass er سير auch an dieser Stelle so aufgefasst hat. Der arabische Übersetzer benutzt سار , سير , مسيّر fast immer um das griechische πορεύομαι, πορεία zu übersetzen.[16] Deshalb bin ich der Meinung, dass er buchstäblich wiedergab, was er in seinem (Minuskel-) Text zu lesen meinte, nämlich eine Form von πορεύ-. Außerdem hat er statt ὥσπερ wahrscheinlich ὥστε (*et propter hoc*)[17] gelesen, wodurch der Satzteil ganz anders angeschlossen wurde, nämlich beim Vorhergehenden statt beim Nachfolgenden.

Das wichtigste Kriterium für die Annahme einer abweichenden Lesart des Griechischen ist also, dass dieser Übersetzer regelmäßig von سار (s-y-r) abgeleitete Ausdrücke für πορεύ- ω/ομαι verwendet. Für eine andere Verwendung im Sinne von „Verlauf", „Ausrichtung" lässt sich kein Beleg finden.

Um zu verstehen, was während der Übersetzung ins Arabische geschehen sein könnte, ist es notwendig zunächst einmal zur Überlieferung des griechischen Textes zurückzukehren. Man kann zum Beispiel versuchen herauszufinden, wie die Vorlage des arabischen Übersetzers an dieser Stelle ausgesehen haben mag.

Aristoteles argumentiert,[18] dass, während bei Vierfüßern und Menschen der Mund, die Speiseröhre, der Magen und das Exkretionsorgan in einer geraden Linie liegen, diese Linie gekrümmt werden muss, um die Lage dieser Organe bei Muschel- und Weichtiere zu beschreiben.[19] Aristoteles beschreibt ausführlich die gerade Linie dieser Organe bei *den Vierfüßern und den Menschen*, die auch für *Krustentiere und Insekten* gilt. Erst in 685a 1 κέκαμπται γὰρ erwähnt er den wesentlichen Unterschied, den man bei *Muschel- und Weichtieren* feststellt.

[16] Andere Beispiele in der Umgebung dieser Textpassage sind 681b 35 πορευτικὰ السيار *ambulanti*; 683b 25 πορευτικὰ السيار *ambulantia*; 684a 18 πορευτικώτερα تسير على رجليها *ambulat super pedes*; 685a 15 πορευτικούς يسير *ambulat*; 686a 35 τοῖς πορευτικοῖς السيار *ambulantibus*; 684a 7 τὴν πορείαν السيار *ambulatio*.

[17] Für die Wiedergabe von ὥστε durch ولذلك (Scotus *et propter hoc*) siehe z. B. 681a 7.

[18] Im Folgenden wird die Interpretation des griechischen Textes wiedergegeben nach Herrn Prof. C.J. Ruijgh, dem 2004 verstorbenen Lehrstuhlinhaber für griechische Sprachwissenschaft an der Universiteit van Amsterdam und Mitglied der Kommission für die Reihe Aristoteles Semitico-Latinus. Herr Ruijgh stellte mir seine Forschungsergebnisse kurz vor seinem Tod freundlicherweise zur Verfügung.

[19] Vgl. die Diagramme auf S. 295 meiner Edition.

Eine mögliche Interpretation des Satzes, der mit ἀμφοτέρων anfängt, lautet: „Denn bei beiden Tierarten ist der natürliche Bau so, wie wenn man in einer geraden Linie—wie es (in Wirklichkeit) bei den vierfüßigen Tieren und den Menschen der Fall ist—zunächst am oberen Ende der geraden Linie den Mund auf der Höhe von Punkt A denken würde, dann, bei Punkt B, die Speiseröhre und bei Punkt Γ den Magen". Τοῦτον... τὸν τρόπον (684b 27, = οὕτως) korreliert mit ὥσπερ (b 22). Anscheinend unter dem Einfluss des vorangehenden τῶν ἀνθρώπων beschreibt Aristoteles die für Menschen charakteristische Lage der vier Organe, bei der die gerade Linie vertikal ist.

Nach τὴν κοιλίαν weicht Aristoteles von der Konstruktion, die von νοήσειεν abhängt, ab und geht zur Konstruktion des Hauptsatzes über: „(Die gerade Linie verläuft weiter) vom Darm bis zum Durchgang für die Exkremente, wo sich Punkt Δ befindet."

Nach einem Exkurs folgt dann erst das, was logischerweise der Schluss sein muss, nach ὥσπερ εἴ τις νοήσειεν ἐπ' εὐθείας, nämlich (685a 2–3) ὥσπερ ἂν εἴ τις τὴν εὐθεῖαν ἐφ' ἧς τὸ Ε κάμψας προσαγάγοι τὸ Δ πρὸς τὸ Α.

Der arabische Übersetzer muss einen korrupten griechischen Text mit einer Form von πορεύεσθαι, πορεία oder πορευτικός vor sich gehabt haben, da von Fortbewegung an dieser Stelle nicht die Rede ist und erst in 684b 29 von ἕνεκα τῆς κινήσεως und in b 33 von κινητικάς. Über die genaue Art dieser Korruption kann man nur spekulieren.

Paläographisch liegt es auf der Hand, anzunehmen, dass ein Kopist einen Unzialtext in *scriptio continua* vor sich hatte, worin er ωcπερειτιc falsch las, nämlich als ωcπορειτιc: Eine Verwechslung von ε und ο kommt oft vor. Aus πορει könnte er dann πορεία gemacht haben. Er (oder ein späterer Kopist) könnte anschließend, angesichts der Konstruktion οὕτως (= τοῦτον τὸν τρόπον)—ὥστε, ωc zu ὥστε verändert haben: „in der Weise, dass". Ein Kopist könnte das νοήσειεν zwischen ὥστε πορεία τις und ἐπ' εὐθείας weggelassen haben, da es dort unverständlich war. (Es ist auch denkbar, dass der Text in der Unzialhandschrift an dieser Stelle durch ein Beschädigung unleserlich war. In diesem Falle wäre es denkbar, dass auch τις zur unleserlichen Stelle gehörte und vom Kopisten weggelassen wurde.) Das auf diese Weise übrig gebliebene ὥστε πορεία τις ἐπ' εὐθείας war interpretierbar als „so dass es eine Art Fortbewegung in einer geraden Linie gibt".

Bei der Interpretation des korrumpierten griechischen Textes wurde τοῦτον... τὸν τρόπον als rückverweisend aufgefasst: بقدر النوع الذى ذكرنا *sicut diximus* (*litt.* „secundum modum quem diximus“) und dadurch wurde ὥστε als Anfang eines neues Satzes interpretiert (وكذلك „so dass“, „und darum“: *et propter hoc*). Außerdem wurde nach ὥστε das nomen actionis πορεία (τις), Arabisch سير , im Lateinischen durch ein verbum finitum (*ambulant*) wiedergegeben.

Das von Scotus gewählte *uniformiter* könnte auf eine Interpretation von „gradlinig“ zu „gleichmäßig“ beruhen, und daher: „einförmig“. Die Änderung in *inuniformiter* (mss. *ACH*) könnte vielleicht von einem Gelehrten stammen, der die Fortbewegung von Muschel- und Weichtieren als ungleichmäßig betrachtete.[20]

Schließlich: Durch diese Korruption zu ὥστε πορεία (τις) ἐπ’ εὐθείας geriet der Satzteil πρῶτον μὲν κτλ in die Schwebe. Um die accusativi στόμα, στόμαχον und κοιλίαν unterzubringen, hat der arabische Übersetzer sie durch ein „der Mensch hat“ فاما الانسان فله (Scotus: *homo vero habet*) ergänzt.

Dies alles ist natürlich nur eine mögliche, keine gesicherte Erklärung für das, was während der komplizierten Überlieferung dieses Fragments geschehen sein könnte.

Trotzdem zeigt dieser Fall, dass ein Herausgeber, der sich auf das Gebiet der arabisch-lateinischen Tradition von Aristoteles-Übersetzungen begibt, sich bei der Suche nach einer Erklärung für bestimmte Übersetzungsprobleme einer Reflektion über den Ausgangstext von Aristoteles nicht entziehen kann.

[20] Da es sich hier um Handschriften handelt, die meistens (nicht immer!) die ursprüngliche Lesung von Scotus erhalten (nämlich *ACH* [*unumformiter W*]), könnte man hier auch auf eine falsche Lesung des Arabischen durch Scotus schließen: غير مستقيم statt سير مستقيم, anschließend in der lateinischen Tradition der Handschriften *BDE* inhaltlich „rückverbessert“ zu *uniformiter* (eine Lesung, die übrigens auch Albertus Magnus hat, weil er eine Handschrift aus letzterer Tradition [*BDEW*] benutzte; *De animalibus*, Ed. H. Stadler, 2 Bde. [Münster, 1916–21], XIV, 15, 958, 6); Aristotle, *De animalibus*, Part Two, ed. van Oppenraay, Introd. S. XVIff.

AL-ʿĀMIRĪ ON VISION AND THE VISIBLE. VARIATIONS ON TRADITIONAL VISUAL THEORIES

Elvira Wakelnig

The main body of Greek scientific and philosophical writings which reached the Arabic-Islamic world from the ninth century onwards included material on visual and optical theories. The most famous Muslim philosophers, namely al-Kindī, al-Fārābī, Ibn Sīnā and Ibn Rushd all dealt with the various Greek theories of vision and added their own views.[1] And so did Abū 'l-Ḥasan al-ʿĀmirī (d. 992), who might be less famous, but not less interesting. He gives his account on vision in a separate work, *Risālat al-qawl fī 'l-ibṣār wa'l-mubṣar*.[2] The treatise has more or less gone unnoticed until now,[3] so the present paper shall be a first step towards altering this situation and shall point out, among other things, a striking textual parallel to Ḥunayn ibn Isḥāq hitherto unnoticed.

As information about al-ʿĀmirī is rather scarce and only few of his writings are known to be extant, we might be at a loss how to assess the position of the *Ibṣār* within al-ʿĀmirī's work. Fortunately, however, the author himself explicitly states the purpose of this treatise: he wants to complete the account of his epistemology by adding what is still missing, namely a presentation of the insight (*baṣar wa-ruʾya*) granted by the senses.[4]

---

[1] Cf. D. Lindberg, *Theories of Vision from al-Kindī to Kepler* (Chicago, 1976). For al-Kindī cf. P. Adamson, 'Vision, Light and Color in al-Kindī, Ptolemy and the Ancient Commentators,' *ArScPhil* 16 (2006), pp. 207–36; and for al-Fārābī cf. B. Eastwood, 'Al-Fārābī on Extramission, Intromission and the Use of Platonic Visual Theory,' *Isis* 70 (1979), pp. 423–5.

[2] The *Ibṣār* is preserved in a single manuscript: ms. Cairo, Taymūriyya, *Ḥikma* 98 and edited by S. Khalīfāt in *Rasāʾil Abī 'l-Ḥasan al-ʿĀmirī wa-shadharātuhū al-falsafiyya* (Amman, 1988), pp. 411–37.

[3] In modern research literature Paul Kraus was the first to acknowledge the existence of the *Ibṣār* in his review of Brockelmann's *Geschichte der arabischen Litteratur* in *Orientalia*, NS 6 (1937), pp. 283–9. In a personal letter to Franz Rosenthal (dated 29 July 1939), he expressed the intention to study the *Ibṣār*, as it "is an interesting treatise directed against the Kalām and dealing with optico-philosophical problems." (F. Rosenthal, 'State and Religion according to Abū l-Ḥasan al-ʿĀmirī,' *IQ* 3 [1956], p. 43). I have presented a preliminary version of the present paper at the *Symposium Graeco-Arabicum Quartum* 'The Arabic Aristotle' at the University of Bochum on 4 February 2005.

[4] *Rasāʾil*, ed. Khalīfāt, p. 413. Additionally al-ʿĀmirī mentions in his preface that the motivation to write the *Ibṣār* was to answer to a "brother praiseworthy for his high

Al-ʿĀmirī discerns four kinds of insight (*aqsām al-baṣar wa'l-ru'ya*):

- The bodily part or kind (*al-qism al-jusdānī*): it is the insight perceived by the five senses and common to man and animal. The broad mass of people knows and uses this kind of insight and its philosophical discussion is the object of al-ʿĀmirī's *Ibṣār*.
- The intellectual one (*al-qism al-ʿaqlī*) is the insight of the sound intellects (*ru'yat al-ʿuqūl al-ṣaḥīḥa*) by which they take hold of the universal concepts (*al-maʿānī 'l-kulliyya*). As al-ʿĀmirī states, he has already dealt with the intellectual insight in his *Tafsīr li-Kitāb al-burhān*, a commentary on Aristotle's *Posterior Analytics*.
- The holy one (*al-qism al-qudsī*) is the insight of the holy spirits (*al-arwāḥ al-muqaddasa*) by which they grasp what can be deduced from the spiritual substances. Among men this insight is granted only to the prophets, and therefore al-ʿĀmirī has discussed it when talking about prophecy in his *Kitāb al-irshād li-taṣḥīḥ al-iʿtiqād*.
- The imaginative one (*al-qism al-wahmī*) is the insight of the rational souls (*al-nufūs al-nāṭiqa*) due to their imaginative powers (*al-quwā 'l-mutakhayyila*). It is perceived while sleeping or being awake, by natural disposition or by a special skill (*ṣināʿa*). Al-ʿĀmirī has explained this part in the fourth chapter of his *Kitāb al-nask al-ʿaqlī wa'l-taṣawwuf al-millī*.

Although al-ʿĀmirī has thus, according to his own account, discussed all the relevant parts of his epistemology, its reconstruction proves nevertheless difficult,[5] since apart from the *Ibṣār* all abovementioned

rank in wisdom" who had asked for an explanation of vision and related subjects (ibid., p. 411).

[5] For instance al-ʿĀmirī's criterion for distinguishing between intellects (*ʿuqūl*) and spirits (*arwāḥ*) remains doubtful. Spirits seem to be an elevated and inspired kind of intellects—the intellects of prophets and angels. In al-ʿĀmirī's *Kitāb al-amad ʿalā 'l-abad* (*A Muslim Philosopher on the Soul and its Fate*, ed. and trans. E. Rowson [New Haven, 1988], pp. 104–5) the holy spirit (*rūḥ muqaddasa*) is said to be the soul in the "extreme degree in elevation" (*al-ghāya fī 'l-rafʿa*), namely when having reached the rank of prophecy by its noblest power, that is "the power of adopting the best of what lies within the limits of possibility" (*quwwat al-īthār li-afḍāl mā fī ṭarafay al-imkān*). In his commentary to this passage (ibid., p. 273) Rowson points out that in al-ʿĀmirī's *Kitāb al-iʿlām bi-manāqib al-Islām* (ed. A. Ghurāb [Cairo, 1967], p. 206) two different kinds of spirits are distinguished: the rational (*nuṭqiyya*) spirits, by which the Intellect is reached, and the holy (*qudsiyya*) ones, which are peculiar to the prophets. Unfortunately the manuscript of the *Iʿlām* is illegible hereafter, so we can not tell which kind of elevation (*iqāma*) is reached by the holy spirits.

works are lost. Even the long and numerous fragments of the *Kitāb al-nask al-ʿaqlī*, which are preserved in al-Tawḥīdī, Miskawayh and the texts of the *Ṣiwān al-ḥikma*-tradition,[6] do not seem to derive from the chapter in question, as they do not deal with imagination at all. It is hence the bodily or sensual kind of al-ʿĀmirī's epistemology, i.e. sense perception, of which we are informed best.

Al-ʿĀmirī restricts his presentation of the senses mainly to the discussion of sight. This restriction is already foreshadowed in the preface of the *Ibṣār*, where he states that he was explicitly asked to explain "the perception of the visibles by the sense of sight".[7] We might also assume that—inspired by Aristotle[8]—al-ʿĀmirī considers the sight as the sense *par excellence* and therefore as being able to serve as an example for all the other senses as well.

The overall structure of al-ʿĀmirī's exposition is shaped by its two main source texts: Aristotle's *De anima* and the *Book of the Ten Treatises on the Eye* ascribed to Ḥunayn ibn Isḥāq.

From the *De anima*, al-ʿĀmirī does not borrow complete passages, but he addresses issues vital to the discussion of sense perception with similar arguments and in a similar order as Aristotle does in *De anima* II, 6–7. The Aristotelian issues he adopts are the following: the three kinds of perceptible or visible objects, namely the specific, the common and the incidental object of sense;[9] colour, light, the transparent, the dark and the fiery or shining.[10] Al-ʿĀmirī's dependence on Aristotle is however not exactly close: he does not follow the Aristotelian text faithfully, slightly changes the order of topics discussed and inserts a

[6] Al-Tawḥīdī, *al-Muqābasāt*, ed. M. Ḥusayn (Baghdad, 1970), pp. 340–54; Miskawayh, *al-Ḥikma al-khālida*, ed. ʿA. Badawī (Cairo, 1952), pp. 347–73; *The Muntakhab Ṣiwān al-ḥikmah of Abū Sulaimān as-Sijistānī*, ed. D. Dunlop (The Hague, 1979), pp. 128–9; *The Mukhtaṣar Ṣiwān al-Ḥikma of ʿUmar b. Sahlān al-Sāwī*, ed. R. Kartanegara (Chicago, 1996), pp. 264–7. Whereas al-Tawḥīdī and the two texts of the *Ṣiwān al-ḥikma*-tradition explicitly indicate that they are quoting from *al-Nask al-ʿaqlī*, we can only assume that Miskawayh does so as well by comparing his quotations with the ones of the three other texts. For the texts of the *Ṣiwān al-ḥikma*-tradition cf. H. Daiber, 'Der Ṣiwān al-Ḥikma und Abū Sulaimān al-Manṭiqī as-Sijistānī in der Forschung,' *Arabica* 31 (1984), pp. 36–68, esp. for al-ʿĀmirī pp. 42–3.

[7] *Rasāʾil*, ed. Khalīfāt, p. 411; cf. also above note 4. However, on p. 413 al-ʿĀmirī claims that he was asked about the sensual part of insight as such, not especially about vision.

[8] Cf. *De anima* III 3, 429a 3: "sight is the most highly developed sense" (trans. J.A. Smith in *The Complete Works of Aristotle I* [Princeton, 1995], p. 682).

[9] *Rasāʾil*, ed. Khalīfāt, pp. 414–22 and *De anima* 418a 7–25.

[10] Cf. *Rasāʾil*, ed. Khalīfāt, pp. 423–5 and Aristotle, *De anima* 418a 26–419a 10.

considerable amount of other philosophical material (e.g. rejections of views of the *mutakallimūn*). Thus it is hard to tell which Arabic rendering of the *De anima* might have been his source text.[11] We may however assume that a complete translation of the Aristotelian work—or at least what al-ʿĀmirī took to be a complete translation—was at his disposal, because in the introduction to the *Ibṣār* he mentions his plan to write a commentary on it.[12]

---

[11] The Arabic transmission of the *De anima* (= *DA*) still poses many unsolved questions. Cf. A. Elamrani-Jamal, '*De Anima*. Tradition Arabe,' in R. Goulet (ed.), *Dictionnaire des Philosophes Antiques, Supplément* (Paris, 2003), pp. 346–58. The main texts of the Arabic *DA*-tradition are the following:

- Ps.-Isḥāq's translation edited by ʿA. Badawī in *Arisṭūṭālīs fī 'l-nafs* (Kuwait, ²1980), pp. 3–88;
- Ibn Sīnā's *Taʿlīqāt*, which contain evidence of two different Arabic translations of the *DA*, the second one—starting at 431a 14—being Ps.-Isḥāq's (cf. H. Gätje, *Studien zur Überlieferung der aristotelischen Psychologie im Islam* [Heidelberg, 1971], p. 35ff). They are edited in *Arisṭū ʿind al-ʿArab*, ed. ʿA. Badawī (Kuwait, ²1979), pp. 75–116;
- Themistius' paraphrase of the *DA* edited by M. Lyons in *An Arabic Translation of Themistus on the 'De Anima'* (Thetford, Norfolk, 1973). The paraphrase seems to have "a strong family resemblance" to the *DA*-translation in the first part of Ibn Sīnā's *Taʿlīqāt* (cf. ibid., p. x; Gätje, *Studien*, p. 64);
- An anonymous paraphrase of the *DA* in the tradition of Philoponus' Commentary edited by R. Arnzen in *Aristoteles'* De Anima. *Eine verlorene spätantike Paraphrase in arabischer und persischer Übersetzung* (Leiden, 1998), pp. 181–351. There is some doubt whether or at least in what form Philoponus' commentary on the *DA* was available in Arabic (cf. Elamrani-Jamal, '*De Anima*,' pp. 353–4 and D. Gutas, 'Avicenna's Marginal Glosses on the *De Anima* and the Greek Commentatorial Tradition,' in P. Adamson et al. [eds.], *Philosophy, Science and Exegesis in Greek, Arabic and Latin Commentaries*, ii [London, 2004], p. 83).
- It is not clear whether Alexander's Commentary on the *De anima*, of which only fragments are preserved in Greek, was translated into Arabic or not (cf. Gätje, *Studien*, p. 72f). The same applies to his own work *On the Soul* (cf. ibid., p. 69f). A second book on the soul, which was compiled by his school, the so-called *Mantissa*, was—at least partly—translated into Arabic. The relevant treatises for our topic are the following: *Maqāla al-Iskandar fī 'l-radd ʿalā man yaqūl inna 'l-ibṣār yakūn bi'l-shuʿāʿāt al-khārija ʿinda khurūjihā min al-baṣar*, edited in *Commentaires sur Aristote perdus en grec et autre épitres*, ed. ʿA. Badawī (Beirut, 1971), pp. 26–30; and *Maqālat Iskandar al-Afrūdīsī fī kayfa yakūn al-ibṣār ʿalā madhhab Arisṭāṭālīs* or *Kitāb al-ibṣār* as Ibn al-Nadīm calls it, which was—according to the manuscript—translated by Isḥāq ibn Ḥunayn (cf. Gätje, *Studien*, p. 140ff and the edition of the text ibid., pp. 147–63). One of Alexander's *Quaestiones* translated into Arabic might also be worth mentioning in the present context: *Fī 'l-lawn wa-ayy shay' huwa ʿalā ra'y Arisṭūṭālīs* edited in H. Gätje, 'Die arabische Übersetzung der Schrift des Alexander von Aphrodisias über die Farbe,' *Nachrichten der Akademie der Wissenschaften in Göttingen* 10 (1967), pp. 364–74. For Alexander cf. also R. Goulet and M. Aouad, 'Alexandros d'Aphrodisias,' in R. Goulet (ed.), *Dictionnaire des Philosophes Antiques*, i (Paris, 1994), pp. 125–39.

[12] *Rasāʾil*, ed. Khalīfāt, p. 414. For the question whether al-ʿĀmirī also used the *Parva naturalia* cf. below.

Having finished his treatment of *De anima* II, 6–7, al-ʿĀmirī turns to a discussion of the instrument of vision, i.e. the eye, and related issues. Herein he closely follows the first three treatises of Ḥunayn ibn Isḥāq's *Kitāb al-ʿashr maqālāt fī 'l-ʿayn*, a compilation of ten physiological studies on the eye in the tradition of Galen.[13]

The final passage of the *Ibṣār* is dedicated to a discussion of the connection between the visible object and the eye, in the course of which al-ʿĀmirī gives clear accounts of the extramission theory and the intromission theory. He himself apparently adheres to the latter.

Having outlined the general structure of the *Ibṣār* I will now present some of al-ʿĀmirī's rather original interpretations of traditional views on vision concerning the following issues:

- the objects of vision
- the transparent and the visible bodies
- the structure of the eye
- the instruments and tools of vision.

(A) The Objects of Vision

The first example of a visible object presented by al-ʿĀmirī is, astonishingly, writing (*kitāba*):[14] He states that the excellence of the art of writing is attested by the intelligent people as it preserves the particular voices which help to elucidate the inner thoughts (*īḍāḥ ʿan al-ḍamīr*), and in reading written accounts of what these voices have said, sight replaces hearing. We may assume that in claiming that, al-ʿĀmirī indirectly

[13] *Rasāʾil*, ed. Khalīfāt, pp. 427–33 and *The Book of the Ten Treatises on the Eye Ascribed to Hunain ibn Is-hâq*, ed. M. Meyerhof (Cairo, 1928), pp. 73–111. For a discussion of the exact title of Ḥunayn's work cf. ibid., pp. XXIX–XXXII; and for a discussion of its contents and its relation to Galen cf. pp. XXXIX–XLVI.

[14] *Raṣāʾil*, ed. Khalīfāt, pp. 417–18:

لسنا نشك أن صناعة الكتابة هي إحدى الصناعات المشهود لها بالفضل عند العقلاءِ ؛ وأنها في الحقيقة حاكيّة للأصوات المجزّأة التي يستعان بها على الايضاح عن الضمير ؛ وأن ثمرتها هي أن يُجعل القول الملفوظ به ذا ثبات على الأزمنة ؛ وأن الأبصار قد تخلف الاسماع في استثباتها ، ....

I take this passage to mean that by the means of writing oral teachings, e.g. Aristotle's philosophy, are preserved for future generations. Peter Adamson has brought to my attention that al-Kindī also uses the example of writing as an object of vision in his *De Aspectibus*, Prop. 11. It is taken from the version of Euclides' *Optics* ascribed to Theon of Alexandria and used to demonstrate the selective focus of vision (cf. Adamson, 'Vision, Light and Color,' p. 216, n. 31).

challenges the view expressed by Aristotle in the *Parva naturalia* (437a 5–15) that hearing "contributes most to the growth of intelligence. For rational discourse is a cause of instruction in virtue of its being audible, which it is, not in its own right, but incidentally; since it is composed of words, and each word is a symbol."[15] However, as al-ʿĀmirī does not mention Aristotle and we do not even know for sure whether he read the *Parva naturalia* or not, this must remain speculation. In the course of the argument in the *Ibṣār*, writing is used to demonstrate yet another aspect: If the perception of writing becomes weak, although there is no damage in the organ of sight and no obstacle from outside, the colour of the writing should be intensified and the letters saturated. Thereby the writing will be more shining and more easily perceived by the eye. What sounds like a practical advice at first sight turns out to have philosophical implications, as al-ʿĀmirī takes it to prove the essential correspondence (*munāsaba dhātiyya*) and natural relation (*ʿalāqa ṭabīʿiyya*) between the visual power (*al-quwwa al-mubṣira*) and the seen colours (*al-alwān al-marʾiyya*).[16] This refers to a lengthy exposition a few pages earlier about teleology: there God is said to create essential correspondences and natural relations between all the different substances and their specific aims, so that every substance is most apt to fulfil the purpose intended for it. Therefore the visual power of the eye is created in such a way that it is best equipped for its task, i.e. the perception of colour. Colour in its turn is created in such a way that its being perceived by the eye is most simple and easy.[17]

Having thus established colour as crucial for the act of vision, al-ʿĀmirī defines colour as bodily quality and concludes that if we see colour, we also see the surface to which the colour belongs and thus the body which is limited by the perceived surface. Along with the body

---

[15] Trans. J.I. Beare in *The Complete Works of Aristotle*, i, p. 694. The Arabic text of the passage can only be quoted as given in Ibn Rushd's Epitome of the *Parva naturalia* (= *PN*; ed. ʿA. Badawī in *Arisṭūṭālīs fī ʾl-nafs* [Beirut, ²1980], p. 206), as the passage in question is missing at the beginning of ms. Rampur 1752, the single manuscript of the Arabic *PN* which is known today (according to a personal communication of Rotraud Hansberger, who is presently preparing an edition of the manuscript in question):
والسمع في الإنسان هو الطريق إلى التعلم ، لأن التعلم إنما يكون بكلام ، والكلام إنما يتأدي إليه من طريق السمع . إلا أن فهم دلالة الألفاظ ليس هو للسمع ، وإنما هو للعقل .

[16] *Rasāʾil*, ed. Khalīfāt, p. 418:
فان كان الاشباع في اللونِ سبباً لسهولة الإبصارِ ، المحوُ منِ اللون علةً لصعوبة الإبصارِ ، ولم نشك أن بين القوة المبصِرة وبين الألوان المرئية مناسبة في معنى الادراك ذاتية ، وعلاقة من جهة الاحساس طبيعية .

[17] *Rasāʾil*, ed. Khalīfāt, p. 414ff. The discussion finally ends in a rejection of the occasionalism of the *mutakallimūn*.

we perceive its qualities, namely its shape and its external form, i.e. the fourth class of qualities describes in Aristotle's *Categories* 10a 10ff.[18] In the course of his further investigation, al-ʿĀmirī finally comes up with the following seven different kinds of visibles: colour (*lawn*), magnitude (*ʿiẓam*), shape (*shakl*), number (*ʿadad*), distance (*masāfa*), movement (*ḥaraka*) or rest (*sukūn*),[19] and substance (*jawhar*). Except for distance, we already find all these objects of sense perception in the *De Anima* (418a 10–25). An equivalent of al-ʿĀmirī's 'distance' might be found in Ḥunayn's 'position of the body' (*waḍʿ al-jism*) and 'the intervening distance' (*al-masāfa baynahā wa-baynahū*).[20]

In accordance with the Aristotelian tradition, al-ʿĀmirī further divides these different kinds of sense objects into what is perceptible in itself (*mudrak bi'l-dhāt*) and what is incidentally perceptible (*mudrak bi'l-ʿaraḍ*). Among the incidental objects of sight, al-ʿĀmirī classifies substance, number and distance. He exemplifies this classification with regard to substance: one can perceive the specific colour, magnitude, shape and movement or rest of a thing, without knowing (*min ghayr an yaʿrif*) what kind of substance it is. That is to say, the perception of substance, as well as the perception of number and distance, requires some kind of knowledge, i.e. some intellectual operation.[21]

The objects or qualities (*maʿānī*) which are visible in themselves are subdivided into two kinds: what is perceived primarily (*mudrak idrākan awwaliyyan*[22] or *bi'l-qaṣd al-awwal*), namely colour, and what is perceived secondarily (*mudrak bi'l-qaṣd al-thānī*), namely magnitude, shape, movement or rest. The distinctive criterion between these two kinds is the same as in the Aristotelian tradition: what is perceived primarily

[18] *Rasāʾil*, ed. Khalīfāt, p. 418:
ثم إذا كان القوام الجسماني غيرَ منفك عن الشكل ، والتخطيط ، وسائر ما عَدَّه المنطقيون نوعاً رابعاً من الكيفيات ، صار التخطيط والشكل ، بل هيئة التركيب والتأليف أيضا ، مدركا بتوسطه .

[19] Al-ʿĀmirī (*Rasāʾil*, ed. Khalīfāt, p. 419) introduces the term 'condition of rest or movement' (*hayʾat al-sukūn wa'l-ḥaraka*) first and afterwards he simply speaks of 'condition' (*hayʾa*).

[20] *Ten Treatises*, ed. Meyerhof, p. 108, where Ḥunayn does not mention number, rest or substance.

[21] *Rasāʾil*, ed. Khalīfāt, p. 421. Al-ʿĀmirī's inclusion of number among the incidental objects of sense contradicts *DA* 418a 18. Philoponus, in his commentary on the *DA*, explicitly states that substance is not perceived by perception, but by intellect (cf. Arnzen, *Aristotles'* DA, p. 397).

[22] According to al-ʿĀmirī's account further down below (*Rasāʾil*, ed. Khalīfāt, p. 423) this expression is widely used as a definition of colour: [*al-lawn*]...*huwa al-mudrak bi-ḥāssat al-baṣar li-dhātihī idrākan awwaliyyan*. However, it can not be an actual definition as it is a relative one (*waṣf iḍāfī*).

is specific to one sense and cannot be perceived by any other sense, whereas what is secondarily perceived, is common to more than one sense and not specific to any of them. However, al-ʿĀmirī also strengthens another aspect and this aspect is evidently of greater importance to him than the one already mentioned: the six qualities, which are visible either incidentally or secondarily, are only seen with the help of colour, whereas colour—being primarily visible—is in no need of any other quality to be seen. Therefore, al-ʿĀmirī concludes, the sages have called colour 'the absolutely visible' (*marʾiyy ʿalā ʾl-iṭlāq/muṭlaqan*) and the other six qualities 'the visibles by colour' (*marʾiyya bi-tawassuṭ al-lawn*).[23] In the *De anima*, Aristotle does not explicitly express this view, he only implies it. His statement in the *Parva naturalia* (437a 6) that "the faculty of seeing, thanks to the fact that all bodies are coloured, brings tidings of multitudes of distinctive qualities of all sorts" is more outspoken.[24] However, al-ʿĀmirī's statement clearly opposes the account of the *De anima* paraphrase, according to which the perceptible specific to only one sense affects the organ of sense more intensively than the perceptible common to more than one sense, but the common sensible still affects the organ of sense in itself. This interpretation of the Aristotelian text is original to the *De anima* paraphrase and partly due to Philoponus, partly due to a misinterpretation of his view. Philoponus himself states that the perceptibles in themselves affect the organ of sense with varying intensity. However, he also clearly states that a common sensible can only affect the organ of sense together with the specific sensible.[25] Herein we may detect the source for al-ʿĀmirī's claim that magnitude, shape, number, distance, movement or rest, and substance are only visible because of the existence of colour in the perceived, that is to say that they are only seen by colour. Then we have to assume that al-ʿĀmirī had access to a more accurate source for the teachings of the Aristotelian commentators than the *De anima* paraphrase.

Regarding terminology, al-ʿĀmirī's text differs completely from the texts of the Arabic *De anima*-tradition. It shows the closest similarity to

---

[23] *Rasāʾil*, ed. Khalīfāt, p. 422:
ولهذا ما وُصِفَ اللونُ عند الحكماء بأنه مرئيٌّ على الاطلاق ؛ ووُصِفَت الستةُ البواقي نأنها مدركةٌ بحاسة البصر ، أو مرئيةٌ بتوسطَ اللون . وقد يجوز أن يُثْبَتَ بالنظر من غير أن يصفوه بأنه مرئيٌّ مطلقا .

[24] Trans. Beare in *The Complete Works of Aristotle*, i, p. 694. As this passage is not extant in the single manuscript of the Arabic rendering of the *P.N* (cf. above, note 15), it is hard to tell if the *P.N* could be al-ʿĀmirī's source here.

[25] Arnzen, *Aristotles'* DA, p. 255; for Philoponus' interpretation and its relation to the text of the paraphrase, cf. ibid., p. 397f.

Ḥunayn's *Ten Treatises*, where colour is said to be sensed primarily (*al-lawn huwa shayʾ yuḥissuhūl-baṣar ḥissan awwaliyyan*) and where movement is said to be recognised by the sense of vision, but not as an absolute perception (*ḥiss muṭlaq*).[26]

(b) The Transparent and the Visible Bodies

The next issue al-ʿĀmirī addresses in his treatise—and herein he follows the Aristotelian course of argument—is the definition of colour. In order to present the definition of the essence (*māʾiyya*)[27] of colour as given by the sages, he first classifies all bodies as belonging to four kinds:[28] the transparent body (*mustashiff* or *mushiff*),[29] the luminous (*muḍīʾ*), the shining (*dhū barīq* or *barrāq*) and the turbid or dark one (*mā lā barīq lahū* or *kadir*). The transparent body conveys or transmits (*tuʾaddī*) colours and lights—to the eye, we may add—,[30] but it is not itself perceived by the sense of sight.[31] The luminous body makes what is potentially transparent an actually transparent body through the rays (of light) (*anwār*) spreading out from it. The shining body is visible in complete darkness because of its shimmer (*lamaʿān*). The turbid body can be seen when it is reached by light (*ḍawʾ*) and might therefore be interpreted as the coloured one. The latter three kinds of bodies are the ones which

---

[26] *Ten Treatises*, ed. Meyerhof, p. 108. Hunayn does not as clearly distinguish between what is secondarily perceived and what is incidentally perceived as al-ʿĀmirī does.

[27] According to Khalīfāt the manuscript reads *māʾiyya* instead of the more common *māhīya*. In one case Khalīfāt reports a scribal corruption *māʾsīya* (*Rasāʾil*, p. 423, note 78), hence we may assume that the copyist was not familiar with the term at all.

[28] *Rasāʾil*, ed. Khalīfāṭ, p. 424:
فإذاً الأجسام كلُّها يجب أن تكون مفتنّة إلى أقسام أربعة ، وهي : المُشفَّة ، والمضيئة ، والبراقة، والكدرة . فأما المستَشفَّة فمن خاصيّتها أنها تؤدي الألوان والأضواء ولنَ [أن *corr.* Khalīfāt] تصير مدرَكةً بحاسة البَصر . وأما المضيئة فمن خاصيّتها أنها – بأنوارِها الساطعة عنها – تجعل المشفَّة بالقوة مشفة بالفعل . وأما البراقة فمن خاصيّتها أنها بقوة لمعانها – تصير مرئية في الظلام المطبق . وأما الكدرة فَمن خاصيّتها أنها تصير مرئية عند حصولها في الضوء .

[29] For the use of *mushiff* and *mustashiff* cf. Arnzen, *Aristotles'* DA, p. 256, note 88.

[30] For *tuʾaddī* cf. a similar expression in Themistius' paraphrase (Lyons, p. 90):
والهواء (..) وسائر الأشياء التي تقدّم قولنا بأنها مشفّة تؤدّي الالوان إلى البصر لا دائما لكن عندما يصير مشفا بالفعل بحضور الضوء إياه .

and in the *DA*-paraphrase (Arnzen, p. 257):

و[الضوء] يؤديه [اللون] إلى الجسم الصافي بتوسط الهواء .

[31] In this case (*Rasāʾil*, ed. Khalīfāt, p. 424) Khalīfāt's correction of *wa-lan* to *wa-an* (cf. ibid., note 88 and above, note 28) has to be rejected. Al-ʿĀmirī's point here is, apparently, that the transparent is not perceptible in itself, as he implicitly expresses further down below (ibid., p. 424) where he says that the luminous, the shining and the turbid bodies share the possibility of being seen.

can be seen, and as colour is what is seen primarily, al-ʿĀmirī combines these two statements to finally define colour: Colour is the rays of a non-transparent body (*shuʿāʿ jism ghayr mustashiff*). Light and colour are therefore to be understood as being of one category (*jins*), as al-ʿĀmirī spells out clearly further down below.[32]

Although al-ʿĀmirī's fourfold classification of bodies is not explicitly stated in Aristotle's *De anima*—or in the commentaries known to have been translated into Arabic—, we already find most of the constituting elements of the classification and the definition there, namely in II, 7.[33] What we do not find in the Aristotelian tradition is the interpretation of the turbid or dark body as the coloured one and the definition of colour as a being a ray. Interestingly, al-Kindī may be a possible fore-runner for both concepts.[34]

---

[32] *Rasāʾil*, ed. Khalīfāt, p. 434 and below, p. 429. For the relation between light and colour cf. Aristotle, *DA* 418a 26: what is visible is colour: *al-mubṣar lawn* (*Fī 'l-nafs*, ed. Badawī, p. 45); *DA* 418b 11f: light is the colour of the transparent (*wa-ammā al-ḍaw' fa-huwa lawn al-ṣafā'*, ibid.) and Themistius' paraphrase (Lyons, p. 89): *fa'l-marʾīy wa'l-lawn ammā bi'l-mawḍūʿ fa-shay' wāḥid bi-ʿaynihī wa-ammā bi'l-maʿnā fa-yakhtalifān*; (ibid., p. 91): *wa-dhalikā anna 'l-ḍaw' bi-manzilat al-lawn li'l-mushiff.* This last statement can be found—nearly identically—in Alexander (Gätje, *Studien*, p. 153).

[33] For the transparent body: cf. Aristotle, *DA* 418b 5: not visible in itself: *wa-hadhā al-ṣafā' fī 'l-jumla shay' manẓūr ilayhī lā bi-dhātihī* (*Fī 'l-nafs*, ed. Badawī, p. 45); cf. also Themistius' paraphrase (Lyons, p. 90, quoted above in note 30). With regard to the transparent, we might also detect the influence of al-Kindī, the teacher of al-ʿĀmirī's teacher al-Balkhī: according to al-Kindī, vision does not see the transparent (*Rasāʾil al-Kindī al-falsafiyya*, M. Abū Rīda [Cairo, 1953], ii, p. 65). For a discussion of this Kindian view, which Peter Adamson terms a "negative" conception of the transparent cf. Adamson, 'Vision, Light and Color,' p. 229 and p. 232. For the luminous body: cf. Aristotle, *DA* 418b 10: light as the activity of the transparent *qua* transparent: *fa'l-ḍaw' fi'l hadhā al-ṣafā'* (*Fī 'l-nafs*, ed. Badawī, p. 45); cf. also Themistius' paraphrase (Lyons, p. 90, quoted above in note 30; and p. 91): *wa'l-ḍaw' huwa istikmāl mā wa-kamāl al-mushiff min ṭarīq mā huwa al-mushiff.* For the shining body cf. Aristotle's "things that appear fiery or shining" in darkness (*DA* 419a 3): the Badawī-translation (*Fī 'l-nafs*, p. 46) has *al-ashyā' al-nāriyya al-muḍī'a* and therefore seems to be an unlikely source for al-ʿĀmirī, who uses *muḍī'* for the luminous bodies, i.e. the sources of light as e.g. the sun. Themistius (Lyons, p. 88) uses the expression *ashyā'... talmaʿ fī 'l-ẓalma* in his introductory remark to section IV, which has no equivalent in the Badawī-translation. Al-ʿĀmirī uses *lamaʿān*, a derivative of the same root *l-m-ʿ*.

[34] In both cases Peter Adamson has brought the similarity of the views expressed by al-Kindī and by al-ʿĀmirī to my attention. If I am right in understanding al-ʿĀmirī's turbid or dark body as the coloured one, that would make al-ʿĀmirī's view fit perfectly with al-Kindī's concept of the element earth being the bearer of colour. Cf. Adamson, 'Vision, Light and Color,' pp. 225–36. In the *De radiis* every coloured thing is said to be seen by the rays which it emits (M.-T. D'Alverny and F. Hudry, 'Al-Kindī. De Radiis,' *AHDLMA* 41 [1974], pp. 139–260, §III, 224: omne coloratum radios suos emittit quibus videtur). Cf P. Adamson, *Al-Kindī* (New York, 2007), p. 191, where he also opts for the authenticity of the *De radiis*. However, Adamson assumes that the

Having presented this first classification, al-ʿĀmirī interestingly adds a second one: it differs from the first one only in so far as now it is no longer the bodies themselves that are focused on, but the rays which are emitted by them. That is to say he explains transparency (*ishfāf*), light (*ḍaw'*) and shimmer (*barīq*) as rays of the bodies classified above.[35] His reason for doing so might be sought in trying to avoid a clear answer to the question whether light is a body or not, and as a clear statement against the extramission theory, according to which it is the eye which emits rays in order to see. Thus we can detect here a first indication that al-ʿĀmirī believes in a variation of the intromission theory, which claims that rays—not images—are emitted from the visible bodies and affect our eyes. The fourth kind of rays which is evidently missing from the account—since al-ʿĀmirī has announced that rays can be divided into four kinds—must be the rays of a turbid or coloured body, i.e. the rays by which the colour of a body is transmitted to the eye. The question of the essence (*dhāt*) of these rays—especially the rays of light—, i.e. whether they are substance or accident, is acknowledged by al-ʿĀmirī as being a crucial one. However, he does not want to focus on it in his *Ibṣār*.

---

work was written in al-Kindī's old age, because of the striking doctrinal divergences to some other works.

[35] *Rasāʾil*, ed. Khalīfāt, p. 424:

فمن الواجب ... أن تكون الأشعة في أنفسها مفتنّةً إلى أقسام أربعة : أحدها الاشفافُ وحِقيقته أنه شعاعٌ يستَعدُّ به الجسمُ لتأدية الألوان والاضواء . والثاني الضوءُ وحقيقته أنه شعاعٌ جاعل المستشفَّ بالقوة مستشفا بالفعل . والثالث البَريقُ وحقيقته أنه شِعاعٌ يؤثر في الأعين لا بمعونة الضوء . ثم كل واحد من هذه الاشعة قد يكون قويا وقد يكون ضعيفا ، وقد يكون طارئا في الجسم ، غريبا ؛ وقد يكون ثابتا فيه ، ذاتيا . وهي كلها في الحقيقة كيفياتٌ جسمانية وهيئاتٌ انفعالية .

Unfortunately al-ʿĀmirī does not expound his concept of rays any further. So we do not know how he pictured, for example, the connection between a body and its rays. In any case it is remarkable that he calls rays a bodily quality. The concept that rays are a bodily quality may be inspired by Aristotle's account of the transparent in the *Parva Naturalia* 439a 20ff: 'But what we call transparent is...capable of no separate existence of its own, but residing in these [the bodies], and subsisting likewise in all other bodies in greater or less degree.' (trans. J.I. Beare in *The Complete Works of Aristotle*, i, p. 697), a passage which—as first mentioned by C. Prüfer and M. Meyerhof ('Die aristotelische Lehre vom Licht bei Ḥunain b. Isḥāq,' *Der Islam* 2 [1911], p. 125, n. 1)—seems to have inspired (Ps.-[?])Ḥunayn ibn Isḥāq's description of light (L. Cheikho, 'Fī 'l-ḍawʾ wa-ḥaqīqatihi, Maqāla li-Ḥunayn ibn Isḥāq,' *al-Mashriq* 2 [1899], p. 111):

ان الضياء كيفية وذلك أنه يقبل والاضعف وهذا من خواص الكيفية.

For the doubtful authorship of the treatise cf. Arnzen, *Aristoteles DA*, pp. 708–17.

For the terminology cf. also Miskawayh in his *Risāla fī 'l-nafs wa'l-ʿaql*, where he speaks of 'the bodily form' (*al-ṣūra al-jismiyya*) which exists only through the existence of the body, as is the case with ray of light (*nūr*), light (*ḍaw'*), heat, and the like (M. Arkoun, 'Deux épîtres de Miskawayh,' *BEO* 17 [1961–62], p. 44 and P. Adamson, 'Miskawayh's Psychology,' in idem (ed.), *Classical Arabic Philosophy. Sources and Reception* [London, 2007], pp. 39–54, whose English translation of the passage I am following).

(c) The Structure of the Eye

What al-ʿĀmirī apparently wants to focus on and what he therefore discusses next is the primary instrument (*al-āla al-awwaliyya*) for obtaining sight, namely the substance of the eye.[36] He introduces this subject by describing the conditions which must be provided when dissecting an (animal) eye and the nerves which connect the eye to the brain. These introductory passages are taken almost literally from Ḥunayn's *Ten Treatises*, and I quote them at some length to show al-ʿĀmirī's close textual dependence on Ḥunayn as well as his eclectic treatment of his source text:

Ḥunayn, *Ten Treatises* (ed. Meyerhof), pp. 90.11–15, 85.8–9, 92.7–12, 95.6–9, 100.10–20, 91.2–4, 91.19–92.2[37]	Al-ʿĀmirī, *al-Ibṣār* (ed. Khalīfāt, *Rasāʾil*) pp. 427.11–428.10
– ومن أراد أن يراه وقت التشريح فانما يتهيأ له رؤيته بان يقصد نحو ثلاثة أشياء :	إنَّ التفتيشَ عن تركيب الأعضاء الحيوانية يتعلق بصناعة التشريح . ثم العينُ لمَّا كانت ذاتَ أجزَاء كثيرة وكان حجمها من بين الأعضاء [موصوفا] بالصغر فمن الواجب إذن أن يكون المُسْتَعْمِلُ لتشريحها متحرِّيا فيه لشرائط ثلاثة :
أحدها أن يجعل تفتيشه عنه في دماغ حيوان عظيم الجثة ، والثاني أن يكون تشريحه لدماغ ذلك الحيوان ساعة يموت ، والثالث يتحرى أن يكون هذا الموضع الذى يشرحه فيه حارا [نيرا] .[38]	أحدها أَن يجعل تفتيشه في جثة حيوان عظِيم الشبح ؛ والثاني أن يستعمل التشَريحَ فيهً ساعةَ موته ؛ والثالث أن يكشفَ عنه في هواءٍ حار.
– وينبت من الدماغ سبعة أزواج عصب: الزوج الأول والثاني منها يأتيان الى العينين-. . . .أن هاتين العصبتين تنبتان من الدماغ من موضعين مختلفين أعني من جانبي آخر بطني الدماغ المقدمين، ثم لا تمضيان على استقامتهما الى العينين لكنهما تتعوجان في جوف عظم الرأس . وتتصل احداهما بالأخرى بالقرب من المنخرين حتى يصير ثقباهما ثقبا	ومهما أتينا من الشرائط الثلاثة ثم استعملَ المُشَرِّحُ فيه الرفقَ واللطافة ظهر له أن الزوجَ الأولِ من أزواج العصبِ – أعني به الأزواج السبعة التي تكون منتهية بالدماغ دون النخاع – ينبتان من بطنيه المقدَّمين من جانبيِّ اليُمنة والشأمة ؛ ثم ينحدران إلى العينين ، لا على استقامة ، بل يتعوَّجان في جوف عَظمِ الرأسِ ؛ ويتصلً أحدُهما

[36] *Rasāʾil*, ed. Khalīfāt, pp. 425 and 427ff.

[37] In the Arabic text every new passage is introduced by "–", whereas in the English translation page and line of the Arabic edition are given.

[38] According to Meyerhof (*Ten Treatises*, p. 21, note 2) both manuscripts on which he bases his edition read „حارا", nevertheless he edits „نيرا" following the Greek source text by Galen. The *Ibṣār* reads the same as the two manuscript of the *Ten Treatises* and this may be taken as a clear indication that Ḥunayn was indeed al-ʿĀmirī's source and not some other Galenic adaptation.

واحدا . ثم تفترقان بعد اتصالهما على المكان وتذهب كل عصبة منهما إلى العين المحاذية لمبدأ منشأها ...

– فأما السبب الخامس في اتصال عصبتي البصر وافتراقهما بعد الاتصال وهو السبب المبنيّ على المنفعة الأولى بذلك ، وهي أشرف المنافع وأجلها خطرا وأعظمها قدرا في فعل البصر . فهو أن يكون الانسان لا يبصر الشيء الواحد شيئين .

– فأما الروح النوري الذي يأتي الى العينين فقد يصل منه اليهما في المجريين النافذين في عصبتي البصر ليس قوّته فقط بل نفس جوهره ، ومقدار ما يصل منه اليهما مقدار يفي بما يحتاج إليه لفعل البصر. والدليل على أن نفس جوهر هذا الروح يصل إلى العينين ...ومما يستدل به أيضا على ذلك أنه متى غمضت إحدى العينين اتسعت حدقة العين الأخرى ... أن ذلك الاتساع انما كان من قبل االطبقة العنبية عندما تمددت بامتلاء الموضع الذي من ورائها داخل منها ،

-....وذلك لأن كل واحد منهما جعل باطنها في الغاية من اللين ليكون حسها أذكى وجعل ظاهرها يميل إلى الصلابة قليلا لما في ذلك من حرزها وبعدها عن قبول الآفات.

- وعُرّض طرف كل واحدة من العصبتين في العين التي تفضي اليها . وانتسج فصار شبيها بالشبكة ، ومن أجل ذلك سمى هذا الطرف من العصبة في العين الطبقة الشبكية...

بالآخر حتى يصير ثقباهما ثقباً واحداً ؛ ثم يفترفانِ بعد اِتصالهما على المكان ويذهبُ كل ثقباهما ثقبا واحداً ؛ ثم يفترفان بعد اتصالهما على المكان ويذهبُ كلّ عَصَب منهما إلى العين المحاذية لمبدأ منشأهما . والحكمة فيِ اتصالهما بعد[39] افتراق مَبدئهما أن يكونَ اجتلابُ المَدَدِ من موضعين من غير أن يُرى الشيءُ الواحِدُ اثنين .

وجُعِلَ العصبان ، من بين العصب كلِّها ، جوفاوين بنحوَين يدركهما الحس لأنه يجري فيهما – من أصل الدماغ – جوهرٌ نوريٌّ على مقدارٍ

يفي بما يُحْتَاجُ إليه في فعل الإبصار . والدليل على أن جريان الروح [هو] فيهما

أنَّا متى غمضنا إحدى العينين اتسعت حدقة العين المفتوحة لأنها تمتدّ بامتلائها من مادة إبصار صاحبتها .

وخُصَّتِ العصبتان بأن جُعِلَ داخلهما في غاية اللين كما جُعِلَ خارجهما مائلاَ إلى الصلابة لما فيه من البعد عِن قبول الآفات.

ثم يَعْرِضُ للعصبتين عند انتهائهما إلى العين أن ينتسج طرفَ [كل منهما] فيصير شبيها بالشبكة وتسميه أرباب الصناعة طبقة شبكية .

(90.11–15) He who wishes to see it [the optic nerve] at the time of the *autopsy*, can only obtain a view of it if he observes *three* things.

Firstly he must conduct his search on a brain *of an animal that has a large* body. *Secondly* the *autopsy* on the brain of the animal must be performed *immediately after* it has died. *Thirdly* it is desirable that the place in which he dissects should be *hot* [bright].

The search for the composition of the animal limbs is related to the art of autopsy. As the eye has many parts and the space between the limbs is small, it is necessary that he who performs its *autopsy* provides for *three conditions*: *Firstly he must conduct his search on* the body of an animal *that has large* size. *Secondly* he must perform the *autopsy* on it *immediately after* its death. *Thirdly* he must uncover it in *hot* air.

Whenever we have provided for the three conditions, the person who

[39] In this case the textual comparison with Ḥunayn's text confirms Khalīfāt's correction of "bi-ghayr" to "baʿd".

(85.8–9) *Seven pairs of nerves* proceed from *the brain: the first* and second *pairs* enter the eyes;
(92.7–12) It is that these nerves *rise* in different places, *viz.* in the posterior parts of the *sides of the anterior ventricles* of the brain; then they do not continue their *way straight to the eyes, but turn aside in the cavity of the skull-bone* and *unite with one another* near the nose, *so that their canal becomes one; after which they separate again immediately after their junction and each nerve continues on its way to the eye which is opposite to its point of origin in the brain*…

performs the autopsy on it must use care and delicacy to see that *the first pair* of the *pairs of nerves*—that is the *seven* pairs that end in *the brain* and not in the spinal chord—*rises* in the inner *sides of the anterior ventricles*, *viz.* in the right and the left side; then it descends *to the eyes*, not in a *straight way*, *but turns aside in the cavity of the skull-bone*; and it *unites with one another, so that their canal becomes one; after which they separate again immediately after their junction and each nerve continues on its way to the eye which is opposite to its point of origin in the brain.*

(95.6–9) Concerning the fifth argument for the *junction* of the two optic nerves and their subsequent *separation*, it is the argument founded on the chief use of it (the vision), on the purpose that is noblest, highest in rank and most powerful in the function of vision, i.e. that man shall not see *one object* double.

The wisdom that is manifest in their *junction* after their *separated* points of origin is that support is brought from the two places without having *one object* appearing as two.

(100.10–20) But as for the *lucid* spirit which goes to the eyes it is not only its power but its actual substance which reaches them by way of two canals piercing the optic nerves, the quantity of it reaching them being (just) the *quantity necessary to fulfil the function of vision.*

Among all nerves, the two were created hollow and perceptible to the senses, because a *lucid* substance—originating in the brain—flows in them in a *quantity* which is *necessary to fulfil the function of vision.*

The argument that the substance of this *spirit* reaches the eyes itself… Another argument is that when one *shuts one eye, the pupil of the other one becomes enlarged,*… this enlargement is caused by the uvea (iris) when it is *distended by the filling up* of the place behind and inside it,….

The argument for the flow of the *spirit* in them is that if we *shut one eye, the pupil of the shut one becomes enlarged,* because it is *distended by its being filled up* by the matter of vision of the other eye.

(91.2–4) This is so because the interior has been created *as soft as possible* in order to render their sensibility finer; and because their external parts *incline* only slightly *to hardness, which* protects and *wards off damages*.

It is characteristic for the two nerves that their interior is *as soft as possible,* as well as that their external parts *incline to hardness, which wards off damages.*

(91.19–92.2) Thereupon the extremity of each of *the two nerves* broadens out inside *the eye* which it has reached and *forms a web resembling a net*, and therefore this extremity of the nerve in the eye is *called the net-like tunic* (retina), ...
(Translation Meyerhof, slightly adapted)

Then at their getting ultimately to *the eye*, it happens to *the two nerves*[40] that the extremity [of both of them] *form a web resembling a net*. The masters of the art *call* it *the net-like tunic*.

(The translation follows Meyerhof's translation of Ḥunayn as closely as possible to clearly show the textual parallels.)

With the mentioning of the net-like tunic or layer al-ʿĀmirī's description of the eye itself begins. Since he omits parts of Ḥunayn's detailed account, the structure of eye described by him turns out to be slightly different than the one described in his source:

The net-like tunic (*ṭabaqa shabakiyya*) encloses the glass-like humour (*ruṭūba zujājiyya*) which in its turn is in close contact with the ice-like humour (*ruṭūba jalīdiyya*),[41] i.e. the lens. This latter humour is protected by the grape-like tunic (*ṭabaqa ʿinabiyya*), which is in its turn protected by a cover or membrane (*ghishʾā*).

The main difference between Ḥunayn's and al-ʿĀmirī's presentation of the eye lies in the assumption of a different number of humours and tunics: Ḥunayn mentions three humours, i.e. the ice-like, glass-like and albuminoïd humour, and seven tunics: three tunics lie behind the glass-like humour, i.e. the net-like, the secondine-like and the sclerotic tunic; three tunics lie in front of the albuminoïd humour, i.e. the grape-like, the horn-like and the connecting tunic; and the cobweb-tunic, which lies between the ice-like and the albuminoïd humour.[42] Al-ʿĀmirī describes only two humours, namely the ice-like and the glass-like humour. It seems probable that he incorporates the albuminoïd humour into the

[40] The comparison of the two texts might hint at a possible corruption in al-ʿĀmirī's text: perhaps the original meaning of "يعرض" was also "broadening" in spite of Khalīfāt's vocalisation.

[41] Al-ʿĀmirī's description of the ice-like humour (*Rasāʾil*, ed. Khalīfāt, pp. 428–9) follows Ḥunayn's account (*Ten Treatises*, ed. Meyerhof, pp. 73–6), but it is less detailed and omits important information.

[42] *Ten Treatises*, ed. Meyerhof, pp. 74–5, 78–81; cf. also Ḥunayn's *Kitāb al-masāʾil fī ʾl-ʿayn*, ed. P. Sbath and M. Meyerhof in *Le Livre des questions sur l'œil de Ḥonaïn ibn Isḥāq* (Cairo, 1938), p. 18ff.

ice-like one. He also acknowledges only two tunics, one tunic in front of the ice-like humour, namely the grape-like tunic, and one behind the glass-like humour, namely the net-like tunic. He further describes a membrane in front of the grape-like tunic. This membrane seems to be identical with Ḥunayn's sclerotic tunic or hard membrane (*al-ghishāʾ al-ṣulb*). However, both Ḥunayn and al-ʿĀmirī explicitly state that the opinions on the numbers of the humours and the tunics vary.[43]

What is strikingly missing from al-ʿĀmirī's description of the eye is any mentioning of a hole in the grape-like tunic, through which a luminous spirit passes to meet the object of vision.[44] This omission is not surprising when we take into account that al-ʿĀmirī adheres to the intromission theory. Therefore the spirit does not have to leave the eye at all, as it does in Galen and Ḥunayn who only claim that it transforms the air in front of the eye. It is, however, interesting that al-ʿĀmirī does not do away with the entire concept of the spirit, but keeps it as providing the connection from the eye to the brain, as we will see next.

(d) The Instruments and Tools of Vision

Al-ʿĀmirī introduces the visual spirit (*al-rūḥ*)[45] as instrument of the brain. He explains the difference between an instrument (*adāh*) and a tool (*āla*) by stating that the instrument is directly joined to the agent, whereas the tool is separate. The whole process of vision only takes place, when the object of vision and the organ of sight are connected by a continuous chain of tools and instruments: colour, which is primarily visible, is transmitted by its tool, i.e. light (*ḍawʾ*), to the eye. The perceiving part of the eye, namely the ice-like humour, i.e. the lens, is

---

[43] *Rasāʾil*, ed. Khalīfāt, p. 429; Ḥunayn (*Ten Treatises*, ed. Meyerhof, p. 81) even reports that some people acknowledge only two tunics, as al-ʿĀmirī does. However, according to Ḥunayn these people omit the net-like tunic saying it is not a protecting tunic and describe the horn-like and the sclerotic tunic as one. In contrast to this description, al-ʿĀmirī does mention the net-like tunic and even mentions a membrane, which may be his equivalent to the horn-like and sclerotic tunic.

[44] Cf. *Ten Treatises*, ed. Meyerhof, p. 79.

[45] For mention of the spirit cf. *Rasāʾil*, ed. Khalīfāt, p. 428 (quoted above), p. 432f. We may assume that al-ʿĀmirī intentionally avoids the term '*al-rūḥ al-bāṣir*': as it is used by Ḥunayn who believes in the spirit's leaving the eye (*Ten Treatises*, ed. Meyerhof, e.g. p. 103), al-ʿĀmirī is perhaps concerned with the possibility of misleading his readers to interpret him as believing the same. For al-ʿĀmirī's discussion of instrument and tool cf. ibid. pp. 430 and 432.

connected to the brain by the visual spirit. The brain is the primary perceiving organ (*al-ḥāss al-awwalī*), the spirit is its instrument. This instrument flows in the hollow nerves which stretch from the brain to the ice-like humour. The latter is tool and instrument at the same time. We may take this to mean that the ice-like humour is a tool for colour as the object of vision and an instrument for the brain as the primary perceiving organ.

The transparent is missing from the described connection between the object of vision and the brain. This omission is understandable when we take into account that according to al-ʿĀmirī the transparent is merely a bearer or transmitter for the rays causing vision, i.e. the rays of the non-transparent bodies. Therefore these rays transmitted by the transparent are understood as a tool for the eye, not the transparent as their transmitter. Additionally the transparent in itself is only potentially transparent. It needs light to be actualised and perfected. Thus with regard to vision there seems to be a mutual dependence of light and the transparent. When discussing the transparent for a second time, al-ʿĀmirī implicitly acknowledges this:[46] the illuminated transparent is the absolute transparent that is suitable for the bearing (*ḥaml*) of colours. However, the transparent, when bearing colours, also bears light at the same time. Its presence thus becomes apparent because of its two layers (*bi-ṭabaqatayn*), namely light and colour. Therefore light and colour must be of one and the same category (*jins*), but the first is fine (*laṭīf*), shining (*mushriq*) and spreading by itself, whereas the other is thick (*kathīf*), turbid (*kadir*) and only spreading by the help of the other.

Another aspect al-ʿĀmirī addresses is the way in which the transparent bears light and colour: he claims that it is a way of transmission (*taʾdiya*) without further explaining this concept. His main focus seems to be the rejection of understanding this bearing as some kind of being affected or acted upon (*infiʿāl*).[47] Herein he is in agreement with, for example, Alexander of Aphrodisias.[48]

---

[46] *Rasāʾil*, ed. Khalīfāt, pp. 433; and 434:
ثم إن سطع الضوءُ عليهِ [المشف] ، صار مستشفّاً على الاطلاق ، فانه حينئذ يَصلحُ لحمل الألوان . ومهما حملها فقد صار في تلك الحالة متجليا بطبقتين : إحدهما الضوء ، والاخرى اللون . غير أن حمله لهما لن يكون على سبيل الانفعال بل يكون على سبيل التأدية . والدليل على أن الجوهر المشف يكون حاملا لهما معاً هو أن البيت المضيءَ متى كسيَ جدرانُه باللون الأسود ضَعُفَ ما فيه من الاضاءة والاشراق . وبهذا يُعْرَفُ أن اللون والضوء جنسهما جنسٌ واحدٌ ، وهو شعاعُ جسم غير مستشفّ .

[47] *Rasāʾil*, ed. Khalīfāt, p. 434.

[48] Gätje, *Studien*, p. 149ff. Cf. also S. Sambursky, 'Philoponus' Interpretation of Aristotle's Theory of Light,' *Osiris* 13 (1958), p. 116.

Having discussed the object of vision, the perceiving organ, the connection between these two and related questions, al-ʿĀmirī is finally dealing with the fundamental problem, namely which theory of vision is the right one: he only mentions two visual theories, i.e. the extramission theory and the intromission theory. As adherents to the first he presents Euclid, Ptolemy and their followers, and also Galen and his followers among the physicians. Al-ʿĀmirī does not make the effort to reject the extramission theory right away. He simply says that the philosophers like Aristotle, Alexander, Themistius and others have taken vision to work in the same way as the other senses, especially hearing. When describing their intromission theory, al-ʿĀmirī also reports their argument against the extramission theory, i.e. that visible things near and far are perceived at the same time. That is to say if we assume that a ray is emitted from the eye, travels to the object of vision and finally comes back to eye, then we have to admit that the ray will take a longer time for its travel the further the perceived object is away.

Concerning the later development of the intromission theory, al-ʿĀmirī knows of two different opinions concerning the question where the differentiation of colour takes place, namely either in the ice-like humour or where the nerves originate from the brain. He does not explicitly opt for either of these two views. However, given the fact that earlier al-ʿĀmirī has presented the ice-like humour as an instrument to the brain, we may assume that he would have preferred the second approach.

Yet, due to the scarcity of information this and other aspects of al-ʿĀmirī's theory of vision must remain doubtful.[49]

[49] I am very grateful to Rotraud Hansberger, Heidrun Eichner and Peter Adamson for reading and commenting on earlier versions of this paper.

CROSS-CULTURAL TRANSMISSIONS OF ARABIC PHILOSOPHY AND SCIENCE

IBN SAB'ĪN AND RAIMUNDUS LULLUS—THE QUESTION OF THE ARABIC SOURCES OF LULLUS' LOGIC REVISITED*

Anna Akasoy and Alexander Fidora

> Narratur, quod quidam Rex Tunicii, qui Miramamolinus vocabatur, peritus in Logica et in Naturalibus, disputavit cum quodam Catholico Religioso valde bene sciente historias et etiam ita bene Arabicum, sed non erat bene fundatus in Logicalibus et in Naturalibus.[1]

In his *Liber de convenientia, quam habent fides et intellectus in obiecto* (quoted above) the Catalan missionary, philosopher and mystic Raimundus Lullus makes no effort to disguise his admiration for Muslim learning in philosophy and logic which he considered more advanced than among his fellow Christians.

The question of the impact of Arabic or Islamic philosophy and culture on the opus of Raimundus Lullus has puzzled researchers for many years. A variety of possible connections have been suggested and discussed, at times quite controversially. These potential Islamic influences range from logic and philosophy over mysticism to literature. Apart from a general reflection of elements of Islamic culture in the œuvre of Lullus, several Muslim authors have been put forward as very specific sources of influence. The most obvious source might be al-Ghazālī. According to his own claim, Lullus abbreviated the part on logic of al-Ghazālī's *Maqāṣid al-falāsifa* in Arabic first, translated this abbreviation subsequently into Latin (the basis of his *Compendium Logicae Algazelis*) (1271–72) and composed in addition to this a Catalan

* Research for this article was partly conducted during a stay of Anna Akasoy at the Warburg Institute thanks to a Frances Yates fellowship. It has received support from the research project 'The Perception of Islam in Medieval Europe' (Project HUM2004-03957-C02-02/FILO of the DGI-MEC and 2005SGR00538 of the AGAUR-Generalitat de Catalunya) at the Universitat Autònoma de Barcelona. We would like to thank Charles Burnett for his helpful suggestions.

[1] Ed. F.P. Wolff (MOG, IV; Mainz, 1729; repr. Frankfurt, 1965), Int. xi, p. 4 (p. 574).

verse version. Emilio de Santiago Simón pointed out the Andalusian mystic and statesman Ibn al-Khaṭīb as a possible source for Lullus' tree motif,[2] and John Dagenais[3] and, in this volume, Josep Puig have called attention to the parallels between the *Rasāʾil* of the Ikhwān al-ṣafāʾ and Lullus' *Fèlix o Llibre de les meravelles*. Hans Daiber addressed the question of Lullus' relationship to Islam in several publications.[4] Many of these discussions have been conveniently described and summarised by Sebastián Trías Mercant.[5]

In this article we set out to examine one of these suggestions, presented for the first time by Charles Lohr in 1976 at the second congress of Raimundus Lullus studies in Mallorca.[6] Lohr pointed out a number of parallels between Lullus' logic and metaphysics on the one hand and the *Budd al-ʿārif*, a philosophico-mystical compendium written by Ibn Sabʿīn of Murcia (*c.* 1217–70), on the other. Since then this suggestion has been discussed several times by Lohr himself, but also by other scholars, most notably Dominique Urvoy, who supported the suggestion

---

[2] E. de Santiago Simón, 'Raimundo Lulio e Ibn Al-Jatib. Notas para un curioso paralelismo,' *Miscelánea de estudios árabes y hebraicos* 29–30 (1980), pp. 189–93. See also M. Cruz Hernández, 'El símbolo del árbol en Ramón Llull e Ibn al-Jatīb,' in *Studia Lullistica. Miscellanea in honorem Sebastiani Garcias Palou* (Palma de Mallorca, 1989), pp. 19–25, p. 25, and D. Urvoy, 'Le symbole de l'arbre chez les auteurs arabes antérieurs à Lull,' in F. Domínguez and J. de Salas (eds.), *Constantes y fragmentos del pensamiento luliano. Actas del simposio sobre Ramon Llull en Trujillo, 17–20 septiembre 1994* (Tübingen, 1996), pp. 91–77.

[3] 'New Considerations on the Date and Composition of Llull's *Llibre de bèsties*,' in M. Duran, A. Porqueras Mayo and J. Roca Pons (eds.), *Actes del segon col·loqui d'estudis catalans a Nord-Amèrica, Yale 1979* (Barcelona, 1982), pp. 131–9, here 134–5.

[4] 'Der Missionar Raimundus Lullus und seine Kritik am Islam,' *Estudios Lulianos* 25 (1981–83), pp. 47–57; 'Raimundus Lullus in der Auseinandersetzung mit dem Islam. Eine philosophiegeschichtliche Analyse des *Liber disputationis Raimundi Christiani et Homeri Saraceni*,' in M. Lutz-Bachmann and A. Fidora (eds.), *Juden, Christen und Muslime* (Darmstadt, 2004), pp. 136–72; and 'Ramundus Lullus' Dispute with Homer Saracenus in the Year 1307. An Inquiry Into Their Theological Positions,' in M.I. Ripoll Perelló (ed.), *Actes de les Jornades Internacionals Lul·lianes. Ramon Llull al s. XXI. Palma, 1, 2 i 3 d'abril de 2004* (Palma, 2005), pp. 259–64.

[5] 'Arabismo e islamología en la obra de Ramon Llull,' *Semitica Escurialensia Augustiniana. Homenaje a Fray Luciano Rubio OSA. La Ciudad de Dios* 208 (1995), pp. 439–52.

[6] Lohr gave a paper with the title 'Ibn Sabʿīn de Murcia y el desarrollo de la *Ars luliana*'. Versions of this paper were published in the following articles: 'Christianus arabicus, cuius nomen Raimundus Lullus,' *Freiburger Zeitschrift für Philosophie und Theologie* 31 (1984), pp. 57–88; 'Arabische Einflüsse in der neuen Logik Lulls,' in *Raymond Lulle, Christianisme, Judaïsme, Islam. Les Actes du Colloque sur R. Lulle, Université de Fribourg, 1984* (Collection « Interdisciplinaire », 12; Fribourg, 1986), pp. 71–91; 'Ramon Llull: « christianus arabicus »,' *Randa* 19 (1986), pp. 7–34; 'Islamic Influences in Lull's Logic,' *Estudi general* 9 (Barcelona, 1989), pp. 147–57; 'Influência árabe na *Nova lógica* de Raimundo Lúlio,' *Leopoldianum* 18 (1991), pp. 5–18.

with additional arguments,[7] and Miguel Cruz Hernández, who rejected it harshly.[8] We will begin our exposition with a short presentation of Ibn Sabʿīn and his context. We will then summarise and review Lohr's suggestion and the additional arguments presented by Urvoy. We will offer a detailed comparison of the relevant passages in Lullus' logical works and the parallels in Ibn Sabʿīn's *Budd al-ʿārif* and discuss how far these parallels might be due to common sources, namely al-Ghazālī's *Maqāṣid al-falāsifa* and the *Rasāʾil* of the Ikhwān al-ṣafāʾ. On a different occasion we intend to address a hypothesis which Lohr presented in his doctoral thesis and which seems to have been overshadowed in later years by the comparisons with Ibn Sabʿīn. According to this hypothesis, the structure of Lullus' *Logica Algazelis* is an imitation of contemporary Arabic texts on logic in general.[9]

1. Ibn Sabīʿn and his Philosophical Works

ʿAbd al-Ḥaqq ibn Sabʿīn is probably one of the most peculiar figures in the history of Arabic or Islamic philosophy. Born in *c.* 1217 in a fortress in the vicinity of Murcia to an almost unknown family he left his home in 1238 or thereabouts and followed the route of many Andalusian émigrés in that time: first to Ceuta, then to Bougie and Tunis, and finally via Egypt to Mecca, where he died in 1270. Ibn Sabʿīn's biography and works as well as his reputation reveal an unusual combination of philosophical and mystical traditions. Biographical sources and catalogues of libraries all over the world record titles of altogether nearly sixty texts by Ibn Sabʿīn.[10] Many of them are not preserved, and the majority of those which are extant have not yet received a full study. The two best explored texts are at the same time the two earliest preserved works of

[7] *Penser l'Islam. Les présupposés islamiques de l'« Art » de Lull* (Paris, 1980), pp. 144–7. With Marie-Thérèse Urvoy, 'Un « penseur de frontière » en Islam: Ibn Sabʿīn,' *Bulletin de littérature ecclésiastique* 98 (1997), pp. 31–55; 'Les thèmes chrétiens chez Ibn Sab'în et la question de la spécificité de sa pensée,' *SI* 44 (1976), pp. 99–119; 'La place de Ramon Llull dans la pensée arabe,' *Catalan Review* 4 (1990), pp. 201–20.

[8] M. Cruz Hernández, 'El símbolo del árbol en Ramón Llull e Ibn al-Jatīb,' p. 25.

[9] C. Lohr, *Raimundus Lullus'* Compendium Logicae Algazelis. *Quellen, Lehre und Stellung in der Geschichte der Logik*, doctoral thesis (Freiburg, 1967), p. 33.

[10] For a full list of Ibn Sabʿīn's writings cf. the article on Ibn Sabʿīn by Anna Akasoy in J. Lirola Delgado and J.M. Puerto (eds.), *Biblioteca de al-Andalus*, v (Almería, 2007), pp. 29–38.

Ibn Sabʿīn, the *Sicilian Questions*[11] and the *Budd al-ʿārif*. Both were written in Ceuta, apparently before Ibn Sabʿīn developed the more radical traits of his mysticism that would bring him into trouble with more orthodox Islamic scholars. Even a more philosophical work such as the *Budd al-ʿārif* turned out to be received rather controversially. Al-Bādisī, for example, claims that Ibn Sabʿīn was expelled from Ceuta after the publication of that text,[12] but modern scholarship has not yet been able to reconstruct the exact circumstances of these events. It seems that Ibn Sabʿīn attracted a certain number, albeit rather small, of disciples with his mystical ideas. His philosophical treatises, however, as far as we can tell, did not have any significant influence.

Even though many questions regarding Ibn Sabʿīn's biography, works and impact on later generations remain, he was certainly a remarkable character and indicative of the intellectual landscape of the Western Mediterranean in the thirteenth century. His philosophical works in particular offer various interesting insights, for instance into the question about which authors were considered the most relevant authorities for which aspects of Aristotelian philosophy.

2. Ibn Sabʿīn as a source of Raimundus Lullus: Lohr's suggestion

In various publications Charles Lohr offered a number of arguments to support his claim that Raimundus Lullus used Ibn Sabʿīn's *Budd al-ʿārif* as a source of inspiration, in particular for his logic. These arguments, as mentioned above, were further developed by Dominique Urvoy who presented some additional examples. In what follows we will deal with seven of these arguments.

2.1. *The two intentions*

Lullus developed a very distinct concept of the 'two intentions', *prima* and *secunda intentio*, which differs substantially from the Scholastic usage. According to the 'traditional' definition, a *prima intentio* is a concept

[11] For an edition and German translation of the *Sicilian Questions* cf. A. Akasoy, *Philosophie und Mystik in der späten Almohadenzeit. Die* Sizilianischen Fragen *des Ibn Sabʿīn* (Leiden, 2006) and the shorter version Ibn Sabʿīn, *Die Sizilianischen Fragen*. Arabisch-Deutsch (Herders Bibliothek der Philosophie des Mittelalters, 2; Freiburg, 2005).

[12] Al-Bādisī, *al-Maqṣad al-sharīf fī dhikr ṣulaḥāʾ al-Rīf*, ed. S.A. Iʿrāb (Rabat, 1402/1982), on Ibn Sabʿīn pp. 32, 34–5 and 69.

of a thing, whereas a *secunda intentio* is a concept of a concept. Lullus' distinction between the first and second *intentio* goes beyond this idea,[13] and needs to be understood within the broader context of his thought which has been characterised by Mark Johnston as having a 'moralizing' character.[14]

The distinction between first and second *intentio* is present in Lullus' writings from early in his career and underlies his entire work. Already from the years 1265–74 onwards, in the *Compendium Logicae Algazelis* as well as the famous *Llibre de contemplació en Déu*, Lullus develops this idea, explaining that the first *intentio* should be understood as the final cause of man, i.e. God, whereas the second *intentio* has to be conceived as a means to this end. When man confuses the two, he uses his freedom against the natural order and becomes a sinner.

This moral dimension has a metaphysical foundation, which Lullus presents in his most important account of first and second *intentio*: the *Liber de prima et secunda intentione* (1276–83), dedicated to his son Domènec. Here Lullus defines *intentio* 'as the act of the natural appetite which requires the perfection, to which he is entitled by nature'.[15] *Intentio* means the striving for perfection not only of man, but of the entire created realm, including 'elements, plants, birds, beasts and so on', which follow 'the order and the rule of the *intentio* for which they have been created'. 'Only man, however, who is at the top of creation, can sin and thus act against the *intentio* he has been created for.'[16]

---

[13] Even though Lullus did not follow the traditional interpretation of *prima* and *secunda intentio*, he was obviously familiar with the scholastic acceptation of the terms as can be seen, among others, from the following quotation from the *Ars generalis ultima*, ed. A. Madre (ROL, XIV; Tunhout, 1986), p. 367: 'The logician deals with secondary intentions (*de secundariis intentionibus*), which are attached to the first.'

[14] 'Llull's procedures or argumentation are 'moralizing'. This term appears throughout this study and is fundamental to understanding its explanations of Llull's methods and doctrines. It embraces two related senses. First, it designates the 'ethico-ontological' duty of every being to acknowledge God, according to Llull's doctrine of 'intention' (...). In so far as this acknowledgement indistinctly embraces both faith in, and understanding of, God, the relationship between these two modes of knowledge becomes one of the great problems in Llull's development of his Art. Second, this 'moralization' designates Llull's effort to explain the status, function, or understanding of any real or rational being according to that duty, and these explanations thus constitute a kind of tropological exegesis of creation.' (M.D. Johnston, *The Spiritual Logic of Ramon Llull* [Oxford, 1987], p. 5.)

[15] Raimundus Lullus, *Liber de prima et secunda intentione*, ed. F.P. Wolff (MOG, VI; Mainz, 1737, repr. Frankfurt, 1965), Int. ix, p. 2 (p. 538).

[16] Ibid., 541 (5).

Charles Lohr dealt in various instances with analogies between Lullus' *prima* and *secunda intentio* and Arabic sources. In his article 'Christianus arabicus, cuius nomen Raimundus Lullus' (published in 1984) he pointed out that the distinction between two intentions was well known in Arabic literature, as Kwame Gyekye had already shown in 1971.[17] In Arabic translations of Aristotle and philosophical works discussing these texts the expressions *ʿalā 'l-qaṣd al-awwal* and *ʿalā 'l-qaṣd al-thānī* render Aristotle's πρώτως/πρῶτον and δεύτερον.[18] In this sense they simply mean 'primarily, in the first sense' and 'in the second place'. This is indeed the meaning these expressions have in many Arabic texts, philosophical or otherwise. There is, however, also another usage, which, according to Lohr, seems to match with Lullus: in the context of creation, as Lohr emphasised, God's first intention can be nothing outside himself, whereas the second intention is directed at the world He creates. Various Arabic texts bear witness to such a cosmological notion of the terms *al-qaṣd al-awwal* and *al-qaṣd al-thānī*,[19] among them the *Rasāʾil* of the Ikhwān al-ṣafāʾ (discussed in further details below).

In an article published two years later Lohr added another Arabic testimony to this list of metaphysical or cosmological usages: Ibn Sabʿīn's *Budd al-ʿārif*.[20] In this text, *al-qaṣd al-awwal* and *al-qaṣd al-thānī* occur in a sense very similar to their appearance in the *Rasāʾil* of the Ikhwān al-ṣafāʾ. Ibn Sabʿīn describes the relationship between the world and its Creator in the following terms:

> With regard to His emanating, He is the first creator *bi'l-qaṣd al-awwal*, and with regard to that which is emanated, He is the first creator *bi'l-qaṣd al-thānī*. It is true that created things have neither creation nor emanation nor being nor truth nor an active being nor a knowing faculty except *bi'l-qaṣd al-awwal*.[21]

Lohr assumed in these two articles that Ibn Sabʿīn and Lullus had been inspired by the same source, i.e. the Ikhwān al-ṣafāʾ ('It was undoubtedly through the encyclopaedia that the distinction of two intentions

[17] 'Christianus arabicus,' pp. 67–9; K. Gyekye, 'The Terms « prima intentio » and « secunda intentio » in Arabic Logic,' *Speculum* 46 (1971), pp. 32–8.

[18] S. Pines, 'Un texte inconnu d'Aristote en version arabe,' *AHDLMA* 23 (1956), pp. 5–43, p. 18f.

[19] Lohr mentions: M. Asín Palacios, *La espiritualidad de Algazel y su sentido cristiano*, iv (Madrid, 1941), p. 127; A. Neuwirth, *ʿAbd al-Laṭīf al-Baġdādī's Bearbeitung von Buch Lambda der Aristotelischen Metaphysik* (Wiesbaden, 1976), pp. 16, 86, 186–90 (Exkurs).

[20] 'Arabische Einflüsse in der neuen Logik Lulls,' p. 78.

[21] Ibn Sabʿīn, *Budd al-ʿārif*, ed. J. Kattūra (Beirut, 1978), p. 28.

came to Lull…'),[22] whereas Dominique Urvoy in his study on Islamic influences on Lullus, *Penser l'Islam*, stressed the role of Ibn Sabʿīn and considered the parallel an argument for Lullus' familiarity with the *Budd al-ʿārif*.[23]

Both authors, Lohr and Urvoy, relied in their discussion of the *Budd al-ʿārif* on the doctoral thesis and an article by Esteban Lator, which include summaries of parts of Ibn Sabʿīn's compendium.[24] Amongst other things, Lator argued that the use of *al-qaṣd al-awwal* and *al-qaṣd al-thānī* in several of Ibn Sabʿīn's writings supported the authenticity of his authorship of the *Sicilian Questions*.[25] This argument might have misled later authors who were concerned with similarities in Lullus' texts. In Urvoy's rendering of the argument, it appears as if it were not only the expression, but the *concept* of the two intentions which appeared prominently in the *Budd al-ʿārif*. This is not the case. As far as the *Sicilian Questions* are concerned, Lator clearly overstated the frequency of the terms there. Similarly in the *Budd al-ʿārif*, the passage quoted above is not followed by any further elaboration of the concept of the two intentions.

Unlike Ibn Sabʿīn, the Ikhwān al-ṣafāʾ discuss the difference between the two terms explicitly and, similar to Lullus, they expound on it in ethico-metaphysical terms:

> There is no doubt that the good things which are related to the good fortune of the heavenly bodies happen through the providence and intention (*qaṣd*) of God the Sublime. As for the bad things which are connected with ill fortunes, they don't happen because of (His) intention. An example for this is that the sun rises and shines over some places and heats the water for some time. Then the sun disappears again and these places cool down. This happens through the providence of God the Sublime and is necessarily decreed by His wisdom, since it is good and useful for everybody. (…) It is not according to the first intention, however, when at some times and in some places animals or plants suffer from excessive heat or when they are damaged by low temperatures.[26]

---

[22] 'Christianus arabicus,' p. 69.

[23] *Penser l'Islam*, p. 147.

[24] S. Lator, *Die Logik des Ibn Sabʿīn von Murcia*, doctoral thesis (University of Munich, 1942) and 'Ibn Sabʿīn de Murcia y su «Budd al-ʿārif»,' *al-Andalus* 9 (1944), pp. 371–417.

[25] 'Ibn Sabʿīn de Murcia y su «Budd al-ʿārif»,' pp. 382–9, n. 4.

[26] *Rasāʾil Ikhwān al-ṣafāʾ*, 4 vols. (Beirut, 1377/1957), iii, pp. 474–5.

Thus, if we were to trace Lullus' doctrine of intention back to Arabic sources, it should rather be seen in the tradition of the Ikhwān al-ṣafāʾ. However, occasional and general similarities in these concepts of first and second intention notwithstanding, Lullus' moralising concept is distinct from the Arabic sources we have referred to. An important peculiarity of his concept is that the theory of first and second *intentio* does not apply to God, in whom there is no difference between intentions: 'In God's *intentio* there is no division and no second *intentio*, in which there would be a more (and less); for his entire *intentio* is infinite and eternal. And therefore, my Son, there is no place for second *intentio* in God's *intentio*.'[27]

In his study of Lullus' doctrine of intentions, Walter W. Artus pointed to the Augustinian background of Lullus' moralising use of the terms. This does of course not exclude the possible Arabic influence on this aspect of Lullus' thought. It rather suggests that his doctrine of intentions should be interpreted as a complex result of different influences and original insights at the same time.[28]

2.2. *Definitions*

Another possible parallel between Lullus and Ibn Sabʿīn which Lohr called attention to is a seemingly peculiar concern with definitions which can be identified in both authors. At the end of the part on logic in the *Budd al-ʿārif* Ibn Sabʿīn presents a discussion of technical terms used by jurists, theologians, philosophers and mystics. According to Lator this section was partly inspired by the Ikhwān al-ṣafāʾ.[29] Lohr recognised here another parallel with Lullus' *Logica nova* (1303), in particular the philosophical dictionary (fourth *distinctio*) and the application of the methodology of the *Logica nova* to the concept of nature and the four faculties of theology, philosophy, law, and medicine (sixth *distinctio*).[30]

With this argument Lohr pointed out one of the most interesting and complex possible parallels between Ibn Sabʿīn and Raimundus Lullus.

---

[27] Raimundus Lullus, *Liber de prima et secunda intentione*, Int. ix, p. 3 (p. 539).

[28] W.W. Artus, 'Ramon Llull on First and Second Intentions: a Basic Ethical Doctrine,' in B.C. Bazán, E. Andújar and L.G. Sbrocchi (eds.), *Les philosophies morales et politiques au Moyen Âge. Moral and Political Philosophies in the Middle Ages* (Actes du IX Congrès international de philosophie médiévale. Ottawa, du 17 au 22 août 1992, S.I.E.P.M.), 3 vols. (New York, 1995), ii, pp. 978–90.

[29] Lator, *Die Logik des Ibn Sabʿīn von Murcia*, pp. xvi–xvii.

[30] 'Arabische Einflüsse in der neuen Logik Lulls,' p. 78.

Even though significant parts of Ibn Sabʿīn's philosophical œuvre are clearly based on the works of other authors,[31] not least the Ikhwān al-ṣafā', there are some distinctive instances of originality, among them Ibn Sabʿīn's methodologically careful usage of definitions as well as his concept of definitions.[32] In his *Sicilian Questions*, for example, he uses the different definitions of philosophical and religious key terms such as *tawḥīd* as a tool for analysing and occasionally harmonising seemingly contradictory approaches by different authors to the same problem. For Ibn Sabʿīn the definitions used in a certain tradition are on a par with the overall status of this tradition among the methods of gaining knowledge. These specific ideas distinguish Ibn Sabʿīn from other medieval Muslim writers who shared his general interest in problems surrounding definitions.

One of the most captivating features of Ibn Sabʿīn is the combination of philosophy and mysticism. Ibn Sabʿīn does not try to harmonise these two systems which are indeed difficult to reconcile. In his various writings he rather makes a clear difference between terms and concepts as they are understood by the philosophers and the mystics respectively, and occasionally he adds explanations concerning other intellectual traditions such as *fiqh* and *kalām*. The technique of using definitions as a tool for explaining the various dimensions, even the very existence, of a problem as well as the diverging approaches of different scholarly traditions is evident in several of Ibn Sabʿīn's writings. The discussion of technical terms in the *Budd al-ʿārif* is another reflection of this concern.

However, as mentioned above, being attentive to definitions in general is not a unique feature of Ibn Sabʿīn, but rather a common attitude among Arabic writers attested to by a number of discussions in a variety of fields. Even though other authors did not approach the problem from exactly the same angle as Ibn Sabʿīn, they shared his concern for a careful use of language. Also in the more specific context of the Western Mediterranean, treatises dealing with the ambiguity of language or even consisting exclusively of definitions were not unfamiliar. Ibn al-Sīd al-Baṭalyawsī, for instance, explained in his *al-Inṣāf fī 'l-tanbīh ʿalā 'l-maʿānī wa'l-asbāb allatī awjabat al-ikhtilāf bayna al-muslimīn fī ārā'ihim* the differences of opinion among the Muslims by their different usages of certain terms. Isaac Israeli compiled a comprehensive

[31] Akasoy, *Philosophie und Mystik*, pp. 177–331.
[32] Ibid., pp. 129–32.

list of definitions of philosophical key concepts which was translated twice into Latin in the twelfth century, by Gerard of Cremona and, most probably, by Dominicus Gundissalinus.[33] Both versions of the so-called *Liber de definitionibus* were extremely popular in Latin medieval philosophy.[34]

The question which concerns us here is whether the specific character of Ibn Sabʿīn's discussion of technical terms at the end of the logical part in the *Budd al-ʿārif* corresponds with Lullus' approach and confirms that he depended on the *Budd al-ʿārif* rather than on these better known texts. A closer look at the *Budd al-ʿārif* reveals that Lohr and other scholars might have been again misled by Lator's presentation of the subject. Lator offered in his doctoral thesis together with a study of Ibn Sabʿīn's logic an extensive German summary of a part of the *Budd al-ʿārif*. This summary is at times quite close to the original Arabic text, but it should not be mistaken for a translation. In Lator's rendering of Ibn Sabʿīn's discussion of the terminology of jurists, Ashʿarites, philosophers and Sufis it does indeed seem as if this was a lexicon. Yet, the complete Arabic text offers a detailed exposition on the methodological and epistemological traditions of these scholars and is hardly a lexicon in the sense of a list of words with brief definitions—the discussion offers thus valuable insights into Ibn Sabʿīn's approach to the issue of definitions and the way he locates himself within the different traditions of Islamic scholarship, but it has little in common with Lullus' approach. The fourth *distinctio* of the *Logica nova* is thus, again, much closer in its structure to other Arabic texts which, again, were far more popular than Ibn Sabʿīn's *Budd al-ʿārif*, such as Ibn Sīnā's *Kitāb al-ḥudūd*, from which Ibn Sabʿīn copies the definitions of 'eternity' or 'creation' in the *Sicilian Questions*.[35]

---

[33] Edited by J.T. Muckle, 'Isaac Israeli, *Liber de definitionibus*,' *AHDLMA* 12–13 (1937–38), pp. 299–340.

[34] Isaac's work is first quoted by Dominicus Gundissalinus in his famous treatise *De divisione philosophiae*, ed. L. Baur (BGPhMA, IV, 2–3; Münster, 1903), pp. 5–9. From there it made its way to the University of Paris: The *Liber de definitionibus* appears very frequently in the so-called introductions to philosophy of the Masters of the Arts faculty in Paris. See C. Lafleur, *Quatre Introductions à la philosophie au XIIIe siècle. Textes critiques et étude historique* (Université de Montréal. Publications de l'Institut d'Études Médiévales, XXIII; Montreal/Paris, 1988), *passim*.

[35] Akasoy, *Philosophie und Mystik*, pp. 301–8.

2.3. *The correlatives*

From the 1290s onwards, Lullus paid much attention to the development of his theory of correlatives. According to this theory, every entity is constituted by a threefold structure, i.e. an agent, a patient and an act.

Divine goodness (*bonitas*), for instance, implies a *bonificativum*, a *bonificabile* and a *bonificare*, and so do the remaining divine attributes.[36] The same holds for the created realm, in so far as creation is an image of the divine. Thus, the 'essential properties' of man—as Lullus sometimes calls the correlatives—are *homificativum, homificabile* and *homificare*,[37] and those of his understanding: *intellectivum, intelligibile* and *intelligere*.[38] These correlatives are central to Lullus' *ars* and constitute what has been described as his 'Trinitarian World Picture'.[39] They are at the very heart of his metaphysics and logic, and play a decisive role in his apologetics insofar as they serve as a proof for the Trinity.[40]

The correlatives occupy a very prominent position in debates around possible Islamic influences on Lullus. Hans Daiber, for example, suggested in his analysis of the *Liber disputationis Raimundi Christiani et Homeri Saraceni* that it was Lullus' intimate knowledge of the Arabic language and its grammar which inspired his theory of the correlatives.[41]

Others were eager to discover specific Arabic sources for Lullus' correlatives. Even though Charles Lohr acknowledged that the idea of an identity or union between *ʿaql, ʿāqil* and *maʿqūl* was not unique to Ibn Sabʿīn in Arabic literature, he clearly favoured, again, the *Budd al-ʿārif* over earlier Arabic texts as a source for Raimundus Lullus:

> Christian controversialists writing in Arabic took up this doctrine as an analogy for the Trinity. For example Yaḥyā ibn ʿAdī († 974), an Oriental Lull, who tried to provide natural arguments for the Trinity, maintained that the intellect (*ʿaql*), intelligens (*āʿqil*), and intellectum (*maʿqūl*) are one. Lull could have found the Sufi version of this doctrine as the definition of

[36] Cf., among others, Raimundus Lullus, *Liber correlativorum innatorum*, ed. H. Riedlinger (ROL, VI; Turnhout, 1978), p. 132.

[37] For instance, Raimundus Lullus, *Ars generalis ultima*, p. 23.

[38] Ibid., pp. 28–9.

[39] R. Pring-Mill, 'The Trinitarian World Picture of Ramon Lull,' *Romanistisches Jahrbuch* 7 (1955–56), pp. 229–56.

[40] For a thourough study of this doctrine see J. Gayà, *La teoría luliana de los correlativos. Historia de su formación conceptual* (Palma de Mallorca, 1979).

[41] 'Raimundus Lullus in der Auseinandersetzung mit dem Islam,' pp. 161–2.

knowledge (*ʿilm, āʿlim, maʿlūm*) among the definitions of the theologians in the *Budd al-āʿrif* of Ibn Sabʿīn.[42]

Yet, a closer look at the passage in the *Budd al-ʿārif* that Lohr refers to, reveals that Ibn Sabʿīn does not discuss any Sufi version of this doctrine here, but rather the definition of knowledge as presented by the theologians (Ashʿarites):

> The definition of knowledge (*ʿilm*) is that something becomes clear as it is. He who knows (*al-ʿālim*) is he who possesses knowledge (*man lahu ʿilm*), and the meaning of what is known (*al-maʿlūm*) is that which he who knows (*al-ʿālim*) knows by way of (speculative) consideration (*naẓar*).[43]

Even though this chapter of the *Budd al-ʿārif* has the title 'The definition of knowledge according to the jurists, the Ashʿarites, the philosophers and the Sufis' (*Ḥadd al-ʿilm bi-naẓar al-fuqahāʾ wa'l-Ashāʿira wa'l-falāsifa wa'l-mutaṣawwifa*), Ibn Sabʿīn deals with the jurists, the theologians and the philosophers, but does not dedicate a separate passage to the Sufis. Lohr's suggestion does not seem to have convinced Dominique Urvoy either. When the latter dealt with this parallel he pointed out the omnipresence of this concept among writers with Neoplatonic tendencies,[44] among them al-Fārābī,[45] who was highly influential in the Muslim West, and also the *Theology* of pseudo-Aristotle,[46] the *Kitāb al-ḥadāʾiq* of the Andalusian author Ibn al-Sīd al-Baṭalyawsī[47] and *Fī ittiṣāl al-ʿaql bi'l-insān* of his compatriot Ibn Bājja.[48]

There are clearly parallels between Raimundus Lullus' correlatives and these Arabic sources. All of them render the notion of diversity within a metaphysical entity. But this might be as far as the comparability goes. Lullus' emphasis on the Trinity must have been alien to Muslim writers, some of whom, like Ibn Sabʿīn, used the terms *ʿilm*, *ʿālim* and *maʿlūm* as a simple grammatical argument. Rather, the close

---

[42] 'Arabische Einflüsse in der neuen Logik Lulls,' pp. 86–7; 'Christianus arabicus,' p. 85.

[43] *Budd al-ʿārif*, ed. Kattūra, p. 105.

[44] 'La place de Ramon Llull,' p. 211ff; 'Les musulmans pouvaient-ils comprendre l'argumentation lullienne?,' *El debat intercultural als segles XIII i XIV. Actes de les I Jornades de filosofia catalana, Girona, 25–27 d'abril de 1988* (= Estudi general 9; Girona, 1989), pp. 159–70, pp. 164–5.

[45] Al-Fārābī, *Philosophische Abhandlungen aus Londoner, Leidener und Berliner Handschriften*, ed. F. Dieterici (Leiden, 1890), p. 43.

[46] Lib. VIII, c. IV.

[47] Ed. M. Asín Palacios in *al-Andalus* 5 (1940), pp. 91 and 93.

[48] Ed. M. Asín Palacios in *al-Andalus* 7 (1942), p. 3.

relation between the theory of the correlatives and Trinitarian motives in Lullus' thought may suggest that he is drawing not on Arabic material alone, but that his theory of correlatives is also based on St. Augustine. In fact, in *De Trinitate* we find strikingly similar formulations, e.g.: 'Love is the love of something, and love loves something. Therefore, there are three: the loving, what is loved and love.' (viii, x, 14).

2.4. *The nine subjects*

Lullus included already in his *Ars inventiva* (1290) the idea of nine 'subjects': *divina, angelica, caelestis, rationativa, imaginativa, sensitiva, vegetativa, elementativa*, and *instrumentativa*. The same idea appears in the first *distinctio* of the *Logica nova*.[49] Charles Lohr interpreted this distinction as another testimony to the influence of Ibn Sab'īn, whose *Budd al-'ārif* contains a concept of emanation in nine steps: God, universal intellect, soul, nature, matter, absolute body, sphere of heavenly bodies, elements (*arkān*), created beings.[50] Lohr conceded that the nine-fold emanation was a common idea in Neoplatonic literature, not least in the *Rasā'il* of the Ikhwān al-ṣafā' and Ibn Sīnā, yet he stated:

> Lull's relation to Ibn Sab'īn appears not so much in the items included in the list, as in the way in which the list is conceived. All of these lists are presented as a means of ascending from the particular things of this world to the nine universal things or of descending from these universals to the particulars.[51]

This is of course more or less the whole point of Neoplatonism, which is also acknowledged by Lohr who quotes a variety of texts that attest to the general prominence of this idea. There is indeed nothing specific about Ibn Sab'īn's list of emanations. In fact, the same distinction appears verbatim in the *Rasā'il* of the Ikhwān al-ṣafā'.[52] Given this prominence of the nine-fold emanation in the *Rasā'il* of the Ikhwān al-ṣafā', one can hardly claim that it is unique to Lullus and Ibn Sab'īn only.

After putting forward this fairly general parallel, however, Lohr proceeds with his argument and regards the distinction in more detail. He

[49] Lohr, 'Christianus arabicus,' pp. 69–70.

[50] Ibid., pp. 74–5. For the passage in the *Budd al-'ārif* cf. ed. Kattūra, p. 122.

[51] Lohr, 'Christianus arabicus,' p. 75.

[52] *Rasā'il Ikhwān al-ṣafā'*, 4 vols. (Beirut, 1377/1957), iii, p. 202. Cf. as well S. Diwald, *Arabische Philosophie und Wissenschaft in der Enzyklopädie. Kitāb Iḫwān aṣ-Ṣafā' (III), Die Lehre von Seele und Intellekt* (Wiesbaden, 1975), p. 107, and I.R. Netton, *Muslim Neoplatonists. An Introduction to the Thought of the Brethren of Purity, Ikhwān al-ṣafā'* (London, 1982), p. 35.

argues that Lullus' 'direct dependence on the logic of Ibn Sabʿīn's *Budd al-ʿārif* may be seen in the derivation he provides for the subjects in his *Logica nova*. Substance, he tells us, is either incorporeal or corporeal'.[53] Lohr thus seems to regard Lullus' division of substance into incorporeal and corporeal a clear hint to Ibn Sabʿīn. Yet, the latter is in this part clearly influenced by the Ikhwān al-ṣafāʾ[54] who employ the same distinction,[55] and further than that the parallel with Lullus cannot be drawn as Lohr admits.[56]

2.5. *Philosophical and mystical ontology*

In addition to the arguments originally presented by Charles Lohr, Dominique Urvoy suggested another possible influence of Ibn Sabʿīn on Raimundus Lullus: the term *wujūd muṭlaq*.[57] Urvoy translated this expression as 'existence absolue = Être Nécessaire' and claimed that it was 'caractéristique de l'enseignement d'Ibn Sab'în'.[58] According to Urvoy, Lullus adopted the term, which appears as *ens necessarium*, or 'as the Arabs say' *hujuden muclach*, in his *Liber disputationis Raimundi Christiani et Homeri Saraceni*, from Ibn Sabʿīn's heretical movement, possibly from his disciple al-Shushtarī (1212–69).

This argument contains a variety of problems. First of all, Urvoy does not explain in which sense *wujūd muṭlaq* should be 'caractéristique de l'enseignement d'Ibn Sabʿîn'. In fact, he quotes not a single text by Ibn Sabʿīn or one of his followers to substantiate that it could have been used in a similar meaning by Lullus. The term *wujūd muṭlaq* was used by Ibn Sabʿīn, but it appears also in a number of other mystical and philosophical writings. Its sense is more than ambiguous. The word *wujūd* might smell more than anything else of Ibn Sīnā's ontology, and Urvoy seems to identify *wujūd muṭlaq* quite rightly with Ibn Sīnā's *wājib al-wujūd*. As Hans Daiber has argued in detail, Lullus could indeed very

[53] Lohr, 'Christianus arabicus,' p. 75.
[54] Akasoy, *Philosophie und Mystik*, pp. 315–18.
[55] Netton, *Muslim Neoplatonists*, p. 46.
[56] In a later article, Lohr does not mention this. Cf. his 'Islamic Influences in Lull's Logic,' p. 150. Here, the differences between Lullus and Ibn Sabʿīn seem hardly worthy of notice. Dominique Urvoy also adopted Lohr's argument without hesitation and took the parallel as proof for Lullus' dependence on the *Budd al-ʿārif*, cf. his 'L'idee de « christianus arabicus »,' *al-Qanṭara* 15 (1994), pp. 497–507, p. 499 and 'La place de Ramon Llull,' p. 205.
[57] 'La place de Ramon Llull,' p. 213.
[58] Ibid., p. 213.

well have borrowed his *ens necessarium* from Ibn Sīnā's philosophy, possibly via al-Ghazālī's *Maqāṣid al-falāsifa*.[59]

Yet, the prominence of the term *wujūd muṭlaq* is not limited to Avicennian philosophy or even to the texts of the *falāsifa*. In this sense, Urvoy made a valid claim when he called attention to the use of the term among Andalusian mystics. The concern with *wujūd* among Sufis of the twelfth and later centuries might very well betray the impact of Ibn Sīnā's metaphysics, but it introduces its own perspective which stems from the idea of the mystical experience. In his study on Ibn ʿArabī's epistemology, William Chittick pointed out that *wujūd* is not only the verbal noun of *wajada* in the passive meaning of 'to be found', i.e. 'to exist', but also in the active meaning, 'to find'.[60] For a Sufi *wujūd muṭlaq* could have two inseparable dimensions at least: God's absolute existence (based on the passive meaning), and the exclusive focus of the Sufi on God, 'to see only God', which is derived from the active meaning of *wajada*.

Urvoy regarded the use of *wujūd muṭlaq* as a term characteristic of Ibn Sabʿīn an unusual case for Lullus, whose adoption of Islamic theology more often than not mirrored Almohad *kalām* which held sway over the Western Mediterranean during the first half of the thirteenth century. Even though Urvoy does not seem to have considered *wujūd muṭlaq* part of Almohad *kalām*, a contemporary of Ibn Sabʿīn did: the notorious polemicist and archenemy of Ibn Sabʿīn, Ibn Taymiyya (1263–1328), condemned the Almohad Mahdi Ibn Tūmart in a *fatwā*, probably written during his stay in Alexandria where he encountered many Sufis from the Arab West.[61] In his refutation of Ibn Tūmart's theology Ibn Taymiyya criticised him for having adopted the idea of a God without attributes. Ibn Taymiyya identifies this idea with the concept of *wujūd muṭlaq*, which he attributes equally to Ibn Sīnā and Ibn Sabʿīn. One can hardly claim that Ibn Taymiyya can serve here as a witness for Ibn Sabʿīn's influence on Lullus (especially considering that he understood the term in yet another sense), but the passage in the *fatwā* clearly demonstrates that *wujūd muṭlaq* can be associated with

[59] 'Raimundus Lullus in der Auseinandersetzung mit dem Islam,' pp. 139–42.

[60] *The Sufi Path of Knowledge. Ibn al-ʿArabi's Metaphysics of Imagination* (New York, 1989), pp. 3–4.

[61] H. Laoust, 'Une *fetwā* d'Ibn Taimīya sur Ibn Tūmart,' *Bulletin de l'Institut Français d'Archéologie Orientale* 59 (1960), pp. 157–84, pp. 166–7. (Arabic) and pp. 177–9 (French translation). Cf. also T. Nagel, *Im Offenkundigen das Verborgene. Die Heilszusage des sunnitischen Islams* (Göttingen, 2002), pp. 167–8.

Almohad *kalām* as well as with Ibn Sabʿīn and Ibn Sīnā. These testimonies suggest that the term *wujūd muṭlaq* with the diverse meanings rendered by it was very much debated in the thirteenth and fourteenth centuries in the Mediterranean. Lullus' usage confirms that he was aware of these debates among Muslim scholars, but it does not clearly point to a specific source, and if any, it was most likely Ibn Sīnā via al-Ghazālī.

2.6. *The thirteen propositions*

In his discussion of the similarities between the divisions of emanations as they appear in the works of Lullus and Ibn Sabʿīn (cf. above, 2.4.), Lohr had admitted that these parallels might be due to a common source, the *Rasāʾil* of the Ikhwān al-ṣafāʾ. Yet, according to Lohr, there is one element common only to Lullus' *Logica nova* and Ibn Sabʿīn's *Budd al-ʿārif* and without equivalent in the *Rasāʾil*, a classification of thirteen types of propositions (cf. the table below).[62] Lohr was right when he claimed that the *Rasāʾil* do not include such a list, but neither does it appear in the *Logica nova* for the first time in Lullus' works: some thirty years earlier he had included a very similar list in his *Logica Algazelis*. There, it was clearly based on his Arabic *Vorlage, Maqāṣid al-falāsifa*. The list can hardly count as al-Ghazālī's invention either. Not surprisingly, it appears in the Persian *Vorlage* of the *Maqāṣid*, the *Dāneshnāme* of Ibn Sīnā, who in turn had borrowed it from a late ancient source, as Dimitri Gutas has elucidated.[63] After Ibn Sīnā included the list in the *Dāneshnāme* it gained significant influence in Islamic theology and philosophy, as shown by Joseph van Ess.[64]

In his doctoral thesis Lohr analysed the list of thirteen propositions as it appears in the *Logica Algazelis* and demonstrated, among other things, that it was not derived from Gundissalinus' earlier Latin translation of the *Maqāṣid*, but rather based on the Arabic text itself. Yet, in his 'Christianus arabicus' Lohr claimed that the version of the thirteen propositions as it appears in the *Logica nova* was not based on the *Maqāṣid* or Lullus' own Arabic abbreviation of it, its Latin translation or the Catalan verse version. According to Lohr, Lullus had copied

[62] 'Christianus arabicus,' pp. 78–9.

[63] 'Paul the Persian on the Classification of the Parts of Aristotle's Philosophy: a Milestone between Alexandria and Baġdād,' *Der Islam* 60 (1983), pp. 231–67.

[64] J. van Ess, *Die Erkenntnislehre des ʿAḍudaddīn al-Īcī. Übersetzung und Kommentar des ersten Buches seiner* Mawāqif (Wiesbaden, 1966), p. 398ff.

the list this time from Ibn Sab'īn's *Budd al-'ārif*. Unfortunately, he did not present any reasons for this conclusion.

Are there then any peculiarities in the *Logica nova*-version of the thirteen propositions which suggest that it was borrowed from the *Budd al-'ārif*? A quick glance at the passage in question reveals that Ibn Sab'īn mentions merely the terms for the propositions—unlike Lullus in both texts as well as his *Vorlage* Ibn Sīnā/al-Ghazālī, all of whom present additionally short explanations and examples, as can be seen in the table in the appendix to our article. The comparative table reveals another peculiarity of the lists in Lullus' works: the terminology in the version of the *Logica nova* is entirely different from that in the *Logica Algazelis*. This astonishing difference requires further explanation, but the assumption of Ibn Sab'īn as a source of the *Logica nova* certainly does not provide a satisfactory explanation. Lullus was an extremely prolific author who did not refrain from mentioning the same idea more than once in his writings, each time with a new choice of words—the same might have happened here.

2.7. *The nine questions*

Another coincidence between Lullus and Ibn Sab'īn alleged by Lohr, are Lullus' nine 'Rules' or general questions. These questions are characteristic for Lullus' *ars* in general.

Thus, according to Lullus, anything can be subject to the following questions: whether something exists (*utrum*), what it is (*quid?*), of what it is (*de quo?*), why it is (*quare?*), how much it is (*quantum?*), which it is (*quale?*), when it is (*quando?*), where it is (*ubi?*), how it is and with what it is (*quomodo et cum quo?*). The last two questions form part of a single rule of dual character, namely the rule represented by the letter K of Lullus' *ars*. Each of the rules is divided in turn into several species, which display the various meanings which can be attributed to each question. Thus, the first rule has three possible meanings: the affirmative, the negative or the dubitative. The second rule has four species, the third three, etc.

In the *Budd al-'ārif* a similar list of nine questions appears very early in the text. Ibn Sab'īn considers them essential for understanding any object:

> There are nine fundamental examinations (*mabāḥith*) and quests (*maṭālib*) in the knowledge of how things really are, and there are also nine expressions which form the respective questions. To each question corresponds

> a particular answer, and they are completely different from each other. They are: is it (*hal huwa*), what is it (*mā huwa*), how much is it (*kam huwa*), which is it (*ayy huwa*), how is it (*kayfa huwa*), where is it (*ayna huwa*), when it is (*matā huwa*), why is it (*li-mā huwa*), and who is it (*man huwa*). Whoever believes that he knows how things really are and talks about their causes and reasons, needs to have previously achieved knowledge in these nine fields and the answers to the individual questions according to its truth and aim.[65]

Yet, this idea was not Ibn Sabʿīn's original invention, but can be traced back to the *Rasāʾil* of the Ikhwān al-ṣafāʾ where this passage had been taken from literally, as already pointed out by Jūrj Kattūra, the editor of the *Budd al-ʿārif*, in a footnote to these sentences.[66] Again, it is not unlikely that Lullus was inspired by Arabic texts on logic when he turned the nine questions into one of the cornerstones of his *ars*, but this particular and highly original development has too little in common with Ibn Sabʿīn or the Ikhwān al-ṣafāʾ to establish a closer link, or even to argue that it was the former rather than the latter which inspired Lullus. This has already been stressed by Anthony Bonner, who, in turn, drew attention on Ibn Sīnā as a possible source for Lullus' rules.[67]

3. Conclusion

Miguel Cruz Hernández was adamant in his rejection of any debate of specific Arabic sources of Raimundus Lullus and ascribed the obvious parallels to a global influence of Islamic culture on the Catalan missionary.[68] Our research into this question has led us to the conclusion that this statement contains both right and wrong elements. On the one hand we share his criticism that many comparisons with specific texts are based on rather scant evidence which is quite often taken out of context. Yet, this should not lead us to abandoning the question of specific influences altogether. The parallels with Ibn Sabʿīn might not be convincing, but the similarities with the Ikhwān al-ṣafāʾ and al-Ghazālī probably point to a closer relationship than simply a com-

[65] *Budd al-ʿārif*, ed. Kattūra, p. 39.

[66] *Rasāʾil Ikhwān al-ṣafāʾ*, 4 vols. (Beirut, 1377/1957), i, p. 262.

[67] A. Bonner, 'Possibles fonts musulmanes de les deu regles i qüestions de Ramon Llull,' in *Pensar en diàleg. Miscel·lània en homenatge al Prof. Dr. Eusebi Colomer* (= *Revista Catalana de Teologia* 19; Barcelona, 1994), pp. 93–8.

[68] 'El símbolo del árbol,' p. 24.

mon milieu—it might have been these very sources that Lullus used, since these texts precisely had an enormous impact on the intellectual tradition in the Western Mediterranean. Thus, already in the twelfth century parts of the encyclopaedia of the Ikhwān al-ṣafāʾ, concerning the theory of demonstration, were translated into Latin, most probably by Dominicus Gundissalinus in Toledo.[69] Yet, we are not in the position to judge whether Lullus had complete versions of these texts at his disposal or excerpts.

At any rate, the previous pages have shown that, far from being closed, the dossier of Raimundus Lullus and his Arabic sources remains still open to discussion. Or, as Hans Daiber put it recently: 'Ramon Llull's concern with Islamic culture has still not been dealt with in a conclusive manner.'[70] Fortunately, it seems that today, with the increasing number of critical editions of medieval Arabic treatises on the one hand and the progress of the edition of Lullus' Latin and Catalan works on the other hand, a thorough study of the question is no longer only desirable but is becoming possible on a much more reliable basis than some decades ago.

The question at stake is of course not only of philological interest: it is crucial for an adequate reconstruction of the most central issues in Lullus' systematic evolution as well as for determining his paradigmatic place in the history of ideas, half way between Islamic philosophy and Christian faith.

---

[69] Cf. pseudo-al-Kindi, *Liber introductorius in artem logicae demonstrationis*, ed. A. Nagy (BGPhMA II, 5; Münster, 1897), pp. 41–64. Friedrich Dieterici had first shown that the *Liber introductorius* stems from the discussion of the *Posterior Analytics* in the encyclopaedia of the Ikhwān al-ṣafāʾ. Henry George Farmer and, more recently, Carmela Baffioni have suggested Muḥammad ibn Maʿshar al-Bustī as its author (H.G. Farmer, 'Who Was the Author of the *Liber introductorius in artem logicae*?,' *JRAS* s.n. [1934], pp. 553–6, and C. Baffioni, 'Il *Liber introductorius in artem logicae*: Problemi storici e filologici,' *Studi filosofici* 17 [1994], pp. 69–90).

[70] 'Raimundus Lullus in der Auseinandersetzung mit dem Islam,' p. 172.

The Thirteen Propositions

مقاصد الفلاسفة (تحقيق دنيا, القاهرة 1961, ص 102–109	*Compendium Logicae Algazelis*, ed. Lohr, p. 100	*Maqāṣid (Gundissalinus)*, ed. Muckle, pp. 274–8	بد العارف (تحقيق كتوره, ص 91)	*Logica nova*, ROL XXI, pp. 98–9
(1) أما الأوليات فهي التي تضطر غريزة العقل بمجردها إلى التصديق بها كقولك الاثنان أكثر من الواحد والكل أعظم من الجزء والأشياء المساوية لشئ واحد متساوية. (...)	(1) Prima enim propositio est de necessario, ut: 'Totum est maius sua parte'.	(1) Primae sunt, quas per se necesse est intellectui naturaliter credere. Ut haec: 'Duo plus sunt quam unum,' et 'Totum maius est sua parte,' et 'Quaecumque aequalia eidem, et inter se.' (...)	(1) الأوليات	(1) Primus modus est sicut dicere: Magis est totum, quam sua pars.
(2) المحسوسات مثل قولنا الشمس مستنيرة وضوء القمر يزيد وينقص.	(2) Secunda est de sensualitate, ut de eo, quod patet alicui quinque sensuum particularium, qui sunt visus, auditus, gustus, odoratus et tactus.	(2) Sensibiles sunt, ut haec: 'Sol est lucidus,' et 'Claritas lunae crescit et decrescit.'	(2) المحسوسات	(2) Secundus est per sensibilitatem, quoniam uisum est, omnem hominem esse animal et habere caput.

(3) التجريبيات ما يحصل من مجموع العقل والحس كعلمنا بأن النار تحرق والسقمونيا تسهل الصفراء والخمر يسكر. (...)	(3) Tertia est de experientia, ut expertum est, quod 'Piper calefacit', et 'Opium infrigidat'.	(3) Experimentales sunt, quas acquirimus intellectu et sensu. Ut hoc quod scimus, quod 'Ignis adurit,' et 'Scamonia ventrem solvit,' et 'Vinum inebriat.'	(3) التجريبات	(4) Quartus modus est per experientiam, sicut probatum est, quod ignis est calidus; et scamonea est laxatiua.
(4) المتواترات ما علم بإخبار جماعة كعلمنا بوجود مصر ومكة وإن لم نبصرهما ومهما استحالِ الشك فيه سمى متواترا. (...)	(4) Quarta est de tavetur, quod est vocabulum arabicum, ut si aliquid ostensum sit in speciali, et in generali sit concessum. Ut quando aliquis credit Parisius esse dato, quod ibi non fuerit, nec ipsum viderit.	(4) Famosae sunt, quas vulgo referente didicimus. Sicut haec quod 'Aegyptus est,' quamvis numquam vidimus. De quo si nullatenus dubitaverimus, vocabitur famosa. (...)	(4) المتواترات	(5) Quintus modus est per communem conceptionem, sicut homo rem non uisam per auditum cognoscit, sicut Sortes, qui numquam Parisius fuit, sed quia audiuit a pluribus, dicit illud esse.

Table (*cont.*)

مقاصد الفلاسفة (تحقيق دنيا, القاهرة 1961, ص 102–109	*Compendium Logicae Algazelis*, ed. Lohr, p. 100	*Maqāṣid (Gundissalinus)*, ed. Muckle, pp. 274–8	بد العارف (تحقيق كتوره, ص 91)	*Logica nova*, ROL XXI, pp. 98–9
(5) أما القضايا التي قياساتها في الطبع معها فهي القضايا التي لا تثبت في النفس إلا بحدودها الوسطى ولكن يعزب عن الذهن الحد الأوسط فيظن أنها مقدمة أولية. (...) مثاله أنك تعلم أن الاثنان نصف الأربعة على البداهة. (...)	(5) Quinta est de re continenti in se ipsa veram demonstrationem. Ut manifestum est, quod 'Duo sunt medietas quattuor'.	(5) Propositiones vero quae secum habent probationem suam naturaliter sunt illae, quibus non acquiescit animus nisi per medium terminum. Quem quia in promptu est intelligere, idcirco putatur propositio illa esse prima quae scitur sine medio termino. (...) Verbi gratia, cum audimus quod 'Binarius est medietas quaternarii,' statim scimus quod quidem non scitur nisi per medium. (...)	(5) القضايا التي لا يخلو الذهن عن حدودها الوسطى وقياساتها الصادقة	(3) Tertius modus est quando subiectum continet in se ea, per quae cognitum est, sicut numerus quinarius, qui in se continet duas et tres unitates; et homo caput et collum.

(6) الوهميات هي مقدمات باطلة ولكنها قويت في النفس قوة تمنع من إمكان الشك فيه وذلك في أثر حكم الوهم في أمور خارجة عن المحسوسات لأن الوهم لا يقبل شيئا إلا على وفق المحسوسات (...)	(6) Sexta est de cogitatione. Ista autem nullius est valoris. Potest enim falsa assidue reperiri.	(6) Opinabiles sunt propositiones falsae quae ita sunt fixae in animo, ut nemo possit dubitare de eis. Quod contingit ex actione aestimationis in ea quae sunt praeter sensibilia. Aestimatio enim non percipit aliquid nisi secundum sensibilia in quibus consuevit. (...)	(6) الوهميات	(9) Nonus est de consensu, eo quod multa assentiuntur, quae non sunt uera.
(7) أما المشهورات فهي القضايا التي لا يعول فيها إلا على مجرد الشهرة ونظر العوام. والظاهر بين أهل العلم أنها أوليات لازمة في غريزة العقل مثل قولك الكذب قبيح (...)	(7) Septima est de publico. Ut: 'Homo mendax vilis debet ab omnibus reputari'.	(7) Maximae sunt propositiones quae non recipiuntur nisi in quantum sunt manifestae. Et putat vulgus et simplices doctores esse primas comitantes naturam intellectus. Ut haec: 'Mendacium est turpe,' (...)	(7) المشهورات	(7) Septimus est de publico, quia de homine mandace publice dicitur, quia non est uerax.

Table (*cont.*)

مقاصد الفلاسفة (تحقيق دنيا, القاهرة 1961, ص 102–109	*Compendium Logicae Algazelis*, ed. Lohr, p. 100	*Maqāṣid (Gundissalinus)*, ed. Muckle, pp. 274–8	بد العارف (تحقيق كتوره, ص 91)	*Logica nova*, ROL XXI, pp. 98–9
(8) أما المقبولات فهي المقبول من أفاضل الناس وأكابر العلماء ومشايخ السلف إذ تكرر نقل ذلك منهم على ذلك الوجه. (...)	(8) Octava est de suppositione nondum habita. Ut plures supponunt multa, quorum veritatem ignorant	(8) Receptibiles sunt illae, quae habentur a sanctis hominibus vel a maioribus sapientium vel ab antiquis et sensibus, cum constiterit eas ab eis fuisse dictas vel in libris suis scriptas (...)	(8) المقبولات	(8) Octauus est de suppositione, quae non est necessaria, quia quandoque est uera, quandoque est falsa; sicut Saraceni, qui praesupponunt quod Mahometus fuerit propheta; quod falsum est.
(9) أما المسلمات فهي التي سلمها الخصم أو كان مشهوراً بين الخصمين فقط فإنه يستعمل معه دون غيره (...)	(9) Nona est de concessione, cui reprehensio non debetur.	(9) Concessae sunt, quas concedit adversarius vel sunt manifestae inter ambos tantum. (...)	(9) الممكنات	(12) Duodecimus est de aestimatione, quoniam multa aliter existimantur, quam sint.

(10) أما المشبهات فهي التي يحتال في تشبيهها بالأوليات والتجريبيات والمشهورات ولا تكون بالحقيقة كذلك ولكنها تقاربها في الظاهر.	(10) Decima est de similitudine, quae quasi assimilatur primae. Assimilatur enim ei impossibilitate; in necessitate autem non.	(10) Simulatoriae sunt, quas studet homo assimilare primis vel experimentalibus vel maximis. Nec sunt ut illae, sed videntur esse ut illae.	(10) المتشابهات	(10) Decimus est de similitudine, quoniam omnis similitudo non est necessaria.
(11) أما المشهورات في الظاهر فهي كل قول يقبله كل من يسمعه كافة ببادئ الرأي وأول النظر (...)	(11) Undecima est de generalitate. Ut generale est quemlibet suum amicum dirigere.	(11) Maximae in apparentia sunt, quas qui audit statim recipit in principio, sed cum diligenter attendit (...)	(11) المشهورات في الظاهر	(11) Vndecimus est de publica fama, quae in aliquibus suis particularibus non est necessaria; sicut dicere: Omnis femina ebria est luxuriosa.
(12) أما المظنونات فما يفيد غلبة الظن مع الشعور بإمكان نقيضه. (...)	(12) Duodecima est de opinione. Ut tempore noctis euntem latronem esse, plures homines opinantur.	(12) Putabiles sunt quae faciunt putare aliquid, quamvis animus percipiat posse esse eius oppositum. (...)	(12) المظنونات	(6) Sextus modus est per opinionem, sicut de homine frequenter eunte de nocte, opinio est, ut sit latro uel adulterator.

Table (*cont.*)

مقاصد الفلاسفة (تحقيق دنيا, القاهرة 1961, ص 102–109	*Compendium Logicae Algazelis*, ed. Lohr, p. 100	*Maqāṣid (Gundissalinus)*, ed. Muckle, pp. 274–8	بد العارف (تحقيق كتوره, ص 91)	*Logica nova*, ROL XXI, pp. 98–9
(13) أما المخيلات فهي مقدمات يعلم أنها كاذبة ولكنها تؤثر في النفس بالترغيب والتنفير.	(13) Tertia decima est de imaginatione. Ut aliqui imaginantur pistem (corr. piscem) morenum, quia serpenti assimilatur, esse totaliter venenosum.	(13) Imaginativae vel transformativae sunt propositiones, quas scimus esse falsas, sed imprimunt in animo vel appetendum vel respuendum. Sicut hoc quod dicitur 'Mel videtur esse stercus.' (…)	(13) المتخيلات	(13) Tertius decimus est de infecta imaginatione per nimium actum imaginationis; sicut homo zelotypus, qui tantum imaginatur suam uxorum esse luxuriosam, quod credit eam luxuriari. Et murenam esse serpentem.

ALEXANDER OF APHRODISIAS, *DE UNITATE*: A PSEUDEPIGRAPHICAL TESTIMONY OF THE *DE UNITATE ET UNO* BY DOMINICUS GUNDISSALINUS*

Cristina D'Ancona

I

Alexander of Aphrodisias was known among the Latin scholars of the thirteenth century both directly and indirectly. They got directly acquainted with some of his treatises and commentaries on Aristotle. Indirectly, his interpretations of the master's doctrines gained wide currency through the Latin translation of Averroes' Great Commentaries on the *De anima* and *Metaphysics*, where his peculiar readings of Aristotle's thinking are mentioned. Some of Alexander's writings were translated from Greek, others from the Arabic version of the Greek original. To make the picture even more complex, for some of the works of the *Alexander Latinus* the Greek original is extant, whereas for others it is lost to us. Luckily, two outstanding surveys on Alexander are nowadays available, by Robert W. Sharples[1] and by Richard Goulet together with Maroun Aouad.[2] Both of these include an exhaustive presentation of Alexander's works, as well as the mention of those among them which have Arabic and/or Latin translations. Before discussing the authorship of a small treatise *De unitate* with which Alexander is credited in the Latin manuscript tradition, it might be of some use to recall here the main data concerning the translations of his works, roughly following the chronology of the translations into Latin from Arabic and Greek.[3]

---

* I would like to express here my warmest thanks to Concetta Luna for her reading of a first draft of this paper and for her many helpful remarks. I would also like to express my most sincere thanks to Anna Akasoy and to Charles Burnett for their help with the English of this paper and for the substantial information added: see especially Appendix ii.

[1] R.W. Sharples, 'Alexander of Aphrodisias: Scholasticism and Innovation,' in W. Haase and H. Temporini (eds.), *Aufstieg und Niedergang der römischen Welt*, Teil ii. *Principat*, xxxvi/2 (Berlin, 1987), pp. 1176–1243.

[2] R. Goulet and M. Aouad, 'Alexandros d'Aphrodisias,' in R. Goulet (ed.), *Dictionnaire des philosophes antiques*, i (Paris, 1989), pp. 125–39.

[3] This survey may count as a partial update of E. Cranz, 'Alexander of Aphrodisias,' in P.O. Kristeller, *Catalogus translationum et commentariorum*, i (Washington, 1960), pp. 77–135.

Unless the fragments of the commentaries on the *Posterior Analytics*, *Topics* and *Sophistici Elenchi* circulating around 1150 under Alexander's name really trace back to him (which seems not to be the case),[4] the first works by Alexander made available to Latin readers belong (i) to the collection of *scripta minora* labelled by its editor Ivo Bruns as the *Mantissa* (i.e., 'supplement')[5] to the *De anima* and (ii) to the so-called *Quaestiones*.[6] In addition, there is (iii) a short writing *On Time*, to which

[4] Alexander's commentary on the *Topics* is extant and edited in the *CAG* series, ii/2 (Wallies, 1891). On the other hand, the commentary on the *Posterior Analytics* is known to us only through the quotations of later commentators (Themistius, Philoponus and Eustratius) and the commentary on the *Sophistici Elenchi* published in the *CAG* ii/3 (Wallies, 1898) is not by Alexander but by Michael of Ephesus: see Sharples, 'Alexander of Aphrodisias,' p. 1184, and Goulet and Aouad, 'Alexandros d'Aphrodisias,' p. 130. As for the Latin commentaries on the *Posterior Analytics*, *Topics* and *Sophistici Elenchi* attributed to Alexander, see L. Minio-Paluello, 'Note sull'Aristotele latino medievale. ix. Gli *Elenchi Sofistici*: redazioni contaminate colla ignota versione di Giacomo Veneto (?); frammenti dello ignoto commento d'Alessandro di Afrodisia tradotti in latino,' *Rivista di filosofia neo-scolastica* 46 (1954), pp. 223–31, also in idem, *Opuscula. The Latin Aristotle* (Amsterdam, 1972), pp. 241–9. Minio-Paluello maintained that the fragments of a commentary on the *Sophistici Elenchi* preserved in two mss., one housed in Oxford and the other in Vienna, traced back to the lost genuine commentary by Alexander. S. Ebbesen, 'Anonymus Aurelianensis II, Aristotle, Alexander, Porphyry and Boethius. Ancient Scholasticism and Twelfth-Century Western Europe,' *Cahiers de l'Institut du Moyen Age Grec et Latin* 16 (1976), pp. 1–128, esp. 108–18, raises doubts about Minio-Paluello's contention and advances Philoponus' authorship. The 'Anonymus Aurelianensis' edited by Ebbesen also bears witness of a commentary by "Alexander" on the *Posterior Analytics*, which Ebbesen tentatively traces back to Philoponus (pp. 98–107). Still the 'Anonymus Aurelianensis' gives slight evidence of a Latin translation by James of Venice of (fragments of) a commentary on the *Topics* attributed to Alexander, and possibly tracing back to Philoponus: see the discussion in Ebbesen, pp. 118–20. On the translations by James of Venice see the survey by L. Minio-Paluello, 'Iacobus Venetus Grecus: Canonist and Translator of Aristotle,' *Traditio* 8 (1952), pp. 265–304 (= *Opuscula*, pp. 189–228).

[5] I. Bruns, *Alexandri De Anima cum mantissa*, in *Alexandri Aphrodisiensis praeter commentaria scripta minora* (Supplementum Aristotelicum, ii; Berlin, 1887), i, *Praefatio*, p. vi: 'de anima libri mantissa titulo vulgato'. The *Mantissa* has been translated into English and Italian: see *Supplement to* On the Soul, trans. R.W. Sharples (London, 2004) and *De anima ii (Mantissa)*, trans. P. Accattino (with P. Cobetto Ghiggia) (Alessandria, 2005). According to Sharples, p. 3, the so-called *Mantissa* 'does seem to constitute a series of texts which have been arranged, without regard to their literary form, approximately in the sequence in which topics are discussed in Aristotle's *De Anima* and, following it, in Alexander's'.

[6] Sharples, 'Alexander of Aphrodisias,' pp. 1194–5, gives an account of the nature of this collection of texts which is worth quoting in full: 'Bruns divided the short Greek texts in the *Mantissa, Quaestiones* and *Ethical Problems* into six classes: (A) problems followed by their solutions, *quaestiones* properly so called, (B) explanations (ἐξηγήσεις) of particular Aristotelian texts, (C) summaries (ἐπιδρομαί) of sections of Aristotle's works or of doctrines on particular topics, (D) fragments, (E) collections of arguments

nothing corresponds in the Greek corpus of Alexander's works. They have all been translated from Arabic in Spain, in all likelihood by Gerard of Cremona (d. 1187), although some of them do not appear in the list of Gerard's translations appended by his *socii* to the biography of the master.[7] This is precisely the case for the *De intellectu*, the most influential among the works of the *Alexander Latinus*.

The Latin *De intellectu* corresponds to the second item of the *Mantissa*, labelled as *Περὶ νοῦ* in the main manuscript of Alexander's works and

to demonstrate a particular point, and (F) discussions of particular topics not differing in character, but only in their shorter length, from the treatises (...). The *Mantissa* texts differ from the *Quaestiones* in that the former seem to relate more closely to Alexander's own longer treatises (...), the latter to Aristotelian texts. It seems improbable that all these texts are by Alexander himself, though they all seem to reflect the work of his school. The assembly of the collections was inexpertly done, texts on similar topics being separated and duplicates not recognised; some of the titles, too, are inept and seem to be the work of editors (...). In the *Quaestiones*, too, there are certain groupings; text on psychology (which mostly follow the order of the text of Aristotle's *De Anima*), on issues raised in Aristotle's *Physics* and in his *De Generatione et corruptione*. Certain preoccupations also recur; with possibility and potentiality, with providence and the motion and influence of the heavens, with questions of form and matter, universal and particular'. Some of the *Quaestiones*, together with parts of the *Mantissa* and other treatises, form a part of the corpus of the Arabic Alexander. A list was given by A. Dietrich, 'Die arabische Version einer unbekannten Schrift des Alexander von Aphrodisias über die *Differentia specifica*,' *Nachrichten der Akademie der Wissenschaften in Göttingen, Philologisch-historische Klasse* (1964), pp. 85–148, and it is customary to indicate the Arabic version with the number in Dietrich's list (pp. 93–100), as D1ff. For a list of the Arabic versions of Alexander's *Scripta minora* see Sharples, 'Alexander of Aphrodisias,' pp. 1192–4, and Goulet and Aouad, 'Alexandros d'Aphrodisias,' pp. 132–5.

[7] The first to draw attention to the bio-bibliography of Gerard of Cremona (d. 1187) made by his students at Toledo was B. Boncompagni, *Della vita e delle opere di Gherardo Cremonese traduttore del secolo duodecimo e di Gherardo da Sabbionetta astronomo del secolo decimoterzo, notizie raccolte da Baldassarre Boncompagni* (Rome, 1851). Later on, the bio-bibliography has been edited three times: F. Wüstenfeld, 'Die Übersetzungen arabischer Werke ins Lateinische seit dem XI. Jahrhundert,' *Abhandlungen der Königlichen Gesellschaft der Wissenschaften zu Göttingen* 22 (1877), esp. pp. 57–81; K. Sudhoff, 'Die kurze Vita und das Verzeichnis der Arbeiten Gerhards von Cremona,' *Archiv für Geschichte der Medizin* 14 (1923), pp. 73–82; C. Burnett, 'The Coherence of the Arabic-Latin Translation Programme in Toledo in the Twelfth Century,' *Science in Context* 14 (2001), pp. 249–88. The bio-bibliography has been translated into English: M. McVaugh, 'The Translations of Greek and Arabic Science into Latin. 7. A List of Translations made from Arabic into Latin in the Twelfth Centrury. Gerard of Cremona (ca 1114–1187),' in E. Grant (ed.), *A Source Book in Medieval Science* (Cambridge, Mass., 1974), pp. 35–41. Further titles have been added to this list: see G. Sarton, *Introduction to the History of Science*, ii. *From Rabbi ben Ezra to Roger Bacon* (Baltimore, 1931), pp. 338–44 and R. Lemay, 'Gerard of Cremona,' in *Dictionary of Scientific Biography* (New York, 1970–90), xv, suppl. 1, pp. 173–92.

in Bruns' edition.[8] It is a short essay on the three meanings allegedly given by Aristotle to the word νοῦς and on the separate status of the third kind of intellect, the ποιητικὸς νοῦς or θύραθεν νοῦς.[9] Alexander's *Περὶ νοῦ* was translated into Arabic by Isḥāq ibn Ḥunayn[10] and played an important role in al-Fārābī's own treatise *Fī 'l-ʿaql*,[11] not without some decisive Neoplatonic reorientations to which M. Geoffroy has recently called attention.[12] The Arabic version was translated into Latin and this translation has been edited by G. Théry in 1926,[13] on the basis of four manuscripts housed in Paris and one in Vendôme.[14] Gerard of Cremona's authorship is based on the title of the text as given by the ms. Paris, Bibliothèque Nationale de France, *lat.* 6325, one of the five basic manuscripts of Théry's edition. However, Alexander's *De intellectu* does not appear in the list of Gerard's translations mentioned above,[15]

---

[8] Venezia, Biblioteca Nazionale Marciana, *gr.* 258. This ms., from the collection of Cardinal Bessarion, belongs to the so-called 'collection philosophique': see L.G. Westerink's account in Damascius, *Traité des premiers principes*, i, ed. L.G. Westerink and trans. J. Combès (Paris, 1986), pp. lxxiii–lxxx, esp. p. lxxiv. The ms. has been studied by L. Labowsky, 'William of Moerbeke's Manuscript of Alexander of Aphrodisias. Bessarion Studies iii,' *Mediaeval and Renaissance Studies* 5 (1961), pp. 155–62. Lotte Labowsky has shown that it was the property of William of Moerbeke, who translated a number of Alexander's works from Greek into Latin (see below, notes 37–48). As Bruns declared in the *Praefatio* of his 1887 edition, he followed this ms. in giving the subtitles to the *Mantissa*. In Bruns' edition the *Περὶ νοῦ* occupies p. 106, l. 18–p. 113, l. 24.

[9] Bibliography on the *Περὶ νοῦ* in Sharples, 'Alexander of Aphrodisias,' p. 1189, and Goulet and Aouad, 'Alexandros d'Aphrodisias,' p. 134. Up-to-date bibliography in Alexander of Aphrodisias, *De intellectu*, ed. and trans. P. Accattino (Turin, 2001).

[10] The Arabic version of Alexander's *De intellectu* has been published twice: J. Finnegan, 'Texte arabe du *Περὶ νοῦ* d'Alexandre d'Aphrodise,' *MUSJ* 33 (1956), pp. 157–202 and ʿA. Badawī (ed.), *Shurūḥ ʿalā Arisṭū mafqūda fī 'l-yūnāniyya wa-rasāʾil ukhrā. Commentaires sur Aristote perdus en grec et autres épîtres* (Beirut, 1972), pp. 31–42.

[11] Abū Naṣr al-Fārābī, *Risāla fī 'l-ʿaql*, ed. M. Bouyges, (Beirut, ²1983). Latin translation edited by E. Gilson, 'Les sources gréco-arabes de l'augustinisme avicennisant,' *AHDLMA* 4 (1929–30), pp. 5–149; ed. of the Latin text, pp. 108–26.

[12] M. Geoffroy, 'La tradition arabe du *Περὶ νοῦ* d'Alexandre d'Aphrodise et les origines de la théorie farabienne des quatre degrés de l'intellect,' in C. D'Ancona and G. Serra (eds.) *Aristotele e Alessandro di Afrodisia nella tradizione araba* (Padova, 2002), pp. 191–231.

[13] G. Théry, *Autour du décret de 1210: ii. Alexandre d'Aphrodise. Aperçu sur l'influence de sa noétique* (Kain, 1926), pp. 74–82.

[14] The fifteen extant Latin mss. of Alexander's *De intellectu* are listed by Théry (*Autour du décret de 1210*, pp. 69–74). To this list another item should be added: see below, *Appendix ii*.

[15] See above, note 7. The title is reproduced by Théry, *Autour du décret de 1210*, p. 74. Two features of the Latin translation are worth mentioning: (i) the Latin gives confirmation to Isḥāq ibn Ḥunayn's authorship of the Arabic version (see p. 181, l. 2 Finnegan; p. 31, l. 13–14 Badawī): in the mss. Città del Vaticano, Biblioteca Apostolica Vaticana, *lat.* 2186, Cambridge, Caius College 996, Paris, Bibliothèque Nationale de

nor has any systematic analysis of the translation style been made, to my knowledge, in order to prove or disprove such an attribution.[16]

France, *lat.* 6325, *lat.* 6443 and *lat.* 16602, the translation is ascribed to 'Isaac filius Iohannici'; Isḥāq ibn Ḥunayn is mentioned also within the translation: see p. 77, l. 16 Théry, 'Dixit Ysaac in hoc loco', p. 81, l. 13: 'Dixit Ysaac in hoc loco'; according to Théry, the entire passage at p. 81, l. 13–19 is a glossa by Isḥāq ibn Ḥunayn. (ii) In the first of the passages mentioning 'Ysaac', there is an allusion to the Greek text of the *De intellectu* which led Théry to think that Gerard of Cremona had at his disposal also a Greek ms. The passage as printed by Théry runs as follows, p. 77, l. 16–18: 'Dixit Ysaac in hoc loco [deficit aliquid de libro graeco per quem correximus istum]: Non autem ob hoc quod nos intelligimus illam, sit ipsa intellectus et intellectum, sed quia hec forma que est sic...'. Théry comments on this passage as follows, p. 77 note (b): 'Ce texte représente une glose du traducteur. Cette remarque est fort intéressante. Elle nous montre que Gérard de Crémone avait deux textes devant lui: le texte d'Ishaq qui lui servait de base et un texte grec qui lui servait à contrôler ou à corriger Ishaq'. Théry also repeats, pp. 82–3, that 'Pour exécuter cette traduction, Gérard avait deux textes devant lui: un texte arabe, celui d'Ishaq ben Honein (...) qui représente une version de l'ouvrage d'Alexandre, glosée en certains endroits; et un texte grec avec lequel il contrôle et corrige la traduction arabe'. In all likelihood this remark by Théry gave rise to the legend, which appears here and there in the accounts on Gerard's translations, that he knew also Greek. However, there is no evidence that Gerard had any Greek nor that there was any Greek text of Alexander available in al-Andalus at that time. On the other hand, it is well known that Isḥāq ibn Ḥunayn checked against other Greek or Syriac mss. the texts he was translating, so that the most natural reading of the quoted sentence is: 'Dixit Ysaac in hoc loco: deficit aliquid de libro graeco per quem correximus istum. Non autem ob hoc quod nos intelligimus illam, sit ipsa intellectus et intellectum, sed quia hec forma que est sic...', giving the glossa to Isḥāq himself instead than to Gerard. What settles the issue is the fact that the glossa appears also in the Arabic (see p. 186, l. 27 Finnegan; p. 35 n. 1 Badawī: the latter eliminates the passage from the text. Finnegan, p. 186, correctly remarks that Théry 'par erreur attribue cette glose à Gérard de Crémone' and Badawī rightly gives it to Isḥāq ibn Ḥunayn). Finally, M. Alonso, 'Traducciones del arcediano Domingo Gundisalvo,' *Al-Andalus* 12 (1947), pp. 295–338, esp. p. 317, remarks that the glossa cannot be by the translator and suggests 'un estudioso posterior' who remarked the failure of the 'pasaje latino comparado con el griego', but this is ruled out by the presence of the note in the Arabic.

[16] On Gerard's typical style of translation see I. Opelt, 'Die Übersetzungstechnik des Gerhard von Cremona,' *Glotta* 38 (1959), pp. 135–70; G. Serra, 'Note sulla traduzione arabo-latina del *De Generatione et corruptione* di Aristotele,' *Giornale critico della filosofia italiana* 53 (1972), pp. 383–427 (criticizing Opelt); idem, 'Alcune osservazioni sulle traduzioni dall'arabo in ebraico e in latino del *De Generatione et corruptione* di Aristotele e dello pseudo-aristotelico *Liber de Causis*,' in *Scritti in onore di Carlo Diano* (Bologna, 1975), pp. 385–433; idem, 'Due studi arabo-latini. i. Note in margine a *anniyya-anitas* ii. Gerardo da Cremona traduttore del *Flos Alfarabii*?,' *Medioevo* 19 (1993), pp. 27–66. See also P.P. Ruffinengo, '*Al-Kindī. Trattato sull'intelletto. Trattato sul sogno e la visione.* Introduzione, traduzione italiana del testo arabo, lessico arabo-latino,' *Medioevo* 23 (1997), pp. 337–94, giving a lexicon of Gerard's translations of the Kindian treatises *On the Intellect* and *On Sleep and Sight*. Alonso, 'Traducciones del arcediano Domingo Gundisalvo,' pp. 315–17, challenged Gerard's authorship in favour of Dominicus Gundissalinus' one. First, he questioned Théry's reading of the title of the ms. Paris, Bibliothèque Nationale de France, *lat.* 6325; then, he argued that (i) Alexander's *De*

Two other short treatises translated from Arabic belong to the so-called *Quaestiones*:[17] *q.* i.5, translated under the title *Quod augmentum et incrementum fiunt in forma et non in yle*,[18] and *q.* iii.3, translated under the title *De sensu*.[19] Both are mentioned in the list of Gerard's translations, together with another treatise which, unlike the first two, does not correspond to any Greek original: *De motu et tempore*.[20]

Q. i.5 is a very short account—half a page in Bruns' edition[21]—on the fact that only form is responsible for the growth of a living being, not matter. The Arabic version is extant and has been published by ʿA. Badawī and by H.-J. Ruland;[22] also the Latin version of the Arabic has been edited by Théry and Ruland.[23] Théry credited Gerard

---

intellectu lacks in the list of Gerard's *socii*; (ii) the translation of *ʿaql* by *intellectus* instead than by *ratio* and of *mustafād* by *adeptus* instead of by *acquisitus* points to Gundissalinus and not to Gerard; (iii) this translation is transmitted by the manuscripts which contain other translations by Gundissalinus. It seems to me that only a systematic analysis of the translation may shed light on the issue.

[17] On the nature of this work see above, note 6.

[18] See below, note 20.

[19] See below, note 27.

[20] In the list by Gerard's *socii* the title runs as *Tractatus unus Alexandri Affrodisii de tempore, et alius de sensu, et alius de eo quod augmentum et incrementum fiunt in forma et non in yle.*

[21] *Alexandri Aphrodisiensis praeter commentaria scripta minora*, ed. I. Bruns, ii (Berlin, 1882), p. 13, l. 9–32. The title runs as διὰ τί ἡ αὔξησις κατὰ τὸ εἶδος μόνον, ἀλλ' οὐχὶ καὶ κατὰ τὴν ὕλην.

[22] The Arabic version, number 19 in the list by Dietrich (see above, note 6), bears the title *Maqālat al-Iskandar al-Afrūdīsī fī anna 'l-nushūʾa wa'l-namāʾa yakunāni fī 'l-ṣūrati lā fī 'l-hayūlā*: see Badawī, p. 51, l. 1 p. 52, l. 8, and H.-J. Ruland, 'Die arabische Übersetzung der Schrift des Alexander von Aphrodisias über das Wachstum (Quaestio i 5),' *Nachrichten der Akademie der Wissenschaften in Göttingen*, *Philologisch-historische Klasse* (1981), pp. 53–74 (Arabic text pp. 62–6).

[23] Théry, *Autour du décret de 1210*, pp. 99–100; Ruland, 'Die arabische Übersetzung der Schrift des Alexander von Aphrodisias über das Wachstum,' pp. 72–4. The mss. of the Latin version are listed by Théry, pp. 97–9: they all overlap with those of the *De intellectu* (which however are much more); as in the case of the *De intellectu*, to the list by Théry another item should be added (see below, *Appendix ii*). An interesting feature of the Arabic and Latin versions is that they make explicit—at variance with the Greek text, which does not—the commonly acknowledged relationship of this short *quaestio* with Aristotle's *De generatione et corruptione*. Compare the beginnings of the the *quaestio* in the Greek and Latin (for the sake of brevity I am omitting the quotation of the Arabic, which is rendered literally by the Latin): εἰ αὔξεται τὸ ὑποκείμενον τῷ αὔξεσθαι λεγομένῳ (ὑπομένει δὲ καὶ ἡ ὕλη, οὐ μόνον τὸ εἶδος· οὐ γὰρ δὴ πᾶσα ἀλλάσσεται), διὰ τί κατὰ τὸ εἶδος μόνον, ἀλλ' οὐχὶ καὶ κατὰ τὴν ὕλην ἡ αὔξησις; οὐ γὰρ πᾶσα ἡ ὕλη ἡ ἐν τοῖς αὐξομένοις ἀλλάσσεται, ἀλλὰ μένοντός τινος ἐξ αὐτῆς ἄλλη προσγίνεται, ed. Bruns, p. 13, l. 11–15; 'Aristoteles dicit in libro De generatione et corruptione quod augmentum et incrementum fiunt in forma et non in yle. Quidam autem negaverunt illud et dixerunt quod corpus et totum vel omne quod suscipit augmentum non crescit nisi forma (Ruland: <in> forma) sua et sua materia. Philosophus dixit quod augmentum fit in forma et non in yle. Alexander. Volo ego absolvere hanc ambiguitatem et dico

of Cremona with the translation—even though there is no evidence in the manuscripts—and he was surely right, because, as we already know, the *De eo quod augmentum* features in the list of Gerard's translations; in addition, the Latin bears some of the typical hallmarks of the *Gerardiana methodus*.[24]

Q. iii.3 is explicitly presented as an exegesis of Aristotle's *De anima* ii 5.[25] Alexander embarks upon a literal explanation of this Aristotelian chapter devoted to sense perception, and appends to the exegesis a summary of the main claims made in it.[26] The Arabic has been edited by H.-J. Ruland,[27] and the Latin version, bearing the title *Tractatus Alexandri Affrodisii de sensu*, has been edited twice: by Théry and, once again, by Ruland himself.[28] Gerard of Cremona's authorship is not explicitly stated in the manuscripts, but, as we have seen before, a treatise *De sensu* by Alexander is mentioned in the list of his translations.[29] The existence also of a Latin translation of Alexander's commentary on Aristotle's *De sensu et sensato*[30] clouded the issue and made Théry raise doubts about the authorship of this translation, but in the end

---

quod yle permutatur paulatim in augmento, scilicet materia rei; quod est quia expirat (Ruland: remanet) pars eius et advenit aliud extrinsecus preter (Ruland: absque) quod pereat yle tota; quoniam si periret tota, non remaneret forma secundum dispositionem suam', Théry, p. 99, l. 1–12, Ruland, p. 72, l. 5–14.

[24] Théry, *Autour du décret de 1210*, p. 100: 'Cette traduction ne porte pas de nom d'auteur. Mais il n'y a pas lieu de douter que le traducteur ne soit le même que celui du traité *De Motu et tempore*, contenu dans les mêmes manuscrits, c'est-à-dire Gérard de Crémone". For the mention within the list of Gerard's translations see above note 7; a typical Gerardian saying, 'redeamus ergo et dicamus', which repeatedly occurs for instance in the Latin *Liber de causis*—one of the most famous translations by Gerard of Cremona—is found also in the *De augmento*, Théry, p. 99, l. 12, Ruland, p. 72, l. 15; see also the typical rendering of *ʿalā ḥālin wāḥidatin* by 'secundum dispositionem unam', Théry, p. 99, l. 26–7, Ruland, p. 73, l. 27; of *kāna qāʾiman ghayra wāqiʿin taḥta ʾl-fasādi* by 'stans non cadens sub corrupcione', Théry, p. 100, l. 5–6, Ruland, p. 74, l. 4.

[25] *Q.* iii.3 occupies pp. 82, l. 21–86, l. 35 of Bruns' edition and its title runs as follows: Λέξεως ἐξήγησις ἐκ τοῦ δευτέρου ὁμοίως Περὶ Ψυχῆς, δι' ἧς περὶ αἰσθήσεως λέγει, ἔστι δὲ ἥδε· διωρισμένων δὲ τούτων λέγωμεν κοινῇ περὶ πάσης αἰσθήσεως (= *De anima* ii 5, 416b 32–3).

[26] Cf. p. 86, l. 4 Bruns: τὰ δὲ κεφάλαια τῶν εἰρημένων.

[27] Number 14 in Dietrich's list (see above, note 6); edited by H.-J. Ruland, 'Die arabische Übersetzung der Schrift des Alexander von Aphrodisias über die Sinneswahrnehmung,' *Nachrichten der Akademie der Wissenschaften in Göttingen, Philologisch-historische Klasse* (1978), pp. 159–225.

[28] Théry, *Autour du décret de 1210*, pp. 86–91, Ruland, 'Die arabische Übersetzung der Schrift des Alexander von Aphrodisias über die Sinneswahrnehmung,' pp. 212–25.

[29] See above, note 20.

[30] See below, note 43.

he was inclined to ascribe it to Gerard, albeit unaware of the witness of Gerard's *socii*.[31]

Finally, there is the treatise *De motu et tempore*, whose importance—despite its shortness and the loss of the Greek text—has been emphasized by R.W. Sharples.[32] Théry surmised that the short writing might have counted as a part of Alexander's lost commentary on the *Physics*,[33] but this hypothesis has been disproved by Sharples, who also discusses the possibility, suggested by F.W. Zimmermann, that the *De motu et tempore* is the treatise that Ibn al-Nadīm alluded to in his *Kitāb al-fihrist*, when he mentioned a *Maqāla* by Alexander *Against Galen on Time and Space*.[34] The Arabic text is extant and has been published by Badawī.[35] The Latin version is attributed to Gerard both in the manuscripts and in the list mentioned above; in the manuscripts it bears the title *Tractatus Alexandri Affrodisii de tempore, translatus a magistro Girardo Cremonensi in Toleto*.[36]

---

[31] Théry, *Autour du décret de 1210*, pp. 83–6, was led astray by the claim of the nineteenth-century scholar F. Wüstenfeld (see above, note 7), who thought that this treatise was the Latin version of Alexander's commentary on the *De sensu et sensato*; Théry correctly remarked that Alexander's treatise is related not to the *De sensu et sensato*, but to *De anima* ii 5, and credited Gerard with the translation. Théry advanced Gerard's authorship also for the Latin version of the commentary on the *De sensu et sensato*, but on this he was surely mistaken: see below, pp. 467–8 and note 45.

[32] R.W. Sharples, 'Alexander of Aphodisias on Time,' *Phronesis* 27 (1982), pp. 58–81 (containing the English translation of the Latin text). The treatise, according to Sharples, 'enables us to see how Aristotle's theory of time was handled and modified by one of the most important of his ancient followers; and it is also of significance as one stage in the series of discussions of time which begins with Plato and continues with the Neoplatonists and beyond' (p. 58). Further bibliography in Sharples, 'Alexander of Aphrodisias,' pp. 1192–3, and Goulet and Aouad, 'Alexandros d'Aphrodisias,' p. 135. See also J. Janssens, 'L'Avicenne Latin: un témoin (indirect) des commentateurs (Alexandre d'Aphrodise—Thémistius—Philopon),' in R. Beyers et al. (eds.), *Tradition et traduction. Les textes philosophiques et scientifiques grecs au Moyen Age latin* (Leuven, 1999), pp. 89–105.

[33] Théry, *Autour du décret de 1210*, p. 97. On the extant traces of Alexander's lost commentary on the *Physics* see M. Rashed, 'Alexandre d'Aphrodise et la *Magna Quaestio*. Rôle et indépendance des scholies dans la tradition byzantine du corpus aristotélicien,' *Les études classiques* 63 (1995), pp. 295–351; idem, 'A "New" Text of Alexander on the Soul's Motion,' in R. Sorabji (ed.), *Aristotle and After* (London, 1997), pp. 181–95.

[34] F.W. Zimmermann and H.W. Brown, 'Neue arabische Übersetzungen aus dem Bereich der spätantiken griechischen Philosophie,' *Der Islam* 50 (1973), pp. 313–24, esp. p. 314; discussion in Sharples, 'Alexander of Aphodisias on Time,' pp. 72–8.

[35] Ed. of the Arabic text in Badawī, pp. 19–24 (title: *Maqālat al-Iskandar al-Afrūdīsī fī 'l-zamān. Tarjamat Ḥunayn ibn Isḥāq*); there is another edition of the Arabic text, with English translation *The Refutation by Alexander of Aphrodisias of Galen's Treatise on the Theory of Motion*, ed. and trans. N. Rescher and M. Marmura (Islamabad, 1965, *non vidi*).

[36] Ed. of the Latin text in Théry, *Autour du décret de 1210*, pp. 92–7; the mss. are the same which served as the basis for the edition of the *De intellectu*: see above, note 23. The quoted title with the mention of Gerard and Toledo appears in two mss., Paris,

Some seventy years later, around 1260 and no later than 1286,[37] other works by Alexander were translated, this time from Greek, by William of Moerbeke: the commentaries on the *Meteorologica*[38] and *De sensu et sensato*,[39] as well as the two writings *On Fate*.[40]

The Latin translation of Alexander's commentary on the *Meteorologica* was completed on 24 April 1260:[41] this is the earliest among the dated translations by William of Moerbeke. The relationship with Moerbeke's translation(s) of Aristotle's *Meteorologica* has been recently studied by G. Vuillemin-Diem.[42]

As for the translation of the commentary on Aristotle's *De sensu et sensato*,[43] William of Moerbeke's authorship is not explicitly mentioned

---

Bibliothèque Nationale de France, *lat.* 16602 and Cambridge, Caius College 996: see Théry, p. 92.

37 The year 1286 is the *terminus ante quem* of all the translations by William of Moerbeke: his death surely antedated 1286 but the exact date remains unknown, whereas a document dated 1286 mentions him as *bone memorie*: see on this and on all the details of William of Moerbeke's bio-bibliography the basic study by M. Grabmann, *Guglielmo di Moerbeke O.P., il traduttore delle opere di Aristotele* (Rome, 1946), esp. pp. 55–6 and A. Paravicini Bagliani, 'Guillaume de Moerbeke et la cour pontificale,' in J. Brams and W. Vanhamel (eds.), *Guillaume de Moerbeke. Recueil d'études à l'occasion du 700ᵉ anniversaire de sa mort (1286)* (Leuven, 1989), pp. 23–52, esp. p. 34.

38 Extant in Greek: ed. Hayduck, *CAG* iii/2.

39 Extant in Greek: ed. Wendland, *CAG* iii/1.

40 Edited by Bruns in the *Supplementum Aristotelicum* (see above, note 21), pp. 164–212; see also *Alexander of Aphrodisias on Fate*, trans. R.W. Sharples (London, 1983). Further bibliography in Sharples, 'Alexander of Aphrodisias,' p. 1187, and Goulet and Aouad, 'Alexandros d'Aphrodisias,' p. 131; also the *Περὶ τύχης* of the *Mantissa* (p. 176, l. 1–p. 179, l. 23) has been translated into Latin: see below, note 48.

41 The title and the colophon of the mss. mentioning the date of the translation and the place where it was made, Nicaea, have been reproduced by the editor of the Latin translation as well as by others: see Alexander of Aphrodisias, *Commentaire sur les Météores d'Aristote. Traduction de Guillaume de Moerbeke*, ed. A.J. Smet (Louvain, 1968), pp. xi–xii. See also Grabmann, *Guglielmo di Moerbeke*, pp. 134–5 and W. Vanhamel, 'Biobibliographie de Guillaume de Moerbeke,' in *Guillaume de Moerbeke*, pp. 301–83, in part. p. 309, 'Aperçu des données biographiques attestées par les documents' and p. 350 (with additional bibliography).

42 G. Vuillemin-Diem, 'Zu Wilhelm von Moerbekes Übersetzung der aristotelischen *Meteorologie*. Drei Redaktionen, ihre griechische Quellen und ihr Verhältnis zum Kommentar des Alexander von Aphrodisias,' in *Tradition et traduction* (quoted above, note 32), pp. 115–66, has shown that William of Moerbeke made a substantial use of his translation of Alexander's commentary on the *Meteorologica* in the third *recensio* of his translation of Aristotle's own work (the so-called *Recensio toletana*, dated by G. Vuillemin-Diem around 1270).

43 The Latin translation has been edited by C. Thurot, 'Alexandre d'Aphrodisias. Commentaire sur le traité d'Aristote *De sensu et sensibili* édité avec la vieille traduction latine,' *Notices et extraits des manuscrits de la Bibliothèque Nationale et autres bibliothèques* 25 (1875).

in the manuscripts, but was advanced by the nineteenth-century editor Ch. Thurot on stylistic grounds; still on stylistic grounds, but the other way round, its attribution to William of Moerbeke was challenged by Théry, who was inclined to attribute the translation to Gerard of Cremona.[44] However, this possibility was ruled out by M. Grabmann in his monograph on William of Moerbeke. Grabmann called attention to the presence of Greek words transliterated, a fact which obviously points to a Graeco-Latin translation.[45] There is a general scholarly consensus on the attribution of this translation to William of Moerbeke,[46] as well as on its importance as a source for Thomas Aquinas' own commentary on this Aristotelian work.[47]

Finally, a Graeco-Latin translation of Alexander's *De fato ad imperatores* and the shorter *De fato* of the *Mantissa* is extant: both translations have been attributed with good reason to William of Moerbeke, even though they are not explicitly ascribed to him.[48]

Between the time of the Arabic-Latin translations of Alexander's small treatises and the appearance of the Greek-Latin translations by William of Moerbeke, Alexander's interpretations of some Aristotelian doctrines—especially his interpretation of the nature and role of intellect according to Aristotle—gained wide currency thanks to the quotations and remarks made by Averroes in his own commentaries on the Aristotelian corpus.[49] To quote but one example, in his 1271 treatise *De unitate intellectus contra Averroistas* Thomas Aquinas openly declares that he owes his knowledge of Alexander's opinion concerning soul and intellect to

---

[44] See above, note 31.

[45] Grabmann, *Guglielmo di Moerbeke*, pp. 132–4; on the Greek words in William of Moerbeke's translations see L.-J. Bataillon, 'L'usage des mots hybrides gréco-latins par Guillaume de Moerbeke,' in *Guillaume de Moerbeke*, pp. 295–9.

[46] See Vanhamel, 'Biobibliographie de Guillaume de Moerbeke,' pp. 350–52.

[47] A. Mansion, 'Le commentaire de saint Thomas sur le *De Sensu et sensato* d'Aristote. Utilisation d'Alexandre d'Aphrodise,' in *Mélanges Mandonnet*, i (Paris, 1930), pp. 83–102; R.A. Gauthier, 'Introduction,' in *S. Thomae Aquinatis opera omnia*, xlv, 2, *Sentencia libri de sensu et sensato* (Rome, 1985), pp. 87–111.

[48] P. Thillet (ed.), *Alexandre d'Aphrodise, De fato ad imperatores. Version de Guillaume de Moerbeke* (Paris, 1963). The Latin translation is anonymous in the four manuscripts on which the edition is based, but Thillet devotes a section of his Introduction (pp. 19–27) to establish William's authorship. Then follows the edition of the Latin version of the *De fato* of the *Mantissa* (pp. 109–16).

[49] The indirect influence of Alexander's doctrines might have been much earlier than this: Théry, *Autour du décret de 1210*, pp. 16–18, points to the indirect knowledge of Alexander's exegesis of the Organon through Boethius.

Averroes,[50] even though the Latin version of Alexander's *De intellectu* was in principle available to him on chronological grounds.

Even from this scanty survey one gets the impression that the *Alexander Latinus* has quite a different profile with respect to the Greek and Arabic ones. I shall not venture to sum up in a short and by the same token questionable account the prominent features of the genuine Alexander of Aphrodisias; but it is apparent that the Arabic and the Latin Alexander each have a particular profile, different from one another and from that of the Greek Alexander. One may be tempted to say that the Arabic Alexander is chiefly the author of the *Principles of the Universe*[51] and the Latin one chiefly the author of the *De intellectu*, were this not a generalization of little or no scientific value. More promising is the remark that our knowledge of the *Alexander Latinus* is incomplete: for instance, the critical edition of the Latin version of the commentary on the *De sensu et sensato* is still a desideratum.[52] Better knowledge of the Latin versions of Alexander's works may also help to solve some

[50] *De unitate intellectus contra Averroistas*, ed. Leonina, p. 302, l. 93–103: 'Quod autem Alexander intellectum possibilem posuerit esse formam corporis, etiam ipse Auerroys confitetur; quamuis, ut arbitror, peruerse uerba Alexandri acceperit, sicut de uerba Themistii preter eius intellectum assumit. Nam quod dicit, Alexandrum dixisse intellectum possibilem non esse aliud quam preparationem que est in natura humana ad intellectum agentem et ad intelligibilia: hanc preparationem nichil aliud intellexit quam potentiam intellectiuam que est in anima ad intelligibilia'. Here, as well as in the *Summa theol.*, i, q. 88, a. 1 *resp.* and in the *Summa contra gentiles* ii, 62, Thomas refers to Alexander's position as presented by Averroes (wrongly, in his opinion: cf., in the quoted passage of the *Summa theol.*, 'ut ipse imponit Alexandro'). Thomas' source is the Great Commentary on the *De anima*, iii, 1, p. 393.196 p. 395.257 Crawford. At first sight one would say that Averroes is referring to Alexander's lost commentary on the *De anima*, since he himself is commenting on the *De anima* and says 'Alexander autem sustentatur *super hunc sermonem*...' (p. 393, l. 196 Crawford, my emphasis); but later on it appears that Averroes' sources are Alexander's *De anima* and *De intellectu*: 'Et hoc aperte et universaliter propalavit in initio libri sui de Anima (...) Et dixit etiam in tractatu quem fecit de Intellectu secundum opinionem Aristotelis' (p. 394, l. 214–15 and 216–17).

[51] The *Treatise by Alexander of Aphrodisias on the Discourse about the Principles of the Universe according to Aristotle, the Philosopher* (*Maqālat al-Iskandar al-Afrūdīsī fī 'l-qawl fī mabādi' al-kull bi-ḥasab ra'y Arisṭāṭālīs al-faylasūf*), lost in Greek, is extant in Arabic in two versions. It has been edited by 'A. Badawī, *Arisṭū 'inda 'l-'Arab. Dirāsa wa-nuṣūṣ ghayr manshūra* (Cairo, 1947), pp. 253–77 and by C. Genequand, *Alexander of Aphrodisias On the Cosmos* (Leiden, 2001). See also G. Endress, 'Alexander Arabus on the First Cause. Aristotle's First Mover in an Arabic Treatise Attributed to Alexander of Aphrodisias,' in *Aristotele e Alessandro di Afrodisia nella tradizione araba* (quoted above, note 12), pp. 19–74.

[52] Vanhamel, 'Biobibliographie de Guillaume de Moerbeke,' p. 351 remarks that 'En attendant l'édition critique, on ne dispose que de l'édition de 1875, basée uniquement sur le manuscrit de Paris, qui est nettement le moins bon des quatre témoins'.

minor, but interesting, questions still open and may shed some light on the intricacies of the transmission of his corpus. A case in point is the treatise *On Unity* whose attribution to Alexander is still under discussion in contemporary scholarship. Reading the Latin text which attributes this treatise to Alexander is equal to undermining such an attribution, to the effect that *On Unity* should definitely be cancelled from the list of Alexander's works: it is to this admittedly minor point that this note is devoted.

II

Albeit not unprecedented,[53] the survey on the Arabic translations of Greek works published between 1889 and 1896 by Moritz Steinschneider[54] was in a sense the starting point of a new story. Following in the footsteps of his predecessors, Steinschneider embarked upon an analysis of the Graeco-Arabic heritage; at variance with them, he paid special attention to the critical evaluation of the sources[55] and, most essential for present purposes, took into account systematically the Hebrew and Latin translations.[56] After a still valuable account of the Arab bio-bibliographical sources, Steinschneider turned to analysing the information given on the Greek philosophers therein. His entry on Alexander of Aphrodisias was based on the *Kitāb al-fihrist* by Ibn al-Nadīm, compared with al-Qifṭī's *Tarīkh al-ḥukamā'* and Ibn Abī Uṣaybiʿa's *'Uyūn*

[53] The seminal works in the field were J.G. Wenrich, *De auctorum graecorum versionibus et commentariis syriacis arabicis armeniacis persicisque commentatio* (Leipzig, 1842) and A. Jourdain, *Recherches critiques sur l'âge et l'origine des traductions latines d'Aristote et sur des commentaires grecs ou arabes employés par les docteurs scolastiques* (Paris, 1819 and 1843; repr. New York, 1960).

[54] M. Steinschneider, *Die arabischen Übersetzungen aus dem Griechischen* (Graz, 1960): this is the reprint of a series of essays published between 1889 and 1896.

[55] Stenschneider, *Die arabischen Übersetzungen aus dem Griechischen*, p. 7: 'Es ist klar, dass diese kritische Frage—i.e., the question whether or not all the Greek works mentioned in the bio-bibliographical sources were actually translated into Arabic—eine Frage nach den Quellen ist. (...) was aber zu wünschen bleibt, ist eine strenge Kritik der Quellen hinsichtlich ihrer Affiliation und eine bequemere Anordnung'.

[56] Steinschneider, *Die arabischen Übersetzungen aus dem Griechischen*, p. 13: 'Nachdem ich so die arabischen Quellen verfolgt habe, gehe ich zu zwei von Wenrich nur sehr wenig benutzten Supplementärquellen über. (...) Ich meine die aus den arabischen stammenden hebraïschen und lateinischen Uebersetzungen'. Steinschneider willingly acknowledged his debt towards the schoolmen who before him had dealt with the Arabic-Latin translations, especially F. Wüstenfeld (see above, note 7), but, at variance with them, took into account also the anonymous translations and made a systematic survey of the catalogues of the mss. in the Western libraries.

al-anbāʾ fī ṭabaqāt al-aṭibbāʾ. Steinschneider also checked the information given by the ancient sources against the accounts of some of his immediate predecessors (Wenrich and Leclerc).[57] In addition, he included an account on seventeen works by Alexander translated into Arabic, listed by Miguel Casiri in his eighteenth-century catalogue of the Arabic mss. of El Escorial.[58] Finally, he called attention to five other works attributed to Alexander in Arabic, Hebrew and Latin translations.[59]

The last item in this series was a treatise *De unitate*. Steinschneider provided the reader with the following information: (i) the work was translated by Gerard of Cremona; (ii) in one of the two manuscripts, housed in Paris, it was attributed to Alexander in the title, but to al-Kindī in the colophon; (iii) Leclerc favoured the Kindian authorship, in consideration of the fact that among al-Kindī's works a treatise *Fī 'l-tawḥīd* is recorded; (iv) in the other manuscript, housed in Rome, the treatise was given to Alexander also in the colophon; (v) in the entry on Alexander of Ibn Abī Uṣaybiʿa's *ʿUyūn al-anbāʾ* two titles appear, each of which might be a candidate to represent the Arabic antecedent of the Latin *De unitate*: a treatise *Fī 'l-tawḥīd* and the so-called *Opinions of the Philosophers on God's Unicity, Ārāʾ al-falāsifa fī 'l-tawḥīd*; (vi) the Latin translation begins as follows: 'Hic liber potest dividi in partes'.[60]

---

[57] For Wenrich see above, note 53; L. Leclerc, *Histoire de la médecine arabe. Exposé complet des traductions du grec, les sciences en Orient, leur transmission à l'Occident par les traductions latines* (Paris, 1876; repr. New York, 1971).

[58] Steinschneider, *Die arabischen Übersetzungen aus dem Griechischen*, pp. 94–6, quoting the contents of ms. El Escorial, Biblioteca del Monasterio de San Lorenzo, ms. 794, from M. Casiri, *Bibliotheca Arabico-Hispana Escurialensis* (...), i–ii (Madrid, 1760–70; repr. Osnabruck, 1969), i, p. 242.

[59] Steinschneider, *Die arabischen Übersetzungen aus dem Griechischen*, pp. 97–8, lists the following works: (i) *Kitāb al-nafs*, which he translated into German on behalf of Ivo Bruns for his edition of Alexander's *De anima* in the *Suppl. Ar.* i; (ii) *Fī 'l-zamān*, namely, the *De tempore* translated into Latin by Gerard of Cremona (see above, note 20); (iii) a historical work; (iv) a commentary by Alexander on the *Physiognomica*, which Steinschneider suspected to have something to do with the *Secretum secretorum*, in which case the Alexander alluded to would have been Alexander the Great, namely, the character of the *Secretum secretorum*; (vi) our *De unitate*.

[60] Steinschneider, *Die arabischen Übersetzungen aus dem Griechischen*, p. 97. Leclerc, *Histoire de la médecine arabe*, p. 494, had pointed to al-Kindī's authorship (as recalled by Steinschneider) as follows: 'L'incipit donne cet opuscule sous le nom d'Alexandre (d'Aphrodisias) tandis que l'explicit le donne sous celui d'Alkindy. Cette dernière attribution nous paraît préférable par la raison que nous trouvons dans la liste bibliographique d'Alkindy un livre intitulé Ettouhid, de l'unité de Dieu'.

The information given by Steinschneider, which I shall return to shortly, is reflected in the two contemporary surveys on Alexander mentioned above. Sharples, referring to Steinschneider, relies on the attribution to al-Kindī and credits Gerard of Cremona with the translation;[61] Goulet-Aouad, albeit warning that the work might be spurious, list the *De unitate* among the works by Alexander which seem to be completely lost and, referring in turn to Steinschneider, advance the hypothesis that the *De unitate* might correspond to the *Opinions of the Philosophers on God's Unicity*.[62]

The manuscripts containing the *De unitate* attributed to Alexander (or to Alexander and al-Kindī) are the following: Paris, Bibliothèque Nationale de France, *lat.* 6443,[63] correctly indicated by Steinschneider, and Rome, Biblioteca Angelica, 242 (*olim* C. 4. 10), mistakenly indicated by Steinschneider as H 10, n. 11.[64] In addition, there is the ms. Paris,

---

[61] Sharples, 'Alexander of Aphrodisias,' p. 1199.

[62] Goulet and Aouad, 'Alexandros d'Aphrodisias,' p. 138, indicating through an asterisk that the work might be spurious. S. Fazzo, 'Alexandros d'Aphrodisias,' in Goulet (ed.), *Dictionnaire des philosophes antiques*, Supplément (Paris, 2003), p. 69, states that Prof. Burnett had indicated to her that the *De unitate* ascribed to Alexander is in fact the *De unitate et uno* by Dominicus Gundissalinus.

[63] This is a well known ms., containing as it does the Avicenna Latinus, the *De processione mundi* by Dominicus Gundissalinus, as well as the Latin al-Kindī, al-Fārābī and al-Ghazālī. It was known to Wüstenfeld and Jourdain (see above, note 7 and 53), to A. Nagy, *Die philosophischen Abhandlungen des Ja'qūb ben Isḥāq al-Kindī* (Münster, 1897), pp. xxx–xxxi; to C. Baeumker, *Alfarabi. Ueber den Ursprung der Wissenschaften (De ortu scientiarum)* (Münster, 1916), to G. Bülow, *Des Dominicus Gundissalinus Schrift Von dem Hervorgange der Welt (De Processione mundi)* (Münster, 1925), pp. x–xi, to Gilson (see above, note 11), pp. 109–10, to Théry, *Autour du décret de 1210*, p. 73 (see also below, note 71), to M.-Th. D'Alverny, 'Avicenna Latinus,' *AHDLMA* 28 (1961), pp. 281–316, esp. p. 310. As for the Latin corpus of Avicenna's writings, it has been examined by S. Van Riet from the viewpoint of the so-called 'lections doubles' of the text of the *De anima*: see Avicenna Latinus, *Liber de anima seu sextus de naturalibus*, ed. S. Van Riet, i–iii (Louvain, 1972), p. 110* (whith the important remark that the text of the *De anima* given by this ms. is not the full text but an *abreviatio*). As for the text of Avicenna's *Metaphysics*, S. Van Riet lists it among the testimonies of the so-called 'texte revu': Avicenna Latinus, *Liber de philosophia prima sive scientia divina*, ed. S. Van Riet, i–iv (Louvain, 1977), p. 129*. As for the Latin version of Ghazālī's *Maqāṣid al-falāsifa*, this ms. is one of the testimonies of the separation between the treatise on logic and those on the metaphysics and physics—a separation which is one of the salient features of the Latin version of Ghazālī's *Maqāṣid*. See D. Salman, 'Algazel et les latins,' *AHDLMA* 10–11 (1935–36), pp. 103–27, esp. p. 121. Even though 'ce ms classique a été maintes fois décrit' (so Salman, p. 121 n. 1), I have found nowhere a full list of its contents, which might be of some use to dress here (even though there are some small treatises which I have not been able to identify): see below, *Appendix ii*.

[64] I owe to the kindness of Dr. Mirella Fidomanzo, Biblioteca Angelica, the indication of the true shelfmark of the ms. May Dr. Fidomanzo find here my most sincere thanks for her decisive help.

Bibliothèque Nationale de France, *lat.* 6325, which labels the treatise as *Tractatus Alexandri de unitate translatus de greco in latinum.*[65] The inspection of two of these manuscripts[66] leaves no doubt: the work is neither by Alexander of Aphrodisias nor by al-Kindī: instead, it is the *De unitate et uno* by Dominicus Gundissalinus, the famous pseudo-Boethian writing edited in 1891 by Paul Correns[67] and then by Manuel Alonso S.J. in 1956.[68] A comparison of the beginning and end of the *De unitate* attributed to Alexander with Gundissalinus' *De unitate et uno* will suffice to settle the issue. Since the mss. Paris, Bibliothèque Nationale de France, *lat.* 6325 and 6443 were known to the editors of the *De unitate et uno* (whereas the ms. Rome, Biblioteca Angelica, 242 was not), I will quote only the Rome ms.

Gundissalinus' *De unitate et uno*, ed. Correns p. 3.2–5	Ms. Rome, Biblioteca Angelica, 242, fol. 27 va
Unitas est, qua unaquaeque res dicitur esse una. Sive enim sit simplex sive composita, sive sit spiritualis sive corporea: res unitate una est; nec potest esse una nisi unitate, sicut nec alba nisi albedine, nec quanta nisi quantitate.	Unitas est <qua> unaqueque res dicitur esse una. Sive enim sit simplex sive composita, sive spiritualis sive corporea, res unitate est una, nec potest esse una nisi unitate, sicut nec alba nisi albedine nec quanta nisi quantitate.

[65] This ms., which I have not seen, is mentioned by P. Correns, *Die dem Boethius fälschlich zugeschriebene Abhandlung des Dominicus Gundisalvi* De unitate (Münster, 1891), p. 12, and by M. Alonso, 'El *Liber de unitate et uno*,' *Pensamiento* 12 (1956), pp. 65–77 and 179–202, p. 65 note 1.

[66] I have read the Paris ms. on a microfilm and the Rome ms. in the Biblioteca Angelica.

[67] See above, note 65.

[68] See above, note 65. See also M. Alonso, 'El *Liber de unitate et uno*. Gundisalvo intérprete de sí mismo,' *Pensamiento* 13 (1957), pp. 159–99. A German translation of the *De unitate et uno* has been recently provided by A. Fidora and A. Niederberger, *Vom Einen zum Vielen. Texte des Neuplatonismus im 12. Jahrhundert* (Frankfurt, 2002), pp. 66–79 (with facing Latin text).

Gundissalinus' *De unitate et uno*, ed. Correns, p. 11.4–14	Ms. Rome, Biblioteca Angelica, 242, fol. 27 rb
Unde aperte datur intelligi, quod discretae et continuae quantitatis radix una est, eo quod composita sunt ex una re et resolvuntur ad unum; et etiam quia partes corporis, quo magis fuerint sibi coniunctae et constrictae, ipsum corpus erit spissius et magis quantum, ut lapis, et, e contrario, quo magis fuerint partes corporis dissolutae et rarae, ipsum erit subtilius et levius et minus quantum, ut aer. Verum est igitur, quod continua quantitas non venit in substantiam nisi ex coniunctione et confluctione unitatum in illa. Unitas igitur est qua unaquaeque res est una et est id quod est.	Unde aperte datur intelligi quod discrete et continue quantitatis radix una est, eo quod composita sunt ex una re et resolvuntur ad unum, et etiam quia partes corporis quo <magis> fuerint sibi coniuncte et constricte, ipsum corpus erit spissius et magis quantum, ut lapis; et e contrario, quo magis fuerint partes corporis dissolute et rare, ipsum erit subtilius et levius et minus quantum, ut aer. Verum est igitur quod continua quantitas non venit in substantiam nisi ex coniunctione et constructione unitatum in illa. Unitas igitur est qua unaqueque res dicitur esse una.

In the subsequent edition by M. Alonso, the sentences 'unitas est qua unaquaeque res dicitur esse una' and 'discretae... unitatum in illa' are put in quotation marks. The reason lies in that Alonso, following in Correns' footsteps, paid much attention to the sources and detected in the two sentences as many quotations.[69] It had been precisely the analysis of the sources which had enabled first Hauréau and then Correns to demonstrate the pseudo-epigraphical nature of the *De unitate et uno*, formerly attributed to Boethius. In all likelihood on the basis of the misattribution in some Latin manuscripts,[70] the *De unitate et uno* had in fact been published within Boethius' works in the two editions by Henry Lorit (Glareanus) at Basle (1546 and 1570) and hence in the nineteenth-century edition within the *Patrologia Latina*.

Correns was aware of the fact that the *De unitate et uno* had been attributed also to Alexander of Aphrodisias, first because the ms. Paris,

[69] Alonso, 'El *Liber de unitate et uno*,' pp. 69 and 77, refers to Euclid's *Elements* and to pseudo-Dionysius the Areopagite's *De divinis nominibus* for the first sentence, to the *Fons vitae* for the second one.

[70] Cf. Correns, *Die dem Boethius fälschlich zugeschriebene Abhandlung*, p. 12.

Bibliothèque Nationale de France, *lat.* 6443 was one of his three basic manuscripts,[71] and second because this attribution had already been discussed by Barthélemy Hauréau, who, in his essay on the origins of the pantheistic doctrine proclaimed by David of Dinant,[72] pointed to the *De unitate et uno* and was the first to substantiate doubts on the Boethian authorship of this writing, authoritatively advanced in the thirteenth century by no less a scholar than Thomas Aquinas.[73] In search of the genuine author of the *De unitate et uno*, Hauréau took into account also the alleged authorship of Alexander of Aphrodisias. It had been easy for him to disprove Alexander's authorship and Correns limited himself to remarking, with Hauréau, that the presence of three quotations from the New Testament[74] and of echoes from Augustine was enough to rule out the possibility for the writing to trace back not only to Alexander, but also to the other candidate of the ms. Paris, Bibliothèque Nationale de France, *lat.* 6443, namely, al-Kindī. It took one and a half pages for Correns to establish that 'weder Alexander noch ein Araber sind Verfasser der Schrift'.[75]

To disprove Boethius' authorship was a bit more complicated, because the *De unitate et uno* contains several distinctive ideas and even verbatim quotations from Boethius' works, a fact which explains the attribution of the treatise to him in some manuscripts, as well as the citation of its opening sentence as Boethius' by Alanus of Lille.[76] However, the presence of quotations from an author much later than Boethius, namely, Salomon ibn Gabirol (*c.* 1021–*c.* 1058), ruled out this authorship too

[71] See below, *Appendix i*. The basic mss. of Correns' edition are Paris, Bibliothèque Nationale de France, *lat.* 16605, *lat.* 14700 and *lat.* 6443. Also, it is worth mentioning that it did not escape to Théry, *Autour du décret de 1210*, p. 73, that the treatise labelled as *Liber Alexandri de unitate* in the ms. Paris, Bibliothèque Nationale de France, *lat.* 6443 was nothing but Gundissalinus' *De unitate et uno*.

[72] B. Hauréau, 'Mémoire sur la vraie source des erreurs attribuées à David de Dinant,' *Mémoires de l'Académie des Inscriptions et Belles Lettres* 29 (1879), pp. 319–30.

[73] Aquinas' disclaimers of Boethius' authorship (*Quodl.* ix a. 6 and *De Spir. creat.*, 1) are quoted by Alonso, 'El *Liber de unitate et uno*,' p. 190. In both passages Thomas says in as many words that the *De unitate et uno* 'non est Boetii'.

[74] The three passages (p. 10, l. 5, p. 10, l. 11 and 12–13 Correns = p. 76, l. 136, 140–41 and 142 Alonso) come respectively from i *Io.* 5, 8, *Act.* 4, 32 and i *Cor.* 6, 16.

[75] Correns, *Die dem Boethius fälschlich zugeschriebene Abhandlung*, pp. 14–15.

[76] Cf. Correns, *Die dem Boethius fälschlich zugeschriebene Abhandlung*, p. 17 and Alonso, 'El *Liber de unitate et uno*,' pp. 185–6.

and left Correns with no other possibility than a writer who had had access to Ibn Gabirol's *Fons vitae*.[77] By means of textual comparisons, Correns demonstrated that the *De unitate et uno* heavily borrowed from the *Fons vitae*, thus narrowing the focus to a Christian writer who was conversant with Boethius and had access to the *Fons vitae*. Relying on one of the manuscripts of the *De unitate et uno* which seems to attribute the treatise to Dominicus Gundissalinus[78]—who translated so many works from Arabic into Latin,[79] and moreover cooperated with John of Spain in the translation of Ibn Gabirol's *Fons vitae*[80]—Correns explored

[77] C. Baeumker, *Avecenbrolis (Ibn Gebirol) Fons vitae ex arabico in latinum translatus ab Iohanne Hispano et Dominico Gundissalino* (Münster, 1892–95). Cf. also J. Schlanger, *La philosophie de Salomon Ibn Gabirol. Étude d'un néoplatonisme* (Leiden, 1968).

[78] According to Correns, *Die dem Boethius fälschlich zugeschriebene Abhandlung*, p. 14, the ms. Oxford, Corpus Christi 86 gives the treatise to Gundissalinus, but Alonso, 'El *Liber de unitate et uno*,' p. 65, n. 1, challenges this claim: 'La atribución aquí a Gundisalvo no parece tan expresa, como afirmaban Hauréau, Correns, Bonilla etc.'. I have not seen the ms., but a note by Correns sheds light on this point: the attribution to Gundisalvi comes from the catalogue of the mss. at Corpus Christi college by Coxe (Correns also declared that 'Genauere Nachrichten über die Oxforder Handschrift zu erlangen war mir nicht möglich') and is grounded on the fact that, as in one of the basic mss. of the *De unitate et uno* (Paris, Bibliothèque Nationale de France, *lat.* 14700), this treatise appears as the last chapter of Gundissalinus' own *De divisione philosophie* (not to be confused with the Latin translation of Farābī's *Iḥṣā' al-ʿulūm*).

[79] M. Alonso, 'Traducciones del arcediano Domingo Gundisalvo,' *Al-Andalus* 12 (1947), pp. 295–338, credited Gundissalinus with the translation of Farābī's *Iḥṣā' al-ʿulūm*, *ʿUyūn al-masā'il*, *Kitāb al-tanbīh ʿalā sabīl al-saʿāda* and *On the Intellect*; al-Kindī's *On the Intellect*, Alexander of Aphrodisias' *On the Intellect*, Isaac Israeli's *Book of Definitions*, the *Liber introductorius in artem logicae demonstrationis* by Muḥammad ibn Maʿshar al-Bustī, Ghazālī's *Maqāṣid al-falāsifa*, Avicenna's *Metaphysics* and the so-called *De convenientia et differentia subiectorum*. Afterwards, the attribution of some of these translations has been discussed: for an up-to-date account see M.-Th. D'Alverny, 'Translations and Translators,' in R.L. Benson and G. Constable (eds.), *Renaissance and Renewal in the Twelfth Century* (Cambridge, Mass., 1982), pp. 421–62 and H. Daiber, 'Lateinische Übersetzungen arabischer Texte zur Philosophie und ihre Bedeutung für die Scholastik des Mittelalters. Stand und Aufgaben der Forschung,' in J. Hamesse and M. Fattori (eds.), *Rencontres de cultures dans la philosophie médiévale. Traductions et traducteurs de l'Antiquité tardive au XIVe siècle* (Louvain, 1990), pp. 203–50. On Gundissalinus see now A. Fidora, *Die Wissenschaftstheorie des Dominicus Gundissalinus. Voraussetzungen und Konsequenzen des zweiten Anfangs der aristotelischen Philosophie im 12. Jahrhundert* (Berlin, 2003). Finally, it is worth mentioning that the fact that the translator and the author of the philosophical treatises are one and the same person has been challenged by A. Rucquoi, 'Gundisalvus ou Dominicus Gundisalvi?,' *Bulletin de philosophie médiévale* 41 (1999), pp. 85–106; see the critical remarks by A. Fidora and M.-J. Soto Bruna, 'Gundisalvus ou Dominicus Gundisalvi? Algunas observaciones sobre un reciente artículo de Adeline Rucquoi,' *Estudios Eclesiásticos* 76 (2001), pp. 467–73.

[80] Correns, *Die dem Boethius fälschlich zugeschriebene Abhandlung*, p. 33, took into account the famous colophon of the ms. Paris, Bibliothèque Mazarine 3472, which assesses the cooperation between Iohannes Ispanus and Dominicus Gundissalinus in the translation

this possibility in depth and came to the conclusion that the author of the *De unitate et uno* was Gundissalinus. He also compared some passages in other works by him with Boethius and the *Fons vitae* (namely, the main sources of the *De unitate et uno*), showing that the composition technique was exactly the same. Finally, he listed some striking parallel passages in the *De unitate et uno* and in other works by Gundissalinus. The latter's authorship, established on such firm grounds, has not been questioned since then.

When Manuel Alonso provided a new edition of Gundissalinus' *De unitate et uno*, he relied on Correns' research and added two main points to the dossier: a detailed analysis of the sources, parallel passages and later citations of the *De unitate et uno*, and a new hypothesis on Gundissalinus' aims in gathering together Boethius' and Ibn Gabirol's formulae on unity and its causal role with respect to all beings. Instead of being a mere compilation, the *De unitate et uno* was, in Alonso's opinion, a strong reaction against the unacceptable emanationist ideas of the *Fons vitae*, which Gundissalinus wanted to counter by inserting in the strategical points of the compilation some doctrinal remarks of his own.[81] As interesting as it may be, to discuss Alonso's interpretation would exceed the scope of this paper; still, it seems to me to be worth mentioning, as the most effective way of giving an idea of the nature and contents of this short treatise without reproducing it in full.

We are now in a position to correct Steinschneider's statements (i), (iii) and (v). The treatise *De unitate* is not a translation by Gerard of Cremona and, although attributed to al-Kindī in the colophon of the

---

of the *Fons vitae* ('non absque iuvante Domingo'). M. Alonso, 'Juan Sevillano, sus obras proprias y sus traducciones,' *Al-Andalus* 18 (1953), pp. 17–49, maintaned that Magister Iohannes Hispanus, John of Seville and Avendauth were one and the same person, but the identity between the two 'Iohannes' has been challenged by L. Thorndike, 'John of Seville,' *Speculum* 34 (1959), pp. 20–38; M.-T. D'Alverny, 'Notes sur les traductions médiévales d'Avicenne,' *AHDLMA* 19 (1952), pp. 337–58, esp. pp. 345–6; Ead., 'Avendauth,' in *Homenaje a Millás-Vallicrosa* (Barcelona, 1954–56), i, pp. 19–43, esp. pp. 23–4; J.F. Rivera Recio, 'Nuevos datos sobre los traductores Gundisalvo y Juan Hispano,' *Al-Andalus* 31 (1966), pp. 267–80 and C. Burnett, '*Magister Iohannes Hispanus*: towards the Identity of a Toledan Translator,' in *Comprendre et maîtriser la nature au Moyen Age. Mélanges d'histoire des sciences offerts à Guy Beaujouan* (Geneva, 1994), pp. 425–36.

[81] Alonso, 'El *Liber de unitate et uno*. Gundisalvo intérprete de sí mismo,' p. 186: 'Por lo dicho se verá que la verdadera finalidad del *De unitate et uno* es la de deshacer las construcciones apriorísticas de Ibn Gabirol. Pero, al mismo tiempo, con los tres géneros primitivos de Gundisalvo con todos sus individuos salidos inmediatamente de Dios, quedaban arruinados el *Liber de causis* con su tríada descendente, el sistema de al-Fārābī y el de Ibn Sīna, refundido por al-Ghazālī y todos los sistemas fundados en el *Ab uno non venit nisi unum*'.

ms. Paris, Bibliothèque Nationale de France, *lat.* 6443 (as correctly stated by Steinschneider in item [ii]),[82] it bears no relationship with al-Kindī's lost treatise(s) *Fī 'l-tawḥīd*,[83] nor does it bear any with the treatise *On the Opinions of the Philosophers on God's Unicity*, with which Alexander is credited (in all likelihood wrongly) by Ibn Abī Uṣaybiʿa. Steinschneider's items (iv) and (vi) are true but require some additional comment. Both refer, one explicitly (even though under an erroneous shelfmark) and the other implicitly, to the ms. Rome, Biblioteca Angelica, 242. Before coming back to Steinschneider's (iv) and (vi), I would like to sum up the main information on this manuscript, based on the catalogue of the library made by the nineteenth-century scholar Enrico Narducci[84] and on my own inspection of it.

The manuscript is ascribed by Narducci to the beginning of the fourteenth century. In very tiny handwriting, it contains the Latin translation of works by Aristotle, Alexander of Aphrodisias, Galen, al-Kindī, Avicenna, al-Ghazālī, plus some writings of Latin authors: a list of the contents will be found below, *Appendix ii*.

On fol. 26rb, lines 32–4, one can read the following colophon of the Latin translation of Alexander's *On the Intellect*: 'Explicit liber Alexandri philosophi de intellectu et intellecto secundum sententiam Aristotelis translatus de greco in arabicum ab Ysaac filio Iohannitii'. This colophon has been cancelled in red. Immediately after, namely, at lines 35–8, the following words can be read, written by the same hand as the main text and marked in the margin as *glosula*:

> Hic liber potest dividi in quattuor partes. In primo (*pro*: prima) describit quod est formaliter unitas et sicut habet esse in re. Secundo ostendit quod

[82] Ms. Paris, Bibliothèque Nationale de France, *lat.* 6443, fol. 194ra lin. 41–2: *Explicit liber de unitate Alquindi.*

[83] In the list by Ibn al-Nadīm in his *Kitāb al-fihrist* (ed. G. Flügel [Leipzig, 1871–72]), al-Kindī is credited with three writings *On Unity*. All of them are lost; they are the *Risāla fī 'l-tawḥīd min jihat al-aʿdād* (*On Unity from the Point of View of Numbers*, ed. Flügel, p. 256, l. 26 = n. 46 in the list by G. Atiyeh, *Al-Kindī. The Philosopher of the Arabs* [Rawalpindi, 1966]), the *Risālat al-tawḥīd bi-tafsīrāt* (*On Unity with Expositions*, ed. Flügel, p. 259, l. 19 = n. 185 Atiyeh), and the *Risāla fī 'l-iftirāq al-milal fī 'l-tawḥīd* (*On the Differences among Religious Communities on Unicity*, ed. Flügel, p. 259, l. 20–21 = n. 189 Atiyeh). Atiyeh also lists as n. 192 a *Kalām maʿa Ibn al-Rāwandī fī 'l-tawḥīd* (*Discourse with Ibn al-Rāwandī on Unity*), which is not mentioned in the *Kitāb al-fihrist.*

[84] *Catalogus codicum manuscriptorum praeter graecos et orientales in Bibliotheca Angelica olim coenobii Sancti Augustini de Urbe* integrum confecit, adnotationibus instruxit, indicibus locupletavit, privatis impensis publicae studiosorum commoditati edidit Henricus Narducci (Rome, 1893), i, pp. 138–40.

> res non solum habent esse ab hac unitate. Tertio a quo omnis res habet esse. Quarto ostendit ad quod tendat motus uniuscuiusque rei.

Still on the same folio 26rb, at lines 39–50, by the same hand and marked in the margin as *alia glosula*, the following account can be read:

> In prima parte determinat auctor diversitatem inter creantem unitatem et unitates creatas et etiam inter creatas unitates inter se secundum maiorem appropinquationem proprie et vere idest creanti (idest creanti *add. s.l.*) unitati et elongationem ab ipsa et maiorem subtilitatem et spissitudinem ipsius materie. In secunda parte ostendit quod propter maximum et optimum effectum unitatis omnino ad unum tendunt et quod omnia etiam similia sunt, unum dici volunt et appetunt, et quod omnis multitudinis radix est unitas. In tertia parte ostendit quod propter diversitatem nature (?), scilicet secundum appropinquationem vere unitati vel elongationem ab ipsa, idest secundum subtilitatem et essitudinem (*pro*: spissitudinem) ipsius nature (?), non uno modo sed pluribus modis dicitur aliquid unitate unum, et hoc debet esse superius. Incipit liber philosophi de unitate.

The words 'Incipit liber philosophi de unitate' are the last line (namely, line 50, as in the entire ms.) of fol. 26rb. At fol. 26va, line 1, Gundissalinus' *De unitate et uno* begins under the label *Incipit liber Alexandri de unitate* (see in the *Appendix i* below the variant readings). At fol. 27rb, line 12, the following colophon can be read: *Explicit liber Alexandri philosophi de unitate.*

This elicits some qualifications on Steinschneider's items (iv) and (vi). As stated by Steinschneider (iv), the ms. of the Biblioteca Angelica contains a treatise *De unitate* attributed to Alexander both in the title and in the colophon, which is but the *De unitate et uno* by Gundissalinus. A misguided doubt on the authorship led someone (maybe the same scholar who wrote the marginal remarks *glosula* and *alia glosula*?) to cancel the correct colophon of the Latin translation of Alexander's *On the Intellect*. Steinschneider's claim (vi), namely, that the translation begins by the words 'Hic liber potest dividi in partes' should be corrected, first because the work is not a translation but a treatise, and second because this is the beginning not of the treatise, but of the first *glosula* quoted above, giving a rough account of its contents.

I must confess that I would have been much happier to find, concealed under the pseudo-Alexander's *De unitate*, some trace in Muslim Spain of one of the *Risālāt* on the *tawḥīd* with which al-Kindī is credited in the Arab bio-bibliographical sources. Unfortunately, this seems not to be the case: the various attributions seem to be nothing but understandable mistakes. The attribution to al-Kindī in the colophon of the ms. Paris, Bibliothèque Nationale de France, *lat.* 6443 can be the reverberation of the proximity of other Kindian treatises in Gerard's translation. In turn, its attribution to Alexander in the title of the same manuscript, as well as in the title and colophon of the Rome manuscript, in all likelihood depends upon the presence in the close vicinity of other treatises by Alexander in Latin translation. The change of authorship between the title and colophon in the Paris manuscript on the one hand, and, on the other, the hesitation in the manuscript of the Angelica (albeit misplaced) may both give a hint in this direction, suggesting that Alexander's alleged authorship of the *De unitate* was far from being unchallenged, even by those who did not recognize in it the *De unitate et uno* by Gundissalinus—another writing whose authorship is extremely uncertain in the manuscripts.

This is a meagre result indeed. However, one may wonder if, notwithstanding all this, the *De unitate et uno* by Gundissalinus does not bear in itself an actual trace of the circulation of al-Kindī's ideas (although not treatises) on the *tawḥīd*. Containing as it does the distinction between the *creans unitas* and the *create unitates*[85] which is so clearly reminiscent of the *Liber de causis*,[86] the pseudo-Alexander may count as an echo of a

[85] *De Unitate et uno*, ed. Correns, p. 5, l. 15–p. 6, l. 13: 'Prima enim et vera unitas, quae est unitas sibi ipsi, creavit aliam unitatem, quae esset infra eam. Sed quia omne creatum omnino diversum est ab eo, a quo creatum est, profecto creata unitas a creante unitate omnino diversa esse debuit et quasi opposita. Sed quia creatrix unitas non habet principium neque finem nec permutationem nec diversitatem, ideo creatae unitati accidit multiplicitas et diversitas et mutabilitas; ita ut in quadam materia sit habens principium et finem, in quadam vero principium et non finem, quia in quibusdam subiacet permutationi et corruptioni, in quibusdam permutationi sed non corruptioni. (…) Quanto enim unaquaeque unitas fuerit propinquior primae et verae unitati, tanto materia formata per illam erit magis una et simplicior; et e contrario, quanto remotior fuerit a prima unitate, tanto erit multiplicior et compositior'.

[86] *Liber de causis*, prop. 31(32), ed. A. Pattin, 'Le *Liber de causis*. Édition établie à l'aide de 90 manuscrits avec introduction et notes,' *Tijdschrift voor Filosofie* 28 (1966), pp. 90–203, esp. p. 201, l. 89–p. 203, l. 16: 'Necessarium est unum faciens adipisci unitates et ipsum non adipiscatur, sed reliquae unitates omnes sunt acquisitae. Et illius quidem significatio est quod dico: si invenitur unum faciens acquirere non acquisitum, tunc quae differentia inter ipsum et primum acquirere faciens? (…) Illud ergo in quo est unitas fixa non inventa ex alio est unum primum verum, sicut ostendimus; et illud in quo est unitas inventa ex alio est praeter unum primum verum. Si ergo est ex alio, est ex uno primo acquisita unitas. Provenit ergo inde ut uni puro vero et reliquis unis sit unitas iterum et non sit unitas nisi propter unum verum quod est causa unitatis. Iam ergo manifestum est et planum quod omnis unitas post unum verum est acquisita,

typical Kindian move.[87] I would like to offer this note as a little token of gratitude to the scholar who, *inter permulta alia*, has established the presence of al-Kindī's *On First Philosophy* in Muslim Spain.[88]

creata; verumtamen unum verum purum est creans unitates, faciens acquirere, non acquisitum, sicut ostendimus".

[87] Compare in particular the following passage from al-Kindī's *First Philosophy*: 'Unity, therefore, when an accident in all things, is not the True One, as we stated previously: the True One being the one *per se* which is never multiple in any way (...). As for all the kinds of one other than the True One, when they occur in whatever they are, it is *per accidens*. (...) Consequently, the first cause uf unity in unified things is the True One which does not acquire unity from another (...). The cause of unity in unified things is accordingly the True One, the First, and everything which receives unity is caused, every one other than the One in truth being one metaphorically and not in truth. (...) Accordingly every multiplicity comes to be through unity, and if there were no unity the multiple would never have being. (...) Therefore the cause of coming to be is due to the True One, which does not acquire unity from a donor but is rather one through its essence' *(Rasā'il al-Kindī al-falsafiyya*, ed. M. Abū Rīda [Cairo, 1950–53], pp. 161–2; *Al-Kindi's Metaphysics. A Translation of Ya'qūb ibn Isḥāq al-Kindī's Treatise On First Philosophy (fī al-Falsafah al-ūlā)*, trans. A.L. Ivry [Albany, 1974], pp. 112–13).

[88] H. Daiber, 'Die Kritik des Ibn Ḥazm an Kindīs Metaphysik,' *Der Islam* 63 (1986), pp. 284–302.

APPENDIX I

Variant Readings of Gundissalinus' *De Unitate et Uno*, Including ms. Rome, Biblioteca Angelica, 242

Pages and lines are given according to the edition by P. Correns (see note 65). This edition is based on the mss. A, B and C. Correns added in an appendix (pp. 50–56) the collation of three other mss., D, E and F. The Rome ms. has been labelled as G.

A Paris, Bibliothèque Nationale de France, *lat.* 16605
B Paris, Bibliothèque Nationale de France, *lat.* 14700
C Paris, Bibliothèque Nationale de France, *lat.* 6443
D Vienna, Österreichische Nationalbibliothek, *lat.* 195
E Vienna, Österreichische Nationalbibliothek, *lat.* 5508
F Munich, Bayerische Staatsbibliotheek, *lat.* 527
G Rome, Biblioteca Angelica, 242
a Edition by Henry Lorit, Basle 1546

Page 3 Correns
Incipit liber Alexandri de unitate C incipit liber de unitate et uno E incipit liber philosophi de unitate G fol. 26rb l. 50 incipit liber Alexandri de unitate G fol. 27ra l. 1 Anitii Manli Severini Boethi de unitate et uno a
2 qua *om.* G **3** sive[2] *om.* AG **4** una est: est una G **7** esse *om.* G **9** est *om.* G **10** in creatis: in rebus creatis Ga || scilicet *om.* Ga **11** materiae: materia G *fortasse recte* **12** nonnisi *om.* G

Page 4 Correns
4 destruitur res: res destruitur Ga **5** unitione: unitate CGa || servatur: conservatur Ga || autem *om.* G **6** unito: unitate CG **8** quare: quia FG **10** unde: unum FGa || concomitantur: committantur G **13** in hoc numero: hoc in munere BDEa hoc immunem G || habeat esse una: habeat esse et una CG **14** quia ex quo: ex quo Ca *fortasse recte* quia G **17** esse naturaliter: naturaliter esse G **18** sint *om.* G **21** unitionem: unionem Ba unitatem EG **22** formam: forma CFGa

Page 5 Correns
2 diffluit: defluit DEFG || sua *om.* G **6** igitur materia non unitur per se *om.* DEG **9** enim: vero G **11** quod: quicquid BEFG **15** vera: una CGa

| | quae est: est que G **17** ab eo *om.* BCFGa **18** creata *om.* G | | creante: creata G **19** opposita: quia G **22** *post* principium *add.* ut in generatis Ga **23** *post* vero *add.* ut in creatis Ga | | *post* quibusdam *add.* non G **24** *post* corruptioni *add.* ut in corporibus terrenis G | | sed: et CG

Page 6 Correns
1 quibus: quibusdam ABCFGa **3** actu: effectu G **4** *post* caelestibus *add.* corporibus FGa corporibus supercaelestibus E **5** *post* sunt *add.* et G | | fuerit: fuit G **7** in uniendo: in humido G | | ob: propter CG **8** essentia eorum: eorum essentia G **9** enim: in G **10** fuerit: fit G **11** e contrario: e converso Ga **14** ad esse materiam: materiam ad esse G **15** non: et G **21** infra: circa G **26** substantiam *om.* G

Page 7 Correns
2 grossitudinem suam: grassitudinem F crassitudinem suam G crassitudine sua a **4** initii: initium FG *fortasse recte* **4** extremitatis: extremitas C *fortasse recte* **6** virtutis: principii unitatis vel virtutis F principii G virtutis principii a **7** superiore: superiori G | | ad: et ad G | | fit degeneratio: sic degeneratione G **9** quae *om.* EG | | nascitur: nascetur G **10** inspissatur: spissatur BEFG **13** est *om.* EG **14** est *om.* G **15** rariores: minores BGa **16** pars: propter G res a | | materiae: hoc G **18** recipit: recepit G | | qua dignior est: digniorem G

Page 8 Correns
3 sensibilis animae: animae sensibilis CGa **7** est quasi lumen: quasi lumen est G | | sic: converso G **8** non: et non G **11** tota *om.* G **13** *post* quo *add.* enim G **15** magis enim : enim magis CG | | *post* iam *add.* supra CG **16** *post* constringitur *add.* et DEGa **17** post ultimas *add.* partes FG | | non enim est: si est enim G **18** *post* tantum *add.* terciam G | | luminis: lux G **19** pervenit: penetrat G **20–21** partem materiae infimam: ultimam materiam G **22** *post* fit *add.* sicut predictum est BGa sicut praedictum est non D sicut nunc dictorum est F

Page 9 Correns
2 cum *om.* EG **3** occultatur: occultatus est G **9** *post* vitreae *add.* etiam G | | et: ut Ga **14** esse unitate: unitate esse G **15** simplicitate: simplicitatis G *fortasse recte* **21** *ante* congeries *add.* et CEG **25–10.1** ut—numero *om. hom.* G

Page 10 Correns
2 sed: et C scilicet G **6** *post* participatione *add.* unius G *fortasse recte* **8** more: in corde G *fortasse recte* **9** quia vel: quod G **14** ut etiam: ut ea D ut in ea G **16** nituntur: videntur G || nituntur^2: mentiuntur B metiuntur D cernunt (?) G **19–21** in—continuae *om. hom.* G **23** ergo non est: non ergo est G **24** intellectus: intuet G **24–25** continuatio disgregatorum: disgregatorum continuatio CG continuatorum disgregatorum D **25–11.2** continua quantitas—ut *om. hom.* G

Page 11 Correns
3 ut dictum est ex unitatibus est: ex unitatibus est ut dictum est G **6** magis *om.* G **11** confluctione: constructio D constrictione E constructione G **13** est una: dicitur una esse C dicitur esse una DG **13–14** et est id quod est *om.* G

explicit liber de divisione philosophie in tres partes et partium in partes suas secundum philosophos B explicit liber de unitate Alquindi C explicit liber de unitate et uno a Boetio editus D explicit liber de unitate et uno E explicit liber Aristotilis de unitate et uno F explicit liber Alexandri philosophi de unitate G

APPENDIX II

Contents of ms. Rome, Biblioteca Angelica, 242

1. Algazelis *Liber Philosophie*: fol. 1r–7v (incomplete; see J.T. Muckle, *Algazel's Metaphysics: a Medieval Translation* [Toronto, 1933]).

2. Aristotelis *De morte et vita*: fol. 8r–v.

3. Aristotelis *De historia animalium*: fol. 9r–17v.

4. Jacob Alchindi, *De ratione*: fol. 18r–v (i.e., the *De intellectu*: cf. Nagy, quoted above, note 63, p. xxxi and see n. **8.** in this list).

5. Abubecri Rhasis *Introductorius*: fol. 18v–19r (i.e., the *Liber introductorius in artem logicae demonstrationis* by Muḥammad ibn Maʿshar al-Bustī: see H.G. Farmer, 'Who was the author of the *Liber introductorius in artem logicae demonstrationis*?,' *JRAS* (1934), pp. 553–6). This work has been identified as a letter of the Ikhwān al-ṣafāʾ. See C. Baffioni, 'Il *Liber Introductorius in artem logicae demonstrationis*: problemi storici e filologici,' *Studi filosofici* 17 (1994), pp. 69–90. I wish to thank Charles Burnett for bringing my attention to this study.

6. Alchindus, *De quinque essentiis*: fol. 19r–20r (see Nagy, p. xxxi; ed. Nagy, pp. 28–40).

7. Liber Alexandri *De augmento*: fol. 20r (see notes 22–24).

8. Alchindus, *De intellectu et intellecto*: fol. 20r (this is another version of n. **4**. Both versions are published and compared by Nagy, pp. 1–11).

9. Isaac, *Liber definitionum et descriptionum*: fol. 20v–24r (see J.T. Muckle, 'Isaac Israeli, *Liber de Definicionibus*,' *AHDLMA* 12–13 [1937–38], pp. 299–344).

10. Anonymous treatise named 'Entendent' by Narducci, p. 139: fol. 24v. *Incipit*: 'usus fuit ennendent cum hec quinque distinguerentur', *explicit*: 'in philosophia prima'.

11. Liber Alexandri *De intellectu*: fol. 25r–26r (see notes 10–15).

12. Pseudo-Alexander *De unitate*: fol. 26r–27r (i.e., Dominicus Gundissalinus *De unitate et uno*).

13. Aristotelis *Sex Principiorum*: fol. 27r–29r (often attributed to Gilbert of Poitiers; see L. Minio-Paluello, 'Magister Sex Principiorum,' *Studi medievali* 6 [1965–66], pp. 123–51; also in *Opuscula*, pp. 536–64; ed.: *Categoriarum supplementa: Porphyrii Isagoge, translatio Boethii et Anonymi fragmentum vulgo vocatum Liber sex principiorum* ed. L. Minio-Paluello adiuvante B.G. Dod [Bruges, 1966]).

14. *Flos Boethii divisionum*: fol. 29r–v (possibly an excerptum from Boethius' *De divisione*, cf. Anici Manlii Boethii *De divisione liber*, ed. J. Magee [Leiden, 1998]).

15. Aristotelis *Meteora*: fol. 29v–33r.

16. *Flos Alfarabii secundum sententiam Aristotelis*: fol. 33r–v (Latin translation of parts of the *'Uyūn al-masā'il* attributed to al-Fārābī: see M. Cruz Hernández, 'El *Fontes questionum* (*'Uyūn al-masā'il*) de Abū Naṣr al-Fārābī,' *AHDLMA* 25–26 (1950–51), pp. 303–23, ed. of the Arabic and Latin texts; Cruz Hernández credits Gundissalinus with the translation; see also Serra 1993, quoted above note 16: ed. of the Latin text; Serra attributes the translation to Gerard of Cremona).

17. Aristotelis *Topicorum* liber tertius: fol. 33v–34v.

18. Augustini *Liber de spiritu et anima*: fol. 35r–40r (pseudo; Alcherus Claravallensis, *De spiritu et anima*, *PL* 40, col. 779–816; cf. Thomas Aquinas, *Q. d. De Anima*, q. 12 ad 1: 'Ad primum ergo dicendum quod liber iste De spiritu et anima non est Augustini, set dicitur cuiusdam Cisterciensis fuisse; nec est multum curandum de hiis que in eo dicuntur').

19. Macrobii *In Somnium Scipionis*: fol. 40v–41r.

20. Boethii *Ad Symmacum De sancta trinitate*: fol. 41v–44r.

21. Aristotelis *De causis et proprietatibus elementorum*: fol. 44v–46v (pseudo; see C.B. Schmitt and D. Knox, *Pseudo-Aristoteles Latinus. A Guide to Latin Works Falsely Attributed to Aristotle before 1500* [London, 1985], n. 14).

22. Aristotelis *Liber pure bonitatis*: fol. 47r–49v (i.e., the *Liber de Causis*).

23. Galeni *De diebus decretoriis liber tertius*: fol. 50r–53v (Latin version of Galen, *De diebus decretoriis*, book III, ed. Kühn, ix, pp. 901–41. I am grateful to Charles Burnett for referring me to Galen's *On Critical Days*.

24. *Tractatus De commixtione elementorum*: fol. 54r–77r (indicated by Narducci, p. 140, as a treatise by Constantinus Afer; in all likelihood Urso of Calabria as suggested by Charles Burnett: see A. Birkenmajer, 'Le role joué par les médecins et les naturalistes dans la reception d'Aristote aux XIIe et XIIIe siècles,' in idem, *Études d'histoire des sciences et de la philosophie du Moyen Âge* [Wroclaw, 1970], pp. 1–15, esp. p. 4).

Contents of the ms. Paris, Bibliothèque Nationale de France, *lat.* 6443

1. Avicenna, *Metaphysica*: fol. 2r–43v (see note 63).
2. Avicenna, *Physica*: fol. 44r–69r.
3. Avicenna, *De Anima* (*Sextus Naturalium*): fol. 70r–90v (see note 63).
4. Avicenna, *De caelo et mundo*: fol. 90v–96v (pseudo; possibly a compilation by Ḥunayn ibn Isḥāq based mostly on Themistius' paraphrasis of Aristotle's *De caelo*, as contended by M. Alonso, 'Ḥunayn traducido al latin por Ibn Dāwūd y Domingo Gundisalvo,' *Al-Andalus* 16 [1951], pp. 37–47; ed.: O. Gutman, *Liber Celi et Mundi. A Critical Edition with Introduction* [Leiden, 2003]; Gutman challenges Alonso's hypothesis).
5. Dominicus Gundissalinus, *De processione mundi*: fol. 96v–100v (see note 63).
6. Avicenna, *De animalibus*: fol. 101r–142v.
7. Algazel, *Metaphysica*: fol. 143r–157v (see note 63).
8. Algazel, *Physica*: fol. 158r–165v (see note 63; fol. 165vb: *explicit Algazel totus*; fol. 166r–v is blank).
9. *De intellectu possibili questio*: fol. 167r–172r (*incipit*: Sicut omnes homines natura scire desiderant).
10. *Philosophia communis*: fol. 172v–180v (*inc.*: humana natura multipliciter est animalis sicut scribitur primo metaphysice).
11. Alfredus Anglicus, *De mixtione*: fol. 181r–182r (not listed among Alfredus' works by J.K. Otte, 'The Life and Writings of Alfredus Anglicus,' *Viator* 3 [1972], pp. 275–91; however, following Otte, one may speculate about the fact that Alfred wrote some exegetical work—glosses or a commentary—on the *De Generatione et corruptione*, which may be alluded to by a title such as *De mixtione*).
12. Alfredus Anglicus, *De motu cordis*: fol. 182r–184v (see C. Baeumker, *Des Alfred von Sareshel [Alfredus Anglicus] Schrift De motu cordis* [Münster, 1923]).
13. *Liber de presagiis tempestatum*: fol. 184v. = Pliny, *Natural History*, bk. 18, ch. 78–90 (probably only a fragment here). I am grateful to Charles Burnett for this information.
14. *Tractatus de ordine universi*: fol. 185r–v.
15. Liber Avicenne *De ortu scientiarum*: fol. 186r–v (pseudo; fol. 186va: *explicit Abucanus de ortu scientiarum*; possibly the Latin translation of Fārābī's lost *Fī ẓūhur al-falsafa*, according to M. Steinschneider, *Al-Farabi [Alpharabius] des arabischen Philosophen Leben und Schriften*

[St. Petersburg, 1869, repr. Amsterdam, 1966], p. 89; edited as Fārābī's by Baeumker, 1916, see note 63; see also H. Bédoret, 'Les premières traductions tolédanes de philosophie. Œuvres d'Alfarabi,' *Revue Néoscolastique de Philosophie* 41 (1938), pp. 80–97, esp. pp. 88–93).

16. Isaac Israeli, *De definitionibus*: fol. 187r–190r (title: Incipit liber Ysaac de diffinitionibus translatus a magistro Gerardo Cremonensi in Toleto. See above, n. **9.** in the list of ms. Rome, Biblioteca Angelica, 242).

17. Averroes, *De substantia orbis*: fol. 190r–193r (fol. 193rb: explicit tractatus Averoys de substantia orbis incipit liber Alexandri de unitate).

18. Pseudo-Alexander *De unitate*: fol. 193r–194r (i.e., Dominicus Gundissalinus, *De unitate et uno*).

19. Liber Alexandri *De augmento*: fol. 194r (see notes 22–4 and n. **7.** in the list of ms. Rome, Biblioteca Angelica, 242).

20. Liber Alexandri *De tempore*: fol. 194r–195r (see notes 32–6).

21. Alquindi, *De intellectu et intellecto*: fol. 195r (incomplete; see above, n. **4.** and **8.** in the list of ms. Rome, Biblioteca Angelica, 242, and below, n. **25** in this list).

22. *Liber introductorius in artem logice demonstrationis collectus a Mahomet discipulo Alquindi*: fol. 195r–197v (see above, n. **5.** in the list of the ms. Rome, Biblioteca Angelica, 242).

23. Alfarabius, *De intellectu et intellecto*: fol. 197v–199v (see note 11).

24. Liber Alquindi *De somno et visione*: fol. 199v–200v (edited by Nagy, pp. 12–27).

25. Liber Alexandri philosophi *De intellectu et intellecto secundum sententiam Aristotelis* translatus de greco in arabicum ab Ysaac filio Ioachim: fol. 200v–202r (see above, n. **21.** and n. **4.** and **8.** in the list of ms. Rome, Biblioteca Angelica, 242).

26. Algazel, *Logica*: fol. 208v–220v (see C. Lohr, '*Logica Algazelis*. Introduction and Critical Text,' *Traditio* 21 [1965], pp. 223–90).

AVERROES' COMMENTARY ON ARISTOTLE'S *DE GENERATIONE ANIMALIUM* AND ITS USE IN TWO THIRTEENTH-CENTURY HEBREW ENCYCLOPEDIAS

Resianne Fontaine

1. Introduction

Among Averroes' commentaries on Aristotle's works the *Kitāb al-ḥayawān* (*De animalibus*), the commentary on Aristotle's zoology, is one of the least studied. Averroes completed it in November 1169 (ṣafar 565) in Seville after he had left Cordoba. The text contains *De partibus animalium* and *De generatione animalium*. Averroes, however, did not use these names, adopting instead the numbering into Books XI–XIX, as was common in the Arabic zoological tradition. Books I–X, covering the *Historia animalium* were not commented upon by Averroes.[1]

Averroes' *Kitāb al-ḥayawān* belongs to the commentaries that have not been preserved in the original Arabic; it is extant only in the Hebrew and Latin translations. If it is true that the Latin version is based on the Hebrew one, as is commonly assumed, the Hebrew translation is the earliest complete testimony of Averroes' zoology. This translation was made by Jacob ibn Machir in 1302 and later served as the basis of Gersonides' super-commentary on *De animalibus* (1323). However, before the Arabic text had been translated into Hebrew, the commentary was already made accessible to Jewish scholars in Christian lands through the surveys provided by two thirteenth-century Hebrew authors, namely Judah ha-Cohen in his *Midrash ha-ḥokhmah* (Hebrew version *c.* 1247) and Shemtov ibn Falaquera in his *De'ot ha-filosofim* (*c.* 1260).[2] In both

---

[1] For the history of the transmission of Aristotle's zoological works in the Arabic tradition, see the editors' introductions in Aristotle, *Generation of Animals. The Arabic Translation Commonly Ascribed to Yaḥyā ibn al-Biṭrīq*, ed. J. Brugman and H.J. Drossaart Lulofs (Leiden, 1971), and R. Kruk, *The Arabic Version of Aristotle's* Parts of Animals (Amsterdam, 1979). See also M. Zonta, 'The Zoological Writings in the Hebrew Tradition: The Hebrew Approach to Aristotle's Zoological Writings and to their Ancient and Medieval Commentators in the Middle Ages,' in C. Steel, G. Guldentops and P. Beullens (eds.), *Aristotle's Animals in the Middle Ages and the Renaissance* (Leuven, 1999), pp. 45–8.

[2] On these and other encyclopedic texts see S. Harvey (ed.), *The Medieval Hebrew Encyclopedias of Science and Philosophy* (Dordrecht, 2000) and the literature mentioned in the 'Selected Bibliography,' pp. 520–21.

'encyclopedias' a section on zoology is found within the framework of a survey of Aristotelian natural philosophy, a survey for which Averroes' commentaries are the most important sources. While Judah ha-Cohen's survey can be called an excerpt, Falaquera's is far more extensive and contains passages that are literal or almost literal translations of the Averroian texts. Through these encyclopedias Jewish savants who did not read Arabic could familiarize themselves with Aristotelian natural philosophy as read by Averroes at a time when only a few of his commentaries in this field were available in Hebrew translation.[3] The aim of this paper is to discuss some features of Averroes' procedure as a commentator in *De generatione animalium* and to examine how the Hebrew encyclopedists in turn transmit his commentary. To this end, I shall select a few topics from Books XV and XIX.

Before investigating these questions, however, it is worthwhile to briefly discuss the category to which this commentary belongs: that of the Epitomes or that of the Middle Commentaries. The *Kitāb al-ḥayawān* is the only commentary that Averroes has written on Aristotle's zoology. Steinschneider tended to consider it to be an Epitome, but it looks more like a Middle Commentary.[4] Not only is it considerably longer than Averroes' other Epitomes, but it also follows the Aristotelian text closer than the Epitomes do. The title of the Hebrew translation is *Sefer ba'alei ḥayyim*, which does not disclose very much. The colophon starts with the words: 'completed is the *be'ur* of all the scientific portions of these books'.[5] The term *be'ur* may render the Arabic *talkhīṣ*, but it is more likely that its meaning here is explanation or 'exposition' in general, since Averroes' use of the term is not consistent.[6] The Latin translation refers to the commentary as a *paraphrasis*. In several Hebrew manuscripts our text is grouped with other commentaries by Averroes. However, no evidence can be deduced from these manuscripts with respect to

[3] For a survey of Hebrew translations of Aristotelian texts and of Averroes' commentaries see M. Zonta, *La filosofia antica nel Medioevo ebraico* (Brescia, 1996), pt. 2, and S. Harvey, 'Arabic into Hebrew: the Hebrew Translation Movement and the Influence of Averroes upon Medieval Jewish Thought,' in D.H. Frank and O. Leaman (eds.), *The Cambridge Companion to Medieval Jewish Philosophy* (Cambridge, 2003), pp. 258–80.

[4] M. Steinschneider, *Die hebräischen Übersetzungen und die Juden als Dolmetscher* (Graz, 1956), p. 144 n. 258.

[5] Ms. Paris BN héb 956, fol. 485r.

[6] D. Gutas, 'Aspects of Literary Form and Genre in Arabic Logical Works,' in C. Burnett (ed.), *Glosses and Commentaries on Aristotelian Logical Texts. The Syriac, Arabic and Medieval Latin Traditions* (London, 1993), pp. 41–3.

the question of the type of commentary: in some cases it is included among Middle Commentaries and in others among the Epitomes.

Although its structure is more akin to a Middle Commentary, Averroes' *De animalibus* does not reveal the neat division into parts, chapters and sections that we know from other Middle Commentaries, such as that on the *Physics* or *On the Heavens*. It should be noted that such a division is also lacking in the Middle Commentary on the *Meteorology* (1172), which is nevertheless classified as a Middle Commentary and is clearly different from the Epitome on this text. It is also absent in Averroes' commentary on the *Parva naturalia*, the only other Aristotelian treatise on natural philosophy on which Averroes composed only one commentary and which was written around the same time as that on the *De animalibus*. The Middle Commentary on the *Physics*, the first 'genuine' Middle Commentary in the field of natural philosophy, dates from the same period. In sum, in the years 1169–72 Averroes composed Middle Commentaries alongside two commentaries that differ from Epitomes on the one hand and from Middle Commentaries on the other. It might well be that he did not feel himself confined to writing one specific type of commentary. Another explanation may be that the format he chose also depended on the state of the text he was commenting on and/or on the many demands on his time. In his colophon of *De animalibus*, Averroes complains about the corrupted state of the Arabic translation he had before him and about the 'troubles of the time'. He states that he completed it in a short time, expressing the hope that God would grant him time to return to it and to go over it.[7] From this it may be inferred that Averroes did not consider this commentary to be definitive.

2. The Origin of Semen

Book XV of *Kitāb al-ḥayawān* comprises the following issues: sexual generation vs. spontaneous generation; the procreative parts of males and females in various species of animals and their respective functions; generation in bloodless animals; the origin and constitution of semen, and the female contribution to generation (cf. *De generatione animalium* i). Here I will discuss Averroes' refutation of the theory that

[7] Ms. Paris, fol. 485r.

semen originates from all the parts of the body, and his discussion of spontaneous generation.

The refutation of what using a modern term is called the pangenesis theory is found within the framework of Aristotle's discussion of the origin of semen. According to this theory, semen is drawn from the whole of the parent's body that is to say from all the bodily parts. In this view, which goes back to Hippocrates, each part of the body reproduces itself, the fingers of the embryo being formed from the fluid that derived from the parent's fingers etc. Averroes' description and refutation of this view is based on Aristotle's expositions in *GA* i.17–18. It can immediately be said that Averroes' account of this issue provides an excellent example of his inclination to systematize and structure the text he is commenting on, a feature that is known from other commentaries as well. In this respect he certainly follows Aristotle himself, but he often goes much further than his model. When dealing with the pangenesis theory the commentator structures his source by changing the order where he deems this fit; by combining passages that are similar in content, and by inserting methodological comments or an occasional personal observation.

To begin with Averroes not only changes the order of Aristotle's description of the *four* 'lines of arguments' put forward by the ancients in support of the pangenesis theory but also reduces them to three. The four arguments are, briefly put: (i) the intense pleasure involved in the emission of semen; (ii) mutilated parents produce mutilated offspring; (iii) the young resemble their parents not only as a whole, but also part for part, and (iv) just as there is an origin for the whole, there must be semen that is the origin for each of the parts, for children sometimes also resemble their parents with respect to so-called 'acquired' characteristics, such as scars and the like. According to the proponents of the pangenesis theory, these phenomena can only be explained by assuming that semen is drawn from all the parts of the body.[8]

Averroes rearranges the order of the arguments listed by Aristotle, by mentioning the proof from resemblance (iii) as the first and strongest proof, whereas Aristotle did not classify them in any order of importance. Averroes' second proof, also the second in Aristotle, is the argument that mutilated parents produce mutilated offspring. His third and last proof, which Aristotle lists as the first, is the intense

[8] Aristotle, *GA* 721b 9–722a 1; Averroes, *De animalibus*, ms. Paris, fol. 453v.

pleasure experienced during sexual intercourse. As for the number of arguments, it is clear from Averroes' listing that he considered the fourth argument to be a mere extension of the third, as both deal with the issue of resemblance. In fact, in Aristotle's formulation there seems to be no substantial difference between the third and fourth arguments, and the examples of offspring with scars that he provides are intended to lend plausibility to (iii) and (iv).[9] Averroes combines them, saying that the third proof is 'strengthened' by the fact that children sometimes resemble their parents with respect to non-natural, accidental things. Moreover, at this point he inserts an observation of his own by adding: 'As for me, I have seen something similar in my eldest son, for the father of his mother had a birthmark on his chest, and my son has one too, but his is small, whereas that of his grandfather is large, and I do not know whether it will grow.'[10] Moreover, after having listed the arguments offered by the proponents of the theory, Averroes inserts a comment on Aristotle's methodology, noting that Aristotle refuted these arguments and also refuted the conclusion itself, as it is fitting in demonstrative refutation not to solely rely on one objection. Another methodological note is that the proponents of the theory under consideration can in fact be divided into two groups and that some of the objections provided by Aristotle invalidate the claim of one group and others those of both groups.[11]

What then follows in the Aristotelian text is a series of objections against the theory that the being drawn of semen from all the parts is the cause of generation and/or similarity.[12] It is only towards the end of his account that the two remaining arguments (i–ii) are briefly dismissed.[13] The emphasis is thus on the argument from similarity which is presumably why Averroes calls it the most important argument and places it ahead of the others. In general Averroes adopts Aristotle's order of objections—designated in the Hebrew text as *re'ayot* (arguments)—against the argument based on resemblance, but neatly numbers them from one to ten, apparently for the sake of clarity. On two occasions he deviates from Aristotle's order: firstly, he inserts as

[9] Aristotle, *GA* 721b 28–34.
[10] Ms. Paris, fol. 453v.
[11] Ibid., fol. 453v–454r.
[12] Aristotle, *GA* 722a 2–723b 33.
[13] Ibid., 723b 3–724a 8.

the third *re'ayah* Aristotle's objection against the second argument of the ancients (that was based on mutilation), namely that it frequently occurs that a parent who misses a certain part produces offspring that has that particular part.[14] Averroes goes on to say that this applies especially to plants and elaborates on this. In the corresponding passage in the *De generatione animalium* Aristotle indeed refers to plants that miss parts without, however, relating it to the more general argument based on mutilation.[15] The same objection appears towards the end of the discussion when he deals with the refutation of this argument, in accordance with the order in the Aristotelian account.[16] The second deviation from Aristotle's order concerns the objection that, if semen would be drawn from all the parts of the body, females would be able to bring forth offspring without a male. This objection is absent in the corresponding discussion in Aristotle, but it is found later on when Aristotle presents his own theory of generation and refers to the pangenesis theory.[17] These deviations thus seem to reflect Averroes' concern for presenting a complete and coherent account, an account that is at times even more systematic than in the original text. On the other hand, it should be noted that Averroes' discussion of *GA* 722a 17–723b 2 is less clear than in the Aristotelian text. Here Aristotle investigates a number of questions to which the pangenesis theory gives rise, such as whether the semen is drawn only from the uniform parts or also from the non-uniform parts and whether or not the parts of the body are scattered within the semen. In this regard he examines views of Anaxagoras and Empedocles that bear on these questions. Averroes summarizes the entire discussion in his fourth, fifth and sixth arguments, without, however, pointing out the interrelation between them and without mentioning Aristotle's predecessors, which to some extent goes at the expense of clarity. Nonetheless, it is generally obvious that his numbering and re-arrangement was intended to underscore the conclusiveness of Aris-totle's refutation of the theory under consideration.

[14] Ms. Paris, fol. 454r (cf. Aristotle, *GA* 724a 8).
[15] Aristotle, *GA* 722a 12–16.
[16] Ms. Paris, fol. 454v.
[17] Aristotle, *GA* 730a 24–32.

How do the Hebrew encyclopedists survey this discussion in their encyclopedias? Judah ben Solomon ha-Cohen refrains from covering it altogether, limiting himself to recording the opinion of the ancients without presenting any arguments for it, and adding that Aristotle refuted them with lengthy arguments.[18] His silence is in accordance with his stated intention of being concise, which implies that in general he deliberately chooses to omit theories that were held by Aristotle's predecessors from his encyclopedia. In other words, Judah did not deem it necessary to inform his readers about the pros and cons of the pangenesis theory.

Ibn Falaquera, in contrast, has more to say on the subject, although he is certainly less extensive than his source. Of the arguments of the ancients he lists only the first, the argument based on resemblance, characterized by Averroes as the most important, but he provides the refutation of the other two towards the end of his account, just as his source does. Falaquera summarizes the first argument as follows: 'One of their arguments (*re'ayot*) is that if [semen] did not come forth from all the parts of the progenitor, the offspring would not resemble its progenitor with respect to the whole or with respect to some of the parts. And he said that this is the cause of resemblance.'[19] He immediately adds to it Averroes' observation on the two classes of proponents of the theory under consideration. Falaquera omits Averroes' personal note on his son, which is not particularly surprising, but also his methodological observation on the need to refute all of the relevant assertions. This may appear surprising in view of the importance of the statement, but it becomes soon clear why Falaquera prefers to skip it, since he himself does not list all of Averroes' objections. While roughly following the commentator's account, he does not take over Averroes' numbering of the objections, introducing a new argument instead by 'furthermore' (*we-'od*), and limiting himself to listing six out of Averroes' ten arguments. The four arguments that he omits are Averroes' fourth, fifth and sixth arguments, namely the arguments in which Aristotle examined some of his predecessors' views, and the very first argument. This argument asserts that the pangenesis theory cannot account for the fact that offspring resembles their parents with respect to parts from which no semen can be drawn such as hair and

[18] Ms. Oxford, Bodleian Poc. 343, fol. 52v.

[19] Ms. Leiden Or. 4758, fol. 237v b2–6.

nails, or matters like voice and movement.[20] All in all then, Falaquera reproduces the following arguments against the pangenesis theory: (i) the argument from resemblance does not hold, for sometimes offspring resemble not their parents, but their grandparents; (ii) parents that lack a certain part may produce offspring that is complete; (iii) we observe that sometimes offspring resembles their parents, but sometimes they do not (as in spontaneous generation); (iv) females would be able to generate on their own; (v) transplanted cuttings of plant can procreate, and (vi) in some insects the female can be observed to insert an non-uniform part into the male upwards from below. These correspond to Averroes' second, third, eight, seventh, ninth and tenth proofs respectively. Aristotle considers the last proof on insects as the weightiest proof, a detail that is omitted by Averroes and Falaquera.[21]

It is regrettable that Falaquera is not as extensive in his treatment of the pangenesis theory as he is in other passages, for a bit more verbosity could have thrown light on a problematic passage in Averroes' text that is found towards the end of the discussion under consideration. Here Aristotle refutes the argument that the intense pleasure experienced during sexual intercourse is due to the fact that semen is drawn from the whole body. Aristotle's objection is that 'the pleasure comes at the end, but according to their theory it should occur (a) in every one of the parts, and (b) not simultaneously, but earlier in some and later in others.'[22] It is obvious from the Greek text that 'some' and 'others' refers to 'parts', and this is exactly how the Arabic translator has understood this sentence.[23] However, the corresponding passage in Averroes reads: 'If the cause of the pleasure were the being drawn of semen from all the parts, the pleasure should occur in both of them (*bi-sheneihem*) in the same manner and at the same time, and not in one of them before the other.'[24] Apparently, the expression 'in both of them' refers to the partners engaged in intercourse. In any event, this is how Gersonides interprets Averroes' words in his super-commentary: 'According to me, [Averroes] means to say that if the cause of the pleasure were the being

[20] Cf. Aristotle, *GA* 722a 3–8.

[21] Ibid., 723b 19–28.

[22] Ibid., 724a 2–3; *Generation of Animals*, trans. A.L. Peck, rev. ed. (Cambridge, Mass., 1963), p. 71.

[23] Ed. Brugman and Drossaart Lulofs, p. 28, lines 9–11. The question of which Arabic translation of the Greek text underlied Averroes' commentary needs further investigation.

[24] Ms. Paris, fol. 454v.

drawn of semen from all the parts during intercourse, the pleasure would of necessity occur in the male and the female in the same manner and at the same time.'[25] Falaquera skips over this passage and limits himself to providing what Averroes, following Aristotle, believed to be the true cause of sexual pleasure; intense stimulation.[26] All this implies that as far as the pangenesis theory is concerned, Falaquera is less systematic and less complete than his source. As we shall soon see, however, this does not apply to all sections of Book XV.

3. Spontaneous Generation

The discussion on spontaneous generation reveals a different picture, both with respect to Averroes' comments on the Aristotelian text and with respect to its coverage by the Hebrew encyclopedists. The main difference between the Aristotelian text and the commentary is that Averroes' discussion of spontaneous generation is largely absent in his source. Both Aristotle and Averroes are concerned to underscore the basic difference between sexual reproduction and spontaneous generation: what has been generated from spontaneous generation, that is, from putrescent matter cannot generate something that is similar in kind, and that which is generated from it likewise cannot produce offspring of the same kind, so that in this manner of reproduction, generation will come to an end. To demonstrate this, however, Averroes considerably expands upon Aristotle's exposition. The discussion is found at the beginning of *De generatione animalium* i.1. Averroes follows the Aristotelian text up to the point where Aristotle says that among the insects we find species that arise from putrescent matter and that produce offspring that are different in kind and that are neither male nor female.[27] At this point, Averroes inserts a digression on spontaneous generation, which we will briefly review.

His first step is to show that what is generated from male and female is invariably and of necessity of the same kind as its progenitors on the ground that is impossible for a species to have more than one matter, unless one accepts that transmigration is possible (both the Hebrew

[25] Ms. Vatican, ebr. 42, fol. 77r.

[26] Aristotle, *GA* 723b 25; Averroes and Falaquera add the sensitivity of the genital parts and titillation as causes. Averroes, ms. Paris, fol. 454v; Falaquera, ms. Leiden, fol. 238ra.

[27] Aristotle, *GA* 715b 2–7.

and Latin version have the Arabic term *tanāsukh* here). Next he refutes the view of the alchemists who held that it is possible for one thing to be produced both by art and by nature, and the view that a certain species is produced from copulation in the majority of cases but in some instances from putrefaction, for instance mice that are generated in dung. Averroes argues that in such cases the senses are misled by the strong similarity between the two products, but that the two cannot be of the same kind. To prove this he goes on to explain that the relation of any species to its four causes must be necessary and eternal and that no species can exist by chance, referring in the ensuing discussion to Aristotle's *De demonstratione*.[28]

Averroes thus goes markedly beyond Aristotle. It should be emphasized that like Aristotle he does not deny the possibility of spontaneous generation as such, but only the conclusions drawn from it by the alchemists. The views that Averroes refers to, such as a kind of mouse that has been observed to be generated from clay and the artificial production of living beings, are to be found in works ascribed to Jābir ibn Ḥayyān.[29] It would seem therefore that Averroes' primary objective in this passage is to refute the alchemists on the basis of sound philosophical reasoning, and in particular the view that within one species both generation from male and female and from putrescent matter can take place. However, Averroes' text presents a problem here, because this is precisely what Aristotle believes to be the case with respect to insects: in 715b 2–16 he explains that some of the insects are generated from matter and that the product of these insects is neither male nor female. This is precisely the passage where Averroes inserts his digression, and therefore his criticism of the alchemists may include this particular view of Aristotle. It is, however, difficult to determine whether or not this is the case, because there is also some terminological confusion. The Hebrew translation of Averroes' commentary renders Aristotle's insects (*ta entoma*) by the rather unusual term *ba'al hayyim ha-qelafi* (meaning something like 'hard-skinned'). On other occasions the Hebrew takes over the Arabic term *al-muḥazziz* to denote insects. At the end of the digression Averroes repeats that in spontaneous generation reproduction will come to an end, as he said just before he left off from

[28] Ms. Paris, fol. 451r–v.

[29] See R. Kruk, 'A Frothy Bubble: Spontaneous Generation in the Medieval Islamic Tradition,' *JSS* 35/2 (1990), pp. 276–8.

the Aristotelian text, but here the term *ḥulyot* (vertebrates) appears to denote insects. In the next sentence he summarizes *GA* 715b 17–30 saying that among the *ba'alei ha-ḥeres ha-qasheh* (hard-skinned animals) none arises from male and female, for their nature resembles that of plants that arise from putrefaction.[30] The Aristotelian text has *ostrakaderma* here, the so-called testacea, another class of bloodless animals that are elsewhere rendered by *ba'al ha-ḥulyot* in Averroes' commentary. The question may thus be raised whether Averroes was aware of exactly which bloodless animals Aristotle was actually dealing with. This implies that the question whether Averroes was indeed criticizing Aristotle can be answered only when the terminology concerning the four classes of bloodless animals throughout *Kitāb al-ḥayawān* has been sorted out. Needless to say, this issue also involves the problem of which Arabic translation underlies Averroes' commentary.

Turning now to the Hebrew encyclopedias, it can be noted that both provide more information on the issue under consideration than on the pangenesis theory. *The Midrash ha-ḥokhmah* provides a concise but adequate description of the essential difference between sexual and spontaneous generation. However, it omits Averroes' digression. The *De'ot ha-filosofim* includes it almost in its entirety and often literally, although Falaquera omits from it Averroes' reference to *De demonstratione* and its explanation. Sometimes his text is clearer than that of his source, and occasionally his version may be useful for the constitution of the Averroian text. He is also more consistent with respect to terminology: Aristotle's insects are in both passages rendered with *ha-me'uzar ha-guf* and in the passage before the digression he wisely refrains from providing names for the various classes of bloodless animals, referring simply to: 'many of the species of bloodless animals'.[31] Moreover, he adds some pieces of information to his source that are taken directly from the Aristotelian text, as has already been observed by M. Zonta.[32] Falaquera then goes on to record the well-known view of Ibn Sīnā,

[30] Ms. Paris, fol. 451v.

[31] Ibid., fol. 235va 21.

[32] M. Zonta, 'Mineralogy, Botany and Zoology in Medieval Hebrew Encyclopedias. "Descriptive" and "Theoretical" Approaches to Arabic Sources,' *ArScPh* 6 (1996), p. 308 n. 136 (ms. Leiden 236ra 22–7 = *GA* 715b 21–5). Another case in point is ibid., 236rb 3–6; these lines quote *GA* 716a 15 on why the earth is sometimes called female and the sun 'male and progenitor'.

according to which at the equator it is possible that a certain mixture arises that can receive the human form, a view that Averroes did not mention here.[33] In all probability Falaquera added it for the sake of completeness. His extensiveness here may indicate that he considered the issue of spontaneous generation to be relevant to his readership. To make the text more comprehensible for his readers, he takes the trouble of explaining to his readers what *tanāsukh* is: 'the belief that the soul migrates from one living being to another that is called in Arabic *tanāsukh.*'[34]

4. Straight and Curly Hair

An interesting digression by Averroes is found in Book XIX where he deals with the cause of straight and curled air. Aristotle mentions the following two causes for it: the kind of exhalation contained in hair, the smoky exhalation making the hair curly, and the amount of fluid in it, for hair with little fluid in it will easily be contracted by the heat and dryness of the environing air.[35] Averroes follows this explanation, but also states that the constitution of hair may be due to both causes. Moreover, he says that the air of the environment has a strong impact (*roshem ḥazaq*), a statement that has no parallel in Aristotle's text.[36] The reason for Averroes' emphasis on the environing air becomes soon clear when he starts to elaborate on a passage where Aristotle contrasted the straight hair of the Scythians and the Thracians to the curly hair of the Ethiopians.[37] Averroes goes much further by saying: 'The inhabitants of the regions in between these two such as the people of Andalusia have hair that is intermediate between straight and curly, and in particular the city that is the most well-balanced (*shaweh*) of all the cities of Andalusia, like Cordoba and of those that are close to it, for the equilibrium of that city is testified to by the qualities of its inhabitants, and their good intelligence, their natural proneness to the best ethical virtues, especially the qualities that are related to peace and

[33] Cf. Ibn Sīnā, *Kitāb al-shifāʾ, al-ṭabīʿiyyāt wa'l-maʿādin wa'l-āthār al-ʿulwiyya*, ed. A. Muntaṣir et al. (Cairo, 1965), p. 27.

[34] Ms. Leiden, fol. 235vb 5.

[35] Aristotle, *GA* 782b 19–783a 2.

[36] Ms. Paris, fol. 482v.

[37] Aristotle, *GA* 782b 33–793a 1.

to proper judgment (*ṭuv ha-maḥshavah*), and these people are different from those of Seville even though these are close'.[38]

This digression is noteworthy not so much because it provides information on Andalusia, for also in other commentaries, for example in those on the *Meteorology*, Averroes adds observations about Andalusia, nor because it expresses adherence to the widely held view that inhabitants of the so-called moderate regions possess the most excellent qualities, but first and foremost because it contrasts the people of Cordoba with those of Seville. One is tempted to relate this utterance to Averroes' personal circumstances—one may recall that he completed the commentary in Seville just after he left Cordoba and that he refers to 'the troubles of the time'.[39]

Remarkably, in regard of this digression the two Hebrew encyclopedias reveal another picture than might be expected on the basis of the two foregoing examples. Although the *De'ot* is as usual much more extensive than the *Midrash ha-ḥokhmah*, it does not include Averroes' elaboration. The *Midrash ha-ḥokhmah*, however, contains at least some reminiscence of it since Judah refers to the constitution of hair of 'the people of Spain' as intermediate between that of the inhabitants of northern and southern regions.[40] This is the more remarkable, since Judah is extremely brief in this section, and in fact does not refer at all to straight and curly hair, but instead to thick and thin hair, a topic that in Aristotle's and Averroes' accounts precedes that of straight and curly hair. An explanation for the fact that despite his brevity he found it worthwhile to provide this piece of information may be that he was a Spaniard (from Toledo) himself, although he resided at the court of Frederick II in Italy at the time he composed the Hebrew version of his work.

As for the *De'ot*, the account of hair deviates from that of his source in that it begins with an observation of the usefulness of hair.[41] Moreover, he mentions Galen's view that hair is drier than bone and that it cannot serve as nourishment whereas 'other people' claim that

[38] Ms. Paris, fol. 482v. The word *mahshavah* can also be translated by 'deliberation', 'thinking' or 'reflection'. The Latin translation of this passage is to be found in M. Alonso, 'Averroes, observador de la naturaleza,' in *al-Andalus* 5 (1940), pp. 215–30, repr. in idem, *Teología de Averroes* (Madrid, 1947; repr. Cordoba, 1998), p. [25]–41 (on p. 28).

[39] Ibid., p. 32.

[40] Ms. Oxford, fol. 59v.

[41] Ms. Leiden, fol. 264vb.

hair is nourishment for the bat.[42] Here again we come across Falaquera supplementing his source. It should be noted, though, that Galen figures prominently in Averroes' text too, since throughout *De animalibus* the commentator notes the contending views of Aristotle and Galen on various biological issues. The sustained discussion of this controversy is but one of the interesting features of Averroes' commentary. A systematic study of Averroes' attitude vis-à-vis the two authorities would greatly contribute to our appreciation of Averroes as commentator and to our knowledge of Galenism in the Arabic tradition.

The examples that I have presented suggest that Averroes was a faithful expositor of Aristotle's zoological thought, but that he also felt free to re-arrange and supplement the text he was commenting on. Likewise, the two Hebrew authors reveal a certain independency vis-à-vis Averroes' *Kitāb al-ḥayawān*, adapting it to their own purposes and using it in a manner they deemed fit for their audiences. Needless to say, more research on these texts is required to determine whether these preliminary conclusions are correct.

[42] Ms. Leiden, fol. 265ra.

RAMON LLULL AND THE ISLAMIC CULTURE OF THE MEDITERRANEAN

Josep Puig Montada

Cultural interchanges have often taken place because of physical occupation of the land. The Islamic presence in the Iberian Peninsula, as is well known, dates from the year 92/711 when Arab and Berber troops landed near Gibraltar. The military invasion reached as far as Poitiers, where in 732 the Muslims were defeated by the Francs. From that year on, the Muslim domination lost ground. The Francs conquered the towns of Girona in 785, Vic in 798, and Barcelona in 801. They created a buffer zone (*Marca Hispanica*) between their French kingdom and the Muslim enemy.

However, without any foreign help, the Christians who lived in the Spanish Asturian mountains fought for the return of their country (*reconquista*), and they advanced even faster than their Christian counterpart in the east, moving the border to the Douro river. By contrast, the Eastern Christians, the Catalans and the Aragonese, did not show such activity. The Catalans, who became independent from the Francs in 878, remained (since the *reconquista* of Barcelona in 801 until 1148) near the border of the Llobregat River. In 1148 they pushed south and took the town of Tortosa on the Ebro River and repopulated the land of Tarragona.

The different nature of the Castilian and the Catalan *reconquista* is obvious.[1] A look at the present map of the Iberian Peninsula shows the results: whereas most of the peninsular area is Spanish speaking, the area where Catalan is spoken extends across the eastern border. (The Aragonese dialect later melted into the Castilian language.) No wonder, as Catalonia lived for almost 350 years in peace with the Muslim state beyond the Llobregat River.

Catalonia suffered sporadic attacks; the most serious was the expedition of Ibn Abī ʿĀmir, known as Almanzor, who entered and sacked Barcelona in 985. During this period, the counts of Barcelona sent their

[1] G. Taravini, *Per a una història de la cultura catalana medieval* (Barcelona, 1996), pp. 6–9.

emissaries to the powerful caliphs of Cordoba, ʿAbd al-Raḥmān III and al-Ḥakam II (912–976). Being on good terms with the Cordoban caliphate helped them keep their independence from the Francs. There was also an interest in the goods, material as well as intellectual, that Cordoba possessed. Arabic culture was superior to the culture in the Northern kingdoms and Arabic science was frequently translated into Latin on both sides of the Pyrenees. Catalonia was, no doubt, also a place where transmission took place. The visit of Gerbert d'Orilhac or d'Aurillac, the future pope Sylvester II, in 967 and the scholarly contacts he established during his stay bear witness to this.[2]

Since 1137, the King of Aragon was also the count of Barcelona, but both territories kept their own institutions and languages. After the territorial gains of 1148, Catalonia reached its modern day borders. There was a movement to expand the realm, but it did not aim at the Muslim south by means of war, but by means of alliances with the culturally and physically close Provence. King Peter II the Catholic, was the brother-in-law of Raymond VI, count of Toulouse, and he aspired to build a confederate state with the Provence, but the project was not realized. Peter II and Raymond VI fought Simon de Montfort, who represented the interests of the Northern Frenchmen, but they were defeated, and Peter II died in the battlefield of Muret (1213). His son, King James I, saw the difficulties of this political strategy oriented towards the north, and went in the other direction. He attacked less dangerous neighbors: the Muslim petty kingdoms of Mallorca, which he conquered in 1229, and of Valencia, conquered between 1236 and 1238.

At this point, Castile had taken for herself the major part of the Iberian Peninsula. Ferdinand III, who reigned from 1217–52, had conquered the rich Andalusia and imposed tribute on the remaining Muslim states, Murcia and Granada. The Catalans and Aragonese had better options than a risky war against the neighboring, now Christian land: the Italian islands and the southern Italian peninsula were more attractive. In 1286 the Catalan-Aragonese conquered Sicily, in 1284 Corsica, in 1326 Sardinia and between 1442 and 1458 the Kingdom of Naples.

The Aragonese crown, which kept a confederated structure throughout all of its territories, became a sea power, a thalassocracy. The

---

[2] See *Vie de Gerbert* by A. Olleris, especially his introduction to *Oeuvres de Gerbert, pape sous le nom de Sylvestre II* (Clermond-Ferrand, 1867); P. Riche, *Gerbert d'Aurillac, le pape de l'an mil* (Paris, 1987); O. Guyotjeannin and E. Poelle (eds.), *Autour de Gerbert d'Aurillac, le pape de l'an mil* (Paris, 1996).

Mediterranean Sea linked its territories as well as its trade centers, many of which were in Islamic countries. Around 1250, Islam and Arabic were present in the Catalan life in two ways: through trade with Arab countries and through the population of the newly conquered territories. Although the church tried to convert the Muslims under Christian rule, their attempts were not very successful; the population continued to speak Arabic and stuck to their Muslim faith.

Even if the Christian countries did not lay as far behind the scientific development of the Arab world around 1250 as they had done in 900, they were always eager to learn from them, and as far as literary creation is concerned, Arabic sources influenced the nascent Romance literature too.

Castile was ruled between 1252 and 1284 by Alphonse X the Wise, who developed a coherent cultural strategy. As F. Márquez Villanueva has shown in his studies,[3] the king known as *el Sabio* was conscious that his expanded kingdom needed its own intellectual backbone, which could not solely consist of the Franco-Latin tradition. Alphonse X made two decisions which affected the success of his *Espanna* project: he made the vernacular language the medium of all communication and he incorporated Arabic science, and to a lesser extent, Arabic literature. They contributed by expanding knowledge and creating a national identity.

Arab civilization influenced Catalan culture to a lesser degree than the Castilian, because the former used to draw heavily on the French Latin civilization. Besides this quantitative difference, there is also a qualitative one. There was no royal project like the one carried out by Alphonse X, and its impact was characterized by a kind of familiarity created by the contact with Muslims who remained in the country after the *reconquista*, as well as with Mediterranean neighbors, whose lives and customs attracted curiosity.

We find great familiarity with the Arabic and Islamic culture in two major Catalan authors: Ramon Llull and Anselm Turmeda (1352–1432?).[4] The Franciscan friar Turmeda immigrated to Tunis and converted to Islam, but he never severed his ties with Catalonia. Llull, on the contrary, tried to convert Muslims, all Muslims, to Christianity.

---

[3] His most recent publication: *El concepto cultural alfonsí*, edición revisada y aumentada (Barcelona, 2004).

[4] His best known works are the *Dispute de l'âne*, ed. A. Llinarès (Paris, 1984) and the *Tuḥfa*: *Fray Anselm Turmeda ('Abdallāh al-Tarjumān) y su polémica islamo-cristiana*, ed. and trans. M. de Epalza (Madrid, [2]1994).

In spite of their diametrically opposite positions, they both shared an understanding of Muslim culture as alive, familiar and interacting with the Catalan world.

Ramon Llull (d. 1316)[5] was born in Palma de Mallorca. His parents moved to the island with the Catalan conquest by James I in 1229 and he must have been born a few years later, around 1232. The basic source for his biography is the *Vita coetanea*,[6] where we learn that he was educated at the royal court, being *seneschal* to Prince James II. Llull loved troubadour poetry and once, at night, when he was composing a poem, the image of Jesus on the cross appeared to him. These revelations recurred. At that time Llull was 30 years old, was married and had children.

After these apparitions Llull went to Barcelona, from where he made the pilgrimage to Santiago de Compostela and to Rocamadour in southern France. He came back to Mallorca in 1265 to study Latin, theology and—this is remarkable—Arabic. He bought an educated slave who taught him Arabic as well as Islamic theology for nine years.[7] He mastered Arabic so well that books such as *Llibre del gentil e tres savis*,[8] the encyclopedic *Llibre de contemplació en Déu*, dated 1272,[9] and maybe *La lògica del Gatzell*[10] were first written in Arabic. In 1270 he composed the *Art abreujada d'atrobar veritat*[11] in which he relies on *rationes necessariae*, 'necessary premises' for the purpose of demonstrating the Christian

[5] Various monographs inform of his life and thought: A. Llinarés, *Raymonde Lulle. Philosophe de l'action* (Grenoble, 1963); L. Sala Molins, *La philosophie de l'amour chez Raymond Lulle* (Paris, 1974); E. Colomer, *De la Edad Media al Renacimiento* (Barcelona, 1975); M. Cruz Hernández, *El pensamiento de R. Llull* (Madrid, 1978); L. Badia and A. Bonner, *Ramon Llull. Vida, pensamiento y obra literaria* (Barcelona, 1993), as well as the monographic volume of *Revista española de filosofía medieval* 5 (1998).

[6] *Vita beati R. Lulli*: ed. H. Harada in *R. Lulli Opera Latina* (= ROL) nº 189 (Turnout, 1980), pp. 259–309. Published also by B. de Gaiffier in *Analecta Bollandiana* 48 (1930), pp. 130–78 and by E.W. Platzeck in *Das Leben des seligen R. Llull* (Düsseldorf, 1964), pp. 145–80.

[7] *Vita beati R. Lullii*, ed. Harada, pp. 279–80.

[8] In *Obras de R. Lull*, ed. J. Rosselló, i (Palma de Mallorca, 1901), pp. 1–305, *la creença dels Sarrahins* appears on pp. 231–9. New ed. A. Bonner in *Nova edició de les obres de Ramon Llull*, n. 2 (Palma de Mallorca, 2001), pp. 159–97.

[9] *Obres originals de R. Llull*, ed. M. Obrador Bennassar, ii–viii (Palma de Mallorca, 1906–14).

[10] Published in *Obres*, xix (Palma de Mallorca, 1936), pp. 1–62. It follows the model of an *urjūza*, Arabic didactical poem. Llull summarizes the logic, from the Eisagoge to the syllogistic.

[11] *Ars compendiosa inveniendi veritatem*, ed. I. Salzinger in *Opera*, i (Mainz, 1721).

truth. The concept of necessary premises is familiar in Arabic theology and Llull always adheres to it.[12]

Llull became known for these writings. In 1276, Llull obtained from James II of Mallorca,[13] the necessary endowment to found a monastery where Arabic would be taught. The place chosen by Llull was the Miramar convent close to Palma of Mallorca, where 13 friars learned Arabic; it lasted for 17 years.

In 1286 Llull moved to Paris and to its university, where he received the highest degree, that of a *Magister artium*. His concern for the conversion of Muslims is apparent in the *Disputatio fidelis et infidelis*, which he composed there.[14]

In 1292, he presented to Pope Nicholas IV (1288–92) two short writings: *Tractatus de modo convertendi infideles*[15] and *Quomodo Terra Sancta recuperari potest.*[16] Both titles tell us enough about their contents. The same year or perhaps in 1293, moved by his desire to apply the theory to the conversion of the Muslims, Llull left Genoa for Tunis, which was ruled by Abū Ḥafṣ ʿUmar (1284–95), Caliph of the Ḥafṣid dynasty.[17] This move was possible because, since 1290, the relations between Abū Ḥafṣ and the Kings of Aragon Alphonse (d. 1291) and James II (d. 1311) had improved. Llull took a provocative attitude by telling the Muslim scholars: 'Argue, and if your reasoning is better than the Christian, I shall convert to Islam'.[18] However, he failed completely in his purpose, and was jailed. Llull was allowed to leave Tunis, but was threatened with stoning if he did not leave the country.

By the end of 1293 he returned to the northern shore of the Mediterranean, namely to Naples, where he gave the *Liber de quinque sapientibus*[19] to Pope Celestine V (1293). In 1296 Llull was again in Paris and unsuccessfully defended his *Ars* among the faculty of Paris University. He

---

[12] J. Stöhr, 'Las "rationes necessariae" de Ramon Llull a la luz de sus últimas obras,' *Estudios Lulianos* 20 (1976), pp. 5–52.

[13] After the death of James I in 1276, the territories were divided into two kingdoms: Majorca and Aragon-Catalonia.

[14] *Opera*, ed. I. Salzinger, iv (Mainz, 1729).

[15] *ROL*, ed. Rambaud-Buhot, iii (Palma de Mallorca, 1954), pp. 99–102. Critical ed. F. Domínguez Reboiras, *Liber de passagio* in *ROL* nº 28 (Turnhout, 2003), pp. 333–53.

[16] *ROL*, ed. Rambaud-Buhot, iii (Palma de Mallorca, 1954), pp. 96–8. Critical ed. F. Domínguez Reboiras, *Liber de passagio* in *ROL* nº 28, pp. 328–31.

[17] R. Brunschvig, *La Berbérie orientale sous les ḥafṣides des origines à la fin du XV siècle*, 2 vols. (Paris, 1940–47), i, pp. 88–110.

[18] *Vita beati R. Lulli*, p. 289.

[19] *Opera*, ed. I. Salzinger, ii (Mainz, 1720).

tried to succeed by adapting the system to pious subjects and composed an *Arbre de filosofia d'amor*.[20]

In 1299 he was in Barcelona, and got from King James II written permission to preach in the synagogues, because his aim was to convert the Jews as well.[21] After an absence of twenty years he went back to Mallorca, and while staying on the island he got the false news that the Tartars[22] had conquered Syria. He knew that they were not Muslims and decided to travel there to meet 'Cassanus the Emperor of the Tartars'. Cassianus must be an alteration of Ghāzān, the most prominent Ilkhan—subordinate Khān—of a Persian kingdom; Maḥmūd Ghāzān, a Buddhist, became Khān in 1295 and declared himself Muslim, compelling other Mongol notables to follow suit.

On his way to Syria in 1301 he went to the island of Cyprus, which was still under Christian rule, and realized that the news was false. He then traveled to Armenia Minor, the last Christian kingdom in continental Asia. The following year, 1302, he left Armenia for Genoa and from there he traveled to Paris and Lyon (1305), where he asked Pope Clemens V to establish monasteries where priests would learn the languages of the countries to which they would go as missionaries. We know that he began his *Ars magna generalis ultima* in Lyon.[23]

Since Mallorca had an active trade with Bougie in Algeria, he embarked for this Mediterranean port in 1306. In the middle of its central square, Llull provoked the Muslims by screaming: 'Christian law is true, holy and accepted by God, but Muslim law is false and erroneous, and I am ready to show it'.[24] The people around almost killed him, but he succeeded in having himself brought before the

---

[20] Available in two editions, one by J. Rosselló in *Obras* (Palma de Mallorca, 1901), and another Barcelona, 1989.

[21] J. Régné, *History of the Jews in Aragon. Regesta and Documents 1213–1327* (Jerusalem, 1978), p. 505, n° 2719, dated Barcelona, 30 October 1299.

[22] The original Tatars probably came from east central Asia or central Siberia; unlike the Mongols, they spoke a Turkic language. After the conquests of the Mongol Genghis Khan, the Mongol and Turkic elements merged, and the invaders became known in Europe as Tatars. Llull must have heard of Hülegü Khān (1217–65) who was the grandson of Genghis Khan. He sacked and burned Baghdad in 1258 (executing the last ʿAbbāsid caliph) and captured Aleppo and Damascus in 1260. Further advances were checked by the Mamlūks, who defeated him (Sepember 1260) at the decisive battle of ʿAyn Jālūt in Syria. Hülegü withdrew to Azerbaijan, adopted Islam, and founded the dynasty of the Ilkhāns. Cf. J. Gayà, 'Ramon Llull en Oriente (1301–1302),' *Studia Lulliana* 37 (1997), pp. 25–78.

[23] *Opera*, ed. I. Salzinger, i (Mainz, 1729). Critical edition A. Madre, *ROL*, n° 128 (Turnhout, 1986).

[24] *Vita beati R. Lulli*, p. 297.

qāḍī, and he argued with him about the Trinity. Llull was jailed and Genovese and Catalans in the town asked the *qāḍī* to mitigate jail conditions. He eventually was put on a ship sailing for Genoa, but a storm destroyed the ship off the shore near Pisa. Llull saved his life and stayed in Pisa, where he finished the *Ars generalis*. From his discussion with the religious authority of Bougie he wrote the *Disputatio Raimundi Christiani et hominis Sarraceni*.[25]

Llull visited Pope Clemens V in Avignon in 1308. In 1309, in Montpellier, Llull composed the *Liber de fine*,[26] which deals in a large measure *contra Sarracenos*. The next year he was in Paris, where he joined the fight against the Averroists with a *Liber reprobationis aliquorum errorum Averrois*,[27] which was to be followed by more anti-Averroistic works.

In 1311, Pope Clemens V assembled the Church Council in Vienne, in the Dauphiné, and before the congregated bishops Llull pleaded for three things: the teaching of Oriental languages to missionaries in Rome, Paris and Toledo; the creation of a military order to conquer the Holy Land; and the fight against the Averroists.

In that year Abū Yaḥyā Zakarīyā al-Liḥyānī (1311–17) ousted the Ḥafṣides from power in Tunis and then forged an alliance with the Aragonese king James II and the Sicilian king Frederick III.[28] The situation became favorable for missionary activity and in 1314 Llull traveled for a second time to Tunis where he could preach without danger. To help the missionary work, he wrote additional texts in Arabic similar to the *Ars consilii*.[29] He did so for over one year until his death, which is estimated to have occurred in March 1316.

These are some details in the life and writings of Ramon Llull that show the depth of his knowledge of the Arabic language and of the Muslim culture. This is the reality that Dominique Urvoy,[30] S. Garcías Palou,[31] and Hans Daiber—see below—have analyzed in detail, and S. Trías

[25] *Opera*, ed. I. Salzinger, iv (Mainz, 1729). Critical edition A. Madre in *ROL* n° 32 (Turnhout, 1959).

[26] Montpellier, 1305, printed Palma 1665, critical edition A. Madre, *ROL* n° 122 (Turnhout, 1981).

[27] *ROL* n° 164, ed. H. Reidlinger (Turnhout, 1978), pp. 288–318.

[28] Brunschvig, *Berbérie*, i, pp. 128–43. After Ibn al-Liḥyānī was dethroned, he died in Egypt in 1326.

[29] *ROL* n° 253, ed. J. Stöhr, ii (Palma de Mallorca, 1960), pp. 213–69.

[30] D. Urvoy, *Penser l'islam. Les presupposées islamiques de l'art de Llull* (Paris, 1980).

[31] *Ramón Llull y el islam* (Palma de Mallorca, 1981).

Mercant has summarized.[32] Llull was divided in his soul between seeking a peaceful conversion or fighting a crusade; the failure of the first may have led him to embrace the second. Llull's permanent concern was to convert Muslims to Christianity from the inside, becoming himself a *christianus arabicus.* This meant, for him, knowing the language as well as Islamic theology. He was convinced that he could demonstrate the truth of the Christian religion by using the same instruments that the Muslim theologians employed to defend Islam.

The analysis of the Islamic influence upon Llull's thought has mainly focused on the doctrine of the divine attributes. Muslim theology has extensively dealt with this subject, and also with that of the divine names, to which Llull devoted a *Libre qui és de cent noms de Déu.*[33] The common interest is obvious, but modern research does not agree upon the exact extent of the influence. Llull must have known some Islamic works on the fundamentals of Islam (*uṣūl al-dīn*) and his *Llibre de contemplació en Déu* deserves to be studied concerning possible influences of this kind of Muslim treatise.

As for the influence of Arabic philosophy, Hans Daiber has shown the clear connection with Avicenna, by the intermediary of al-Ghazālī, in his analysis of the *Liber disputationis Raimundi Christiani et Homeri Saraceni.*[34] Knowledge of Arabic literature falls within Llull's universal knowledge of Islam. Although it plays a minor role compared to theology, it has its importance because of Llull's original adaptations of some of its elements.

The presence of Arabic literary sources is most obvious in a book called *Fèlix* or *Book of the World Marvels*, which Llull wrote in Paris,[35] which deplores the fact that men know and love God so little. The main figure of the book is a young man called Fèlix who learns from a philosopher the science of nature. The philosopher guides Fèlix through the natural realms: starting from the heavens to living beings.

---

[32] 'Arabismo e islamología en la obra de Ramon Llull,' *Semitica Escurialensia Augustiniana. Homenaje a Fray Luciano Rubio OSA. La Ciudad de Dios* 208 (1995), pp. 439–52.

[33] *Obres de Ramon Llull*, ed. S. Galmés, xix (Palma, 1936), pp. 79–170.

[34] 'Raimundus Lullus in der Auseinandersetzung mit dem Islam. Eine philosophiegeschichtliche Analyse des *Liber disputationis Raimundi Christiani et Homeri Saraceni*,' in M. Lutz-Bachmann and A. Fidora (eds.), *Juden, Christen und Muslime* (Darmstadt, 2004), pp. 136–72, and especially pp. 138–56.

[35] 'En tristícia e en llanguiment estava un home en estranya terra. Fortment se meravellava de les gents d'aquesta món com tan poc coneixien e amave Déu, qui aquest món ha creat,' *Fèlix*, ed. S. Galmés (Barcelona, 1932), p. 25; ed. M. Batllori et al., in *Obres essencials*, i (Barcelona, 1957), p. 319.

After teaching Fèlix the composition of the metals, the philosopher bids farewell to him, and goes his own way.

Fèlix then meets two beggars, who tell him that they belong to the 'Order of the Apostles', a religious order that was founded in 1260 by Gherardo Segalelli. The positive, though brief, discussion of the Order of the Apostles leads John Dagenais to the conclusion that 'the first six books of the *Fèlix* and the prologue to Book VII, the *Llibre de les bèsties* were written before the spring of 1286,'[36] when Pope Honorius condemned the order. Most scholars—S. Galmés, M. Batllori, and A. Llinarès—entertain the view that Book VII, i.e., the *Llibre de les bèsties* was written independently and before the rest of the books of the *Fèlix* that were written between 1288 and 1289 during Llull's stay in Paris that had began in 1286.

The fictitious meeting with members of the Order of the Apostles serves as a bridge to a new story. The beggars are just coming from a meadow where the animals have gathered to elect their new king. In the middle of an encyclopedic work, fiction is inserted and the characters of the fiction are animals who talk and have human intelligence.[37] This kind of literary construction is not a novelty, and Llull could have been inspired by the *Epistles* of the Brethren of Purity, a group of scholars who lived in Baghdad towards the end of the tenth century. They were linked by friendship and common activities, and we know some names. There are 52 *Epistles*, dealing mainly with philosophical and scientific subjects but the chapter on animals is different. It tells of a dispute between animals and man about who is superior.[38] The insertion of the fable in an encyclopedia is the only motif common to the Brethren of Purity and to Llull, who develops a different argument.

The animals meeting in the meadow are divided into two factions: those living upon grass and those devouring other animals. The carnivores win with the election of the lion as the king, and the lion allows them to nourish themselves by eating the herbivore animals.

---

[36] 'New Considerations on the Date and Composition of Llull's *Llibre de bèsties*,' in M. Duran, A. Porqueras Mayo and J. Roca Pons (eds.), *Actes del segon colloqui d'estudis catalans a Nord-Amèrica*, Yale 1979 (Publicacions de l'Abadia de Montserrat, 1982), pp. 134–5.

[37] The chapter on the animals was translated into English by E. Allison Peers, *The Book of the Beasts* (London, 1927; repr. Westport, Conn., 1978).

[38] For a content of the *Epistles*, see Y. Marquet, *La philosophie des Ikhwân aṣ-ṣafâ'* (Algiers, 1975). The Arabic text was edited by K. Bustānī, *Rasā'il Ikhwān al-ṣafā'* (Beirut, 1927; repr. 1957).

One night, the king and his advisers, all carnivores, feel hungry and follow the advice of the fox, who tells them to eat a calf that is the son of the ox, and a foal that is the son of the horse. The bull and the horse, hurt by the action, flee the kingdom and place themselves at the service of man. But, their fate under this new master is not better, because man puts them to work. Even worse, the ox finds out that the master wants to sell him to the butcher.

The episode has introduced the fox, the protagonist of the story. The fox is not strong enough to become king and he knows it, but wants to become the main adviser to the king. However, the lion chooses the bear, the leopard, the ounce, the snake and the wolf as his counselors. The herbivores are not happy, and the king chooses the cock too. The cock and the snake represent the herbivores, but the lion does not trust the fox and does not appoint him. From then on, the fox takes revenge against the lion, using 'cunning speech and his skill'.

Llull calls the fox Na Renard, or Na Renart, and borrows the name from the fox who is the protagonist of a famous series of *fabliaux*, dating to the thirteenth century,[39] but Llull's 'dame Renart' is of the feminine gender because the fox is feminine in Latin as well in Catalan. The French Renart became the paradigm of cunning and trickery, but he remains a very amusing figure and, if we compare him with Llull's Renard, he is more naive, he is teasing, stealing cheese from a crow and similar things, whereas Llull's Renard is a master of conspiracy:

The fox of Llull sees that the elephant, in spite of being an herbivore, is the only animal strong enough to replace the lion. So, the fox employs all of his art to persuade the elephant to join his campaign. The elephant is reluctant because he eats grass and the fox eats meat so that, by nature, they are opposite. The arguments they employ are mainly stories, such as the following which the elephant tells to the fox:

> In a certain land it came to pass that a kite was holding in its claws a rat, and a hermit prayed to God that the rat might fall into his robe. Through the prayers of this holy man God caused the rat to fall into the robe of that hermit, who thereupon prayed that God would turn it into a fair damsel. 'Fair child, said the hermit, wilt thou have the sun for a husband?' 'Nay, sir,' she answered, 'for the clouds rob the sun of all its brightness.' Then the hermit asked her if she herself have the moon for a husband,

[39] The classical edition is by E. Martin, *Le Roman de Renart*, 3 vols. (Strasbourg, 1882–87). Also: J. Dufournet et A. Méline, 2 vols. (Paris, 1985). Cf. K. Varty, *The Roman de Renart. A Guide to Scholarly Work* (Lamhan, Md., 1998).

> and she answered that it is not brightness itself, but that it rather gets its light from the sun. 'Fair child, said the hermit, wilt thou have a cloud for a husband?' 'Nay, sir, she answered, for the wind chases the cloud whithersoever it will.' Neither did she desire the wind for a husband since the mountains impede its motion, nor did she desire the mountains, since men make tunnels therein; nor did she desire a man for husband, for men slay rats. So in the end, the damsel begged the hermit to pray God to turn her into a rat, as she was before, and give her a fair rat as a husband.[40]

The source of this story is found in the book of *Kalīla wa-Dimna*, a collection of Hindu tales that Ibn al-Muqaffaʿ translated from the Pahlevi version into Arabic by the middle of the eighth century.[41] The book was translated in Christian Spain in 1251, when Alphons X the Wise was still a prince.[42] Llull knew *Kalīla wa-Dimna* and refers to it in this and other instances;[43] Armand Llinarès, who published the old French translation of Llull's *Llibre de les bèsties*, stated that most of the citations of Arabic literature found in the book belong to *Kalīla wa-Dimna*.[44]

The fox understands very well the message of the elephant, but he does not give up and tells him another story to neutralize the elephant's suspicion.[45] Once the elephant is no longer suspicious of him, the fox wants to persuade him that they are able to kill the lion using craftiness and tricks. The story is about animals which ate grass and were hunted by the lion. To spare themselves all the stress of running from the lion, they made an agreement with him to sacrifice themselves in this way: every day they drew lots and that one that lost went to the lion to be eaten.

One day the lot fell upon a hare. She was afraid of dying and delayed going to the lion until noon. The lion was by that time very hungry and furious, and asked the hare for the reason for her delay. She gave as an excuse that another lion wanted to eat her, pretending that he was the king of the same country and that she had to run away from that lion. The king believed her and asked her to show him where the other lion was. The hare brought him to a pool surrounded by a high

[40] *The Book of the Beasts*, p. 15.

[41] First edited by A. Silvestre de Sacy, *Calila et Dimna ou Fables de Bidpai* (Paris, 1816).

[42] The Old Spanish version was edited by J.E. Keller and R. White Linker, *El libro de Calila e Digna* (Madrid, 1961).

[43] See here *Kalilah and Dimnah. An English Version of Bidpai's Fables Based upon Ancient and Spanish Manuscripts*, trans. T.B. Irving (Newark, Del., 1980), pp. 107–8.

[44] *Le livre des bêtes* (Paris, 1964), pp. 18–27.

[45] *Book of the Beasts*, pp. 16–17.

wall and the lion saw himself reflected on the water, but the hare made him believe that the image was the other lion who disputed his kingdom. The lion wanted to kill him and jumped into the water, where he drowned. The source of the story is again *Kalīla wa-Dimna*.[46]

The elephant counteracts with a story of opposite purport. He eventually agrees to join the conspiracy, when the fox relates the example of Eve and the snake. The fox has the necessary skill and wisdom to kill the lion.

The ox and the horse are not forgotten by Llull: they decide to flee from the servitude of man and go back to the court of the lion king. The ox meets the fox on his way to the court and relates his fate with man, but the ox tells him that the king of the animals is evil too and that the ox can bring the king of the animals into a better state if he follows the fox's counsel. The ox hesitates, but after the fox tells him the story of a hermit who was able to redress the behavior of a bad king, he accepts to follow the fox's counsel.

The ox goes to a meadow, not far away from the lion and his court. Once he has recovered from the weakness caused by working for man, he bellows three times a day and three times a night following the fox's counsel. The lion king trembles out of fear, as the king of the animals does in the tales of *Kalīla wa-Dimna* when the ox Sençeba bellows.[47]

The fox appears before the lion, who asks him if he knows who is bellowing; it seems to be a strong animal. Instead of telling him the truth, the fox tells him a story. A juggler 'had hung his drum in a tree and the wind blew upon the drum and caused it to strike the branches of the tree,'[48] and the sound was increased by the echo. A monkey heard the sound, looked for the place where it came from and found the drum. Since the sound was so loud, he thought the drum was filled with butter or something good to eat, and he broke it, but the drum was empty. We read again a story found originally in the *Kalīla wa-Dimna*.[49]

The lion understands the meaning of the story, but he cannot stop trembling every time the ox bellows. His court is also full of fear, except for Na Renard. The lion asks him again, why he is not afraid of the powerful voice? The fox does not tell the truth, but answers with another

[46] *Kalilah*, trans. Irving, pp. 20–21.
[47] *Kalilah*, trans. Irving, p. 3.
[48] *Book of the Beasts*, p. 27.
[49] *Kalilah*, trans. Irving, p. 10.

story that we know comes from the *Kalīla wa-Dimna*:[50] a snake used to go into the nest of a crow and eat her young. The crow could not fight against a snake and had to develop a strategy: she took the garland of the king's daughter and flew with it to the place where the den was. The servants of the princess followed the crow to pick up the garland. When they saw that it was close to the snake, they killed it.[51] The crow employed 'subtlety and art' to slay the snake as the fox does, and for this reason he has no fear of those who are stronger than he is.

The snake, which is a counselor to the king, is not pleased with the answer and tells them a story that also comes from the *Kalīla wa-Dimna*.[52] A goose promised to save the fish in a pond from the fishermen by flying them to another pond, but what she did instead was to bring them to a place where she could eat them 'without the labor of fishing.'[53] The crab asked her to be flown out, and seized the neck of the goose with both pincers. The crab did not see any pond but the bones of the fish the goose had eaten: it understood what had happened and squeezed the neck of the goose, taking revenge for all his friends in the pond. The fox understands that he is the crab of the story, avenging the animals that live upon grass, but goes on with another story, to prove that the snake is not to be trusted; 'all the evils which are in the world have come through the serpent.'[54]

The lion is persuaded by the fox to go to the animal with the powerful voice and bring him before the king. The ox begs forgiveness from the king, who grants it to him and asks about the kingdom of man. For the ox 'the most false and evil beast in this world is man,'[55] and to prove it, he relates a story of a hermit who rescued from a well a bear, a crow, a snake and a man who had fallen into it. The animals all show gratitude toward their rescuer but not the man, who accuses the hermit of stealing a garland from the king's daughter. This is a story known again from *Kalīla wa-Dimna*.[56] As a result, the lion is now afraid of the kingdom of man and sends the leopard and the ounce as emissaries to the king of men and gives him the dog and the cat

[50] *Kalilah*, trans. Irving, p. 17 and p. 19.
[51] *Book of the Beasts*, pp. 29–30.
[52] *Kalilah*, trans. Irving, pp. 17–19.
[53] *Book of the Beasts*, pp. 30–33.
[54] *Book of Beasts*, p. 33.
[55] *Book of Beasts*, p. 37. 'La pus mala bèstia e la pus falça que sia en est món, és lo hom,' ed. S. Galmés, p. 117; ed. Batllori, p. 378.
[56] *Kalilah*, trans. Irving, pp. 172–5.

as presents. More importantly, the fox has gained his confidence and the lion king appoints him as his chamberlain, instead of the dog, and later makes him one of his counselors.

The fox is the protagonist of Llull's story—one main difference from the *Kalīla wa-Dimna*—and he relentlessly pursues the plan to remove the lion, install the elephant as king, and acquire the real power as his only counselor. The fox incites the king to take advantage of the leopard's absence and to seduce the leopard's wife. The leopard returns home from his embassy and the members of his household recount to him 'the dishonor that the king had done to him by corrupting his wife.'[57]

The leopard denounces the king and challenges him, but for reasons of his stature, the ounce has to defend the honor of the king. He battles against the leopard, who kills him. The truth of the king's treason is made known as a result of the battle and the king becomes so furious that he slays the exhausted leopard. The situation in the court of the lion is deplorable, and the smart snake points to the cause: 'Since the fox and the ox had been admitted to his court it has never been without anxiety and tribulation.'[58]

The ox defends himself against the accusations of the snake and declares to the king that the fox persuaded him to bellow in order to scare him. But the fox knows how to get out of the trap and the lion eventually eats the ox. Then the fox has reached the summit of his power in the court: the snake, the leopard, the ounce and the wolf have left the council, and the fox has made the rabbit the chamberlain, 'for the cock and the peacock and the rabbit feared the fox equally, and the lion believed all that the fox said to him.'[59]

However the lion thinks he needs more counselors, a view that the cock shares in opposition to the fox, who praises the cock as being so wise that the lion does not need any other counselor, and to illustrate his assertion tells another story:

> There was a man who could understand the language of the animals but God had given him this capacity on the condition that he not tell anybody what he overheard; otherwise he would die. The man owned a donkey and an ox, and the ox had a very hard time drawing water from a well. He complained to the donkey, who counseled him not to eat his oats in

[57] *Book of the Beasts*, p. 62.

[58] *Book of Beasts*, p. 69, *depuys que lo bou e na Ranart foren de sa cort, no fo sa cort sens treball e tribulació*, ed. Galmés, p. 142; ed. Batllori, p. 385.

[59] *Book of Beasts*, p. 75.

> the evening and to act as if he were ill, so that the farmer will let him rest in the stall. The farmer, indeed, let the ox rest, but led the ass to the well in his place. The ass learned his lesson. To avoid the hard labor, he told the ox that the master wanted to sell him to the butcher.

The master who had listened to the talk laughed at it. His curious wife inquired about the reason for his laughing. He could not tell her anything; otherwise he would die, but the woman insisted and did not want to drink and to eat anything. She was going to die. The master who loved his wife was ready to tell what he had heard, and started writing his will.

His dog and the cock of his coop saw all this. The cock told the dog that the master deserved to die, 'for he was an evil man and unfit to be the lord over his wife.'[60] If he were in his place, he would beat his wife until she would eat and drink. The master, who had heard the cock's advice, beat his wife and she ate and drank.

We know the story from its Arabic source: from the Arabian Nights as represented in their first version.[61] Llull puts the story into the mouth of the fox to praise the wisdom of the roosters in order to make additional counselors in the court superfluous. But the cock in Llull's story is not as smart and he will recount a story that will trigger the wrath of the lion king. The story comes from *Kalīla wa-Dimna* and is about a parrot, a crow and a monkey.[62]

> The parrot and the crow were sitting in a tree and at the foot of the tree there was a monkey who had placed wood upon a glow-worm, thinking that it was a fire, and he blew upon the wood with the intent to make a fire whereby he might warm himself. The parrot cried to the monkey and told him that it was not a fire but glow-worm; and the crow said to the parrot that he should instruct or correct any who would not receive counsel or correction.[63]

However the parrot repeated his warning many times and eventually went down from the tree to make the monkey understand the situation, but the monkey just slew the parrot. The cock wanted to persuade the lion to have more counselors, but the lion understood that he was a monkey and became very angry at the cock. The fox killed and ate him.

---

[60] *Book of Beasts*, p. 79.

[61] *Kitāb alf layla wa-layla*, ed. M. Mahdi, 3 vols. (Leiden, 1984–94). English translation by Husain Haddawy, *The Arabian Nights* (New York, 1990), pp. 11–15.

[62] *Kalilah*, trans. Irving, p. 41.

[63] *Book of the Beasts*, p. 80.

Na Renard was now in a very strong position being the king's only counselor, and actually did not need to overthrow the king, but he feared the elephant could betray him and reveal the plot to make the elephant the king. He decided to go on with the conspiracy and ended up getting caught in his own net. The elephant remained faithful to his king and denounced the fox who paid for his treason with his life. Llull has constructed the narrative with the purpose of warning the kings of his time about the dangers of their counselors and about the intrigues inside the royal court.

Who was the king to whom he was addressing the book? Llull was writing the book in France, and the king was Philip the Fair (1268–1314), whose mother was Isabelle of Aragon, sister of king James II of Mallorca. Philip the Fair was enthroned after the death of his father, Philip the Bold on the 5th of October 1285. He was very young—17 years old—and could easily be misled by his counselors. And the narrative is based on the rise and fall of Na Renard as an instance of an ambitious courtesan. It is likely that Llull was thinking of a real person. Llinarés suggests two names of counselors who could possibly be identified with the wicked Renard of Llull.[64]

To some extent Llull found his inspiration in the main personage of the *Roman du Renart*. John Flinn, however, did not believe in an effective influence of the *Roman du Renart* upon the *Llibre de les bèsties*.[65] Although there are some common elements, such as the frame of the lion's kingdom—the king Noble of the *Roman du Renart* has been identified with Philip Augustus (reg. 1180–1223)[66]—and the name of Renart, they do not build strong evidence for a relevant influence. In contrast the evidence showing the Arabic influence is undeniable: Llull used many stories taken from *Kalīla wa-Dimna* and from the Arabian Nights but he also shared their belief that story-telling can be more effective than dry reasoning.

Talking animals are an old literary device by means of which the author can tell truths that he is afraid to speak about. Although we can-

---

[64] A. Llinarès suggests two courtiers: Enguerran de Marigny (executed 1315) and, most likely, Pierre de la Brosse (executed 1278), cf. 'Introduction' (nt. 45), pp. 32–4.

[65] *Le roman de Renart dans la littérature française et dans les littératures étrangères au Moyen Age*, ([Toronto], 1963), p. 146.

[66] J. Subrenat, 'Un point de vue sur la fonction royale sous Philippe-Auguste; le roi Noble dans *Le Roman de Renart*,' in *Histoire et Société. Mélanges offerts à Georges Duby*, iii (Aix-en-Provence, 1992), pp. 167–77.

not be certain where Llull got the idea for using this device, we know that Ibn al-Muqaffaʿ, the translator of the *Kalīla wa-Dimna* into Arabic, had written an introduction to it where he said that slipping words in animals' mouths was an artifice (*ḥīla*) that allowed wise authors of such stories 'to use cunning in their sayings'.[67] The old Spanish translation is far more explicit in this point: talking animals allowed these wise people 'to tell the truth in a veiled manner'.[68]

---

[67] *Wajadū mutaṣarrafan fī 'l-qawl*, in ʿAbdallāh ibn al-Muqaffaʿ, *Kalīlah wa-Dimnah*, ed. ʿA. ʿAzzām (Algiers, 1973), p. 5.

[68] 'E ayuntaronseles para esto tres cosas buenas: la prymera, que los fallara usados en rrazonar, e trobaronlas, segun que los usavan, para dezir encobiertamente lo que querian,' *El libro de Calila e Digna* (nt. 43), p. 3.

LANGUAGE AND LITERATURE

NOTES SUR QUELQUES-UNS DES TÉMOIGNAGES MÉDIÉVAUX RELATIFS À L'*HISTOIRE PHILOSOPHIQUE* (*ἡ φιλόσοφος ἱστορία*) DE PORPHYRE

Emily Cottrell

I – August Müller et les fragments de l'*Histoire philosophique*

Friedrich August Müller (n. 3/12/1848–m. 12/9/1892) publia au Caire en 1882 sous le pseudonyme arabisé d'Imrū' al-Qays ibn al-Ṭahhān[1] les *Sources des informations sur les Générations des médecins* (*ʿUyūn al-anbāʾ fī ṭabaqāt al-aṭibbāʾ*), par Ibn Abī Uṣaybiʿa (590/1194 ou 600/1203–668/1270), en deux tomes reliés en un, dépourvus des introductions et de l'apparat critique qui avaient pourtant été préparés.[2] Ceux-ci furent publiés à Königsberg en 1884 sous la forme d'un deuxième volume accompagnant une réimpression du texte édité en 1882. La dimension de l'ouvrage et sa richesse en firent l'un des classiques de l'orientalisme. L'*Encyclopédie de l'Islam* et l'*Encyclopaedia Iranica* l'utilisent systématiquement pour établir la biographie et la bibliographie des personnages qui y sont mentionnés.[3] Ibn Abī Uṣaybiʿa naquit dans une famille de médecins qui travaillèrent au service de Saladin puis des Ayyoubides et de leurs vassaux à Damas, à Alep et au Caire à la fin du XIIe siècle et pendant tout le XIIIe siècle (fin VIe–VIIe A.H.).[4]

---

[1] Comme A. Nallino le remarque, c'est une simple traduction de son nom allemand, *Raccolta di Scritti*, éd. M. Nallino (Roma, 1944), v, p. 143, n. 1. Bien qu'il signât du nom d'August Müller, c'est sans doute son autre prénom, Friedrich, qui avait été traduit Imrū 'l-Qays, en hommage au poète Friedrich Schiller. Ni Georges Anawati ('Tawārīkh al-ʿulūm al-ṭibbiyya fī usrat Ibn Abī Uṣaybiʿa al-Dimashqī wa-kitābihi *ʿUyūn al-anbāʾ fī ṭabaqāt al-aṭibbāʾ* al-muʾallaf sanat 640,' in N. al-ʿAṭṭār [éd.], *Al-Nadwa al-rābiʿa li'l-samāt al-insāniyya li'l-ʿilm wa'l-ʿamal fī bilād al-Shām fī 'l-bimaristān al-Nūrī* [Damas, 4–11 oct. 1985], cahier n° 12, pp. 1–41), ni Nallino ne croient à une conversion éventuelle d'August Müller.

[2] Une réimpression de l'édition du Caire accompagné du volume de corrections publiées à Königsberg a été publié par F. Sezgin et al. dans la collection Islamic Medicine, 2 vols. (Frankfurt, 1995).

[3] Les ouvrages d'Ibn al-Qifṭī et de Bar Hebraeus, repris tout autant par les encyclopédies, sont à utiliser avec plus de prudence. A. Nallino relève néanmoins plusieurs erreurs et confusions chez Ibn Abī Uṣaybiʿa, cf. Nallino, *Raccolta*, p. 144.

[4] *EI*[2], art. Ibn Abī Uṣaybiʿa (J. Vernet); A. Nallino, *Raccolta*, pp. 137–43.

Des sources qu'Ibn Abī Uṣaybiʿa utilisa pour composer son histoire de la médecine, laquelle recouvre aussi bien le monde arabo-persan que l'Inde, la Grèce et l'Andalousie, on sait peu, aucune étude d'ensemble de l'ouvrage n'ayant été entreprise.[5] Sa proximité avec la cour lui permettait sans aucun doute l'accès aux meilleures bibliothèques de son époque. Il cite fréquemment les ouvrages qu'il utilise. Outre son père, médecin lui-même, son oncle, Rashīd al-Dīn ʿAlī (m. 616/1219), joua un rôle important dans sa formation et Ibn Abī Uṣaybiʿa donne le détail du cursus (médecine, philosophie, astronomie, musique) que celui-ci avait suivi, ainsi que le fait qu'il connaissait l'arabe, le persan et le turc.[6] Il étudia avec lui la médecine, l'ophtalmologie, et la philosophie (Rashīd al-Dīn ʿAlī avait étudié les commentaires d'Aristote avec ʿAbd al-Laṭīf al-Baghdādī, qui resta proche d'Ibn Abī Uṣaybiʿa et de son père aussi bien en Egypte qu'en Syrie).[7] Les informations dont il fait état sur les médecins andalous pourraient avoir été obtenues auprès des nombreux voyageurs qui visitaient le Caire où il résida pendant quelques années, ainsi que de l'herboriste Ibn al-Bayṭār (m. 646/1248) qui séjourna à Damas à partir de 633/1235.[8] Ibn Abī Uṣaybiʿa étudia avec lui la botanique dans les ouvrages de Dioscoride, de Galien et d'al-Ghāfiqī. Quant aux médecins indiens, il utilise les ouvrages d'al-Yaʿqūbī (m. après 292/905), d'Abū Bakr al-Rāzī (m. 313/925 ou 323/935), d'Ibn al-Nadīm (m. 385/995 ou 388/998), et d'Ibn al-Qifṭī (m. 646/1248).[9] Quant à ses sources grecques, qu'il cite généralement avec précision, ce sont les ouvrages classiques et leurs commentaires datés principalement de l'époque romaine et de l'antiquité tardive. En tant que médecin, sa connaissance intime des écrits de Galien (m. 216 ap. J.-C.) et d'Hippocrate (*c.* 460–375 av. J.-C.) lui donnait accès à tout un pan de la littérature antique que nous ignorons.

[5] Les études anciennes ont été rassemblées sous le titre *Studies on Ibn Abī Uṣaybiʿa (d. 1270) and his ʿUyūn al-anbāʾ fī ṭabaqāt al-aṭibbāʾ*, ed. F. Sezgin et al. (Frankfurt, 1995).

[6] Anawati, p. 7. L'oncle d'Ibn Abī Uṣaybiʿa connaissait le persan, y compris sa métrique en poésie, et le turc (Anawati, 'Tawārīkh al-ʿulūm,' p. 7).

[7] Anawati, 'Tawārīkh al-ʿulūm,' pp. 14–15.

[8] Ibn Abī Uṣaybiʿa (éd. Müller, ii, 133) raconte les promenades pendant lesquelles ils herborisaient dans les environs de Damas. Il précise qu'Ibn al-Bayṭar commençait toujours par lui lire le texte grec de Dioscoride (Anawati, 'Tawārīkh al-ʿulūm,' pp. 17–18).

[9] Müller réalisa une étude du chapitre sur les médecins indiens ('Arabische Quellen zur Geschichte der indischer Medizin,' *ZDMG* 34 [1880], pp. 465–556; réimpr. in *Islamic Medicine*, iv [Frankfurt, 1995]). S.M. Khan, 'An Arabic Source for the History of Ancient Indian Medicine,' *Indian Journal for the History of Science* 161 (1981), pp. 47–56.

Parmi d'autres philosophes grecs, la présence de Pythagore (*c.* 569–494 av. J.-C.) au sein des médecins est en soi révélatrice d'une certaine *Weltanschauung* d'Ibn Abī Uṣaybiʿa. La figure d'un Pythagore médecin n'étonnera pas quiconque a lu la *Vie de Pythagore* que composa Porphyre (n. *c.* 232–233 – m. avant 304 ap. J.-C.). August Müller put identifier de nombreux parallèles entre ce texte et le chapitre d'Ibn Abī Uṣaybiʿa consacré à Pythagore. La mention de Porphyre s'y trouvait avec celle du premier livre des *Histoires des philosophes*, titre derrière lequel Müller reconnut l'*Histoire philosophique*. C'est en effet au livre premier de cet ouvrage perdu pour l'essentiel que la *Vie de Pythagore* appartenait, selon le témoignage de Cyrille d'Alexandrie (376–444 ap. J.-C.), qui le cite dans son *Contre Julien*. Il revient à Hans Daiber, auquel j'ai la joie de dédier ces notes, d'avoir le premier insisté sur l'importance du chapitre consacré par Ibn Abī Uṣaybiʿa à Pythagore.[10] Ses recherches ont montré, grâce à l'édition du *Commentaire des Vers d'or* attribué à Jamblique, qui venait s'ajouter aux deux textes fondamentaux que sont la traduction des *Vers d'or* par Ḥunayn ibn Isḥāq et un deuxième *Commentaire des Vers d'or* attribué à Proclus, que l'influence des (néo-)pythagoriciens n'est pas à négliger dans l'étude de la philosophie médiévale composée en langue arabe.[11]

1. *August Müller et le Catalogue d'Ibn al-Nadīm*

August Müller avait travaillé avec Julius Lippert à terminer l'édition du célèbre *Catalogue (Kitāb al-fihrist)* d'Ibn al-Nadīm par Gustav Flügel, qui mourut avant de terminer son entreprise. L'ouvrage parut à Leipzig en 1871–1872.[12] On commençait alors à peine à entrevoir la richesse inépuisable du *Catalogue* achevé par le libraire bagdadien Ibn al-Nadīm

[10] H. Daiber, 'Hellenistisch-kaiserzeitliche Doxographie und philosophischer Synkretismus in islamischer Zeit,' *Aufstieg und Niedergang der römischen Welt*, ii/36.7 (Berlin, 1994), pp. 4974–92, voir en particulier pp. 4984–5. Nous espérons bientôt pouvoir proposer une étude du chapitre dans son entier.

[11] H. Daiber, *Neuplatonische Pythagorica in arabischem Gewande. Der Kommentar des Iamblichus zu den Carmina aurea* (Amsterdam, 1995). La traduction des *Vers d'or* appartient à un compendium attribué à Ḥunayn ibn Isḥāq, *Ādāb al-falāsifa*, éd. ʿA. Badawī (Koweit, 1406/1985) (nombreuses traductions et études relevées par H. Daiber, *Bibliography of Islamic Philosophy*, i [Leyde, 1999], s.v. Pythagoras (Ps.): *Carmina aurea*). Ibn al-Ṭayyib, *Proclus' Commentary on the Pythagorean Golden Verses*, éd. et trad. N. Linley (New York, 1984).

[12] Ibn al-Nadīm, *Kitāb al-Fihrist*, éd. G. Flügel et al. (Leipzig, 1871–72); réimpr. dans *Historiography and Classification of Science in Islam* (Frankfurt, 2005). Une nouvelle édition intégrant un chapitre retrouvé de l'ouvrage fut publiée par R. Tajaddod (Ibn Nadīm, *al-Fihrist* [Téhéran, 1971]). La traduction de B. Dodge, *The Fihrist of al-Nadīm. A Tenth-Century Survey of Muslim Culture* (New York, 1970) est à utiliser avec précaution.

en 377/987. Dans la liste des ouvrages de Porphyre qui étaient en circulation au IVe/Xe siècle, on pouvait lire ceci:

T1 Porphyre, *Histoire philosophique* = Smith[13] 194aT: Ibn al-Nadīm, *Catalogue (al-Fihrist)* VII, 1 [ch. Furfuriyūs], éd. Tajaddod, p. 313 ult / éd. Flügel, p. 253, l. 18 (cf. Ibn al-Qifṭī, éd. Müller-Lippert, p. 257, l. 7–9; Bar Hebraeus, éd. Saliḥānī, p. 78, l. 19)

Livre des *Informations (akhbār: 'nouvelles', 'récits')*[14] *sur les philosophes*. J'en ai vu le quatrième livre *(al-maqāla)* en syriaque.	كتاب أخبار الفلاسفة و رأيت منه المقالة الرابعة سريانيّ.

J. Wenrich avait noté déjà en 1842 que ce titre devait renvoyer à l'*Histoire philosophique* de Porphyre mais aucune édition des fragments n'était encore disponible.[15] Le terme *maqāla*, dérivé du verbe *qāla*, 'dire' peut signifier 'chapitre' ou même 'traité'. Il traduit ici le grec *'logos'* par lequel Eusèbe désigne les livres de l'*Histoire philosophique* dans sa *Chronique*.[16]

Une seconde mention par Ibn al-Nadīm de l'*Histoire philosophique* n'attestait pas en revanche de sa connaissance directe du texte. Il s'agit du témoignage du médecin et philosophe nestorien Ibn Suwār (Abū 'l-Khayr ibn al-Khammār)[17] (m. *c*. 421/1030) à propos d'un extrait dans lequel Porphyre affirmait que Thalès avait été chronologiquement le premier philosophe. Pour Flügel, l'ouvrage mentionné sous le titre *al-Ta'rīkh ('L'Histoire')* devait être la *Vie de Pythagore* de Porphyre, plutôt que son *Contre les Chrétiens* ou l'un de ses ouvrages d'histoire de la philosophie.[18] La version grecque conservée de la *Vie de Pythagore* étant incomplète, rien n'empêchait qu'elle ait contenu un extrait sur Thalès et les sept Sages.

---

[13] Porphyrius, *Fragmenta*, éd. A. Smith (Leipzig, 1993).

[14] Sur les sens techniques de ce terme, F. Rosenthal, *A History of Muslim Historiography* (Leyde, 1952), pp. 67–71 et A.A. Duri, *The Rise of Historical Writing among the Arabs* (Princeton, 1983), pp. 42–75.

[15] J.G. Wenrich, *De auctorum graecorum versionibus et commentariis syriacis, arabicis, armeniacis percicisque commentatio* (Leipzig, 1842), p. 281.

[16] Eusèbe, *Chronique*, in Porphyrius, *Fragmenta*, éd. Smith, frag. 200F et 200aF (F. Jacoby, *Die Fragmente der griechischen Historiker* [Berlin, 1923–] [= *F. Gr. H.*], 260F4): *'apo tou prôtou logou tès philosophou historias'*. Paul Kunitzsch discute l'usage de ce terme de Isḥāq ibn Ḥunayn (IIIe/IXe s.) à Naṣīr al-Dīn al-Ṭūsī (VIIe/XIIIe) pour désigner les livres de l'*Almageste* dans son *Der Almagest. Die Syntaxis Mathematica des Claudius Ptolemaüs in arabisch-lateinischer Überlieferung* (Wiesbaden, 1974), pp. 130–31.

[17] *EIr*, i, pp. 330–31 (W. Madelung).

[18] Ibn al-Nadīm, *Kitāb al-Fihrist*, éd. Flügel et al., ii, p. 111, n. 7.

T2 Porphyre, *Histoire philosophique* = Smith 194bT = *F. Gr. H.* 260F1a:[19] Ibn al-Nadīm, *Catalogue (al-Fihrist)* VII,1, éd. Tajaddod, p. 305 ult.–p. 306, l. 2 / éd. Flügel, p. 245, l. 13–16

Abū 'l-Khayr ibn al-Khammār m'a dit en présence d'Abū 'l-Qāsim ʿĪsā ibn ʿAlī alors que je lui avais demandé qui avait été le premier à discuter de philosophie que Porphyre de Tyr affirmait dans son livre *L'Histoire (al-ta'rīkh)*, qui était en syriaque, que le premier des sept philosophes *(falāsifa)* fut Thalès, fils de Mālis al-Amalīsī.[20] Deux livres [ou 'chapitres', *maqāla*] de cet ouvrage ont été traduits en arabe. Abū al-Qāsim confirma cela. Selon d'autres, le premier à avoir discuté de philosophie fut Pythagore *(Būthāghūras)* fils de Mnésarque du peuple de Samos *(Sāmiyā)*.	قال لي أبو الخير بن الخمّار بحضرة أبي القاسم عيسى بن علي و قد سألته عن أوّل من تكلّم في الفلسفة فقال زعم فرفوريوس الصوري في كتابه التاريخ و هو سريانيّ إنّ أوّل الفلاسفة السبعة ثالس بن مالس الأمليسى* و قد نُقل من هذا الكتاب مقالتيْن إلى العربيّ فقال أبو القاسم كذا هو و ما أنكره. فقال آخرون : إنّ أوّل من تكلّم في الفلسفة بوثاغورس بن ميسارخس [= منيسارخس] من أهل ساميا.** * Flügel : الأمليسي ** Tajaddod : سامينا

August Müller réalisa en 1873 une traduction commentée du chapitre VII du *Catalogue* d'Ibn al-Nadīm, consacré à la philosophie. Pour le passage ci-dessus (T2), Müller donna à propos de la mention du titre la traduction suivante: «…in seinem buche 'die Chronik', das *ich in einem* Syrischen *exemplar studiert habe*…» (orthographe allemande XIXe). Par souci d'exactitude, il avait indiqué en italique ses propres gloses. Cette précision n'était pas inutile en ce qu'elle insistait sur le fait qu'Ibn Suwār devait avoir lu l'ouvrage qu'il mentionnait, puisqu'il était en mesure de donner la langue dans laquelle il circulait. En note à ce passage, Müller précise : « Gemeint ist die *philosophos historia* die auch [*Fihrist*] 253,18 als *akhbār al-falāsifa* erwähnt wird, und deren erstes buch also ausser dem leben des Pythagoras auch die früheren philosophen behandelt haben

[19] Le témoignage de Ḥājjī Khalīfa (m. 1657), relevé par F. Jacoby, *F. Gr. H.*, 260F1c, est trop tardif pour être considéré. Il reproduit en l'abrégeant Ibn al-Nadīm (260F1a).

[20] A. Müller, *Die griechischen Philosophen in der arabischen Überlieferung (abgedruckt aus der Festschrift der Franckischen Stiftungen zu dem fünfzigjährigen Doctorjubiläum des Herrn Geheimen Rats Professor Bernhardy)* (Halle, 1873); réimpr. Ibn al-Nadīm, *Kitāb al-Fihrist* (Frankfurt, 2005), p. 33, n. 1 considère qu'il s'agit d'une dittographie. Le *alif* initial dans *al-Amalīsī* sert de voyelle d'appui à deux consonnes, comme le veut la règle pour les noms étrangers en arabe classique. Les graphies *al-Malisī* (Pseudo-Ammonius) et *al-Malaṭī* (Bar Hebraeus) existent aussi, montrant une tendance à transformer le *ṭā'* en *ṣād* puis en *sīn*.

muss...».[21] Moritz Steinschneider accepta cette identification : « Die Biographien der Philosophen (oder der *Taarikh*)...».[22] Selon Müller, la mention de Thalès devait donc renvoyer à un passage du livre I de l'*Histoire philosophique* dans lequel les sept Sages étaient mentionnés. Un tel extrait était par ailleurs connu grâce au *Contre Julien* (I, 38) de Cyrille d'Alexandrie. Néanmoins Ibn Suwār ne précise pas s'il a eu connaissance de la totalité de l'*'Histoire'* de Porphyre, ou s'il a lu cela dans un ouvrage composé par un auteur de langue syriaque, ou bien encore s'il a reçu cette information d'un collègue. Si Ibn Suwār avait été le traducteur[23] des deux livres de l'*Histoire philosophique* traduits en arabe, Ibn al-Nadīm aurait sans doute connu cet ouvrage, et il n'aurait pas manqué de le mentionner dans la notice qu'il consacre à son contemporain.

En 1860, August Nauck avait publié l'édition de trois opuscules philosophiques de Porphyre : les fragments de l'*Histoire philosophique* (y compris la *Vie de Pythagore*), le quatrième livre du *De abstinentia*, et la *Lettre à Marcella*.[24] La rareté des textes arabes édités en Europe empêchaient toute collecte sérieuse des fragments dans cette langue. En 1856,

[21] Cf. Müller, *Die griechischen Philosophen*, p. 42, n. 1 rejoignant Flügel (cf. n. 18 supra). Par la suite, plusieurs auteurs furent amenés à citer cet extrait sans mentionner la note qui l'accompagnait. L'hypothèse d'une *Chronique* composée par Porphyre de façon indépendante trouvait là un soutien involontaire.

[22] M. Steinschneider, *Die arabischen Übersetzungen aus dem Griechischen* (Leipzig, 1889–93; réimpr. Graz, 1960), p. 64 (= p. 16). Bar Hebraeus cite à propos de la datation d'Homère et d'Hésiode «Porphyre l'historien *(Fūrfūriyūṣ al-mu'arrikh)*», *Ta'rīkh mukhtaṣar al-duwal*, éd. A. Ṣaliḥānī (Beyrouth, 1958), p. 36.

[23] Ainsi que l'affirme de façon erronée Franz Rosenthal, 'Arabische Nachrichten über Zenon den Eleaten,' *Orientalia* 6 (1937), pp. 21–67; réimpr. *Greek Philosophy in the Arab World. A Collection of Essays* (Hampshire, 1990), p. 39, n. 2, relisant sans doute trop rapidement le *Catalogue (al-Fihrist)* d'Ibn al-Nadīm. L'erreur est reprise par R. Walzer dans *EI*², art. Furfūriyūs. Pour ajouter à la catastrophe, Walzer ira jusqu'à affirmer dans un article très influent sur 'Porphyry and the Arabic Tradition', paru dans la série des Entretiens sur l'antiquité classique (*Porphyre: Huit exposés suivis de discussions* [Vandœuvres, 1966], p. 283), que les quatre livres de l'*Histoire philosophique* avaient été traduits en syriaque et que deux livres circulaient aussi en arabe. Il reprend la date erronée donnée par Rosenthal, ibid., p. 21, n. 3, pour l'édition Müller de Ibn Abī Uṣaybi'a (1892–94 au lieu de 1882–84).

[24] A. Nauck (éd.), *Porphyrii Opuscula tria* (Leipzig, 1860); idem, *Porphyrii Opuscula Selecta* (Leipzig, 1886). La deuxième édition utilise de nouveaux manuscrits et rend obsolète la première. Nauck était lui-même l'élève de 'Godofredus Bernahardy' (cf. supra n. 20), avec lequel il avait étudié dans les années 1840 et auquel il dédia les deux éditions. Une traduction augmentée des fragments de *L'Histoire de la Philosophie* a été proposée par A.-P. Segonds en appendice à Porphyre, *Vie de Pythagore – Lettre à Marcella*, éd. et trad. E. Des Places (Paris, 1982), pp. 163–97.

D. Chwolsohn, un orientaliste russe de Saint-Petersbourg qui avait étudié à Leipzig utilisa pourtant les fragments de Porphyre conservés par Cyrille d'Alexandrie la théologie des Sabéens qu'il présentait à partir de nombreaux ouvrages parmi lesquels le *Livre des religions et des sectes (Kitāb al-milal wa'l-niḥal)* de Shahrastānī (m. *c*. 548/1153).[25] Nauck cita l'ouvrage sur lequel Chwolsohn avait attiré son attention mais remarqua contre la valeur du témoignage de Shahrastānī, que les symboles pythagoriciens – des énigmes qui connurent un succès certain dès l'Antiquité et dont atteste par ex. *Vie de Pythagore* (que nous indiquerons sous le sigle *VP* dans les références abrégées), ch. 42 – étaient malheureusement attribués à Socrate dans le *Livre des religions et des sectes*.[26]

En 1886, peu de temps après la publication des ouvrages d'Ibn al-Nadīm et d'Ibn Abī Uṣaybiʿa, les fragments grecs de l'*Histoire philosophique* furent réédités par Nauck. Il ajouta à sa précédente collection un nouveau texte de Porphyre *(L'Antre des nymphes)*, ainsi qu'un fragment supplémentaire à la collection d'extraits de l'*Histoire philosophique*, pris cette fois-ci dans un témoignage en langue arabe, celui d'Ibn al-Nadīm (ci-dessus T2). Mais le court extrait d'Ibn al-Nadīm rapportant le témoignage d'Ibn Suwār fut donné selon la traduction de Müller en 1873 sans que Nauck s'aperçoive que la traduction de l'arabe avait été glosée pour lui donner plus d'intelligibilité, et il ne reproduisit pas les italiques.[27] Pour ajouter à la confusion, Nauck joignit à ce fragment (frag. V Nauck) la citation par Shahrastānī de la date de Thalès selon Porphyre, sans doute parce qu'elle était connue aussi de Bar Hebraeus dont l'édition d'Oxford de la *Chronique abrégée des nations (Ta'rīkh mukhtaṣar al-duwal)* était lue depuis le XVIIe siècle.[28]

---

[25] D. Chwolsohn, *Die Ssabier und der Ssabismus* (Saint-Pétersbourg, 1856; réimpr. Amsterdam, 1965), p. 719. Abū-'l-Fatḥ Muḥammad al-Shahrastānī, *Religionspartheien und Philosophen-Schulen*, trad. T. Haarbrücker (Halle, 1850–51).

[26] Nauck mentionne Chwolsohn tant en 1860 (introduction, p. IX) qu'en 1886 (pp. VI–VII). Il acceptera en revanche la date de Thalès selon Shahrastānī citant Porphyre (frag. V Nauck).

[27] Cette erreur fut reprise à la suite de Nauck par F. Jacoby, *F. Gr. H.*, 260F1 et par Segonds, p. 169, n. 4.

[28] Bar Hebraeus, *Historia compendiosa dynastiarum*, éd. et trad. E. Pococke (Oxford, 1663), p. 33; *Ta'rīkh mukhtaṣar al-duwal*, éd. Ṣaliḥānī, p. 36.

2. *August Müller et Ibn Abī Uṣaybiʿa*

Dans le chapitre qu'il consacre à Pythagore dans ses *Générations des médecins* (p. 37, l. 8–p. 43, l. 10 Müller), Ibn Abī Uṣaybiʿa cite Porphyre à deux reprises et mentionne dans un cas le titre de l'ouvrage dont il tire la citation. Müller reconnut là deux fragments arabes inédits de l'*Histoire philosophique* de Porphyre, ainsi qu'il l'annonça triomphalement au congrès des orientalistes tenu à Leyde en 1883.[29] Ces fragments, ainsi que les parallèles entre la *Vie de Pythagore* de Porphyre et les fragments d'Ibn Abī Uṣaybiʿa qui avaient été relevés par Müller ne sont pas mentionnés par Nauck qui pourrait ne pas avoir eu connaissance de l'ouvrage d'Ibn Abī Uṣaybiʿa, le volume consacré aux notes et aux parallèles venant alors seulement de paraître à Königsberg en 1884.

Une partie des paragraphes pour lesquels A. Müller indique dans ses notes des parallèles avec la *Vie de Pythagore*[30] sont cités par Ibn Abī Uṣaybiʿa d'après un ouvrage de Mubashshir ibn Fātik composé en 440/1048, les *Sentences choisies et meilleures maximes (Mukhtār al-ḥikam wa-maḥāsin al-kilam* ou *al-kalim)* mais il apparaît qu'Ibn Abī Uṣaybiʿa a un texte de Porphyre plus étendu que celui de Mubashshir.[31] De ce livre, seule une traduction espagnole médiévale avait alors été publiée par Hermann Knust en 1879, lequel donnait un grand nombre de sources et de parallèles dans la littérature classique grecque et latine.[32] Un manuscrit de Mubashshir fut utilisé par Müller pour proposer certaines émendations au texte, ainsi que l'ouvrage lui aussi inédit d'un autre lecteur de Mubashshir, Shams al-Dīn al-Shahrazūrī (m. entre 688/1287 et 704/1305).[33] Alors qu'Ibn Abī Uṣaybiʿa cite non seule-

---

[29] Müller mentionne ('Über Ibn Abi Oseibi'a und seine Geschichte der Ärzte,' in *Actes du 6ème congrès International des Orientalistes tenu à Leyde en 1883*, deuxième partie, section 1, pp. 257–80, p. 270 [réimpr. *Islamic Medicine*, iv, p. 20, cf. supra n. 5]) 'deux petits extraits' *(mit zwei kleinen Bruchstücken)* sans plus de précision.

[30] Éd. Müller, *Lesarten* aux pages 37 à 43 du vol. I, in Ibn Abī Uṣaybiʿa, *ʿUyūn al-anbāʾ*, ii, pp. 6–8.

[31] Mubashshir ibn Fātik, *Los Bocados de Oro (Mukhtār al-ḥikam wa maḥasin al-kalim*, ou *al-kilam)*, éd. ʿA. Badawī (Madrid, 1958).

[32] H. Knust, *Mitteilungen aus dem Eskurial* (Tübingen, 1897); F. Rosenthal, 'Al-Mubashshir ibn Fātik. Prolegomena to an Abortive Edition,' *Oriens* 13–14 (1960–61), pp. 132–58.

[33] Il existe trois éditions concurrentes de ce texte: *Nuzhat al-arwāḥ wa-rawḍat al-afrāḥ fī taʾrīkh al-ḥukamāʾ wa'l-falāsifa*, éd. S. Khūrshīd Aḥmad (Hyderabad, 1396/1976); *Tāʾrīkh al-ḥukamāʾ—Nuzhat al-arwāḥ wa-rawḍat al-afrāḥ*, éd. ʿA. Abū Shuwayrib (Tripoli, 1398 h. lyb./1988); *Kitāb nuzhat al-arwāḥ wa-rawḍat al-afrāḥ: «Tawārīkh al-ḥukamāʾ»*, éd. M.ʿA. Abū Rayyān (Alexandrie, 1414/1993). Une comparaison des éditions et des manuscrits disponibles est donnée dans E. Cottrell, 'Shams al-Dīn al-Shahrazūrī et les manuscrits

ment Mubashshir, mais d'autres sources à l'intérieur de son chapitre 'Pythagore' (nommément : Ṣāʿid al-Andalūsī, Porphyre, Mubashshir, Plutarque). Mubashshir en revanche ne mentionne pas ses sources et les deux fragments dans lesquels Ibn Abī Uṣaybiʿa mentionne Porphyre n'apparaissent pas chez lui.

F1a Porphyre, *Histoire philosophique*: Ibn Abī Uṣaybiʿa, *Générations des médecins*, éd. Müller, p. 38, l. 9–11

Porphyre rapporte, au premier traité (*maqāla*) de son *Livre des Récits (akhbār) sur les philosophes, anecdotes [les concernant], et leurs opinions*, des récits étranges qui circulèrent à propos de Pythagore, comme le fait qu'il prophétisait et ce qu'on lui attribuait de miracles qu'il avait annoncés ou dont on avait été le témoin, comme il les avait prévus.	و ذكر فُرْفُوريُوس في المقالة الاولى من كتابه في أخبار الفلَاسفة و قصصهم و آرائهم حكايات عجيبة ظهرتْ عن فيثاغورس ممّا تكهِّن به و من إخباره بمغيّبات سمعتْ منه و شُوهدتْ كما قاله.[34] De حكايات عجيبة (« des récits étranges ») à la fin, cf. *VP* 28–29 (éd. Des Places, p. 49, l. 12–16).

Ce premier fragment dans lequel Porphyre est mentionné est l'un des 'deux petits extraits' dont Müller avait annoncé la découverte au congrès des orientalistes de 1883 sans donner d'indications quant aux limites qu'il leur donnait. Il s'insère dans le texte d'Ibn Abī Uṣaybiʿa entre un extrait de *VP* 39 (éd. Des Places, p. 53, l. 22–24) sur la promesse et la confiance et des extraits de *VP* 42 reprenant en partie les symboles pythagoriciens.

F1b Porphyre, *Histoire philosophique*: Shams al-Dīn al-Shahrazūrī, *La Promenade des âmes et le Jardin des réjouissances (Kitāb nuzhat al-arwāḥ wa-rawḍat al-afrāḥ)*, éd. Khūrshīd Aḥmad, p. 92, l. 5–7 / éd. Abū Rayyān, p. 185, l. 4–5

Porphyre a rapporté dans son *Histoire* des récits étranges au sujet de Pythagore…	قدِ ذكر فرفريوس في تاريخه حكايات عجيبة ظهرتْ عن فيثاغورس…

de *La Promenade des âmes et le Jardin des réjouissances*: *Histoire des philosophes* (*Nuzhat al-arwāḥ wa-rawḍat al-afrāḥ fī Taʾrīkh al-ḥukamāʾ*),' *BEO* 56 (2004–2005), pp. 225–60.

[34] Le style assez redondant porte la trace de syriacismes: *mimmā takahhana bihi wa min ikhbārihi… kamā qālahu*. On remarque aussi le peu d'importance attaché au fait que ces pronoms de rappel soient mis au masculin ou au féminin et les passifs improprement utilisés (le sujet est exprimé).

En revanche, chez al-Shahrazūrī ce paragraphe s'insère entre les symboles et la biographie (tandis que dans Ibn Abī Uṣaybiʿa la source qui est citée pour la biographie est Mubashshir ibn Fātik). Bien que Shahrazūrī utilise de façon systématique dans son ouvrage les *Sentences choisies* de Mubashshir,[35] il n'est pas possible à ce stade de déterminer s'il a eu accès à un manuscrit de Mubashshir plus complet que ceux qui nous ont été conservés et dont Ibn Abī Uṣaybiʿa aurait connu la copie ou un parent de celle-ci, ou bien s'il transcrit directement Ibn Abī Uṣaybiʿa.[36] D'une façon générale, al-Shahrazūrī n'hésite pas à abréger ses sources (qu'il cite assez peu). Il semble utiliser ici le même texte qu'Ibn Abī Uṣaybiʿa en donnant pour la source un titre abrégé, ce qui pourrait confirmer l'équivalence proposée par A. Müller et M. Steinschneider entre l'*Histoire des philosophes (Akhbār al-falāsifa)* et l'*Histoire* ou *Chronique (al-Taʾrīkh)* de Porphyre.

Le deuxième fragment considéré par A. Müller comme appartenant à l'*Histoire philosophique* ne peut certainement pas être qualifié de 'petit extrait' comme le laissait entendre sa formule. Porphyre y mentionne le fait qu'une partie seulement des ouvrages attribués à Pythagore alors en circulation étaient authentiquement pythagoriciens. Il cite les noms des principaux faussaires. Ce passage, remarqué à la fin du XIXe siècle par Müller et par Steinschneider, a été résumé et étudié par B.L. van den Waerden dans un article sur les ouvrages attribués à Pythagore, publié dans le supplément de la *Real-Encyklopädie* paru en 1965.[37] Thesleff, qui avait édité une collection de fragments d'ouvrages pseudo-pythagoriciens cités par Stobée (*c.* 400 ap. J.-C.), proposa que les deux cent quatre-vingts livres mentionnés par Porphyre dans cet extrait aient appartenu à ce corpus pseudoépigraphe.[38] L'extrait a récemment été donné en traduction anglaise dans un livre consacré au

[35] R.P. Dozy et al., *Catalogus Codicum Orientalium* (Leyde, 1851), iii, pp. 342–3.

[36] Pour un autre cas d'utilisation possible d'Ibn Abī Uṣaybiʿa par al-Shahrazūrī, comparer la notice Ibn al-Aʿlām al-Baghdādī dans les deux ouvrages.

[37] Müller, *Die griechischen Philosophen*, pp. 33–4; Steinschneider, *Die arabischen Übersetzungen aus dem Griechischen* (1893), p. 4, n. 10. B.L. van den Waerden, art. 'Pythagoras,' in A. Pauly, G. Wissowa, W. Kroll, *Real-Encyklopädie der classischen Altertumwissenschaft*, Suppl. x (Stuttgart, 1965), pp. 862–4.

[38] M. Steinschneider, *Zur pseudepigraphischen literatur des Mittelalters* (Berlin, 1862; réimpr. Amsterdam, 1965), p. 46, n. 29 et p. 83; Müller, *Die griechischen Philosophen*, p. 33, n. 7 et p. 35 ; H. Thesleff, *An Introduction to the Pythagorean Writings of the Hellenistic Period* (Abo, 1961). Dans 'On the Problem of the Doric Pseudo-Pythagorica,' in *Pseudepigrapha I: Huit exposés et discussions* (Vandeuvres, 1972), pp. 59–87, Thesleff identifie ainsi les faussaires: Archippos, Nearchos, Kleinias, Megilos, Proros, et suggère de lier Kleinias et Proros pour lire le nom de Kleemporos, cité par Pline (NH XXIV 159).

philosophe pythagoricien Archytas de Tarente, mais là encore sans le texte arabe. L'auteur de l'ouvrage, Carl Huffman, admet l'hypothèse qu'il provienne de l'*Histoire philosophique*.[39]

F2 Porphyre, *Histoire Philosophique*: Ibn Abī Uṣaybiʿa, *Génération des médecins*, éd. Müller, p. 42, l. 20–p. 43, l. 4

J'ai copié [ou 'traduit', *naqaltu*][40] du livre de Porphyre *Informations sur les philosophes, anecdotes [les concernant], et leurs opinions* [ceci] : Il dit : Les livres du sage Pythagore *(Fūthāghūras)* qu'Archytas *(Arkhūṭas)* le philosophe de Tarente a rassemblé à lui seul sont [au nombre de] quatre-vingts. Quant aux livres qu'il avait de tout son effort recherchés, édités et rassemblés auprès des hommes matures qui étaient de la famille *(jins)* de Pythagore le philosophe, des hommes de son parti, et de ceux qui dépendaient directement de l'héritage de ses sciences, il y en a deux-cent. Mais celui [= Archytas (?)][41] qui se distingue par la pureté de son intellect et en retire les faux qui sont dits être de la bouche du sage et [qui circulent] sous son nom [mais] ont été fabriqués par des gens malhonnêtes et qui sont: [1] le *Livre de la prière (Kitāb al-munājāh)* ; [2] le *Livre de la description des professions viles*; [3] le *Livre de la science des ruses* ; [4] le *Livre des règles de l'organisation (taṣwīr) des banquets* (*majālis al-khumūr*) ; [5] le *Livre de la préparation des tambours, des cymbales, et des lyres* ; [6] le

و نقلتُ من كتاب فرفوريوس في أخبار الفلاسفة و قصصهم و آرائهم : قال و أمّا كتب فوثاغورس الحكيم التي انفرد بجمعها أرخوطس الفيلسوف الطارنطيني فتكون ثمانين كتاباً.

فأمّا التي اجتهد بكلية جهده في التقاطها و تأليفها و جمعها من جميع الكهول الذين كانوا من جنس فوثاغورس الفيلسوف و حزبه و ورثة علومه رجل فرجل فتكون مائتي كتاب عددا. فمن انفرد بصفوة عقله و عزل منها الكتب الكذبية المقولة على لسان الحكيم و اسمه التي اختلقها أناس فَجَرَة و هي كتاب المناجاة و كتاب وصف المهن السيّئة و كتاب علم المخاريق و كتاب أحكام تصوير مجالس الخمور وكتاب تهيئة الطبول و الصنوج و المعازف و كتاب الميامر الكهنوتية و كتاب بذر الزروع و كتاب الآلات و كتاب القصائد و كتاب تكوين العالم و كتاب الأيادي و كتاب المروءة و كتبٍ أخَر كثيرة تشاكل هذه الكتب مما اخترق حديثا فسيسعد سعادة الأبد.

[39] C.A. Huffman, *Archytas of Tarentum. Pythagorean, Philosopher and Mathematician King* (Cambridge, 2005), p. 616.

[40] Le verbe *naqala* en arabe peut avoir le sens technique de 'recevoir par transmission orale' dans le cadre de la jurisprudence religieuse, ou encore de 'traduire' (Ḥunayn l'emploie) ou de 'recopier'. La possibilité qu'Ibn Abī Uṣaybiʿa ait su assez de syriaque n'est pas à écarter, mais notons qu'il n'est pas possible au Moyen Âge de déterminer si un auteur a traduit lui-même ou a fait traduire un texte qu'il cite.

[41] M. Rashed, dans sa traduction du texte pour C. Huffmann (cf. n. précédente) considère qu'il s'agit d'Archytas, mais le syntaxe ne permet pas de l'affirmer avec certitude.

Livre des discours hiératiques (al-mayāmir al-kahanūtiyya, gr. *ieroi logoi ?)* ; [7] le *Livre des semences* (*bidhr al-zurūʿ*) ; [8] le *Livre des instruments* ; [9] le *Livre des poèmes* , [10] le *Livre de la genèse du monde* ; [11] le *Livre des mains* ; [12] le *Livre de la vertu*[42] et de nombreux autres livres qui ressemblent aux livres forgés récemment, il jouira d'une félicité éternelle.

Il dit : Et ces hommes malhonnêtes ces criminels qui fabriquèrent ces faux livres que nous avons mentionnés, ils sont d'après ce que la tradition nous a rapporté : Aristippe[43] le rhéteur (*al-muḥaddith*), et Nikos Monophtalmos [ou traduction d'un jeu de mot sur 'Prôros[44] qu'on appelait l'Aveugle (gr. *Pôros ?*)'], et un Crétois *(min iqrīṭiyya)* du nom de Qūniyūs [Kainias ? Phrontidas ? Kallicratidès ?][45] ainsi que Megillos

و قال و أمّا الرجال الآثمة الذين اختلقوا هذه الكتب الكاذبة التي ذكرناها فانّهم على ما أدّت الينا الروايات ارسطيبوس المحدّث و نقوس الذي كان يكنّى عين الناقص و رجل من أهل أقريطية يقال له قونيوس و ماغيالوس و فوخجواقا مع آخرين أطغى منهم و كان الذي دعاهم إلى اختلاق هذه الكتب الكاذبة على لسان

[42] Certains de ces titres en rappellent d'autres, mentionnés par dans Diogène Laërce (= D. L.) (avec la latitude que permet la traduction), attribués à Archytas ([1], [3], [7], [8], [10] (sur lesquels voir Huffman, pp. 30–32 ; 607), à Aristippe ([4], [12], cf. D. L., II, 84 *'Au préposé à la coupe'* et II 85 *'Sur la vertu'*), et à Démocrite ([3], [7], [9], [10], [11], [12], cf. D. L., IX, 46 : on attribue à Démocrite un traité intitulé *'Pythagore'* qui précède un *'Sur la Vertu'*). Le *'Livre des mains'* pourrait être un ouvrage d'astronomie (traduction littérale de *Prokheiroi kanones*, 'tables manuelles'?). On connaît d'Okkelos (ou Okellos) des fragments d'un ouvrage correpondant au titre [10].

[43] B.L. van den Waerden donne la forme 'Aristaios' et Thesleff ('On the Problem of the Doric Pseudo-Pythagorica') 'Archippe', qui ne correspondent ni au texte de Müller ni aux manuscrits. De plus, Archippe de Tarente est considéré par Porphyre comme un véritable pythagoricien (cf. *VP* 56–7). Le commentaire qui suit sur la raison de cette forgerie (être bien reçu parmi les jeunes gens) pourrait circuler à l'encontre d'Aristippe de Cyrène, que Diogène Laërce classe parmi les Socratiques. Rappelons que Porphyre donnait de Socrate, dans le IIIe livre de l'*Histoire philosophique*, un portrait peu flatteur (fragments X–XII Nauck, à compléter par de nouveaux fragments issus de Théodoret de Cyr, édités par A. Smith [cf. supra n. 13], 210 F–217 F). Jamblique mentionne par ailleurs un pythagoricien du nom d'Aristippe de Tarente (*De Vita pythagoricae*, §267).

[44] Prôros est cité par Jamblique (*De Vita Pythagoricae*, §267). Diogène Laërce (II, 83) mentionne une lettre d'Aristippe de Cyrène à Prôros.

[45] Si l'on accepte cette reconstruction [*yu-QĀL-LAH(U)-KRTYDS] avec chute du yu-initial et transformation du ductus de QŪNYŪS et QRTYDS (confusion du *wāw* en *rāʾ* et du *wāw* en *dāl*) il faut considérer que 'et un Crétois' est une traduction calque.

(*Māghiyālūs*),[46] et *Fūkhjwāqā ainsi que d'autres pires encore que ceux-là. Et ce qui les avait poussé à fabriquer ces ouvrages mensongers au nom de Pythagore le philosophe et en les lui attribuant c'était le désir d'être reçus par les jeunes gens grâce à lui et qu'ainsi [ceux-ci] les honorent, les révèrent et les assistent.

Les livres du sage sur lesquels il n'y a pas de doute sont au nombre de deux cent quatre-vingts.[47] Ils avaient été oubliés, jusqu'à ce qu'ils réapparussent auprès d'un groupe de sages doté d'une intention pure et de piété, qui les trouvèrent, les rassemblèrent et les éditèrent. Avant cela [ces ouvrages] n'étaient pas connus en Hellas,[48] car ils étaient conservés en Italie.

فوثاغورس الفيلسوف و اسمه كي يقبلوا عند الاحداث بسببه فيكرموا و يؤثروا و يواسوا.
فأمّا كتب الحكيم التي لا ريب فيها فهي مائتان و ثمانون كتابا و قد كانت منسية حتى جاء الكيان بقوم حكماء ذوي نية و ودع فحصلوها و جمعوها و ألفوها و لم تكن قبل ذلك مشهورة ببلدة الاذا لكنها كانت مخزونة في إيطاليا .

L'extrait s'insère chez Ibn Abī Uṣaybiʿa entre une série d'anecdotes et une citation de Plutarque indiquant que Pythagore serait l'inventeur du mot 'philosophie'. Cette citation, que l'on trouve aussi chez Ibn al-Nadīm en introduction à une bibliographie des ouvrages pythagoriciens disponibles à son époque et que reproduit en partie Ibn Abī Uṣaybiʿa sans donner de source, est connue par son parallèle dans la traduction

---

Sur ce personnage, cf. Thesleff, 'On the Problem of the Doric Pseudo-Pythagoric,' p. 86, n. 1 et p. 95. La traduction de van den Waerden, col. 863, par 'trois Crétois' est erronée.

[46] Un néo-pythagoricien du nom de Magillos est connu de Jamblique, et l'ouvrage qui lui est attribué est un faux (Thesleff, *Introduction to the Pythagorean Writings*, p. 115). L'identification de Magillos et Clinias fut proposée par M. Steinschneider (*Die arabischen Übersetzungen aus dem Griechischen*, 1893), p. 4, n. 10. Cette dernière semble plus difficile à expliquer à partir du ductus (*QWNYWS). La seule justification à l'hypothèse 'Clinias' pourrait être le fait qu'il était crétois et qu'il apparaissait avec un Spartiate du nom de Magillos dans les *Lois* et l'*Epinomis* de Platon. Un Keinias de Tarente est mentionné par Jamblique (*De Vita Pythagoricae*, §267). Pour le dernier personnage, on pourrait deviner derrière le ductus corrompu les noms d'Okkelos, Apollophane, ou Polemarchos, tous liés au pythagorisme.

[47] Il y a ici une contradiction avec ce qui précède, comme le remarquent van den Waerden et Huffman. Il faut donc considérer ce paragraphe soit : 1) comme un résumé de ce qui précède par le traducteur (double-traduction); 2) comme une reprise du texte de Porphyre après une incise provenant d'un autre auteur ayant noté en marge une liste de faux (scolie).

[48] *Ilādhā*, nom syriaque de la Grèce (Hellas).

arabe des *Opinions des philosophes (Arā' al-falāsifa)* par Qusṭā ibn Lūqā (m. 300/912), que connaissait Ibn al-Nadīm.[49] Quant à Shahrazūrī, il a, semble-t-il, comme pour le fragment F1 ci-dessus, eu connaissance du texte d'Ibn Abī Uṣaybiʿa ou de sa source qu'il réduit à une simple mention des deux cent quatre-vingts livres composés par Pythagore conservés en Italie, soit le dernier paragraphe introduit par la mention de Porphyre.[50]

Ce fragment ne provient pas nécessairement du livre I de l'*Histoire philosophique*, puisque le livre dont l'extrait est tiré n'est pas indiqué. Le fait qu'Archytas soit mentionné dans le texte laisse envisager la possibilité qu'il ait fait partie du IVe livre de l'*Histoire philosophique*, dont la tradition (en fait, le seul témoignage d'un néoplatonicien du IVe siècle ap. J.-C., Eunape de Sardes) rapporte qu'il contenait la vie de Platon. L'éventualité qu'il provienne d'une vie de Platon n'est pas à exclure. Les *Lettres* de Platon (dont la fameuse VIIe Lettre) contiennent en effet un échange (pseudépigraphe) entre celui-ci et Archytas où les ouvrages de Pythagore sont évoqués.[51] Il paraît en outre impossible de déterminer si l'ensemble du fragment est attribuable à Porphyre et à son *Histoire philosophique* en particulier ou si des additions ultérieures ont pu lui être ajoutés au cours de sa transmission.[52] Outre des tournures dont les résonnances chrétienne et/ou syriaque pourraient être imputables au traducteur (*al-mayāmir al-kahanūtiyya*[53]; *jaʾā al-kiyān; fa-sayasʿadu saʿādat al-ʿabad*; enfin la forme syriaque *Ilādhā* pour Hellas), on remarque que les ductus des noms propres transmis semblent corrompus, ce qui va dans le sens d'un original syriaque.

---

[49] H. Daiber, *Aetius Arabus. Die Vorsokratiker in arabischen Überlieferung* (Wiesbaden, 1980), texte arabe I, 3, 8 (p. 5.13) (= Pseudo-Plutarque I, 3, 8).

[50] Shahrazūrī, *Kitāb nuzhat al-arwāḥ*, éd. Khūrshīd Aḥmad, p. 104, l. 9–11; éd. Abū Rayyān, p. 197, l. 1–3 (il faut lire 'en Italie' et non 'à Antioche': confusion fréquente due à la proximité des ductus de ces deux noms). Une autre mention des deux cent quatre-vingts ouvrages, qui précède les maximes que portait Pythagore sur sa bague et sa ceinture est conservée par ailleurs dans Mubashshir, *Mukhtār al-ḥikam*, éd. Badawī, p. 61, l. 9 et Shahrazūrī, éd. Khūrshīd Aḥmad, p. 103, l. 9; éd. Abū Rayyān, p. 196, l. 4, ainsi que dans Ibn Abī Uṣaybiʿa, éd. Müller, p. 41, l. 2–4.

[51] Platon, *Lettres*, trad. L. Brisson (Paris, 1987), voir aussi Diogène Laërce, VIII, 80–81.

[52] Eusèbe, *Préparation évangélique*, X, 3, 24–6, cite un passage de la *Leçon de philologie* de Porphyre sur les faussaires.

[53] J.-B. Belot, *Dictionnaire Arabe-Français* (Beyrouth, 1890; rééd. 1986) connaît le mot *kahanūt*, 'sacerdoce, prêtrise', p. 711 (absent du dictionnaire de M. Ullmann [*Wörterbuch der klassischen arabischen Sprache* (Wiesbaden, 1970–)]).

II – L'une des sources d'Ibn Abī Uṣaybiʿa: Mubashshir ibn Fātik

Franz Rosenthal publia en 1937 un article intitulé 'Arabische Nachrichten über Zenon den Eleaten' dans lequel il présentait des extraits des *Sentences choisies et meilleures maximes (Mukhtār al-ḥikam wa-maḥāsin al-kalim* ou *al-kilam)* de Mubashshir ibn Fātik, livre qu'Ibn Abī Uṣaybiʿa cite à plusieurs reprises pour ses informations sur les médecins et philosophes grecs de l'Antiquité. L'auteur, mécène et bibliophile d'origine syrienne, avait vécu à la cour fatimide du Caire au milieu du XIe siècle ; il fut proche de deux grands scientifiques de son époque : Ibn al-Haytham (m. 430/1039) et Ibn Riḍwān (m. *c.* 443/1061). Rosenthal, au fait des découvertes de Knust et de Franceschini, qui avaient relevé dans leurs éditions des traductions espagnole et latine médiévales de Mubashshir ibn Fātik d'intéressants parallèles, connaissait bien sûr l'édition des *Générations des médecins* d'A. Müller.[54] Il reprit les recherches de ce dernier en s'intéressant à la tradition manuscrite de Mubashshir ibn Fātik, ainsi qu'à celle de la *Promenade des âmes et le Jardin des réjouissances (Nuzhat al-arwāḥ wa-rawḍat al-afrāḥ)* d'al-Shahrazūrī qui, comme Ibn Abī Uṣaybiʿa, avait utilisé les *Sentences choisies* au milieu du XIIIe siècle.

Rosenthal avait travaillé plusieurs années à une édition critique des *Sentences choisies et meilleures maximes* à laquelle il renonça lorsque ʿAbd al-Raḥmān Badawī publia la sienne. Badawī, qui connaissait les travaux de Rosenthal, ne les utilisa pas. Il priva ainsi les lecteurs des propositions de lectures de ce dernier. Pire encore, sans fournir un appareil critique complet, il laissa de côté la tradition représentée par le manuscrit de Berlin et les manuscrits turcs.[55] Celle-ci comporte de nombreuses traces d'arabe vulgaire ou 'moyen' qui pourraient avoir été corrigées dans l'autre recension, caractérisée en outre par de nombreuses lacunes, ce qui rend l'édition donnée par Badawī inutilisable en tout cas pour la reconstruction des sources grecques.[56]

---

[54] H. Knust, cf. ci-dessus n. 32. E. Franceschini, 'Il *"Liber philosophorum moralium antiquorum"*,' *Atti del Reale Istituto Veneto di scienze, lettere ed arti* 91/2 (1931–32), pp. 393–597 et éd. séparée Venise 1932.

[55] Rosenthal, 'Al-Mubashshir ibn Fātik,' pp. 139–43; G. Schoeler, *Arabische Handschriften*, ii (Stuttgart, 1990), s.v. n° 327 = Ms. or. quart. 785, pp. 391–4.

[56] Sur l'arabe moyen comme trace écrite du phénomène de diglossie de l'arabe, cf. P. Larcher, 'Moyen arabe et arabe moyen,' *Arabica* 48 (2001), pp. 578–609. Rosenthal, 'Al-Mubashshir ibn Fātik,' est un compte-rendu de l'édition Badawī à laquelle il propose de nombreuses corrections.

Grâce à des extraits issus des chapitres consacrés par Mubashshir à Zénon (*c.* 495–430 av. J.-C.), Solon (*c.* 633–561 av. J.-C.), Pythagore (*c.* 569–494 av. J.-C.), les fragments présentés par Rosenthal montrent que des parallèles pertinents pouvaient être découverts dans les textes arabes qui étaient donc à prendre comme témoins historiques d'originaux grecs.[57] Pour deux des trois chapitres (Solon et Pythagore), on connaissait en grec des parallèles expressément attribués à Porphyre.[58] En comparant les trois chapitres, Franz Rosenthal remarqua un plan commun à toutes les notices[59] ainsi que des excursus consacrés à des personnages contemporains. Quelques remarques s'imposent cependant à la lecture de cet article dans lequel l'auteur utilise parfois la tradition transmise par Mubashshir, parfois celle que transmet Ibn Abī Uṣaybiʿa, et parfois encore, celle de Shahrazūrī. La date de publication (1937) explique les difficultés pour Rosenthal de voyager en Allemagne, où il travaillait, et de se rendre aux Pays-Bas pour y vérifier les manuscrits qu'il avait utilisés.

1. *Zénon (et Diagoras)*

A partir de la traduction allemande donnée par Rosenthal, Mario Untersteiner fut en mesure de proposer en 1961 une analyse du fragment conservé par Mubashshir sur Zénon. D'autres parallèles ont été relevés par Angelo Sodano, dans une édition récente des fragments de l'*Histoire philosophique*, qui s'appuie à nouveau sur la traduction allemande.[60] Nous nous bornerons donc à quelques points sur lesquels ces auteurs n'ont pas proposé d'éclairage. Au début du chapitre sur Zénon, Mubashshir entremêle des informations relatives à Zénon d'Elée d'une part, et à Zénon de Kition le stoïcien d'autre part et le même imbroglio se retrouve dans le portrait qui clôt la première moitié de la notice avant l'amorce des sentences que Rosenthal traite à la fin de son article (pp. 57–63). La courte description du début de la notice de Mubashshir a son parallèle dans Diogène Laërce IX, 25, comme le remarque Untersteiner, mais ce portrait reprend aussi les traits que Platon attribue à Zénon dans le *Parménide*, ainsi qu'une partie de la description fournie par Diogène Laërce

[57] Rosenthal, 'Arabische Nachrichten,' pp. 22–3.

[58] Cyrille d'Alexandrie, *Contre Julien*, P. G. 76, 532 A-B; 961 A-B. Cf. fragment VII Nauck, trad. A. Segonds, in Porphyre, *Vie de Pythagore*, éd. Des Places, p. 182.

[59] Rosenthal, 'Arabische Nachrichten,' p. 38.

[60] M. Untersteiner, *Zenone. Testimonianza e Frammenti* (Florence, 1970), pp. 17–21; A.R. Sodano, *Storia della filosofia* (Milan, 1997), frag. 30 et notes p. 151s.

(VII, 1) à propos de Zénon de Kition. Une remarque sur Empédocle est rapprochée par Rosenthal d'un extrait de l'*Histoire philosophique* de Porphyre[61] et il cherche à comprendre le texte comme une information sur Zénon en suggérant une lacune,[62] alors qu'il pourrait s'agir d'une simple glose sur Empédocle ou Parménide. L'extrait de Porphyre nous dit en effet qu'Empédocle fut le favori de Parménide, à propos duquel Diogène Laërce (IX, 25) rapporte que si Téleutagoras était le père biologique de Zénon (éd. Badawī, p. 40, l. 2 / éd. Rosenthal, p. 30, l. 2), c'est Parménide qui était son père adoptif. La mention de Téleutagoras apparaît aussi chez Mubashshir. On peut également comprendre comme une remarque portant sur Parménide les deux phrases qui suivent et selon lesquelles Empédocle fut le disciple de Zénon (éd. Badawī, p. 40, l. 2–4); il faudrait alors rapprocher cet élément de Diogène Laërce VIII, 56 (ce que fait Untersteiner): Parménide eut pour élèves Empédocle et Zénon.

Rosenthal découvrit en outre, dans le récit consacré à Zénon, un paragraphe entier sur Diagoras de Mélos (éd. Rosenthal, p. 30, l. 10–p. 31, l. 4 = éd. Badawī, p. 40, l. 10–p. 41, l. 6).[63] Celui-ci fut étudié en détail par F. Jacoby qui accepta l'hypothèse de Rosenthal: bien qu'on n'aie pas d'argument décisif pour l'avancer, il n'était pas invraisemblable que les éléments rapportés par Mubashshir aient pour provenance l'*Histoire philosophique* de Porphyre.[64] La remarque qui précède l'excursus sur Diagoras d'après laquelle Parménide et Zénon auraient appartenu à l'école des 'Obscurs' *(al-ghawāmiḍ)* a été interprétée par Badawī et Untersteiner comme renvoyant à l'école dialectique et Untersteiner note qu'un extrait de la *Chronique* d'Eusèbe de Césarée (perdue en grande partie, mais dans laquelle Eusèbe utilisait les ouvrages de son contemporain Porphyre) donne Zénon et Héraclite 'l'obscur' *(skoteinos)* comme ayant vécu à la même époque, ce qui pourrait avoir entraîné une confusion.[65] En revanche, lorsque Rosenthal note que le texte ne semble pas avoir été glosé par les chrétiens car Diagoras 'l'Athée' *(al-māriq)* (ou *'l'impie'*) meurt de sa belle mort sans qu'aucune remarque d'ordre

---

[61] Nauck, frag. VIII, trad. Segonds p. 182; cf. Untersteiner, *Zenone*, p. 22.

[62] Rosenthal, 'Arabische Nachrichten,' p. 36, n. 3 et n. 4.

[63] Rosenthal, 'Arabische Nachrichten,' p. 30 et pp. 35–6.

[64] F. Jacoby, 'Diagoras ὁ ἄθεος,' *Abhandlungen der deutschen Akademie der Wissenschaften zu Berlin, Klasse für Sprächen, Literatur und Kunst* (1959), Abh. n° 3.

[65] Mubashshir, *Mukhtār al-ḥikam*, éd. Badawī, p. 40, n. 6; Untersteiner, *Zenone*, pp. 28–9.

moral ne soit à relever, il faut replacer l'anecdote dans son contexte : Diagoras était connu pour avoir rejeté les rites de la religion grecque, ce qui n'était certainement pas blâmable aux yeux des chrétiens.[66] Quant au jugement porté sur le livre de Diagoras qui aurait été découvert à sa mort et écrit en phrygien : « [Il était] rempli d'horreurs sur les choses divines (*mamlūʾ mafāḥish fī ʾl-umūr al-ilāhiyya*, éd. Rosenthal, p. 31, l. 3–4 = éd. Badawī, p. 41, l. 6)[67] », il pourrait être celui d'un intermédiaire chrétien à moins que ce ne soit l'effet d'une traduction biaisée.[68]

On a par ailleurs proposé qu'un texte consacré à un Zénon 'sophiste' et connu déjà de Rosenthal ait fait partie de l'*Histoire philosophique*. François Lasserre et Charles Genequand ont présenté, à la suite de M. Klamroth à la fin du XIXe siècle, un extrait consacré aux philosophes dans l'*Histoire (al-Taʾrīkh)* d'al-Yaʿqūbī.[69] L'exposé qui dépeint les opinions du 'groupe de Zénon' et que les auteurs rapprochent des théories de Protagoras (autre 'athée' notoire, si l'on devait le rattacher aux chapitres consacrés à Diagoras et à Socrate dans une histoire 'philosophique') donne un exemple de raisonnement dialectique, mêlé de commentaires à la première personne. Une comparaison (non fournie par les auteurs) du vocabulaire utilisé par al-Yaʿqūbī avec les chapitres de Mubashshir étudiés par Rosenthal montre quelques similitudes qui restent cependant assez mineures, et le style de Mubashshir dépasse de loin celui de l'extrait d'al-Yaʿqūbī. La mention de l'appartenance de Zénon au groupe des Sophistes (al-Yaʿqūbī, *Histoire*, éd. Houtsma, p. 166, l. 7–8) est accompagnée de précisions sur le sens de ce mot en

[66] Cf. Rosenthal, 'Arabische Nachrichten,' p. 39. R. Janko, 'The Derveni Papyrus,' *Classical Philology* 96/1 (Jan. 2001), pp. 1–32, resitue l'extrait découvert par Rosenthal dans le contexte des procès d'impiété du IVe siècle av. J.-C. Janko donne p. 6, n. 22 une bibliographie actualisée des travaux sur Diagoras à laquelle on peut ajouter R. Goulet (éd.), *Dictionnaire des Philosophes Antiques*, i (Paris, 1989), s.v. 'Diagoras' (M. Aouad et L. Brisson).

[67] Il faut corriger Badawī avec Rosenthal, 'Arabische Nachrichten,' p. 36, citant Tatien, *Orationes ad Graecos*, 27. Badawī lit dans ses manuscrits 'écrit en phénicien'.

[68] Si l'on accepte l'hypothèse de Janko selon laquelle l'interprétation du poème orphique retrouvée dans le papyrus de Deverni est à attribuer à Diagoras, on peut lire dans sa traduction (provisoire) des fragments ce qui pouvait choquer les lecteurs de l'antiquité tardive. Janko accepte l'hypothèse de Rosenthal selon laquelle le passage remonte à l'*Histoire philosophique* de Porphyre.

[69] Al-Yaʿqūbī, *Taʾrīkh*, éd. M.T. Houtsma (Leyde, 1883); C. Genequand et F. Lasserre, 'Chapitres d'une histoire de la philosophie grecque chez al-Yaʿqūbī,' *Museum Helveticum* 42 (1985), pp. 191–204; M. Klamroth, 'Über die Auszüge aus griechischen Schriftstellern bei al-Jaʿqūbī,' *ZDMG* 42 (1888), pp. 1–44.

grec (serait l'équivalent de *mughāliṭa*, *'mélangé'*) et en arabe.[70] Enfin, une correction proposée par Houtsma et suivie par Lasserre-Genequand ne semble pas légitime : il n'est pas nécessaire de chercher à retrouver sous le ductus السالقطة*] la lecture التناقضية *al-tanāqūḍiyya*, *'ceux qui se contredisent mutuellement'* ; on lit en fait très bien السافسطة *(al-sāfisṭa)* lequel est attesté en syriaque et sous la forme السفسطة en arabe.[71] Ceci entraîne la possibilité que le texte utilisé par al-Yaʿqūbī ou la source de celui-ci ait été en syriaque.

2. *Solon (et les sept Sages)*[72]

Rosenthal présente ce chapitre très partiellement sans pour autant s'expliquer sur les raisons de ce choix, qui tient sans doute à une tradition manuscrite complexe. Il coupe les cinq premières lignes du chapitre (éd. Badawī, p. 34, l. 1–6) ainsi que les deux derniers paragraphes (éd. Badawī, p. 36, l. 3–11) avant les sentences, qu'il ne traite pas non plus. Les deux paragraphes omis sont donnés à des emplacements différents selon les manuscrits et Rosenthal semble vouloir s'être limité au fragment du texte pour lequel on avait un parallèle chez Porphyre. Mubashshir commence par préciser que Solon avait aboli les lois de Dracon l'impie *(al-māriq)*, utilisant la même épithète que dans l'excursus sur Diagoras.[73] Plusieurs éléments peuvent être rapprochés des informations rapportées sur Solon dans le *Timée* de Platon, la *Constitution d'Athènes* d'Aristote, et la *Vie de Solon* par Plutarque. Le manque d'anecdotes sur la vie de Solon (une partie des éléments biographiques concernant les lois ou des maximes qui lui sont attribuées se trouvent dans la partie de la

[70] *Al-mughāliṭ* (dans le sens de «quelqu'un qui induit en erreur») est utilise par Ḥunayn et son école dans la traduction du *Commentaire au serment d'Hippocrate*. Cf. F. Rosenthal, 'An Ancient Commentary on the Hippocratic Oath,' *Bulletin of the History of Medicine* 30 (1956); réimpr. idem, *Science and Medicine in Islam. A Collection of Essays* (Aldershot, 1991), p. 5, n. 4: 'Thessalus the Sophist (*Al-mughāliṭ*)'. Rosenthal note que Galien est a l'origine de ce portrait peu flatteur, par exemple dans le *De atra bile*, où Galien accuse Thessalus d'user de *sophismata*.

[71] C. Brockelmann, *Lexicon Syriacum* (Halle, ²1928), p. 491. Le dictionnaire arabe-allemand (et arabe-anglais en traduction) de Hans Wehr connaît *al-safisṭa*.

[72] Smith (supra n. 13) reprend le fragment publié par Rosenthal dans sa collection (203aF). En revanche il ne donne pas les extraits sur Zénon et Diagoras, qui ne comportent pas la mention de Porphyre et ne sont connus par aucun parallèle.

[73] Mubashshir, éd. Badawī, p. 34, l. 4. La phrase arabe est une traduction-calque: au lieu de comprendre que Solon avait abrogé les lois de Dracon sauf celles concernant les meurtres *(tôn phonikôn)*, le traducteur a compris 'sauf les lois des Phéniciens' *(illā al-latī jaʾathum min Fuwīnīks)*!

notice consacrée aux sentences) a pu inciter le traducteur à rechercher ailleurs des informations destinées à combler cette lacune. Le récit de la légende de l'heptade des sept Sages, dans laquelle le nom de Solon figure en bonne place, joue ce rôle.

Comme l'a remarqué Alain-Philippe Segonds, le texte de Mubashshir présente une version plus longue et plus détaillée de la légende des sept Sages que celle qui nous est rapportée par Cyrille d'Alexandrie, citant Porphyre.[74] A la suite de Rosenthal, il note que seul Porphyre relève le fait que les Sages auraient été à l'origine neuf, ce qui donne du poids selon lui à l'attribution à Porphyre.[75] Le texte qu'édite Badawī ne reprend pas les corrections proposées par Rosenthal, lesquelles correspondent à la tradition rapportée par les manuscrits de Shahrazūrī: en suivant le manuscrit de Berlin, Badawī édite au pluriel (lequel recouvre ici de l'arabe dialectal plus encore que 'moyen' : *wa-ammā al-akhirīn*... que Badawī corrige en *wa-ammā al-akhirūn*, p. 36, l. 1) ce qui pour Rosenthal doit cacher un duel qui devrait être lu *al-akhirayn* (même ductus que *al-akhirīn*), donné improprement au cas oblique (la particule *ammā* aurait dû entraîner le cas sujet).[76] Selon cette interprétation, Anacharsis et Myson ne furent pas comptés dans l'heptade car ils avaient suggéré de donner le trépied à Thalès, tandis que les sept vrais Sages avaient chacun pu désigner autre plus sage qu'eux-mêmes.[77] En fait, le sixième sage avait lui aussi suggéré de le donner à Thalès, avant que ce dernier ne propose qu'on le place dans le temple d'Apollon, ce qui rend difficile d'accepter l'interprétation de Rosenthal. Il est fort

---

[74] Cyrille d'Alexandrie, *Contre Julien*, frag. IV Nauck (trad. Segonds, in Porphyre, *Vie de Pythagore*, éd. Des Places, p. 181); PG 76, 544D1–545A8 (= I, 38 de la traduction des *Sources Chrétiennes* par Paul Bruguière et Pierre Evieux [Paris, 1985], t. I = livres I et II). Porphyre est l'auteur d'un traité sur le *'Connais-toi toi-même'*, qu'il dédia à Jamblique et qui pourrait être lié à cet intérêt pour les sept sages. Il cite par ailleurs le *Du Trépied*, d'Andrôn d'Ephèse, dans les extraits de sa *Leçon de philologie* conservés chez Eusèbe, *Préparation évangélique*, X, 3.

[75] Segonds in Porphyre, *Vie de Pythagore*, éd. Des Places, p. 171. En fait, Diogène Laërce (I,41) connaît aussi une liste de neuf sages.

[76] Rosenthal, 'Arabische Nachrichten'. Les manuscrits de Shahrazūrī ont la phrase au duel, ainsi que l'éditent Khūrshīd Aḥmad et Abū Rayyān. Je n'ai pas consulté l'édition Abū Shuwayrib, souvent fautive, dans le cadre de cet article.

[77] Mubashshir, éd. Badawī, p. 35, l. 13–p. 36, l. 2: *'fa-ja'alūhu fī haykal Afūlūn al-ladhī fī Dālfis fa-ṣārat sābi'at* ['l'heptade'] *al-ṭaranbūdh* (c'est la lecture des manuscrits mais Badawī corrige en *aṭranbūdh*) *li'l-sab'at al-ḥukamā' al-ladhīna marra 'alā aydīhim. Wa-ammā al-akhīrūn al-ladhīna lam yataffaqū ma'ahum fī hadhā al-ma'nā fa-aqarrū* [ou *afradū*, avec Rosenthal?] *fī faḍīlat Thālis* [Badawī: *fī faḍīlat al-sinn*, sic!]. Comparer Rosenthal, 'Arabische Nachrichten,' p. 41, l. 18–19 (trad., p. 43). La dernière phrase peut être rapprochée du témoignage de Diogène Laërce (I, 32).

vraisemblable que le texte arabe conservé soit tronqué et ait contenu l'évocation d'autres listes de sages, comme le fait par exemple Diogène Laërce dans l'excursus qu'il consacre à la légende dans son chapitre sur Thalès. (I, 40–42). Il faudrait alors comprendre la fin du récit (« Quant aux autres qui ne s'accordent pas avec ceux-là sur ce point...» *Wa-ammā al-akhīrūn alladhīna lam yataffaqū maʿahum fī hadhā al-maʿnā)* comme se rapportant aux historiens qui discutent du nombre de sages, et non pas aux Sages qui viennent d'être mentionnés.

3. *Pythagore (et Phérécyde)*

Confronté à un texte souvent fautif dans les deux manuscrits de Mubashshir qu'il avait consultés, Rosenthal utilise ici le chapitre édité par Müller d'après Ibn Abī Uṣaybiʿa (dont il adopte l'ordre des paragraphes sans mentionner qu'il diffère de celui de Mubashshir). Mais son édition du texte arabe d'Ibn Abī Uṣaybiʿa est incomplète comme le montre la comparaison. Rosenthal omet d'étudier : une série de conseils éthiques dont certains peuvent être rapprochés des *Vers d'or* ; une partie des symboles pythagoriciens de *VP* 42 ; les sentences ; la bibliographie et l'anecdote sur les faussaires pourtant explicitement citées d'après Porphyre ; la citation de 'Plutarque' ainsi que la courte bibliographie qui suit. La mention explicite du titre de l'ouvrage de Porphyre (cf. supra F1) comme source de la biographie, laquelle n'est pas conservée dans le texte de Mubashshir, est laissée de côté sans aucune explication.

En présentant sur deux colonnes, en grec et en arabe, plusieurs paragraphes pour lesquels on disposait du texte grec de Porphyre, Rosenthal montre dans le détail que malgré les erreurs de traduction et un enchaînement différent des paragraphes, on peut retrouver le texte de Porphyre sous le texte arabe.[78] Son hypothèse quant aux changements d'ordre dans la notice Pythagore, selon qu'on consulte le texte grec conservé ou Mubashshir, est qu'ils sont intervenus par addition de notes prises ailleurs dans l'*Histoire philosophique* lorsqu'un trop grand nombre de sources mentionnées pourrait avoir eu comme effet de perdre le lecteur au milieu de noms inconnus.[79] Les comparaisons

[78] Rosenthal, 'Arabische Nachrichten,' pp. 41–56.

[79] Rosenthal, 'Arabische Nachrichten,' p. 44, n. 3. Cette hypothèse est démentie dans un cas où le nom de la source (Eudoxe) est présent dans le texte sous la forme d'une traduction-calque. C'est ainsi qu'il faut comprendre le début mystérieux du chapitre de Mubashshir: «Ils rapportent que [Pythagore] avait pour théorie le voyage... *(dhakarū*

avec le grec montrent que le chercheur est nécessairement confronté à un problème méthodologique insoluble : surcorriger le texte transmis ou donner un texte parfois aberrant.[80]

Le texte conserve pour certains noms propres la trace de la déclinaison du grec, mais cela ne peut être la preuve de ce que le texte-source ait été dans cette langue, car il est fréquent en syriaque de fixer un substantif emprunté au grec non pas au nominatif mais à l'un des cas obliques.[81] Rosenthal note à propos de la transcription des noms Lèmnon, Imbron (les manuscrits portent *imkron*, qui s'explique par la ressemblance en syriaque entre le Bēth et le Kōph) et Skuron, que ceux-ci ont pu être interprétés comme des pluriels, ce qui explique la traduction par 'tribus' au lieu de 'îles'.[82] Ayant soumis son étude à un spécialiste d'histoire grecque, ce dernier lui fit remarquer que le grec *tèn Lèmnon* avait été confondu quelques lignes plus loin avec *tèn limnèn*, 'le lac' (éd. Badawī, p. 53, l. 2 = Ibn Abī Uṣaybiʿa, éd. Müller, p. 38, l. 22). Le nom 'al-Buḥayra',[83] donné au dème d'Alexandrie en Egypte, dans lequel se trouvaient de nombreux monastères syriaques, pourrait avoir été plus familier aux traducteurs que l'île de Lemnos. Rosenthal remarque par ailleurs un excursus auquel on ne peut trouver de parallèle dans la *Vie de Pythagore* et qui se rapporte cette fois-ci à Phérécyde, considéré par certains comme l'un des sept Sages.[84] La mention de Phérécyde dans la *Vie de Pythagore* pourrait, dit-il, avoir servi d'amorce et entraîné l'addition de notes tirées d'autres chapitres. Le rapprochement qu'il propose avec Diogène Laërce n'est pas textuel (il faut rassembler deux passages de I, 117–18 pour retrouver l'anecdote rapportée par Mubashshir).

Comme A. Müller lors du Congrès international des orientalistes en 1883, Rosenthal arrive à la conclusion que le texte arabe est presque incompréhensible sans l'original grec, tant les erreurs de traduction sont nombreuses. Il précisera dans la notice *Fīthāghūras* ('Pythagore') de

---

annahu kāna yarā al-siyāḥa…)» (éd. Badawī, p. 52, l. 11) qui cache la source et le titre mentionnés au début de *VP* 7 : « Eudoxe (eu-dokeô, sc. *yarā*, 'considérer', cf. Daiber, *Aetius Arabus*, I, 3, 3) dans son [livre]: *Voyage* (*siyāḥa*, plutôt de l'ordre de la 'déambulation') *autour de la terre…*» Il revient à Constantinos Macris (Université de Crète), que je remercie, d'avoir élucidé cette tournure.

[80] Par exemple, Rosenthal, 'Arabische Nachrichten,' p. 51, n. 4, p. 52, n. 1 et n. 3.

[81] F. Altheim et R. Stiehl, *Porphyrios und Empedokles* (Tübingen, 1954), pp. 21–2 citant Nöldeke et Brockelmann.

[82] Rosenthal, 'Arabische Nachrichten,' p. 44, n. 2.

[83] *EI*², s.v. Buḥayra (G. Wiet).

[84] Rosenthal, 'Arabische Nachrichten,' pp. 46–7 et 47 n. 3.

l'*Encyclopédie de l'Islam*: « Les circonstances[85] en partie historiques, mais pour la plupart légendaires de sa vie ont été connues très en détail par un long [résumé] de sa biographie tiré de la *Philosophos historia* de Porphyre conservé chez al-Mubashshir (52 sqq) et Ibn Abī Uṣaybiʿa (I 38 sqq) ». Par ailleurs, il estime que plusieurs éléments des notices 'Solon', 'Pythagore' et 'Zénon' pourraient provenir de l'*Histoire philosophique*, peut-être via un chroniqueur chrétien.[86] Ce dernier élément vient sans doute de sa connaissance des travaux de A. Baumstark et de la lecture de Bar Hebraeus (cf. infra, IIIe partie). Baumstark avait en effet remarqué que Théodore bar Konai (m. début IVe/Xe s.) utilisait une source qui connaissait l'*Histoire philosophique*, et qu'Eusèbe, qui cite abondamment Porphyre, était largement répandu parmi les chrétiens syriaques.[87] L'hypothèse d'un intermédiaire 'gréco-syriaque' fut admise tant par F. Altheim et R. Stiehl, que par H. Daiber.[88]

En comparant la tradition des manuscrits de la *Vie de Pythagore* à celle de Cyrille d'Alexandrie, les éditeurs s'aperçoivent de nombreuses lectures divergentes, desquelles Segonds conclut que la *Vie de Pythagore* a dû se détacher assez tôt de l'*Histoire philosophique*. Deux des trois fragments de la *Vie de Pythagore* pour lesquels on possède un parallèle dans le *Contre Julien* de Cyrille d'Alexandrie sont cités comme appartenant au « livre *(bibliô)* premier de [l']*Histoire de la philosophie* ». Ils concernent tous deux 'les Sages' ou 'les sept Sages', ce qui pourrait indiquer que le chapitre sur Pythagore se trouvait lié aux notices consacrées aux sept Sages et à leur légende. Diogène Laërce nous rapporte que certains auteurs donnaient Pythagore comme faisant lui aussi partie des sept Sages et Aude Busine a noté le lien privilégié entre les cercles pythagoriciens et cette légende.[89] Selon le témoignage d'Eusèbe, le premier livre de l'*Histoire philosophique* contenait en outre la chronologie des événements depuis la chute de Troie [1181 av. J.-C. selon Eusèbe] jusqu'à la première olympiade [776

[85] *EI*², ii, 950a (F. Rosenthal) de la traduction française, qui donne improprement l'anglicisme 'sommaire' pour 'summary' (résumé).

[86] Rosenthal, 'Arabische Nachrichten,' p. 39, et avec moins de certitude p. 40 (les éléments donnés comme pouvant provenir de l'*Histoire philosophique* sont donnés p. 35, p. 36 et p. 37, n. 1 et p. 42).

[87] A. Baumstark, 'Griechische Philosophen und ihre Lehren in syrischer Ueberlieferung,' *Oriens Christianus* 5/1 (1905), pp. 1–25, en particulier 6, 7, et 10. Rosenthal, *History of Muslim Historiography*, p. 78.

[88] Rosenthal, 'Arabische Nachrichten,' p. 40; Altheim et Stiehl, p. 16, n. 6; Daiber, 'Hellenistisch-kaiserzeitliche Doxographie,' pp. 49–84.

[89] A. Busine, *Les Sept Sages de la Grèce antique* (Bruxelles, 2002), pp. 75–8.

av. J.-C.].[90] Les chapitres 'Solon' et 'Phérécyde' devraient en principe, comme le chapitre 'Pythagore', avoir appartenu au livre I, puisqu'ils sont liés à la légende des sept Sages.

III – La date de Thalès selon Porphyre: témoignages arabes

1. *Ibn Suwār et Mubashshir*

Dans la Ire partie, nous avons vu que, dans le *Catalogue* d'Ibn al-Nadīm, le témoignage de Porphyre était évoqué dans le cadre d'une discussion sur les origines de la philosophie. Ibn al-Nadīm expose à ses lecteurs le fait que deux traditions concurrentes existaient : l'une donnant Thalès et d'autres sages contemporains[91] comme premiers philosophes, et l'autre lui donnant Pythagore pour inventeur.[92] La discussion rapportée par Ibn al-Nadīm montre qu'Ibn Suwār situait Thalès comme le plus ancien des sept Sages,[93] et ceci d'après un ouvrage syriaque dont il avait connaissance. Cependant, la mention de la source de cette information comme étant l'*Histoire (al-Ta'rīkh)* de Porphyre engendre une ambiguïté quant au but visé : Ibn Suwār souhaite-t-il affirmer que Thalès fut le premier chronologiquement, ou n'est-ce là qu'une allusion à la tradition des sept Sages ? La suite du texte d'Ibn al-Nadīm, qui donne l'opinion alternative selon laquelle c'est Pythagore qui était le premier philosophe, donne du poids au fait que l'information sur Thalès avait une valeur chronologique. Dans les *Sentences choisies*, Mubashshir rapporte la légende des sept Sages dans un chapitre consacré à Solon (il n'a pas de chapitre sur Thalès), peut-être dans le but de situer l'époque de Solon. Le fait que Mubashshir ait puisé à des sources antérieures à l'époque d'Ibn al-Nadīm a été établi et documenté par les recherches de Franz Rosenthal

[90] *F. Gr. H.*, 260F4, trad. Segonds, pp. 178–9.

[91] Voir aussi Diogène Laërce I, 12 ; I, 40.

[92] La tournure quelque peu familière employée par Ibn al-Nadīm engendre une ambiguïté quant au sens : *'man takallama'* signifie 'discuter de' tandis que dans les sources grecques il s'agit de l'invention du mot 'philosophie' (traduction correcte dans le Pseudo-Plutarque arabe, cf. Daiber, *Aetius Arabus*, I, 3, 8).

[93] Il n'est pas possible de projeter le débat sur les différentes acceptions du mot 'sage' ou 'philosophe', dont atteste par exemple Diogène Laërce (I, 12), sur notre fragment, comme souhaitaient le faire Altheim et Stiehl, 'Porphyrios und Empedokles,' pp. 9–10, tout au plus peut-on voir là la trace du mot utilisé dans la source.

et de Dimitri Gutas.[94] Plus récemment, Mohsen Zakeri a retrouvé dans les collections de sentences que transmet Mubashshir bon nombre de parallèles attestés par ailleurs dans les fragments conservés des ouvrages de ʿAlī ibn ʿUbayda al-Rayḥānī (m. 219/834).[95] On ne peut donc exclure qu'Ibn Suwār ait connu le passage de l'*Histoire philosophique* conservé dans les *Sentences choisies* de Mubashshir dans lequel la liste des sept Sages 'contemporains' était donnée.

F1 [Thalès] : Mubashshir, *Sentences choisies*, éd. Badawī, p. 34, l. 6–8 (émendé avec éd. Rosenthal, 'Arabische Nachrichten,' p. 40)

Solon était l'un des sept Sages qui vécurent tous à la même époque et qui sont : Thalès *(Thālīs)*, Solon *(Sūlūn)*, Pittacos *(Biṭāqūs)*, Périandre *(Bāriyāndrūs)*, Chilon *(Khīlūn)*, Cléobule *(Qlāwubūlūs)* et Bias *(Biyās)*.	و كان سولون أحد الحكماء السبعة الذين كانوا في وقت واحد و هم ثاليس و سولون و بيطاقوس[1] و بارياندروس و خيلون و قلاوبولوس و بياس. [1] بطاقوس Badawī

2. *Les témoignages de Porphyre et de Cyrille sur Thalès*

La datation de Thalès est l'objet d'un fragment souvent repris par les auteurs de langue arabe. On en trouve la première attestation dans le *Florilège du Coffret de sagesse (Muntakhab Ṣiwān al-ḥikma)*.[96] Si l'on peut dater cette recension du VIIe/XIIIe siècle en raison de la mention de personnages ayant vécu à cette époque (ʿAbd al-Laṭīf al-Baghdādī et Ibn Hubal), l'original dont elle est issue et qui devrait porter le titre de *Coffret de sagesse (Ṣiwān al-ḥikma)* est perdu. Son attribution à Abū Sulaymān al-Sijistānī (m. ca 375/985) apparaît sous la plume de Ẓahīr al-Dīn al-Bayhaqī (m. 565/1169), lequel vécut non loin de Nishapour et composa avant 549/1154 un *Complément au Coffret de sagesse (Tatimmat*

[94] Rosenthal, 'Al-Mubashshir ibn Fātik,' pp. 135–6; idem, 'Some Pythagorean Documents Transmitted in Arabic,' *Orientalia*, NS, 10 (1941), p. 104–15; 383–95; D. Gutas, *Greek Wisdom Literature in Arabic Translation. A Study of the Graeco-Arabic Gnomologia* (New Haven, 1975), index s.v. Mubashshir.

[95] M. Zakeri, *Persian Wisdom in Arabic Garb* (Leyde, 2007), index, s.v. Mubashshir.

[96] *The Muntakhab Ṣiwān al-ḥikma*, éd. D.M. Dunlop (La Haye, 1979); *Ṣiwān al-ḥikma*, éd. ʿA. Badawī (Téhéran, 1974). Sur les textes issus du *Ṣiwān al-ḥikma* (auxquels il faut ajouter *La Promenade des Âmes, Kitāb nuzhat al-arwāḥ*, d'al-Shahrazūrī), voir D. Gutas, 'The Ṣiwān al-Ḥikma Cycle of textes,' *JAOS* 102 (1982), pp. 645–50 et E. Cottrell (supra n. 33).

Ṣiwān al-ḥikma) dans la tradition littéraire des *dhayl*-s.[97] Il actualisa la succession des philosophes que donnait le *Ṣiwān* jusqu'à sa propre époque, et surtout jusqu'à sa propre région (le lointain Khorassan). L'attribution à Abū Sulaymān al-Sijistānī a été discutée mais il est d'autant plus difficile de trancher que le texte dont il est question est perdu. Les critères d'attribution des textes médiévaux résistent en général mal aux exercices d'hypercritique, et la nature composite des textes contenus dans les deux abrégés, le *Florilège (Muntakhab)* et l'*Epitomé (Mukhtaṣar)* du *Coffret de Sagesse (Ṣiwān al-ḥikma)*, incite plutôt à suspendre tout jugement.[98]

L'introduction du *Florilège (Muntakhab)* comporte plusieurs extraits juxtaposés dans lesquels différentes versions des débuts de l'histoire de la philosophie sont exposées (par Thalès ayant voyagé en Egypte; par Pythagore ; par le prophète coranique Luqmān[99]). La présentation doxographique générale (*Florilège*, éd. Dunlop, §§1–13) est suivie d'éléments de chronologie et d'une nouvelle histoire des origines dans laquelle la philosophie est située par rapport à d'autres inventions telles que l'écriture, le calcul, la géométrie, l'astronomie, la musique, et la médecine (*Florilège*, éd. Dunlop, §§13–19). Ces extraits sont principalement issus d'ouvrages du IIIe–IVe/IXe–Xe s., lorsque les sources sont citées.[100] Il paraît cependant difficile de déterminer s'ils appartiennent au *Coffret de Sagesse (Ṣiwān al-ḥikma)* original ou s'ils ont été ajoutés par l'auteur du *Florilège (Muntakhab)* au XIIIe s. La première mention d'un élément à visée chronologique situe l'apparition de la philosophie « à l'époque de Bukhtnaṣar » (éd. Dunlop, §13 = l. 176). C'est en effet sous ce règne qu'aurait eu lieu l'éclipse dont Thalès avait pu calculer la date et être ainsi à même de l'annoncer à ses compatriotes. Cet événement est rapporté dans le *Florilège du Coffret de*

---

[97] C.F. Farah, *The Dhayl in Medieval Arabic Historiography* (New Haven, 1967).

[98] Remise en cause de l'attribution à al-Sijistānī par W. al-Qāḍī, 'Kitāb Ṣiwān al-Ḥikma: Structure, Composition, Authorship and Sources,' *Der Islam* 58 (1981), pp. 87–124.

[99] C'est sans doute al-ʿAmirī qui est à l'origine de la légende selon laquelle un prophète coranique fut le maître d'Empédocle (plus ancienne attestation datée *Amad*, III, 1 [*A Muslim Philosopher on the Soul and its Fate. Al-ʿAmirī's Kitāb al-Amad ʿalā l-abad*, éd. et trad. E.K. Rowson (New Haven, 1988), éd. p. 70, trad. p. 71]). Au contraire, Abū Bakr al-Rāzī identifie correctement Alcmène dans les passages de son grand traité médical (*al-Ḥāwī*) où celui-ci est cité avec Empédocle. Nous espérons revenir sur ce point dans une prochaine publication.

[100] Sources et parallèles en partie élucidés par H. Daiber, 'Der Ṣiwān al-ḥikma von Abū Sulaimān al-Manṭiqī as-Siǧistānī in der Forschung,' *Arabica* 31 (1984), pp. 26–68.

sagesse et sous une forme résumée dans *L'Abrégé de la Chronique des nations (Ta'rīkh mukhtaṣar al-duwal)* de Bar Hebraeus (comparer *Florilège*, éd. Dunlop, §13–14, l. 176–180 et *Abrégé*, éd. Ṣaliḥānī, p. 37, l. 17–21). Dans les sources arabes, il s'agit d'une éclipse lunaire, tandis que les sources grecques mentionnent la prévision des éclipses solaires.[101] Pline situe l'éclipse prévue par Thalès durant la quatrième année de la 48e olympiade (= 585–584 av. J.-C.), ce qui correspond effectivement à l'époque du règne de Nabuchodonosor II (r. 605–562), habituellement identifié à Bukhtnaṣar.[102]

Le *Florilège du Coffret de sagesse* contient d'autres informations sur l'époque de Thalès qu'il partage en partie avec Shahrastānī, l'*Abrégé* de Bar Hebraeus et Shahrazūrī.[103] Selon le *Florilège*, Thalès aurait vécu 382 ans après Homère, et il y aurait un intervalle de 28 ans entre l'existence *(kawn)* de Thalès et le début du règne de Bukhtnaṣar. Les Grecs (ou 'les Ioniens', *ummat al-yūnāniyyīn*) auraient inventé la poésie 200 ans avant la philosophie, et leur premier philosophe aurait vécu 951 ans après la mort de Moïse. La source donnée à la suite de ces éléments par le *Florilège* est « Cyrille [d'Alexandrie], dans son livre en réponse [au livre] de Julien dans lequel celui-ci avait critiqué les Evangiles [= le *Contre les Galiléens*] » (*Florilège*, §13, l. 181–6). Le texte se poursuit immédiatement avec la date alternative rapportée d'après Porphyre selon laquelle Thalès aurait vécu 123 ans après Bukhtnaṣar (*Florilège*, §14, l. 187), rapprochée cette fois-ci (par l'auteur du *Florilège* ou par sa source ?) de la prise d'Athènes par Khusraw [= Xerxès] ibn Dārā [en 480 av. J.-C. selon Eusèbe] et de la prédication du prophète biblique Malachie, à l'époque duquel auraient vécu les philosophes Démocrite et Anaxagore (*Florilège*, §14, l. 187–9). Comme le remarque S.M. Stern, Cyrille n'utilise pas la chronologie des rois de Babylone que présente ici le *Florilège*, mais

---

[101] Diogène Laërce (= *D. L.*), *Vies et doctrines des philosophes illustres*, trad. sous la direction de M.-O. Goulet-Cazé (Paris, 1999), p. 81, n. 1 et n. 3 (R. Goulet). Voir aussi *D. L.* I, 24; Pline, *Histoire Naturelle*, II, 53; Hérodote, *Histoire*, I, 74.

[102] Par exemple al-Bīrūnī, *al-Āthār al-bāqiya* = *The Chronology of Ancient Nations*, éd. et trad. C.E. Sachau (London, 1879), p. 89, trad. p. 101; Bar Hebraeus, *Mukhtaṣar al-duwal*, p. 41. G. Vajda, 'Bukhtnaṣ(ṣ)ar,' *EI*^{2}, est inutilisable sans l'article de B. Carra de Vaux dans la 1re édition.

[103] Shahrastānī, *Kitāb al-milal wa'l-niḥal*, éd. M.F. Badrān, 2 vols. (Cairo, 1366–75/1947–55), ii, p. 936, l. 1–7 Badrān; *Livre des Religions et des Sectes*, trad. J. Jolivet (Paris, 1995), ii, p. 259, cf. Shahrazūrī, *Kitāb nuzhat al-arwāḥ*, éd. Khūrshīd Aḥmad, p. 29, l. 13–p. 30, l. 1; éd. Abū Rayyān, p. 105, l. 6–8. Le fragment est donné tronqué par F. Jacoby, *F. Gr. H.*, 260F1b, et à sa suite par A. Sodano (éd.), *Porphyrius. Storia della filosofia* (Milan, 1998), frag. VIII.

celle des olympiades des Grecs, lesquelles se déroulaient tous les quatre ans. Selon Cyrille (*Contre Julien*, I, 14), Thalès serait né durant la 35e olympiade[104] (640–637 av. J.-C.), date qui correspond à notre texte si l'on considère que l'intronisation de Nabuchodonosor II eut lieu en 610 av. J.-C., comme le fait Eusèbe, dont les tables chronologiques sont largement citées par Cyrille. Le témoignage de Cyrille est donc cité ici à travers un ou plusieurs intermédiaires.[105] Shahrastānī a exactement les mêmes éléments que le *Florilège* qu'il reformule et intègre en conclusion de son chapitre 'Homère' dans le but de préciser la chronologie de ce dernier.

F2 [Thalès] : *Florilège*, §14, l. 187 (trad. Stern, p. 451) = Shahrastānī, éd. Badrān, ii, p. 936 (trad. Jolivet, p. 259) = Shahrazūrī, *Kitāb Nuzhat al-Arwāḥ*, éd. Khūrshīd Aḥmad, p. 29, l. 13–p. 30, l. 1 / éd. Abū Rayyān, p. 105, l. 6–8 (cf. Bar Hebraeus, *Abrégé*, éd. Ṣaliḥānī, p. 31, l. 8–9)

Porphyre dit que Thalès apparut en 123 de Bukhtnaṣar.	و ذكر فرفوريوس أنّ ثاليس ظهر في سنة ثالث و عشرين و مائة من ملك بختنصر.

En revanche, la date que donne Porphyre doit provenir d'une source différente car, comme l'a remarqué Röper,[106] le 'Bukhtnaṣar' ici mentionné correspond en fait à Nebonassar (r. 747–737 av. J.-C.). Ptolémée avait donné en introduction à ses *Tables manuelles (Prokheiroi kanones)* la chronologie de la dynastie des rois babyloniens (le *Kanôn basileiôn*), à partir de laquelle les plus anciens événements astronomiques étaient datés, mais il ne semble pas que Porphyre l'ait utilisée. La confusion dans les sources arabes autour des deux Bukhtnaṣar est évidente à la lecture de Bīrūnī qui donne comme premier roi de la dynastie des Chaldéens (vassaux des rois de Babylone dont il présente aussi la chronologie et dont le premier serait Nemrod) 'Bukhtnaṣar I,

[104] Cette date apparaît aussi dans les manuscrits de Diogène Laërce (I, 37) comme étant celle de la naissance de Thalès selon la *Chronique* d'Apollodore. Mais pour faire correspondre le texte à l'acmé proposée par le même Apollodore en 585, Diels et Jacoby ont corrigé le texte transmis en '39e olympiade' (624–621 av. J.-C.). Voir Diogène Laërce, p. 90, n. 4 (R. Goulet).

[105] S.M. Stern, 'Abū 'Īsā Ibn al-Munajjim's Chronography,' in S.M. Stern, A. Hourani et V. Brown (éds.), *Islamic Philosophy and the Classical Tradition* (Oxford, 1972), pp. 437–66 (réimpr. S.M. Stern, *Medieval Arabic and Hebrew Thought* [Londres, 1983]), p. 439.

[106] F. Röper, *Lectiones Abulpharagiae* (Danzig, 1844), que cite Stern, 'Abū 'Īsā Ibn al-Munajjim's Chronography,' p. 464, n. 1.

d'après lequel est calculée la chronologie dans l'*Almageste (wa-minhu mabdā᾿ al-ta᾿rīkh fī 'l-Majisṭī)*' et 'Bukhtnaṣar II', comme étant le 'conquérant de Jérusalem' (*Āthār*, p. 89, trad. Sachau, pp. 100–101). Le premier est, toujours selon Bīrūnī, le père de Nabūkhadnāṣar et le second, le fils d'un Nabūkhadnāṣar qui n'est pas désigné comme second et serait le fils de Nabopolassar ! Historiquement, Nabuchodonosor I règne ca 1128–1106 av. J.-C. et il n'est semble-t-il pas connu des sources grecques et arabes, tandis que Nabuchodonosor II est, selon les sources akkadiennes, le fils de Nabopolassar.[107] Il semblerait que Bukhtnaṣar (ou *N-BKHTNṢR ?) et Nabūkhadnāsar soient des doublets.

Stern a pu donner dans un article posthume plusieurs fragments de ce qu'il considère être la source du *Florilège* pour ce passage.[108] Grâce à des extraits parallèles dans la *Chronographie universelle (al-Mukhtaṣar fī akhbār al-bashar)* d'Abū al-Fidā᾿ (m. 732/1331), il propose que la citation de Cyrille provienne de *L'Exposé de chronographie universelle selon la méthode apodictique (al-Bayān ʿan ta᾿rīkh sinī al-ʿālam ʿalā sabīl al-ḥujja wa'l-burhān)* d'Abū ʿĪsā ibn al-Munajjim (début IVe/Xe s.), lequel aurait utilisé une source chrétienne dans laquelle figurent Eusèbe, Cyrille et Andronicus. Il remarque que les dates d'Eusèbe semblent, pour certaines, avoir été corrigées à partir des astronomes. D. Serruys a émis une hypothèse similaire sur la *Chronique* d'Eusèbe à partir d'éléments des deux versions syriaques et de la traduction arménienne qu'il faudrait approfondir.[109] Les extraits parallèles de Bar Hebraeus n'apportent pas d'éclairage sur les sources

[107] *Der Kleine Pauly. Lexicon der Antike*, éd. K. Ziegler et W. Sontheimer, iv (Munich, 1972), col. 36, s.v. Nebukadnezar. Dans le *Qānūn al-Masʿūdī* (éd. Hyderabad, i, p. 155) en revanche, Bīrūnī donne correctement comme père Nebopolassar à Nabuchodonosor II (mais c'est à Nebopolassar qu'il attribue un très long règne de 43 ans, tandis qu'il n'accorde qu'un règne de deux ans à son fils). Elie de Nisibe, dans sa table du 'Nombre des années des rois de Babylone et d'Egypte selon la computation du sage Ptolémée', connaît trois Nabuchodonosor : le premier de la dynastie: 'Nabuchodonosor Ier ou Nabonassar', un deuxième: 'Nabuchodonosor ou Nabopolassar', père du troisième: 'Nabuchodonosor, son fils, celui qui incendia Jérusalem' (L.J. Delaporte, *La Chronographie d'Elie bar-Ṣinaya* [Paris, 1910], pp. 30–31).

[108] L'article de S.M. Stern, 'Abū ʿĪsā Ibn al-Munajjim's Chronography,' ayant été publié de façon posthume, plusieurs coquilles apparaissent (par ex. l'intervalle entre Moïse et 'Nebuchadnezzar' donné p. 441 est absent du tableau résumant cette chronologie p. 440).

[109] Stern, 'Abū ʿĪsā Ibn al-Munajjim's Chronography,' p. 441; sur Andronicus, chronographe copte du VIe s., voir D. Serruys, 'Les Canons d'Eusèbe, d'Annianos et d'Andronicos d'après Elie de Nisibe,' *Byzantinische Zeitschrift* 22 (1913), pp. 1–36. L'auteur remarque qu'Andronicus a 284 années supplémentaires par rapport au total d'Eusèbe pour l'époque qui précède la mort d'Alexandre [323 av. J.-C.]. Sur l'hypothèse de corrections effectuées à partir des tables des astronomes, voir Serruys, p. 7.

dont pourraient provenir ces éléments disparates. Quelques lignes après avoir cité Cyrille et Porphyre dans l'*Abrégé* (éd. Ṣaliḥānī, p. 31, l. 6–9), dans le même ordre que le *Florilège* (qu'il reformule) et Shahrastānī, il précise que ce qui vient d'être dit sur les époques des anciens philosophes est tiré des livres d'Eusèbe et d'Andronicus, lorsque ceux-ci concordaient avec Jacob d'Edesse (*c.* 640–708) « qui maîtrisait aussi bien l'hébreu que le grec et le syriaque » (éd. Ṣaliḥānī, p. 31, l. 15–18).[110] L'*Abrégé* de Bar Hebraeus est, comme on le sait, un résumé arabe, qu'il réalisa quelques mois avant sa mort, de sa plus vaste *Chronique* syriaque ; il y ajouta des éléments empruntés aux livres arabes et persans qu'il avait trouvés dans la bibliothèque de l'observatoire de Marāgha.[111] On peut trouver dans tout ouvrage classique des emboîtements de citations, ce qui semble être le cas lorsque Porphyre est cité après Cyrille, qu'il précède de plus d'un siècle. Le mode de présentation des chronographies engendre lui aussi des erreurs. Il peut être modifié par les copistes pour des raisons pratiques : les tables dynastiques sont le plus souvent copiées en colonne dans un but synoptique et les notices peuvent être réparties différemment. Ainsi, Bar Hebraeus cite de nouveau Porphyre à propos de la date d'Homère et d'Hésiode (*Abrégé*, éd. Ṣaliḥānī, p. 36, l. 13–14), qu'il synchronise avec le roi Joatham, fils d'Osée. Cependant, dans la *Chronique* syriaque, il donne la datation par Porphyre d'Homère et d'Hésiode comme concomitante avec la 48e année d'Osée, lorsque celui-ci fut attaqué par le roi d'Assyrie, Tighlath-Pilser [r. 745–727 av. J.-C.] en représailles, à la suite de la guerre déclenchée par Osée contre les Arabes de Petra. Après quelques commentaires sur Joatham, Bar Hebraeus ajoute (seulement dans l'*Abrégé*) que c'est à cette époque que Thalès aurait été connu comme philosophe selon la *Chronique* d'Eusèbe (*ʿalā mā dhakarahu Ūsābiyūs al-Qayṣarī fī taʾrīkhihi al-musammā Khrūnīqūn*, éd. Ṣaliḥānī, p. 37, l. 16–17). Selon ce passage, la *Chronique* d'Eusèbe aurait situé Thalès moins d'une quarantaine d'années après Homère et Hésiode.[112] Ceci correspond semble-t-il à la plus haute des trois dates

[110] Jacob d'Edesse avait corrigé les canons chronologiques de la *Chronique* d'Eusèbe, qui comportait une erreur de trois ans, et composé une suite de celle-ci. Ed. et trad. E.W. Brooks, *Chronicon Jacobi Edesseni* in *Chronica minora*, CSCO, Scriptores Syri, IIIe sér., t. IV.

[111] E.A. Wallis Budge, *The Chronography of Abû'l-Faraj Bar Hebraeus* (Oxford, 1932), p. 1.

[112] Selon les dates données par Elie de Nisibe dans sa table des 'Années de la maison d'Adam, selon l'opinion d'Andronicus, le chronographe', l'addition des périodes d'Osée (Ozias), Joatham et Achaz s'élève à 44 ans (Bar Hebraeus donne 52 ans pour

donnée par la *Chronique* de Jérôme (qui traduisit la *Chronique* d'Eusèbe en latin) pour Thalès, qui fait du philosophe un contemporain de Osée, à la 5e olympiade. En revanche, dans sa *Chronique* syriaque, Bar Hebraeus écrit que l'acmé de Thalès aurait eu lieu à la 25e année de Menasseh, faisant de lui un contemporain de Senacherib, roi des Assyriens, et du prophète Isaïe, une cinquantaine d'années avant Nabopolassar. Selon Cyrille, Homère et Hésiode naissent 164 ans après la prise de Troie, ce qui est pour Homère une date plus haute encore que celle donnée par Porphyre selon le témoignage tardif du dictionnaire byzantin de Soudas, pour qui Homère aurait vécu 275 ans après la prise de Troie, soit vers 908 av. J.-C. (*F. Gr. H.*, 260F19, trad. Segonds, p. 179), en contradiction avec le témoignage de Bar Hebraeus. Quant à Hésiode, il aurait vécu cent ans après Homère, toujours selon Porphyre cité par Soudas (*F. Gr. H.*, 260F20, trad. Segonds p. 180).

Pour conclure, la datation de Thalès selon Porphyre dans le *Florilège*, telle qu'elle fut transmise par Shahrastānī et par Bar Hebraeus fut considérée par Jacoby comme un extrait d'un ouvrage de Porphyre dont l'existence n'est pas avéré et qui aurait été intitulé *Chronique*. Il regroupe en un seul fragment attestant de l'existence de cet ouvrage [*F. Gr. H.*, 260F1] la mention de Thalès par Ibn Suwār conservée par Ibn al-Nadīm [*F. Gr. H.*, 260F1a, cf. ci-dessus Ire partie, T2 Porphyre], celle de sa date selon Porphyre citée chez Bar Hebraeus et Shahrastānī, [*F. Gr. H.* 260F1b, cf. ci-dessus IIIe partie, F2 Thalès] lesquels pourraient dépendre ici du *Florilège* et enfin un résumé d'Ibn al-Nadīm par le libraire ottoman Kātib Çelebi (= Ḥājjī Khalīfa, m. 1657) [*F. Gr. H.*, 260F1c]. Jacoby émet lui-même des doutes sur l'existence d'une *Chronique* de Porphyre, doutes relevés par Stern[113] qui précise qu'il considère le fragment comme provenant de l'*Histoire philosophique*, de même que c'est à cet ouvrage selon lui qu'Ibn al-Nadīm fait allusion. L'idée d'une *Chronique* composée par Porphyre apparaît en 1849 dans le troisième volume des *Fragmenta Historicorum Graecorum* de Karl et Theodor Müller.[114] Brian Crocke, qui remarque cette filiation, a donné tous les arguments

---

le règne d'Osée (durée parallèle chez Jérôme) tandis qu'Elie de Nisibe lit 12 ans). La disposition en tables de la traduction française par L. J. Delaporte rend ce travail bien utile, comme le note Serruys qui considère que la traduction latine de Brooks et Chabot est cependant meilleure (*Opus Chronologicum*, CSCO, Scriptores Syri, IIIe sér., t. VII et t. VIII).

113 Stern, 'Abū 'Īsā Ibn al-Munajjim's Chronography,' p. 464, n. 14.

114 K. et T. Müller, *Fragmenta Historicorum Graecorum*, iii (Paris, 1849), p. 688b (= *F. H. G.*) lesquels renvoient à Niebuhr, *Kleine Schriften* (Bonn, 1828), i, p. 188, qui ne fait que

qui permettent de réfuter l'existence d'une *Chronique*, mais n'étant pas arabisant, il retient le témoignage d'Ibn al-Nadīm comme seule attestation de cet ouvrage, ainsi que le faisait avant lui Jacoby (*F. Gr. H.*, comm. p. 854 et p. 856).[115]

Nous avons vu dans ce qui précède que la mention par Ibn Suwār d'une '*Histoire*' de Porphyre ne peut en aucun cas être considérée comme univoque. En revanche, comme le note Brian Croke à propos de l'*Histoire philosophique*, la possibilité demeure que des tables chronographiques aient circulé de façon séparée. Outre les présentations complexes des ouvrages de chronographie, la raison pour laquelle une partie de ces tables pourrait avoir eu une existence indépendante vient de ce que certaines d'entres elles devaient provenir du *Contre les Chrétiens* de Porphyre, ouvrage banni et qui subit plusieurs autodafés. Franz Rosenthal notait déjà que « l'*Histoire philosophique* contenait des matériaux chronologiques qui nous sont attestés en grec dans la *Chronique* d'Eusèbe », visant sans doute le fragment dans lequel l'évêque de Césarée précisait que le livre premier de l'*Histoire philosophique* donnait la chronologie des événements depuis la chute de Troie à la première olympiade.[116] Ce premier livre comportait donc outre une vie de Pythagore et des informations sur les sept Sages (que Porphyre ne pouvait croire contemporains), des éléments de chronologie sur Homère et Hésiode. Avec autant de matériaux réunis en un seul volume, comment interpréter le fait que Socrate ait figuré dans le volume III et Platon dans le volume IV ? Il faut, soit rejeter le témoignage de Cyrille, qui attribue au volume I la *Vie de Pythagore* et l'anecdote sur la légende de l'heptade des Sages, soit considérer que l'ouvrage contenait une présentation par école. Dans ce cas, le volume IV contenait sans doute des éléments sur l'école néoplatonicienne, puisque, selon Eusèbe (*F. Gr. H.*, 260T2), les éléments chronologiques donnés par Porphyre s'étendaient jusqu'à l'époque de Claude le Gothique (r. 268–270), sous le règne duquel mourut Plotin. S'il en va pour les auteurs grecs de l'antiquité tardive comme des auteurs arabes médiévaux, il semblerait que ces ouvrages dans lesquels se mêlaient chronologies, biographies, doxographies, et bibliographies

citer Georges le Syncelle, dont Serruys montre qu'il utilise Annianos (que connaît aussi Bar Hebraeus).

[115] B. Croke, 'Porphyry's Anti-Christian Chronology,' *Journal of Theological Studies* 34 (1983), pp. 168–85, à la p. 178.

[116] Rosenthal, *History of Muslim Historiography*, pp. 77–8 (réf. supra n. 14). Cf. *F. Gr. H.*, 260F4.

servaient avant tout de vade-mecum à leurs auteurs et aux savants qui les recopiaient, ajoutant leurs propres notes et corrections au fur et à mesure de leurs lectures. On connaît en partie les méthodes des traducteurs grâce au témoignage de Ḥunayn ibn Isḥāq qui commençait par la collation des manuscrits disponibles et l'étude des traductions existantes.[117] Dans le cas des ouvrages historiques, le découpage en tables dynastiques, canons chronologiques et notices entraînait là aussi une infinité de possibilités quant à la transmission du texte.

[117] G. Bergstrasser, *Ḥunain Ibn Isḥāq über die syrischen und arabischen Galen-Übersetzungen* (Leipzig, 1925); idem, *Neue Materialien zu Ḥunain Ibn Isḥāq's Galen-Bibliographie* (Leipzig, 1932); réimpr., éditions récentes et comptes-rendus in Daiber, *Bibliography of Islamic Philosophy*, ii, n° 1382–3]; F. Rosenthal, *The Technique and Approach of Muslim Scholarship* (Rome, 1947); S. Brock, 'The Syriac Background to Ḥunayn's Translation Techniques,' *ARAM* 3 (1991), pp. 139–62; D. Gutas, *Greek Thought Arabic Culture* (Londres, 1998).

ARABIC PARTICLES AND GRAECO-ARABIC TRANSLATIONS: ON THE USES OF *GALEX* I

Dimitri Gutas

In a 1982 review of Hans Daiber's *Aetius Arabus*, a work in the vanguard of Graeco-Arabic studies as I called it then, I noted that the glossaries it contained were 'exact and reliable and of potential benefit to an eventual Graeco-Arabic lexicon.'[1] That eventuality actually did materialize, and to this day Daiber's book has continued to be among the most valuable resources of *GALex*, edited by Gerhard Endress and myself.[2]

One decision that had to be taken during the preparatory stages of *GALex* was the extent of coverage to be allocated to what Arab grammarians classify as *ḥurūf*—particles, conjunctions, prepositions, etc. These are 'service' parts of speech, inextricably linked to the mechanics of the operation of a specific language, and a historical lexicon of Greek and Arabic, the argument could be made, had nothing to gain by listing numerous instances where *idhā*, for example, translates Greek ὅταν, 'when', or εἰ, 'if'. After looking closely at the material, however, we decided, despite the eventual cost in terms of the ultimate length of the completed *GALex*, to treat the *ḥurūf* extensively but illustratively, not exhaustively. In retrospect, the benefits of detailed analysis of the *ḥurūf* for an understanding of how sense is reflected in Arabic style and syntax have indeed been substantial and very much worth the effort.

Study of the particles in the light of the Greek sentences which they help translate allows, most significantly, the classification of their various uses in new ways—i.e. according to the variety of Greek structures that lie behind them. This is very useful for Arabic, for two reasons. First, classical Arabic syntax is still in need of detailed and sustained investigation, despite the significant advances in the field since the

---

[1] 'The Present State and Future Tasks of Graeco-Arabic Studies: Remarks apropos H. Daiber's *Aetius Arabus*,' *JRAS* (1982), pp. 113–23, 118.

[2] *A Greek and Arabic Lexicon (GALex). Materials for a Dictionary of the Medieval Translations from Greek into Arabic* (Leiden, 2002), i; work continues on the second volume. It is a privilege and a pleasure to express, however briefly through these lines, my appreciation and good wishes for a happy and long retirement to a friend and colleague whose work has been a source of inspiration to all of us working on *GALex*.

publication of Reckendorf's studies over a hundred years ago.[3] A new look at Arabic syntax in comparison with Greek should prove fruitful. Second, the comparison with Greek is particularly helpful because in Greek, the complexity and sophistication of the language rests for the most part on its morphology and wealth of 'service' vocabulary—the highly complex mood and tense system and the inordinately rich set of particles and conjunctions—which are immediately visible and therefore allow easy classification of the various levels of sense reflected in the corresponding structures; in the case of classical Arabic, however, its complexity and sophistication depend much more on syntax and style and on the extremely variegated use of a very limited number of particles, virtues which are, for the most part, hidden and have therefore remained relatively unappreciated. One still reads, and most students of Arabic still believe, that *inna* means 'verily'. To illustrate the point just made, in what follows I will present the uses of two Arabic particles that can be elucidated through comparison with the Greek, as they are listed in the corresponding entries of the first volume of *GALex*.

First, let me take *idhā* again: on the basis of qualitatively distinct categories of Greek structures that it helps translate, its use can be divided into three (*GALex* pp. 154–5; the fourth category listed on p. 155 depends on inaccurate renderings and is therefore not immediately relevant to this discussion). It translates either Greek words, or the meaning of Greek syntactic constructions, or the semantic implications of Greek syntax and vocabulary. The first needs little comment, if any; *idhā* naturally translates Greek temporal conjunctions and conditional particles. However, what is not clear is the extent to which the bewildering variety of Greek constructions in this area, all slightly nuanced and thus distinct from each other, is semantically leveled in Arabic—since *idhā* clauses allow of little variation—or whether there are sufficient and consistent variations in the syntax of *idhā* clauses and their immediate environment that do, in fact, reflect these nuances. The question has yet to be studied, and *GALex* contributes nine pages of illustrative material to that end (pp. 156–64).

In the second category, *idhā* introduces clauses that translate, among others, the Greek syntactic structures known as the genitive absolute

[3] For a recent listing of works on Arabic syntax see the bibliographies in the English translation by J. Rodgers of W. Fischer's *A Grammar of Classical Arabic* (New Haven, 2002), especially pp. 290–301.

and the participium coniunctum. These are essentially circumstantial clauses in Greek—'the case being what it is...', 'seeing that...', as is obvious through their translation by means of *idh* (*GALex* pp. 150–53), but are here rendered with an *idhā* clause. Two questions might be raised. The first is whether, if *idhā* introduces essentially a temporal/conditional clause, there is semantic overlap between circumstantial and temporal clauses, or *idhā* also has circumstantial extensions. The second is the extent to which *idhā* shares functions with *idh* and *wa-* of *ḥāl*, or to put it the other way around, what discriminates among and determines the use of *idh*, *idhā*, and *wa-* of *ḥāl*.

The third category is by far the most interesting and the most unexpected; here *idhā* clauses help translate Greek adjectives, adverbs, and nouns where a corresponding Arabic adjective, prepositional phrase or accu-sative, or a noun or *maṣdar* might have been thought appropriate. Thus, for example, the translator of the Artemidorus dream book renders ὁ μάντις ὅ γε ἀληθής with *fa-idhā kāna 'l-ʿarrāfu ṣādiqan* (*GALex* p. 172, §5.1.1).[4] The nature of attributive adjectives (like ἀληθής, 'veracious' here) is such that they tend to generate the impression that the quality attributed is permanent; in this example, this impression, in the Greek, is blunted by the particle γε: 'the seer—well, at least the veracious one.' The Arabic admirably renders this qualification, i.e., the modality of the veraciousness attributed to the seer, by employing, instead of an adjective, the *idhā* clause: '*when* the seer happens to be veracious...'. This sensitivity to modality is to be expected from Arabic, a language actually built on it: the difference among the *faʿala*, *faʿila* and *faʿula* forms of verbs directly depends on it, as does that among *fāʿil*, *faʿīl*, *fiʿʿīl*, and *faʿʿāla* nouns with their direct implications of durability and intensity of attributions.[5]

This categorization of the function of *idhā* clauses in Arabic deepens and refines our understanding of them and ought to lead, in future grammars, in explicit differentiation among the strictly temporal/conditional, circumstantial, and modal uses of this particle.

---

[4] *Artemidori Daldiani Onirocroticon Libri v*, ed. R.A. Pack (Leipzig, 1963), p. 212, l. 7; *Le livre des songes, Traduit du grec en arabe par Ḥunayn b. Isḥāq*, ed. T. Fahd (Damascus, 1964), p. 384, l. 12.

[5] Slightly to extrapolate, Avicenna's refinement of Aristotle's modal logic may owe not a little to this inherent quality of Arabic. See D. Gutas, *Avicenna and the Aristotelian Tradition* (Leiden, 1988), pp. 179–80.

Second, the decision by the editors of *GALex* to accord extensive illustrative treatment to 'service' parts of speech has also led to serious attention being paid to substantives which, through usage, have themselves assumed 'service' functions. A case in point is *amr*. It has two basic meanings, (1) 'command' (pl. *awāmir*), which will no longer occupy us here, and (2) 'matter, affair' (pl. *umūr*), which has additional semantic extensions that lend it (2.1) a grammatical and syntactic use, as well as (2.2) a conceptual use (*GALex* p. 369–93, §§20–46). Its major grammatical or syntactical use (2.1) is its employment, first, (2.1.1) as a substitute for pronouns, and second (2.1.2), as a generic or dummy substantive added in Arabic whenever a noun is omitted or implied in Greek but needs to appear in Arabic due to the requirements of Arabic style or syntax. As an example for the first (2.1.1), cf. the passage in Aristotle's *Physics* (197a 26), τύχη δὲ ἀγαθὴ μὲν λέγεται ὅταν ἀγαθόν τι ἀποβῇ, 'chance is called good when something good results' = *wa-yuqālu fī 'l-bakhti innahū jayyidun matā kānat ʿuqbāhu amran jayyidan maḥmūdan* (*GALex* p. 370, §21.1),[6] with *amr* directly translating the indefinite pronoun, 'something'. For the second (2.1.2) there are numerous varieties; one that is common is for *amr* to stand in the place of the substantive implied when the Greek uses the nominal formation of the article followed by adjectives, adverbs, etc.; e.g., τὰ ἐκτός (the [things that are] outside) = *al-umūru llatī min khārijin* (*GALex* p. 379, §34.2)[7]; τὸ ἐξ ἀνάγκης (that [which is] by necessity) = *al-amru lladhī yakūnu bi'ḍṭirārin* (*GALex* p. 380, §34.6).[8] Another is for *amr* to stand in the place of the object of a verb which in Greek was expressed absolutely (i.e., without an object), e.g., οὐ κελεύει δὲ φοβεῖσθαι (does not enjoin to have fear) = *fa-lā yakūnu 'l-khawfu ʿan amrin* (*GALex* p. 380, §35.5),[9] etc.

The conceptual use of *amr* (2.2), however, is most significant and little noted. It consists in *amr* being used redundantly in association with abstract and general nouns—to paraphrase the Arab grammarians, an *amr al-zāʾida*, so to speak—in order to specify and individuate the sense of the Arabic word which translates the Greek substantive. This use

[6] Arisṭūṭālīs, *al-Ṭabīʿa*, ed. ʿA. Badawī (Cairo, 1964–65), p. 125, l. 5.

[7] Alexander Aphrodisiensis, *De anima cum mantissa*, ed. I. Bruns (Berlin, 1887), p. 174, l. 6; H.-J. Ruland, *Die arabischen Fassungen von zwei Schriften des Alexander von Aphrodisias*, doctoral thesis (Saarbrücken, 1975 [1976]), p. 201, l. 70).

[8] Aristotle, *De partibus animalium* 639b 21; R. Kruk, *The Arabic Version of Aristotle's* Parts of Animals (Amsterdam, 1979), p. 6, l. 24.

[9] Aristotle, *De anima* 432b 31; Arisṭūṭālīs, *Fī 'l-nafs*, ed. ʿA. Badawī (Cairo, 1945), p. 81, l. 12.

manifests itself in two ways. In the first (2.2.1), in what may be called the partitive use of *amr*, the expression consists of a phrase including *amr* with a following genitive, in which the word in the genitive is actually the one that translates the Greek abstract noun; thus what is expressed is not the abstract thing itself but its *amr*. For example, when Aristotle says, ἐπισκοποῦντες δὲ περὶ ψυχῆς, 'as we study the soul', the translator renders it by *fa-idhā naẓarnā wa-fak(k)arnā fī amri 'l-nafsi* (*GALex* p. 391, §45)[10] instead of just *fī 'l-nafsi*, which, after all, is the subject of the entire book. An extreme case of this usage is encountered in the translation of Themistius, when he is talking about the olfactory sense and its object: ἡ δὲ ὀσμὴ καὶ ὀσφραντὸν τί ποτέ ἐστιν, οὐχ οὕτω ῥᾴδιον ἀποδοῦναι ὥσπερ καὶ τὸ χρῶμα καὶ τὸ διαφανὲς καὶ τὸν ψόφον, 'it is not as easy to give an account of what smell and the smellable are as it was for color and the transparent and sound' = *fa-ammā 'l-rā'iḥatu wa'l-mashmūmu mā humā fa-laysa ta'diyatu dhālika bi'l-sahlati mithla ta'diyati amri 'l-lawni wa-amri 'l-mushiffi wa-amri 'l-ṣawti* (*GALex* p. 392, §45),[11] instead of just *al-lawn*, etc., without the *amr*, as indeed the translator does with ὀσμή, 'smell', which he renders merely with *al-rā'iḥa* and not *amru 'l-rā'iḥa*! However, the abstract noun need not be 'abstract' only in the conventional sense; indefinite substantives or even substantivized adjectives are also treated similarly by the translators, as, e.g., in the following: καὶ πολλὰ ἡμῖν δῆλά ἐστιν οὕτως ἔχοντα, 'many [things] are obvious to us as being such' = *wa-hāhunā ashyā'u kathīratun bayyinun min amrihā annahā bi-hādhihi 'l-ḥāli* (*GALex* p. 391, §45),[12] instead of *minhā*, or, in this case, just *bayyinun annahā*.

In all these examples, the function of *amr* would seem to be to make an abstract or indefinite concept more specific and as a result to direct attention not to its entirety but only either to its essential core, its 'nature,' or to the particular issue about it that is under investigation. The limiting sense of this partitive use is thus that what is under discussion is not everything concerning the abstract or indefinite object but only its nature or a specific problem that is at issue. This sense is consistent with the regular definition of *amr* (pl. *umūr*) given by Arab lexicographers,

---

[10] *De anima* 403b 20; *Fī 'l-nafs*, p. 7, l. 18.

[11] *Themistii In libros Aristotelis De anima paraphrasis*, ed. R. Heinze (Berlin, 1899), p. 67, l. 27; M.C. Lyons, *An Arabic Translation of Themistius' Commentary on Aristoteles* De anima (Oxford, 1973), p. 109, l. 11.

[12] Aristotle, *De interpretatione*, 19a 12; *Manṭiq Arisṭū*, ed. 'A. Badawī (Kuwait, 1980), p. 112, l. 1.

i.e., *sha'n, ḥāl*, etc. There is, however, nothing tangible in the Greek text to elicit such a rendering other than the translator's sense of the Greek sentence in its semantic context. This is clear also from the fact that, naturally, not every instance of a Greek abstract noun is rendered with *amr* followed by a genitive. Whether the presence of *amr* in these cases is arbitrary or there is an underlying pattern is a problem that is well worth investigating; and it would seem that a productive way of analysis is through a systematic juxtaposition of Arabic sentences containing the *amr al-zā'ida* with the Greek sentences they translate.

The second way (2.2.2) in which an apparently redundant *amr* is used in the translation of abstract nouns is individuating: *amr* functions as a generic substantive, 'thing,' to which the word that actually translates the Greek abstract noun is added as an adjective—i.e., 'goodness' becomes 'a good *amr*, a good thing'—thus effectively transforming the abstract noun into a concrete manifestation of itself. This phenomenon is easier to understand than the preceding one, and it has at least two major reasons. One is the discomfort with abstract nouns, when used by themselves, which can be observed in the old *ʿarabiyya*. A good example can be found in one of the earlier translations, the Aristotelian zoological treatises. Aristotle here speaks of menstrual fluids as impure semen because they lack the principle of soul, imported by semen; hence the inability of an unfertilized egg to conceive: τὸ γὰρ θῆλυ ὥσπερ ἄρρεν ἐστὶ πεπηρωμένον καὶ τὰ καταμήνια σπέρμα, οὐ καθαρὸν δέ· ἓν γὰρ οὐκ ἔχει μόνον, τὴν τῆς ψυχῆς ἀρχήν. καὶ διὰ τοῦτο... τὸ συνιστάμενον ᾠὸν... τὴν ἀρχὴν οὐκ ἔχει = 'For the female is, as it were, a mutilated male, and the menstrual fluids [are] semen, only not pure; for there is only one [thing] they have not [in them], the principle of soul. For this reason... the egg that is formed... does not have the principle [in question].'[13] This is all rendered almost literally as, *li-anna 'l-unthā mithla dhakarin madrūrin wa'l-ṭamtha zarʿun laysa bi-naqiyyin li-anna laysa fīhi shay'un wāḥidun, aʿnī awwala 'l-ṭibāʿi; wa-min ajli hādhihi 'l-ʿillati bayḍu 'l-rīḥi... ʿādama 'l-amra 'l-awwala* (*GALex* p. 392–3, §46).[14] Leaving aside the rendering of ψυχῆς, 'soul' as *ṭibāʿ*,[15] it is to be noted that in the first instance, where

[13] Translation adapted from that by A. Platt. See note 16 below.

[14] *De generatione animalium* 737a 27–32; *Generation of Animals. The Arabic Translation Commonly Ascribed to Yaḥyā ibn al-Biṭrīq*, ed. J. Brugman and H.J. Drossaart Lulofs (Leiden, 1971), p. 65, l. 8–10.

[15] This may be due to the interpretation of the translator or, more likely, to a textual variant, or marginal gloss, of φύσεως, 'nature' for ψυχῆς, 'soul'. This has not been remarked upon by Drossaart Lulofs, who both edited the Greek text in the Oxford

ἀρχή (principle) is attached to a genitive (ἀρχὴ ψυχῆς, 'principle of soul'), i.e., it is the ἀρχή, 'principle' of *something*, the Arabic has no problem translating literally, *awwal* followed by a genitive, *awwala 'l-ṭibāʿi*; in the second, however, ἀρχή, 'principle' is used by itself, and the translator feels that he has to add *al-amr* to concretize and individuate *al-awwal*.[16] In the usage of later translations, ἀρχή, 'principle' would have been rendered by *mabdaʾ* instead, an *ism* which by its very morphology has more concrete implications than *awwal*.

Second, Greek uses abstract nouns even when it wishes to express individual manifestations of the quality intended. In these cases the individuation effected by the use of *amr* and the concomitant rendering in Arabic of the abstract quality as an adjective, or a verb in a relative clause, is unavoidable. E.g., τὸ φίλους ὁρᾶν...παρέχει...ἡδονάς, 'seeing friends [in dreams]...provides...pleasures' = *idhā raʾā fī manāmihī...aṣdiqāʾahū...dalla dhālika ʿalā amrin ladhīdhin* (*GALex* p. 393, §46); similarly, πλουσίῳ δὲ πρὸς τὸ ἄρχειν ἤδη καθεστῶτι...βαρύνσεις... προαγορεύει, 'for the rich man who has assumed a ruling [function]...it augurs...burdens' = *wa-ammā 'l-ghaniyyu idhā kāna fī riʾāsatin ... fa-inna dhālika yadullu ʿalā annahū yanāluhū...amrun yathqulu ʿalayhi* (*GALex* p. 393, §46).[17] In these cases, both ἡδονάς, 'pleasures' and βαρύνσεις, 'burdens' in Greek, used as they are in the plural, do indeed indicate individual instances of pleasure and oppression, and the Arabic renderings with *amr* (*amrun ladhīdhun* and *amrun yathqulu ʿalayhi* respectively) are necessary; *ladhdha* and *thiql* by themselves would have been impossible here.

series (1965), in which he took fully into account the Arabic translation with the help of S. A. Bonebakker, and assisted in the edition of the Arabic text prepared by J. Brugman (1971). It is a pity that, as he mentions in the Preface of the Arabic edition (p. xi), he was forced to leave out all references to deviations from the Greek text.

[16] It appears that few languages can sustain intelligibility at the same level of abstraction as classical Greek—or at least Aristotle's stenographic style in his lecture notes; interestingly enough, the English translation of this very passage by A. Platt (*The Complete Works of Aristotle*, ed. J. Barnes [Princeton, 1984], p. 1144) also adds a prepositional phrase for the sake of specificity: 'the egg...has not the principle *in question*' (emphasis added).

[17] Both citations are from Artemidorus' *Onirocriticon* (as in n. 4), p. 8, l. 14 = 17, l. 4–5 and 26, l. 11 = 50, l. 8 respectively).

RECLAIMING BABYLON: THE MULTIPLE LANGUAGES OF THE QURʾĀN

Angelika Neuwirth

1. The Project

The provocative title of this paper has been chosen deliberately: the Qurʾān in Islamic tradition has been read as a rather monolithic entity consummate in itself; if related at all to a cultural and historical origin, it is more often than not conceived as born into the ethnically and culturally 'pure' environment of *al-Jāhiliyya*, of pre-Islamic Arab culture, represented by the Bedouin and sedentary population of the Arabian peninsula. Though contacts of the Qurʾān's main protagonists, the prophet Muhammad and his community, with other cultural groups have been conceded, these encounters are perceived as episodic rather than functional in terms of serious exchanges of ideas. The Qurʾān in its traditional reading is a text of one language.

Oriental studies have questioned that perception in various ways. The earliest initiative to contextualize the Qurʾān grew out of a Western intellectual movement that targeted the historization of religious traditions as such, though that of Jewish traditions in particular, namely the *Wissenschaft des Judentums*.[1] Abraham Geiger, one of its founding fathers, as early as 1833 published a groundbreaking work with the telling title 'What did Mohammed borrow from Judaism?'[2] Though Geiger was imagining the process of proto-Islamic exchange with Judaism in

[1] See for the *Wissenschaft des Judentums*, M. Kramer (ed.), *The Jewish Discovery of Islam. Studies in Honor of Bernard Lewis* (Tel Aviv, 1999), pp. 103–35; R. Firestone, 'The Qur'an and the Bible. Some Modern Studies of their Relationship,' in J.C. Reeves (ed.), *Bible and Qurʾān. Essays in Scriptural Intertextuality* (Leiden, 2004), pp. 1–22; C. Schulte, 'Die Wissenschaft des Judentums,' in E.-V. Kotowski, J.H. Schoeps and H. Wallenborn (eds.), *Handbuch zur Geschichte der Juden in Europa*, ii: *Religion, Kultur, Alltag* (Darmstadt, 2001), pp. 269–84; C. Schulte, 'Religion in der Wissenschaft des Judentums. Ein historischer Abriss in methodologischer Absicht,' *REJ* 161 (2002), pp. 411–29 and D. Hartwig et al. (eds.), *Im vollen Licht der Geschichte. Die Wissenschaft des Judentums und die Anfänge der Koranforschung* (Würzburg, 2008).

[2] A. Geiger, *Was hat Mohammed aus dem Judenthume aufgenommen?* (Berlin, 1833), repr. with an introduction by Friedrich Niewöhner (Berlin, 2005), English translation: *Judaism and Islám. A Prize Essay by Abraham Geiger*, translated from the German by a member of the Ladies' League in aid of the Delhi Mission (Madras, 1898).

an all too mechanistic fashion, he did unearth significant intertexts of the Qurʾān, showing that particular qurʾānic discussions, once projected against a backdrop of sectarian debates, reveal themselves as responses to relevant theological questions. Whereas Geiger was exclusively concerned with Jewish traditions reflected in the Qurʾān, in the later extensions of his work presented by Hartwig Hischfeld,[3] Joachim Wilhelm Hirschberg,[4] Isaac Gastfreund,[5] Israel Shapiro[6] and Heinrich Speyer,[7] a 'Babylon of cultural languages'—Jewish, Judaeo-Christian and Christian, in addition to the pagan Arab—comes to the fore: Geiger's insistence on a sectarian backdrop was corroborated by the qurʾānic scholarship after him—mostly originating from Geiger's own scholarly tradition, the *Wissenschaft des Judentums*. These studies, conducted in various fields such as the linguistic features of the Qurʾān, the foreign vocabulary and the relations between qurʾānic proper names and those of neighboring traditions, and—most significantly—the Jewish intertexts of the qurʾānic discourse, were to shed ample light on the intertwined worlds of paganism, Judaism and Christianity in the Arabian peninsula contemporary to the genesis of the Qurʾān. The numerous contributions—all targeting a joint project of critical qurʾānic studies—would arguably have resulted in a complete historico-critical analysis of the Qurʾān and its contextualization in a comparative horizon after the model of Biblical studies, had they been allowed to live up to our days of methodologically more diversified scholarship.

After the disappearance of the *Wissenschaft des Judentums*, and the elimination of Jewish scholars from German universities during the early 1930s by the Nazi regime, scholars—no longer equipped with linguistic and hermeneutic tools required for the study of pre-Islamic Qurʾānic intertexts—lost interest in the Qurʾān's setting in Late Antiquity. Instead, they turned their focus to the *persona* of the Prophet whose psychological development seemed to promise a direct insight into the peculiar genesis

[3] H. Hirschfeld, *Beiträge zur Erklärung des Ḳorân* (Leipzig, 1886); idem, *Jüdische Elemente im Korân. Ein Beitrag zur Korânforschung* (Berlin, 1878); idem, *New Researches into the Composition of the Qoran* (London, 1902).

[4] J.W. Hirschberg, *Jüdische und christliche Lehren im vor- und frühislamischen Arabien. Ein Beitrag zur Entstehungsgeschichte des Islams* (Krakow, 1939).

[5] I. Gastfreund, *Mohamed nach Talmud und Midrasch: kritisch-historisch bearbeitet*, 3 vols. (Berlin, 1875–80).

[6] I. Schapiro, *Die haggadischen Elemente im erzählenden Teil des Korans* (Berlin, 1907; originally Ph.D. dissertation, Strassburg, 1906).

[7] H. Speyer, *Die biblischen Erzählungen im Koran* (Gräfenhainichen, 1931; repr. Hildesheim, 1971); idem, 'Mohammed und die Aggada,' *Der Jude* 7 (1923), pp. 202–6.

of the Qurʾān.[8] They thus maintained and even underscored the premise of the earlier established 'authorial paradigm' featuring Muhammad as the author of the Qurʾān who deliberately conceptionalized and composed the text corpus. This paradigm continued to be upheld in Post-World-War II scholarship by scholars like Rudi Paret and Montgomery Watt who, in their reading of Muhammad's message, made ample and often somewhat naïve use of Islamic tradition. It eventually provoked the skeptical response of John Wansbrough and his school in the late seventies. Wansbrough's *Qurʾānic Studies*, by advocating a wholesale dismissal of the Islamic tradition and with it of the Qurʾān's chronological and geographical frame, induced a turn in qurʾānic studies. Even though in the meantime manuscript evidence has been discovered that disproves a later emergence of the Qurʾān,[9] Anglo-Saxon qurʾānic scholarship is still largely skeptical in orientation. In Germany, more 'positivist' revisionist attempts at reconstructing the qurʾānic genesis have been put forward by Günther Lüling[10] and Christoph Luxenberg,[11] reclaiming the Qurʾān as a re-writing of earlier Christian texts—approaches that have further widened the hermeneutic gap that has long been dividing qurʾānic scholarship into Islamic and Western.

What needs to be resumed is the study of the cultural environment of the Qurʾān.[12] This is a task that is presently pursued in the project *Corpus Coranicum* based at the Berlin-Brandenburgische Akademie der Wissenschaften in Berlin.[13] Some principal considerations formulated in that context that also inform the present paper may be worth mentioning here:

[8] See J. Fück, 'Die Originalität des arabischen Propheten,' *ZDMG* 90 (1936), pp. 509–25.

[9] Cf. G.-R. Puin, 'Observations on Early Qurʾan Manuscripts in Ṣanʿāʾ,' in S. Wild (ed.), *The Qurʾan as Text* (Leiden, 1996), pp. 107–11. Wansbrough's hypothesis of a milieu temporally and geographically different from the traditionally assumed Meccan/Medinan environment has not stood up to historical evaluation.

[10] G. Lüling, *Über den Ur-Qurʾan. Ansätze zur Rekonstruktion vorislamischer christlicher Strophenlieder im Qurʾān* (Erlangen, 1974).

[11] C. Luxenberg, *Die syro-aramäische Lesart des Koran. Ein Beitrag zur Entschlüsselung der Koransprache* (Berlin, 2000, 2004), cf. the review by S. Hopkins in *JSAI* 28 (2003), pp. 377–80.

[12] See M. Marx, A. Neuwirth and N. Sinai (eds.), *The Qurʾān in Context* (forthcoming).

[13] The project *Corpus Coranicum*, Textdokumentation und historisch-kritischer Kommentar was inaugurated at the Berlin-Brandenburgische Akademie der Wissenschaften in 2007, cf. M. Marx, 'Ein Koran-Forschungsprojekt in der Tradition der Wissenschaft des Judentuns: zur Programmatik des Akademienvorhabens *Corpus Coranicum*,' in Hartwig (ed.), *Im vollen Licht der Geschichte*.

> Is it really plausible to model Muhammad's relationship to his followers on the relationship between a modern novelist and his readers (...). Is it not more plausible to envisage the qurʾānic texts as having emerged from the complex interaction of a multitude of discursive agents, among whom Muhammad functioned more like a moderator whose revelatory promulgations were able to translate the ideas, motives, concepts, questions and expectations circulating among his followers into an authoritative literary form suitable for liturgical use? Hence, even within the traditional account of the Qurʾān's genesis, to which I largely subscribe, one must not underestimate the extent to which the community of followers that had gathered around Muhammad were involved in negotiating and articulating the various stages of religious consensus that we find reflected in the Qurʾān. Of course, the process of the Qurʾān's genesis could not have gotten under way if there had not been an original nucleus of texts promulgated by the religious charismatic Muhammad. But once a small qurʾānic community had crystallized around these earliest revelations, it must have engaged in some kind of discussion on the meaning of existing revelations, on their significance for the community's present situation, on their implications for the community's social and cultic life, on their compatibility or incompatibility with ancient Arabian, Jewish or Christian conceptions, narratives, and practices, etc. It is also likely that any answer that subsequent qurʾānic revelations will draw to some extent on the conceptual and narrative resources that pre-exist within the community. (...) The alternative to the authorial that is here proposed would be to soften the rigidity with which Muhammad, assumed to have been in total and all-embracing control over qurʾānic discourse, is marked off from his supposedly passive recipients'.[14]

Identifying extra-qurʾānic traditions should, however, not be expected to result in more than a delineation of the range of traditions known among the contemporaries of the early community—certainly, *pace* Luxenberg, not in the discovery of the 'sources' of the Qurʾān. John Wansbrough and his school[15] have justly maintained that not only the fully developed Islamic culture, but the Qurʾān itself emerged from a sectarian milieu, alerting us again to the polemical and apologetic framework[16] of the qurʾānic communications. The Qurʾān thus should

---

[14] N. Sinai, 'Orientalism, Authorship, and the Onset of Revelation' (unpublished paper).

[15] J. Wansbrough, *Quranic Studies. Sources and Methods of Scriptural Interpretation* (Oxford, 1977; repr. Amherst, NY, 2004). P. Crone and M. Cook, *Hagarism. The Making of the Islamic World* (Cambridge, 1977), cf. G.R. Hawting, *The Idea of Idolatry and the Emergence of Islam. From Polemic to History* (Cambridge, 1999), pp. 10–19.

[16] Polemic against idolatry is the topic of the monograph by G. Hawting (*Idea of Idolatry*); see for a critical appraisal L. Ammann, *Die Geburt des Islam. Historische Innovation durch Offenbarung* (Göttingen, 2001).

be understood as a reading of a plethora of earlier traditions.[17] This observation, however—*pace* Wansbrough—does not contradict, let alone exclude, the assumption that the qurʾānic corpus and the early community developed synchronously in Mecca and later in Medina. Why not heuristically consider the Qurʾān as a 'documentation' of that crucial socio-political process that eventually resulted in the emergence of a community and relocate the focus from the notion of a text authored by the prophet (or later compilers) to the communication process taking place between the prophet and the collective of the earliest listeners? In order to investigate the community's possible religious background, and to understand their debates reflected in the Qurʾān, extra-qurʾānic traditions of late antiquity have to be reconsidered.

Approaching qurʾānic texts primarily through extra-qurʾānic material, would, however, fail to acknowledge their intra-qurʾānic referentiality and blur the *Sitz im Leben* of individual communications within the process of the emergence of a community. The neglect of the intra-qurʾānic context, due to the dismissal of qurʾānic chronology, gravely impairs the scholarly significance of even the most substantial contributions to the reflections of Jewish and Christian traditions in the Qurʾān.[18] However, it has to be kept in mind that the qurʾānic texts, before they were integrated into the text corpus and arranged irrespectively of their chronological sequence, were communicated to the early listeners in response to particular discourses that the community was engaging in, later communications presupposing the memory of earlier ones. Only a close consideration of these intra-qurʾānic contexts allows the multi-layered structure of the qurʾānic text, its multiple voices and 'languages' to become apparent. This micro-structural reading is a step that has been overlept in the scholarship in the tradition of Wansbrough, who stripped the qurʾānic text of its temporary and cultural coordinates, thus rashly projecting it into the realm of literary myth that entails no clues regarding its historical positioning. Literary evidence—which remained

---

[17] Cf. F. de Blois, '*Naṣrānī* (Ναζωραῖος) and *ḥanīf* (ἐθνικός). Studies on the Religious Vocabulary of Christianity and of Islam,' *BSOAS* 65 (2002), pp. 1–30; S. Krone, *Die altarabische Gottheit al-Lāt* (Frankfurt, 1992); Ammann, *Die Geburt des Islam*.

[18] See e.g. H. Busse, *Die theologischen Beziehungen des Islam zu Judentum und Christentum* (Darmstadt, 1994) and M. Bauschke, *Jesus im Koran* (Cologne, 2001). See also N. Robinson, *Christ in Islam and Christianity. The Representation of Jesus in the Qurʾān and the Classical Muslim Commentaries* (London, 1991). Since these studies are oriented towards an ecumenical dialogue, they focus on the canonical text rather than the development of the community reflected in the Qurʾān.

excluded from 'revisionist' scholarship,—strongly supports the idea that the Qur'ān emerged from an oral communication process involving the historical figure of the prophet Muhammad and his listeners.[19]

The following article presupposes the Qur'ān not as the fixed corpus it had become after the death of the prophet, in Mohamed Arkoun's words 'the Closed Official Corpus', but as a chain of communications conveyed to the Meccan and later the Medinan community, whose expectations and religious background are implied in the qur'ānic texts. Two cases of a later re-composition of an individual qur'ānic text in a new 'key', inviting multiple voices, will be presented. Recent studies have corroborated the assumption that Meccan texts have often been re-read in Medina under new auspices, with Jews among the listeners,[20] and the present paper will try to show that these re-readings did not take shape isolated from Jewish exegetical pre-texts. What was a consensus of believing listeners in Mecca vis-à-vis their pagan opponents was challenged by more complex positions once Jewish tradition, a new cultural 'language,' was introduced into the qur'ānic discourse. How does this new language make itself audible? The first case study will focus on the Medinan expansion of a Meccan reading of a Biblical story that had to be revised at Medina in view of a new, Jewish audience involved (part 2), the second case study addresses the Medinan re-reading of the Meccan story of Mary and Jesus (part 3). Here again, extra-qur'ānic traditions, not least Jewish theologumena become apparent that will allow us to draw some conclusions about the Meccan and the Medinan communities' position towards Mary and Jesus and thus Christianity itself.

2. Judaism as a Challenge:[21] the Language of the Jewish Tradition

More than twenty years ago Andrew Rippin emphasised the need of qur'ānic scholarship to extend its reading scope beyond the trans-

[19] Cf. A. Neuwirth, 'Form and Structure in the Qur'ān,' in *EQ* ii, pp. 245–66.

[20] See D. Marshall, *God, Muhammad and the Unbelievers. A Qur'anic Study* (Richmond, 1999).

[21] The following discussion is a summary of an earlier article, A. Neuwirth, 'Meccan Texts—Medinan Additions? Politics and the Re-reading of Liturgical Communications,' in R. Arnzen and J. Thielmann (eds.), *Words, Texts and Concepts Cruising the Mediterranean Sea. Studies on the Sources, Contents and Influences of Islamic Civilization and Arabic Philosophy and Science. Dedicated to Gerhard Endress on his sixty-fifth birthday* (Leuven, 2004), pp. 71–93.

mitted qurʾānic text, and to take into consideration those religious discourses among the Near Eastern communities which the Qurʾān time and again appears to interact with.[22] Whereas Rippin intends to apply this exegetical principle not to the Qurʾān itself, but to the history of reader response, i.e. *tafsīr*,[23] the present study tries to trace the qurʾānic reflections of encounters between the community and the heirs of the earlier traditions. Assuming that the community in Medina would have encountered adherents of earlier religious traditions, particularly Jews, and have been confronted with their beliefs, not only in the form of closed Scriptures but also—as Madigan has lucidly argued[24]—in oral communications, ritual practices and the like, we may reasonably expect particular rabbinical discourses about Scriptural issues to be reflected in the Qurʾān. In what follows the qurʾānic story of the Golden Calf will be discussed, whose re-reading in a Medinan text is closely related to the imposition of fasting and the theological reflection on the divine attributes of wrath and mercy. In the Jewish tradition, these ideas are closely associated with the Biblical episode of the Golden Calf, a story of extraordinary theological significance that even provides the etiology of one of the highest holidays, the Day of Atonement. It is therefore unsurprising that the community's discovery of the theological implications of the biblical episode in the Jewish tradition triggered a reconsideration of the Biblical story as it had been related earlier at Mecca. In the light of the 'new knowledge' this version seemed to invite a more complex reading.

---

[22] See A. Rippin, 'The Qurʾān as Literature. Perils, Pitfalls and Prospects,' in *British Society for Middle Eastern Studies* 10 (1983), pp. 38–47. For an earlier attempt to sound out the qurʾānic reflections of interactions between the early Islamic community and the representatives of other monotheistic religions focusing common discourses, see A. Neuwirth's discussion of *sūrat al-Raḥmān* and its intertext, Psalm 136, celebrated in Jewish tradition as the 'Major Laudes,' *He-hallel ha-gadol* in: 'Qurʾānic Literary Structure Revisited: *Sūrat al-Raḥmān* between Mythic Account and Decodation of Myth,' in S. Leder (ed.), *Story-telling in the Framework of Non-fictional Arabic Literature* (Wiesbaden, 1998), pp. 388–420.

[23] See the critical evaluations of Rippin's claim that scholars cannot expect to understand the original meaning of the Qurʾān and therefore should study the history of reader response in Marshall, *God, Muhammad and the Unbelievers*, pp. 8–15.

[24] D.A. Madigan, *The Qurʾān's Self-image. Writing and Authority in Islam's Scripture* (Princeton, 2001), pp. 193–213. See the review by G.S. Reynolds in *Al-Abḥāth* 50–51 (2002–2003), pp. 225–8.

2.1 *The Story of the Golden Calf in the Meccan suras*

2.1.1 *The earliest evidence: Q 20:83–99*

The story of Moses as a messenger and leader of his people is told eight times in the Meccan suras,[25] without much attention being paid to the episode of the Golden Calf (*Exodus* 32). The event figures prominently in only two of these texts. It appears for the first time in a rather early Meccan sura, Ṭāhā (Q 20), filling the last part (Q 20:87–99) of a long and detailed vita of Moses that makes up the bulk of the sura (Q 20:10–99). The story as such is a particularly empathetic account of Moses' career that stresses the close relation between the biblical prophet and his God. The story in this early text is still purely narrative; it provides a qur'ānic version of the biblical narrative, giving equal attention to the diverse events in the life of Moses. It refrains completely from adopting any particular theological position that might be deduced from the biblical text.

The account of the Golden Calf in particular (Q 20:87–99)[26] is limited to the rough facts: during Moses' temporary absence from his people, the Israelites were subjected to a divine trial they failed to withstand. At the beginning of the episode God informs Moses about this trial (Q 20:83–5, cf. *Exodus* 32:7). This is a strikingly undramatic start of the narrative, one which also does without any reference to the momentous event of the giving of the tablets (*mattan torah*), a founding event of Judaism, nor mentions the unique location where the encounter takes place. Moses returns to his people and learns about the idol worship that has taken place. The blame for this disastrous sin is, however, not placed on members of the community—Aaron is soon exculpated—but on a figure unknown from the biblical account, al-Sāmirī, 'the Samarian'. The tendency to mitigate the monstrosity of the transgression by introducing external agents as culprits is reminiscent of midrashic renarrations of the story, albeit the figure of the Samarian is not employed

[25] It is related in Q 79:15–26 (early Meccan), Q 37:114–22; 20:10–99; 26:10–67 (middle Meccan), Q 40:21–55; 28:1–46; 10:75–93, 7:103–56 (late Meccan) and Q 2:54ff (Medinan). With the exception of Q 7:103–56 and 2:54ff, these stories have been discussed by the present author in 'Erzählen als kanonischer Prozess. Die Mose-Erzählung im Wandel der koranischen Geschichte,' in R. Brunner et al. (eds.), *Islamstudien ohne Ende. Festschrift für Werner Ende zu seinem 65. Geburtstag* (Wiesbaden, 2002), pp. 323–44.

[26] For the exegetical comments on some obscure details of the story, see Hawting, 'Calf of Gold,' in *EQ* i, pp. 273–6.

there.[27] He is accused of having seduced the people into producing the idol, which, in the shape of a calf, has been venerated in spite of Aaron's warning. Accepting the excuses of his brother, Moses turns his anger on al-Sāmirī, who is—very much in analogy to the fate of the seducer *par excellence*, Iblīs—cursed and expelled. The entire story of Moses, ending in this unexpected scenario of reconciliation instead of retaliation, is concluded by a praise of God's uniqueness (Q 20:98). The story could be classified as a hagiographic account that bears no relation to any particular theological discourse.

It is therefore all the more surprising that theological issues were introduced in a later period, when the story was extended. Because of its particular rhetorical shape the addition can be easily identified as Medinan. In the middle of the story the narrator directly addresses the Banū Isrāʾīl, admonishing them, no longer in an empathetic but rather in a severe tone to avoid God's wrath. The addition (in italics) is placed in the caesura between the very short account of the Exodus (Q 20:77–9) and the story of the Golden Calf (Q 20:83–99).

20:77 And We revealed to Moses (saying), 'Set out at night with My servants and strike for them a dry path in the sea, not fearing to be overtaken nor dreading anything.'[28]	وَلَقَدْ أَوْحَيْنَا إِلَى مُوسَى أَنْ أَسْرِ بِعِبَادِي فَاضْرِبْ لَهُمْ طَرِيقًا فِي الْبَحْرِ يَبَسًا لَا تَخَافُ دَرَكًا وَلَا تَخْشَى
20:78 Then Pharaoh pursued them with his troops and so they were overwhelmed by the water;	فَأَتْبَعَهُمْ فِرْعَوْنُ بِجُنُودِهِ فَغَشِيَهُم مِّنَ الْيَمِّ مَا غَشِيَهُمْ
20:79 Pharaoh thus led his people astray and did not guide them rightly.	وَأَضَلَّ فِرْعَوْنُ قَوْمَهُ وَمَا هَدَى

[27] Speyer, *Die biblischen Erzählungen im Koran*, pp. 229–32, cf. J. Horovitz, *Koranische Untersuchungen* (Berlin, 1926), p. 144f.

[28] The translations are those of M. Fakhry, *An Interpretation of the Qur'an. English Translation of the Meanings. A Bilingual Edition* (New York, 2004). Fakhry's use of the divine name 'Allāh' has been changed into the more universal 'God'. Further modifications occasionally deemed necessary have been marked as such.

20:80 'O Children of Israel! We have delivered you from your enemy and We made a covenant with you on the right side of the Mount, and sent down to you the manna and quails.	يَا بَنِي إِسْرَائِيلَ قَدْ أَنْجَيْنَاكُمْ مِنْ عَدُوِّكُمْ وَوَاعَدْنَاكُمْ جَانِبَ الطُّورِ الأَيْمَنَ وَنَزَّلْنَا عَلَيْكُمُ الْمَنَّ وَالسَّلْوَى
20:81 Eat of the good things We have provided you with, but do not be excessive therein, lest My wrath descend upon you.' Those upon whom My Wrath descends shall be ruined.	كُلُوا مِنْ طَيِّبَاتِ مَا رَزَقْنَاكُمْ وَلا تَطْغَوْا فِيهِ فَيَحِلَّ عَلَيْكُمْ غَضَبِي وَمَنْ يَحْلِلْ عَلَيْهِ غَضَبِي فَقَدْ هَوَى
20:82 I am, indeed, All-Forgiving unto him who repents, does the righteous deed and is well-guided.'	وَإِنِّي لَغَفَّارٌ لِّمَن تَابَ وَآمَنَ وَعَمِلَ صَالِحًا ثُمَّ اهْتَدَى
20:83 'What has led you to go ahead of your people, O Moses?'	وَمَا أَعْجَلَكَ عَن قَوْمِكَ يَا مُوسَى

Verses 20:80–82 shift the speech temporarily from a report to a direct address targeting particular addressees or at least virtual listeners to the Prophet. Although such appeals to listeners are frequent in qurʾānic narrative, this case is special insofar as the group addressed is specified. They are the Medinan Jews, who had not yet been part of the Prophet's listeners when the story was first composed at Mecca, but entered the stage only at a later date, thus provoking an adaptation of the text to fit the new situation. Although the addition is directly connected with a reminder of episodes previously neglected in the plain narrative—God's covenant and His providing the Israelites with heavenly food—its exhortative comment refers to an event yet to be accounted: the sin of the Golden Calf. The extension—using the issue of food as a vantage point for an admonition not to exaggerate (possibly in the respecting of dietary laws)—culminates in a threat of divine wrath that can only be avoided through repentance, belief and good deeds. A closer look, however, reveals that verses Q 20:82b–83 allude to a particularly momentous Scriptural verse, one which is the *locus classicus* of the idea of guilt and atonement in Judaism and belongs to the immediate context of the biblical Golden Calf story. The image of a both wrathful and forgiving God becomes most expressively apparent in a divine self-description revealed to Moses as he was given the new set of tablets.

It contains what Jewish tradition has labeled the 'Thirteen Attributes', *shelosh esre ha-middot*, all connected with wrath or mercy and which were to figure prominently in Jewish liturgy from early times onwards,[29] see *Exodus* 34:6–7.

> The Lord, the Lord, God, merciful and gracious, longsuffering and abundant in goodness and truth, keeping mercy for thousands, forgiving sin and transgression, seeking the iniquity of the fathers on the children and upon the children's children until the third and fourth generation.

In the qur'ānic allusion to this text, which focuses on the notions of forgiveness (*ghafūr* Q 20:82) and wrathfulness (*ghaḍab*,[30] Q 20:81b, c), the biblical threat of a divine retaliation over generations is alleviated; yet the menace prevails and it can only be averted by repentance (*man tāba* Q 20:82). The insertion as such is unique in sura 20; its particular significance will become more evident in the following discussion.

2.1.2 *The later account: Q 7:142–56*

A more extensive account of the episode of the Golden Calf is presented in the late Meccan sura *al-Aʿrāf* (Q 7:142–56). This story differs from the earlier version in that it focuses on Moses' encounter with God on the mountain, cf. *Exodus* 24:1. The text then switches to a later development, Moses' desire to see God. Though Moses is denied the vision of the ensuing theophany, he is given, in the qur'ānic text, a demonstration of God's power, which surpasses human perception, Q 7:143. In the biblical version he is compensated with the communication of the Thirteen Attributes, *Exodus* 34:6–7, not explicitly quoted in the Qur'ān. The qur'ānic story then turns to the Golden Calf episode, which—unlike the version in Q 20—focuses on Moses' complex function in achieving a divine-human reconciliation. Attention

[29] The significance of the Thirteen Attributes for the act of atonement in Judaism is expounded in *Babylonian Talmud Rosh ha-Shana* 17b: It is written: 'And the Lord passed by before him and proclaimed'. R. Johanan said: 'Had this Passage not been written, it would have been impossible to have said it, for it teaches us that the Holy One, blessed be He, wrapped Himself, as does a minister who recites the prayers for a congregation, and pointing out to Moses the regular order of prayer, said to him: "Whenever Israel sins, let him pray to Me, after this manner, and I shall pardon him".' (kind communication by Dirk Hartwig).

[30] The attitude of *ghaḍab*, 'wrath', as ascribed to God is not found in Meccan contexts, except in Q 20:81b,c and 20:86, with the same phrase *(fa/an) yaḥilla ʿalaykum ghaḍabī/ghaḍabun min rabbikum)*. Since Q 20:86 is overlong, the phrase is very likely to have been introduced at a later stage with the other additions that focus on *ghaḍab*.

is given to both the first and second granting of the tablets. But the story is not only more self-reflective, its mood, too, is different from the previous account: whereas in the first version perfect harmony prevailed between Moses and God, in this text there is dissent. Moses' desire to see God is met with disapproval and has to be redeemed by an act of repentance. Further divine discontent is expressed in Q 7:145–7, where God, who had first spoken to Moses, abruptly turns to Muhammad, admonishing him to respect the commands contained in the tablets given to Moses, and threatens those obstinate skeptics among his listeners who have 'always been rejecting the divine signs'. The polemical tone is further underscored by a shift of the responsibility for making the calf from al-Sāmirī, who no longer figures in this version, to the people themselves (Q 7:149f) and to some degree to Aaron, who had represented Moses (Q 7:151).[31] Even though the main strain of the story is in accordance with late Meccan storytelling, which focuses on the trials endured in prophetic experience, the paraenetic addresses to the Prophet and the menacing response to the Jews' refusal to accept the new message, clearly point to a Medinan *Sitz im Leben* of the revised version.

There is, furthermore, another layer of confrontation, which, similar to that observed in the addition to sura 20, manifests itself in explicit references to divine wrath and mercy as well as human repentance. From a narrative viewpoint, these references appear again as interruptions, translating the situation of the historical story into the time of the Prophet's recitation, not however the recitation in Mecca, but its later resumption in Medina. Thus, verses Q 7:152–3 should be considered as a later addition for formal reasons.[32] They present a comment on the sin of the Golden Calf, predicting divine wrath to those who did not repent, while promising mercy to those who had. The concept of a lasting guilt going back as far as the transgression of the Golden Calf that occurred in Biblical times attests to a new interpretation of the story. Also part of this exegetical annex to the story is the concluding verse Q 7:157. It is less easy to determine the localization of Q 7:155–6, which contain Moses' prayer for mercy on behalf of his people as well as God's response, insisting on His absolute freedom to show wrath or mercy

[31] See for the entire account Speyer, *Die biblischen Erzählungen im Koran*, pp. 329–32.

[32] They are, again, not listed among the Medinan additions collected by Tilman Nagel, 'Medinensische Einschübe in Mekkanischen Suren. Ein Arbeitsbericht,' in Wild (ed.), *The Qur'ān as Text*, pp. 59–68. See also idem, *Medinensische Einschübe in Mekkanischen Suren* (Göttingen, 1995).

to whomever He wills, obviously an echo of the Thirteen Attributes. Since the three verses Q 7:155–7 are syntactically closely connected, they may—in spite of their close association with the Golden Calf story[33]—be equally considered part of the extended Medinan annex, in which various incidents of Moses' career are recollected which are no longer related to the Golden Calf episode, which begins with 7:157. Whereas the narrative as such—consisting of Q 7:142–5, 148–51, and 154—ends with Moses taking upon himself the Law and accepting the second set of the Tablets (Q 7:154), the extended version, addressed to an extended public of both believers and skeptical Jewish listeners, ends with the appeal to follow the 'gentile' prophet (Q 7:157):

7:142 And We appointed to Moses thirty nights to which We added ten, and thus the term appointed by his Lord was forty nights. And Moses said to his brother Aaron: 'Succeed me at the head of my people, set matters right and do not follow the path of the mischief-makers.'	وَوَاعَدْنَا مُوسَى ثَلاثِينَ لَيْلَةً وَأَتْمَمْنَاهَا بِعَشْرٍ فَتَمَّ مِيقَاتُ رَبِّهِ أَرْبَعِينَ لَيْلَةً وَقَالَ مُوسَى لأَخِيهِ هَارُونَ اخْلُفْنِي فِي قَوْمِي وَأَصْلِحْ وَلا تَتَّبِعْ سَبِيلَ الْمُفْسِدِينَ
7:143 And when Moses came on Our appointed time and his Lord spoke to him, he said: 'Lord, show me [Yourself] so that I may look at You.' He said: 'You will not see Me; but look at the mountain. If it stays in its place, you shall see Me.' But when his Lord revealed Himself to the Mountain, He leveled it to the ground, and Moses fell down unconscious. When he woke up,	وَلَمَّا جَاءَ مُوسَى لِمِيقَاتِنَا وَكَلَّمَهُ رَبُّهُ قَالَ رَبِّ أَرِنِي أَنْظُرْ إِلَيْكَ قَالَ لَنْ تَرَانِي وَلَكِنِ انظُرْ إِلَى الْجَبَلِ فَإِنِ اسْتَقَرَّ مَكَانَهُ فَسَوْفَ تَرَانِي فَلَمَّا تَجَلَّى رَبُّهُ لِلْجَبَلِ جَعَلَهُ دَكًّا وَخَرَّ مُوسَى صَعِقًا فَلَمَّا أَفَاقَ قَالَ سُبْحَانَكَ تُبْتُ إِلَيْكَ وَأَنَا أَوَّلُ الْمُؤْمِنِينَ

[33] Though the beginning of Q 7:155 seems to relate to the episode of the punishment of the horde of Koraḥ, Moses' bargaining with God to spare the Israelites a most severe punishment, is part of the events following the disaster of the Golden Calf, see *Exodus* 32:31–2.

he said: 'Glory be to You, I repent unto You and I am the first of the believers.'

7:144 Then He said: 'O Moses, I have chosen you above all men for My Messages and My words; so take what I have given you and be one of the thankful.'

قَالَ يَا مُوسَى إِنِّي اصْطَفَيْتُكَ عَلَى النَّاسِ بِرِسَالاَتِي وَبِكَلاَمِي فَخُذْ مَا آتَيْتُكَ وَكُن مِّنَ الشَّاكِرِينَ

7:145 And We wrote for him on the Tablets about everything, providing exhortation and a clear exposition of everything. So take it resolutely and bid your people to take the best part of it. I will show you the abode of the wicked.

وَكَتَبْنَا لَهُ فِي الأَلْوَاحِ مِن كُلِّ شَيْءٍ مَّوْعِظَةً وَتَفْصِيلاً لِّكُلِّ شَيْءٍ فَخُذْهَا بِقُوَّةٍ وَأْمُرْ قَوْمَكَ يَأْخُذُواْ بِأَحْسَنِهَا سَأُرِيكُمْ دَارَ الْفَاسِقِينَ

7:146 'I will turn away from My Signs those who are unjustifiably arrogant on earth, and who, if they see each sign, will not believe in it; and if they see the path of rectitude will not follow it, but if they see the path of error, will follow it as their path. That is because they denied the truth of Our Signs and failed to pay heed.

سَأَصْرِفُ عَنْ آيَاتِيَ الَّذِينَ يَتَكَبَّرُونَ فِي الأَرْضِ بِغَيْرِ الْحَقِّ وَإِن يَرَوْاْ كُلَّ آيَةٍ لاَّ يُؤْمِنُواْ بِهَا وَإِن يَرَوْاْ سَبِيلَ الرُّشْدِ لاَ يَتَّخِذُوهُ سَبِيلاً وَإِن يَرَوْاْ سَبِيلَ الْغَيِّ يَتَّخِذُوهُ سَبِيلاً ذَلِكَ بِأَنَّهُمْ كَذَّبُواْ بِآيَاتِنَا وَكَانُواْ عَنْهَا غَافِلِينَ

7:147 'Those who deny Our Signs and the Meeting of the Hereafter—their works are nullified. Will they be rewarded except according to what they do?'

وَالَّذِينَ كَذَّبُواْ بِآيَاتِنَا وَلِقَاء الآخِرَةِ حَبِطَتْ أَعْمَالُهُمْ هَلْ يُجْزَوْنَ إِلاَّ مَا كَانُواْ يَعْمَلُونَ

7:148 And the people of Moses took, after he went away, a calf made of their jewellery—a mere body which lowed. Did they not

وَاتَّخَذَ قَوْمُ مُوسَى مِن بَعْدِهِ مِنْ حُلِيِّهِمْ عِجْلاً جَسَدًا لَّهُ خُوَارٌ أَلَمْ يَرَوْاْ أَنَّهُ لاَ يُكَلِّمُهُمْ وَلاَ يَهْدِيهِمْ سَبِيلاً اتَّخَذُوهُ وَكَانُواْ ظَالِمِينَ

see that it did not speak to them and could not guide them to any path? They took it [for worship] and were wrongdoers.

7:149 And when they felt deep regret, and saw that they had gone astray, they said: 'If our Lord will not have mercy on us and will not forgive us, we will certainly be among the losers.'

وَلَمَّا سُقِطَ فِي أَيْدِيهِمْ وَرَأَوْاْ أَنَّهُمْ قَدْ ضَلُّواْ
قَالُواْ لَئِن لَّمْ يَرْحَمْنَا رَبُّنَا وَيَغْفِرْ لَنَا لَنَكُونَنَّ
مِنَ الْخَاسِرِينَ

7:150 And when Moses returned to his people, angry and very sad, he said: 'Evil is what you did following my departure. Were you in a hurry regarding your Lord's Commandment?' He cast the Tablets down and took hold of his brother's head, dragging him towards him. He said: 'Son of my mother, the people deemed me weak and were about to kill me, so do not let the enemies rejoice at my plight, and do not reckon me one of the unjust people.'

وَلَمَّا رَجَعَ مُوسَى إِلَى قَوْمِهِ غَضْبَانَ أَسِفًا
قَالَ بِئْسَمَا خَلَفْتُمُونِي مِن بَعْدِي أَعَجِلْتُمْ أَمْرَ
رَبِّكُمْ وَأَلْقَى الأَلْوَاحَ وَأَخَذَ بِرَأْسِ أَخِيهِ يَجُرُّهُ
إِلَيْهِ قَالَ ابْنَ أُمَّ إِنَّ الْقَوْمَ اسْتَضْعَفُونِي وَكَادُواْ
يَقْتُلُونَنِي فَلاَ تُشْمِتْ بِيَ الأَعْدَاء وَلاَ تَجْعَلْنِي
مَعَ الْقَوْمِ الظَّالِمِينَ

7:151 He said: 'Lord, forgive me and my brother and admit us into Your Mercy; for you are the Most Merciful of all.'

قَالَ رَبِّ اغْفِرْ لِي وَلأَخِي وَأَدْخِلْنَا فِي
رَحْمَتِكَ وَأَنتَ أَرْحَمُ الرَّاحِمِينَ

7:152 surely, those who worshipped the calf will be visited by wrath from their Lord and humiliation in the present life; and thus We recompense those who invent lies.

إِنَّ الَّذِينَ اتَّخَذُواْ الْعِجْلَ سَيَنَالُهُمْ غَضَبٌ مِّن
رَّبِّهِمْ وَذِلَّةٌ فِي الْحَيَاةِ الدُّنْيَا وَكَذَلِكَ نَجْزِي
الْمُفْتَرِينَ

7:153 And those who perpetrated the evil deeds and repented thereafter and believed—surely your Lord is thereafter All-Forgiving and Merciful.

وَالَّذِينَ عَمِلُواْ السَّيِّئَاتِ ثُمَّ تَابُواْ مِن بَعْدِهَا
وَآمَنُواْ إِنَّ رَبَّكَ مِن بَعْدِهَا لَغَفُورٌ رَّحِيمٌ

7:154 And when the anger of Moses abated, he took up the Tablets, in the text of which are guidance and mercy for those who fear their Lord.

وَلَمَّا سَكَتَ عَن مُّوسَى الْغَضَبُ أَخَذَ الأَلْوَاحَ
وَفِي نُسْخَتِهَا هُدًى وَرَحْمَةٌ لِّلَّذِينَ هُمْ لِرَبِّهِمْ
يَرْهَبُونَ

7:155 And Moses chose from his people seventy men for Our appointed time. When the earthquake overtook them, he said: 'Lord, had You pleased, You would have destroyed them, together with me, before this time. Will You destroy us for what the fools among us have done? This is only Your Trial, with which You lead astray whomever You will and guide whomever You will. You are our Protector; so forgive us and have mercy upon us. You are the Best Forgiver of all.

وَاخْتَارَ مُوسَى قَوْمَهُ سَبْعِينَ رَجُلاً لِّمِيقَاتِنَا
فَلَمَّا أَخَذَتْهُمُ الرَّجْفَةُ قَالَ رَبِّ لَوْ شِئْتَ
أَهْلَكْتَهُم مِّن قَبْلُ وَإِيَّايَ أَتُهْلِكُنَا بِمَا فَعَلَ
السُّفَهَاء مِنَّا إِنْ هِيَ إِلاَّ فِتْنَتُكَ تُضِلُّ بِهَا مَن
تَشَاء وَتَهْدِي مَن تَشَاء أَنتَ وَلِيُّنَا فَاغْفِرْ
لَنَا وَارْحَمْنَا وَأَنتَ خَيْرُ الْغَافِرِينَ

7:156 'And ordain for us good in this world and in the Hereafter; we turn repentant to You.' He said: 'I smite with My punishment whomsoever I please, and My Mercy encompasses all things; and I will ordain it to those who are pious and give the alms, and to those who believe in Our Signs.

وَاكْتُبْ لَنَا فِي هَـذِهِ الدُّنْيَا حَسَنَةً وَفِي
الآخِرَةِ إِنَّا هُدْنَـا إِلَيْكَ قَالَ عَذَابِي أُصِيبُ
بِهِ مَنْ أَشَاء وَرَحْمَتِي وَسِعَتْ كُلَّ شَيْءٍ
فَسَأَكْتُبُهَا لِلَّذِينَ يَتَّقُونَ وَيُؤْتُونَ الزَّكَـاةَ وَالَّذِينَ
هُم بِآيَاتِنَا يُؤْمِنُونَ

7:157 'And to those who follow the Messenger, the unlettered Prophet whom they find mentioned in their Torah and

الَّذِينَ يَتَّبِعُونَ الرَّسُولَ النَّبِيَّ الأُمِّيَّ الَّذِي
يَجِدُونَهُ مَكْتُوبًا عِندَهُمْ فِي التَّوْرَاةِ وَالإِنْجِيلِ
يَأْمُرُهُم بِالْمَعْرُوفِ وَيَنْهَاهُمْ عَنِ الْمُنكَرِ

Gospel. He enjoins them to do good and to forbid evil, and makes lawful to them the good things and unlawful the impure things. And He relieves them of their heavy burden and the shackles that were upon them. Thus those who believe in him and who honour and support him and follow the light which has been sent down with him—those are the prosperous.'	وَيُحِلُّ لَهُمُ الطَّيِّبَاتِ وَيُحَرِّمُ عَلَيْهِمُ الْخَبَآئِثَ وَيَضَعُ عَنْهُمْ إِصْرَهُمْ وَالأَغْلاَلَ الَّتِي كَانَتْ عَلَيْهِمْ فَالَّذِينَ آمَنُواْ بِهِ وَعَزَّرُوهُ وَنَصَرُوهُ وَاتَّبَعُواْ النُّورَ الَّذِيَ أُنزِلَ مَعَهُ أُوْلَـئِكَ هُمُ الْمُفْلِحُونَ

Verses 7:152–3 and 7:155–7 introduce into the story a discursive dimension very much like the additions to sura 20. They focus on wrath (*ghaḍab* Q 7:152, *ʿadhāb* Q 7:156) as well as mercy and forgiveness (*raḥma* Q 7:154, 155, 156, *ghufrān* Q 7:153, 155), the self-denying role of Moses in obtaining God's mercy (Q 7:151), and—on the human side—repentance (*tawba* Q 7:153). It resembles the discourse of atonement so intimately connected with the biblical story of the Golden Calf, which, however, with its underscoring of the still prevailing divine wrath is alien to the corresponding Meccan qurʾānic narrative. We may thus assume that verses Q 7:145–7, 7:152–3 and 7:157 (and presumably also 7:155–6) belong into a Medinan context, where the discourse of atonement enshrined in the notions of wrath/mercy and repentance surfaced in the debates between the earlier and the later 'heirs' of the biblical tradition. Even though these verses do not address the Medinan Jews directly, the unambiguous appeal to them in Q 7:157, which promises them divine mercy if they follow the Prophet (*fa-lladhīna āmanū bihi wa-ʿazzarūhu wa-naṣarūhu wa-ttabaʿū ʾl-nūra lladhī unzila maʿahu ulāʾika humu ʾl-mufliḥūn*) suggests that the verses Q 7:145–7 and 7:152–3 are also addressed to that group in particular. The notion of unpredictable divine wrath and mercy in Q 7:153 and the penitential attitude in Q 7:156 once more recall the biblical verse on God's 'Thirteen attributes' (*Exodus* 34:6), a reflection of which we have already encountered in the additions to the earlier account of sura 20. The narrative of the Golden Calf in this text appears to have undergone a revision—communicated in a modified recitation before a more diversified Medinan public—, one that integrates the discourse of atonement, which had not yet been significant in the Meccan story.

2.2 *The public in the text*

It is worthwhile to reflect briefly on this public. The presence or at least virtual presence of listeners educated in the Jewish tradition, who had still been absent during the period when the two versions of the story were first presented, looms through the additions. The new readings of the story focus on the idea of God's wrath that has to be calmed by repentance, projecting it on the Jews in particular. The idea of human guilt and divine forgiveness as such (to be attained after the perpetrators' repentance), that is familiar already from early sections of the Qurʾān, figures prominently in both of suras 20 and 7. It is exemplified in both texts by an identical narrative context: the story of Adam's disobedience and his immediate reconciliation with God. The drama of humans sinning against God and being reconciled again, is presented as a universal drama in the earlier, Meccan, versions of both Moses stories. But it is only in Medina, that God's wrath is adduced. The notion of divine wrath in these texts serves exclusively to threaten the Jews among the listeners, whose own traditions—which had been regarded until then as universal monotheist knowledge—are now turned against them. To charge the story of the Golden Calf with a dimension of not yet wiped out guilt is a new and fateful step in the community's interaction with Jewish listeners. The idea seems to have been inspired by discussions about the Thirteen Attributes, a biblical text that owes its significance to its particular *Sitz im Leben* in Jewish religious life.

3. Christianity as a Counter-Paradigm: the Christian vs. the Jewish Cultural Language

At approximately the same time—the texts discussed above and those that are to follow in this article belong to the early Medinan developments—Christian self-expression manifested itself prominently in the qurʾānic debates. One particularly striking case of the use of Christian cultural language in response to a Jewish challenge will be presented in what follows.[34] In an earlier study by the present author on the images of Mary and Jesus in *sūrat Maryam* and related Meccan texts, the Meccan testimonies of Mary and Jesus had been discussed extensively.[35]

---

[34] See for a more extensive version of this discussion, A. Neuwirth, 'Mary and Jesus—Counterbalancing the Biblical Patriarchs. A Re-reading of *sūrat Maryam* in *sūrat Āl ʿImrān* (3:1–62),' in *Parole de l'Orient* 30 (2005), pp. 231–60.

[35] A. Neuwirth, 'Imaging Mary—Disputing Jesus. The Images of Mary and Jesus in

In contrast to this earlier analysis the present study[36] portrays a political development. It attempts to demonstrate the substantially new reading of the Christian stories during the Prophet's Medinan activities, when earlier texts were frequently remodeled to fit the polemical and apologetic needs of the emerging community, who by then were challenged by learned representatives of the older monotheistic traditions. The fate of *sūrat Maryam* is no exception to the rule. It was subjected to a re-reading that served a 'political' purpose: to disempower the predominant Jewish tradition represented by Āl Ibrāhīm, whose weighty superiority in terms of scriptural authority had to be counter-balanced by a non-patriarchal tradition, that of the Āl ʿImrān.

3.1. *The narrative (Q 3:34–62)*

The third sura of the Qurʾān, Āl ʿImrān, a Medinan text, documents a revision of the stories of Mary and Jesus[37] in *sūrat Maryam* (Q 19:1–33),[38] presenting a later re-reading of the earlier text which reflects new perspectives.

The following observations will go beyond a narrative analysis,[39] trying to trace the *Sitz im Leben* of the text, i.e. its sociopolitical function within the historical development of the genesis of the Qurʾān. For this

sūrat Maryam and Related Meccan Texts' that will appear in J. Butler and T. Hoffmann (eds.), *Qurʾānic Studies. Literary and Hermeneutical Dimensions* (with a comprehensive bibliography). For the specifically Medinan portrayal of Christians, see D. Marshall, 'Christianity in the Qur'an,' in L. Ridgeon (ed.), *Islamic Interpretations of Christianity* (Richmond, 2001), pp. 3–29. This work, though differentiating between Meccan and Medinan texts, is interested in the interaction with or the image of real Christians in the Qurʾān rather than the perusal of Christian tradition in apologetic-polemical qurʾānic texts.

[36] It is a résumé of a more extensive study, A. Neuwirth, 'Icon of Piety—Trigger of Controversity. Re-reading the Meccan Stories of Mary and Jesus in Q 3:1–62,' to appear as a contribution to the Festschrift for Hartmut Bobzin.

[37] See for Jesus in the Qurʾān: N. Robinson, 'Jesus,' in *EQ* iii, pp. 7–21, and idem, *Discovering the Qur'an. A Contemporary Approach to a Veiled Text* (London, 1996); J. Bouman, *Das Wort vom Kreuz und das Bekenntnis zu Allah* (Frankfurt, 1980); G. Parrinder, *Jesus in the Qur'an* (London, 1965). None of these authors, however, takes interest in the qurʾānic texts as re-reading of earlier texts, though Robinson proceeds chronologically.

[38] See for the image of Mary in the Qurʾān: B. Freyer-Stowasser, 'Mary,' in *EQ* iii, p. 288–96. The article is not interested, however, in historical developments within the Qurʾān and, again, does not present the stories as readings of earlier intra- and extra-qurʾānic texts. Theological dimensions have been uncovered by A. Neuwirth, 'The House of Abraham and the House of Amram: Genealogy, Patriarchal Authority, and Exegetical Professionalism,' in A. Neuwirth et al. (eds), *The Qurʾān in Context. Textual Palimpsests and Historical Entanglements* (forthcoming) and by M. Marx, 'Glimpses of Mariology in the Qurʾān,' ibid.

[39] A.H. Mathias Zahniser, 'The Word of God and the Apostleship of ʿĪsā: a Narrative Analysis of Āl ʿImrān (3):33–62,' *JSS* 36 (1991), pp. 77–112.

purpose we will discuss not only the story of the family of Mary and Jesus, referred to as Āl ʿImrān[40] (Q 3:33–62), hence the sura's name, but also the introductory section (Q 3:1–32). The text which combines an address to and a report about the Medinan Jews,[41] gives evidence of a changed attitude towards the hitherto unquestioned rank of the Israelites (Āl Ibrāhīm) as the sole elects and transmitters of Scripture. The argument in Q 3:1–33 suggests that the intention behind the revised account of the story of Mary and Jesus was to establish a counter-tradition to the predominant Jewish patriarchal tradition. The story had already been presented in *sūrat Maryam* in an artistically highly sophisticated way which would not have required a 'reproduction' if not for religious-political reasons. In the new version the story receives in *sūrat Āl ʿImrān*, the sole female protagonist of *sūrat Maryam* is further empowered by a second, more active figure, her mother, who provides her with a—matrilineary—genealogy of her own. Furthermore, Jesus is presented for the first time as an apostle, who somehow reverses the public self-representations of earlier prophets and reveals explicit non-patriarchal traits. Together with her mother and her son, Mary constitutes the Āl ʿImrān, the Family of Amram, that is for the first time and uniquely in the Qurʾān presented in sura 3 as equal in rank with the patriarchal family of Abraham.

	3:33
3:33 God chose Adam, Noah,	إِنَّ اللَّهَ اصْطَفَى آدَمَ وَنُوحًا
the family of Abraham	وَآلَ إِبْرَاهِيمَ وَآلَ عِمْرَانَ
and the family of ʿImrān	عَلَى الْعَالَمِينَ
above all mankind,	

[40] See R. Tottoli, "Imrān,' in *EQ* ii, p. 509. Tottoli regards the attribution of Mary and Jesus to the family of ʿImrān as a confusion due to 'a Christian tendency to utilize earlier Biblical figures as "types" for later ones.' Indeed, the Qurʾān—being a re-reading of both Biblical and post-biblical lore—reflects exegetical processes that have reshaped the Biblical texts. The presentation of Biblical memory enriched by exegetical tradition is to be considered as a characteristic of the Qurʾān rather than the result of a confusion.

[41] For the origin of the Medinan Jews, see M. Gil, 'The Origin of the Jews of Yathrib,' *JSAI* 4 (1984), pp. 203–24; W.M. Watt, *Muhammad at Medina* (Oxford, 1956), ch. iv: 'Muḥammad and the Jews,' pp. 192–220.

	3:34
3:34 Descending one from another; above all mankind,	ذُرِّيَّةً بَعْضُهَا مِن بَعْضٍ
God is the All-Hearing, All-Knowing!	وَاللّٰهُ سَمِيعٌ عَلِيمٌ

The story of Mary's unnamed mother who vows to give her child to the service of God (following the narrative of the Protevangelium of James) is strikingly explicit in its gender-specific physical details, talking overtly about the female womb and child delivery (*fī baṭnī, waḍaʿtu, waḍaʿat, unthā*). Having born a daughter instead of the expected son the woman is still determined to fulfill her vow. She herself names the child Maryam/Mary—the role of the father being unoccupied in the story—and hands her over to the temple, asking God for guardianship for the child and her future offspring. Thus, contrary to the scenario in the Protevangelium, in the qurʾānic story no male human protagonist is involved, if we disregard for a moment the parallel story of Zechariah, who, at any rate does not play an active role in Mary's context.— The ensuing story of a male figure, Mary's son Jesus, does not counterbalance the female dominance, since his submissive appearance in public, failing to exert any patriarchal authority, is more a reversal than a recapture of the public self-representations of earlier male prophets. The Āl ʿImrān family, renouncing patriarchal power, appears as a counter-model to the firmly established patriarchal family of Abraham, the Āl Ibrāhīm.

3.2. *The prologue (Q 3:1–33) and the hermeneutical locus classicus: Q 3:7–8*

The immediate context of the verses under discussion, the prologue to these stories, starts out evoking the main reference of the sura, the Gospel,[42] *al-injīl*, a Scripture otherwise nowhere mentioned in introductory sections:[43]

[42] See S.H. Griffith, 'Gospel,' in *EQ* ii, p. 342f. Griffith does not attempt to locate the Gospel as part of the qurʾānic discourse about revelation within the process of the genesis of the Qurʾān.

[43] The Gospel, *injīl*, which in the Qurʾān denotes one single Scripture—not a multiplicity of writings—is mentioned twelve times in the Qurʾān, exclusively in Medinan verses, with the exception of our text always within qurʾānic debates, never, except in sura 3, in the programmatic initial verses of a sura. *Injīl* is most frequently contextualized with *tawrāt* (Q 3:3, 3:18, 3:74, 5:66, 5:68, 5:110, 7:57); once it appears with *tawrāt* and *qurʾān*: Q 9:111.

3:3 He has revealed the Book to you in truth confirming what came before it; and he has revealed the Torah and the Gospel.

نَزَّلَ عَلَيْكَ الْكِتَابَ بِالْحَقِّ مُصَدِّقاً لِّمَا بَيْنَ
يَدَيْهِ وَأَنزَلَ التَّوْرَاةَ وَالإِنجِيل

It continues as a seemingly conventional debate about revelation and prophethood and their acceptance by diverse groups, until with verse 7, the Qurʾān's so-called hermeneutical *locus classicus* is evoked:

3:7 It is He Who has revealed to you the Book with verses which are precise in meaning and which are the Mother of the Book and others which are ambiguous. As to those in whose hearts there is swerving, they follow what is ambiguous in it, desiring sedition (seduction) and desiring to interpret it. However no one except God knows its interpretation. Those well-grounded in knowledge say: 'We believe in it: all is from our Lord'; yet none remembers save those possessed of understanding![44]

هُوَ الَّذِيَ أَنزَلَ عَلَيْكَ الْكِتَابَ مِنْهُ آيَاتٌ
مُّحْكَمَاتٌ هُنَّ أُمُّ الْكِتَابِ وَأُخَرُ مُتَشَابِهَاتٌ
فَأَمَّا الَّذِينَ فِي قُلُوبِهِمْ زَيْغٌ فَيَتَّبِعُونَ مَا تَشَابَهَ
مِنْهُ ابْتِغَاءَ الْفِتْنَةِ وَابْتِغَاءَ تَأْوِيلِهِ وَمَا يَعْلَمُ
تَأْوِيلَهُ إِلاَّ اللّهُ وَالرَّاسِخُونَ فِي الْعِلْمِ يَقُولُونَ
آمَنَّا بِهِ كُلٌّ مِّنْ عِندِ رَبِّنَا وَمَا يَذَّكَّرُ إِلاَّ أُوْلُواْ
الألْبَابِ

3:8 Lord, do not cause our hearts to swerve after you have guided us and grant us your mercy. You are indeed the munificent Giver.

رَبَّنَا لاَ تُزِغْ قُلُوبَنَا بَعْدَ إِذْ هَدَيْتَنَا وَهَبْ
لَنَا مِن لَّدُنكَ رَحْمَةً إِنَّكَ أَنتَ الْوَهَّابُ

Verse 3:7 is a *crux interpretum*, that has inspired numerous studies,[45] since the expression *al-rāsikhūna fī ʾl-ʿilm*, 'those firmly rooted in knowledge',

[44] The verse may also be read: *However no one except God and the well-grounded in knowledge* know its interpretation. They say: 'We believe in it…'

[45] See C. Gilliot, 'Exegesis of the Qur'an, Classical and Medieval,' in *EQ* ii, pp. 99–124, particularly pp. 99–100. See also J. McAuliffe, 'Text and Textuality. Q 3:7 as a Point of Intersection,' in I. Boullata (ed.), *Literary Structures of Religious Meaning in the Qurʾān* (Richmond, 2000), pp. 56–76, discusses the inner-Islamic exegetical positions.

can be construed both as the end of the sentence preceding it, and as the beginning of the one following it, thus making the prerogative of exegesis either be reserved to God or shared by the learned. In view of the fact that desire of interpretation, *taʾwīl*, of the ambiguous passages is equated with desire of dissension, the first option seems to convey the originally intended meaning: only God is entitled to interpret.—But is it really, as current scholarship suggests, the prerogative of interpretation that is at stake here? Or is it not rather the nature of revealed texts that is reconsidered in Q 3:7?

3.2.1 *Qurʾānic scripturality and situatedness*

The concession of a hermeneutic ambiguity in Scripture[46] comes as a surprise considering the numerous previous qurʾānic self-declarations as emanating from a particularly clear (*mubīn*) text (cf. Q 26:2 *tilka ayātu ʾl-kitābi ʾl-mubīn*, 'those are the signs of the clear scripture', and often). Why should there be verses that are ambiguous? The problem remains unsolved as long as the qurʾānic situation in an ongoing debate with adherents of the older religions is ignored. Q 3:7 certainly is not a theoretical statement, but its harsh rebuke of the exegesis of certain people seems to respond to existing hermeneutic practices. Intertexts, indeed, loom behind the verses: scriptural text units admitting more than one interpretation are an acknowledged reality in the Jewish reading of the Bible, and their identification as such is part of exegetical practice. Jewish tradition since the Tannaitic period, distinguishes different 'faces', *panim*, of the Torah.[47] This exegetical perception of scriptural texts as

See for the hermeneutic implications of the verse Madigan, *The Qurʾān's Self-image*, and in response: U. Rubin, review of Madigan, in *JSAI* 28 (2003), p. 381–6. See further S. Wild, 'The Self-referentiality of the Qur'an. Sura 3:7 as an Exegetical Challenge,' in J. Dammen McAuliffe, B.D. Walfish, J.W. Goering (eds.), *With Reverence for the Word. Medieval Scriptural Exegesis in Judaism, Christianity, and Islam* (Oxford, 2003), pp. 422–36.

[46] The admission of ambiguity in the text may be related to the concept of the 'faces' of Scripture, *panim shel ha-torah*, that according to Jewish hermeneutics coexist beside each other. Early qurʾānic exegesis refers to the *mutashābihāt al-Qurʾān* as *wujūh al-Qurʾān*. The verse would then be the outcome of a process of differentiating textual qualities that was encouraged through the exchange with the older religions.

[47] There is an early genre of commentary covering the polysemy caused by lexemes that admit more than one meaning, called *mutashābih al-Qurʾān* and *al-wujūh* (= *panim*) *wa-ʾl-naẓāʾir* respectively. In my view, the concept of a positively connotated ambiguousness of Scriptural verses, so prominent in Jewish tradition with its concept of *panim shel ha-torah*, did not, as Gilliot holds, originate only in early exegesis but seems to be reflected already in the Qurʾān itself.

being *per se* liable to more than one understanding, being ambiguous, *mutashābih*, seems to be reflected already in our qurʾānic section. Furthermore, the term *taʾwīl* points to professional exegetical expertise: *taʾwīl* etymologically sounds like an Arabic rendering of *reductio ad primum*, obviously a technique practiced in the circles of learned scriptural exegetes of that time. Equally, the two terms *mutashābih* and *muḥkam* resound categories of Hellenistic rhetoric, *mutashābih* matching the Aristotelian *amphibolos*, *muḥkam* coming close to its opposite, *pithanos*.[48]

It is interesting to see—and now we return to the narrative context of the story of Mary—that to discredit such a professional exegesis, the text refers to an imagery that is unfamiliar from earlier qurʾānic discussions of revelation. It strikes the reader as virtually sexual, or at least charged with gender associations, in contrasting the desire (*ibtighāʾ*) of sedition (*fitna*, a word that also denotes seduction)[49] through scriptural exegesis with the believers' respect for the 'mother text', *umm al-kitāb*. In striking contrast to the skeptics' 'unchaste' exploitation of the text in search for *fitna*, the believers 'firmly rooted in knowledge' immediately reconnect the ambiguous verses to the 'mother of Scripture', *umm al-kitāb*, thus legitimizing their 'otherness'. The qualification of the core of the Scripture as *umm*, 'mother', deserves attention: it initiates the conflation of two discourses, the power-informed discourse of the vertical revelation on the one hand (*kitāb, tanzīl*), and the more submissive female discourse of procreation on the other. The scriptural image of *umm al-kitāb* again points to the Qurʾān's place in a debate. It seems to reflect, albeit not precisely in that way, a hermeneutic category in rabbinic scholarship, where a reading according to the securely transmitted scriptural text, is

---

[48] Cf. H. Bonitz, *Index Aristotelicus* (Berlin, 1870; repr. Graz, 1955).

[49] Though *ibtighāʾ* in the Qurʾān is mostly oriented towards spiritual targets, such as the face of God, or His content it may denote less noble desires such as covetousness Q 13:17. In view of the presence of the root *b-gh-y* in the sense of 'whore' in the context of the Mary story, however, the sexual connotation seems not to be alien to the Qurʾān. Against the backdrop of the frequent spiritual targets of *ibtighāʾ* in the Qurʾān, the use in context of *fitna* should appear particularly subversive. *Fitna*, 'temptation' (though its meanings extend to encompass trial, straying from the right path, intracommunal strife), in the Qurʾān often is a divine strategy to test human belief. An agent of *fitna* par excellence, though not figuring explicitly in the Qurʾān, is, however, woman; cf. the *ḥadīth mā taraktu baʿdī fitna aḍarr ʿalā ʾl-rijāl min al-nisāʾ*, see A.J. Wensinck, *Concordance et indices de la tradition musulmane* (Leiden, 1937–88), v, p. 63. The *ḥadīth* is adduced by W. Saleh, 'The Woman as a Locus of Apocalyptic Anxiety,' in A. Neuwirth et al. (eds.), *Myths, Historical Archetypes and Symbolic Figures in Arabic Literature. Towards a New Hermeneutic Approach* (Beirut, 1999), pp. 123–45, 128. Fakhry translates *fitna* with 'sedition', thus blurring the virtual erotic implication.

a reading 'that has a mother', namely in scripture itself, *yesh em la-miqra*, in contrast to a reading that freely diverges from the canonical shape only relying on the transmitted consonantal structure, that is qualified as *yesh em la-masoret*.[50]

The gendered image of *umm al-kitāb* is further underscored by the reference to procreation in the preceding verse, Q 3:6, that preludes the hermeneutic debate about the *āyāt muḥkamāt* and the *āyāt mutashābihāt*:

3:6 It is He Who forms you in the wombs as He pleases, There is no God but He, The Mighty, the Wise.	هُوَ الَّذِي يُصَوِّرُكُمْ فِي الأَرْحَامِ كَيْفَ يَشَاء لا إِلَهَ إِلا هُوَ الْعَزِيزُ الْحَكِيمُ

Outwardly a predication of God's omniscience, this verse can as well be read as a statement about conception and motherhood, where the period before giving birth appears to mirror the antagonism inherent in the two kinds of verses, God creating the child in his/her mother's womb without the child's gender being clearly recognizable (*muḥkam*), the unborn child remaining *mutashābih* for human observers until his/her birth and only God knows about his/her nature. The divine purpose behind this ambiguity is the focus of the story of Mary's birth. Mary's mother is unaware that she is bearing a female child and thus dedicates the unborn to the temple. It is to God's discretion in what shape the hidden will come to the fore—in procreation equally as in revelation. It is important however, that God makes, in both his acts of procreation and revelation, use of a female agency, i.e. the womb of the mother, *raḥim*, for procreation and the core of the Scripture, *umm al-kitāb*, for revelation. Since a positive female agency, the *umm al-kitāb*, is established in the revelation process, and respected by the believing listeners, the aberration of the unbelievers, logically, again appears in gendered terms: they cling to the ambiguous, *mutashābih*, for their 'desire' of disorder, *fitna*, a term that conveys a conspicuous allusion to the female power of seduction. Taking this gender-oriented subtext of the verses into consideration, it is difficult not to be reminded of the image of the 'jealous' God known from the Hebrew Bible who laments the

[50] See *Babylonian Talmud, bSukka 6b* (and often), and cf. the discussion in W. Bacher, *Die exegetische Terminologie der jüdischen Traditionsliteratur* (Leipzig, 1899; repr. Hildesheim, 1990), i, p. 119f; Horovitz, *Koranische Untersuchungen*, p. 65, (unjustifiedly) rejects the Jewish intertext, arguing that it does not fit with the Meccan mentions of *umm al-kitāb*.

adulterous leanings of his elect people. More directly, however, the section evokes the ambiguity inherent in the image of Mary.

In this context a look at a central liturgical text of Eastern Christianity, the *Akathistos hymnos*, may be rewarding. In this 'Praise of the Virgin' recited during the period of lent, Mary herself is the subject of diverse perceptions, being recognized as an unambiguous symbol of faith by the believers and an agent of disturbance for the unbelievers: χαίρε των απίστων αμφίβολον άκουσμα, χαίρε των πιστών αναμφίβολον καύχημα, 'Hail, doubtful rumour of the faithless, Hail, undoubtful pride of the faithful!' (*Ifraḥi yā khabarā yaltabisu ʿalā 'l-kuffār. Ifraḥi yā fakhrā li'l-muʾminīna la yushawwihuhu 'ltibās*; literally: 'Rejoice, you who is ambiguous news to the unbelievers, rejoice, you who personifies the pride of the believers.' In this text, it is not the 'mother of the Book',) but the 'mother of the Word' that is praised as an object of faith unambiguous to the believers—ambiguity arising only with the disbelievers.[51]

3.3 *Some methodological conclusions*

What matters in *sūrat Āl ʿImrān* (Q 3:1–62) primarily is the politicization of the earlier purely didactic and edifying story of sura 19. Mary's story is retold to support a female-dominated genealogy of elects, Mary being staged as one of two women who entertain a particularly close relation to the divine. Their empowerment serves as a dialogical response and counterbalance to the overweight of the one hitherto solely acknowledged Jewish tradition that relies on a male genealogy going back to Abraham. Furthermore, the entire realm of revelation is affected by this turn in perspective: female metaphors infiltrate the purely male power discourse. Revelation is—as Q 3:7 highlights—no purely male transfer, there is also a female source of power. Claiming affinity with the female-dominated Āl ʿImrān, the early community is able to cope with the prerogatives of the dominant patriarchal tradition that also holds the more professional tools of exegesis. This attempt to embrace a non-

[51] A more extensive discussion of the Christological implications that might be entailed in the qurʾānic section is presented in A. Neuwirth, 'The House of Abraham and the House of ʿAmram. Female Loci of Divine Empowerment versus Patriarchal Authority and Exegetical Professionality,' in eadem et al. (eds.) *The Qur'an in Context. Textual Palimpsests and Historical Entanglements* (forthcoming).

patriarchal tradition imagined as imprinted by female[52] no less than male experience is, in my view, a relevant stage in the shaping of the qurʾānic message, that is all too easily overlooked since it eventually had to cede its prominence again to the finally re-established Abrahamian paradigm.

No such dynamic is realizable as long as the Qurʾān is read as the fixed, close corpus (*muṣḥaf*) that it became after its canonization. Focused, however, in its pre-canonical shape, as *qurʾān*, i.e. read diachronically and relocated into its culturally complex context, the Qurʾān clearly reveals intrinsic diversity, polyphony, a dialogical structure and a multiplicity of cultural languages.

[52] Cf. Michael Sells, who to my knowledge was the first to investigate female subtexts of the qurʾānic texts, though without attempting to trace a development; see M. Sells, 'A Literary Approach to the Hymnic Suras of the Qur'an. Spirit, Gender, and Aural Intertextuality,' in Boullata (ed.), *Literary Structures*, pp. 3–25. Feminist reading of the Qurʾān is still in its beginning; it is less interested in the gender aspects of the text than in the text's reformist potential, cf. M. Badran, 'Feminism and the Qurʾān,' in *EQ* ii, pp. 199–203.

THE CHEW STICK OF THE PROPHET IN *SĪRA* AND *ḤADĪTH*[1]

Wim Raven

A chew stick (Arabic *siwāk* or *miswāk*) is a twig, mostly of the *arāk* tree (Salvadora persica). The end is chewed until it is fibrous and it is moved to and fro in the mouth to clean the teeth and strengthen the gums. The twig can be green and succulent, or dried.

This article is about the chew stick as a motif,[2] whose development will be followed in one *sīra* text and a number of *ḥadīths*. I have placed the texts loosely in what I think is the chronological order, on the basis of their contents. Admittedly, the arrangement is impressionistic, but a better result may never be possible.

1. The Chew Stick in the Sīra ('Biography of the Prophet')

The earliest Islamic text in which the chew stick occurs is the narrative about the deathbed of the Prophet, in the *sīra* of Ibn Isḥāq (d. 767):

> T1 Yaʿqūb b. ʿUtba, from al-Zuhrī, from ʿUrwa: ʿĀʾisha said, The apostle came back to me from the mosque that day and lay in my bosom. A man of Abū Bakr's family came in to me with a chew stick in his hand and the apostle looked at it in such a way that I knew he wanted it, and when I asked him if he wanted me to give it to him he said Yes; so I took it and chewed it for him to soften it and gave it to him. He rubbed his teeth with it more energetically than I had ever seen him rub before; then he laid it down. I found him heavy in my bosom and as I looked into his face, lo his eyes were fixed and he was saying: 'No, rather *the

---

[1] This paper was presented at the 23th UEAI-conference in Sassari, Sept. 27th–Oct. 1st, 2006.

[2] See on its medical aspects: G. Bos, 'The *Miswāk*, an Aspect of Dental Care in Islam,' *Medical History* 37 (1993), pp. 68–79; E. Wiedemann, 'Über Charlatane unter den arabischen Zahnärzten und über die Wertschätzung des Zahnstochers bei den muslimischen Völkern,' in idem, *Gesammelte Schriften zur arabisch-islamischen Wissenschaftsgeschichte* (Frankfurt, 1984), ii, pp. 764–9; idem, 'Zur Zahnheilkunde bei den muslimischen Völkern,' ibid., 811; idem, 'Über Zahnpflege bei den muslimischen Völkern,' ibid., 874–8.

> highest companions (*al-rafīq al-aʿlā*) in Paradise*.' I said, 'You were given the choice and you have chosen, by Him Who sent you with the truth!' And so the apostle was taken.[3]

What part does the chew stick play in this story? Did the narrator want to make the Prophet purify himself or at least freshen up in view of his meeting with the high companions in paradise? That seems unlikely.

Or did the narrator intend to present the Prophet as establishing a *sunna*? At first sight, that may seem somewhat more feasible. The later a *sunna* is placed in the Prophet's life, the smaller the chance that it could be changed or abolished. Therefore we see *sunna*s established during the conquest of Mekka and the farewell pilgrimage; and why not also during the final illness and on the Prophet's deathbed? An example is the protest by the Prophet on his deathbed against *ladd*, the forced administration of a medicine through the mouth corners of a sleeping or unconscious patient. Indeed this became the precedent for a ban.[4]

But if our text had intended to establish a *sunna*, of what kind would it be? The use of the chew stick was so widespread, that Arabs hardly needed a prophetic example for it. It was a pre-Islamic habit and indeed non-Muslim pastoral peoples of Eastern Africa use such twigs for cleaning their teeth until today. If non-Arab Muslims, e.g. Persians, had other habits of dental care, they may have needed encouragement to clean their mouth the Arab way, or with a twig of that specific plant, but our text does not look like establishing any *sunna* at all. There is no obvious intention. Neither legal nor pious interest in purity are present here, and how and when to apply the *siwāk* is neither discussed nor demonstrated. Moreover, it would have been tactless to stress the importance of that little twig against the setting of the very last moments of the prophet.

---

[3] *Das Leben Muhammeds nach Muhammed ibn Isḥāḳ* [...], *bearbeitet von ʿAbd el-Malik ibn Hischām* [...], ed. F. Wüstenfeld, 2 vols. (Göttingen, 1858–60; henceforth Ibn Isḥāq), p. 1011. The translation is that of A. Guillaume (*The Life of Muhammad. A Translation of Isḥāq's* [sic!] *Sīrat Rasūl Allāh* [Oxford, 1955]). Asterisks mark my own modifications. Guillaume translated as follows: 'Nay, the most Exalted Companion is of paradise'. Given the capitals, he had God in view. But who would ever refer to God as a companion? It may be better to take *rafīq* as a plural, and think of the company that is meant in Qurʾān 4:69: 'the company of those whom God has favoured: the prophets, the veracious, the witnesses and the righteous people. What excellent companions (*rafīq*) they are!' Furthermore, I do not read the phrase as a nominal sentence, as Guillaume does, but as the object to a presupposed 'I have not chosen this world, but rather...'. On the choice that prophets have when they are about to die cf. Ibn Isḥāq, p. 1006.

[4] Ibn Isḥāq, p. 1007, and see A.J. Wensinck et al., *Concordance et indices de la tradition musulmane*, 8 vols. (Leiden, 1936–88) s.v. *ldd*.

The deathbed story as a whole may have the underlying intention to enhance the status of Abū Bakr's family. According to one well-known *sīra* narrative, Abū Bakr himself had put up a poor show during the Prophet's final illness, since at the crucial moment he was having some quality time with his wife in al-Sunḥ, at the outskirts of Medina.[5] In the *sīra*, it is always important which companions are present or absent during important events, and what part they play.[6]

In the deathbed narrative, the prophet's intimacy with ʿĀʾisha may be emphasised to compensate for Abū Bakr's absence. The prophet did not only die on her lap after she had nursed him; she also pre-chewed the chew stick for him. ʿĀʾisha was his favourite wife, but she was also the daughter of Abū Bakr. The person who carried the chew stick was a relative of hers. He serves to underline once more the presence of Abū Bakr's family in the Prophet's final hour. The message seems to be that even though Abū Bakr was absent, his relatives were there.

Yet the main *raison d'être* of the chew stick motif in Ibn Isḥāq's *sīra* seems to be of a purely narrative nature. It is a universal topos in deathbed stories to recall what the deceased did just before he died, so as to emphasise the liveliness he had shortly before. We may recount such stories ourselves, after a relative died. The deceased was still so much alive: he or she asked about the garden, or showed an interest in today's newspaper, and as a contrast, one hour later that lively person was dead. In the case of Muḥammad, the sign of liveliness was that he was still interested in dental hygiene, as on a normal day.

2. From *Sīra* to *Ḥadīth*

The above *sīra* narrative was saved into some *ḥādīth* collections. That was not a great step, for Ibn Isḥāq's story has a chain of transmitters and therefore was already a *ḥadīth*. But it did mean an upgrade. *Qiṣṣa*

[5] Ibn Isḥāq, p. 1010.

[6] In a widespread story about the *hijra*, e.g. (Ibn Isḥāq, pp. 323, 327–9; al-Ṭabarī, [*Taʾrīkh al-rusul waʾl mulūk* =] *Annales*, ed. M.J. de Goeje et al., 15 vols. (Leiden, 1879–1901), i, pp. 1237–40), Abū Bakr is the travelling companion of the Prophet, a motif suggested by Qurʾān 9:40. No less than three of his children are emphatically present and undertake creditable tasks in connection with the *hijra*, whereas ʿAlī stays behind and plays a subordinate part. This version contrasts with a story in which Abū Bakr is nearly absent, or present in a bothersome way, whereas ʿAlī carries off the palm (R.G. Khoury, *Wahb b. Munabbih*, 1. *The Heidelberg Papyrus PSR Heid Arab 23* [Wiesbaden, 1972], pp. 140–42).

and *sīra* represent early types of Muslim scholarship, which in the Islam of the *ʿulamāʾ* have a lower status than *ḥadīth*.

Ḥadīth scholars of course were above all interested in transferring legal materials from the *sīra* to *ḥadīth*, but took over narratives as well, when they contained an ethical lesson or a behavioural example, or were simply dear to them. Among these was also the deathbed story.

In the standard *ḥadīth* collections,[7] the deathbed story is found in various versions. I quote only the relevant parts.

> T2 is virtually identical with T1; it has the same chronological framework and the same *isnād*. (AbH vi, 274)

> T3 ʿĀʾisha said, When the Prophet died, he was leaning his back against me. Then ʿAbd al-Raḥmān ibn Abī Bakr entered with a chew stick in his hand. The Prophet called for it (*daʿā bi*). I took it, prepared it and handed it over to him, and he began to rub his teeth with it. Then his hand became heavy, and he became heavy on me, while he said, 'Oh God, the highest companions,' twice. Then he passed away. ʿĀʾisha said, The Prophet passed away on my lap. (AbH vi, 200/25698)

ʿĀʾisha's relative is called by name here: it is her brother. The Prophet is a bit livelier, for it is he who takes the initiative and calls for (*daʿā bi*) the stick. But his collapse is more dramatic as well; he dies during the very use of it. The main intention of this version seems to be the telling of an endearing story with some drama; secondly, there is a definite emphasis on the family of Abū Bakr.

> T4 ʿĀʾisha: The Prophet died in my house, on my day, and in my lap. ʿAbd al-Raḥmān ibn Abī Bakr entered with a succulent chew stick in his hand. The Prophet looked at it, and I thought he might want it. So I took it and chewed it, shook it out and chewed it for him, prepared it and handed it over to him. He rubbed his teeth with it better than I had ever seen him rub before; then he held it up to me, but it fell out of his hand. I began to pray to God for him with the prayer that Gabriel used to say (*daʿā li bi*), for so he prayed when he was ill, but he did not do so during this illness. He raised his eyes to heaven and said, 'The highest companions, the highest companions,' and passed away. Praise be to God, who mixed my saliva with his in the last of his days in this world! (AbH vi, 48/24271)

---

[7] The nine works which are dealt with in Wensinck, *Concordance*, i.e. the *Ṣaḥīḥ*s by al-Bukhārī and Muslim and the *Sunan* works, which together are known as 'the six books', plus three more, among which Aḥmad b. Ḥanbal's *Musnad*. I quote these works according to the system and the abbreviations in A.J. Wensinck, *A Handbook of Early Muhammadan Tradition* (Leiden, 1927), p. xvii. In some cases, I added *ḥadīth* numbers from modern editions.

In T4, the stick is called 'succulent': a slight contamination from a legal *ḥadīth*, on which see T10 below. The story is somewhat dramatised. The word *daʿā* was given an extra preposition and generated a new motif: that of the prayer. ʿĀʾisha's prayer is not enough to prevent the Prophet's decease; Gabriel's not praying for him this time may serve to explain that the Prophet had to die at all, which certain believers found difficult to accept.[8] ʿĀʾisha's preparation of the chew stick is expanded upon. The mixing of her saliva with his emphasises her intimacy with the Prophet, as does the first sentence as well.

> T5 ʿĀʾisha said, The Prophet passed away in my house, on my day, and in my lap. One of us [women] used to pronounce incantations over him with a prayer (*tuʿawwidhuhu bi-duʿāʾ*) when he was ill. So I was about to do that, but then he raised his head to heaven and said, 'Among the highest companions, among the highest companions.' ʿAbd al-Raḥmān ibn Abī Bakr passed by with a succulent piece of palm branch in his hand. The Prophet looked at it and I thought he might want it. So I took it and chewed the end of it, shook it out and handed it over to him. He rubbed his teeth with it better than I had ever seen him rub before. Then he wanted to give it to me, but his hand fell (or: it fell out of his hand). God mixed my saliva with his on the last of his days in this world and the first in the later world. (Bu 64, 83w)

T5 is an offspring of T4. The prayer has become an incantation here and the arrangement is slightly different. ʿĀʾisha's initiatives are even more numerous than in T4. Al-Bukhārī included this text in a chapter on 'The illness and death of the Prophet' and seems to have seen no legal or behavioural tenor in it.

> T6 ʿĀʾisha said, The Prophet died in my house when it was my turn, and in my lap. God mixed my saliva with his. ʿAbd al-Raḥmān ibn Abī Bakr entered with a chew stick. The Prophet was too weak to use it, so I took it, chewed it and rubbed his teeth with it.[9]

T6 is very short and not at all keen on story-telling. The Prophet is very weak and ʿĀʾisha does virtually everything, but that is not what matters here. The motif has been processed for the *sharīʿa*. The Muslim practice, also in our days, of helping children and old and sick people with the *siwāk* in this way has its explicit precedent here.

---

[8] Notably ʿUmar; Ibn Isḥāq, *Sīra*, p. 1012.

[9] Bu 57, 4b, in the chapter: 'About the houses of the wives of the Prophet'.

> T7 ʿAbd al-Raḥmān ibn Abī Bakr entered with a *siwāk*, rubbing his teeth with it. The Prophet looked at him and I said, 'Give me that *siwāk*, ʿAbd al-Raḥmān!' So he gave it to me and I gnawed and chewed it and I gave it to the Prophet, and he rubbed his teeth with it, leaning against my bosom. (Bu 11, 9)

The last Tradition also modifies a part of the original story for a legal purpose. The broader context has been left out. This time ʿAbd al-Raḥmān has the *siwāk* not in his hand, but in his mouth, and ʿĀ'isha took it into her mouth after that. A *sīra* fragment was adapted to fit a legal question: is it permitted to use someone else's chew stick? At least this is what al-Bukhārī found important, for he presented this *ḥadīth* under the chapter heading, 'Using someone else's *siwāk*'. T4 and T5 already spoke of 'mixing the saliva', but had no legal focus yet.

The notion that the Prophet's *siwāk* was prepared for him also found its way into another Tradition, which has no connection with the deathbed scene:

> T8 When the Prophet wanted to use the *siwāk*, he gave it to me, so that I could wash it; then I used it myself first, then I washed it and handed it to him. (AD 1, 28)

The matter of sharing or passing on a *siwāk* had already been dealt with in an older legal Tradition:

> T9 The Prophet once used a *siwāk*, while two men were with him. Then he was inspired (*ūḥiya ilayhi*) to pass it on to the elder of them, and so he did.[10]

3. A Legal Discussion: The Use of the Chew Stick While Fasting

Two relatively early *ḥadīth* collections contain a number of *ḥadīth*s and reports that bear witness to an early legal discussion about the chew stick: the *Muṣannaf* of ʿAbd al-Razzāq ibn Hammām al-Ṣanʿānī (126–211/724–827) and that of Ibn Abī Shayba (159–235/775–849). These texts are unaware of the small detail from the deathbed scene. They simply take the *siwāk* for granted and concentrate on the question, whether and how it should be applied during the fast.

---

[10] AR x, 19603, cf. Mu 42, 19; Mu 53, 70 '...in a dream'; AD 1, 27/50 (mentions it as a specialty of the Medinese).

One may assume that the early Muslims used the chew stick naively, also during Ramaḍān. Then some legal scholars developed scruples, in practice or only in theory: is the use of the *siwāk* a breaking of the fast or not?

I have not found texts that forbid the use of the chew stick in Ramaḍān in general terms, but a number of Traditions focus on the saliva that is produced in the mouth. Does swallowing that saliva amount to 'drinking', which is forbidden during the fast? One can imagine three types of qualms. Chewing the *siwāk* stimulates the production of saliva; a succulent twig may bring its own humidity into the mouth; and when the twig is soaked in water there is even more external liquid. Reasons enough for the pious to invoke the sayings and examples of previous authorities, who insisted on a dry *siwāk*, i. e. a withered twig that was not soaked in water to soften it.

T10 Al-Ḥakam[11] considered a succulent *siwāk* for a fasting person reprehensible.

T11 Ziyād ibn Ḥudayr al-Asadī said, 'I never saw a man who was more assiduous with the *siwāk* than 'Umar ibn al-Khaṭṭāb while he was fasting, but it was a withered piece of wood,' i.e. dry. (AR iv, 7485)

T12 Abū Hurayra said, 'I made my mouth bleed twice today with a *siwāk* while I was fasting.'[12]

Others thought of the few nutrients present in the twig:

T13 Qatāda [b. Di'āma; 66–117/680–735; *EI*²] considered it reprehensible for a fasting person to use a succulent [piece of] palm branch as a *siwāk*, because it counts as eating food (*ṭa'm*). (AR iv, 7494)

T14 'Amr ibn Shuraḥbīl [Abū Maysara] said, Do not use a succulent *siwāk* while you are fasting, for something of its food may enter your throat.' (AR iv, 7493)

T15 Al-Ḍaḥḥāk [b. Muzāḥim; gest. *c.* 106/724; TT iv, 453/784] considered it reprehensible. He said, It is sweet and bitter.[13]

---

[11] Al-Ḥakam b. 'Abdallāh, Successor, TT ii, 428/748, IAS xxx, 37/5, and 'Amr b. Shuraḥbīl Abū Maysara, d. *c.* 66/685; TT viii, 47/78; IAS xxx, 37/6.

[12] AR iv, 7486. This implies a dry *siwāk*, for a succulent, pre-chewed or well-soaked *siwāk* would not have caused bleeding. 'Twice' no doubt refers to the beginning and the end of the day; see below, T39.

[13] IAS xxx, 37, 4. 'Sweet and bitter' sounds like a circumscription of *ṭa'm* in the sense of 'taste, flavour'; cf. Lane, *Lexicon* p. 1854a.

So at least two legal scholars, living around 700, and perhaps some Successors and Companions were against succulent *siwāk*s during the fast. Others reacted by explicitly permitting or even recommending it:

> T16–T19 Ibrāhīm [al-Nakhaʿī] said, There is no harm in the [use of the] *siwāk* for a fasting person.[14]

> T20 Ziyād ibn Jarīr said, I never saw a man who was more assiduous with the *siwāk* than ʿUmar ibn al-Khaṭṭāb while he was fasting. (IAS xxx, 35/3, 4)

> T21 Khabīsha said, I came to ʿĀʾisha and asked her about the [use of the] *siwāk* for a fasting person. She said, This is my chew stick; I have it [always] at hand while I am fasting. (IAS xxx, 35/5)

> T22 [ʿAbdallāh] ibn ʿAbbās [3BH–68/619–687; *EI*²] was asked after about the *siwāk* for a fasting person. He said, Purity is good, use the *siwāk* in any case. (IAS xxx, 35/6)

In these texts, the nature or the *siwāk* is not specified (dry or wet). But saying, 'there is no harm in…' as well as the ostentatious behaviour ascribed to Companions of the Prophet imply an opposition against the older opinion that there *is* harm.

A text that tries to play down the problem of nutrients in the chew stick:

> T23 A man came to [Muḥammad] b. Sīrīn [TT ix, 214/336; 33–110/653–728] and asked, What about the *siwāk* for a fasting person? He answered, There is no harm in it. Then [that man] said: [But] it is a palm branch that contains food (*ṭaʿm*)! [Ibn Sīrīn] said, The water [in your mouth] has a taste (*ṭaʿm*) while you rinse.[15]

3.1. *Dry or wet chew sticks during the fast*

Other texts explicitly approve of a *succulent* chew stick during the fast.

> T24–T32 ʿAṭāʾ [b. abī Rabāḥ; d. 114/732; *EI*²] said, There is no harm in using a succulent (or: green) *siwāk* for a fasting person.[16]

---

[14] IAS xxx, 36/1. Saʿīd ibn al-Musayyab (IAS xxx, 36/8) and Sālim [b. ʿAbdallāh ibn ʿUmar] (IAS xxx, 36/2) and ʿAbdallāh ibn ʿUmar (IAS xxx, 35/2) similarly.

[15] IAS xxx, 37/1. If I understand the text properly, there is a play on the two meanings of *ṭaʿm*, cf. note 13.

[16] Mujāhid, ʿAbdallāh ibn ʿAbbās, al-Ḥasan [al-Baṣrī], Ibrāhīm [b. Yazīd al-Nakhaʿī], ʿUrwa [ibn al-Zubayr] and [Sufyān] al-Thawrī similarly (AR iv, pp. 7491, 7492, 7497; IAS xxx, 36/9–13; 37/2).

One might not go so far as moistening one's chew stick with water from outside:

> T33 Al-Shaʿbī[, ʿĀmir b. Sharāḥīl, *c.* 40–105/660–725; *EI²*] used the *siwāk* without moistening it. (IAS xxx, 37/8)

But ʿAṭāʾ did moisten it:

> T34 ʿAṭāʾ moistened it when it was dry. (IAS xxx, 37/7)

ʿAṭāʾ did not only approve of a succulent or wet *siwāk* during the fast, he even forbade dry ones:

> T35 Ibn Jurayj [80–150/699–767; *EI² Suppl.*] asked ʿAṭāʾ, What sort of *siwāk* is forbidden?
>
> – 'When the *siwāk* is dry and no water comes from it.'
> – 'What is the so-called water of the chew stick [*māʾ al-siwāk*]?'
> – 'The saliva that comes onto it from the head and the mouth.'
> – 'So when the *siwāk* is dry and has no juice (*ʿuṣāra*) [it is forbidden]?'
> – 'Yes.'[17]

But why would ʿAṭāʾ forbid dry *siwāks*? Possibly since they make the gums bleed, as in T12, and/or because he was against bigotry and zealotry. But maybe ʿAṭāʾ was not quite so relaxed about this subject matter after all, as a somewhat peculiar text shows:

> T36 Ibn Jurayj asked ʿAṭāʾ, May a person who is fasting use a *siwāk*?
>
> – 'Yes.'
> – 'And may he swallow his saliva? And when he does so, does he break his fast then?'
> – 'No, but it is forbidden to do so.'
> – 'And when he swallows it after he was informed that it is forbidden?'
> – 'Then he has broken his fast.'
> ʿAṭāʾ said this more than once. (AR iv, p. 7487)

The consequence of ʿAṭāʾ's opinions would be that one may use a succulent or even wet *siwāk* during the fast, but has to spit out the saliva caused by it.

---

[17] AR iv, p. 7490. Fasting is not mentioned in this text, but it is in the chapter heading and the surrounding Traditions. Moreover, the text makes sense only in a context of fasting.

The following text sounds a bit tired of the whole issue:

> T37 Al-Shaʿbī said, There is no harm in the use of a succulent or a dry *siwāk* for a fasting person. (IAS xxx, 37/3)

3.2. *The chew stick allowed once or twice a day during the fast*

People who wanted to make the prohibition less severe or—depending on the starting point—wanted to limit the general permission, discussed the possibility of using it at least twice a day during the fast:

> T38 ʿUrwa [b. al-Zubayr, 23–93/643–712; *EI*²] used the *siwāk* twice a day: in the early morning and in the late afternoon. (IAS xxx, 35/7)

> T39 Abū Hurayra was asked about the *siwāk* for a fasting person. He said, I threw it into my mouth twice today. (IAS xxx, 36/6)

Some said, Not at the end of the day:

> T40–T42 Mujāhid [b. Jabr al-Makkī, *c.* 21–102/642–720; *EI*²]...[18] considered the use of the *siwāk* at the end of the day reprehensible for a fasting person.

This is probably because using a *siwāk* shortly before the evening indicates a lack of stamina. But other reasons came up as well:

> T43 ʿAṭāʾ...answered, To God, the breath of a fasting person is more fragrant than musk. (IAS xxx, 35/8)

> T44 Al-Ḥakam [b. ʿAbdallāh]...considered it reprehensible at the end of [the day], because he considered it desirable that the smell of his breath returned into his stomach. (IAS xxx, 36/7)

Others had definitely nothing against it:

> T45 Sālim [b. ʿAbdallāh b. ʿUmar (?)] saw no harm in a *siwāk* for a fasting person when the sun faded. (IAS xxx, 36/3; cf. 36/2)

> T46 Ḥasan [al-Baṣrī, 21–110/642–728; *EI*²] said, 'There is no harm in it at the end of the day; it is a thing that cleans, so let one use it at the beginning and the end of the day.' (AR iv, 7489)

> T47 Ibrāhim [b. Yazīd al-Nakhaʿī, 50–96/670–715; TT i, 177/325] found no harm for a fasting person in using the *siwāk* at the beginning of the day and at the end of it. (AR iv, 7496)

[18] Maymūn ibn Mihrān (d. *c.* 735) and Muḥammad [b. Sīrīn] similarly (AR iv, 7495; IAS xxx, 35/9, 36/4).

And others wanted to widen the permission, or were just tired of the fuss about this mini-problem:

> T48 [ʿAbdallāh] b. ʿUmar used the *siwāk* while fasting when he went to the midday *ṣalāt*. (AR iv, 7488; IAS xxx, 35/10)
>
> T49 ʿĀmir [b. Sharāḥīl al-Shaʿbī] said, A fasting person may use the *siwāk* at any time of the day he wants. (IAS xxx, 36/5)

Finally the Prophet is invoked for a laid-back attitude:

> T50 ʿAbdallāh ibn ʿAmir ibn Rabʿīa from his father, I saw the Prophet using a *siwāk* while he was fasting (var.: innumerable times). (AR iv, 7484; IAS xxx, 35/1)

This is the only prophetic Tradition on the chew stick during the fast in the *Muṣannafs* of ʿAbd al-Razzāq and Ibn Abī Shayba. It is also the only one that survived in the standard collections.[19] The discussion with all its ramifications apparently had come to an early end, and only a broad prophetic *sunna* remained.

In what follows, I arrange the other Traditions and reports about the chew stick by subject matter, without any claim to completeness.[20] Some motifs already occurred in the earlier collections.

4. Praise of the Chew Stick

> T51 ʿUbayd ibn ʿUmayr said, The *siwāk* gives the mouth a sweet smell and satisfies the Lord. (AR x, 19603)
>
> T52 ʿAlī ibn Abī Ṭālib said, Your mouths are roads for the Qurāʾn, so make them smell good with the chew stick. (IM 1, 7/291)
>
> T53 The Prophet said, Apply the *siwāk*, for it gives the mouth a sweet smell and satisfies the Lord. (AbH ii, 108)
>
> T54 The Prophet said, The *siwāk* cleans the mouth and satisfies the Lord.[21]

---

[19] Bu 30, 27; AD 14, 26. I start from the assumption that prophetic Traditions tend to be younger than such as go back to a Companion or a legal scholar of the 8th century (cf. J. Schacht, *Origins of Muhammadan Jurisprudence* [Oxford, 1950], e.g. p. 5; G.H.A. Juynboll, *Muslim Tradition. Studies in Chronology, Provenance and Authorship of Early Ḥadīth* [Cambridge, 1983], *Index* s.v. *isnād*).

[20] I left out, for instance, all *ḥadīth*s that have multiple topics.

[21] IAS i, 169/10; Bu 30, 27; Nas 1, 5; Dā 1, 19; AbH i, 3, 10; vi, 47, 62, 124, 146, 238.

T55 Al-Shaʿbī said, The *siwāk* cleans the mouth and clears the eyes. (IAS i, 169/15, 171/27)

4.1. *The sunnas of the prophets, the fiṭra*

T56 The Prophet said, Four things belong to the *sunnas* of the prophets: perfuming oneself, marrying, the use of the chew stick and henna.[22]

T57 The Prophet found Jibrīl slow, but [the latter] said, How can [a revelation] come to you [plur.], while you do not cut your nails, do not clean your knuckles and do not apply the chew stick?[23]

T58 Ten things belong to the *fiṭra*: . . . the chew stick. . . .[24]

5. Strict or Moderate Rules?

As we saw in the texts about the fast, some Muslims used the chew stick with a vengeance. This also becomes manifest where the fast context is lacking. A Successor, a Companion and the Prophet himself serve as examples for such behaviour.

T59 Al-Ḥakam [b. ʿAbdallāh] said, I stayed with Mujāhid and he used the *siwāk* assiduously. (IAS i, 170/19)

T60 Jābir ibn ʿAbdallāh [d. 78/697; *EI*² Suppl.] used the *siwāk* when he went to bed, when he got up during the night and when he went to the morning prayer. I said to him, You bother yourself a lot with that *siwāk*. He answered, Usāma reported to me that the Prophet used to do it like this. (IAS i, 169/6)

T61 The chew stick had abraded the gums of the Prophet. (Ibn Saʿd i, 169)

Of course the zealots knew that there was neither a qurʾānic verse nor an explicit prophetic command about the chew stick. The best they could do was to suggest that there had *nearly* been a revelation about it.

T62 [ʿAbdallāh] b. ʿAbbās said, [The prophet] ordered us so continuously to use the chewing stick that we were afraid a revelation would be sent down on him about it. (IAS i, 169/11, 171/28; Ṭay 2739)

---

22 IAS i, 170/21, AbH v, 421 *bis*; Tir 9, 1.

23 I.e., practise the *ādāb* given with the *fiṭra*. IAS, 171/24.

24 AD 1, 29 *bis*; Nas 48, 1 *ter*; AbH iv, 264; vi, 137, and many other such enumerations, which I left out of my corpus.

> T63 From [ʿAbdallāh] b. ʿAbbās, The Prophet said, I was ordered [so often] to use the chewing stick that I thought a *qurʾān* or a revelation would be sent down on me. (AbH i, 337)

The second best was a near-command from the part of the Prophet:

> T64 The prophet said, Were it not that I might overburden my community, I would order use of the chew stick.[25]

> T65 The prophet said, Were it not that I might overburden my community, I would order use of the chew stick at every *ṣalāt* (or: *wuḍūʾ*).[26]

T65 may well have been at the centre of the zealots' attempts to enforce the *siwāk*; cf. the connection with *ṣalāt* below (T79).

The anti-zealot faction could use a description of the behaviour of the Prophet's wife Maymūna to explicitly unlink the *siwāk* from prayer and to promote a laid-back attitude in general,

> T66 The *siwāk* of Maymūna bint al-Ḥārith, the wife of the Prophet, was left to soak in water. When work or prayer distracted her [she forgot about it]; otherwise she took it and used it. (IAS i, 170/20)

They more succesfully mobilised a prophetic dictum—if I understand it correctly—to play down the importance of the whole subject:

> T67 The prophet said, I bothered you (*akthartu ʿalaykum*) a lot about the *siwāk*.[27]

6. Using the Chew Stick at Certain Moments

The Companions of the Prophet walked around with a chew stick behind their ears, 'where a secretary has his pen' (T68).[28] This implies that it was always ready for use. ʿAbdallāh ibn ʿUmar is said to have used one at every meal (T69).[29] The Prophet was said to have ordered the use of the *siwāk* on Friday, among other measures of purity and

---

[25] IAS i, 169/13; AbH i, 80 *bis*; ii, 287 bis, 530–1; Bu 94, 9 *bis*; Mu 2, 42; Tir, 1 18 *bis*; Nas 1, 7.

[26] AR x, 19605; IAS i, 168/5, 170/16; Bu 2, 8a; 30, 27; Mu 2, 42; Nas 1, 7/287; Dā 1, 18c/686; Mā 2 115; AbH i, 214; ii, 399, 429, 460, 517; iii, 442; iv, 225; v, 410; vi, 80, 429.

[27] IAS i, 170/30; Bu 11, 8b; Nas 1, 6; Dā 1, 18a,b/684, 685; AbH iii, 143, 249.

[28] IAS i, 168/12, 171/29; AD 1, 25/48.

[29] IAS i, 170/23.

hygiene (T70).[30] He had one with him while traveling, among other items of toiletry (T71).[31] A number of Traditions report the use the *siwāk* by the Prophet at various moments in the house of ʿĀʾisha. One may, but need not, think of the deathbed story (T1) here, for ʿĀʾisha is often the authority for intimate details in the Prophet's life.

> T72 ʿĀʾisha said, When the Prophet entered his house, the first thing he did was using the chew stick.[32]
>
> T73 When the Prophet woke up with his wife he called a slave girl named Barīra[33] to bring him a *siwāk*.
>
> T74 ʿĀʾisha said, After every sleep, at night or during the day, the Prophet used the *siwāk* when he woke up. (AbH vi, 121)
>
> T75 ʿĀʾisha said, After every sleep, at night or during the day, the Prophet used the *siwāk* when he woke up, before he performed the ablution. (IAS i, 169/9; AD 1, 30/57; AbH vi, 160)

6.1. *Chew stick and prayer*

The ablution mentioned in T75 links the *siwāk* with the *ṣalāt*, as some separate Traditions do as well. It may be applied before or afterwards.

> T76 The Prophet said, A *ṣalāt* with *siwāk* is seventy times more excellent than one without. (AbH vi, 272)
>
> T77 The Prophet used to perform two *rakʿa*s and then use the *siwāk*. (IAS, 169/7)
>
> T78 When the Prophet woke up after sleep at night or during the day he used (or: cleaned his mouth with) the *siwāk* before he performed the ablution.[34]

The following Tradition presents the *siwāk* as an obligation at every *ṣalāt*, but one that is less burdensome than the ablution:

> T79 [...] The Prophet had been ordered to perform the ablution at every *ṣalāt*, pure or not. When that was too burdensome for him, he was ordered to use the *siwāk* for every *ṣalāt* [...]. (AD 1, 25/48; Dā 1, 3b)

---

[30] AR iii, 5301, 5318; Mu 7, 7; Mā 2, 113.

[31] Ibn Saʿd, [*K. al-Ṭabaqāt al kabīr*] *Biographien Muhammeds, seiner Gefährten* [...], ed. E. Sachau, 9 vols. (Leiden, 1904–40), i, 170.

[32] IAS i, 168/3; Mu 2:43, 44; AD 1, 27/51; Nas 1/8; IM 1, 7/290; AbH vi, 182, 188, 192.

[33] She was ʿĀʾisha's slave girl. IAS i, 171/26.

[34] Ibn Saʿd i, 169; AD i, 30/57; AbH vi, 160.

6.2. *The chew stick at night*

Using the *siwāk* when getting up at night became one of the most discussed topics:

> T80 [Abū Wāʾil] Shaqīq said, We were ordered to clean our mouth when we rose at night. (Nas 20, 11b)

> T81 When the Prophet rose at night he cleaned his mouth with the *siwāk*.[35]

> T82 The Prophet used the *siwāk* several times at night. (IAS i, 170/17)

But why would one get up at night if not to pray?

> T83 ʿAlī said, When one of you rises at night, he should use the *siwāk*. For when a man rises at night, uses the *siwāk* and then performs the ablution and the *ṣalāt*, an angel comes and stands behind him, listening to the Qurāʾn, and he comes closer and closer until his mouth is on that man's mouth, and he does not recite a verse that does not enter his belly. (IAS i, 170/18)

> T84 ʿĀʾisha said, The *siwāk* and the water for the ablution were prepared for the Prophet; when he got up at night he went to the privy and then used the *siwāk*. (AD Ṭahāra 30/56)

> T85 The Prophet used to perform at night twice two *rakʿa*s; then he went away and applied the *siwāk*. (IM 1, 7/288; AbH i, 218)

The nightly prayer may be specified as the *tahajjud*, a *ṣālāt* with an even number of *rakʿa*s,

> T86 When the Prophet rose for the *tahajjud*, he cleaned his mouth with the *siwāk*.[36]

Alternatively, it may be the *witr*, a *ṣalāt* with an odd number of *rakʿa*s (T87–T89),[37] or as a combination of *tahajjud* and *witr* (T90),[38] each with varying numbers of *rakʿa*s.

---

[35] IAS i, 169/8; Bu 4, 73b; Bu 11, 8; Bu 19, 9b; Mu 2:46b, 47; AD 1, 30/55; Nas 1, 2; Nas 20, 10; IM 1, 7; AbH v, 382, 397, 402, 407. AbH v, 402 adds as a variant: *liʾl-tahajjud*, cf. T86.

[36] IAS i, 167/1, 168/2; Ṭay 409; Bu 19, 9b; Mu 2:9, 46a; IM 1, 7/286; Dā 1, 20; AbH v, 390. On this type of *ṣalāt*, see A.J. Wensinck, 'Tahadjdjud,' in *EI*^{2}.

[37] Mu 6, 139; AD 1, 30/58; Nas 20, 43. On this type of *ṣalāt*, see A.J. Wensinck, 'Witr,' in *EI*^{2}.

[38] Mu 6:182, 183, 191; AbH vi, 123; cf. Mu 2, 48.

I refrain from presenting the Traditions that deal with the number of *rakʿa*s in the various *ṣalāts*, or how often the *siwāk* is used in between, or whether the *siwāk* should be used before or after going to the privy, etcetera. For the moment it may suffice to observe that the *siwāk* became linked to the nightly prayer and developed intricate sub-topics there.

7. Handling the *Siwāk*

As we have seen, a chew stick can be succulent or dried. It can be pre-chewed, moistened with water, prepared for someone else and shared with someone else. A number of texts give yet other details about handling the *siwāk*. When a man did not actually use the *siwāk*, he wore it behind his ears, as we saw in T68. When the Prophet used the *siwāk*, he placed the end of it on his tongue (T91)[39] and produced the sound *aʿ aʿ* (or *ih ih*), as if he were vomiting (T92).[40] Or he put it under his upper lip, when he interrupted his chewing on account of astonishment or anger (T93).[41]

8. Final Remarks

Some readers may think, in the wording of T67, that they were 'bothered a lot about the chew stick'. Why look into such tedious texts about a tiny subject?

Both Muslims and non-Muslims always claim that *ḥadīth* is an extremely important genre, second only after the Qurʾān as a source of Islamic Law. Yet it is never studied systematically and in its entirety,[42] and even decent editions are lacking. If *ḥadīth* is really so important, one should study it, and it is better to start with a small subject. The chew stick requires a monograph rather than an article; more sizeable topics would require volumes.

Apart from the difficulty of handling masses of texts in innumerable variants, every *ḥadīth* raises the following questions: What is it about?

---

[39] Mu 2, 45; AbH iv, 417.

[40] Bu 4, 73a; AD 1, 26 (*ih ih*); Nas 1, 3; Ibn Saʿd, *Ṭabaqāt*, i, p. 170.

[41] Mu 33, 15; Nas 1, 4.

[42] This will change with the publication of G.H.A. Juynboll, *Encyclopedia of Canonical Ḥadīth* (Leiden, forthcoming).

where and in what time did it originate and who may held responsible for bringing it into circulation?[43]

- What is a Tradition about? What does it want to convey? What part does it play in which discussion? Often enough it is not about what it seems to be about. It may contain a reaction to a contrary point of view, it may be an argument in a dispute, it may represent a phase in the growth of an opinion. Therefore it should not be isolated and then summarised under a simple heading, as Wensinck had done in his *Handbook* of 1927. It must be read within the context of *all* related Traditions. Hence a text corpus like that above is necessary.

- How can a Tradition be dated? One dating method is the Common Link Method. Invented by Schacht, developed by Juynboll, disputed but then taken over by several other scholars, this method now seems to be accepted among non-Muslim scholars.[44] It focuses on the youngest person in a series of transmitters that all *isnād*s of a given Tradition have in common. He is the common link, who is supposed to have proliferated the Tradition. The biographical lexicons will tell us *where* he did so, who he was and where his authorities came from.

However, this method can be applied only in selected cases. Three conditions must be fulfilled: the text must go back to the Prophet; it has to be transmitted with a number of *isnād*s; and these have to partially resemble each other. Only then a common link may become manifest. Of all the texts in our corpus, only four fulfil these criteria. This is not the place to draw *isnād*-trees; whoever may want to do so, may find the *isnād*s via the footnotes to the texts. Here I simply give my results.

- Out of the seven *isnād*s of T67 ('I bothered you a lot...), six have the following part in common: ʿAbd al-Wārith—Shuʿayb ibn Ḥabḥāb—Anas ibn Malik—Prophet. In this Baṣran *isnād*, ʿAbd al-Wārith ibn Saʿīd (TT vi, 441/923; lived *c*. 102–180/720–796) is the common link.

[43] Most of these questions were formulated by Juynboll, *Muslim Tradition*, p. 7.

[44] Schacht, *Origins*, pp. 163–75; Juynboll, *Muslim Tradition*, pp. 206–17. The most encompassing description of the method will no doubt be found in Juynboll, *Encyclopedia*. The easiest introduction hitherto may be Juynboll, 'Ḥadīth and the Qurʾān,' in *EQ* ii, pp. 378–81. See also H. Motzki, 'Dating Muslim Traditions. A Survey,' *Arabica* 52 (2005), pp. 204–53.

- In five out of six *isnāds* of the Basran Tradition T91–T92 (about the Prophet wearing the *siwāk* on his tongue), Ḥammād ibn Zayd (TT iii, 9/14; 98–179/717–795) is the common link.
- T72[45] ('When the Prophet entered his, or: 'Ā'isha's, house...'), has in all variants of its Kufan *isnād* the following transmitters in common: al-Miqdām—Shurayḥ—'Ā'isha—Prophet. Al-Miqdām ibn Shurayḥ ibn Hāni' al-Ḥārithī (TT x, 287/504, died *c.* 100/720) is the common link.
- Of T81 and T86 (about the *siwāk* while rising at night), the transmission is quite complicated. But all *isnāds* have in common: Abū Wā'il—Ḥudhayfa ibn al-Yamān—Prophet. The Kufan Successor Abū Wā'il Shaqīq ibn Salama (TT iv, 361/609; 1–82/622–701) seems to be the common link, then, and an unusually early one too.

It is good to have a few Traditions dated. Yet it is obvious that the Common Link Method is not of great help for understanding the rise and decline of our topic as a whole.

Just going through the texts with some common sense will at least provide an outline. The deathbed story in Ibn Isḥāq's *sīra* text (T1) is older than 150/767. Independent of that, the texts about the *siwāk* during the fast (T10–T50) came into being between the dates of the authorities quoted in them and the appearance of the *Muṣannafs* of 'Abd al-Razzāq and Ibn Abī Shayba; say between 100–205/720–820. The 'Strict or moderate rules'-discussion (T59–T67) was already there in those early days; it must have started as a branch of the '*siwāk*-and-fast'-topic. Witness the number of quotations in the standard collections, it gained momentum some decades later, and correspondingly the Prophet himself is the authority in the often quoted T64–65 and T67. One may date them roughly between 185–265/800 and 875. *Within* this group I would not yet venture any chronological arrangement.

Maybe the *siwāk* during the nocturnal *ṣalāt* is old, given that there is an early common link. But the intricate ramifications of this topic (T86–T90 and more!) seem to be very late. Of course studying the topic 'Rising at night (for prayer)' as a whole would help increase knowledge about the use of the *siwāk* before, during and after the night *ṣalāts*, to get a clearer picture of Abū Wā'il as a common link, and to understand

[45] In this case we can even link textual variants with transmitters in younger parts of the *isnāds* ('partial common links').

the Traditionist mind better. The question 'strict or moderate?' can no doubt be studied in connection with innumerable other subjects. And this goes for most topics: the associated fields have to be studied as well. Many such text corpora, then, will have to be gleaned together and impressionistic reading will remain indispensable for both understanding and dating *ḥadīth*.

Abbreviations (apart from those for the standard *ḥadīth* collections, on which see note 7):

AR = ʿAbd al-Razzāq ibn Hammām al-Ṣanʿānī, *al-Muṣannaf*, ed. Ḥabīb al-Raḥmān al-Aʿẓamī, 11 vols. (Beirut, 1970–72).
IAS = ʿAbdallāh ibn Muḥammad ibn Abī Shayba, *al-Kitāb al-muṣannaf fī 'l-aḥādīth wa'l-āthār*, 15 vols. (Bombay, 1966–83).
TT = Ibn Ḥajar al-ʿAsqalāni, *Tahdhīb al-tahdhīb*, 12 vols. (Hyderabad, 1325–27).

AL-JĀḤIẒ ON ʿABBĀSID CALIPHS AND PEOPLE IN BASRA

*Jens O. Schmitt**

Even though most of the writings of the great Arabic prose-writer al-Jāḥiẓ (d. 255–868 or 869) are lost, many have been preserved. The most comprehensive list of these writings was compiled by Charles Pellat in 1984.[1]

In 1954, Charles Pellat published a note on a manuscript in Berlin (*Al-Mukhtār min kalām Abī ʿUthmān al-Jāḥiẓ*, Ahlwardt 5032), which contains excerpts of works of al-Jāḥiẓ,[2] some of them unknown, others slightly different from the published versions. Already before publication of this note, the Berlin manuscript had been used several times for editions.[3] At first, Pellat doubted the trustworthiness of this manuscript

* First of all, I would like to thank Professor Daiber for providing me with copies of the manuscripts, for encouraging me to edit the texts, and for many helpful ideas. It is a pleasure to offer him these texts in the present festive volume. I am grateful to Atik Akhouaji for several discussions and valuable suggestions on the Arabic text and for pointing out errors in the manuscript. I am indebted to Miriam Kaatz for sharing with me her discovery of traces of this text in al-ʿAwtabī. Almuth Rhode and my mother Inge Schmitt commented on the introduction. I am grateful to the editors for their inclusion of this article in the volume, their patience, and their many helpful suggestions.

[1] C. Pellat, 'Nouvel essai d'inventaire de l'œuvre ǧāḥiẓienne,' *Arabica* 31 (1984), pp. 117–64. This list is an updated version of his first list of 1956, C. Pellat, 'Essai d'inventaire de l'œuvre ǧāḥiẓienne,' *Arabica* 3 (1956), pp. 147–80. The list of Bū Mulḥim is based on this earlier one (*Rasāʾil al-Jāḥiẓ*, ed. ʿA. Bū Mulḥim, 3 vols. [Beirut, 2004], *Rasāʾil kalāmiyya/Kashshāf āthār al-Jāḥiẓ*, pp. 343–415). Note that in the shorter list of J. Jabar an edition of the *Burṣān* is attributed mistakenly to Pellat (J. Jabar, *Al-Jāḥiẓ fī ḥayātihi wa-adabihi wa-fikrihi* [Beirut, 1999], p. 59).

[2] C. Pellat, 'Notice sur un manuscrit arabe de Berlin,' *Oriens* 7 (1954), pp. 85–6.

[3] Van Vloten seems to have used it without indicating it precisely in his edition of the *Tria opuscula* (Leiden, 1903) under the siglum 'Berol.' (edited posthumously). Later, T.H. Ḥājirī published a *risāla* concerning the death of Abū Ḥarb al-Ṣaffār al-Baṣrī 1946 in *al-Kātib al-Miṣrī* (3/9, pp. 38–44) and one year later a *risāla* on Muḥammad ibn al-Jahm al-Barmakī in the same journal (5, pp. 55–62). These texts are also published in his edition of the *Majmūʿ rasāʾil al-Jāḥiẓ* (Beirut, 1983), as *Risālat rithāʾ wa-tabyīn*, pp. 20–27, and as *Fuṣūl fī ʾl-hijāʾ*, pp. 33–40. The latter *risāla* was also published by ʿUmar Abū ʾl-Naṣr in his edition of the *Āthār al-Jāḥiẓ* (Beirut, 1969) as *Fuṣūl lam tunshar*, pp. 33–40. Furthermore, Pellat published a *risāla* on anthropomorphism in 1953 using this manuscript as 'Risāla li'l-Jāḥiẓ,' *Mashriq* 47 (1953), pp. 281–303. He used some parts of the Berlin manuscript for his edition of the *Kitāb al-tarbīʿ wa'l-tadwīr* (Damascus, 1955). Later, he published a biographical portrait of al-Aḥnaf ibn Qays in 1969 ('Al-Jāḥiẓ

and explained these deviations as insertions and abridgments by a later person. Subsequently, however, Pellat revised this judgment.[4] In any case, his description still requires some corrections.[5]

In 1986, Professor Daiber published a note on a manuscript he had discovered in Hyderabad in India entitled *Al-Fuṣūl al-mukhtāra min kutub Abī ʿUthmān ibn Baḥr al-Jāḥiẓ* (Andhra Pradesh Government Oriental Manuscripts Library and Research Institute, ms. Muḥaḍarāt 33, Arab. 137).[6] It comprises the same material as the Berlin manuscript. In this manuscript, which—like the Berlin manuscript—were written in the seventeenth century, the name of the compiler is given as Ḥamza al-Iṣfahānī (d. before 360/970 or 971). According to Daiber, this substantiates the reliability of the collection, as Ḥamza al-Iṣfahānī is otherwise known as a collector of poetry, i.e. someone who is expected

---

wa'l-Aḥnaf ibn Qays,' *Mashriq* 63, pp. 661–72). Pellat mentions the Berlin manuscript as a source of information on Ibrāhīm ibn al-Sindī in his article on him in *EI*².

[4] Pellat, 'Nouvel essai', p. 121.

[5] Fol. 97a–98b, described by Pellat as 'épître contre l'obstination', is the beginning of the *Burṣān* (ed. Hārūn [Beirut, 1990], p. 27, l. 3–7; p. 28, l. 12–p. 29, l. 7; p. 31, l. 3–p. 32, l. 11; p. 33, l. 10–11). Fol. 100b–101b are not a part from *Ḥujaj al-nubuwwa*, but the beginning of *Khalq al-Qur'ān* e.g. in *Rasā'il kalāmiyya*, ed. Bū Mulḥim, p. 165–6, l. 13. Fol. 101b–103a, identified as 'extrait d'une épître sur la colère', are the beginning of the *Wukalā'* in *Rasā'il adabiyya*, ed. Bū Mulḥim, p. 227–8, l. 6; p. 228, l. 7–10, 13–6. There is an insertion between 228, 6 and 228, 7 (*Rasā'il al-Jāḥiẓ*, ed. M. Bāsil ʿUyūn al-Sūd, 4 vols. [Beirut, 2000], iv, p. 70) not included in the printed text which runs (H47/B102b): في الأيام الكثيرة. وقد قال بعض السلف أقرب ما يكون العبد من غضب الله إذا غضب. وذكروا عن أبي هريرة رضى الله عنه إنه قال إن رسول الله صلى الله عليه وسلم قال: ليس الشديد بالصرعة. إنما الشديد من يملك نفسه عند الغضب. واعلم أن

Fol. 104b, 'extrait d'une lettre sur l'orgueil', is part of the *Kitāb fī 'l-nubl wa'l-tanabbul*, ed. Pellat in *Arabica* 14 (1967), pp. 259–83, at p. 283, l. 11–4, and also as *Dhamm al-kubr* in *Rasā'il*, ed. al-Sūd, iv, p. 130, l. 1–3. Pellat, who edited the text himself, neither used nor mentioned the Berlin manuscript. It also seems to be published in part by A.F. Rifāʿī, *ʿAṣr al-Ma'mūn*, 3 vols. [Cairo, ³1928], iii, p. 82. Already in 1953 Pellat mentioned that he edited some otherwise unknown texts in this work (C. Pellat, *Le Milieu Baṣrien et la Formation de Ǧāḥiẓ* [Paris, 1953], xxxiii), but later did not mention him again. Fol. 105–106b, 'louange de Quraish': the beginning is almost identical with the text given by Rifāʿī, iii, p. 82, l. 10–18. In two cases, there is *kayfa* instead of *hal*. There are also some other deviations. Rifāʿī must have used another manuscript.

Fol. 106b–111a, 'court fragments': these are first sentences of book openings with pious wishes. Some of them are identical with those in edited works, some slightly different, some unknown. Maybe these are studies on these opening sentences. I will give an edition of the set soon.

Thus, also Pellat's list from 1984 should be updated in numbers 57, 169, 183, 237.

[6] No. 70 (p. 32) in H. Daiber, 'New Manuscript Findings from Indian Libraries,' *Manuscripts of the Middle East* 1 (1986), pp. 26–48.

to render texts in their authentic wording. Unfortunately, the order of the pages in this manuscript is confused. Several consecutive pages do not belong together, and other parts are completely missing.[7] It seems that the two manuscripts were not copied from a common source, since some of the variants cannot be explained as scribal errors.

Therefore, this collection of excerpts may have been distributed as an entire set. This is also suggested by the fact that later Arabic texts, namely Abū Ḥayyān al-Tawḥīdī[8] and al-ʿAwtabī,[9] include several quotations from al-Jāḥiẓ, which are extant only in this collection. Al-Dīnawarī quotes a number of similar sentences without attributing them explicitly to al-Jāḥiẓ, and further parallels occur in the writings of Abū 'l-Faraj al-Iṣfahānī and Ibn Qutayba. Furthermore, the collection of excerpts contains several sentences and poems which appear also in the writings of al-Mubarrad, a disciple of al-Jāḥiẓ, but which are not included in the surviving writings of al-Jāḥiẓ himself. One of the peculiarities of the collection is the presence of a passage from the beginning of Book Seven of al-Jāḥiẓ's *Kitāb al-ḥayawān*, in which al-Mubarrad and his rival Thaʿlab appear opposite each other and which is not included in the

[7] The order compared with the Berlin ms. seems to be as follows (correspondences at page breaks, otherwise progressing the same order): H4-B3a;
H5-B66b; H20-B78a; H21-B7b; H26-B11a; H27-B59b; H30-B62b; H31-B54a;
H38-B59b; H39-B51b; H42-B54a; H43-B62b; H46-B65a; H47-B3b; H115-B79a;
H124/125: after the Aḥnaf-article, there are the first three lines of B89a, then follows B86b, the whole rest of B89a is found again on H129; H126-B87b; H167-B116a. There are two numberings, I refer to the European numbering.

As some folios are missing completely, the following edition is in some cases based on the Berlin manuscript only. Both manuscripts are vocalized, but in difficult or misleading cases vowels or diacritics seem to be missing. Regarding *hamza* conventions, they are written in a kind of Middle Arabic.

In what follows, the Berlin and Hyderabad manuscripts are indicated by ب and ه. س shall indicate ساقط من الأصل.

[8] Abū Ḥayyān al-Tawḥīdī, *al-Baṣā'ir wa'l-dhakhā'ir*, ed. W. al-Qāḍī (Beirut, ^{4}1999). The portrait of the pretended philosopher is in iii, p. 133f, no. 454; some *fuṣūl* appear in iii, p. 139, no. 473 and viii, p. 168, no. 582.

[9] In M. Hinds, *An Early Islamic Family from Oman. Al-ʿAwtabī's Account of the Muhallabids* (Manchester, 1991), Miriam Kaatz discovered that the article on al-Muhallab ibn Abī Ṣufra appears, sometimes verbatim, in the text of al-ʿAwtabī al-Ṣuḥārī, *Al-Ansāb* (2 vols. [Masqat, 1981–84]) (personal communication). Unfortunately, since I did not have access to the relevant second volume, I have to refer to the translation here. Hinds stated that he had not found the sentences ascribed to al-Jāḥiẓ in his edited writings and therefore even doubted the accuracy of this ascription. The edition here demonstrates that al-ʿAwtabī was right. Also, the text on al-Aḥnaf ibn Qays, edited by Pellat, contains one verse which can also be found in al-ʿAwtabī (Hinds, p. 34). Furthermore, some parts of the 'réponse à un adversaire' (F. 86b–87b) must have taken their way into al-ʿAwtabī (Hinds, p. 41).

edited version of the *Kitāb al-ḥayawān*.[10] This suggests that it would be worthwhile investigating the role al-Mubarrad played in compiling the collection of excerpts. Likewise, the date of its composition is of great significance. Were parts taken out of the complete works, or does the selection contain material not originally intended for publication by al-Jāḥiẓ?

Among the as yet unedited material are biographic descriptions of ʿAbbāsid caliphs. The beginning of this section of the collection is identical with the wording of a passage in the *Kitāb al-bayān*. This section might stem from al-Jāḥiẓ's *Kitāb imāmat Banī ʾl-ʿAbbās*,[11] but it does not accord with another text of al-Jāḥiẓ published under the title *Kitāb al-ʿAbbāsiyya*. The editor, Sandūbī, suggested that these two books were identical, but it seems quite possible that al-Jāḥiẓ wrote two independent works on the ʿAbbāsids and that entries 1 (*Kitāb fī masāʾil al-ʿAbbāsiyya*) and 99 (*Kitāb imāmat Banī ʾl-ʿAbbās*) in Pellat's list are not identical. Furthermore, the section contains several biographies of famous people of Basra, which might have been taken from the lost *Kitāb akhlāq al-wuzarāʾ*.[12] These biographies are edited below.

Among the remaining, unedited, parts of the manuscripts are brief satirical portraits, several short *rasāʾil*, and quotations from the pious introductory passages of books. Some of these can be traced back to known works, from which they occasionally deviate slightly; others may have been taken from unknown writings. It seems quite peculiar that nothing more than these first sentences of the beginnings of books or *rasāʾil* were selected for the collection. Finally, there are a number of isolated sentences which appear under the heading *Fuṣūl qiṣār* in the Hyderabad, and under *Fuṣūl* in the Berlin manuscript. Most of these sentences belong to the genre of *ḥikam wa-amthāl*, and some of them appear also in different contexts in some of al-Jāḥiẓ's edited works.[13] Pellat suggested that these *fuṣūl* stemmed from the *Fuṣūl fī ʾl-adab*, a work by al-Jāḥiẓ which has not been preserved. They might also have

[10] After ابن الحر in the *Kitāb al-ḥayawān*, ed. Hārūn, 8 vols. (Beirut, 1996), vii, p. 7, l. 10, the insertion is (B 100a) ولا تفضيل المبرد على ثعلب.

[11] Pellat, 'Nouvel essai', no. 99.

[12] Pellat, 'Nouvel essai', no. 238.

[13] *Ḥayawān*, v, p. 225, l. 11–p. 226, l. 4 and *Fakhr al-sūdān*, ed. Bū Mulḥim (= *Rasāʾil siyāsiyya*), p. 552, l. 13–15.

been taken from the *Kitāb al-amthāl*, provided this was not identical with the *Kitāb amthāl ʿAlī*.

The biographies are characterized by subtle descriptions and references to psychological motives, which is typical of al-Jāḥiẓ. As far as I can tell, their language and style are that of al-Jāḥiẓ.

Among the ʿAbbāsid caliphs who are portrayed here, are al-Saffāḥ, al-Manṣūr, al-Maʾmūn, al-Muʿtaṣim, and al-Wāthiq. The absence of four other consecutively ruling caliphs might be due to a decision of al-Jāḥiẓ, or to a later selection by the compiler. Likewise, the varying length might have been a decision of al-Jāḥiẓ or of a later compiler.

The passages offer some details about al-Maʾmūn and the wars of, and revolts against, al-Muʿtaṣim. Particularly noteworthy is a peculiarity in the temporal order given by al-Jāḥiẓ. According to earlier historians such as al-Yaʿqūbī and al-Ṭabarī, the head of the Khurramī sect, Bābak, was defeated by the troops of al-Muʿtaṣim, killed and crucified and only later, Māzyār, the leader of the Qārinids, was crucified next to him.[14] This order is also given by later historians such as Ibn al-Athīr and al-Dhahabī.[15] Ibn Khayyāṭ and Ibn Kathīr only mention Bābak's execution.[16] Al-Jāḥiẓ, however, reports that Māzyār was the first to be crucified and only later was Bābak crucified next to him. This is a striking difference, especially since these executions happened during al-Jāḥiẓ's lifetime. Another testimony to this event is a verse of the poet Abū Tammām, who accompanied al-Muʿtaṣim in at least one other battle. At first sight the verse confirms the order given by al-Jāḥiẓ, saying that Bābak became a neighbor of Māzyār.[17] Yet, when Ibn Khallikān quotes this verse he repeats the chronological order of al-Ṭabarī and Yaʿqūbī in his own text.[18] Al-Masʿūdī quotes the verse

[14] Al-Yaʿqūbī, *Tārīkh*, 2 vols. (Beirut: Dār Ṣādir, s.a.), ii, p. 477; al-Ṭabarī, *Tārīkh al-umam wa'l-mulūk*, 6 vols. (Beirut, 2003), v, p. 259 (ed. J. Barth, M.J. De Goeje et al., 15 vols. [Leiden, 1879–1901], iii, p. 1298).

[15] Ibn al-Athīr, *al-Kāmil fī 'l-tārīkh*, 13 vols. (Beirut, 1965–66), vi, p. 504; al-Dhahabī, *Tārīkh al-Islām*, ed. Tadmurī (Beirut, 1987–99), years 221–30, p. 24. A discussion of the connections between these historians exceeds the scope of this article.

[16] Ibn Khayyāṭ, *Tārīkh*, ed. S. Zakkār, 2 vols. (Damascus, 1967–68), ii, p. 788; Ibn Kathīr, *al-Bidāya wa'l-nihāya fī 'l-tārīkh*, 14 vols. (Cairo, 1348–58 A.H.), x, p. 286.

[17] Abū Tammām, *Dīwān, Sharḥ al-Khaṭīb al-Tabrīzī*, ed. M. ʿAbduh ʿAzzām, 4 vols. (Cairo, 1952–65), ii, p. 207: ولقد شفى الأحشاء من برحائها أن صار بابك جار مازيار.

[18] Ibn Khallikān, *Wafayāt al-aʿyān*, ed. I. ʿAbbās, 8 vols. (Beirut, 1969–72), v, p. 123.

in a description of the execution of Yāṭis and mentions that Māzyār's cross inclined to that of Bābak.[19] Thus, one might understand the verse of Abū Tammām as describing a later movement of the crosses, not a temporal order of the executions. Yet, one should also pay attention to the fact that there is a variant of this verse which does not mention Māzyār.[20] Here, al-Jāḥiẓ compares Bābak with a Khārijite because of his hostility, but he does not say that Bābak was a Khārijite. Even though it does not seem very likely, it is possible that this Bābak was another person with the same name, who was crucified after the execution of Māzyār.[21] Another possibility is that the names were later exchanged by a scribe. As Māzyār became a Muslim, the Khārijite comparison would become more obvious, and *wa-man qara'a minhu*[22] would refer to Māzyār's reading of the Koran.

Al-Jāḥiẓ also reports the killings of two other people, Bāṭīn and Jaʿfar al-Kurdī. Bāṭīn is said to be crucified next to Māzyār and Bābak. This Bāṭīn cannot be identical with Baṭin al-Kharijī who appears much earlier in al-Ṭabarī,[23] since al-Ṭabarī related that the above-mentioned Yāṭis al-Rūmī was crucified next to Bābak.[24] Since both manuscripts contain the same name, even though they were not copied from a common source, it might be that the names of Bāṭīn and Yāṭis were confused at an earlier stage. Other authors offer further variants for the name Yāṭis, among them Bāṭis. In a list of variants given by Pellat and Marin,[25] Bāṭīn is not included. Ibn al-Athīr mentions a certain Nāṭis (variants: Māṭis and Māṭir), who was crucified in Samarra. Ibn Kathīr mentions the name Manāṭis. According to other authors, Yāṭis is a kind of military deputy in Amorion. Our Bāṭīn is said to be ruler of the waste lands or of the high-lands of the Byzantine Empire (*al-ḍawāḥī*). In Ibn Isfandiyār, Māzyār is crucified opposite Nāṭis who is said to be

[19] Al-Masʿūdī, *Murūj al-dhahab wa-maʿādin al-jawhar*, ed. C. Pellat, 7 vols. (Beirut, 1965–73), nos. 2820 and 2821.

[20] *Sharḥ al-Ṣūlī li-Dīwān Abī Tammām*, ed. Kh.R. Nuʿmān, 2 vols. (Baghdad, 1977–78), i, p. 546: إن صار بابك جاره بالنار.

[21] On first sight, the name could be read in the Hyderabad manuscript as Bābal, but such an interpretation does not lead to any further results.

[22] Said with reference to Bābak's hostility toward those who read the Qur'ān (see below).

[23] III, p. 572; ed. de Goeje, ii, p. 881

[24] V, p. 260; ed. de Goeje, iii, p. 1302.

[25] Al-Masʿūdī, *Murūj al-dhahab*, ed. Pellat, no. 2818; al-Ṭabarī, *The Reign of al-Muʿtaṣim (833–842)*, trans. E. Marin (New Haven, 1951), p. 68.

lord of Amorion.[26] Thus this offers yet another temporal variant, with Māzyār crucified after Yāṭis. One possibility is that Bāṭīn is identical with Yāṭis, in which case we are dealing here with just another variant. Another possibility is that al-Jāḥiẓ had an entirely different person in mind, whose name 'Bāṭīn' might be related to the Bāṭiniyya? The latter does not seem very likely since he appears just after the conquest of Amorion.

The second person, Jaʿfar al-Kurdī, cannot be identical with the person of that name mentioned by al-Ṭabarī,[27] since he appears after the death of al-Muʿtaṣim. Rather, he seems identical with Jaʿfar ibn Mihrajash al-Kurdī.[28]

One of the most striking peculiarities around the history of this collection of excerpts is a short passage quoted by al-Dīnawarī which is almost identical with that offered by al-Jāḥiẓ.[29] This surprising similarity notwithstanding, according to the text, Bābak is crucified first, then Māzyār alongside him, and finally Jaʿfar al-Kurdī next to Māzyār. According to al-Jāḥiẓ's version, Jaʿfar al-Kurdī was simply 'killed', but Bāṭīn was crucified next to Māzyār and Bābak. However, Bāṭīn is not mentioned by al-Dīnawarī. Two possible conclusions may be drawn: either that al-Dīnawarī copied the passage from al-Jāḥiẓ and corrected the text according to the order of events, or that both shared a common source. Al-Dīnawarī died a few years after al-Jāḥiẓ and is reported to have met al-Mubarrad,[30] who could have provided the missing link. Yet, one should bear in mind that al-Jāḥiẓ is clearly the earlier author and had personal connections with the caliph, at least with al-Mutawakkil. One might understand these difficulties as a reflection of al-Jāḥiẓ's critical attitude toward the reliability of early historians.

---

[26] Ibn Isfandiyār, *History of Ṭabaristān*, trans. E.G. Browne (Leiden, 1905), p. 156.

[27] V, p. 410; ed. De Goeje, iii, p. 1659.

[28] V, p. 270; ed. De Goeje, iii, p. 1322.

[29] Abū Ḥanīfa al-Dīnawarī, *al-Akhbār al-ṭiwāl*, ed. ʿĀmir and al-Shayyāl (Cairo, 1960), p. 402; ed. ʿAlī (Beirut, 2001), p. 587f. ʿAlī identifies (p. 588, note 4) Jaʿfar al-Kurdī with Jaʿfar ibn Fahrajas in Ibn al-Athīr. This person must be indentical with Mihrajash (see note above).

[30] ʿAbd al-Qādir ibn al-Baghdādī, *Khizānat al-adab*, ed. Hārūn, 13 vols. (Cairo, 1967–86), i, p. 55. For further references, see Lewin's article in *EI*².

The descriptions of people in Basra also vary in length. They seem to be remnants of the *Kitāb akhlāq al-wuzarā'*. The part on Ibrāhīm ibn al-Sindī is nearly identical with a passage in the *Kitāb al-bayān*, except that in our text his death is mentioned. The sketch of al-Naẓẓām contains many elements from the *Kitāb al-ḥayawān*. The three following biographies on al-Ḥajjāj ibn Yūsuf, al-Muhallab ibn Abī Ṣufra and Qays ibn ʿĀṣim share sentences and verses with writings of al-Mubarrad, and there are further similarities with a passage on Qays in the *Kitāb al-aghānī*. Substantial parts of the passage on al-Muhallab are quoted by ʿAwtabī, and some details are mentioned by al-Ṭabarī.

Next there is one person whose identity cannot be established. Daiber suggested to me that he might be identical with the secretary Fatḥ ibn Khāqān. One might also assume that this text provided a kind of sample intended as a model for biographical sketches.

According to our text, al-Ḥasan al-Baṣrī meets al-Farazdaq and is asked to pray for his wife and remarkably refuses; most probably, a divorce rather than a funeral is meant there.

Parts of some sayings of al-Muhallab, which, according to Daiber, are more coherent here than in the *Bayān*, are also mentioned as a kind of legacy by al-Ṭabarī and al-Mubarrad, but nothing of this kind is mentioned in our collection or in the *Bayān*. The Muhallabī family is presented as being very keen on maintaining and extending its dominion, which could be interpreted as a subtle criticism, when, for example, Yazīd is mocked for his sitting on the throne of Sulaymān. Noteworthy in this part are repetitions of formulations containing 'on earth'.

The name of one of al-Muhallab's companions is given here as Ḥarīsh ibn Halīl, whereas al-Mubarrad and Ibn Khallikān mention the name as Ḥarīsh ibn Hilāl. Variants are also found in some verses. Qays ibn ʿĀṣim is presented as having a bad character; a verse of Mālik ibn Nuwayra is preserved;[31] Qays is said to be a witness of the wedding between Musaylima and Sajāḥ. In the manuscripts, the entry on al-Aḥnaf ibn Qays is the last biography in the part on the *wuzarā'*. Since the passage has already been edited by Pellat, I shall only present

---

[31] Verses by him are rare. Ṣaffār, *Mālik wa-Mutammim Ibnā Nuwayra* (Baghdad, 1968), does not mention it.

the variants here. A verse in this passage is also quoted by al-ʿAwtabī, who attributes it to al-Muhallab.[32]

According to Daiber, the date of the composition of these biographies should be estimated as between 847 and 852, since he assumes these texts preceded the composition of the *Kitāb al-bayān*.[33] Yet, the death of Ibrāhīm ibn al-Sindī is mentioned only in this collection and not in the similar part in the *Bayān*. This might suggest that our text was written after the *Bayān*, or at least shortly after Ibrāhīm's death. Furthermore, since al-Mubarrad (826–98) is mentioned in opposition to Thaʿlab, the excerpts in part or in whole were probably written after his rivalry with Thaʿlab had developed.

Further investigation into the date and circumstances of the composition of this collection will have to wait until after completion of the publication of the remaining parts of the two manuscripts.[34]

To sum up, the following passages were most probably written by al-Jāḥiẓ. They may very well be remnants of the lost *Kitāb imāmat Banī 'l-ʿAbbās* and the *Kitāb akhlāq al-wuzarāʾ*, but we might also be dealing with versions of al-Jāḥiẓ's writings which the author did not intend for publication. Even though the origins of these passages are not completely clear, some of them have clearly influenced later authors and should be taken into consideration as a new source for some chronological problems of early ʿAbbāsid history.

من كتاب إمامة بني العبّاس (؟)[35]

[B64b/H45] لا نعلم أحداً بعد الأنبياء، صلوات الله وسلامه(1) عليهم، أفضل من الخلفاء ولا أحقّ بالسناء والكرامة والرفعة والفضيلة وإنما غاية الناس ومنتهى شرفٍ المتشرف أن يخدمهم ويتّصل بهم. فكل علم لا يرفعونه متّضعٌ وكل حكمة لا ينبهونها(2) خاملة

---

[32] Pellat, 'Al-Jāḥiẓ wa'l-Aḥnaf ibn Qays', p. 670, l. 9; Hinds, *Early Islamic Family*, p. 34.

[33] H. Daiber, lecture given at the conference 'Al-Jāḥiẓ, a Muslim Humanist for our Time?,' Beirut, 2005. Proceedings are forthcoming.

[34] An edition of all these remnants is in preparation. Furthermore, I will compare the known parts of these two manuscripts with the printed texts in order to detect some useful textual variants.

[35] The Arabic numbers between brackets refer to the critical apparatus below.

وكل سوق لا ينفق عندهم كاسدةٌ. ولو[36] لا أن دولة بني العباس صارت عجمية خراسانية وكانت دولة بني مروان عربية أعرابية في أجناد شامية(3). والعرب أوعى لما تحفظ وتصنع(4) وأحفظ لما تأتى أولا بالأشعار(5) التي تقيد عليها مآثرها وتخلد بها(6) محاسنها. فثبت بذلك لبني مروان شرف كبير(7) ومجد تليد وتدابير لا تحصى لأربت(8) مناقب ملك من ملوك بني العباس على مناقب جميع من وُلد بني (9) مروان وأبي (10)سفين.

ولو أن أهل خراسان حفظوا على أنفسهم وقائعهم في أهل الشام وتدبير ملوكهم وسياسات كبرائهم وما جرى في ذلك من فوائد الكلام ومن شريف المعاني كان فيما قال المنصور وفعل في أيامه وما أسّس لمن بعده ما يفي بجماعة(11) ملوك بني مروان.

ولقد تتبّع أبو عبيدة النحوي وأبو الحسن المدائني وهشام بن الكلبي والهيثم بن عدي أخباراً قد اختلّت وأحاديث قد تقطعت فلم يدركوا إلا قليلا من كثير وممزوجاً من خالص. وعلى كل(12) حال فإنا إذا صرنا إلى بقية(13) [B65a] مما رواه العباس بن محمد وعبد الملك بن صالح والعباس بن موسى وإسحاق بن عيسى وإسحاق بن سليمان وأيوب بن جعفر وما رواه إبراهيم بن السندي عن السندي وعن صالح بن عبد المصلى وعبد القدوس وعن مشيخة بني هاشم وعن مواليهم عرفنا بتلك البقية كثرة ما فات من ذلك وبذلك الصحيح أين موضع الفساد مما صنعه الهيثم بن عدي وتكلفه هشام بن الكلبي.

[B65b] وأما السفّاح فأول ما نذكره منه أن طهره الله تعالى بالعفاف وليداً وناشئاً ورجلا وزينه بالبسطة في العلم والجسم والهمّة والقصد والقناعة. فنشأ بخير ما ينشأ به الرجال أديباً عفيفاً نزيهاً. لم ير له قط صبوة ولا غرام لشهوة الجاهلية ولا ملابسة لظنين، ولم ير منتجعاً قط ولا راجلاً إلى ذي سلطان ولا مخاصما إلى قاض. ألبسه الله رداء الحلم ووقار السكينة وألقى عليه محبّة التواضع وبرّأه من الطمع الموقع وحلاه بحلية الجود والسجدة وأتاه الفقه في الدين والأصالة في الرأي وجعله بصيرا ذكيا زمينا أديبا تفهيما إذا قال، فهما إذا استمع، يزين صمته إذا صمت، بيانه إذا نطق، ويزين بيانه إذا نطق، صمته إذا صمت من غير عي. يجود ويعطي فلا يبلغ جوده أحد. ويحلم فلا يضطرّ صاحبه إلى الحلم، أشدّ الناس إلجاما لنفسه عن هوى وأعظمهم عليها سلطانا في حملها على ما تكره وكفرها عمّا تنازع إليه. وأحذر الناس بالطريقة الواسطة العادلة من السنّة بين الخفاء والغلو والجود بين البخل والسرف والأناة بين التثبّط والتسرّع وأصبرهم لنفسه عمّا تهوى وعمّا لا تهوى.

---

[36] ولو... تكلّفه هشام بن الكلبي: cf. al-Jāḥiẓ, al-Bayān, ed. I. Shams al-Raʾī, 2 vols. (Beirut, 2003), ii, p. 248, *Bayān*, ed. Hārūn, 4 vols. (Cairo, 1948–50; Beirut, ³1968), iii, p. 366f. Note that our text is most similar to that of Hārūn's manuscript "ه" which he added for the third edition, located in Istanbul, see i, p. 24 there. There is also a French translation of this passage by Charles Pellat ('Milieu Basrien,' p. 143), who included the passage because of what it revealed with regard to al-Jāḥiẓ's attitude toward the historian Abū ʿUbayda.

[B66a] وأما المنصور فهو الذي لو عدل به جميع الملوك حزماً وعزماً وحلماً لرجح بهم رجحان الثقيل بالخفيف والكثير بالقليل والكبير بالصغير.[37]

[B66a] وأما المأمون فكان واحد عصره وخطيب دهره، أبين الناس بياناً وأبسطهم لساناً وأجودهم سخاء وأفخمهم لفظاً وأكثرهم أدباً وأتمّهم منقباً وأبعدهم في العلوم مذهباً وأكثرهم فيها تصرّفاً وأقلّهم تكلّفاً وأنداءهم راحةً وأعظمهم عفواً وأوصلهم رأياً وأبعدهم غوراً، ظاهراً برّه، مبايناً قدره، عظيماً شأنه، واضحة أعلامه، وهذا عند تخوّن الزمان وذهاب الأعلام وعموم الفساد وتبدّل الناس.

ولو لم يكن من عجب شأنه وقوة سلطانه إلا أنه ما امتحن منذ كانت الدنيا إمامٌ مهديٌّ ولا غير إمام بمثل ما امتحن به[38] من العنوق العظام ومن اضطراب العوامّ ولا صادف من فساد الزمان [B66b/H5] وإكداء الصواب وإلحاح الخطأ(14) مثل الذي صادف منه، كل ذلك يرجع منصوراً مظفراً. ثم لم يتعقّب له في جميع ذلك رأي إلا ازداد على التكشيف حسناً وعلى الأيام جدّة وظهوراً. فلما كان الزمان الذي وافق سلطانه مخصوصاً من الفساد بغايته ومن خطأ الرأي بأشنعه(15) وكان الله بمنّه وفضله يريد الاستنقاذ لهم على يده وكشف حيرتهم بإرشاده وتعهّد جرائهم بصفحه. قدر لطبائعهم المخصوص بغاية الاستصلاح لهم، كانوا مخصوصين بغاية الاستنقاذ لأنفسهم، (16) فهيّأ لهم المفصّل بالكمال في الحلم والسعة في العلم المعروف بإيضاح المشكل وتسهيل المتوّعر.

قصد إلى الداء وقد أعضل بأهله وصبر على معالجة العسير بفضل عزمه فكلما (17) ازدادوا علي العلاج نَبوَة[39] ازداد عند نَبوَتهم رأفة. فمن ذلك إنه بلغ من تأتّيه ورفقه وعلمه وحسن تخلصه أن ألّف بين الثلج والنار وجمّع بين الضب والنون لأنّ الذي ألّف بين المعتزلي والنائي[40] وجمع بين الأزرقي والرافضي قد ألّف بين الثلج والنار وبين(18) الضب والنون.[41]

وقد رأينا حذّاق (19) الأطبّاء يداوون الأبدان ولم نر طبيبا (20) يداوي القلوب ويعالج الأهواء ويثرى من سقم سوء الاختيار.

ولو لا أن العيان اضطرّ العقول إلى بديع رأيه وعظيم حلمه وعجيب عفوه لم نر أن طبع البشر يحتمل مثل صفحه ولا يتسع لمثل(21) تجاوزه ولا يبلغ كنه رأيه فبدّ (22) الحلماء(23) حلمه وغضّ من(24) الأجواد جوده.

وقد كان المثل جرى بغيره في غير عصره(25) فتطلّبته الأمثال وتنازعت إليه الأقوال وحنّ إليه قلب الزاهد وتاقت إليه نفس الراغب. فهو جماع الخير ومفتاحه ودفاع (26) الشرّ ومغلاقه.

---

[37] Cf. *Ḥayawān*, vii, p. 182.

[38] A similar formulation occurs in al-Jāḥiẓ, al-*ʿUthmāniyya*, ed. Bū Mulḥim (*Rasāʾil siyāsiyya*), p. 257.

[39] Severity, according to the *Tāj*.

[40] مانوي ؟

[41] Cf. *Ḥayawān*, vii, p. 236.

فهل رأيتم كعقد أمانه وثبات عهده ودوام وفائه على بعد مداه وتقادم عصره وثقل موؤنته والصبر على مكروهه؟
وهل رأيتم أفعالا أشبه بأخلاق وأخلاقاً أشبه بأعراق من أعراقه(27) بأخلاقه وأخلاقه بأعراقه؟
وهل سمعتم بأعدل منه في حكمه(28) ولا أقصد في فعله ولا أشدّ في قوله على غناء طرفه وذكاء عينه ودوامِ طريقته وحكايةِ آخر أمره لأوّلِه؟
وهل(29) وضع أصلا لم يفرّعه وركنا لم يشيّده وأمراً فلم يستمّه؟

[B68a/H6] وأما المعتصم فلو شئنا أن نطيل الذكر ونطيب في الوصف لوجدنا إلى ذلك أنهج السبل وأسهل الطرق وأتمّ الأسباب وأكبر(30) الأعوان وأظهر الحجج، ولذكرنا المعروف غير المجهول والظاهر دون الباطن.
أما جماله وبهاؤه وقوامه وتمامه ومركبه ونصابه فقد كشف لكم العيان وأغنتكم المشاهدة عن الامتحان.
وأما عشرته وإنصافه وقربه وحلمه وصبره وقلّة تلوّنه وتلقّي(31) الحالات به وثبات عقده واستقَامة طريقته وتشابه أفعاله وتناسب أخلاقه وكثرة اعتقاده وكثرة تغافله والعناية بأمر العثرة والتفقّد لحال الصغير والكبير فقد باشرتموه بأبصاركم وجاءكم به من تتابع الأخبار وقرب الأسانيد ما يُثلج(32) صدوركم وينفى الشبه عن قلوبكم.
وأما أيامه الغرّ المشهورة وفتوحه العظام المذكورة التي لو أن واحداً منّا تهيّأ(33) لملك مستضعف لصار به مهيباً ولواهي الركن لصيّره قويّاً ولمشنوّ السلطان لجعله محبباً ولمحدود يجعله مظفراً.
فليس الشمس بأنور من برهانه ولا القمر بأضوى من دلائله.
وهي الفتوح التي تسكت الأزرقي وتخرس الرافضي وتجدل السنّي الجماعي وتعمّ الأمّة بالسرور والرعية بالمحبّة والحبور.
وهي الفتوح التي خصّت وعمّت واستفاضت وتشعّبت.
وهي التي تصغر معها كبار الفتوح وتدقّ مع بهائها جسام النعم، وما لها عيب إلا أنها تضع من كل رفيع وتصغر من كل جسيم.
وما ظنّك بأيام أطلقت ألسنة المفحمين وحوّلت المستعجمين في طباع الناطقين واستوى في معرفتها العالم والجاهل(34)[42] والأقصى والأدنى؟
وأما حزمه وعزمه فقد عرفتموها بآثار التدبير ومخارج الأمور.
وأما الأيد والبطش وشجاعة القلب والبصر بالحروب فقد أربى فيه على كل بطل(35) وغمر كل مدبر.
وهذا شيء لم يقتبسه من العلماء ولم تختلف فيه(36) الرواة وقد شاهدتموه(37) كما شاهدنا(38) وعلمتموه كما علمنا.

[42] Cf. Qurʾān 39:9.

وأما العلم بالخراج وعمارة البلاد(39) وما يُحمل من الوظائف وأبواب المال ومصلحة الثغور فقد علم ذلك وزراؤه وكتّابه والمطيفون به وكل من اتّصل بالسلطان وعرف أمور الخلفاء.

وأما بركته ويمنه فقد عرفتموه لما رأيتم من النجح وأبصرتم من غرائب الظفر.

وأما اجتهاده في أمر بيضتكم وضبط أطرافكم وتقوية(40) سلطانكم فقد رأيتموه كيف يبذل ما لا يبذل مثله من الأموال وسمحت(41) نفسه بما لا تجود(42) به نفوس الأجواد[43] وأنها لهائلة في السمع فكيف في العمل!

ولو لا أنا عاينّا لاحتجنا من تتابع الأخبار وترادف الدلائل إلى ما لم يحتج(43) إليه في جليل الاسم ولا في صغيره ولا إلى العتاد والعُدّة وما أعدّ لعدوّكم من رباط الخيل[44] ونجب الرجال. فقد رأيتم خيوله وسلاحه على أنه إن كان قد أعذر(44) ذلك فإن أحبّ الأمور أليه إن تكون عدّته وافرة(45) وقوته نامية وقاطعة لأسباب الطمع ومانعة من خواطر الشيطان يرى ذلك أدعى إلى السلامة وأهنأ للنعمة وأجمع لشمل الأمّة. ألا تراه كيف يتوقّى الدماء وكيف يستصلح بالرغبة دون الرهبة؟ ألا تراه لا يعاقب حتى يكون ترك العقاب فساداً وتعوّد(46) التغافل عجزاً؟

وهل علمتم أحداً نصب له في خاصّة نفسه حرباً وواجه(47) نحوه جنداً؟ وهل نصب له إلا من نصب للإسلام؟ وهل عاداه إلا من عادى القرآن؟ وهل رأيتم الحق والقول بالحق(48) في زمان قط أقوى ولا أهل الاختلاف في دهر قط أسكن ولا أهل السنّة والجماعة فيه أكثر ولا أرفع ولا العامّة فيه أهدى ولا الثغور فيه أحصن منه في زمانه وفي دولته وفي أيامه؟

ومن شأن العوامّ أن يملّ طول الولاية مع العدل وأن تسأم السلطان مع حسن النظر وإن كان ذلك شأنهم وعلية طبائعهم فلكل حال من الأحوال(49) من الملالة نصيب ولكل شهر من السآمة حظ، وعلى حساب ذلك يكون اليوم والساعة واللحظة والطرفة. ووجدنا المعتصم بالله على خلاف ذلك وضدّه، ووجدناهم في كل حال فيه أرغب وعليه أحدب وإليه أميل وبه أكلف. وحسباني لك دليلا وكفاك به شاهداً ولقد(50) كافوا.

وغاية مناهم أن ينفق (51) في مصالحهم من بيت مال (52) عامّتهم وأن يردّ عليهم ما أخذ من حواشي أموالهم. فلم يرض(53) المعتصم بالله مبلغ مناهم ومنتهى آمالهم حتى وفى(54) بيت مالهم بماله وأنفق على عوامّهم من خاصّة ملك يده.

ولم يمتحن إلا صاحبُ ظنّة ولم يوقع إلا بعد زوال الشبهة. يوفي الأشراف حقوق أقدارهم ويزيدهم فوق استحقاقهم ويؤلف بين قلوب المختلفين ويزيد في بصيرة المتفقين.

وهذا كلّه عيان يغني عن الإسناد وظاهر يغنيك(55) عن السؤال.

وقد كانت للخلفاء فتوح ولكنّ(56) لم يتّفق لأحد مثل ما اتّفق(57) للمأمون وعبد الملك بن مروان ومحاربتهما إنما كانت لمن قصد إلى(58) مُلكهما فلقد(59) بلغا لعمري في ذلك مبلغاً لم يبلغه

[43] Cf. al-Jāḥiẓ, *al-Jidd wa'l-hazl*, ed. Ḥājirī in *Majmūʿ rasāʾil*, p. 94, l. 10.
[44] Cf. Qurʾān 8:60.

أحد من ملوك الإسلام. وللمعتصم بالله ستّة فتوح عظام جليلة، لم يحارب في واحدة(60) منهنّ إلا من قصد الإسلام والمسلمين دون ملكه خاصّة.

فمن ذلك مازيار ملك طبرستان بعد أن تغلّب وقتل وتهضّم وسبى وتمكّن من تلك القلاع والجبال والمضائق المنيعة والسبل(61) الوعرة حتى ظفر به وقتله وصلبه.

ومن ذلك بابك(62) فإنا لا نعلم خارجياً في الأرض كان أشدّ عداوةً للإسلام وأهله والقرآن(63) ومن قرأ(64) منه بعد أن اتّسقت(65) له العساكر وقتل القوّاد وأخرب البلاد وبعد ما أودع القلوب من الهيبة والمخافة وتجرّد له حتى أخذه أسيرا فقتله وصلبه حيث(66) مازيار.

ومن ذلك فتح عمورية وهي الثانية بعد(67) قسطنطينية،[45] ثم هزيمة الطاغية،[46] ثم أسر باطين صاحب الضواحي بعد أن كان لا يعدّ الغزاة والمطوعية[47] شيئاً فأسره وصلبه إلى جنب بابك ومازيار.

ومن ذلك استباحة الزطّ حتى اجتثّ أصلهم وأباد خضرائهم بعد أن منعوا بغداد الميرة وقتلوا القوّاد وأسروا وغلبوا على البلاد واتّسق لهم من قتل الأجناد وقوّادهم(68) وأسرهم ما لم يتّسق لأحد بعد أن رامهم خليفة بعد خليفة.

ثم كان من أمر(69) جعفر الكردي وتغلّبه وإخافته السبل وقطع سبل المسلمين وجرائته على السلطان ومحاربته الأجناد حتى قتله الله على يده.[48]

ثم الذي كان منه في ضائقة البصرة وشقّ الهند كله حتى عُدموا(70) الخوارجُ من رؤساء(71) ومن إبطال المقاتلة ومن قتل الغُزاة وإخراب(72) السواحل على يدي عمرو بن الفضل الشيرازي أحد بني ربيعة بن حنظلة.

وهذه كلّها إسلامية جماعية لا تنازع فيها ولا اختلاف.

[B71b/H12] وأما الواثق بالله فهو الذي جمع بين المهابة والمحبّة وإيثار الحق وحسن النية والشغف بالعدل والقول به وقمع الظالم وقلة الرخصة وأعمال اليقظة والمسألة في كل حال مع إعطاء كل خصلة من خصال الخير نصيبها من العمل وكل خصلة من خصال الشرّ حقّها من الاجتناب حتى تكاملت فيه خلال الفضل وتتامّت عنده خصال المجد حتى لا تجد خصلة ترجح على أختها ولا ساعة تنقص عن مثلها.

ومن صفاته أصالة الرأي وصحّة العقل والحُسن اللطيف والفهم العجيب، ثم الجود بكل علق ولزوم ذلك في كل حال، ثم طيب العشرة وحسن الملكة(73) وتعهّد المولى وتفقّد حال العشيرة ثم إيثار(74) العلم على كل لذة والبيان على كل صناعة مع المعرفة بما جمع شمل العوامّ وكيف قسمة النعم بين الخواصّ وما يليق بكل مرتبة ويصلح لكل زمان مع شدّة التعقّب وجودة التصفّح ومع ذلك حسن الاختيار وصواب الإيراد والإصدار. إذا اعترم لم يعجز عن

[45] Cf. al-Dīnawarī, *al-Akhbār al-ṭiwāl*, p. 402; ed. ʿAlī, p. 587.

[46] Theophilus.

[47] This form is in the manuscripts. I take it to mean something like exceeding in warring.

[48] Cf. al-Dīnawarī, *al-Akhbār al-ṭiwāl*, p. 402; ed. ʿAlī, p. 588.

الرجوع ولم يملّك عليه اللجاج ولم ير اليد أعجز إلا عن صواب ولا إنفاد العزم لجاجاً إلا في الخطأ ولا يرى الصواب إلا ما وافق الحق. وأحبّ الحقوق إليه ما جانب الهو ى وأزينها في عينه ما زاد في المروة وآثرها عنده أمرّها عاجلة وأحلاها آجلة وأبغض الباطل إليه ما أشبه السخف وناسب الفواحش. يحوط الصوابَ حياطة من قد عرف فضله وينصب للخطأ نصب(75) من قد عرف ضرره. وقد ذلل السبيل إليه حتى سهلت وزاد في أسبابه حتى اتّصلت وتعرّف ما فيه حتى استقصاه وعجمه حتى أفضاه، وكان في طلبه وحزم في التقدّم فيه فبدأ به قبل حدوث الأشغال وقبل أن تأخذ منه الأيام فنظر بعقل سليم من الأسقام وبمعزل من الأهواء وبذهن حديد وقوة وافرة وهيئة جامعة وبغرب غير مغلول وعزم غير مدخول أيام اجتماع قوته وشباب آرائه(76) وحدّة طرفه وثقوب حسّه ودقّة مدخله وانفتاح الأبواب لقرعه فطلبه طلب من يشتهيه ويبصر جوهره ويعرف عاقبته وفضيلته والتمسه بطبيعة مناسبة وعزيزة مشاكلة والأمور لاحقة بعناصرها تابعة لجواهرها. وصادف زماناً جمّ العجائب كثير الغرائب فعرف في أيام يسيرة ما لم تعرفه الملوك في السنين الكثيرة إلا ما خصّ الله به آباءه المنتجبين ومنحه أجداده المكرمين، صلوات الله عليهم أجمعين.

من كتاب أخلاق الوزراء (؟)

[B73a/H13] كان يحيى بن خالد بن برمك وأبوه وحاشيته وبنوه من أشراف العجم(77) وهناك حلّ العلم والحزم(78) ومدّ رواقه.

[B73a/H14] وكان أحمد بن أبي دواد ذا الحلم الفاضل واللسان البين والفقه العجيب والرأي السديد والصدر الرحيب والقول الفصل والجود الغمر والرأي الجزيل والعشرة الكريمة والأخلاق المحمودة والعطايا السنية والقسمة بالسوية وشيخِ العرب وسيد الحضر وغيثِ البدو وقاضي القضاة ومقوّم الولاة ومن قد طبّق الأرض عُرفاً وملأ صدور الأولياء عزاً ومن قد جرّد القول بالعدل وكشف القناع في التوحيد وأقام لكل حالة سوقها حتى عرف الحق من كان يجهله وأقرّ به من كان ينكره وأحبّه من كان يبغضه وأنس به من كان يستوحش منه ودعا إليه من كان ينهى عنه.

[B73a/H14] وكان محمد بن عبد الملك من الكفاة والولاة والحماة ومن أهل النزاهة عن الأمور الوضيعة(79) والطهارة عن الأدناس وصدق اللهجة وسلامة الصدر وبذل البشر مع رفع الحجاب وحضور الذي لم يقل قط بعد طول المنازعة وبعد كثرة المناقلة واستغراق الألفاظ واستنفاد(80) المعاني. لو كنتُ قلتُ كذا وكذا لكان أربح ولو لم(81) أكن قلتُ كذا وكذا لكان أسلم[49] الذي فضل لسانه على لسان البليغ كفضل قلبه على لسانه وفضل علمه على عقله كفضل عقله على علمه وتركيبه أجود تركيب وصيغته أعدل صيغة وبنيانه(82)

[49] Cf. al-Jāḥiẓ, *al-Jidd wa-'l-hazl*, ed. Ḥājirī, p. 103, l. 13.

أوثق بنيان(83) وأ كثرما يرى البنيان من قواعده والمؤمن من إعراقه(84) فإذا كرم العرق قهر لؤم المنشأ وإذا أحكم(85) الركن ثبت الفرع. فما ظنّك بعرق لم تخنه العادة ولم تحكمه القريحة وبقريحة لم يعقدها العرق؟

[B73b/H15] وكان إبراهيم بن السندي خطيباً، فقيهاً، راويةً للشعر،(86) نحوياً. وكان فخم الألفاظ، نبيل المعاني، شريف الأحاديث، كريم المجالسة، وكان كاتب القلم، كاتب اللسان، كاتب العمل. وكان إذا تكلم حسبتَه رؤبة بن العجاج أو مطرفا الغنوى، وإذا عمل الخراج قلتَ هذا نبطي سوادي. كان مرة استاذاً(87) ومرة مريداً(88) ومرة عامل رستاق ومرة عامل الطسوج(89) ومرة كاتب ديوان ومرة صاحب الديوان الأعظم ومرة وزيرا. وكان عالما بالنجوم وبالطبّ وبالمنطق وكتب الحكماء، وكان فرضيا، عروضيا، وكان من كبار(90) علماء المتكلمين ومن كبار المقايسين في الفتوي، وله كتب جياد. وكان أحفظ من أعمى وأفصح من أعرابي وأسمع من فرس[50] وكان أقلّ الناس نوما[51] وأسرعهم انتباها مع غلظ رواية وكثرة لحمة، ومات مع هذا بالمرة الصغرى.

[B74b/H15] وكان فلان خطيبا لسناً وعلامة ناسباً وراويةً للحديث وفقيهاً. وكان يعرف رأي البصريين والكوفيين وعلمائهم،وكان أحفظ الناس لما يسمع من غير معاناة وكان فخما فخم الألفاظ، جيد المعاني، دقيق المسالك، لطيف المذاهب،(91) كثير المخارج. وكان يستعمل الغريب غير مكره(92) له وكانت له أحاديث قليلة الألفاظ والحروف، قليل الأخذ من القرطاس وهي كثيرة المعاني، بعيدة(93) المذاهب. وكان كاتب القلم، جيد الخط والقول، كاتب اللسان، عالما بالحجج، حاضر الجواب، وكاتب العمل. يعرف حجج الديوان من حجج الأحكام، وكان حاسبا. وكان إذا تكلم وتحدث حسبتَه رؤبة بن العجاج وإذا تكلم في الخراج حسبتَه زاذان(94) فرّوخ الأعور.

[B74b/H15] وكان إبراهيم بن سيار فرضياً، عروضياً، وكان حاسباً ومنجماً وكان نسابا وكان حافظا للقرآن العظيم(95) وتفسيره وللتورية والإنجيل والزبور وكتب الأنبياء وقد(96) كان عالج الكيمياء وعرف مذاهبه، وكان أروى الناس لكلام الأوائل ولصنوف(97) نحل الإسلام، [52] وأحسن الناس إخراجا وأبلغهم عند الاحتجاج لسانا. ولم يكتب علما قط ولم يدوّنه. وكان صاحب حديث عالما وكان له نسك وخالط السادة(98) الصوفية وأصحاب المضمار وعرف اختلافهم وكان يقول الشعر إذا أراده وكان يستخرج المعمى وكان حسن العلم بالنحو.

[50] See *Ḥayawān*, i, p. 221 and *passim*.
[51] Cf. *Bayān*, i, p. 277.
[52] أروى لصنوف: *Bayān* i, p. 217.

وقال أبو عبيدة: ما ينبغي أن يكون في الدنيا مثل إبرهيم بن سيار. سألته وهو صبي عن عيب الزجاج فقال: سريع الكسر، بطيء الجبر.[53] ومدحوا النخلة فقال: صعبة المرتقى، بعيدة في الهواء، خشنة المسّ، قليلة الظلّ.[54] وقال: يوماً كنّا نلهو بالأماني ونطيب أنفسنا بالمواعيد فذهب من يعد وقطعتنا الهمومُ عن فضول المنى.[55]

وذكر الخليل بن أحمد فقال: توحّد به العجب فأهلكه. وصوّر له الاستبداد صواب رأيه، ورام إن لا يناله، وفتنته دوائره التي لا يحتاج إليها غيره. كان إبراهيم إذا ذكر الوهم لم يشكّ في جنونه واختلاط عقله. وهكذا كان الخليل بن أحمد وإن(99) كان قد أحسن في شيء.[56] وكان أصحابه يرون إنهم لو ماتوا لتبدّل الدين ولفسدت الحكمة ولاستولى على الناس الجهل ولأ كلتهم علماء الروم والهند[57] فضلا عن علماء الدهرية والزنادقة والخوارج والرافضة(100) وكذلك كان إبراهيم أكثر أهل الأرض تنقّلا وأسرعهم اعتقاداً وأقلّهم على ما اجتنى ثباتا لأنه كان لا يخلو بقوله ولا يجيد انتحاله ويعجل على التصديق ويعمل حسن الظنّ ويسأم طول الروية(101) ويغلط في حق الإنصاف فيعطيه ما ليس له. وكان يجعل(102) قصور خصمه عنه ونقصان من قبل عنه عن مرتبه سبباً للنصرة(103) في مذهبه وحجّة فيما بينه وبين ربّه. وكان كثير الخواطر، قليل الصبر على التحصيل، معجباً بالتفرّد، شديد الجرأة على اعتقاد ما يخرج من طبائع الأمّة، ومن ذلك قال بالمداخلة والطفرة.[58]

وكان أضيق الناس صدراً بحمل سرّ، وكان شرّ ما يكون إذ أكّد عليه صاحب السرّ وكان إذا لم يؤكد عليه، نسي القصّة فيسلم صاحب السرّ.[59]

[B76a/H18] كان الحجاج بن يوسف من أهل الانتقام والسطوة وأصحاب التشفّي والقسوة والصولة عند القدرة. وقال له(104) عبد الملك بن مروان إن الرجل لا يكون عاقلا حتى يعرف نفسه. وأمير المؤمنين يقسم عليك لتخبرنه عن نفسك فقال: أنا حديد حقود ذو قسوة وحسود.[60] فانتحل الشرّ بحذافيره والمروق من جميع الخير بزوبره.(105) ولقد تأنّق في ذمّ نفسه وتجرّد(106) في الدلالة على لؤم طبعه(107) وفي إقامة البرهان على إفراط كفره والخروج من كنف ربّه وشدّة المشاكلة لشيطانه الذي أغواه وقرينه[61] الذي أغراه، هذا مع

---

[53] *Ḥayawān*, iii, p. 471.

[54] *Ḥayawān*, vii, p. 165.

[55] *Ḥayawān*, vii, p. 153. قطعنا الهمومَ عن فصول هـ، قطعنا الهموم عن فضول ب

[56] *Ḥayawān*, vii, p. 165f.

[57] Cf. *Ḥayawān*, iv, p. 206.

[58] These are technical terms of his philosophy.

[59] *Ḥayawān*, v, p. 187.

[60] Similar in wording: *Ḥayawān*, iii, p. 470 and v, p. 592; *Bayān*, ii, p. 177; similar in motivation: Ibn Qutayba, *ʿUyūn al-akhbār*, 4 vols. (Cairo, 1925–30), ii, p. 8.

[61] Companion of the devil: *Bayān*, i, p. 536.

عتوه وطغيانه وشدّة صولته وقسوة قلبه، والعجلة شعبة من الحدّة، وصاحب العجلة إن أصاب فُرصته لم يك محموداً، وإن أخطأها كان مذموما.[62]
الهيثم بن عدي: إن رجالا كانوا إذا التقيا الصفان في الحرب ذهب تدبيرهم غير أنهم كانوا لا يبرحون منهم الحجاج بن يوسف. كان[(108)] أبو كعب مولاه يدبّر له الحرب عند حيرته فيها. وكان أخيفش[(109)] منسلق الأجفان. ألا ترى أن عبد الملك كتب إليه في شأن أنس بن ملك: فلعنك الله أخيفش العينين أصك الرجلين أسوء الجاعرتين[(110)].[63] وكان في صغره يسمّى كليباً.[64] فلما عظم شأنه بالعراق وطغى وعتى سقط عنه هذا الاسم. ولما اتّصل بالحسن خبر موته قال اللّهم أنت قتلته فاقطع عنّا سنّته.[65] فإنه أتانا أخيفش أعميش مقيتا له جميمة يرجّلها. صعد المنبر فأخرج إلينا كفا قصير[(111)] البنان ما عُرف فيها[(112)] عنان في سبيل الله. فقال بايعونا فبايعناه. يصعد إلى هذه الأعواد[66] فينظر إلينا بالتصغير وننظر إليه بالتعظيم. يأمرنا بالمعروف ويتجنّبه وينهانا عن المنكر ويرتكبه.[(113)]

[B77a/H19] وكان الحسن بن أبي الحسن ممن تزوّج نساء عشيرته ورهطه وهو مولاهم على أن الحسن قد جاوز قدر كل ذي قدر. وكان[(114)] يصلّي على كل جنازة شهدها فإذا علم الوالي أنه في جنازته[(115)] تجافاها حتى إذا صلى عليها أقبل فعزى أهلها إكبارا للحسن وقد صلى على أم عبد الأعلى علي[(116)] بن عبد الله بن عامر وهو يومئذ سيّد أهل البصرة فسمع صراخا فالتفت كالمنكر لذلك فعاد إليه عبد الأعلى فقال: جعلني الله فداءك. والله ما علّمته ولا اشتهيتُه حتى سمعتُه. وأتاه الفرزدق يسأله أن يصلي على النوار امرأته[67] فأبى واعتلّ عليه فقال: يا أبا سعيد إذاً تجللني وإياها عار الأبد فأجابه إلى ذلك.
وكان عند أهل البصرة في مستثنى الغاية. كان يقال هو أزهد الناس إلا الحسن وأبين الناس إلا الحسن وأفقه الناس إلا الحسن. وقال أبو شعيب: الحسن خير لأهل البصرة من الجزر والمدّ والمدّ هو حياتهم. يأتيهم فيقف على أبوابهم فإن شاؤوا حجبوه وإن شاؤوا أذنوا له، والله أعلم.[(117)]

---

[62] Cf. al-Jāḥiẓ, *al-Jidd wa'l-hazl*, ed. Ḥājirī, p. 86, l. 5f; al-Maydānī, *Majmaʿ al-amthāl*, ed. A. Ibrāhīm, 4 vols. (Cairo, 1978–79), iii, p. 208, no. 3690: لا يوجد العجول محمودا.

[63] *Bayān*, i, p. 311.

[64] Al-Mubarrad, *al-Kāmil fī 'l-lugha wa'l-adab*, ed. Hindawī, 2 vols. (Beirut, 2003), i, p. 402.

[65] *Bayān*, ii, p. 188.

[66] Al-Ḥajjāj on boards: *Bayān*, i, p. 457.

[67] Most probably as a witness for a divorce, not a funeral, cf. *Kāmil*, i, p. 124 note 6 (from the commentary *Raghbat al-āmil*); Abū 'l-Faraj al-Iṣfahānī, *Kitāb al-aghānī*, 25 vols. (Beirut, 1955–64), xxi, p. 315; Ibn Sallām al-Jumaḥī, *Ṭabaqāt fuḥūl al-shuʿarā'*, ed. M.M. Shākir, 2 vols. (Cairo, s.a.), i, p. 334ff.

[B771b/H20] كان المهلّب بن أبي صفرة رجلاً لا يفي به أحد في الأرض في الحزم والعزم والصدق والأمانة والوفاء والسنّة وحاجة الناس إليه[68] واستغنائه عنهم مع الولايات الكثيرة والعفاف والتوفير والعلم بالمصلحة.(118) وكان أجمع الناس لخصال الرجال.(119) لم يخضّ في فتنة قط[69] ولا عُرف بجاهلية ولا قذف بمجنة. وكان ممن يحمل عنه الأثر ويصدق على الخبر مع مكانه من السلطان وقتله للأقران وكان يقال: بصرة المهلب[70]،لنفيه الخوارج عنها حين كعّ[71] عنهم الأحنف وجمع(120) الوجوه وقلدوه الأمر وعظموا عليه الحق وكل شيء [B78a] بالبصرة فائق جيد قاسمٌ المهلبي عليه مثل السوذق[72] والدقوف[73] والجخال[74] والمقانع،[75] وهو أكثر من أن نحصيه. وأهل البصرة يقولون: جاءِ الناس وقريش والمهالبة. ويقال إن المهلب لم يسابّ أحدا قط في شبيبة(121) ولم يسبّ أحداً في كهولته إلا مرة واحدة فإنه قال لخالد بن عتاب بن ورقاء: يا بن اللخناء![76] وكان مع حزمه وبأسه من أبين الناس وأجودهم كلاما. فمن كلامه: صلة الرحم مثراة في المال، محبّة في الأهل منسأة في الأجل ومنه، تحابّوا. فإن بني الأم يختلفون فكيف بني العلات؟ إن البرّ ينسيء الأجل ويكثر العدد، تبارّوا، تحابّوا. فإن القطيعة تورث القلّة وتعقب الحسرة واتّقوا ذلّة اللسان[77] فإن الرَجل تزّل رجله فيعيش، ويزّل لسانه فهلك. وعليكم بالمكيدة فأنها أبلغ من الشجاعة[78] فإن اللقاء إذا وقعَ وقع القصداء فإن ظفر سعد وإن ظفر به لم نفرّط.[79]

ومن كلامه: خير الولاة من إذا كان في رعيته فكأنه غائب عنهم وهو شاهد لهم ومن إذا خرج من داره خرج في زي سوقة وإذا رجع رجع في زي ملك.[80]

---

68 Need of the people for him: cf. Hinds, *Early Islamic Family*, p. 30, § 28 (ʿAwtabī, ii, p. 130).

69 Cf. Hinds, *Early Islamic Family*, p. 28f, § 27 (ʿAwtabī, ii, p. 129f).

70 Al-Jāḥiẓ, *Jawārī*, in *Rasāʾil adabiyya*, ed. Bū Mulḥim, p. 180, also al-Mubarrad, *Kāmil*, ii, p. 223.

71 Unable to cope with (*Lisān*)?

72 Maybe companies of men (soldiers) journeying by night are meant, or a bird, cf. al-Jawālīqī, *al-Muʿarrab*, ed. Kh. ʿImrān al-Manṣūr (Beirut, 1998), s.v. *sūdhāniq*. These four introductions are unclear, since the text is partly illegible. Explanations given are those of *Lisān* and *Tāj*.

73 Someone considered as blameworthy, esp. as a sodomite? This seems to make little sense here.

74 Effective poison? A hint to *makīda* and the story of the poisoned arrows (Hinds, *Early Islamic Family*, p. 39)?

75 Men in armament.

76 *Kāmil*, ii, p. 250; al-Ṭabarī, iii, p. 553, ed. De Goeje, ii, p. 878; Hinds, *Early Islamic Family*, p. 29, § 27 (ʿAwtabī, ii, p. 129f).

77 Cf. al-Mubarrad, *Kitāb al-taʿāzī wa'l-marāthī*, ed. Kh. al-Manṣūr (Beirut, 1996), p. 82f.

78 See S.M. Yusuf, 'Al-Muhallab b. abī Ṣufra: his Strategy and Qualities of Generalship,' *IC* 17 (1943), pp. 1–15; p. 4 on Muhallab's introducing of *al-makīda* into battles.

79 *Bayān*, i, p. 453; Mubarrad, *Taʿāzī*, p. 82f, Ṭabarī, iii, p. 633f; ed. De Goeje, ii, p. 1083.

80 Remotely similar (modesty): Mubarrad, *Taʿāzī*, p. 83.

ومن كلامه: يعجبني أن أرى عقل(122) الرجل يزيد على لسانه ولا يعجبني أن أرى لسانه يزيد على عقله.[81]

ومن كلامه: في بعض الأيام عجب ؤلاء القوم والله لكان ما ينقص منهم يزاد فيهم إني كلما ظننت أنهم قد ملوا عادوا كأنهم لم يصنعوا شيئا.[82]

وقالٍ الشاعر يريد هجاء المهلب فمدحه لأنه لا يستطيع أن يقول في المهلب هجاء إلا وقع مدحا ومن يقدر على وهن الذهب الإبريز والكوكب الوقاد وهو قوله:

فقدتُك يا مهلب من أمير
أما تبدي بنانك للفقير
أكل الدهر أنت لِدار حرب
أمام القوم في السَلف المغير
إذا صاح السُراة أبا سعيد
تقدّم كنتَ في أولي النفير[83]

وليس في الأرض موضع قَدَم إلاّ وليه مهلبي إلا مكة وفيهم خمسة، أمير بن أمير، وليس يشرع إلى رحبة المهلب إلا باب أمير.

ولي يزيد بنِ المهلب العراق وسارت العرب كلّها تحت رايته ودعا إلى الخلافة وأجابته القبائل طوعا وكرها. فجلس على سرير سليمن بن عبد الملك وفي مقعده حتى إذا جلس عن يمينه فإذا نهض عاد إلى مكانه. وإليه أمر جميع الناس من جهة الكفاية والمكانة والقدر والسياسة وتمليك أعنّة(123) الخيل[84] والأمر والعهن إلا من جهة الكتابة والخدمة. وكان أجود الناس وأشجع النَاس. وهمّه الحجاج بالحسد والعصبية محبه(124) [B79a/H115] أن يسمع أنينه فما(125) قال حسّ[85] حتى دلّوه على موضع نصل قد كان بقي في ساقه فلما وُضع الدهق(126)

---

[81] Remotely similar Ṭabarī, iii, p. 633.; ed. De Goeje, ii, p. 1083; Mubarrad, *Taʿāzī*, p. 83 (لسان/فعل). An expression with a closer, but not identical, wording, but not attributed to al-Muhallab, has Ibn Qutayba, *ʿUyūn*, ii, p. 168. It is followed by a quotation from Yazīd ibn Muhallab.

[82] Cf. Mubarrad, *Kāmil*, ii, p. 252; ii, p. 248 (وكان لم يصنعوا شيئا).

[83] Al-Mubarrad, *Kāmil*, ii, p. 245ff. See ii, p. 246, note 3 for similar versions. The poem as it appears here seems to be a patchwork of different material, or alternatively the verses mentioned there have incorporated this verse. The verses cited in al-Mubarrad have variants which sometimes have the same sense as e.g. ندى and صاح, but others introduce a different sense. Since the verses as they are given here are possible, I decided to keep with the manuscript السُراة instead of الشراة and لدار حرب for لزاز حرب. Al-Mubarrad ascribes it to Abū Ḥarmala al-ʿAbdī. S.M. Yusuf, 'Al-Muhallab and the Poets,' *IC* 24 (1950), pp. 197–9, mentions p. 197 Abū Ḥarmala (with regard to his appearance in the *Kāmil*) as 'typical example of lively discourses with war-poets in his army'. This statement is in accordance with the teaching of war-poems sung about the Muhallabides mentioned later, see below. Meter: *wāfir*.

[84] Hinds, *Early Islamic Family*, p. 66, § 77 (ʿAwtabī, p. 149).

[85] An expression that is possible according to the *Lisān*, suggested by Akhouaji.

على ذلك الموضع أن أنينِ الكِرام[86] وهو الذي قال في ابن الأشعث: غلب على النصر فغلب على الصبرِ، [87] خرج كريماً وقتل كريماً.[88]

ونظر يوماً الحريش بن هليل[89] إلى ولد المهلب وإلى شدّاتهم في تلك(127) الحروب فقال لو لا شفقة أبيهم عليهم ما احتاج إلينا.

وقال بعض الناس ما يخاف آل المهلب إلا الله وهم قوم ليس في الأرض مثلهم. لم يقف مهلبي قط إلا على زرّاد أو ورّاق،(128) وبذلك أوصى المهلب وجرت بهِ العادة.

ولم تر أمة قط من إمائهم في حوضٍ كإماء العرب عندنا(129) ولا غلاما قط مع الصبيان[90] ولا قال مهلبي قط: فعلتُ وفعلتُ، وأنا وأنا، حتى إذا استوى على ظهر فرسهِ رأيت أمراً يخلع القلب ويجوز الصف(130) ولم يكلموا منهزماً ولا ذادوه ولا قضوا له حقّاً حتى يموت وحشةً وكمداً.

وليس في الأرض مع هذه الشدّةِ والصرامة والشكيمة أحسن جواراً منهم، لم يسألوا نازلاً فيهم قط ممن أنت إن كان عربياً أمسكوا عن ذكر جميع العرب ما كان حاضراً مخافة أن يسوءه بعض ما كان(131) منهم.[91]

يتواصون بالصبر ويدرسون أولادهم أشعار الحرب ويخرّون على من يعلمهم الكتابة ويعلمونهم مآثر(132) آبائهم والأشعار التي قيلت فيهم كما يعلمونهم التشهّد.

والبيوتات في الإسلام ثلاثة: بيت المسامعة[92] في ربيعة ثم في بني قيس بن ثعلبة وبيت مسلم بن عمرو[93] في قيس عيلان ثم في باهلة(133) وبيت المهلب في الأزد ثم في عتيك. ولم

---

[86] Al-Ḥajjāj is said to have captured and tortured Yazīd and his brothers. Yazīd's sister Hind was al-Ḥajjāj's wife (Yaqūʿbī, ii, p. 276). Hearing his brother screaming, she left al-Ḥajjāj. See Hinds, *Early Islamic Family*, p. 61, § 69 (ʿAwtabī, ii, p. 146) and al-Ṭabarī, iii, p. 684; ed. De Goeje, ii, p. 1209.

[87] Cf. Yaqūʿbī, ii, p. 310.

[88] Cf. *Bayān*, i, p. 389.

[89] Note the different form حريش بن هلال in Mubarrad, *Kāmil*, ii, pp. 212, 215, 243, also in Ibn Khallikān, vi, p. 284.

[90] Women not sent for pouring and carrying heavy water pots, not very young boys in service as marks of gentle behavior against slaves? Maybe the satirical poems and narrations which claim that al-Muhallab was not of Arab origin or that he at least behaved in a 'non-Arab' manner (cf. S.M. Yusuf, 'Al-Muhallab b. Abī Ṣufra,' *IC* 18 [1944], pp. 131–44, at 131–3), form the background here.

[91] See Yusuf, 'Al-Muhallab b. abī Ṣufra: his Strategy and Qualities of Generalship', p. 13 on Muhallab's treatment of his sons, soldiers, and *mawālī* in equal manner.

[92] Al-Mubarrad, *Kāmil*, i, p. 162: المهالبة mentioned with المَسامعة. On the tribes see ʿU.R. Kaḥḥāla, *Muʿjam qabāʾil al-ʿArab al-qadīma wa'l-ḥadītha*, 5 vols. (Damascus, 1949–75), iii, p. 1088; Ibn Qutayba, *Maʿārif*, ed. ʿUkāsha (Cairo, 1960), p. 419.

[93] See Ibn Qutayba, *Maʿārif*, p. 406.

يبلغ بيت الجارود[94] وبيت بني بدر[95] في الإسلام(134) ما بلغته هذه البيوت ولم تكن(135) مثل المهلب.[96]

[B80a/H116] وكان قيس بن عاصم أغدر الناس وأبخلهم. أسره عباد بن مرثد بن عمرو بن مرثد وسبى أمه يوم أبرق الكثيب فمنّ عليهم وردّهم بغير فداء فأحقره(136) ولم يشكر يده.[97] وكان يسمّى الكذاب وفي ذلك يقول زيد الخيل

فلست(137) بوقاف إذا الخيل أحجمت

ولست بكذّاب كقيس بن عاصم[98]

وكان أيضاً يلقب الغرّ وكان يقال(138) له البدع والمبدع المتلطح بخرئه. وغدر بجار له وجاوره ليتعزز(139) خمار فأقبل يضربه وشرب خمره وقد سكر وهو يقول

وتاجر فاجر جاء الإله به

كأن عثنونه أذناب أجمال[99]

وسكر فوثب على ابنته فاقتضّها[100] فلما أصبح آلى أن لا يشرب الخمر فنعى عليه ذلك مالك بن نويرة فقال

مجوسية سعدُ بن زيد وينتهي

إلى بنت قيس غدرها وفجورها[101]

وهذا الشعر مثل قول أوس بن حجر

والفارسية فيهم غير منكرة

فكلهم لأبيه ضيزن سلف[102]

وقال أبو عبيدة قال قيس بن عاصم وئدت(140) بنتاً في الجاهلية[103] فلكأني أسمع صوتها يا أبه يا أبه.(141) وأراد أن يئد(142) بنتاً له أخرى فأخذها منه جدّ الفرزدق صعصعة بن ناجية[104] ورووا أنه ارتدّ عن الإسلام وكان على الصدقات فعمد إلى ما قبض فقسمه في الردة بين بني منقر، وكان يهجو أبا بكر وعمر والمهاجرين والأنصار رضى الله عنهم فقال(143)

---

[94] See Kaḥḥāla, *Muʿjam*, iv, p. 52.

[95] See Ibn Qutayba, *Maʿārif*, p. 83; Kaḥḥāla, *Muʿjam*, i, p. 68.

[96] Very remote resemblance: al-Jāḥiẓ, *Jawārī*, ed. Bū Mulḥim (in *Rasāʾil adabiyya*), p. 180/181.

[97] *Aghānī*, xiv, p. 84; there: عبادة، يوم أبرق الكبريت.

[98] *Shiʿr Zayd al-Khayl al-Ṭāʾī*, ed. al-Barzat (Beirut, 1988), p. 153, l. 2. Meter: *ṭawīl*.

[99] Cf. Mubarrad, *Kāmil*, i, p. 445, *Aghānī*, xiv, p. 71, 80.

[100] Cf. *Aghānī*, xiv, p. 79.

[101] On this verse see the introduction. The meter seems to be *ṭawīl*. The person named is a pre-Islamic ancestor of Qays, his religion is the one which desires Qays' daughter (Akhouaji).

[102] *Dīwān Aws ibn Ḥajar*, ed. al-Ṭabbāʾī (Beirut, 1996), p. 63, no. 32, 2; see the explanation given there (note 4). Meter: *basīṭ*.

[103] Cf. *Aghānī*, xiv, p. 66f.

[104] Al-Mubarrad, *Kāmil*, i, p. 387 claims that he was opposed to burying daughters alive, cf. *Aghānī* xxi, p. 300.

حَبَوت بما صدّقتُ في العام منقرا
وأيأستُ منها كل أطلس طامع[105]

ولما تنبأت(144) سجاح إليه بوعية آمن بها قيس بن عاصم وخرج معها إلى اليمامة إلى مسيلمة الكذاب(145) فآمن به أيضا وصدقه[106] وكان ممن شهد نكاح مسيلمة لسجاح. وتزعم(146) تميم مع هذا كله أن رسول الله صلى الله عليه وسلم قال لما رأى قيسا: هذا سيّد أهل الوبر.[107] فإن كانوا قد صدقوا على قيس وصدقوا عن النبي، صلى الله عليه وسلم، فيه فهم شرّ الأشرار إذ كان قيس شرّ الخلق إذ كان من صفاته الكذب والغدر واللؤم والفجور والسفه. فإذا كان قيس شرّ الخلق فكيف يكون من قيس سيّده إن كان النبي، صلى الله عليه وعلى آله وسلم،(147) قد قال ما حكوا؟

Textcritical apparatus

ا 6 ولاِ بالشعراء ب: أولا بالشعر ه 5 لما تصنع ه 4 أخبار سامية ب 3 ينهونها ب 2 س. ه 1
مَن وَلَد بني مروان perhaps read ,وأبو ه،ب 10 س. ب 9 لأربى ه،ب 8 كثير ه 7 ب
بغية 13 ه corrected in the margin ,س. ب 12 وما أبقى لجماعة ب 11 .وأبو سفين
فلما 17 الاستنفاد بأنفسهم ه 16 باسنعه ب ,.s.p. vel sim باسنعه ه 15 الخطاب ب 14 ه
عن ه 24 الحكماء ب 23 فند ه؟ 22 مثل ه 21 لطبيبا ب 20 من حذاق ب 19 و ه 18 ب
وأكبر السبل ب 30 هل ب 29 صنعه ه 28 أعلاقة ب 27 ورفاع ب 26 عنصره ب 25
31 ه تكفّى 32 ه يثلج به 33 ه منّا, corrective sign sup. lin. A very problematic passage. Akhouaji proposed reading منها. 34 ب الجاهل والهالم 35 ب بطال 36
وعمادة البلاد ب:وعمارة للبلادھ39 شاهدونا؟ 38 شاهدوه ه؟ 37 يختلف فيه إلى ب
46 راحرة ه؟ 45 ه:أعد ب قد أعذر 44 نحتج ه 43 يجود ب 42 به ب 41 ويقويه ه 40 ه
وقد ه 50 فلكل حول ه 49 القول بالحق ه:الحق والقول بالحق ب 48 وجه ب 47 يعود ه
58 أنفق ه 57 لكنه ه 56 هيغني 55 وقى ه 54 يرض لهم ب 53 المال ب 52 يتفق ه 51
قرأه ه 64 للقرآن ه 63 بابل ه؟ 62 والسبيل ه 61 .corr ,واحد ب 60 فقد ب 59 س. ب
عل... vel عد... ه 70 شأن ه 69 وقواد الأجناد ه 68 من ب 67 جنب ه 66 اتسق ه 65
وأخرب ه 72 كله... رؤساء س. ب 71 .عُدَ Akhouaji proposed reading ,عتل vel
قد عرف فضله وينصب للحطأ نصب س. ه من 75 إيسار ب 74 المالكة ه؟ 73 ومن قد قتل الغزاة, sed sign of correction. 76 ه وسباب الأتة 77 ه. 78 س. ه والعلم الحزم 79 الوصعية
corr. sup. lin. et interlin. 80 ه واستنفاذ 81 ه؟ ب ألم 82 ه وبيانه 83 ه بيان 84
س. ه 90 السطوج ه 89 برمذا ه؟ 88 استفانا ه؟ 87 الشعر ه 86 حكم ه 85 أعرقه ه

[105] Al-Mubarrad, *Kāmil*, i, p. 328, also al-Ṭabarī, ii, p. 287; ed. De Goeje, i, p. 1965 (there as part of a longer poem). Meter: *ṭawīl*.
[106] Cf. *Aghānī*, xiv, p. 83ff.
[107] Cf. *Aghānī*, xiv, p. 71.

س. ه 95 راذا ب:رادا ه 94 كثيرة ب 93 العرب عن فكرة ب 92 المسالك والمذاهب ب 91 96 ب. س قد 97 ب لصنوف 98 الساد ب, ة sup. lin., س. ه, sign of correction 99 ه. س إن 100 والرفاضة ب 101 الروية فيه 102ه يجعل ه, corr. 103 add. sup. lin. ه 104 ه يوما له corr. in marg. 105 وبره ب (or written together: الخيروبره) Al-Ḥajjāj's speech might continue until here. In this case one should read فأنتحل. 106 ب؟ وتجردا 107 ب طبيعته 108 ب وكان 109 ب أحفش 110 ما ما عرف فيها، (corr. by later hand), عرق فيه ه 112 قصيرة ب 111 الحالين ب 113 ه،ب ويرتكبئه 114 ه كان 115 ه جناز 116 ه. س 117 أعلم ٍس. والله 118 شبيبته. Perhaps read شبيبة. ب 121 جميع ب 120 الخير الرجال ب 119 والمصلحة ب 122 add. sup. lin. ب 123 ب تمليكنا vel sim.; corrected with Hind's comment on al-ʿAwtabī (Hinds p. 66, note 178), who seems to have taken it from al-Jāḥiẓ 124 illegible 125 ه فما 126 ه الدهن 127 corr. ب sup. lin. 128 رواق ب 129 عند ماء ب 130 الوصف ه 131 يكون ه 132 ما أثر ه 133 ناهلة ب 134 add. in marg. ه 135 ه؟ تكن ريا ه 136 فأحقرهم ه 137 فليس ه 138 س. ه 139 ليتعرر ب, perhaps read ليتعزز. I take it to mean that either the headache gets stronger or the wine-merchant should be punished by taking away and drinking his wine. Either way, خمار is indefinite. Note: *Aghānī* xiv, p. 71: فازداد سكرا. 140 ب ولدت ه:وادت (Daiber wrote the correct form in his copy). 141 يابه يابه 142 ب ينيك 143 ه. س فقال عنهم الله رضى 144 تنبت 145 ه آله ب:آله وأصحابه وسلم ه 147 يزعم ه 146. س

الأحنف بن قيس

Pellat's edition of al-Aḥnaf ibn Qays, based on ms. B, 81b–86b, is an integral part of the *Wuzarāʾ*. As it is accessible, only the variants in the Hyderabad manuscript (118–124) shall be given here (س: ساقط من الأصل):

663, 5 لك س.; 6 واللفظ القبيح الساقط; 7 كان س.; منظوماً مجموعاً; 9 تميم; 12 عن الإغراء; 18 رايه; 664,3 وعثمان; 4 وبعد ب:بعد ه; 5 فخذلهما وعائشة; 7 مع:في, but على علي في الحكمين corrected in marg. 8/9; 10/11 بنات شحاج وبنات عوج; 11 أخرى; 665,1 كان; 3 عجب; 4 قالت ب:قال ه; 7 فعلت ذلك أن corr. in marg.; حيله ب:حلمه ه; 9 زبّه وقال; أسألكم بالله; 10 خرشة الضبي من نرى; له س.; 11 كل في; 666,4 رجل ناساً; 7 حارثة; 9 حسد إياس; 10 حتى هجاه: فهجاه; 667,1 ممن; 3 بعلم في; 4 الخلال ب:الخصال ه; 5/6 لهم ... الناس س. (corr. in marg., starting with ولم, but most of it illeg. Note that the insertion of this addition is in 6 between بالحلم and عادا.); 668,1 خليفة; 2 وما با:ما هـ; 3 الأحنف بن قيس ولا رأينا ما رأينا; 7 ولا; 9 وذكروه; من; 10 قال س. (corr. in marg.); 11 حليم ب:ظليم هـ; 669,2 عنها ب:منها هـ; 4 صاحبتها; 5 فيه; 6 صلوات الله عليهم فالفسمية; 7 لهم س.; 10 وإنك لورهاء، أبو س.; 11 ولد الأحنف من شق حتار الإست حتى فتق وعوج (corr. in marg. من شق حتار مرتشق vel sim.); 12 قد س.; 14 غالبة; 670, 2 عورة ب:غرزة هـ; 7 وهو; ثلث; 10 الأوزجان; 13 14 lacuna deest; المصعب; الرسول; 15 لم يفرنا لم نفزه وإنا; 16 يعم; 671,1 يخص; 2 ما; 3 أ لا أنزل; 4 معوية وتوكيد; 5 تكلم; 8 أي والله ما فعلت; 9 رهبة فإنه لكذا وإن ابنه لكذا; 12 فقال له; 13 من كثير س.; رجأت ب:جأت هـ; 14 كبير; 672,1 الله corr. sup. lin.; 4 وقال له بعض; 5 صدق بمتنبن; 6 عمر وتنازعوا; 7 بالكلام وعمر وذكروا لسان (اسان); 8 مجلسهم; 9 رحمه الله تعالى س.

Addenda

i) Ms. B, fol. 103a–b: *Wukalāʾ*, ed. Bū Mulḥim, p. 229, l. 3–5; p. 229, l. 10–p. 230, l. 1; fol. 104b, 'vanité': *Tanabbul*, ed. Bū Mulḥim, *Rasāʾil adabiyya*, p. 131, l. 6–8.

ii) For similar expressions, cf. *Ḥakamayn*, ed. Bū Mulḥim, *Rasāʾil siyāsiyya*, p. 340, l. 24–p. 341, l. 2; p. 357, l. 15f.; p. 358, l. 24f.; p. 360, l. 12; p. 367, l. 19f.; p. 384, l. 10.

iii) Among the ca. 352 *fuṣūl*, many are identical with or similar to edited works. Some seem to have been written prior to the edited works and probably stem from personal notes.

iv) Also Abū Fidāʾ, *Tārīkh*, ed. M. Dayyūb, 2 vols. (Beirut, 1997), mentions only Bābak's execution (i, p. 345).

v) For the common order of the executions see also Ibn al-Jawzī, *al-Muntaẓam fī tārīkh al-mulūk wa'l-umam*, ed. M. and M. ʿAṭā, 18 vols. (Beirut, s.a.), xi, p. 100; Ibn Khaldūn, *Tārīkh*, 8 vols. (Beirut, [2]2003), iii, p. 328; ʿAbd al-Qādir al-Baghdādī, *al-Farq bayn al-firaq* (Beirut, [3]2005), p. 202.

DIE ENTSTEHUNGSGESCHICHTE DER ANTHOLOGIE IM *SHARḤ AL-ASHʿĀR AL-SITTA AL-JĀHILIYYA* DES BAṬALYAWSĪ

Lotfi Toumi

Mit *Sharḥ al-ashʿār al-sitta al-jāhiliyya* betitelten einige Quellen den Kommentar des Baṭalyawsī zu den Dīwānen („Gesamtdichtung")[1] der sechs arabischen vorislamischen Dichter Imruʾu l-Qays[2] ibn Ḥujr, al-Nābigha al-Dhubyānī, ʿAlqama ibn ʿAbada, Zuhayr ibn Abī Sulmā, ʿAntara ibn Shaddād und Ṭarafa ibn al-ʿAbd.

Abū Bakr ʿĀṣim ibn Ayyūb al-Baṭalyawsī (gest. 494/1100 oder 1101), ein Literat und zeitweise ein Justizminister (*ṣāḥib al-maẓālim*)[3] am Hofe der Banū 'l-Afṭas in Baṭalyaws (dem heutigen Badajoz an der spanisch-portugiesischen Grenze), unterteilte sein Buch in zwei Teile;[4] im ersten

---

[1] Zur genauen Bedeutung des Wortes „Dīwān" siehe A.A. Duri u.a., 'Dīwān' in *EI*²; vgl. dazu auch A. Müller, *Der Islam in Morgen- und Abendland*, 2 Bde. (Berlin, 1885–87), i, S. 42 (n. 1) und 273 und M. van Berchem, *La propriété territoriale et l'impôt foncier sous les premiers califes* (Genf, 1886), S. 45 (n. 2).

[2] Dieser altarabische Name ist eine Genitivverbindung, die aus folgenden zwei Wörtern besteht: „Imruʾ" und „al-Qays". Wie alle arabischen Namen, die aus Genitivverbindungen bestehen, wird er unterschiedlich dekliniert je nach Kasus; so lautet dieser Name im Nominativ Imruʾu l-Qays, im Akkusativ Imraʾa l-Qays und im Genitiv Imriʾi l-Qays. Aus diesem Grund erscheint er in westlichen orientalistischen Quellen in unterschiedlichen Formen, wie z.B. „Imruulqays" (siehe Ahlwardts Einleitung in al-Aʿlam al-Shantamarī, *al-ʿIqd al-thamīn fī dawāwīn al-shuʿarāʾ al-sitta al-jāhiliyya*, ed. W. Ahlwardt [London, 1870], S. II), oder: „Imraʾalqays" (siehe R. Jacobi, *Studien zur Poetik der altarabischen Qaṣide* [Wiesbaden, 1971], S. vii; E. Wagner, *Grundzüge der klassischen arabischen Dichtung*, 2 Bde. [Darmstadt, 1987–88] i, S. 208; T. Bauer, *Altarabische Dichtkunst. Eine Untersuchung ihrer Struktur und Entwicklung am Beispiel der Onagerepisode*, 2 Bde. [Wiesbaden, 1992] i, S. 280 und ders., *Liebe und Liebesdichtung in der arabischen Welt des 9. und 10. Jahrhunderts* [Wiesbaden, 1998], S. 149). In diesem Artikel transkribiere ich diesen Namen nicht als Kompositum, sondern als Genitivverbindung—wie er im arabischen Original lautet—immer im Nominativ.

[3] Zu dem Amt des *ṣāḥib al-maẓālim* in der islamischen Jurisprudenz siehe al-Māwardī, *al-Aḥkām al-sulṭāniyya*, ed. M. al-Sarjānī (Kairo, 1978), S. 86–107.

[4] Zu dieser Aufteilung schreibt al-Baṭalyawsī am Ende des Dīwāns von ʿAlqama Folgendes (*Sharḥ al-ashʿār al-sitta* Teil 1, ed. N.S. ʿAwwād [Baghdad, 1979], S. 618):

تمّ شعر علقمة بحمد اللّه وعونه وحسن توفيقه، وهو آخر الجزء الأوّل. يتلوه شعر زهير، أوّل الجزء الثّاني إن شاء اللّه تعالى.

> [Der Kommentar zur] Dichtung des ʿAlqama ist [hiermit]—mit Gottes Hilfe und dessen großer Unterstützung—beendet. Gott sei gepriesen dafür. [Dieser Kommentar] ist der letzte Abschnitt im ersten Teil. Es folgt—so Gott der Erhabene

kommentierte er die Dīwāne der ersten drei Dichter, im zweiten die Dīwāne der anderen Dichter.

Sein Kommentar wurde in den Quellen unterschiedlich betitelt. Ibn Khayr[5] (gest. 575/1179)[6] und Ibn Khaldūn[7] (gest. 808/1406)[8] nennen ihn *Sharḥ al-ashʿār al-sitta al-jāhiliyya*, „Kommentar zu den sechs vorislamischen Dichtungen", während al-Marrākushī[9] (gest. 703/1303)[10] ihn unter dem Titel *Sharḥ ashʿār al-sitta*, „Kommentar zu der Dichtung der sechs [vorislamischen Poeten]" kennt.

Obwohl die zweite Version des Titels besser formuliert und daher wahrscheinlicher ist, bevorzuge ich hier den Titel, welchen Ibn Khayr und Ibn Khaldūn nennen, weil diese Version, die von Ibn Khayr stammt, die älteste ist; zumal die Informationen des Ibn Khayr über dieses Buch direkt von Ibn al-Milḥ, dem Schüler des Baṭalyawsī, stammen.[11]

Wie kommt aber al-Baṭalyawsī auf diese Anthologie, die er in seinem Buch kommentiert hat?

Wichtig zu erwähnen ist, dass dieselbe Anthologie, bis auf wenige Abweichungen, in der Epoche Baṭalyawsīs auch von einem anderem Andalusier namens al-Aʿlam al-Shantamarī[12] (gest. 476/1084)[13] kom-

will—[der Kommentar zur] Dichtung des Zuhayr, welcher den ersten Abschnitt des zweiten Teils ausmacht.

5 Ibn Khayr, *Fahrasa*, ed. F. Codera und J. Ribera Tarrago, 2 Bde. (Zaragoza, 1894–95), i, S. 389.

6 Ibn al-Abbār, *al-Takmila li-Kitāb al-ṣila*, ed. ʿI. al-ʿAṭṭār al-Ḥusaynī (Kairo, 1955), ii, S. 525 und H. Derenbourg, *Les manuscrits arabes d'Escurial*, 4 Bde. (Paris, 1884–1941), iii, S. 200.

7 Ibn Khaldūn, *al-Taʿrīf bi'bn Khaldūn wa-riḥlatuhu gharban wa-sharqan*, ed. M. ibn Tāwīt al-Ṭanjī (Kairo, 1951), S. 17; dort meint Ibn Khaldūn eigentlich das Buch des Aʿlam al-Shantamarī, aber sowohl sein Kommentar als auch der Kommentar des Baṭalyawsī tragen in einigen Quellen denselben Titel, wie ich später näher erläutern werde.

8 ʿU. Kaḥḥāla, *Muʿjam al-muʾallifīn* (Damaskus, 1957–61), v, S. 188.

9 Al-Marrākushī, *al-Dhayl wa'l-takmila*, ed. M. ibn Sharīfa und I. ʿAbbās (Beirut, 1964–), v/1, S. 232.

10 Siehe die Einleitung des Herausgebers in al-Marrākushī, *al-Dhayl wa'l-takmila*, iv, S. ـه.

11 Ibn Khayr, *Fahrasa*, i, S. 389.

12 Die *nisba* „Al-Shantamarī" bezieht sich auf seine Geburtsstadt Shantamariyyat al-Gharb (Santa Maria de Algarve); vgl. die Einleitung des Herausgebers in Ṭarafa ibn al-ʿAbd, *Dīwān*, ed. M. Seligsohn (Paris, 1901), S. xi (n. 1).

13 Zu seinem Todesjahr siehe Ibn Khallikān, *Wafayāt al-aʿyān*, ed. I. ʿAbbās (Beirut, 1968–72), vii, S. 82 und Seligsohn in Ṭarafa ibn al-ʿAbd, *Dīwān*; vgl. dazu auch Ibn Bashkuwāl, *Kitāb al-ṣila*, ed. ʿI. al-ʿAṭṭār al-Ḥusaynī (Kairo, 1955), ii, S. 644.

mentiert wurde. Hier stellt sich die Frage: wer von den beiden und ob überhaupt einer von ihnen zuerst diese Anthologie zusammengestellt hat?

Im Gegensatz zu al-Baṭalyawsī gibt al-Aʿlam in der Einleitung seines Kommentars bekannt, dass er die Anthologie selbst zusammengestellt hat; dazu schreibt er:

رأيتُ أن أجمعَ من أشعار العرب ديوانًا يُعين على التّصرّف في جُملة المنظوم والمنثور، وأن أقتصرَ منها على القليل؛ إذ كان شعرُ العرب كلّه متشابهَ الأغراض متجانسَ المعاني والألفاظ، وأن أوثِرَ بذلك من الشّعر ما أجمعَ الرُّواةُ على تفضيله، وآثر النّاسُ استعماله على غيره. فجعلتُ الدّيوانَ متضمّنا لشعر امرئ القيس بن حُجْر الكنْدي، وشعرَ النّابغة زياد بن عَمْرو الذّبياني، وشعر علقمة بن عَبَدة التّميمي، وشعر زهير بن أبي سُلمىَ المُزَني، وشعر طرفة بن العَبْد البَكري، وشعر عنترة بن شدّاد العبسي.[14]

> Ich beschloss, aus der [vorislamischen] Dichtung der Araber eine Anthologie zu erstellen, die das Verständnis der gesamten [klassischen] Poesie und Prosa erleichtern soll. Dabei beschränkte ich mich auf einen kleinen Anteil [dieser Poesie], weil sie im allgemeinen die gleichen Motive, Semantik und Lexikographie hat. [Bei diesem Vorhaben] bevorzugte ich die Dichtung, über deren Vorzüglichkeit ein Konsens unter den Tradenten herrscht, und die von den Menschen am meisten gelesen wird. So erstellte ich also diese Anthologie aus den Dīwānen des Imruʾu ʾl-Qays ibn Ḥujr al-Kindī, des Nābigha Ziyād ibn ʿAmr al-Dhubyānī, des ʿAlqama ibn ʿAbada al-Tamīmī, des Zuhayr ibn Abī Sulmā al-Muzanī, des Ṭarafa ibn al-ʿAbd al-Bakrī und des ʿAntara ibn Shaddād al-ʿAbsī.

Diese Behauptung wird aber von dem Bibliographen Ibn Khayr, der für seine wissenschaftliche Genauigkeit bekannt war, nicht bestätigt, denn dieser schreibt über die Anthologie in dem Kommentar des Aʿlam Folgendes:

> ... يرويها الأستاذُ أبو الحجّاج الأعلم المذكور، عن الوزير أبي سهل يونس بن أحمد الحرّاني، عن شيوخه أبي مروان عبيد الله بن فرج الطوطالقي وأبي الحجّاج يوسف بن فضالة وأبي عمر بن أبي الحبّاب،كلّهم يرويها عن أبي علي [القالي][15] البغدادي، عن أبي بكر عن أبي حاتم، عن الأصمعي رحمه الله.[16]

> ... überliefert hat sie der [bereits] erwähnte Gelehrte Abū ʾl-Ḥajjāj al-Aʿlam, und zwar über den Wesir Abū Sahl Yūnus ibn Aḥmad al-Ḥarrānī, der sie seinerseits über seine Lehrer Abū Marwān ʿUbayd Allāh ibn Faraj al-Ṭūṭāliqī, Abū ʾl-Ḥajjāj Yūsuf ibn Fadhāla und Abū

---

[14] Imruʾu ʾl-Qays, *Dīwān*, ed. M. Abū ʾl-Faḍl Ibrāhīm (Kairo, 1958), S. 3f.

[15] Diese Ergänzung ist notwendig, weil dieser berühmte Tradent und Literat eher unter dieser *nisba* bekannt ist.

[16] Ibn Khayr, *Fahrasa*, i, S. 389.

> ʿUmar ibn Abī ʾl-Ḥabbāb überliefert hat. Sie alle[17] tradierten diese Dichtung über Abū ʿAlī [al-Qālī] al-Baghdādī, der sie seinerseits über Abū Bakr ibn Durayd tradierte; und dieser überlieferte sie über Abū Ḥātim, der sie bei al-Aṣmaʿī—Gott erbarme sich seiner—gehört hat.

Aus dieser Information geht noch nicht hervor, ob die von al-Aʿlam kommentierte Anthologie von ihm selbst zusammengestellt wurde, und zwar aus der übrigen Dichtung, die ihm überliefert wurde, oder ob er sie von den alten Tradenten zusammengestellt überliefert erhalten hat und nun beschloss, sie zu kommentieren. Aber folgende Stelle in derselben Bibliographie des Ibn Khayr lässt keinen Zweifel mehr daran, dass al-Aʿlam die Anthologie der sechs Dīwāne überliefert bekommen und nicht selbst erstellt hat, wie er behauptet:

> ...وممّا ذِكره أبو الحجّاج الأعلم ممّا أخذه عن أبي سهل الحرّاني ما لم يتقدّم ذكره قبل: شعر السُّلَيْك بن السَّلَكة وقصيدة عمرو بن كلثوم وقصيدة لقيط بن مَعْمَر[18] الإيادي وشعر الأسود بن يعفُر[19] وشعر حاتم بن عبد الله الطائي وشعر زيد الخيل والأشعار السّتّة الجاهليّة التي شرحها.[20]

> ...und unter den Werken, die Abū ʾl-Ḥajjāj al-Aʿlam bei Abū Sahl al-Ḥarrānī gehört und [später] erwähnt hat, welche bis jetzt [in diesem Buch] noch nicht aufgezählt wurden: die Poesie des Sulayk ibn al-Salaka, die Qaṣīde[21] des ʿAmr ibn Kalthūm, die Qaṣīde des Laqīṭ ibn Maʿmar al-Iyādī, die Dichtung des Aswad ibn Yaʿfur, des Ḥātim ibn ʿAbd Allāh al-Ṭāʾī, des Zayd al-Khayl und *al-Ashʿār al-sitta al-jāhiliyya*, die er [al-Aʿlam] kommentiert hat.

Nach Ibn Khayr war also das Buch *al-Ashʿār al-sitta al-jāhiliyya* eindeutig eines unter vielen anderen Werken, die al-Aʿlam bei seinem Lehrer Abū Sahl al-Ḥarrānī studiert hat.

Dies beantwortet auch die Frage, warum sowohl al-Baṭalyawsī als auch al-Aʿlam fast die gleiche Anthologie in derselben Epoche kommentiert haben; und zwar, weil diese Anthologie zu ihrer Zeit in Andalusien kursiert und die Aufmerksamkeit der Gelehrten erregt hat,

---

[17] Gemeint sind die zuletzt erwähnten drei Tradenten.

[18] Oder auch: Yaʿmar, wie er in anderen Quellen genannt wird; siehe C. Pellat, 'Laḳīṭ al-Iyādī,' in *EI*².

[19] Oder auch: Yaʿfir, wie er in manchen Quellen genannt wird; siehe C. Pellat, 'Al-Aswad b. Yaʿfur,' in *EI*².

[20] Ibn Khayr, *Fahrasa*, i, S. 398.

[21] *Qaṣīda* bedeutet ein langes arabisches (auch türkisches, persisches usw.) polythematisches Gedicht. Vgl. die Definition der *qaṣīda* bei Jacobi, *Studien*, S. 6; Wagner, *Grundzüge*, i, S. 61ff und im G. Lecomte, 'Ḳaṣīda,' in *EI*².

so dass al-Baṭalyawsī und al-Aʿlam es beide interessant fanden, sie zu kommentieren.

Bestimmte Stellen in al-Baṭalyawsīs *Sharḥ al-ashʿār al-sitta al-jāhiliyya* bestätigen diese These. Er spricht zuweilen von unterschiedlichen Varianten in den *Abschriften*, die er als Grundlage seines Kommentars benutzt hat, so schreibt er an einer Stelle:

وجدتُه في بعض النسخ الصحاح: «أشذّ»، بالذال المُعجمة...[22]

...ich las ihn [den Begriff] in bestimmten authentischen Abschriften: *ashadhdh*, mit dem punktierten *dhāl*.

Ein paar Seiten weiter schreibt er wieder:

...وجدتُه في النّسخة الصّحيحة: «ويمنعها».[23]

...ich las ihn [den Begriff] in der authentischen Abschrift: *wa-yamnaʿuhā*.

Auch an folgender Stelle heißt es:

وجدتُ هذا البيتَ في النسخة...[24]

Ich fand folgenden Vers in der Abschrift...

ʿAwwād, der Herausgeber des ersten Teils dieses Kommentars, meint, dass diese Anthologie, auch wenn sie *mit an Sicherheit grenzender Wahrscheinlichkeit*[25] weder von al-Baṭalyawsī noch von al-Aʿlam erstellt wurde, so aber doch eine *rein andalusische Auswahl* sei. Diese Schlussfolgerung, die er aus den oben zitierten Sätzen des Aʿlam (S. 641) zieht, lautet folgendermaßen:

من هذا القول ندرك أنّ اختيار هؤلاء الشعراء لم يكن من عمل المشارقة ولم يكن من عمل القدامى دون تحديد [...] وإنّما هو اختيار أندلسي بحت، أراد به الأندلسيّون أن يجمعوا ما اعتقدوا أنّه أفضل الشعر الجاهلي وخير نماذجه في كتاب واحد، يكون في متناول المتأدّبين، فيدرسونه ويرجعون إليه، بدلا من الرجوع إلى عدّة دواوين متفرّقة.[26]

[22] Al-Baṭalyawsī, *Sharḥ al-ashʿār al-sitta*, Teil 1, S. 319.
[23] Ebd. S. 322.
[24] Meine Edition (im Erscheinen), iii, S. 360.
[25] Siehe die Einleitung des Herausgebers in: al-Baṭalyawsī, *Sharḥ al-ashʿār al-sitta*, Teil 1, S. 21.
[26] Ebd. S. 20.

> Durch diese Sätze [des Aʿlam] erkennen wir, dass die Auswahl dieser Dichter nie das Werk der „Mashriqīten“[27] oder einiger bestimmter „frühislamischer“ Philologen war, sondern sie ist eine rein andalusische Auswahl, die die Andalusier aus bestimmten Exemplaren der—ihrer Meinung nach—besten vorislamischen Poesie in einem einzigen Buch zusammenstellen wollten. Dieses Buch sollte den Lernenden zur Verfügung gestellt werden, so dass sie es studieren und als eine Quelle benutzen können, anstatt dass sie [immer] auf verschiedene verstreute Dīwāne zurückgreifen [müssen].

ʿAwwāds Argument für seine vermeintliche Erkenntnis ist widersprüchlich. Denn auf der einen Seite betrachtet er al-Aʿlams Behauptung, die besagt, dass dieser die Anthologie selbst erstellt habe, als Beweis für den andalusischen Ursprung der Anthologie; auf der anderen Seite aber vermutet er stark, dass weder al-Aʿlam noch al-Baṭalyawsī die Anthologie zusammengestellt, sondern sie *höchstwahrscheinlich* überliefert bekommen haben; dazu schreibt ʿAwwād:

أغلب الظنّ أنّ هذه الأشعار وصلت الأعلم كما وصلت أبا بكر مجموعة في كتاب.[28]

> Es ist sehr wahrscheinlich, dass diese Dichtung sowohl zu al-Aʿlam als auch zu al-Baṭalyawsī bereits in einem Buch zusammengestellt in ihre Hände gelangt ist.

Nicht nur wegen dieser Widersprüchlichkeit, sondern auch durch eine Information von Ibn al-Nadīm (gest. 438/1047),[29] lässt sich ʿAwwāds These über den andalusischen Ursprung der Anthologie leicht widerlegen. Ibn al-Nadīm erwähnt unter den vielen Büchern, die al-Aṣmaʿī (gest. um 216/831)[30] im Irak verfasst hat, *Kitāb al-qaṣāʾid al-sitta*.[31] Dieses Buch bildet meiner Meinung nach den Hauptteil der Anthologie in den Kommentaren des Baṭalyawsī und des Aʿlam. Dass Ibn al-Nadīm die Anthologie des Aṣmaʿī *Kitāb al-qaṣāʾid al-sitta* und nicht *Kitāb al-ashʿār al-sitta* nennt, dürfte uns nicht allzu sehr irritieren, da mit *qaṣāʾid* hier

---

[27] „Mashriqīten“ wird aus dem arabischen Wort *mashriq* (Osten) abgeleitet und bezeichnet—im Gegenteil zu „Maghrebiner“—die Araber aus dem östlichen Teil der arabischen Welt. Dieser östliche Teil umfasst die gesamte arabische Halbinsel und Ägypten.

[28] Siehe die Einleitung des Herausgebers in: al-Baṭalyawsī, *Sharḥ al-ashʿār al-sitta*, Teil 1, S. 21.

[29] Kaḥḥāla, *Muʿjam al-muʾallifīn*, ix, S. 41.

[30] Ebd., vi, S. 187.

[31] Ibn al-Nadīm, *al-Fihrist*, ed. M. al-Shuwaymī (Tunis, 1985), S. 251.

sicherlich die *muʿallaqāt*[32] gemeint sind; und weil die *muʿallaqāt* die wichtigsten und berühmtesten Gedichte in den sechs Dīwānen der jeweiligen Dichter sind, hat Ibn al-Nadīm meiner Vermutung nach den Begriff *ashʿār* in dem Titel der Anthologie mit dem Begriff *qaṣāʾid* entweder irrtümlich oder sogar absichtlich vertauscht, um dem Leser klar zu machen, um wessen Dichtung es sich in dieser Anthologie handelt, nämlich um die Dichtung der *muʿallaqāt*-Verfasser. Genau dasselbe Phänomen erleben wir auch bei al-Suyūṭī[33] (gest. 911/1505),[34] der den Kommentar des Baṭalyawsī zu den sechs vorislamischen Dīwānen kurz *Kommentar zu den Muʿallaqāt* nennt.

Wenn es nun verschiedene Indizien gibt, die darauf hinweisen, dass die Anthologie in den Kommentaren des Baṭalyawsī und des Aʿlam nicht von ihnen selbst zusammengestellt, ja sogar nicht einmal andalusischen Ursprungs ist, wie ist dann die Behauptung des Aʿlam, dass er beschlossen habe, *aus der Dichtung der Araber eine Anthologie zu erstellen*,[35] zu verstehen, besonders wenn man bedenkt, wie wichtig wissenschaftliche Ehrlichkeit und Genauigkeit traditionsgemäß für seriöse arabische Gelehrte war? Meine Antwort darauf ist, dass al-Aʿlam zwar die Anthologie des Aṣmaʿī als Hauptquelle für seinen Kommentar genommen, sie aber mit weiteren Gedichten der sechs vorislamischen Poeten bereichert hat, die al-Aṣmaʿī entweder nicht kannte, oder aus bestimmten Gründen nicht berücksichtigen wollte, und die al-Aʿlam anderen alten Poesiequellen entnommen hat. Dies ist keine von mir aufgestellte These, sondern eine Information, die al-Aʿlam klar seinen Lesern mitteilt; in der Einleitung seines Kommentars schreibt er:

---

32 *Muʿallaqāt*, auf Deutsch die „Angehängten", ist der Titel einer im 8. Jahrhundert erstellten Auswahl von 7, 9 oder 10 (vgl. dazu Jacobi, 'Allgemeine Charakteristik der arabischen Dichtung,' in H. Gätje [ed.], *Grundriss der arabischen Philologie*, ii [Wiesbaden, 1987], S. 7–31, hier S. 12) arabischen Gedichten, u.a. auch Oden berühmter Dichter aus vorislamischer Zeit (6.–7. Jahrhundert). In den frühen Quellen—wie z.B. Ibn Qutayba: *al-Shiʿr waʾl-shuʿarāʾ*—wurde diese Auswahl zuerst als *al-sabʿ* („die sieben") bzw. *al-sabʿ al-ṭiwāl* („die sieben Langen") bezeichnet. Den Titel *al-Muʿallaqāt* scheint diese Anthologie jedoch erst im 4./10. Jahrhundert von Abū Zayd Muḥammad ibn Abī ʾl-Khaṭṭāb al-Qurashī (*Jamharat ashʿār al-ʿarab*) erhalten zu haben (so Jacobi, *Allgemeine Charakteristik*, S. 12). Über die Bedeutung des Wortes *al-muʿallaqāt* herrscht allerdings Uneinigkeit. Eine mögliche Deutung ist die von C.J. Lyall, *Ancient Arabian Poetry* (London, 1930), S. xliv, der den Namen von *ʿilq* (Kleinod, kostbarer Gegenstand) ableitet. Vgl. dazu T. Nöldeke, *Beiträge zur Kentniss der Poesie der alten Araber* (Hannover, 1864), S. xviiff, ders., *Fünf Moʿallaqāt*, (Wien, 1899–1901), S. 8ff und G. Lecomte, 'Al-Muʿallaḳāt,' in *EI*².

33 Al-Suyūṭī, *Bughyat al-wuʿāt*, ed. M. Abū ʾl-Faḍl Ibrāhīm (Kairo, 1964–65), ii, S. 24.

34 Siehe die Einleitung des Herausgebers in: al-Suyūṭī, *Bughyat al-wuʿāt*, i, S. 15.

35 Siehe oben S. 641.

واعتمدت فيما جلبته من هذه الأشعار على أصحّ رواياتها وأوضح طرقاتها، وهي رواية عبد الملك بن قُريب الأصمعي، لتواطؤ الناس عليها واعتيادهم لها واتّفاق الجمهور على تفضيلها. وأتبعت ما صحّ من رواياته قصائدَ متخيّرة من رواية غيره.[36]

> Bei der Sammlung dieser Dichtung habe ich mich an die authentischste Überlieferung gehalten, deren Weg [zum Ursprung] am klarsten ist, nämlich an die Überlieferung des ʿAbd al-Malik ibn Qurayb al-Aṣmaʿī. [Dies tat ich,] weil diese Überlieferung die Zustimmung der Gelehrten[37] gefunden hat, die sich an sie gewöhnt haben und die sich untereinander über deren Vorzüglichkeit einig sind. Ferner habe ich der Dichtung, die sich als authentisch in al-Aṣmaʿīs Überlieferung erwiesen hat, hervorragende Qaṣīden aus anderen Überlieferungen hinzugefügt.

Weil al-Aʿlam sich mit der Anthologie des Aṣmaʿī, die bei Ibn al-Nadīm *Kitāb al-qaṣāʾid al-sitta* heißt, nicht begnügt, sondern sie durch weitere Gedichte aus anderen alten Poesiesammlungen vergrößert hat, betrachtete er die neu entstandene Anthologie in seinem Kommentar als seine Auswahl.

Auf dieselbe Art und Weise dürfte auch die Anthologie in al-Baṭalyawsīs Kommentar entstanden sein, auch wenn dieser im Gegensatz zu al-Aʿlam keine Details über seine Quellen in der Einleitung seines Buches verraten will, sondern darüber lediglich folgende vage Aussage macht:

...كلّ ما ذكرته في هذا الشرح فمن كتب العلماء أخذته ومن مكنون أقوالهم استخرجته.[38]

> ...alles, was ich in diesem Kommentar erwähnt habe, entnahm ich entweder den Werken der Wissenschaftler, oder ich leitete es aus dem Kern ihrer Erklärungen ab.

Auf die Frage, warum gerade diese sechs Dichter für die Anthologie ausgewählt wurden, schreibt al-Aʿlam—wie oben auf Seite 641 im Original bereits gezeigt wurde—dass er die Dichtung bevorzugt hat, „über deren Vorzüglichkeit ein Konsens unter den Tradenten herrscht, und die von den Menschen am meisten gelesen wird".[39] Wie es zu diesem *Konsens* überhaupt gekommen ist, glaubt Ahlwardt durch die Tatsache erklärt, dass gerade diese sechs Dichter

[36] Imruʾu ʾl-Qays, *Dīwān*, S. 4.
[37] Im Original: „der Menschen", anstatt: „der Gelehrten".
[38] Al-Baṭalyawsī, *Sharḥ al-ashʿār as-sitta*, Teil 1, S. 37.
[39] Imruʾu ʾl-Qays, *Dīwān*, S. 3f.

> Exercised a regulative and permanent influence on the literature of the succeeding centuries. Even though they found a certain form, so to speak, a certain fashion of composition already in vogue, yet they enriched it by elevation and splendour of diction, by variety and novelty of thoughts and images, and in part by the art of transition from one subject of description to another, and thus as it were re-constituted it a model of style.[40]

Als einen weiteren Grund dafür führt Ahlwardt auf, dass „the Compass of their compositions may be called large, in comparison with that of their contemporaries."[41] Darüber hinaus vertritt Ahlwardt die Ansicht, dass man diese sechs Dichter seit jeher vorgezogen hat, weil „their life was not so much implicated with petty local incidents as that of many of the earlier poets, but with memorable events and eminent historical personages, and therefore lent a higher interest to their poetry."[42]

Brockelmanns These für die auszeichnende Bewertung dieser sechs Dichter ist allerdings eine andere, dazu schreibt er:

> Unter der grossen Zahl der vorislamischen Dichter nehmen sechs als die berühmtesten die erste Stelle ein. Sie verdanken ihren Ruhm den Philologen, vermutlich einfach deswegen, weil sie von ihnen allein noch umfänglichere Diwane zusammenbringen konnten.[43]

---

[40] Ahlwardt in: al-Aʿlam ash-Shantamarī, *al-ʿIqd al-thamīn*, S. ii.
[41] Ebd.
[42] Ebd.
[43] *GAL S* i, S. 44.

BEFORE ARISTOTLE BECAME ARISTOTLE: PSEUDO-ARISTOTELIAN APHORISMS IN *ĀDĀB AL-FALĀSIFA*

Mohsen Zakeri

Much has been said about the origin, content and authorship of *Ādāb al-falāsifa* (= ĀF), one of oldest available collections of gnomologia in Arabic.[1] Since the early decades of the nineteenth century this book is wrongly assumed to be an abridgement of an original written or translated by Ḥunayn ibn Isḥāq (d. 260/873). Elsewhere I believe to have shown that ĀF is neither a work of Ḥunayn ibn Isḥāq nor a shorter recension of a previously existing text. Rather, it is an independent book which the fourth–fifth/tenth–eleventh-century author Muḥammad ibn ʿAlī ibn Ibrāhīm al-Anṣārī prepared by using several smaller tracts of a number of earlier authors, among them ʿAlī ibn ʿUbayda al-Rayḥānī (d. 219/834), al-Kindī (d. after 252/865), Ḥunayn ibn Isḥāq, Isḥāq ibn Ḥunayn (d. 296/908) and others.[2] At this stage it is not possible to ascertain whether anybody else prior to al-Anṣārī had compiled a book of similar content in the beginning of the fourth/tenth century. However, the anonymous Istanbul manuscript Köprülü 1608, a comparable but much larger collection, offers itself remotely as a possible model for al-Anṣārī.[3]

[1] ĀF, cf. Lo (the abbreviations are explained at the end of the article), and A. Loewenthal, *Honein Ibn Ishak, Sinnsprüche der Philosophen. Nach der hebräischen Übersetzung Charisi's ins Deutsche übertragen und erläutert* (Berlin, 1896). For studies of this book consult A. Müller, 'Über einige arabische Sentenzensammlungen,' *ZDMG* 31 (1877), pp. 506–28; A. Baumstark, *Syrisch-Arabische Biographien des Aristoteles* (Leipzig, 1898); K. Merkle, *Die Sittensprüche der Philosophen: Kitāb Ādāb al-Falāsifa* (Leipzig, 1921); and D. Gutas, *Greek Wisdom Literature in Arabic Translation. A Study of the Graeco-Arabic Gnomologia* (New Haven, 1975). For a new synopsis of the structure and contents of ĀF, see now O. Overwien, 'Ḥunayn ibn Isḥāq, Ādāb al-falāsifa. Griechische Inhalte in einer arabischen Spruchsammlung,' in R.M. Piccione and M. Perkams (eds.), *Selecta colligere, i. Akten des Kolloquiums, Jena, 21–23. November 2002* (Alessandria, 2003), pp. 95–115; and D. Gutas, 'The Spurious and the Authentic in Arabic Lives of Aristotle,' in J. Kraye (ed.), *Pseudo-Aristotle in the Middle Ages* (London, 1986), pp. 15–43.

[2] M. Zakeri, "ʿAlī ibn ʿUbaida ar-Raiḥānī: a Forgotten Belletrist (*adīb*) and Pahlavi Translator,' *Oriens* 34 (1994), pp. 76–102; idem, 'Ādāb al-falāsifa: the Persian Content of an Arabic Collection of Aphorisms,' *MUSJ* 57 (2004), pp. 173–90.

[3] Zakeri, 'Ādāb al-falāsifa,' pp. 185–90; Z i, pp. 59–73.

Ādāb al-falāsifa is clearly a composite work consisting of several distinct texts, of Greek, Persian, and early Islamic origin, the most conspicuous among them *Ādāb al-faylasūf Mahādharjīs al-muʿallim.*[4] Here as a token of my gratitude to Hans Daiber, who first encouraged me to work on the Arabic gnomologia many years ago, and always readily and cordially placed his vast erudition at my disposal, I offer a new edition and translation of another distinct unit in the ĀF, namely the so-called *Ḥikmat Arisṭūṭālīs*, 'Aristotle's Wisdom', the legendary circumstances of its creation I have discussed in detail at another occasion.[5] The content of this piece is nicely framed in a fabulous story related to the legendary 'gatherings of philosophers' in which the orphan Aristotle is to serve Plato who is teaching a good-for-nothing prince in one of the Houses of Wisdom (*buyūt al-ḥikma*). On the day of examination in the presence of the learned and the dignitaries of the empire, as the prince fails to demonstrate the fruits of Plato's teachings, Aristotle, who has secretly learned everything by heart, steps on the podium and with a brilliant public oration displays the Teacher's fruitful lessons and so rescues him. Deeply impressed by the ingenuity of the young boy, Plato now adopts him as his pupil to teach him all the sciences.

The framework story of 'Aristotle's Wisdom' does not properly fit Overwien's proposal of a preconceived methodical structure of the ĀF, though it could be taken as an example for illustrating the educational procedure in the houses of wisdom described in earlier chapters of the book.[6] Overwien interprets this story as part of the attempts by Alexandrian biographers of Aristotle in late antiquity to harmonize between the two great ancient philosophers.[7]

In the printed text the dicta of the delivered oration are not numbered and the divisions among the items are not always sharp and clear. Sometimes the sentences are attached to one another with a simple conjunctive 'and' without an apparent or inherent relationship between them. Consequently they have been divided differently in different editions. They are numbered here from 1 to 75 for the purpose of easy reference (Loewenthal has done the same in his German translation

[4] Mahādharjīs = Mihr Ādharjushnas. Consult Zakeri, "Alī ibn 'Ubaida ar-Raiḥānī,' pp. 97–102; and see a new edition and translation of Mihr Ādhar's *Ādāb* in Z, pp. 1010–28.

[5] Zakeri, 'Ādāb al-falāsifa,' pp. 185–90.

[6] Overwien, 'Ḥunayn ibn Isḥāq,' pp. 102, 110.

[7] Overwien, 'Ḥunayn ibn Isḥāq,' p. 110.

of the Hebrew version, but his division is slightly different from mine). No alphabetic or any other ordering principle is detectable in the text. However, sentences ns. 4, 7–25, 29, 30, 32 all start with the preposition *bi*, 'With…', and ns. 28, 54–5, 58–65, 68, 71, 74 all start with *man*, 'He who…'. The first four in praise of God and Wisdom are introductory comments by Aristotle himself, and the rest what he had purportedly learned secretly from Plato's lectures.

In addition to al-Anṣārī's ĀF, the following sources have the sermon more or less completely: the anonymous Köprülü 1608 (K folios 14v–15v),[8] al-Mubashshir ibn Fātik, *Mukhtār al-ḥikam* (Mb, pp. 199–201), al-Shahrāzūrī, *Nuzhat al-arwāḥ* (Sh i, 201–2), Ibn Abī Uṣaybiʿa, *ʿUyūn al-anbāʾ* (IAU, pp. 97–8), al-ʿĀmilī, *al-Mikhlāt* (ĀM, pp. 158–9), the Old Spanish version of ĀF, *El libro de los buenos proverbios* (*Libro*, pp. 58–61), and the Hebrew-German translations of ĀF by Loewenthal (Lo, pp. 64–8). Although these all reproduce the same original text, they are not fully identical with one another. Some lack several sayings (K has the frame story but misses ns. 30, 34, 36, 47–8, 50, 56, 59–60, 63, 66, 68–72, 74; the *Libro* includes the frame story but has left out a few apparently corrupt pieces in its original: ns. 28, 31, 34, 36, 63), or have additional ones (i.e. Mb); others contain divergent readings for certain words or phrases. Clusters of several sentences together have found their way into many works including Ibn Durayd, *al-Mujtanā* (p. 47; = MJ), al-Sharīf al-Raḍī, *al-Nahj al-balāgha* (p. 398; = NB), Miskawayh, *Jāwīdān khirad* (p. 12; = J), al-Ṭurṭūshī, *Sirāj al-mulūk* (pp. 50–51), al-ʿĀmilī, *al-Mikhlāt* (p. 69), al-Ibshīhī, *al-Mustaṭraf* (p. 53), and others. The text without the frame story is taken over fully by al-Mubashshir, but he reproduces about one-third of it again among Socrates' sayings (p. 118) without noticing their repetition, and some others independently and anonymously throughout his work (ns. 8, 15, 41, 67, 73).

All the sources just outlined are posthumous to Ḥunayn ibn Isḥāq. Another author who seems to have been familiar with our text even prior to the time of Ḥunayn is ʿAlī ibn ʿUbayda al-Rayḥānī, who lived and worked at the court of Caliph al-Maʾmūn (d. 218/833) and died not long after the caliph.[9] Al-Rayḥānī's *Jawāhir al-kilam* (= Z), now edited and translated, is a large compendium of over 2000 ancient proverbs and proverbial phrases alphabetically arranged as *ḥikam* and without

[8] See Gutas, *Greek Wisdom Literature*, pp. 42–7; Z, pp. 59–73.
[9] See Zakeri, "ʿAlī ibn ʿUbaida ar-Raiḥānī,' pp. 76–102.

any attribution to persons. This has at least half of the content of our text, some identical (cf. ns. 1, 4, 12–15, 17, 19–22, 27, 38–41, 55, 57, 62, 64–6, 72–4), some with editorial modifications (cf. ns. 9, 18, 24, 31, 44–5, 49–51, 54, 58, 67). The existence of these sentences already in the *Jawāhir al-kilam* is a good index to the fact that our sermon had been available to the Arabic reading public prior to the time of its alleged translation or composition by Ḥunayn ibn Isḥāq. In the studies already mentioned, I have shown that several other chapters in the ĀF have close affinity with al-Rayḥānī's work. Whether the text under review here had been one of this author's own previous compilations or translations from Middle Persian, which he also exploited for his *Jawāhir al-kilam*, remains still to be decided.

The works of the above-mentioned anthologists and the further documentation of the variants and parallels of the sayings offered in the following pages present a good testimony to the tremendous popularity of these dicta in the annals of classical Arabic literature. However, the reason for this great success is hard to perceive for a modern day reader. In accordance with its supposed origin, this short tract consists of didactic precepts of a universal nature. The contents and the message they convey are rather ordinary common sense without any surprising turn of thought or delightful ambiguity of formulation and cannot claim any uniqueness or exceptionality. They are not real proverbs with general applicability, but are certainly proverbial, mostly consisting of terse statements of a truth or dogma, hence true aphorisms. Their conciseness and melodious composition should in fact explain their success to some extent.

Nothing in the sayings can be taken specifically as Arab-Islamic, Persian or Greek. The presence of a semi-reference to the famous Socratic saying, 'I know that I do not know,' (cf. n. 40) cannot be—because it is so common and widespread in classical Arab literature—evaluated as a piece of evidence in determining its origin. This tendency may be contrasted with maxims having parallels in Pahlavi sources (cf. ns. 4, 37, 49). These aphorisms are as notorious as all other of their kinds in constantly being attributed to different authorities in accordance with the attitude of the source where we find them. Those educated in the Greek tradition of *paideia* had no difficulty to assign them to the renowned philosophers of the past; those coming from the Iranian background of *farhanq* saw them fit to be spoken only by their ancient great kings and sages; the more orthodox Muslims claimed to have heard them from the Prophet himself; and the Shiites had naturally Imām ʿAlī and his successors as their original composers. Since they were not engraved in stone or written down in other unchallengeable documents, we should be satisfied for now to ascertain some stages of their development and transmission. At any rate, the language of the presentation is solid, plain Arabic and shows no trace of translation.

No systematic line of thought or efficient argumentation is detectable in the *Ḥikmat Arisṭūṭālīs*. Its content may be summarized in general terms as follows: man's primal goal in life is to live a meaningful and praiseworthy existence (*ṭīb al-ʿaysh*) with honor and respect. The key to recognizing this goal and materializing it is *ḥikma*, 'wisdom', which is not defined any closer here, but it is replaced with *ʿaql* (intellect, wisdom), *ʿilm* (science, knowledge), or *adab* (etiquette, decorum) (ns. 1–7, 40, 44, 57–8, 74). Whoever achieves this Wisdom is regarded with reverence in the community (n. 68), and whoever misses it is exposed to the harm of ignorance (n. 58). The path of Wisdom is further characterized in terms of a set of vices to be avoided and virtues to be cultivated. Faculty of speech, its dangers and benefits, and the impact of man's outward behavior on social relations receive the biggest share.

Gentle words win affection in people's hearts and so enable man to fulfill his goals. Openheartedness perfects a joyful and pleasant life. While hasty answers cause stumbling (n. 41), keeping silence at the right time and place wins respect in people's eyes, though speaking the right words opportunely increases prestige and honor (ns. 9–13). Hence when not well-versed in a field or on a topic, one should keep silence (n. 57), because picking up more than one can carry becomes embarrassing (n. 65), rushing causes affliction, and acting hurriedly leads into catastrophe, whereas contemplation brings safety, and thinking before acting prevents regret (ns. 59–62). When not knowing something, it is better to say so (n. 40), and ask questions to learn (n. 64), because the well-informed is content not to dispute when having questions, but the uninformed is plunged into ignorance, beguiled by the arrogance of self-judgment, and withdrawn to fancy from the gate of ascertainment (n. 71).

Nothing is more apt in changing a prosperous life than injustice (n. 73), but justice wins the enemy over, equity begets friendship, modesty helps love grow, chastity purifies deeds, rendering favors brings respect, forbearance adds to one's supporters, and kindness subjugates hearts (ns. 14–20). These gains are further consolidated by benefaction, altruism, fidelity, and veracity (ns. 21–4).

Words of wisdom are parables used to convey good lessons to face the time and the hours that bear harm (ns. 25–8). Experiences are endless and the wise gathers more of them, because the experienced is wiser than the learned doctor. However, if knowledge exceeds understanding it becomes harmful (ns. 55–6, 66). In well-being, the good life can be cherished, but in adversity life becomes turbid (ns. 29–30). It is to remember that refusing to acknowledge favors received causes bereavement, and reminding a favor done annuls it (ns. 31–2); and that the niggard is despised even if he were rich, and the generous is loved even if he were poor (n. 37). Ill-natured is a peril to his companion, and hard-fisted is dull-sighted (ns. 35–6). Avidity is constant poverty, renunciation is visible wealth (ns. 38–9), and ardent desire is the cause of grief (n. 51). Having to endure the company of an imbecile is torture to the spirit; infatuation with women is stupidity; and sorrow over that which is passed away is a waste of time. The friend of an ignorant is exposed to danger; the risk-taker is a wrongdoer, and the one who takes risks endangers himself (ns. 47–50, 53). Patience confirms determination, and its fruits are relief and effacement of tribulation. However, to be impatient in the misfortunes of one's friends is better than to be patient, and to be patient in one's misfortune is better than to be impatient (ns. 52, 72). Reflection on such issues brings forth discernment, training sharpens talent, and acquired virtue makes ancestry irrelevant. All in all, righteousness is the attire of the learned, hypocrisy is the mantle of the ignorant (ns. 42–6).

قال حنين: هذا ما وجدت من حكمة أرسطو. أيها الأشهاد!

Ḥunayn said: This is what I have obtained from Aristotle's Wisdom. He announced to the public: [On the margin is written in Persian: أى حاضران!]

1- العلم موهبة الباري والحكمة عطية من يعطي ويمنع ويحط ويرفع.

1- Knowledge is God's blessing, and Wisdom is the gift of Him who grants and withholds, lowers and raises ranks. (ĀF 53; K 15r; Mb 199; IAU 97; Lo 64–5).

1- فإن الحكمة أعظم المواهب التي وهب الله لعباده.

1- Wisdom is a gift from God on High, who gives it to whom He wills. (Z 2032).

2- أفضل ما أعطي العبد في الدنيا الحكمة.
(Mj 62; J 6: 'Awshahanj'; Usāma 422.)
3- أفضل ما أعطي الرجل العقل والحلم. (Mj 40).
4- الدين أفضل المواهب التي وصلت من الله إلى خلقه وأعظمها منفعة وأحمدها في كل حكمة. (AṢ 33; J 190).
5- أفضل ما رزقهم الله تعالى ومن به عليهم العقل الذي هو الدعامة لجميع الأشياء والذي لا يقدر أحد في الدنيا على إصلاح معيشته ولا إحراز نفع ولا دفع ضرر إلا به.
(ImuqKal 42–3).
6- "ومن يؤت الحكمة فقد أوتى خيرا كثيرا." (Qurʾān 2:269; MawAd 51)
7- احمد الله على تدبيره * قدر الرزق وأعطى ومنع. (AAt 256)
8- الله هو المعطي وهو المانع. (TB xii, 93)

2- والتفاضل في الدنيا والتفاخر هما الحكمة التي هي روح الحياة ومادة العقل الرباني العلوي.

2- Rivalry for precedence and boasting among people should be based on Wisdom, which is the spirit of life and the substance of the high godly intelligence. (ĀF 53; K 15r: Mb 199; IAU 97; Lo 65)

1- الحكمة التي هي حياة للقلب الميت.

1- It is Wisdom that is like life to the heart of the dead (NB 132).

2- الحكمة حياة النفس. (ĀF 50)
3- الأدب حياة القلب. (Bayh 438)

3- التسبيح والتقديس لمعلم الصواب ومسبب الأسباب.

3- Glorification and worship are due the Master of truth and the Cause of all causes. (ĀF 53; K 15r; Mb 199; IAU 97; Lo 65).

4- بالعقول تفاضل الناس لا بالأصول.

4- Superiority of people to one another depends on intelligence not on pedigree. (ĀF 53; Mb 199; IAU 97; Sh i, 201; Lo 65:1; Z 1207; cf. with n. 44 below. On the margin of the Munich ms. 13b is written in Persian: نه بپدران ومادران).

1- الفضل بالعقل والأدب لا بالأصل والحسب لأن من ساء أدبه ضاع نسبه ومن قل عقله ضل أصله.

1- Merit depends on reason and education, not on origin and noble descent, that is because he whose education is defective his descent is of no avail, and he whose reason is insufficient his origin is in vain (MawAd 211; Ibsh 51).

2- لا يكون الشرف بالحسب والنسب ولكن من قبل الأفعال لأن الشرف إنما هو فيه لا في النسب. (Bayh 106).
3– لا يكون الشرف بالنسب لأن الشرف إنما هو بالفضل.

3- Honor does not come from birth, verily honor is in acquired merits. (*Maḥāsin* 163).

4- الشرف بالفضل والأدب لا بالأصل والنسب.

4- Nobility is due to merits and accomplishments, not to origin and descent. (Waṭwāt n. 79; Ikhtiyār 129).

5- الفضل بالعقل والأدب لا بالأصل والنسب.

5- Merit is due to intelligence and education, not to origin and noble birth. (Sagh 50).

6- الشرف بالعقل والأدب لا بالأصل والنسب.

6- Honor lies in the mind and in acquired worth, not in origins and noble birth. (Qābūs 27, trans. 22).
7- Be not reliant on kindred [*payvand*] and great ancestry [*tukhmak*], since in the end, dependence is on one's own deeds. (*Mainyo-i-khard* 133; 2:108–9).

8- بزرگی و ارز در خرد باشد نه در بزرگواری تبار وخاندان. (Dihkhudā i, 253).

8- Honor and eminence are in wisdom, not in noble family and ancestry.

9- گوهر تن از گوهر اصل بهتر است.

9- Personal quality in the individual is nobler than noble descent. (Qābūs 27, trans. 22).

10- أرفع منازل الشرف لأهله العلم والأدب.

10- The highest ranks of nobility to noblemen are knowledge and refined culture. (IAB i, 110: 'Buzurjmihr').

11- الشرف بالفضل والأدب.

11- Nobility comes from merit and cultured manners. (Dihkhudā i, 253).

12- شرف الحسب يحتاج إلى شرف الأدب.

12- The honor of noble descent is in need of the honor of education. (Rāghib i, 31)

13- لا شرف مع سوء الأدب.

13- There is no honor with ill manners. (IQut i, 111: '*fī kitābin li'l-Hind*' = *Kalīla wa-Dimna*; Mb 326; Ṭurṭ 53, 172; Waṭwāt n. 17).

14- سوء الأدب يهدم ما بنى الأسلاف.

14- Ill manners destroy what the ancestors have founded. (Ibn Hindū 353, n. 254: 'Aristotle'; Mb p. 195; IAU 100; Mustaʿṣimī 126).

15- من لم يكن له علم ولا أدب لم يكن له حسب ولا نسب. (Rāghib i, 31)
16- عز الشريف أدبه وحصنه صيانته ونسب الحكيم حكمته.

16- The noble man's honor is education, his fortress is chastity; and the noble lineage of the wise man is wisdom. (Z 1207; cf. also ns. 28 and 547; MJ 45; Ābī v, 191; Usāma 229; Mustaʿṣimī 167)

17- شرف الأدب مستغن عن شرف الحسب.

17- The honor of education is not in need of the honor of noble descent. (Rāghib i, 31)

18- تو را پرسند هنرت چیست و نه گویند پدرت کیست. (سعدی)

18- You will be asked what your virtues are, not who your father was. (Haim 120).
19- Manners make the man. (CDP 176)

20- لا خير فيمن له أصل بلا أدب * حتى يكون على ما زانه حدبا. (Marzub 12).
21- لكل شيء حسن زينة * وزينة العالم حسن الأدب.
قد يشرف المرء بآدابه * فينا وإن كان وضيع النسب. (Yāqūt 20).
22- من كان مفتخرا بالمال والنسب * فإنما فخرنا بالعلم والأدب
لا خير في رجل حر بلا أدب * لا لا و إن كان منسوبا الى العرب. (Yāqūt 20).

وعيت عن أفلاطون الحكيم

I have memorized these by heart from the wise Plato:

5- الحكمة رأس العلوم.

5- Wisdom is the head of all sciences. (ĀF 53; K 15r, + وبها تفاضلت العقول; Mb 191, 199; IAU 97; Sh i, 201; *Libro* 58; Lo 65:2).

1- الحكمة رأس التدبير.

1- Wisdom is the basis of planning.[10]

2- رأس الحكمة حسن الخلق.

2- The principle of wisdom is good character. (Mb 110: 'Socrates'; Sh i, 152; Alon 75 n. 512)

3- رأس العلم الرفق وآفته الخرق.

3- Gentleness is the basis of knowledge, roughness is its bane. (Qud-Dus 20).

4- الرفق رأس الحكمة.
4- Gentleness is the apex of wisdom. (QudSh 4; Ibsh 305).
5- Wisdom is better than strength. (Bible, *Eccl.* 9:16)
6- By wisdom one attains the top in affairs. (MaxAli 19).

---

[10] 'A.R. Badawī, *Sirr al-asrār*, in his *al-Uṣūl al-yūnāniyya li'l-naẓariyyāt al-siyāsiyya* (Cairo, 1956), p. 75.

7- God has given His creatures nothing to place higher than reason. (MaxAli 19).
8- The chief of the talents is knowledge. (MaxAli 14).
9- Knowledge leads to wisdom; accordingly the educated man is the wise one. (MaxAli 14). [11]

6- والآداب تلقيح الأفهام ونتائج الأذهان.

6- Acquired virtues are the pollination of brains, and the fruits of minds. (ĀF 53; K 15r; Mb 199; IAU 97; Sh i, 202; Lo 65:2.)[12]

1- الأدب لقاح العقول وغذاؤها.

1- Wisdom is the pollination and nutrition of intellects. (Thaʿāl Tam 159).

2- ليس غذاء الطعام بأسرع في نبات الجسد من غذاء الأدب في نبات العقل. (AṢ 15).
3- غذاى خرد حكمت است.
3- The food of wisdom is derived from philosophy. (Qābūs 263, trans. 261).

4- توتياى چشم خرد حكمت است.

4- Philosophy is the antimony and collyrium of the eye of the mind. (Qābūs 49, trans. 44).
Wisdom is the antimony of the mind.
5- التجارب لقاح العقول.

5- Experiences are the pollination of minds. (Abū al-Maʿālī 28).

6- المشورة لقاح العقول ورائد الصواب. (Thaʿāl Tam 417; Sh/NB xviii, 383).

[11] D.M. Donaldson, 'Aphorisms in Islamic Ethics,' *MW* 36 (1946), p. 240, has a sentence of similar structure that reads, 'The head of wisdom is the necessity of the truth.' Quoted by Alon, p. 135.

[12] The idea that the intelligence can grow and flourish only with *adab* is developed in *al-Adab al-ṣaghīr*, pp. 12–15. The Old Spanish, following perhaps the Hebrew version, combines the above two sentences as an indirect comment on wisdom and offers a better reading: 'Apris de Platon el philosoph la philosophia que es cabeca de todos los saberes e los ensenamientos buenos; es el ffructo de los entendimientos y conclusion de los sesos.' (*Libro* 58).

7- المشورة لقاح العقل ورائد العقل وحزم التدبير. المشاورة قبل المساورة والمشورة عين الهداية. (Ḥuṣrī 824).
8- النظر في العواقب تلقيح العقول.

8- Thinking about the consequences is the pollination of minds. (AUAm 217; AUGhar i, 243; AskariAmth ii, 332; Fārābī ii, 347; Mayd iii, 119, 131; ZamAm i, 353; Freytag iii, 512).

9- العقول تلقيح العقول. Minds pollinate minds. (Mb 323).

7- بالفكر الثاقب يدرك الرأي العازب.

7- With a sound and penetrating mind, good judgment is reached at. (ĀF 53; K 15r; Mb 199; IAU 97; Sh i, 202; ĀM 158; *Libro* 58; Lo 65:3; IR 193).

1- بإجالة الفكر يستدرك الرأي المصيب. (Mj 47; Ṭurṭ 72; IAB i, 450).
2- تود عدوي ثم تزعم أنني * صديقك إن الرأي عنك لعازب.

2- You befriend my enemy and still think that I am your friend? Verily, good judgment has escaped your mind! (Bashshār 23; IQut iii, 4; TawḥBas i, 32; Sh/NB xx, 15).

8- وبالتأني تدرك المطالب.

8- With circumspection, demands are fulfilled. (ĀF 53; K 15r; Mb 199; IAU 97; Sh i, 202, 160: 'Socrates'; ĀM 158; *Libro* 58; Lo 65:4; Sh/NB xix, 228).
All things come to those who wait. (CDP 4).

1- بالتأني تسهل المطالب.

1- Deliberateness facilitates achieving one's goals. (Alon 74 n. 537; Mj 47; K. 39v; Mb 118; Ṭurṭ 50; Sh i, 202; IAU 97; ĀM 158; MawAd 263; IR 193; Ibsh 53).

2- بالرفق تنال الحاجة وبحسن التأني تسهل المطالب. (Sh/NB xx, 263).
3- بحسن التأني تدرك المطالب وبالنصفة يكثر المتواصلون. (Ṭurṭ 172–3).
4- بالله تدرك المطالب وإليه تبذل الراغب. (Murādī 48).

5- بالتأني تدرك الفرص. (Tha'ālTam 420).
6- بالتأني يدرك الغرض.

6- He who goes slowly goes surely. (Tha'ālTam 43: '*min amthāl Furs*').

7- بالتأني تدرك الفرص والغرض. (Freytag iii, 23)
8- عليك باللين والرفق والحلم والتأني. (IMuqKal 208).
9- من تأنى أدرك ما تمنى.

9- He who is patient achieves what he desires. (Mayd iii, 362).

10- كفى بالإنسان سعادة أن تسهل عليه المطالب فيدرك مراده بأهون سعي وأقل عناء. (al-Māwardī *Tashīl* 68).

9- وبلين الكلمة تدوم المودة في الصدور.

9- With gentle words affection endures in the hearts. (ĀF 53; K 15r; Mb 199; IAU 97; Sh i, 202, 160: 'Socrates'; ĀM 69, 158; *Libro* 58; Lo 65:5).

1- من لانت كلمته وجبت محبته.

1- He whose words are gentle, man must love him. (Z 2442; with more examples; ĀF 153; IAR ii, 279; Murādī 79).

2- وجبت محبة من لانت كلمته.

2- He who speaks kindly deserves to be loved. (JahB ii, 174; ĀF 153; al-MubarKam i, 64; IAR ii, 279, 310; IWahb 291; Rāghib i, 277; Mb 281; Mayd iii, 364; IḤamd iv, 359; Sh/NB xix, 35; Sh i, 317; Z 1852, with more examples).
He whose word is gentle, to love him is a duty. (Kassis 131)

3- هرك سخن وى نرم گشت دوستى وى واجب شد. (*Khirad-nāma* 103).
4- من لانت كلمته استحق من الجميع المحبة.

4- The soft-spoken is worthy of being loved by all. (Mb 281; Ishkiwarī 374).

5- بلين كنف المعاشرة تدوم المودة.

5- Flexibility in relationship prolongs friendship. (Mb 118: 'Socrates'; Mj 47; Ṭurṭ 50).

6- بحسن المعاشرة تدوم المحبة.

6- Good companionship prolongs love. (Ibsh 53).

7- بسلامة الصدر تتأكد المحبة في القلب. (J 18: 'Awshahanj').
8- الكلمة اللينة تلين من القلوب ما هو أشد من الحديد.

8- Gentle words soften hearts harder than iron. (Z 284).

10- وبخفض الجناح تتم الأمور.

10- With unbending oneself towards people all affairs will end well. (ĀF 53; K 15r; Mb 199; IAU 97; ĀM 158; *Libro* 58; Lo 65:6).

1- بخفض الجناح تأنس النفوس.

1- By means of graciousness, souls become more humane. (Alon 74; IQut i, 266; Mj 47; Mb 118; Ṭurṭ 50; IR 264, 266).

2- بخفض الجانب تأمن النفوس. (Sh i, 160: 'Socrates'; Ibsh 53).
3- وألن جناحك تعتقد * في الناس محمدة بلينه
فلربما احتقر الفتى * من ليس في شرف بدونه. (Wash 29; AAt 449).
4- "واخفض لهما جناح الذل." (Qur'ān 17:24; IMutBad 3).[13]

4- And make soft to them (thy two parents) the side of gentleness; *meaning* treat them with gentleness. (trans. Lane, *Lexicon* 973).

11- وبسعة الأخلاق يطيب العيش ويكمل السرور.

[13] Ibn al-Muʿtazz lists this among the metaphors of the Qurʾān.

11- With openheartedness life becomes pleasant and joy becomes perfect. (ĀF 53; K 15r; Mb 118: 'Socrates'; 199: 'Aristotle'; IAU 97; Sh i, 202, 160: 'Socrates'; ĀM 69, 158; *Libro* 58; Lo 65:7; IR 264: 'Plato'; Ṭurṭ 50).

Generous character improves life. (Alon 74).

1- بسعة خلق المرء يطيب عيشه. (Ṭurṭ 50; *Aqwāl* 29; Ibsh 53).

1- With courtesy, life becomes peasant. (Ṭurṭ 50; *Aqwāl* 29; Ibsh 53).

2- مع الرضاء يطيب العيش. (J 16: 'Awshahanj').
3- في سعة الأخلاق كنوز الأرزاق.

3- Treasures of livelihood are in good morality. (Kassis 168; *Risāla* 70; Mj 46; Rāghib i, 274; J 177; MawAd 221; QudDus 22; Thaʿāl Tam 14; Mayd ii, 472; Sh/NB xx, 339).
All doors are open to courtesy. (An English proverb).

12- وبحسن الصمت جلالة الهيبة.

12- With keeping silence at the right time, one's majesty of awe grows. (ĀF 53; K 15r: بالصمت تجب الهيبة; Mb 199: جلال; Sh i, 202; Lo 65:8; *Libro* 58).[14]

1- بكثرة الصمت تكون الهيبة.

1- Reverence is gained by extensive silence. (Z 2187; MJ 47; NB 398 n. 224; see also: *Aqwāl* 53; Mb 118; Ṭurṭ 51; ZamRab i, 782; IR 119; IḤamd i 360; Sh/NB xix, 48 n. 220; Sh i, 160; Ibsh 53, 108).

2- قالوا: استكثر من الهيبة صامت. (JahB i, 270; IAR iii, 82).
3- قال يحيى بن خالد: ما رأيت أحدا قط صامتا إلا هبته حتى يتكلم فإما أن تزداد ذلك الهيبة أو تنقص. (Sh/NB xix, 48).

[14] *Al-hay'a* in the text is a misprint. The Munich ms. has the correct reading. This sentence is used often as part of a cluster; cf. Z 2187–90.

4- Never speak when it is not the time for speech. (MaxAli 23).

13- وبإصابة المنطق يعظم القدر ويرتقى الشرف.

13- With speaking the right words at the right time, prestige grows and honor increases. (ĀF 53; K 15r; Mb 199; IAU 97; Sh i, 202; ĀM 69, 158; Lo 65:9; *Libro* 58; cf. Z 2188).

1- بعدل المنطق تجب الجلالة. (Mj 47).
2- عدل المنظق يوجب الجلالة. (Ṭurṭ 50; Ibsh 53).
3- بالعدل تجب الجلالة. (Mb 118; Sh i, 160).

3- By means of justice, excellence is necessarily brought about. (Alon 74).

4- بعذب المنطق تجب الجلالة. (IR 119).

14- بالإنصاف يجب التواصل.

14- With equity comes friendship. (ĀF 53; K 15r; Mb 199; IAU 97; Sh i, 202, 160: 'Socrates'; ĀM 159; *Libro* 58; Lo 65:10; cf. Z 552).

1- الإنصاف يؤلف القلوب.

1- Equity tames the hearts. (Z 2210; MJ 53; J 181).

2- بالنصفة تكون المواصلة.

2- Fairness occasions comradeship. (Mb 118: 'Socrates'; Ṭurṭ 50; Ibsh 53; Sh i. 160; cf. Alon 74 n. 506).

3- بالنصفة يكثر المواصلون.

3- With fairness, one's friends increase. (NB 398 n. 224; Sh/NB xix, 48 n. 220; Mj 47; Ṭurṭ 50, 172–73; IḤamd i, 360).

4- باعتزالك الشر يعتزلك وبالنصفة يكثر الواصلون.

(MawQaw 182; MawAd 310).

15- بالتواضع تكثر المحبة.

15- With modesty, love increases. (ĀF 53; K 15r; Mb 199; IAU 97; Sh i, 202; ĀM 158; J 50; *Libro* 58; Lo 65:11).

1- بالتواضع تتم النعم.

1- Humility brings kindness to perfection. (Alon 74 n. 506; Mb 118: 'Socrates').

2- ثمرة التواضع المحبة والأمن.

2- Modesty bears love and security. (IQut i, 277; *Tha'lab* i, 257; Askar-iDiw ii, 91; Rāghib i, 519; Ābī iv, 173, 191; J 50; Mb 335; IAR i, 444; Sh/NB xx, 286; Nuwayrī iii, 245, 246; Freytag iii, 63; Z 421).

3- التواضع يورث المحبة.

3- Modesty bears love. (IMuqHik 178).

4- التواضع يوجب المقة.

4- Modesty wins love. Or, Humility involves love. (JahR i, 110; NadīmFihr, 209; Ṭurṭ 50).

5- التودد يوجب المحبة. Affection necessitates love. (IR 268).

16- بالعفاف تزكو الأعمال.

16- With chastity, deeds become pure. (ĀF 53; K 15r; Mb 199; IAU 97; Sh i, 202; *Libro* 58; Lo 65:12).

1- بصالح الأخلاق تزكو الأعمال.

1- Good character purifies one's deeds. (Mj 47; J 12: 'Awshahanj'; Mb 118: 'Socrates'; Ṭurṭ 51; Ibsh 53; Sh i, 160 'Socrates'; cf. Alon 74).

2- بالأدب تنمي العقول وتزكو.

2- With refined education, minds grow and purify. (AṢ 12; cf. with n. 6 above).

3- بالأدب تعمر القلوب.

3- With refined education, hearts become prosperous. (AṢ 36; IḤibb 19).

4- عند كل بلية تزكو الأعمال. (ʿĀmirī 496).

17- بالإفضال يكون السؤدد.

17- By rendering favors, one wins respect. (ĀF 53; K 15r; Mb 199; IAU 97; Sh i, 202; ĀM 158; *Libro* 59; Lo 65:13).

1- بالإفضال تعظم الأقدار.

1- It is through personal merits that the dignity becomes great. (Z 2188; NB 398 n. 224; ThaʿālTam 137; Mb p. 118: 'Socrates'; 203: 'Aristotle'; IḤamd i, 360; Sh i, 160 'Socrates'; Dihkhudā iii, 1560).

2- بالإفضال يكون تعظم الأخطار. (Mj 47; Masʿūdī i, 303; *Kalimāt* 40; J 12).
3- بالإفضال يعظم القدر. (Ṭurṭ 50; Murtaḍā Ṣarrāf 23).

In al-Rayḥānī, this sentence is part of a cluster. Ibn Riḍwān mixes two distinct elements of the text (ns. 13 and 17) together and gives:

4- بالإفضال يعظم القدر ويرتقى الشرف. (IR 119; Ibsh 53).
5- بالإفضال على الناس تعظم الأقدار. (IR 236; IḤamd i, 360).
6- بالإفضال تعلو الأقدار. (Thaʿālak 12; ĀM 164).
7- بالإفضال تشرف الأقدار. (Ṭurṭ 173).
8- بالإفضال تعظم الأقدار. (Sh i, 160).
9- باحتمال المؤن يجب السؤدد.

9- Bearing up under difficulties inevitably produces dominion. (Alon 74; Mj 47; Mb 118, 203; ʿĀmirī 496; Ṭurṭ 51; Murādī 172; Sh/NB xix, 48; Sh i, 160; Ibsh 53).

18- وبالعدل يقهر العدو.

18- With justice, the enemy is conquered. (ĀF 53; K 15r; Mb 199; IAU 97; Sh i, 202; *Libro* 59; Lo 65:14).

1- بالسيرة العادلة يقهر المناوئ.

1- With just conduct, man overpowers his opponents. (MJ 47; NB 398 n. 224; Mb 118: 'Socrates'; 203: 'Aristotle'; Sh i, 160: 'Socrates'; MawAd 311; IAB i, 605; ZamRab ii, 51; iii, 71; IḤamd i, 360; Sh/NB xix, 48 n. 220; cf. Alon 74).

2- أبذل لعدوك عدلك. Spend your justice on your enemy. (Z 172).

19- بالحلم يكثر الأنصار.

19- With forbearance, one's supporters increase. (ĀF 53; K 15r; Mb 199; IAU 97; Sh i, 202; ĀM 69, 159; *Libro* 59; Lo 65:15; ZamRab ii, 51). The good-tempered has many friends.

1- حلمك عن السفيه يكثر أنصارك عليه.

1- Your tolerating the impudent increases your supporters against him. (Z 552).

2- بالحلم عن السفيه يكثر أنصارك عليه.

2- By tolerating the fool, your helpers against him increase. (Mj 47; IAR ii, 279, 281; NB 398 n. 224; Mb 118 'Socrates'; IAB i, 605; Murādī 174; Ṭurṭ 51; IḤamd i, 360; Sh/NB xix, 48 n. 220; Sh i, 160; Ibsh 53).

3- حسب الحليم أن الناس من أنصاره. (Tha'ālTam 413).

4- حسب الحليم أن الناس أنصاره على الجاهل. (Mayd i, 408; ZamRab ii, 21, 51).

5- من عاجل نفع الحلم كثرة أعوان الحليم على الجاهل.

(Qudāma 127; IWahb 258; IAB 616; IR 104; IKhall ii, 501; Ibsh 197).

19.1- [ما أغفل الحساد عن سلامة الأجساد.]

19.1- How neglectful are the envious of the health of bodies! (Z 2190).

1- العجب لغفلة الحساد عن سلامة الأجساد.[15]
(Mj 47; NB 398 n. 225; Ābī iv, 192; MawAd 247; Sh/NB xix, 49; xx, 302; ZamRab ii, 618).

20- بالرفق تستخدم القلوب.

20- With kindness, hearts are subjugated. Or: Kindness wins hearts. (ĀF 54; Mb 199; IAU 97; Sh i, 202; ĀM 69, 159; *Libro* 59; Lo 65–66:16; cf. Z 669, 792).

1- في الرفق تكون السلامة. Kindness occasions safety. (Z 1301).
2- الرفق سبب القدرة. Kindness is the cause of power. (IQut iv, 137).

21- بالإيثار يستوجب اسم الجود.

21- By being altruistic, one is entitled to be called The Altruist. (ĀF 54; K 15r; Mb 199; IAU 97; Sh i, 202; ĀM 159; *Libro* 59; Lo 66:17; cf. Z 473).

1- واعلم أن الذي يوجب لك اسم الجود القيام بواجب الحقوق عند النوائب مع بعض الراغبين وإذا أوجب لك اسم الجود زال عنك اسم البخل. (JahR i, 112).

22- بالإنعام يستحق اسم الكرم.

22- By rendering benefaction, one deserves to be called The Benefactor.[16] (ĀF 54; K 15r; Mb 199; IAU 97; Sh i, 202; *Libro* 59; Lo 66:18; cf. Z 290).

1- والإيثار على النفس موجب لاسم الكرم. (Nuwayrī iii, 204).
2- بإيثارك على نفسك تستحق اسم الكرم. (Mj 47).
3- بالرفق والتودد تستحق [يستحق] اسم الكرم.

3- With gentleness and affection, you deserve to be called The nobleman. (Mb 118: 'Socrates'; Ṭurṭ 51; Sh i, 160 'Socrates'; Ibsh 53).

---

[15] This is missing in ĀF, K and Mb, but Z, MJ and the others who quote this cluster have it in addition.
[16] The numbers 20–22 are combined in various versions.

The attribute of generosity befits you due to your kindness and friendliness. (Alon 74 n. 506).

23- بالوفاء يدوم الإخاء.

23- With fidelity, brotherliness (friendship) endures. (ĀF 54; K 15r; Mb 199; IAU 97; Sh i, 202; ĀM 159; *Libro* 59; Lo 66:19).

1- من الوفاء دوام المواصلة. (J 16: 'Awshajanj').
2- بالصدق والوفاء يلاحظك بالجلالة الأكفاء.

2- By being honest and sincere, your peers regard you with respect. (Mb 118 'Socrates'; Sh i, 160–61).

24- بالصدق يتم الفضل.

24- With veracity, merit becomes complete. (ĀF 54; K 15r; Mb 200; IAU 97; Sh i, 202; ĀM 159; *Libro* 59; Lo 66:20).

1- صدق الرجل أوضح دلائل الفضل.

1- Veracity is the most obvious sign of merit. (Z 988).

25- بحسن الاعتبار تضرب الأمثال.

25- Parables are conveyed for good lessons. (ĀF 54; K 15r; Mb 200; IAU 97; Sh i, 202; *Libro* 59; Lo 66:21).

1- لأهل الاعتبار تضرب الأمثال.

1- Words of wisdom are imparted for those who take warning. (Mj 47; Freytag iii, 23).

2- لأهل الاعتبار في صروف الدهر كفاية وكل يوم يأتي عليك منه علم جديد. (Mb 118: 'Socrates'; Sh i, 161).
3- "وللعقول تضرب الأمثال." (Rāzī 151).
4- العبرة عن أحوال الضرر والأمثال والأقران. (Sh i, 202).
5- ما أكثر العبر وأقل الاعتبار!

5- How abundant are the lessons, and how few are those who listen! (NB 416 n. 297; Sh/NB xix, 203; IḤamd i, 77).

26- الأيام تفيد الأحكام.

26- Time teaches the rules of life. (ĀF 54; K 15r; Mb 200: الحلم, الحكم; IAU 97; *Libro* 59; Lo 66:22. Not understanding this sentence and its relation with the next, the copyist or editor of Mb has confused both).

1- إن أمور العالم تعلمك العلم. (Shahrastānī 933).

27- يستوجب الزيادة من عرف نقص الدنيا.

27- He who recognizes the detriments of this world deserves its benefits. (ĀF 54; K 15r; Mb 200; IAU 97; *Libro* 59; Lo 66:23; *Risāla* 71).

1- كل يوم يفيدك علما، والصدق والوفاء فاجعلهما لك حصنا، وأحق الناس بالرضاء من عرف نقص الدنيا.

1- Each day teaches you a lesson; make truth and fidelity a fortress for yourself; and he is more content who knows the shortcomings of this world. (Z 1456; numbers 26–27 are combined here, to which numbers 23–24 form a logical consequence: truth completes your merit and fidelity wins you friends).

28- من الساعات تتولد الآفات. (ĀF: 54 "التباعات"، مونيخ: الساعات).

28- The hours bear harms. (ĀF 54; K 15r; Mb 200; IAU 97; Lo 66:24).

29- بالعافية يوجد طيب الطعام والشراب.

29- In good fortune, the taste of food and beverage is evoked. (ĀF 54; K 15r; Mb 200 طعم الطعام; IAU 97; *Libro* 59; Lo 66:25).

1- بالعافية توجد عذوبة كل مطعم. (J 99).
2- كل شيء مع العافية طيب.

2- With good health, everything tastes delicious.[17]

30- بحلول المكاره يتنغص العيش ويتكدر.

30- In misfortune, life becomes disturbed and turbid. (ĀF 54; Mb 200: تتكدر النعم; IAU 97; *Libro* 59; Lo 66:26).

1- بعوارض الآفات تكدر النعم على المتنعمين. (Mb 119 'Socrates').

31- النعم بالمن تكفر.

31- Reminding a favor done, annuls it. (ĀF 54; K 15r: Mb 200; IAU 97:يكفر الإحسان; Lo 66:27).

1- لا تشن وجه المعروف بالمن.

1- Do not mar the essence of a favor you confer by reminding it. (Z 1977).

2- الاستطالة على المنعم عليه تهدم الصنيعة وتكدر المعروف.

2- Presumptuousness towards the beneficiary ruins the benefit and spoils the beneficial. (Z 80).

3- من امتن بمعروفه افسده. (Ābī iv, 222).

32- بالجحد للإنعام يجب الحرمان.

32- Being ungrateful for a gift necessitates bereavement. (ĀF 54; K 15r; Mb 200; IAU 97; *Libro* 59; Lo 66:28).

1- من لم يشكر الإحسان لم يعدم الحرمان.

1- He who does not show gratitude for a favor shall not miss bereavement. (Sagh 31; Ibn ʿArabī ii, 480).

---

[17] A. Furayḥa [Frayha], *Lebanese Proverbs*, 2 vols. (Beirut, 1953), ii, p. 529.

2- من لم يشكر ما أنعم به عليه أوشك أن لا يزيد نعمته.

2- He who does not express gratitude for favors bestowed upon him is unlikely to receive any more. (Alon 83 n. 676; Mb 120: 'Socrates'; IAU 79; see Z 1333).

33- ضيق الملول زائل عنه.

33- Restraint of the depressed is ephemeral. Or, Friend of a depressed leaves him. (ĀF 54; Munich: صديق الملول; K 15r; Mb 200 صديق الملك; IAU 97; *Libro* 59; Lo 66:29: 'Fliehe die Freundschaft der Thoren.' The reading is uncertain).

34- الملل من كواذب الأخلاق ولا فعل لملول.

34- Depression is a falsity of character and there is no act for the depressed? (ĀF 54).

35- السيئ الخلق مخاطر بصاحبه.

35- The ill-natured is dangerous to his companion. (ĀF 54; K 15r بنفسه; Mb 200; IAU 97; *Libro* 59; Lo 66:30).

36- الضيق الباع حسير النظر.

36- Hard-fisted is dull-sighted. (ĀF 54; Mb 200; IAU 97; Lo 66:31 reads: 'Wer schnell vorwärts schreitet, schädigt seinen Blick').

37- البخيل ذليل وإن كان غنيا، والجواد عزيز وإن كان مقلا.

37- The niggard is despised even if he were rich; the generous is loved even if he were poor. (ĀF 54; K 15r; Mb 200; IAU 97; ĀM 69, 159; *Libro* 59; Lo 66:32).

1- البخيل أبدا ذليل. (Thaʿ ālTam 440; Sijistānī 293).
2- إن من قنع كان غنيا وإن كان مقترا ومن لم يقنع كان فقيرا وإن كان مكثرا. (MawAd 207–208).

3- Though a man be very poor in the goods of this world, he is (nevertheless) rich if there is moderation in his character.[18]

4- فقير كل من يطمع * غني كل من يقنع. (DiwAli 77).
5- واقنع بقوتك فالقناع هو الغنى * والفقر مقرون بمن لا يقنع. (DiwAli 79).
6- كفى حزنا أن الغنى متعذر * على وأنى بالمكارم مغرم
والله ما قصرت في نيل غاية * ولكنني أسعى إليها فأحرم. (IAB i, 193).
7- كفى حزنا أن الجواد مقتر * عليه ولا معروف عند بخيل.

7- Sufficiently grievous is the liberal man when he becomes too poor to be generous, for no benefaction is to expect in the greedy. (AskariDiw ii, 209; AskariAmth i, 176: 'Abū Nuwās'; Raqīq 661; IḤamd viii, 107).

38- الطمع الفقر الحاضر.

38- Avidity is hidden poverty. (ĀF 54; K 15r; Mb 200; IAU 97; *Libro* 59; Lo 66:33; MubarFad 16; Sh/NB xviii, 85; Nuwayrī iii, 376; Ibsh 94. For more examples see Z 1086).

39- اليأس الغنى الظاهر.

39- Renunciation is tangible wealth. (ĀF 54; K 15v; Mb 200; IAU 97; *Libro* 59; Lo 66:33. More examples Z 1086).

1- التعب مع الطمع، والراحة مع اليأس.

1- Discomfort is with avidity, comfort is with renunciation. (Z 1480).

40- "لا أدري" نصف العلم.

40- 'I do not know' is the half of science. (ĀF 54; Mb 200; IAU 97; Sh i, 202; ĀM 159; *Libro* 59; Lo 66:34; Rāghib i, 50. More examples Z 444).

[18] R.C. Zaehner, *The Teachings of the Magi* (London, 1956), pp. 110–11: 'Adharbād, n. 9'.

Saying, 'I do not know' is half the knowledge. (Haim 356).
Who knows he knows nothing knows much. (An English proverb)

1- "لا أعلم" نصف العلم. (Sh/NB xviii, 236).
2- من قال: "لا أدري" فقد أحرز نصف العلم. (JahB i, 398; ii, 91).
3- إذا ترك العالم قول لا أدري، أصيبت مقاتله.

3- The learned who abandons to say: 'I do not know' is afflicted by his destruction. (IQut ii, 125; MawAd 66; NB 374 n. 85; Sh/NB, xviii, 236).

4- هلك من ترك "لا أدري". (MawAd 66).
5- ما معي من فضيلة العلم إلا علمي بأني لست أعلم.

5- I have no virtue in matters of knowledge except the knowledge that I know not.[19] (Fakhry 161; ĀF 74: 'Plato'; Mb 50: 'Hippocrates'; JahTar 98; IQut ii, 126; AskariSin 371; MawAd 67).

6- لولا أن قولي لا أعلم تثبيتا أني أعلم لقلت إني لا أعلم.

6- Were it not for that my saying, 'I do not know' would be a confirmation that I do know, then I would say, 'I do not know'. (ĀF 74: 'Plato'; JahTar 98; IQut ii, 126; AskariSin 410; Mb 125: 'Socrates'; Sh i, 166: 'Socrates').
Plato said: 'Were it not that my statement, "I do not know," established the fact that I know, I would have said that I do not know.' (Gutas 131).
The truly learned man is he who understands that what he knows is but little in comparison with what he does not know. (MaxAli 13).

7- تا بدانجا رسید دانش من * که بدانم همی که نادانم.

7- To such a pitch has my wisdom come, that now I know how much I know not. (Qābūs 39; trans. 33).
My knowledge has reached to the point to realize I am ignorant.

---

[19] This is the famous Socratic saying, 'I know that I do not know'; see Plato, *Apology* 21E.

8- زسر عقل واقف شد روانم * بدانستم كه من چيزى ندانم. (Nāṣir Khusraw 542).
9- قال بعض الفضلاء: إذا قال لنا إنسان: "لا أدري" علمناه حتى يدري، وإن كان قال: أدري، امتحناه حتى لا يدري.

9- If someone says, 'I do not know,' we teach him until he knows. If he says, 'I do know,' we question him until he does not know. (Sh/NB xviii, 236).

41- السرعة في الجواب توجب العثار.

41- Haste in answering causes stumbling. (ĀF 54; Mb 200; IAU 97; Sh i, 202 يورث; *Libro* 59; Lo 67:35. Cf. Z 125, 791, 2060).

1- من أسرع في الجواب أخطأ في الصواب. (Sagh 51; Ibsh 53).
2- من أسرع في الأشياء يوشك أن يكثر عثاره.

2- He who rushes things is likely to stumble often. (Sh i, 139; Alon 65 n. 342).

3- من أسرع كثر عثاره والتؤدة تؤمن العثار.

3- The hurried stumbles often; the circumspect is secure from stumbling. (ĀF 63: 'Socrates'; IMutAd 120; Mb 100: 'Socrates'; Ḥuṣrī 1009; IAU 77; Usāma 463).
He who hurries stumbles much, and slowness is a guarantee against stumbling. (Alon 83 n. 659).

4- من ركب العجلة لم يأمن العثار.

4- The hasty is not secure from stumbling. (Murādī 172).

5- من ركب العجل أدرك الزلل.

5- He who rides haste ends in waste. (Sagh 51; Ibsh 53).

6- من ركب البغي لم يأمن مغبته. (Ibn 'Arabī ii, 362).
7- بئس المركب العجلة.

7- Very bad is, as a mount, the haste. (Usāma 436).

8- الزلل مع العجل. من أسرع كثر عثاره.

8- Whoso hurries, trips. (IMutAd 120; Thaʿālīam 455).

9- راكب العجل لا يأمن العثار. The hasty is doomed to slip. (IḤibb 274).

10- راكب العجلة مشرف على الكبوة.

10- The rider of haste is doomed to slip. (*Risāla* 69).

11- من ركب العجلة لم يأمن الكبوة.

(TawḥImt ii, 150; *Kalimāt* 25, 40; Ābī iv, 219, 224; Sh/NB xx 311; *Aqwāl* 53).

42- التروي في الأمور يبعث على البصائر.

42- Reflection on affairs brings forth discernment. (ĀF 54; Mb 200; IAU 97; *Libro* 59; Lo 67:36).

1- Matters obscure become clear on reflection. (MaxAli 21).

2- Act only after reflection, and all your affairs will work out well. (MaxAli 21).

3- رو قبل الفعل كيما لا تعاب في فعلك. (Mb 63: 'Pythagoras').

43- الرياضة تشحذ القريحة.

43- Training sharpens talent. (ĀF 54; Mb 200; IAU 97; Sh i, 202; *Libro* 59; Lo 67:37).

Practice makes perfect. (CDP 220).

1- بالأدب تشحذ الفطن.

1- Education sharpens the intelligence. (ĀF 125: 'Ptolemy'; Thaʿālīam 455).

2- الشطرنج يشحذ.

2- Chess sharpens the mind. (Suyūṭī 324).

3- الفقر يخرس الفطن عن حجته.

3- Poverty silences the astute from reasoning. (NB 361 n. 3; QudDus 23; Sh/NB xviii, 87, 88; IḤamd i, 250).

44- الأدب يغني عن الحسب.

44- Acquired virtue makes ancestry irrelevant. (ĀF 54; K 15v; Mb 200: عن النسب; IAU 97; Lo 67:38; *hubb* in the text is a misreading or misprint for *ḥasab*; cf. with n. 4 above).

1- الأدب يصفي الحسب.

1- Education refines noble descent. (*Kalimāt* 39).

2- الأدب أملك بالفتى من نسبه.

2- Good education rules the young man more than his noble descent. (Z 28; cf. 547, 1207).

3- لا ينفع الحسب بغير أدب.

3- Noble descent is useless without a good education. (AṢ 44; J 76; IḤamd i, 250).

4- لا يتم الحسب إلا بالأدب.

4- Noble descent will not be complete other than by education. (Ābī iv, 179; Ibsh 51).

5- من فاته الأدب لم ينفعه الحسب.

5- He who lacks education, a noble descent is of no use to him. (ThaʿālTam 163; Ikhtiyār 129).

6- الحسب محتاج إلى الأدب.

6- Noble descent stands in need of education. (IQut iv, 32; IAR ii, 42; Ābī vii, 48; IḤamd i, 288; Sh/NB xx, 41).

7- الآداب خير من الأنساب، والأعمال خير من الأموال.

7- Acquired virtues are better than noble lineages; good deeds are better than riches. (IḤamd i, 261).

45- التقوى شعار العالم.

45- Piety is the attire of the learned.[20] (ĀF 54; K 15v; Mb 200; IAU 97; ĀM 159; *Libro* 59; Lo 67:39. Cf. Z 1072).

1- التقوى أفضل لباس. Piety is the best garb. (Z 427).
2- أحسن اللباس الورع. Righteousness is the best garb of virtue. (Z 5).

46- الرياء لبوس الجاهل.

46- Hypocrisy is the mantle of the ignorant. (ĀF 54; K 15v; Mb 200; IAU 97; ĀM 159; *Libro* 59; Lo 67:39.

On the margin of the Munich manuscript of *Ādāb al-falāsifa* some old hand gives a Persian version as خود نمودن جامه نادانست. The above two sentences can be made into one).

47- مقاساة الأحمق عذاب الروح.

47- Company of the imbecile is torture of the spirit. (ĀF 54; Mb 200; IAU 97; Sh i, 202; ĀM 159; *Libro* 60; Lo 67:40).

1- مجالسة الثقيل حمى الروح.

1- Companionship of a distasteful person is a pain to the spirit. (ThaʿālTam 180).

2- "روح را صحبت ناجنس عذابى است اليم."

[20] *Shiʿār*, 'undergarment' is a synonym for *labūs*, 'mantle'.

2- Ill-assorted companionship is a torment to the spirit. (Haim 237: 'Ḥāfiẓ').

3- العقل يضجر عند محاورة الجاهل.

3- Conversation with the ignorant grieves the intellect. (SijistSiw 309).

48- الاستهتار بالنساء حلس النوكى.

48- Fondness of women is the saddle blanket of the stupid. (ĀF 54; Mb 200; IAU 97; *Libro* 60; Lo 67:41).

49- الاشتغال بالفائت تضييع للأوقات.

49- To attend what is passed away is a waste of time. (ĀF 54; K 15v; Mb 200; IAU 97; *Libro* 60; Lo 67:42)

1- حزن المرء على ما فات جهل.

1- To feel sorrow for what is lost is ignorance. (Freytag ii, 98).

2- حزن المرء على ما فاته يضره ولا ينفعه.

2- To be sad for what is lost is harmful and of no use. (Z 504).

3- على العاقل أن لا يحزن على شيء فاته من الدنيا. (AṢ 21; J 71, 163)
4- لا تأسفن على ما فاتك من الثراء.

4- Do not feel sorry for the fortune that has escaped you. (J 27: 'Adhurbād').

5- آنچه گذشت فراموش کن و برای آنچه نیامده است تیمار ورنج مبر!

5- Forget that which has departed, and feel no grief and pain on account of that which has not come.[21]

6- هر کاری از تو فائت گشت و فرصت از دست شد در طلب آن تعب بیفایده تحمل مکن و ترك آن گیر. (ṬūsīAd 26).
7- الندم على ما فات من الفشل.

7- Remorse for what is lost is a failure. (Sijistānī 111: 'Thālis al-Malaṭī'; 253: 'Aumānūs'; Ibn Hindū 461 n. 653).

50- المتعرض للبلاء مخاطر بنفسه.

50- He who takes risks endangers himself. (ĀF 54; Mb 200; IAU 97; *Libro* 60; Lo 67:43).

1- من تجلد على النوى فقد تعرض للبلاء. (IDaw i, 178).
2- حذرك من ركوب الخطر مسلم لك من عظيم الزلل.

2- Your precaution in taking risks safeguards you from great mistakes. (Z 556).

3- التغرير بالنفس أحد الخطرين.

3- Taking risks is dangerous. (Ḥamza 2, 513)[22]

51- التمني سبب الحسرة.

51- Desire is the cause of grief. (ĀF 54; K 15v; Mb 200; IAU 97; *Libro* 60; Lo 67:44).

1- عظم الهمة سبب لطول الحسرة.

1- High-aiming ambition is a cause of lengthy grief. (Z 1222).

---

[21] F. de Blois, 'The Admonitions of Ādurbād and their Relationship to the Aḥīqar Legend,' *JRAS* (1984), p. 46.
[22] Cf. A. Spitaler, *Al-Qalamu aḥadu l-lisānaini* (München, 1989), p. 24, n. 48.

High aspirations occasion hardship. (AṢ 27). 2- في بعد الهمة يكون النصب.
3- الرغبة مفتاح النصب ومطية التعب.

3- Ardent desire is the key to distress and the mount of trouble. (NB 427 n. 381; QudDus 23; Mayd iv, 54).

4- الرغبة مفتاح الطلب ومطية الحسرة. (Ṭurṭ 172).

52- الصبر تأييد العزم، وثمرة الفرج وتمحيق المحنة.

52- Patience strengthens determination, and its fruits are relief and effacement of tribulation. (ĀF 54: ثمرة; Munich: ثمرة الفرج; K 15v; Mb 200; IAU 97; *Libro* 60; Lo 67:45).

53- صديق الجاهل مغرور، والمخاطر خائب.

53- The friend of an ignorant is exposed to danger, and the risk-taker is a wrongdoer. (ĀF 54; K 15v; Mb 200; IAU 97; *Libro* 60; Lo 67:46).

1- صديق الجاهل تعب، ونديم العاقل مغتبط. (Usāma 28; Sh i, 161: 'Socrates').
2- مصاحب العاقل مغتبط، ومصاحب الجاهل تعب. (Mb 119: 'Socrates').
3- صاحب العاقل مغبوط، صديق الجاهل تعب. (Ṭurṭ 51).

54- من عرف نفسه لم يضع بين الناس.

54- He who knows himself will not be humiliated among people. (ĀF 54; K 15v; Mb 200; Sh i, 202; ĀM 69, 159; IAU 97; *Libro* 60; Lo 67:47; IAR ii, 422).

He is a good orator who convinces himself. (An English proverb)

1- ما ضاع من عرف نفسه وما أضيع من جهل نفسه!

1- He who knows himself does not perish, but how prone to perdition is he who does not know himself! (Mb 93; Sh i, 135; IAU 76; cf. Alon 88 n. 757).

2- ما ضاع امرؤ عرف قدره.
(IShams 59; ĀK 828; al-ʿĀmilī, Kashkul, p. 828).

3- ما هلك امرؤ عرف قدره.

3- He who is cognizant of his worth will not perish. (AUAm 294; JahB ii, 23; SijistMu 9; Balādhurī vii.1, 362; al-Yaʿqūbī, *Taʾrīkh*, ii 113; Mufaḍḍal 262; Wash 143; IAR ii, 419; Masʿūdī iii, 37; AskariAw i, 95; AskariAmth ii, 318; Rāghib i, 91, 263; J 104; ThaʿālTam 28; Rādūyānī 120; Maʿarrī 34; MawQaw 237; MawAd 292; Wāḥidī 148; Mayd iii, 96; ZamAm ii 295; Waṭwāṭ n. 4).

4- من عرف نفسه فقد أمن الهلاك.

4- He who knows himself is already saved from perdition. (J 265).

5- هرك قدر خود بشناخت از هلاك ايمن شد. (*Khiradnāma* 103).
6- هر كه دانست قدر و قيمت خويش * از هلاكش هميشه بيزارست. (Rādūyānī 120).
7- من عرف قدره لم يهلك. (Qālī ii 172; JahB ii, 339).
8- هلك امرؤ لم يعرف قدره.

8- He who does not know his worth perishes. (NB 389 n. 149; Sh/NB xviii, 355; IḤamd ii, 246).

9- من عرف قدره قل إفراطه.

9- He who knows his worth, his excess is less. (IMuqHik 181).

10- العالم من عرف قدره. He is the learned who knows his own worth. (NB 95).
11- رحم الله امرءا عرف قدره ولم يتعد طوره.

(ṬūsīAd 74; Qushayrī (Persian) 222).

12- أرفع الأشياء أن يعرف الرجل قدر منزلته، ومبلغ علمه وعقله. (ĀF 59).
13- أفضل العقل معرفة المرء بنفسه. (Rāghib i, 19).
14- إن الخير كله فيمن عرف قدر نفسه، وكفى بالمرء جهلا أن لا يعرف قدر نفسه.

14- The good, all of it, is in him who knows his worth. It is sufficient ignorance for a man not to know himself. (J 111: "Alī"; NB 18, 95. More examples Z 684).

15- من جهل قدر نفسه كان للقدر غيره أجهل.

15- He who is ignorant of his own worth, is even more ignorant of people's worth. (ʿĀmirī 496; Ābī iv, 226; J 15; ThaʿālTam 439; Mb 335; Freytag iii, 77).

16- ومن جهلت نفسه قدره * رأى غيره منه ما لا يرى. (Sh/NB xviii, 392).
17- كفى بالمرء نقصا أن يبدو له من عيب أخيه ما يعنا عنه من عيب نفسه. (K 41r).
18- ينبغي لنا أن نبتدئ بمعرفة أنفسنا من قبل أن ننصرف إلى معرفة غيرنا. (ĀF 56).
19- النفس جامعة لكل شيء فمن عرف نفسه عرف كل شيء ومن جهل نفسه جهل كل شيء.

19- The soul encompasses everything, so whoever knows his soul knows everything, and whoever is ignorant of his soul is ignorant of everything. (Mb 93: 'Socrates').

20- الجاهل بنفسه جاهل بكل شيء.

20- He who is ignorant of himself is ignorant of everything else. (Mb 227).

21- من عجز عن علم نفسه عجز عن علم غيره. (ĀF 60).

22- How know another, if one does not know oneself? (MaxAli 16).

23- من عجز عن معرفة نفسه فهو عن معرفة خالقه أعجز.

23- He who is incapable of knowing himself is even more incapable of knowing his Creator. (Sh/NB xx, 292; Mayd iv, 53).

24- من عرف نفسه فقد عرف ربه.

24- He who knows himself already knows his Lord.[23] (Ibn ʿArabī ii, 155; Hujwīrī 247, 353; J 23; Niẓāmī 9; Rādūyānī 120; Waṭwāṭ n. 6; ṬūsīAkh 8; Najm Rāzī 3, 174, 185, 413, 535).

[23] This is a Socratic maxim often attributed to the Prophet Muhammad. See Z 684 for more examples.

25- Know thyself.[24] (CDP 152).

-26 چنین گفتند رو بشناس خود را * طریق کفر و دین و نیك و بد را
کزین ره سوی یزدانست راهت * ترا بس باشد این معنی گواهت.
زدانش زنده مانی جاودانی * زنادانی نیابی زندگانی. (Nāṣir Khusraw 511).
-27 بدان خود را که گر خود را بدانی * زخود هم نیك وهم بد را بدانی
شناسای وجود خویشتن شو * پس آنگه سر فراز انجمن شو
چو خود دانی همه دانسته باشی * چو دانستی زهر بد رسته باشی
ندانی قدر خود زیرا چنینی * خدا بینی اگر خود را ببینی.
(Nāṣir Khusraw 528).
-28 چون گوهر خویش ندانستی * مر خالق خویش را کجا دانی.
(Nāṣir Khusraw 414).
-29 خویشتن خویش را بدان بدرستی * تا ملك خویش را درست بدانی.
(Rādūyānī 120).

For more examples of this famous saying in Persian literature see Dihkhudā iv, 1744–45.

-55 من زاد علمه عقله كان وبالا عليه.

55- When knowledge exceeds understanding, it becomes harmful. (ĀF 54; K 15v; Mb 200; Sh i, 202; IAU 97: على عقله ; *Libro* 60; Lo 67:48; cf. Z 517).

-1 إذا كان علم الرجل أكثر من عقله كان قمينا أن يضره علمه. (IAB i, 533).

1- When a man's knowledge is greater than his intelligence, he deserves to be hurt by his knowledge.

-2 قيل لبعض الحكماء: متى يكون الأدب شرا من عدمه؟ قال: إذا كثر الأدب ونقص العقل. (IQut ii, 40).

-56 المجرب أحكم من الطبيب.

[24] This is the motto inscribed on the sixth-century B.C. temple of Apollo at Delphi.

56- The experienced is wiser than the doctor. (ĀF 54; Mb 200; ĀM 159; IAU 97; *Libro* 60; Lo 67:49. A colloquial proverb says:

1- أسأل مجرب ولا تسأل حكيم.

1- Ask the experienced not the wise. (IQayy 112).

57- إذا فاتك الأدب فالزم الصمت.

57- When not well-versed in a field, keep silence. (ĀF 54; K 15v; Mb 200; IAU 97; *Libro* 60; Lo 67:50; IQut ii, 176; Ṭurṭ 170; ZamRab iii, 262; Ibsh 51; ĀK 727; ĀM 69; cf. Z 1446).

1- إذا فاتك الجواب فالزم الصمت.

1- When not having the right answer, keep silence. (IMutAd 120).

2- إذا فاتك المنطق، فلا يفتك الصمت. (IWahb 307).
3- من أخافه الكلام أجاره الصمت.
(ĀF 124; IMutAd 120; ThaʿālTam 425; Thaʿālibī/Maqdisī 52).
4- "الصمت إن ضاق الكلام أوسع." (AAt 495; Agh iv, 37; Rāzī 160).

58- من لم ينفعه العلم لم يأمن ضرر الجهل.

58- He whom knowledge benefits not, is not secure against the harm of ignorance. (ĀF 54; K 15v; Mb 200; IAU 97; *Libro* 60; AAt 495; Agh iv, 37; Rāzī 160).

1- من لم ينفعه علمه ضره جهله.

1- He whom his knowledge does not benefit, his ignorance harms. (QudSh, 13).

2- ويل لمن علم فلم ينفعه العلم.

2- Woe unto whoso learns, but his learning benefits him not. (Z 1864).

59- من اتأد لم يندم.

59- He who proceeds deliberately shall not regret. (ĀF 54; Munich: تأنى; Mb 200; *Libro* 60; Lo 67:51).

1- من تأنى لم يندم.

1- He who deliberates shall not regret. (IAU 97).

2- من ملك التأني ملك الندم. (Mb 31: 'Homer').
3- مع العجلة الندامة ومع التأني السلامة.

3- Regret comes with haste, safety comes with deliberation. (Ṭurṭ 51; Ibsh 53).

4- التأني أفضل من العجلة. (J 9: 'Awshahanj')
5- أقل التأني أجدى من أكثر العجلة. (J 9 'Awshahanj').

60- من اقتحم ارتطم.

60- He who rushes plunges. (ĀF 54; Mb 200; IAU 97; *Libro* 60; Lo 67:52).

1- من اقتحم اللجج غرق.

1- He who enters raging waves recklessly is drowned. (IAR ii 420; NB 423 n. 349; QudDus 28: "Alī'; Mayd iv, 63; IḤamd i, 254; *Aqwāl* 125; Sh/NB xix, 264).

2- من اقتحم اللجة أتلف المهجة. (Sagh 53).
3- من اقتحم الأمور لقي المحذور. (Sagh 53–54; Usāma 69).
4- من اقتحم الهزل، إرتطم في الجهل. (Sijistānī 294).

61- من عجل تورط.

61- He who acts hurriedly gets into trouble. (ĀF 54; K 15v; Mb 200: "أعجل ; IAU 97; *Libro* 60; Lo 67:53[25]).

62- من تفكر سلم.

---

[25] The printed text has عمل, for which the Munich ms. has *ʾajal*, the better reading.

62- He who contemplates much is safe. (ĀF 54; K 15v; Mb 200; IAU 98; *Libro* 60; Lo 67:54; Z 93. This has a parallel in:

1- من سكت سلم.

1- He who keeps silence is safe. (Mb 119; Ṭurṭ 51; Ibsh 53, 54).

2- المؤمن إذا انظر اعتبر، وإذا سكت تفكر، وإذا تكلم ذكر. (Sh/NB xx, 280).
3- من تفكر أبصر.

3- He who contemplates becomes wise. (ĀF 137; NB 305; Ābī iv, 222; QudDus 28; MawAd 22).

4- من تفكر اعتبر. (IAR iii, 79; IAB ii, 194).
5- من أكثر الفكر اعتبر. He who thinks much learns. (ĀF 158).
6- من عمل سلم. He who acts is safe. (QudDus 28).

63- من روى غنم.

63- He who irrigates the land becomes rich. (ĀF 54; Mb 200; IAU 98; Lo 67:55).

1- من صبر غنم.

1- He who is patient becomes rich. (Mb 119: 'Socrates'; Ṭurṭ 51; Ibsh 53).

2- من صبر نال المنى.

2- He who is patient attains wishes. (Sagh 8; MawAd 263).

64- من سأل علم.

64- He who asks learns. (ĀF 54; K 15v; Mb 200; IAU 98; *Libro* 60; Lo 67:56; cf. Z 492).

1- من فهم علم.

1- He who understands learns. (Mb 119 'Socrates'; Ṭurṭ 51; Ibsh 53).

65- من حمل ما لا يطيق إرتبك.

65- He who picks up more than he can carry will be embarrassed. (ĀF 54; K 15v; Mb 200; IAU 98; ĀM 69, 159; *Libro* 60; Lo 67:57).

1- من حمل ما لا يطيق عجز. (QudDus 28; Freytag iii, 120).
2- من حمل أكثر مما يطيق عجز.
2- He who carries more than he can take will fail. (Z 645).

3- العاقل لا يتكلف ما لا يطيق.

3- The intelligent person does not undertake what he cannot bear. (Z 177).

4- من أمر العاقل ألا يتكلف ما لا يطيق.

4- It is a prerogative of the wise man not to burden himself with more than he can bear. (Mb 347).

5- العاقل لا يضمن ما لا يثق بالقدرة عليه. (J 11).
6- لا تحمل بطنك ما لا يطيق.

(Ṭurṭ 51; IAR iii, 148; Ibsh 54).

7- لا تقدم على أمر تخاف العجز منه. (ĀF 130).

66- التجارب ليس لها غاية، والعاقل منها في زيادة.

66- Experiences have no limit, and the wise gathers always more. (ĀF 54; Mb 200; IAU 98; *Libro* 60; Lo 67:5; AUAm 106; IAR ii, 246; AskariAmth i, 225; Rāghib i, 24; MawAd 274; *Aqwāl* 27; Mayd i, 259; ZamAm i, 305; cf. Z 1087)

1- التجارب ليس لها غاية، والعاقل يستزيد منها إلى غير نهاية. (IAB ii, 187).
2- التجارب لا تنقضي والعاقل منها في زيادة. (Ābī i, 279; IAR ii, 246).
3- كل شيء محتاج إلى العقل، والعقل محتاج إلى التجارب. (IQut i, 34, 281).

67- للعادة على كل شيء سلطان، وكل شيء يستطاع نقله إلا الطباع، وكل شيء تتهيأ فيه حيلة إلا القضاء.

67- Habit exercises power over everything; everything can be changed except natures, and everything can be tricked except destiny. (ĀF 54, 74: 'Plato'; K 15v; Mb 131, 200; ĀM 159; IAU 98; *Libro* 60; Lo 67–68:59–61; IQut i, 275; Shahrastānī 762: 'Hermes'; Usāma 237).

1- كل شيء يستطاع قلبه إلا الطبيعة ويقدر على رده إلا القضاء.
(Tha'ālIJ 1844, 44).

2- فقد العادة أشد من فقد المادة.
2- Giving up habits is harder than giving up wealth. (Z 1792).

3- غلبة القضاء لا يطمع فيه عاقل.

3- No intelligent man longs for overcoming destiny. (Z 1791).

68- من عرف بالحكمة لحظته العيون بالوقار.

68- He who is reputed to be wise, people regard him with reverence. (ĀF 55; Mb 200; IAU 98; *Libro* 60; Lo 68:62; Ibn al-Mudabbir, *al-'Adhrā'*, 43; Sh/NB xx, 323; QudDus 29; Ābī iv, 226).

1- من عرف بالحكمة لاحظته العيون بالهيبة. (TawḥImt ii, 148; *Kalimāt* 22).
2- من عرف بفصاحة اللسان لحظته العيون بالوقار. (Ibsh 67).
3- من تكلم بالحكمة لاحظته العيون بالوقار.

(ZamRab iii, 221; IAB i, 590).

4- من انتشر له الصوت بفضل أدب ونظر إليه العيون بالإجلال فليكن... (Mj 53)
5- إن من عرف بالصدق صار الناس له أتباعا. (JahR 125).

69- قد يكتفي من حظ البلاغة بالإيجاز.

69- One should be content with precision in eloquence. (ĀF 55; Mb 200; IAU 98; *Libro* 61; Lo 68:63).

70- لا يؤتى الناطق من سوء فهم السامع.

70- The speaker is not undermined by the listeners' misunderstanding. (ĀF 55; Mb 201; IAU 98; *Libro* 61; Lo 68:64: 'Der Redner darf den Hörer nicht beschuldigen, wenn dieser ihn schlecht verstanden hat.')

1- يكفي من حظ البلاغة أن لا يؤتى السامع من سوء إفهام الناطق ولا يؤتى الناطق من سوء فهم السامع.
(JahB i, 87; IMud 47; IAR ii, 261; Bayh 427; AskariSin 16; Ḥuṣrī 117; TawḥBas i, 362; Nuwayrī vii, 7)

Al-Jāḥiẓ has the better reading and al-Tawḥīdī, who also combines the two sentences, explains: It is sufficient of eloquence that the audience does not notice the misinstruction of the speaker (that the speaker does not become obscure when speaking eloquently), and that the speaker does not notice the misunderstanding of the audience (that the audience comfortably understands everything he says). To instruct is the responsibility of the speaker, and to understand is the responsibility of the listener. Nobody reproaches an eloquent speaker just because the audience does not have the means to understand.

Sentences 69–70 belong together as in the quoted example. Both seem to go back to Ibn al-Muqaffaʿ's definition of eloquence:

2- ...والإيجاز هو البلاغة. Precision is eloquence. (AskariSin 14).
3- قال الأرسطو: ما البلاغة: فقال: إقلال في إيجاز. (ĀF 82; Mb 205).

71- من وجد برد اليقين أغناه عن المنازعة في السؤال ومن عدم درك ذلك كان مغمورا بالجهل ومفتونا بعجب الرأي ومعدولا بالهوى عن باب التثبت ومصروفا بسوء العادة عن تفضيل التعليم.

71- He who has experienced the comfort of certitude is satisfied not to dispute when having questions, and he who misses this is plunged in ignorance, beguiled by the arrogance of self-judgment, withdrawn to fancy from the gate of ascertainment, turning away by bad habits from the merit of learning. (ĀF 55; Mb 201; Sh i, 202; IAU 98; *Libro* 61; Lo 68:65).

1- قال علي بن الحسين بن على رحمة الله: لو كان الناس يعرفون جملة الحال في فضل الاستبانة وجملة الحال في صواب التبيين لأعربوا على كل ما تخلج في صدورهم ولوجدوا من برد اليقين ما يغنيهم عن المنازعة إلى كل حال سوى حالهم وعلى أن درك ذلك كان لا يعدمهم

في الأيام القليلة العدة والفكرة القصيرة المدة ولكنهم من بين مغمور بالجهل ومفتون بالعجب ومعدول بالهوى عن باب التثبت ومصروف بسوء العادة عن فضل التعلم.

(JahB i, 84; Ḥuṣrī 64).

72- الجزع عند مصائب الإخوان أحمد من الصبر والصبر المرء على مصيبته أحمد من جزعه.

72- Impatience in the misfortunes of friends is better than patience, and patience in one's misfortune is better than impatience. (ĀF 55; Mb 201, 325; ĀM 159; IAU 98; *Libro* 61; Lo 68:66; Karkhī 93).

1- جزعك في مصيبة أخيك أحمد من صبرك وصبرك في مصيبتك أحمد من جزعك.

1- Your impatience in the misfortune of your friend is better than your patience, and your patience in your misfortune is better than impatience. (Z 450; Marzub 200; Rāghib ii, 507; TawḥSad 30; Murādī 102; ZamRab iv, 181; Sh/NB xx, 344).

2- إن تكن مصيبتك في أخيك أحدثت لك خشية فنعم المصيبة مصيبتك وإن تكن مصيبتك بأخيك أحدثت لك جزعا فبئس المصيبة مصيبتك.
(JahB iii, 171–72).

3- متى يكون الجزع أحمد من الصبر؟ قال: في مصيبة أخيك. (Mb 325; Sijistānī 183).

73- ليس شيء اقرب إلى تغيير النعم من الإقامة على الظلم.

73- Nothing is nearer in changing a prosperous life than persistence on injustice. (ĀF 55; K 15v; Mb 201; IAU 98; *Libro* 61; Lo 68:67).

1- ليس شيء أعجل من تغيير نعمة وتعجيل نقمة من الإقامة على الظلم.

1- Nothing is quicker in altering blessings and accelerating punishment than an act of injustice. (Rāghib i, 215; Z 1521; see also Z 1127, 1135).

2- ليس شيء أدعى إلى تغيير نعمة الله وتعجيل نقمته من إقامة على ظلم.

Here the simple and neutral comment on life in the above sentence is turned into a pious and religiously colored statement. (NB 327).

3- ليس شيء لتغيير نعمة وتعجيل نقمة أقرب من الإقامة على الظلم. (J 15: 'Awshahanj').
4- الظلم أدعى شي ء إلى تغيير النعمة وتعجيل النقمة. (Mj 45; *Kalimāt* 39; Ṭurṭ 47).
5- الظلم أسرع شيء إلى تعجيل نقمة وتبديل نعمة. (Thaʿālṭam 452).

74- من طلب خدمة السلطان بغير أدب خرج من السلامة إلى العطب.

74- He who seeks to serve the sovereign without proper decorum leaves safety for danger. (ĀF 55; Mb 201; ĀM 159; IAU 98; *Libro* 61; Lo 68:68; Mustaʿṣimī 116: 'Ibn al-Muqaffaʿ').

1- طلب صحبة السلطان بغير أدب مخاطرة بالنفس.

1- Seeking the company of the king without proper decorum is risky. (Z 1091).

2- من جالس الملوك بغير أدب فقد خاطر بنفسه.

2- He who sits with kings without proper education endangers himself. (ThaʿālAM 52; ThaʿālTam 142: 'Buzurjmihr'; Balādhurī vii.1, 373: 'Aktham'; Nuwayrī vi, 13).

3- من جالس الملوك بغير أدب جلسة فإنه خاطر بروحه وعرض للبلاء نفسه. (Freytag iii, 73).
4- من صاحب الملوك بغير أدب أسلمه الجهل إلى القتل.

(Mustaʿsimī 91; ThaʿālAM 52; Qushayrī 129).

5- من صحب السلطان قبل أن يتأدب فقد غرر بنفسه. (ZamRab iv, 225; Ibsh 115).
6- صحبة السلطان بلا أدب كركوب البرية بغير ماء. (Ābī iv, 238; Sh/NB xix, 150).
7- کسی کو ندارد هنر با خرد * سزد گر در پادشاه نسپرد.

(Firdawsī viii, 2401).

75- الارتقاء إلى السؤدد صعب والانحطاط إلى الدناءة سهل.
75- Rising to dignity is difficult, falling to lowliness is easy. (ĀF 55; K 15v; Mb 201; IAU 98; *Libro* 61; Lo 68:69).

Bibliography of the Sources (A Fuller Bibliography is Provided in Z)

AAt = Abū 'l-ʿAtāhiya, *Dīwān* (Beirut, 1964).
Ābī = Abū Saʿīd Manṣūr ibn al-Ḥusayn al-Ābī, *Nathr al-durr*, 7 vols. (Cairo, 1980–91).
Abū al-Maʿālī Naṣr Allāh ibn Muḥammad Munshī, *Kalīla wa-Dimna* [Persian], ed. M. Minovi (Tehran, 1964).
ĀF = Muḥammad ibn ʿAlī al-Anṣārī, *Ādāb al-falāsifa*, published as: Ḥunayn ibn Isḥāq, *Ādāb al-falāsifa*, ikhtaṣarahu Muḥammad ibn ʿAlī ibn Ibrāhīm ibn Aḥmad ibn Muḥammad al-Anṣārī, ed. ʿA. Badawī (Kuwayt, 1985).
Agh = Abū 'l-Faraj al-Iṣfahānī, *Kitāb al-aghānī*, 24 vols. (Cairo 1924–74).
ĀK = al-ʿĀmilī, *al-Kashkūl* (Beirut, 1983).
Alon = I. Alon, *Socrates Arabus. Life and Teachings* (Jerusalem, 1995).
ĀM = al-ʿĀmilī, *Kitāb al-mikhlāt* (Cairo, 1957).
ʿĀmirī = al-ʿĀmirī, *al-Nask al-ʿaqlī wa'l-taṣawwuf al-mallī*. In S. Khalīfāt, *Rasāʾil Abī 'l-Ḥasan al-ʿĀmirī* (Amman, 1988).
Aqwāl = Aqwāl al-ḥukamāʾ ʿan kalām Jālīnūs wa-Suqrāṭ wa-Nuqrāṭ wa-Arisṭāṭālīs, ms. Baghdad, Maktabat al-Awqāf 9763.
AṢ = Ibn al-Muqaffaʿ, *al-Adab al-ṣaghīr wa'l-adab al-kabīr* (Beirut, 1964).
AskariAmth = Abū Hilāl al-ʿAskarī, *Jamharat al-amthāl*, ed. A. ʿAbd al-Salām, 2 vols. (Beirut, 1988).
AskariAw = Abū Hilāl, *Awāʾil*, 2 vols. (Damascus, 1975).
AskariDiw =Abū Hilāl al-ʿAskarī, *Dīwān al-maʿānī*, 2 vols. (Cairo, 1352).
AskariSin =Abū Hilāl, *al-Ṣināʿatayn, al-kitāba wa'l-shiʿr*, Istanbul 1952.
AUAm = Abū ʿUbayd al-Harawī, *Al-Amthāl*, ed. ʿA. Qutāmish (Damascus, 1980).
AUGhar = Abū ʿUbayd al-Harawī, *Gharīb al-ḥadīth*, 4 vols. (Hyderabad, 1964–67).
Balādhurī = al-Balādhurī, *Ansāb al-ashrāf*, vii, ed. M. al-Yaʿlawī (Berlin, 2002).
Bashshār ibn Burd, *Dīwān shiʿr*, ed. M. Badr al-Dīn al-ʿAlawī (Beirut, 1963).
Bayh = al-Bayhaqī, *al-Maḥāsin wa'l-masāwī*, ed. F. Schwally (Giessen, 1902).
CDP = J. Simpson and J. Speake (eds.), *The Concise Oxford Dictionary of Proverbs* (Oxford, 1998).
Dihkhudā = ʿAlī Akbar Dihkhudā, *Amthāl wa-ḥikam*, 4 vols. (Tehran, 1310).
DiwAli = *Dīwān al-imām ʿAlī ibn Abī Ṭālib*, ed. Y. al-Faraḥāt (Beirut, 1991).
Fakhry = M. Fakhry, *Ethical Theories in Islam* (Leiden, 1994).
Fārābī = al-Fārābī, Abū Ibrāhīm Isḥāq ibn Ibrāhīm, *Dīwān al-adab*, 5 vols. (Cairo, 1974–78).
Firdawsī, *Shāhnāma*, ed. S. Nafīsī, 10 vols. (Tehran, 1314).
Freytag = G.W. Freytag, *Arabum Proverbia*, 3 vols. (Bonn, 1838–43).
Haim = S. Haim, *Persian-English Proverbs* (Tehran, 1956).
Ḥamza al-Iṣfahānī, *al-Durra al-fākhira fī 'l amthāl al-sāʾira*, ed. ʿA.M. Qutāmish, 2 vols. (Cairo, 1972).
Hujwīrī = al-Hujwīrī, *Kashf al-maḥjūb*, ed. V.A. Žukovsky (Leningrad, 1926).
Ḥuṣrī = al-Ḥuṣrī, *Zahr al-ādāb wa-thamar al-albāb*, 2 vols. (Cairo, 1969–70).
IAB = Ibn ʿAbd al-Barr, *Bahjat al-majālis wa-uns al-mujālis*, 2 vols. (Cairo, 1962).
IAR = Ibn ʿAbd Rabbih, *al-ʿIqd al-farīd*, ed. A. Amīn, 7 vols. (Cairo, 1940–53).
IAU = Ibn Abī Uṣaybiʿa, *ʿUyūn al-anbāʾ fī ṭabaqāt al-aṭibbāʾ* (Beirut, n.d.).
Ibn ʿArabī = Ps-Muḥyī al-Dīn al-ʿArabī, *Muḥāḍarāt al-abrār wa-musāmarāt al-akhyār*, 2 vols. (Beirut, 1968).
Ibn Hindū = Abū 'l-Faraj ibn Hindū, *Al-Kilam al-rūḥāniyya fī ḥikam al-yūnāniyya*, ed. S. Khalīfāt in *Ibn Hindū. Sīratuhu, ārāʾuhu al-falsafiyya, muʾallafātuhu*, 2 vols. (Amman, 1995).

Ibsh = al-Ibshīhī, *al-Mustaṭraf min kull fann mustaẓraf* (Beirut, 1982).
IDaw = Ibn Dāwūd al-Iṣfahānī, *K. al-Zahra. The First Half*, ed. A.R. Nykl and I. Tūqān (Chicago, 1932).
IḤamd = Ibn Ḥamdūn, *al-Tadhkira al-Ḥamdūniyya*, ed. I. ʿAbbās and B. ʿAbbās (Beirut, 1996).
IḤibb = Ibn Ḥibbān al-Bustī, *Rawḍat al-ʿuqalāʾ wa-nuzhat al-fuḍalāʾ*, ed. M. ʿAbd al-Ḥamīd (Beirut 1977).
IKhall = Ibn Khallikān, *Wafayāt al-aʿyān wa-anbāʾ abnāʾ al-zamān*, ed. I ʿAbbās, 8 vols. (Beirut, 1968–72).
Ikhtiyār = Ikhtiyār al-Dīn, *Asās al-iqtibās* (Istanbul, 1298).
IMud = Ps. Ibrāhīm ibn al-Mudabbir, *al-Risāla al-ʿadhrāʾ*, ed. Z. Mubārak (Cairo, 1931).
IMuqHik = 'Ḥikam Ibn al-Muqaffaʿ', in ʿU. al-Ṭabbāʿ (ed.), *Āthār Ibn al-Muqaffaʿ* (Beirut, 1978), pp. 177–81.
IMuqKal = Ibn al-Muqaffaʿ, *Kalīla wa-Dimna* (Cairo, 1937).
IMutAd = Ibn al-Muʿtazz, *Kitāb al-ādāb*, ed. I. Kratchkovsky in *Le monde oriental* 18 (1924), pp. 56–121.
IMutBad = Ibn al-Muʿtazz, *al-Badīʿ*, ed. I. Kratchkovsky (London, 1935).
IQayy = Ibn Qayyim al-Jawziyya, *al-Amthāl fī 'l-Qurʾān al-karīm* (Beirut, 1983).
IQut = Ibn Qutayba al-Dīnawarī, *ʿUyūn al-akhbār*, 4 vols. (Cairo, 1925–30).
IR = Ibn Riḍwān, *Al-Shuhub al-lāmiʿ fī 'l-siyāsa al-nāfiʿa*, ed. A.S. al-Nashshār (Casablanca, 1984).
IShams = Ibn Shams al-Khilāfa, *Al-Ādāb* (Cairo, 1931).
Ishkiwarī = al-Ishkiwarī, Quṭb al-Dīn Muḥammad ibn al-Shaykh ʿAlī, *Maḥbūb al-qulūb*, ed. I. al-Dībājī and Ḥ. Ṣadaqī (Tehran, 1999).
IWahb =Ibn Wahb al-Kātib, *al-Burhān fī wujūh al-bayān*, ed. A. Maṭlūb (Baghdad, 1967).
J = *Jāwīdān-khirad.*
JahB = al-Jāḥiẓ, *Al-Bayān wa'l-tabyīn*, ed. A.S.M. Hārūn, 4 vols. (Cairo, 1968).
JahR = *al-Rasāʾil*, ed. A.S.M. Hārūn, 2 vols. (Cairo, 1964).
JahTar = al-Jāḥiẓ, *al-Tarbīʿ wa'l-tadwīr*, ed. C. Pellat (Damascus, 1955).
K = *Ādāb al-falāsifa*, ms. Köprülü 1608 (*majmūʿa*).
Kalimāt = [Anonymous], 'Fī kalimāt mukhtāra min durar al-kalim,' in *al-Tuḥfa al-bahiyya*, Al-Jawāʾib 11884, (repr. Beirut, 1981, pp. 21–48).
Karkhī = al-Karkhī, *Al-Muntahā fī 'l-kamāl*, ed. R. Bostan (Giessen, 1986).
Kassis = R.A. Kassis, *The Book of Proverbs and Arabic Proverbial Works* (Leiden, 1999).
Khiradnāma, ed. M. Tharwat (Tehran, 1372).
Libro = *The Libro de los buenos proverbios*, ed. H. Sturm (Lexington, 1971).
Lo = A. Loewenthal, *Sefer musre ha-philosophim, 'Sinnsprüche der Philosophen,' aus dem Arabischen des Honein ibn Ishak ins Hebräische übersetzt von Jehuda ben Salomo Alcharizi* (Frankfurt, 1896).
Maʿarrī = Abū 'l-ʿAlāʾ al-Maʿarrī, *Rasāʾil Abī 'l-ʿAlāʾ al-Maʿarrī* (Beirut, 1894).
Maḥāsin = *Al-Maḥāsin wa'l-aḍdād*, ed. G. van Vloten (Leiden, 1898).
Mainyo-i-khard. The Pazand and Sanskrit Texts, ed. and trans. E.W. West (Stuttgart, 1871).
Marzub = Al-Marzubānī, *Nūr al-qabas al-mukhtaṣar min al-Muqtabas*, ed. R. Sellheim (Wiesbaden, 1964).
Masʿūdī = al-Masʿūdī, *Murūj al-dhahab wa-maʿādin al-jawhar*, ed. Ch. Pellat, 7 vols., (Beirut, 1965–79).
MawAd = al-Māwardī, *Adab al-dunyā wa'l-dīn*, ed. M. al-Saqqāʾ (Cairo, 1956).
MawQaw = al-Māwardī, *Qawānīn al-wizāra wa-siyāsat al-mulūk*, ed. R. al-Sayyid (Beirut, 1979).
MawTas = al-Māwardī, *Tashīl al-naẓar wa-taʿjīl al-ẓafar fī akhlāq al-malik wa-siyāsat al-mulk*, ed. R. al-Sayyid (Beirut, 1987).

MaxAli = *Maxims of 'Ali*, trans. J.A. Chapman (Oxford, 1937).
Mayd = al-Maydānī, *Majma' al-amthāl*, ed. A. Ibrāhīm, 4 vols. (Cairo, 1977–79).
Mb = Mubashshir ibn Fātik, *Mukhtār al-ḥikam wa-maḥāsin al-kilam*, ed. A. Badawī (Madrid, 1958).
MJ = Ibn Durayd, *al-Mujtanā* (Hyderabad, 1923).
MubarFad = al-Mubarrad, *al-Fāḍil*, ed. A.A. al-Maymanī (Cairo, 1956).
MubarKam = al-Mubarrad, *al-Kāmil*, 4 vols. (Cairo, n. d.).
Mufaḍḍal = al-Mufaḍḍal, *al-Fākhir*, ed. A.A al-Ṭaḥāwī and M.A. al-Najjār (Cairo, 1960).
Murādī = Muḥammad ibn al-Ḥasan al-Murādī, *al-Ishāra ilā adab al-imāra* (Beirut, 1981).
Murtaḍā Ṣarrāf, *Rasā'il Jawānmardān* (Tehran, 1973).
Musta'ṣimī = Yāqūt ibn 'Abd Allāh al-Musta'ṣimī, *Asrār al-ḥukamā'* (Damascus, 1994).
NadīmFihr = Ibn al-Nadīm, *al-Fihrist*, ed. R. Tajaddud (Tehran, 1971).
Najm Rāzī, *Mirṣād al-'ibād*, ed. M.A. Riyāḥī (Tehran, 1366).
Nāṣir Khusraw, *Dīwān*, ed. N. Taqwī (Tehran, 1335).
NB = al-Sharīf al-Raḍī, *al-Nahj al-balāgha*, ed. Sayyid Ja'far Shahīdī (Tehran, [3]1371).
Niẓāmī = al-Niẓāmī al-'arūḍī, *Čahār maqāla*, ed. M. ibn 'Abd al-Wahhāb Qazwīnī (Leiden, 1909–10).
Nuwayrī = al-Nuwayrī, *Nihāyat al-arab fī funūn al-adab*, 27 vols. (Cairo 1923–85).
Qābūs = Qābūs ibn Wushamgīr, *Qābūsnāma*, ed. Gh.Ḥ. Yūsufī (Tehran, 1378); *A Mirror for Princes. The Qābūs-nāma by Kai Kā'ūs Ibn Iskandar*, trans. R. Levy (London, 1951).
Qālī = al-Qālī, *al-Amālī*, ed. M. 'Abd al-Jawād al-Aṣma'ī (Beirut, 1926).
Qudāma = Ps.-Qudāma ibn Ja'far, *Naqd al-nathr* (Cairo, 1933).
QudDus = al-Quḍā'ī, *Dustūr ma'ālim al-ḥikam* (Beirut, 1981).
QudSh = al-Quḍā'ī, *al-Shihāb fī 'l-ḥikam wa'l-ādāb* (Baghdad, 1327).
Qushayrī = al-Qushayrī, *al-Risāla* (Cairo, 1927); Persian trans. Badī' al-Zamān Furūzānfar (Tehran, 1345/1967).
Rādūyānī = al-Rādūyānī, Muḥammad ibn 'Umar, *Tarjumān al-balāgha*, ed. A. Ātash (Istanbul, 1949).
Rāghib = al-Rāghib al-Iṣfahānī, *Muḥāḍarāt al-udabā'*, 2 vols. (Beirut, 1961).
Raqīq = al-Raqīq al-Nadīm, *Quṭb al-surūr fī awṣāf al-khumūr* (Damascus, 1969).
Rāzī = al-Rāzī, *al-Amthāl wa'l-ḥikam*, ed. F. Ḥarīrchi (Damascus, 1987).
Risāla = *Risāla tataḍammanu ādāb wa-ḥikma*. Anonymous. Published together with Sinān ibn Thābit ibn Qurra al-Ḥarrānī's *Siyāsat al-nufūs*, ed. 'A.A. al-Fāwī (Cairo, 1992).
Sagh = al-Ṣaghānī = Ps-al-Tha'ālibī, *Kitāb al-amthāl* (Cairo, 1327).
Sh = al-Shahrāzūrī, *Nuzhat al-arwāḥ wa-rawḍat al-afrāḥ fī ta'rīkh al-ḥukamā' wa'l-falāsifa*, ed. Kh.A. Fārūq, 2 vols. (Hyderabad, 1976).
Sh/Nb = Ibn Abī 'l-Ḥadīd, *Sharḥ Nahj al-balāgha*, 20 vols. (Cairo, 1965–67).
Shahrastani = al-Shahrastānī, *al-Milal wa'l-nihal*, 2 vols. (Cairo, 1910).
SijistMu = Abū Ḥātim al-Sijistānī, *Kitāb al-mu'ammarīn*, ed. I. Goldziher (Leiden, 1899).
SijistSiw = al-Sijistānī, *Ṣiwān al-ḥikma wa-thalāth rasā'il*, ed. A.R. Badawī (Tehran, 1974).
Suyūṭī = al-Suyūṭī, *Ta'rīkh al-khulafā'* (Cairo, 1964).
TawḥBas = al-Tawḥīdī, *al-Baṣā'ir wa'l-dhakā'ir*, ed. I. al-Kaylānī, 4 vols. (Damascus, 1964–69).
TawḥImt = al-Tawḥīdī, *al-Imtā' wa'l mu'ānasa*, ed. A. Amīn and A. al-Zayn, 3 vols. (Cairo, 1939).
TawḥSad = al-Tawḥīdī, *al-Ṣadāqa wa'l-ṣadīq* (Damascus, 1964).
TB = al-Khaṭīb al-Baghdādī, *Ta'rīkh Baghdād*, 14 vols. (Cairo, 1931).
Tha'ālAK = al-Tha'ālibī, *Aḥāsin kalim al-nabī wa'l-ṣaḥāb* [...], ed. J.P. Valeton (Leiden, 1844).
Tha'ālAM = al-Tha'ālibī, *Ādāb al-mulūk* (Beirut, 1990).

Tha'ālibī/Maqdisī = al-Tha'ālibī and Abū Naṣr Aḥmad al-Maqdisī, *al-Ẓarā'if wa'l-laṭa'if fī 'l-maḥāsin wa'l-aḍdād* (Cairo, 1307).
Tha'ālIJ, = al-Tha'ālibī, *al-I'jāz fī al-ījāz*, in *Khams rasā'il*, (Istanbul 1301), pp. 1–100.
Tha'ālTam = al-Tha'ālibī, *al-Tamthīl wa'l-muḥāḍara* (Cairo, 1961).
Tha'lab = *Majālis Tha'lab*, ed. A.S.M. Hārūn, 2 vols. (Cairo, 1948–49).
Ṭurṭ = al-Ṭurṭūshī, *Sirāj al-mulūk* (Cairo, 1319).
ṬūsīAd = Naṣīr al-Dīn al-Ṭūsī, *al-Adab al-wajīz*, ed. Gh.Ḥ. Āhanī (Iṣfahān, 1961).
Ṭūsī Akhl = Naṣīr al-Dīn Ṭūsī, *Akhlāq-i Nāṣirī*, ed. M. Minovi (Tehran, 1369).
Usāma ibn Munqidh, *Lubāb al-ādāb* (Cairo, 1935).
Wāḥidī = al-Wāḥidī, *al-Wasīṭ fī 'l-amthāl*, ed. A.M. 'Abd al-Raḥmān (Kuweit, 1975).
Washshā'= al-Washshā', *Kitāb al-muwashshā aw al-ẓarf wa'l-ẓurafā'*, ed. M. Kamāl (Cairo, 1953).
Waṭwāṭ, [Ṣad kalama] = *Maṭlūb kull ṭālib min kalām 'Alī b. abī Ṭālib*, ed. and trans. H.L. Fleischer (Leipzig, 1837).
Yāqūt al-Ḥamawī, *Mu'jam al-udabā'*, ed. I. 'Abbās, 7 vols. (Beirut, 1993).
Z = M. Zakeri, *Persian Wisdom in Arabic Garb. 'Alī b. 'Ubayda al-Rayḥānī (d. 219/834) and his Jawāhir al-kilam wa-farā'iḍ al-ḥikam* (Leiden, 2007).
ZamAm = al-Zamakhsharī, *Al-Mustaqṣā fī al-amthāl*, 2 vols. (Hyderabad, 1962).
ZamRab = al-Zamakhsharī, *Rabī' al-abrār wa-nuṣūṣ al-akhbār*, 4 vols. (Baghdad, 1976–82).

INDEX OF PERSONS

INDEX OF BOOK TITLES

1. Works of Aristoteles

2. Works by other authors